REPRODUCTIVE BIOLOGY
AND
EARLY LIFE HISTORY
OF
FISHES
IN THE
OHIO RIVER DRAINAGE

Elassomatidae and Centrarchidae

VOLUME 6

REPRODUCTIVE BIOLOGY
AND
EARLY LIFE HISTORY
OF
FISHES
IN THE
OHIO RIVER DRAINAGE

Elassomatidae and Centrarchidae

VOLUME 6

Robert Wallus
Thomas P. Simon

CRC Press
Taylor & Francis Group
Boca Raton London New York

CRC Press is an imprint of the
Taylor & Francis Group, an **informa** business

CITATION GUIDE, Volume 6:

Wallus, R. and T.P. Simon. 2008. *Reproductive Biology and Early Life History of Fishes in the Ohio River Drainage. Volume 6: Elassomatidae and Centrarchidae.* CRC Press, Boca Raton, Florida, USA.

or:

Mettee, M.F. 2008. Spring Pygmy Sunfish. Pages 35–46 in R. Wallus and T.P. Simon, editors. *Reproductive Biology and Early Life History of Fishes in the Ohio River Drainage. Volume 6: Elassomatidae and Centrarchidae.* CRC Press, Boca Raton, Florida, USA.

CRC Press
Taylor & Francis Group
6000 Broken Sound Parkway NW, Suite 300
Boca Raton, FL 33487-2742

First issued in paperback 2019

© 2008 by Taylor & Francis Group, LLC
CRC Press is an imprint of Taylor & Francis Group, an Informa business

No claim to original U.S. Government works

ISBN-13: 978-0-8493-1922-8 (hbk)
ISBN-13: 978-0-367-38716-7 (pbk)

Visit the Taylor & Francis Web site at
http://www.taylorandfrancis.com

and the CRC Press Web site at
http://www.crcpress.com

SPONSORED BY

Allegheny Energy Supply Company

American Electric Power Company

American Municipal Power – Ohio

Buckeye Power

Dayton Power & Light

Duke Energy

Electric Power Research Institute

Engineering Research and Development Center, U.S. Army Corps of Engineers

FirstEnergy Services Company

Indiana Biological Survey, Aquatic Research Center

Louisville Gas & Electric, a Subsidiary of E.ON-US

Nashville District, U.S. Army Corps of Engineers

Ohio Valley Electric Corporation/Indiana–Kentucky Electric Corporation

Owensboro Municipal Utilities

Tennessee Valley Authority

Western Kentucky Energy Corporation, a Subsidiary of E.ON-US

"sunfishes are fun fishes"

by

R. Wallus

sunfishes are fun fishes
this I do declare
it's easy to go and catch some
'cause they're everywhere

you can catch 'em in the river
and the pond and bayou too
in the lake or in the creek
it's really up to you

a cane pole and a bobber
will work most ever time
or get yerself a Zebco
and some monofilament line

you can catch 'em from a boat
or sittin' on the bank
simply drop your bait and wait
then give a big ole yank

bait? you ask
well, what does it take?
'bout anything will do
except your birthday cake

they like worms and minners
and crickets they love to chew
and a Green Giant niblet
might get one fer you

they dislike Rapalas
and will smack 'em up on top
and they'll nibble a lil' plastic grub
soon as you let 'er drop

tiny crawfish crank baits
will bring 'em off the bottom
so will any spinner bait
so fling 'em if ya got 'em

if havin' fun while fishin'
is one of yer many wishes
then fun will be had and you'll be glad
you caught a mess of sunfishes!

Supplementary Resources Disclaimer

Additional resources were previously made available for this title on CD. However, as CD has become a less accessible format, all resources have been moved to a more convenient online download option.

You can find these resources available here: http://resourcecentre.routledge.com/books/9780849319228

Please note: Where this title mentions the associated disc, please use the downloadable resources instead.

FOREWORD

This comprehensive, multivolume series, which originated more than a decade ago, is a much needed taxonomic aid, complete with keys, diagnostic criteria, and illustrated descriptions for identification of the eggs, larvae, and early juveniles of most of the about 285 fishes in the Ohio River Basin. It is also an equally needed compendium of information on the ecology of those early life stages, as well as a summary of the distribution, habitat, and reproductive biology of their parents. The descriptive and early life history information in this single series complements a multitude of state and regional guides that emphasize only adult descriptions, distributions, and biology. Each volume has been anxiously awaited by many fish biologists throughout the central United States, and wherever else the covered species are found.

The early life stages of most fishes represent developmental intervals that are ecologically distinct from each other and especially from their later juvenile and adult counterparts. Knowledge of their changing ecological requirements and limitations, population dynamics, and behavior facilitates more effective monitoring and management of fish populations and habitats. It is also crucial to the evaluation of environmental impacts and recovery of endangered species.

Early life history investigations in fresh waters of the United States received their greatest boost in the 1960s and 1970s in response to federal laws that require assessments or monitoring of adverse environmental impacts on the country's waters, aquatic communities, and endangered fishes. The effects of chemical discharges from industry, thermal effluents from and entrainment in power-plant cooling systems, transport through hydroelectric and pumped-storage turbines, impoundments, water diversions, other habitat changes, and introductions of non-native species on the early life stages of fish were, and in many cases remain, significant concerns across the country.

However, field research on fish eggs, larvae, and early juveniles depends on accurate identification of at least the targeted species among collected specimens. Morphological identification requires knowledge of the appearance of not only the targeted species, but all potentially similar-looking species in the waters sampled and the diagnostic criteria for segregating them. For the early life stages of most species, morphological criteria for identification change dramatically as the fish grow and develop, making diagnosis especially difficult and complicated.

This series will prove invaluable as research on, and management of, the fishes and aquatic ecosystems of the Ohio River Basin (and the rest of the Mississippi River System) continue in the new millennium. The authors and many of the contributors have dedicated much of their lives to advancing our knowledge of the eggs, larvae, and early juveniles of North America's freshwater fishes. As a result of their effort, the original, report-embedded, and previously published information compiled in each volume of this guide goes a long way toward filling immense gaps in our knowledge for future research and management. But, as evidenced by the information that is still missing, much remains to be learned. Even with the completion of this guide, the vast majority of North America's approximately 800 species of freshwater and anadromous fishes (perhaps two-thirds of them, but only about one-sixth of those in the Ohio River Basin) remain inadequately described as larvae for identification purposes. It must be impressed upon the sponsors of early life history research that descriptive biology and the development of taxonomic aids remain a vital part of that research and need to be funded and published accordingly.

Darrel E. Snyder

Larval Fish Laboratory
Colorado State University

PREFACE

Knowledge of early developmental stages of fishes is obviously fundamental to proper understanding of many aspects of fishery biology and ichthyology. It is paradoxical, then, that eggs, larvae, and juveniles of so many species of fishes remain completely or essentially unknown and undescribed.

— **Mansueti and Hardy, 1967**

Prior to the present-day environmental movement, which began in the late 1960s, scientific attention to early life histories of fishes was limited to a handful of investigators possessing the insight, patience, and occupational privilege to pursue such an important but little known aspect of fisheries science. In recent decades, however, water resource issues have emerged as a top priority worldwide. It has been in this atmosphere of public concern and environmental enlightenment that the number of scientific voids has become apparent. In the arena of reproductive biology and early life ecology of fishes, so little had been investigated that larvae of most fish species could not be identified. Even less was known about behavior, ecology, and habitat requirements of young fish.

Regulatory requirements in the 1960s and 1970s resulted in the advancement of scientists' abilities to collect, identify, and quantify larval and juvenile fishes. However, knowledge of such important matters as spawning habitat requirements, reproductive behavior, and ecological relationships during the first few months of life, for even the most common species of fishes, lagged well behind. New information has been and is currently being collected, but because so much research is a direct result of regulatory requirements, it is often necessary to use the knowledge gained only to fulfill reporting requirements. The importance to environmental biologists of disseminating advances in the state of the art to the remainder of the scientific community often becomes secondary to getting on with the next challenges at hand.

It is against this backdrop that the need for a compendium of acquired information was recognized and this particular project was spawned. What first formed an updated guide to identification of early life stages of fishes in the Tennessee River ultimately developed into a resource document on the reproductive biology and early life ecology of the fishes of the Ohio River Basin. The persistence and dedication of the authors, contributors, researchers, and supporters of this project cannot be overstated. Certainly there is more information in existence than has been discovered and incorporated herein. Unfortunately, that will always be the case. What has been provided, however, is the most complete treatise on early life histories of freshwater fishes in North America to date.

The information in this treatise is based on thorough field collections of early life phases and propagation and culture activities throughout the study area. We have added new information on the reproductive biology and early life histories for many species in the Ohio River drainage that previously was unknown. The Ohio River drainage contains a diverse fauna. Approximately 285 species are recognized from the system, including 54 endemic species. Currently, 9 species are federally listed as endangered, 7 are listed as threatened.

This series is divided into seven volumes that represent the inland ichthyofauna of the majority of eastern North America. Each volume contains distinguishing characteristics and a pictorial guide to the families of fishes present in the Ohio River drainage followed by family chapters. Family chapters are organized into species accounts arranged alphabetically within genus and sometimes higher taxonomic groupings (e.g., subgenera). The level of taxonomy presented is dependent on larval diagnostic traits within the family. Where possible, dichotomous keys to species or higher taxa within families are provided, and, when useful, schematic drawings of characters supplement key couplets. Each species account is divided into a variety of subtopics that discuss all aspects of reproductive biology, early life history, and ecology of the young.

The information contained in this series will be invaluable to fisheries managers. They will be able to use the information to better protect and restore fishery resources. Resource planners and environmental scientists will use the information to validate the predicted effects of their decisions and to aid them in mitigating the impacts of their decisions. It is our intention to present the information in a format that will facilitate wide use. Our goal is to produce a resource document that will help a biologist identify a single larval fish, as well as provide a resource for the environmental manager concerned with the health and condition of his watershed jurisdiction. This series is the current state-of-the-art resource for reproductive biology and early life history of North American freshwater fishes.

Robert Wallus
Thomas P. Simon

ACKNOWLEDGMENTS

The authors thank the Tennessee Valley Authority (TVA), the Nashville District of the U.S. Army Corps of Engineers (USACE), the Engineering Research and Development Center (ERDC) of USACE, American Electric Power (AEP), the Electric Power Research Institute (EPRI) and its members, and the Indiana Biological Survey, Aquatic Research Center for sponsoring this work. Agency support for a project of this magnitude is gained only through a commitment of support from individuals. We would be remiss if we did not acknowledge the commitment to this work of W.L. Poppe, W.B. Wrenn, S. Robertson, R.J. Pryor, C. Massey, and W.G. Ruffner of TVA; H.J. Cathey and C.T. Swor of the Nashville District USACE; J. Killgore, ERDC-USACE; R. Reash of AEP; and D. Dixon of EPRI. We are also indebted to J.P. Buchanan, H.J. Cathey, G.E. Hall, C.T. Swor, C.W. Voigtlander, and W.B. Wrenn as editors for their critical and constructive review of original manuscripts and for their insights during the conceptualization of the project and development of the format.

Original photographs of larvae in this volume were taken by T.P. Simon. Murrie V. Graser was the layout and design artist during conceptualization of this project and her excellent contributions are appreciated. We are also grateful to Beth Simon for artistic support of this project.

We would like to express our thanks to the following colleagues who provided specimens, data, technical assistance, publications, reports, manuscript review, or other professional courtesies helpful in completing this volume: N.A. Auer, N.H. Douglas, D.A. Etnier, J.F. Heitman, L.K. Kay, J. Killgore, G. McGowan, and C.A. Taber. Thanks to J. Yarborough for his help in accessing spatio-temporal data for young rock bass, smallmouth bass, and crappies from TVA data bases. Special appreciation is due to J.F. Heitman for his review of most of the text and for the use of his laboratory, technical library, and many other professional courtesies. A special thanks to M.A. Colvin for reviewing and editing the *Pomoxis* chapter. Thanks are due to C.E. Saylor and many other TVA aquatic biologists whose interest in this project was demonstrated by their moral support and by their invaluable assistance in field and laboratory work. Without the assistance of staff from TVA's Technical Library, completion of this project would have been much more difficult. For their attention to detail, timeliness in handling volumes of requests, and positive and supportive attitudes toward this project, we are greatly indebted.

LIST OF ABBREVIATIONS

ABD	Air bladder depth	mm	Millimeter
ADFL	Adipose fin length	MPosAD	Mid-postanal depth
AF	Anal fin	N	Number
AFL	Anal fin length	ORM	Ohio river mile
BDA	Body depth at anus	P1	Pectoral fin
BDE	Body depth at eyes	P1L	Pectoral fin length
BDP1	Body depth at pectoral fin	P2	Pelvic fin
CFL	Caudal fin length	P2L	Pelvic fin length
ChiBL	Chin barbel length	PosAL	Postanal length
cm	Centimeter	Ppt	Parts per thousand
CPD	Caudal peduncle depth	PreAFI	Preadipose fin insertion length
CPL	Caudal peduncle length	PreAFO	Preadipose fin origin length
CPUE	Catch per unit effort	PreAL	Preanal length
DF	Dorsal fin	PreDFFL	Predorsal finfold length
DFL	Dorsal fin length	PreDFL	Predorsal fin length
ED	Eye diameter	REB	Redeye bass
FL	Fork length	RM	River mile
g	Gram	s	Second
GD	Greatest depth	SD	Shoulder depth
GSI	Gonadosomatic index	SL	Standard length
ha	Hectare	SMB	Smallmouth bass
HD	Head depth	SnL	Snout length
HL	Head length	SPB	Spotted bass
HW	Head width	sq	Square
IOD	Interorbital distance	TL	Total length
kg	Kilogram	TRM	Tennessee river mile
km	Kilometer	TVA	Tennessee Valley Authority
m	Meter	UJL	Upper jaw length
LMB	Largemouth bass	YSD	Yolk-sac depth
MaxBL	Maxillary barbel length	YSL	Yolk-sac length
MBL	Mandibular barbel length	YOY	Young of the year

GLOSSARY OF TERMS

Abbreviate heterocercal Tail in which the vertebral axis is prominently flexed upward, only partly invading upper lobe of caudal fin; fin fairly symmetrical externally.

Actinotrichia Fin supports that are precursors of fin rays or spines; also called lepidotrichia.

Adherent Attached or joined together, at least at one point.

Adhesive egg An egg that adheres on contact to substrate material or other eggs; adhesiveness of entire egg capsule may or may not persist after attachment.

Adipose fin A fleshy, rayless median dorsal structure, located posterior to the true dorsal fin.

Adnate Congenitally united; conjoined; keel-like.

Adnexed Flaglike.

Adult Sexually mature as indicated by production of gametes.

Alevin A term applied to juvenile catfish, trout, and salmon after yolk absorption; exhibiting no post yolk-sac larval phase.

Allopatric Having separate and mutually exclusive areas of geographical distribution.

Anadromous Fishes which ascend rivers from the sea to spawn.

Anal Pertaining to the anus or vent.

Anal fin Unpaired median fin immediately behind anus or vent.

Anlage Rudimentary form of an anatomical structure; primordium; incipient.

Antero-hyal Anterior bone to which branchiostegal rays attach; formerly ceratohyl.

Anus External orifice of the intestine; vent.

Auditory vesicle Sensory anlage from which the ear develops; clearly visible during early development.

Axillary process Enlarged accessory scale attached to the upper or anterior base of pectoral or pelvic fins.

Band Continuous, curved line of pigment, usually developed on the fins.

Bar Dorsoventrally elongate line of pigment, usually straight.

Barbel Tactile structure arising from the head of various fishes.

Basibranchials Three median bones on the floor of the gill chamber, joined to the ventral ends of the five gill arches.

Blastula A hollow ball of cells formed early in embryonic development.

Blotch Irregular marking, usually rounded or squarish in outline, on more than one adjacent scale when developed on body.

Body depth at anus Vertical depth of body at anus, not including finfolds.

Branched ray Soft fin ray with two or more branches distally.

Branchial arches Bony or cartilaginous structures supporting the gills, filaments, and rakers; gill arches.

Branchial region The pharyngeal region where branchial arches and gills develop.

Branchiostegals Struts of bone inserting on the hyoid arch and supporting, in a fanwise fashion, the branchiostegal membrane; branchiostegal rays.

Buoyant egg An egg that floats free within the water column; pelagic.

Caeca Finger-like outpouchings at boundary of stomach and intestine.

Calcareous Composed of, containing, or characteristic of calcium carbonate.

Catadromous Fishes that go to sea from rivers to spawn.

Caudal fin Tail fin.

Caudal peduncle Area lying between posterior end of anal fin base and base of caudal fin.

Caudal spot A blotch, short stripe, or distinct spot developed over the terminus of the hypural plate and anterior part of caudal fin rays.

Cement glands Discrete or diffuse structures that permit a larva to adhere to a substrate.

Cephalic Pertaining to the head.

Ceratohyal See antero-hyal.

Cheek Lateral surface of head between eye and opercle, usually excluding preopercle.

Chorion Outer covering of egg; egg capsule.

Choroid fissure Line of juncture of invaginating borders of optic cup; apparent in young fish as a trough-like area below lens.

Chromatophores Pigment-bearing cells; frequently capable of expansions and contractions which change their size, shape, and color.

Cleavage stages Initial stages in embryonic development where divisions of blastomeres are clearly marked; usually include 1st through 6th cleavages (2–64 cells).

Cleithrum Prominent bone of pectoral girdle, clearly visible in many fish larvae.

Coelomic Pertaining to the body cavity.

Confluent Coming together to form one.

Ctenoid scale Scale with comb-like margin; bearing cteni or needle-like projections.

Cycloid scale Scale with evenly curved, free border, without cteni.

Demersal egg An egg that remains on the bottom, either free or attached to substrate.

Dentary Major bony element of the lower jaw, usually bearing teeth.

Dorsal fins Median, longitudinal, vertical fins located on the back.

Early embryo Stage in embryonic development characterized by formation of embryonic axis.

Egg capsule Outermost, encapsulating structure of the egg, consisting of one or more membranes; the protective shell.

Egg diameter In nearly spherical eggs, greatest diameter; in elliptical eggs given as two measurements, the greatest diameter or major axis and the least diameter or minor axis.

Egg pit The pit or pocket in a redd (nest) into which a trout female deposits one batch of eggs.

Emarginate Notched but not definitely forked, as in the shallowly notched caudal fins of some fishes.

Emergence The act of leaving the substrate and beginning to swim; swim-up.

Epaxial Portion of the body dorsal to the horizontal or median myoseptum.

Epurals Modified vertebrae elements which lie above the vertebrae and support part of the caudal fin.

Erythrophores Red or orange chromatophores.

Esophagus Alimentary tract between pharynx and stomach.

Eye diameter Horizontal measurement of the iris of the eye.

Falcate Deeply concave as a fin with middle rays much shorter than anterior and posterior rays.

Fin insertion Posterior-most point at which the fin attaches to the body.

Fin origin Anterior-most point at which the fin attaches to the body.

Finfold Median fold of integument which extends along body of developing fishes and from which median fins arise.

Focal point Location of a fish maintaining a stationary position on or off the substrate for at least a 10-second period.

Fork length Distance measured from the anterior-most point of the head to the end of the central caudal rays.

Frenum A fold of skin that limits movement of the upper jaw.

Ganoid scales Diamond- or rhombic-shaped scales consisting of bone covered with enamel.

Gas bladder Membranous, gas-filled organ located between the kidneys and alimentary canal in teleosts; air bladder or swim bladder.

Gastrula Stage in embryonic development between blastula and embryonic axis.

Gill arches See branchial arches.

Gill rakers Variously shaped bony projections on anterior edge of the gill arches.

Granular yolk Yolk consisting of discrete units of finely to coarsely granular material.

Greatest body depth Greatest vertical depth of the body excluding fins and finfolds.

Guanophores White chromatophores; characterized by presence of iridescent crystals of guanine.

Gular fold Transverse membrane across throat.

Gular plate Ventral bony plate on throat, as in *Amia calva*.

Gular region Throat.

Haemal Relating to or situated on the side of the spinal cord where the heart and chief blood vessels are placed.

Head length Distance from anterior-most tip of head to posterior-most part of opercular membrane, excluding spine; prior to development of operculum, measured to posterior end of auditory vesicle.

Head width Greatest dimension between opercles.

Heterocercal Tail in which the vertebral axis is flexed upward and extends nearly to the tip of the upper lobe of the caudal fin; fin typically asymmetrical externally, upper lobe much longer than lower.

Homocercal Tail in which the vertebral axis terminates in a penultimate vertebra followed by a urostyle (the fusion product of several vertebral elements); fin perfectly symmetrical externally.

Horizontal myoseptum Connective tissue dividing epaxial and hypaxial regions of the body; median myoseptum.

Hypaxial That portion of the body ventral to the horizontal myoseptum.

Hypochord A transitional rod of cells that develops under the notochord in the trunk region of some embryos.

Hypochordal Below the notochord; referring to the lower lobe of the caudal fin.

Hypurals Expanded, fused, haemal spines of last few vertebrae that support the caudal fin.

Incipient Becoming apparent.

Incubation period Time from fertilization of egg to hatching.

Inferior mouth Snout projecting beyond the lower jaw.

Integument An enveloping layer or membrane.

Interorbital Space between eyes over top of head.

Interradial Area between the fin rays.

Interspaces Spaces between parr marks of salmonids.

Iridocytes Crystals of guanine having reflective and iridescent qualities.

Isocercal Tail in which vertebral axis terminates in median line of fin, as in Gadiformes.

Isthmus The narrow area of flesh in the jugular region between gill openings.

Jugular Pertaining to the throat; gular.

Juvenile Young fish after attainment of minimum adult fin-ray counts and complete absorption of the median finfold and before sexual maturation.

Keeled With a ridge or ridges.

Kupffer's vesicle A small, vesicular, ventro-caudal pocketing which forms as blastopore narrows.

Larva Young fish between time of hatching and attainment of juvenile characteristics.

Late embryo Stage prior to hatching in which the embryo has developed external characteristics of its hatching stage.

Lateral line Series of sensory pores and/or tubes extending backward from head along sides.

Lateral line scales Pored or notched scales associated with the lateral line.

Lecithotrophic Pertaining to embryos nourished by nutrients of the egg yolk.

Lepidotrichia See actinotrichia.

Mandible Lower jaw, comprising three bones: dentary, angular, and articular.

Maxillary The dorsal-most of the two bones in the upper jaw.

Meckel's cartilage Embryonic cartilaginous axis of the lower jaw in bony fishes; forms the area of jaw articulation in adults.

Melanophores Black chromatophores.

Mental Pertaining to the chin.

Myomeres Serial muscle bundles of the body.

Myosepta Connective tissue partitions separating myomeres.

Nares Nostrils, openings leading to the olfactory organs.

Narial Pertaining to the nares.

Nasal Pertaining to region of the nostrils, or to the specific bone in that region.

Notochord Longitudinal supporting axis of body which is eventually replaced by the vertebral column in teleostean fishes.

Notochord length Straight-line distance from anterior-most part of head to posterior tip of notochord; used prior to and during notochord flexion.

Obtuse With a blunt or rounded end; an angle greater than 90°.

Occipital region Area on dorsal surface of head, beginning above or immediately behind eyes and extending backward to end of head; occiput.

Oil globules Discrete spheres of fatty material within the yolk.

Olfactory buds Incipient olfactory organs.

Ontogenetic Related to biological development.

Opercle Large posterior bone of the operculum.

Operculum Gill cover.

Optic vesicles Embryonic vesicular structures which give rise to the eyes.

Otoliths Small, calcareous, secreted bodies within the inner ear.

Over yearling Fish having spent at least one winter in a stream; applies to trout and salmon.

Palatine teeth Teeth on the paired palatine bones in the roof of the mouth of some fishes.

Parapatric Distribution of species or other taxa that meet in a very narrow zone of overlap.

Parturition The act or process of birth.

Pectoral fins Paired fins behind head, articulating with pectoral girdle.

Peduncle Portion of body between anal and caudal fins.

Pelagic Floating free in the water column; not necessarily near the surface.

Pelvic bud Swelling at site of future pelvic fin; anlage of pelvic fin.

Pelvic fins Paired fins articulating with pelvic girdle; ventral fins.

Pericardium Cavity in which the heart lies.

Peritoneum Membranous lining of abdominal cavity.

Perivitelline space Fluid-filled space between egg proper and egg capsule.

Pharyngeal teeth Teeth on the pharyngeal bones of the branchial skeleton.

Physoclistic Having no connection between the esophagus and the pneumatic duct; typical of perciform fishes.

Physostomus Having the swim bladder connected to the esophagus by the pneumatic duct; typical of cypriniform fishes.

Plicae Wrinkle-like folds found on the lips of some catostomids.

Postanal length Distance from posterior margin of anus to the tip of the caudal fin.

Postanal myomeres Myomeres posterior to an imaginary vertical line through the body at the posterior margin of the anus; the first postanal myomere is the first myomere behind and not touched by the imaginary line.

Postero-hyal Posterior bone to which branchiostegal rays attach; formerly epihyal.

Postorbital length Distance from posterior margin of eye to posterior edge of opercular membrane.

Preanal length Distance from anterior-most part of head to posterior margin of anus.

Preanal myomeres The number of myomeres between the anterior-most myoseptum and an imaginary vertical line drawn at the posterior margin of anus, including any bisected by the line.

Predorsal scales Scales along dorsal ridge from occiput to origin of dorsal fin.

Prejuvenile Developmental stage immediately following acquisition of minimum fin ray complement of adult and before assumption of adult-like body form; used only where strikingly different from juvenile.

Premaxillary The ventral-most of the two bones included in the upper jaw.

Primordium Rudimentary form of an anatomical structure; anlage.

Principal caudal rays Caudal rays inserting on hypural elements; the number of principal rays is generally defined as the number of branched rays plus two.

Procurrent caudal rays A series of much shorter rays anterior to the principal caudal rays, dorsally and ventrally, not typically included in the margin of the caudal fin.

Pronephic ducts Ducts of pronephic kidney of early development stages.

Pterygiophores Bones of the internal skeleton supporting the dorsal and anal fins.

Redd An excavated area or nest into which trout spawn.

Retrorse Pointing backward.

Rostrum Snout.

Scute A modified, thickened scale, often spiny or keeled.

Semibuoyant Referring to eggs that neither float nor sink, but remain suspended in the water column.

Sigmoid heart The S-shaped heart that develops from the primitive heart tube.

Soft rays Bilaterally paired, usually segmented fin supports.

Spines Unpaired, unsegmented, unbranched fin supports, usually (but not always) stiff and pungent.

Squamation Covering of scales.

Standard length In larvae, straight-line distance from anterior-most part of head to the most posterior point of the notochord or hypural complex.

Stellate Referring to a melanophore which is expanded into a starlike shape.

Stomodeum Primitive invagination of the ectoderm that eventually gives rise to the mouth.

Streak A narrow, faint line of pigment along a single scale row.

Stripe Horizontly elongate line.

Superior mouth Condition when the lower jaw extends upward and the mouth opens dorsally.

Sympatric Species inhabiting the same or overlapping geographic areas.

Teleosts Bony fishes.

Terminal mouth Condition when lower and upper jaws are equal in length and the mouth opens terminally.

Total length Straight-line distance from anteriormost part of head to tip of tail; all older literature references not stated differently are assumed to be total length.

Truncate Terminate abruptly as if the end were cut off.

Urostyle Terminal vertebral element in higher teleosts, derived from the fusion and loss of several of the most posterior centra of the more primitive forms; usually modified for caudal fin support.

Vent Anus.

Vermiculate Having wormlike markings.

Vitelline vessels Arteries and veins of yolk region.

Vitellogenesis Yolk formation.

Viviparous Bearing young alive, regardless of the dependency on yolk stores.

Water-hardening Expansion and toughening of egg capsule due to absorption of water into the perivitelline space.

Weberian apparatus First four vertebrae of cypriniform fishes modified for sound amplification.

Width of perivitelline space Distance between yolk and outer margin of egg capsule.

Xanthophores Yellow chromatophores.

Yearling A fish in its second year.

Yolk Food reserve of embryonic and early larval stages, usually seen as a yellowish sphere diminishing in size as development proceeds.

Yolk diameter Greatest diameter of yolk; more accurately measurable prior to embryo formation.

Yolk sac A baglike ventral extension of the primitive gut containing the yolk.

Yolk-sac depth Greatest vertical depth of yolk sac.

Yolk-sac larva A larval fish characterized by the presence of a yolk sac.

Yolk-sac length Horizontal distance from most anterior to most posterior margin of yolk sac.

CONTENTS

Reproductive Biology and Early Life History of Fishes in the Ohio River Drainage:

An Introduction to the Series

Thomas P. Simon and Robert Wallus

Although numerous descriptions of the ontogeny of individual fish species have been published and a few studies have summarized the existing knowledge of the early life histories of fishes present in particular areas or regions, information is still lacking for many species (Mansueti and Hardy, 1967; Simon, 1985). Important geographical works on the early life histories of fishes have come from coastal regions (Mansueti and Hardy, 1967; Jones et al., 1978; Wang and Kernehan, 1979) and the Great Lakes (Auer, 1982). However, no resource document of this type exists for the large, inland freshwater drainages of the United States.

Fisheries biologists have become acutely aware of this void with their increased need for reproductive biology and early life history information in their conduct of required ecological studies. Information on distribution and abundance of eggs and larvae is useful in determining spawning and nursery areas, spawning seasons, reproductive success, year-class strength, and in some instances relative abundance of adult populations. The conditions and behaviors associated with spawning, as well as the sensitivity of fish eggs and larvae to environmental impacts, are a concern and the cause for assessments and monitoring programs, i.e., 316(b) demonstrations, now required of most industries and utilities.

SERIES OBJECTIVES

The principal objective of this book is to provide an illustrated resource document for biologists who study the reproductive biology and early life history of fishes that occur in the Ohio River or its tributaries. Comprehensive reviews of the literature, as well as presentations of original data, are included. This text has three primary purposes: the advancement and evaluation of larval fish taxonomy, the identification of gaps in the knowledge of reproductive biology and early life history of fishes within the study area, and the stimulation of further research in areas lacking information. The diversity of species in the Ohio River drainage should make this document useful to fisheries biologists throughout the eastern and central United States.

STUDY AREA

The Ohio River originates at the confluence of the Allegheny and Monogahela Rivers at Pittsburgh (ORM 0) and generally flows southwest for 981 miles (1578 km) before entering the Mississippi River near Cairo, IL (ORM 981). After flowing from PA, the Ohio River delineates the geographical boundaries between OH and WV, OH and KY, IN and KY, and IL and KY (Figure 1). Most of the tributaries in the system drain water from these states including headwater tributaries, which flow from NY, MD, and VA.

The southern portion of the Ohio River system is drained by two of its largest tributaries, the Cumberland and Tennessee Rivers (Table 1). The mouth of the Cumberland River enters the Ohio River at ORM 925. Its tributaries are confined to KY and TN. The Tennessee River is the largest tributary system in the Ohio River accounting for approximately 20% of the watershed. The drainage lies mostly in the state of TN, but its headwaters are in the mountains of VA, western NC, eastern TN, and northern GA (Figure 1). From the confluence of the Holston and French Broad Rivers near Knoxville, TN, the Tennessee River flows approximately 652 miles (1049 km) before entering the Ohio River. Its course takes it southwest across TN into AL and then west across northern AL; it turns north at the northeast corner of the state of MS and flows back across TN and western KY to enter the Ohio River near Paducah, KY (ORM 940).

The Wabash River is the second largest Ohio River tributary system (Table 1) and the largest northern tributary (Figure 1). It encompasses approximately 16% of the total watershed and drains most of IN and portions of southeastern IL before its confluence with the Ohio River (ORM 850). The Wabash River is the largest free-flowing tributary of the Ohio River.

The Ohio River drainage contains one of the world's greatest coal-producing regions, several large metropolitan areas (e.g., Pittsburgh, Cincinnati, Louisville, Lexington, Knoxville, Chattanooga, and Nashville), and numerous power plants and large industries. Dams have been built on most of the larger rivers, including the mainstem Ohio, to provide flood control, navigation, hydroelectric power, water supply, and recreation.

The Ohio River system contains a diverse ichthyofauna (Pearson and Krumholz, 1984). Approximately 285 species are recognized from the system (Lee et al., 1980), including 54 endemic species (Table 2). This represents about 40% of the North American fauna. Currently, nine species are federally listed as endangered and seven are listed as threatened (Table 3).

FORMAT

This document is presented as a series of volumes containing family chapters. Information is not presented in phylogenetic sequence. Volume 1 includes the families Acipenseridae through Esocidae (Wallus et al., 1990); Volume 2 represents the single family Catostomidae (Kay et al., 1994); Volume 3 contains information on the catfishes, family Ictaluridae (Simon and Wallus, 2003); Volume 4 covers the single family Percidae (Simon and Wallus, 2006); and Volume 5 includes the families Aphredoderidae through Cottidae, Moronidae, and Sciaenidae (Wallus and Simon, 2006). This volume (6) includes the families Elassomatidae and Centrarchidae. The final volume of the series (7) will compile information for the family Cyprinidae. Common and scientific names generally follow American Fisheries Society Special Publication 29 (Nelson et al., 2004). Exceptions are noted in the introduction to each volume.

Each volume contains distinguishing characteristics and a pictorial guide to the families of fishes present in the Ohio River drainage followed by family chapters. Family chapters are organized into species accounts arranged alphabetically within genus and sometimes higher taxonomic groupings, i.e., subgenera and subfamilies. The level of taxonomy is dependent on larval diagnostic traits within the family. Where possible, dichotomous keys to species or higher taxa within families are provided.

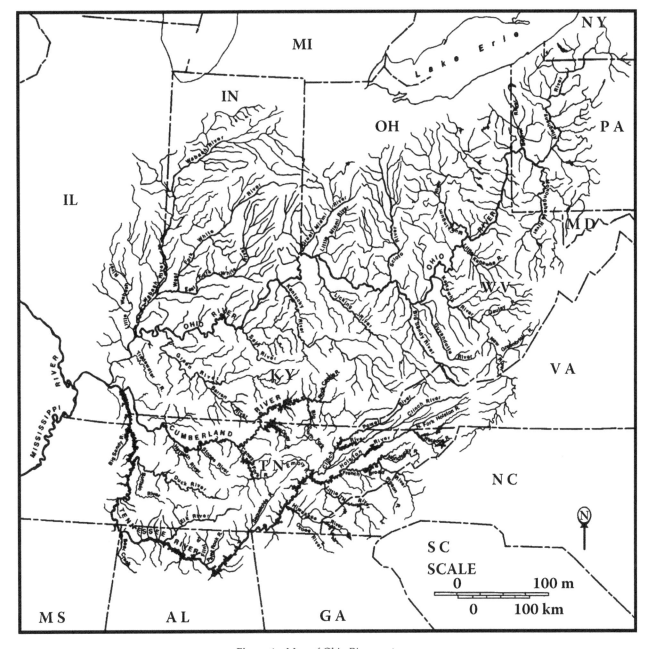

Figure 1 Map of Ohio River system.

When useful, schematic drawings of characters supplement key couplets. Each species account is divided into the following major divisions.

Range

A description of the reported distribution of the species is presented; the principal source for this information is Lee et al. (1980), although more recent references are used, if appropriate.

Habitat and Movement

A description of the habitats with which adults of the species are most often associated and a descrip-

tion of any movement patterns (e.g., diel, seasonal, prespawning, and postspawning) associated with the life history of the species are provided.

Distribution and Occurrence in the Ohio River System

Information about the relative occurrence of the species within the study area comes from state or regional publications such as Gerking (1945), Burr and Warren (1986), Etnier and Starnes (1993), Pearson and Krumholz (1984), Smith (1985), Smith (1979), Jenkins and Burkhead (1994), Mettee et al. (2001), and Boschung and Mayden (2004).

Table 1

Physical characteristics of the Ohio River system.

River Basin	Ohio River Mile	Approximate Drainage Area (km²)
Allegheny	0	30,300
Monongahela	0	19,200
Beaver	25	8,100
Muskingum	172	20,800
Little Kanawha	185	6,000
Kanawha	266	31,600
Guyandotte	305	4,300
Big Sandy	317	11,100
Scioto	356	16,900
Little Miami	464	4,600
Licking	470	9,500
Great Miami	491	14,000
Kentucky	546	18,000
Salt	630	7,500
Wabash	848	85,700
Cumberland	920	46,400
Tennessee	940	106,000
Mainstem Ohio and smaller tributaries	—	64,100
		528,000

Spawning

A description of reproductive characteristics is organized into sections including information on location (habitat), season, temperature, fecundity, sexual maturity (age and size), and spawning act.

Eggs

Information is given on the following:

Description — Characteristics of fertilized eggs, including shape, adhesiveness, buoyancy, color, diameter, and sometimes internal characteristics; information on ovarian eggs may be provided if little information is available for fertilized eggs.

Incubation — Time period in days or hours with associated temperatures.

Development — Reference is made to important embryological studies but little information is provided other than brief comments pertaining to embryonic distinctiveness. Drawings and descriptions of embryology are limited to the presentation of new information.

Development

Descriptions of development within each life phase (yolk-sac larvae, post yolk-sac larvae, and juvenile) arranged into the following subdivisions:

Size Range — Size encompassed by phase, if known.

Myomeres — Usually includes total, preanal, and postanal counts.

Morphology — Information further presented under length or length-range subheadings.

Morphometry — Where available, measurements are presented as percent total length or as percent head length.

Fin Development — Information usually presented under length or length-range subheadings, although individual fins may be used as subheadings with dynamic descriptions of development provided; finfold absorption and median and paired fin development are discussed.

Pigmentation — Information presented under length or length-range subheadings; emphasis placed on patterns of diagnostic importance.

Taxonomic Diagnosis — Fishes most likely to be confused with the species under discussion are listed and, if possible, taxonomic differences described for all life phases. Diagnostic discussions may be presented at the beginning of a family chapter along with keys.

Table 2

Diversity of fish populations in the Ohio River
and its tributaries.

River Basin	No. Native spp.	No. Introduced spp.	Total	No. Endemic spp.
Ohio River proper	102	9	111	0
Allegheny River	97	11	108	0
Monongahela River	93	12	105	0
Little Kanawha River	75	5	80	0
Kanawha River	125	10	135	6
Muskingum River	114	19	133	0
Guyandotte River	68	3	71	0
Big Sandy River	98	5	103	0
Scioto River	114	9	123	1
Little Miami River	95	4	99	0
Great Miami River	103	9	112	0
Licking River	98	5	103	0
Kentucky River	117	10	127	0
Salt River-Rolling Fork	81	2	83	0
Green River	146	5	151	5
Wabash River	151	2	153	0
Cumberland River	175	7	182	10
Tennessee River	220	11	231	32

Source: From C.H. Hocutt and E.O. Wiley, 1986.

Table 3

Listing of endangered and threatened fish species (as of October 16, 2007)
occurring in the Ohio River system.

Endangered	Threatened
Etheostoma (= *Catonotus*) *percnurum* duskytail darter	*Erimystax* (= *Hybopsis*) *cahni* slender chub
Etheostoma sp. bluemask darter (= jewel darter)	*Etheostoma boschungi* slackwater darter
Etheostoma chienense relict darter	*Cyprinella* (= *Notropis*) *caerulea* blue shiner
Etheostoma wapiti boulder darter (= Elk River darter)	*Cyprinella* (= *Hybopsis*) *monacha* spotfin chub (= turquoise shiner)
Notropis albizonatus palezone shiner	*Noturus flavipinnis* yellowfin madtom
Noturus baileyi smoky madtom	*Phoxinus cumberlandensis* blackside dace
Noturus stanauli pygmy madtom	*Percina tanasi* snail darter
Noturus trautmani Scioto madtom	
Speoplatyrhinus poulsoni Alabama cavefish	

Source: From the U.S. Fish and Wildlife Service at http://www.fws.gov/endangered.

ECOLOGY OF EARLY LIFE PHASES

Occurrence and Distribution

Spatial-temporal and other ecological information from the open and gray literature and original data are presented under egg, yolk-sac larval, post yolk-sac larval, larval, and juvenile subheadings.

Early Growth

Preadult growth information.

Feeding Habits

Focus is on preadult feeding habits.

Literature Cited

These include abbreviated citations to literature consulted for that species account. Complete citations appear in the Bibliography and Appendix at the end of each volume. Occasionally, we became aware of important literature after a species or family account had been completed. Such articles are listed in abbreviated citation form as "Other Important Literature" at the end of the appropriate species account and fully referenced in the master Bibliography or Appendix.

TERMINOLOGY

Key morphological attributes and examples of yolk-sac and post yolk-sac larval phases and anatomy are illustrated in Figure 2. Definitions and terms for the early development of fishes vary considerably. We have adopted the following developmental terminology based on Hubbs (1943); however, other terminology exists including Balon (1975, 1981) and Snyder (1976). We choose to use a simple approach that any fish biologist could quickly identify. Since the presence of yolk and fin rays is easily identified, we have only slightly modified Hubbs' (1943) terminology:

Yolk-sac larvae — Phase of development from the moment of hatching to complete absorption of the yolk.

Post yolk-sac larvae — Phase beginning with complete absorption of the yolk and ending when a minimum adult complement of rays is present in all fins and the median finfold is completely absorbed.

Larvae — Includes both yolk-sac and post yolk-sac phases of development.

Juvenile — Phase beginning when an adult complement of rays is present in all fins and the median finfold is completely absorbed, and ending with the attainment of sexual maturity.

GENERAL COMMENTS ABOUT THE TEXT

Superscript numbers in each species account refer to the abbreviated literature citations at the end of each account. In some instances, a numbered, abbreviated citation is preceded by a capital A, denoting the referenced work as gray literature, e.g., internal agency reports, incomplete Dingel-Johnson (D-J) or other project reports, and, generally, unrefereed publications that contain useful information but are not widely circulated or available. Complete citations for journal and other refereed literature are in the Bibliography at the end of each volume; complete citations for gray (A) literature are in the Appendix. In family, genera, or subgenera introductions, taxonomic accounts higher than species (i.e., genus and family), and in some tables encompassing information for more than one species, citations are given by author and date, rather than superscript. Each volume has its own Bibliography; no cumulative bibliography will be attempted.

Throughout the text, original data are indicated by an asterisk superscript. Sources of original data are described at the end of the abbreviated literature list for each species. Reference material used for the description of species development was obtained from a variety of sources, including individual researchers, universities, and agencies. The location of specimens utilized for documentation of morphometric and meristic data and other developmental information is noted. The Tennessee Valley Authority provided many developmental series of eggs and larvae. Most of this material, along with many other specimens from this study, are deposited at the Indiana Biological Survey, Aquatic Research Center, Division of Fishes, Bloomington, IN.

When available, illustrations of development are presented as part of each species description. They vary in quality and source. Some have been reprinted from the literature; others have been redrawn from previously published figures or plates; and many are original illustrations. Illustrators of original drawings are listed in the acknowledgments for each volume and have initialed their work. In instances where more than one source of illustration was available, we used only those that best illustrated important developmental features.

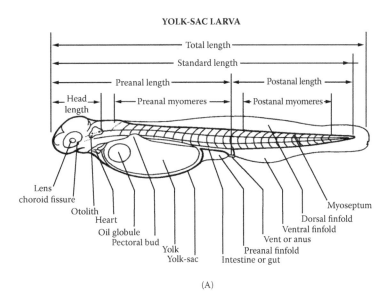

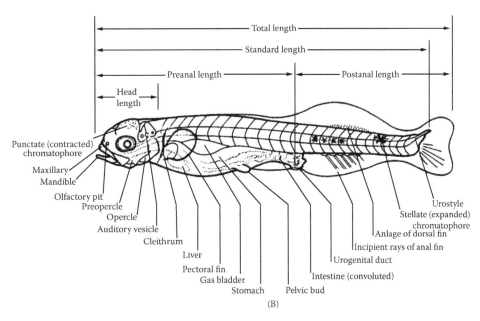

Figure 2 Diagrammatic representation of typical (A) yolk-sac larva and (B) post yolk-sac larva. (Redrawn from original drawing by Alice J. Mansueti in Mansueti and Hardy, 1967.)

Maps provided with each species account are most often used to indicate distribution of the species within the study area and to document reproduction by showing collection localities of early life history phases of that species. However, if the species is rare or has limited distribution, the maps may only show localities of recent adult collections. We have noted this situation in the figure caption.

References to body length are presented as found in the literature, i.e., standard length (SL), fork length (FL), or total length (TL). No conversions to TL were attempted. If body length was presented as length only with no further definition, we presented the information in a similar manner.

LITERATURE CITED

Auer, N.A. 1982.

Balon, E.K. 1975.

Balon, E.K. 1981.

Boschung, H.T., Jr. and R.L. Mayden. 2004.

Burr, B.M. and M.L. Warren, Jr. 1986.

Etnier, D.A. and W.C. Starnes. 1993.

Gerking, S.D. 1945.

Hocutt, C.H. and E.O. Wiley (editors). 1986.

Hubbs, C.L. 1943.

Jenkins, R.E. and N.M. Burkhead. 1994.

Jones, P.W. et al. 1978.

Kay, L.K. et al. 1994.

Lee, D.S. et al. 1980.

Mansueti, A.J. and J.D. Hardy, Jr. 1967.

Mettee, M.F. et al. 2001.

Nelson, J.S. et al. 2004.

Pearson, W.D. and L.A. Krumholz. 1984.

Simon, T.P. 1985.

Simon, T.P. and R. Wallus. 2003.

Simon, T.P. and R. Wallus. 2006.

Smith, C.L. 1985.

Smith, P.W. 1979.

Snyder, D.E. 1976.

Wallus, R. and T.P. Simon. 2006.

Wallus, R. et al. 1990.

Wang, J.C.S. and R.J. Kernehan. 1979.

Distinguishing Characteristics

and Pictorial Guide to the Families of Fishes in the Ohio River Drainage

Robert Wallus and Thomas P. Simon

In all, 27 families of fishes occur in the Ohio River drainage. The following pictorial guide is based on distinguishing characteristics that are diagnostic to separate each of the families. Diagnostic characters for yolk-sac and post yolk-sac stages of development are highlighted for each family.

YOLK-SAC LARVAE

POST YOLK-SAC LARVAE

ACIPENSERIDAE — sturgeons
- Hatching size 7–12 mm TL
- No adhesive organ
- Large, dark yolk sac
- Anus posterior to midbody
- More than 50 total myomeres
- Preanal length of early yolk-sac larvae about 65% TL
- Length from tip of snout to dorsal finfold origin about 25% TL for early yolk-sac larvae

- Extended snout with four ventral barbels
- Ventral mouth
- Heterocercal tail

POLYODONTIDAE — paddlefishes
- Hatching size 8–9.5 mm TL
- Large, dark yolk sac
- More than 50 total myomeres
- No adhesive organ
- Anus posterior to midbody
- Small eye
- Preanal length of early, yolk-sac larvae about 60% TL
- Length from tip of snout to dorsal finfold origin about 35% TL for early, yolk-sac larvae

- Rostrum develops with two ventral barbels
- Numerous sensory patches present on head and operculum
- Heterocercal tail

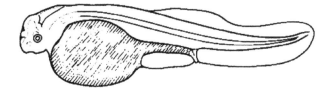

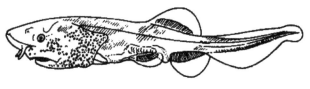

LEPISOSTEIDAE — gars
- Adhesive organ present
- Large, oval yolk sac
- More than 50 total myomeres

- Elongate body
- Extended snout
- Anal fin origin anterior to dorsal fin origin
- Heterocercal tail

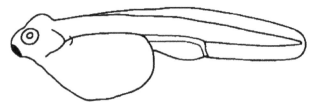

YOLK-SAC LARVAE	POST YOLK-SAC LARVAE

AMIIDAE — bowfins
- Hatching size 3–7 mm TL
- Adhesive organ present
- Total myomeres 60 or more

- Round, robust head
- Gular plate
- Long dorsal fin, origin above pectoral fins

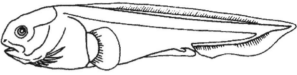

ANGUILLIDAE — freshwater eels
Larvae are absent from the Ohio River drainage, but elvers with adult characteristics occur.

CLUPEIDAE — herrings
- Slender, little pigment, transparent
- Oil may or may not be visible
- Large oil globule, if present, will be located posteriorly
- Posterior vent
- Fewer than ten postanal myomeres
- Dorsal finfold origin anterior, at mid-yolk sac early and just behind head later

- Slender, little pigment
- Posterior vent
- Anal fin posterior to dorsal fin

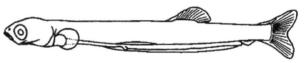

HIODONTIDAE — mooneyes
- Hatch at about 7 mm TL
- Large yolk sac
- Anterior oil globule
- Dorsal finfold origin near midbody

- Robust
- Large eye
- 17 or more postanal myomeres
- Dorsal fin insertion over anal fin

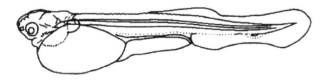

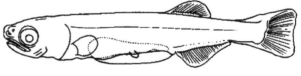

YOLK-SAC LARVAE	POST YOLK-SAC LARVAE

SALMONIDAE — trouts

- Large, greater than 11 mm TL at hatching
- Large yolk, initially pendulus
- Advanced fin development prior to complete yolk absorption
- Vent about two thirds back on body

- Robust
- Large, rounded head
- Adipose fin

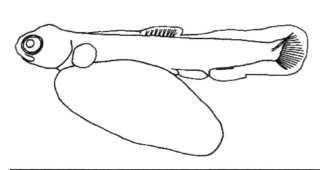 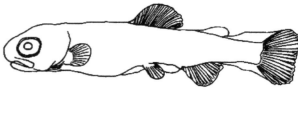

OSMERIDAE — smelts

- Long, slender, herring-like
- Small head
- Yolk positioned well posterior to pectoral fins
- Single, anterior oil globule
- Vent about three quarters back on body

- Elongate, slender, herring-like
- Adipose fin
- Anal fin posterior to dorsal fin

UMBRIDAE — mudminnows

- Yolk with many oil globules
- Vent slightly posterior to midbody
- Urostyle extends to posterior margin of caudal finfold

- Robust
- Darkly pigmented
- Urostyle extends beyond margin of developing caudal fin

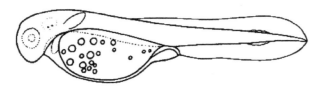 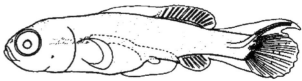

ESOCIDAE — pikes

- Darkly pigmented
- Vent about two thirds back on body

- Elongate
- Extended, depressed, duck-like snout
- Posterior dorsal fin

YOLK-SAC LARVAE	POST YOLK-SAC LARVAE

CYPRINIDAE — carps and minnows

- Yolk long, cylindrical, initially bulbous anteriorly
- Pigmentation varies from light to heavy
- Vent usually slightly beyond midbody

- Pigmentation often in rows; dorsolaterally, midlaterally, along ventral margin of myomeres, and midventrally
- Air bladder obvious, becoming two-chambered, usually pigmented dorsally
- Single dorsal fin

CATOSTOMIDAE — suckers

- Yolk long, cylindrical, initially more bulbous anteriorly
- Vent posterior, two thirds to three fourths back on body

- Mouth shape and position varies from inferior (later in development) to terminal and oblique
- Pigment variable but often in three rows, dorsally, ventrally, and midlaterally, dorsal pigment may also be in one to three rows
- Air bladder obvious
- Single dorsal fin

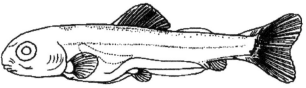

ICTALURIDAE — catfishes

- Large bulbous yolk
- Barbels evident at hatching
- Advanced fin development before complete yolk absorption

- No post yolk-sac larval phase

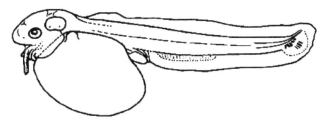

YOLK-SAC LARVAE

POST YOLK-SAC LARVAE

AMBLYOPSIDAE — cavefishes

- No information

- Caudal fin rounded
- Pelvic fins lacking in all but one species (*Amblyopsis spelaeu*)
- Eyes and pigment may be reduced or lacking in all genera except *Chologaster*

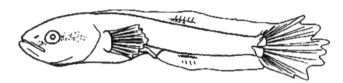

APHREDODERIDAE — pirate perches

- Small, about 3 mm TL at hatching, yolk absorbed between 4–5 mm TL
- Usually fewer than 30 total myomeres
- Anterior oil globule

- Head and body robust
- Usually fewer than 30 total myomeres
- Anus begins to migrate toward gular region at about 9 mm TL

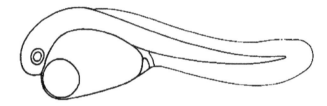

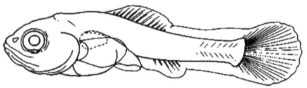

PERCOPSIDAE — trout-perches

- More than 30 total myomeres
- Hatching size 5.3–6 mm TL
- Large head
- Pointed snout with inferior mouth
- Vent slightly anterior

- Large head
- Adipose fin
- Long snout
- Air bladder obvious

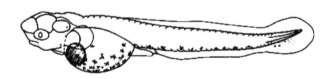

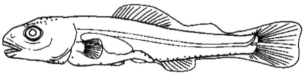

GADIDAE — codfishes

- More than 50 total myomeres
- Large head
- Short gut
- Anterior vent opens laterally on finfold

- Single barbel on chin
- Second dorsal fin and anal fin long
- Isocercal tail
- Pelvic fins positioned under pectoral fins

YOLK-SAC LARVAE	POST YOLK-SAC LARVAE

FUNDULIDAE — killifishes

- Stubby, robust
- Caudal fin with rays at hatching
- Vent anterior, near posterior margin of yolk

- Large head
- Superior mouth
- Rounded caudal fin
- Stocky caudal peduncle
- Ten or more dorsal rays

POECILIIDAE — livebearers

- Inside female

- Scales present at birth
- Rays in all fins at birth
- Superior mouth
- Dorsal fin short, seven to eight rays

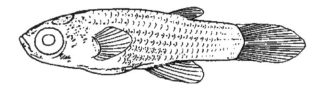

ATHERINOPSIDAE — silversides

- Elongate, slender
- Anterior vent (about one quarter back on body), immediately behind yolk sac
- Preanal myomeres, six to nine
- Preanal finfold absent or vestigal

- Elongate, slender
- Mouth small, terminal
- Two dorsal fins
- Anterior vent

GASTEROSTEIDAE — sticklebacks

- Short (5–6 mm TL), stubby
- Vent at midbody or slightly posterior
- Vitelline vessel over yolk
- Small oil globules present

- Sloping head, superior mouth
- Narrow caudal peduncle

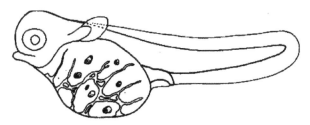 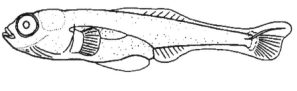

YOLK-SAC LARVAE	POST YOLK-SAC LARVAE

MORONIDAE — temperate basses

- Vent slightly posterior to midbody
- Single, large, anterior oil globule
- Low total myomere count, 25–26 or fewer

- "S" shaped gut
- Low myomere count
- Late larvae with well-developed mouth with teeth
- Spinous dorsal fin develops secondarily

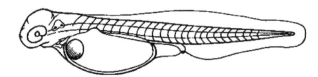
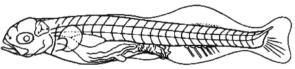

CENTRARCHIDAE — sunfishes

- Large, oval yolk sac at hatching
- Position of oil globule variable, but usually posterior
- Vent anterior to midbody

- Usually robust with large head
- Air bladder distinct
- Gut short, coils with growth
- Spinous and soft dorsal fins continuous

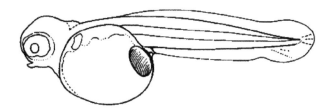
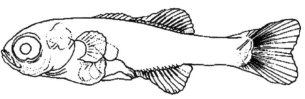

PERCIDAE — perches

- Vent near midbody
- Large anterior oil globule
- Pectoral fins usually well developed at hatching
- Total myomere counts higher than in moronids or centrarchids

- Large pectoral fins
- Spinous dorsal separate from soft dorsal fin

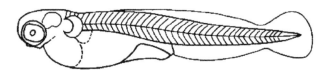
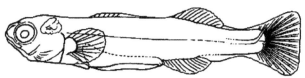

SCIAENIDAE — drums

- Small, 3–5 mm TL
- Large posterior oil globule
- About 25 total myomeres

- Heavy, truncate body
- Large, deep head
- Spinous and soft dorsal fins continuous
- Soft dorsal fin long, 24+ rays

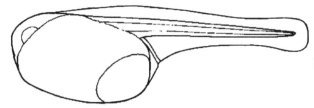
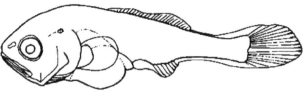

YOLK-SAC LARVAE

COTTIDAE — sculpins

- Robust with large head and large round yolk sac
- Fins well developed before yolk absorption is complete
- Anterior vent

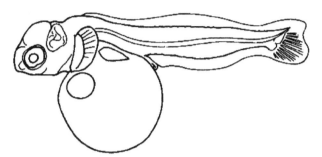

POST YOLK-SAC LARVAE

- Large pectoral fins
- Two dorsal fins
- Second dorsal fin and anal fin long
- Caudal fin spatulate

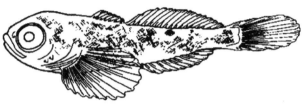

Reproductive Biology

and Early Life History Accounts for Ohio River Drainage Fishes of the Families Elassomatidae and Centrarchidae: An Introduction to the Volume

Robert Wallus and Thomas P. Simon

This is the sixth volume in this series that comprises accounts concerning the reproductive biology and early life history of fishes in the Ohio River drainage. The following 21 species are discussed representing the families Elassomatidae and Centrarchidae:

Elassomatidae
Genus *Elassoma*
E. *alabamae*, spring pygmy sunfish
E. *zonatum*, banded pygmy sunfish

Centrarchidae
Genus *Ambloplites*
A. *rupestris*, rock bass
Genus *Centrarchus*
C. *macropterus*, flier
Genus *Lepomis*
L. *auritus*, redbreast sunfish
L. *cyanellus*, green sunfish
L. *gibbosus*, pumpkinseed
L. *gulosus*, warmouth

L. humilis, orangespotted sunfish
L. macrochirus, bluegill
L. marginatus, dollar sunfish
L. megalotis, longear sunfish
L. microlophus, redear sunfish
L. miniatus, redspotted sunfish
L. symmetricus, bantam sunfish
Genus *Micropterus*
M. coosae, redeye bass
M. dolomieu, smallmouth bass
M. punctulatus, spotted bass
M. salmoides, largemouth bass
Genus *Pomoxis*
P. annularis, white crappie
P. nigromaculatus, black crappie

Systematic hypotheses for this group of fishes are presented and discussed in the following section entitled "Taxonomic Diagnosis of Young Pygmy Sunfishes (Elassomatidae), and Sunfishes (Centrarchidae) in the Ohio River Drainage." Common and scientific names generally follow AFS Special Publication Number 29 (Nelson et al., 2004).

Within this volume is an assimilation of previously published information and original work. Original morphometric and meristic data were taken from developmental series of banded pygmy sunfish, rock bass, redbreast sunfish, green sunfish, warmouth, bluegill, longear sunfish, redear sunfish, and the four basses. Original observations concerning early development, reproductive biology, and early life ecology are presented for rock bass, redbreast sunfish, warmouth, orangespotted sunfish, bluegill, longear sunfish, redspotted sunfish, redeye bass, smallmouth bass, and spotted bass. Noteworthy are new taxonomic keys for distinguishing the genera of young elassomatids and centrarchids, a key to the young stages of basses that occur in the Ohio River drainage, and original descriptions of developing larvae and juveniles of redeye bass and spotted bass.

This work also identifies areas that require additional research. Identification of the early life forms of all elassomatid and centrarchid species is not possible at this time. Currently no reproductive biology or developmental information is known for the dollar sunfish and additional descriptions are needed of early life forms of pumpkinseed, redbreast sunfish, redear sunfish, redspotted sunfish, bantam sunfish, redeye bass, and spotted bass. Hopefully, the material presented herein and the information that is obviously missing will stimulate others to begin filling in the gaps in our knowledge about the reproductive biology and early life ecology of this interesting group of fishes.

Species accounts are authored by a variety of individuals. When citing a species account, the individual author should be cited rather than the genus author. Authorship is also acknowledged for family and genus introductions. For example, the *Ambloplites* genus account should be cited as

> Simon, T.P. IV. 2008. Genus *Ambloplites* Rafinesque. Pages 61–62 in R. Wallus and T.P. Simon, editors. *Reproductive Biology and Early Life History of Fishes in the Ohio River Drainage. Volume 6: Elassomatidae and Centrarchidae.* CRC Press, Boca Raton, Florida, USA.

Species accounts should be cited as

> Simon, T.P. IV and J.V. Conner. 2008. Rock bass *Ambloplites rupestris* (Rafinesque). Pages 63–88 in R. Wallus and T.P. Simon, editors. *Reproductive Biology and Early Life History of Fishes in the Ohio River Drainage. Volume 6: Elassomatidae and Centrarchidae.* CRC Press, Boca Raton, Florida, USA.

Note: Because of production restrictions, the illustrations in some of the figures presented in this volume are not as large as those presented in previous volumes. To aid those who will use this document as a laboratory reference for identifying young fishes, a CD of enlarged figures showing the development of young elassomatids and centrarchids is also provided.

Taxonomic Diagnosis of Young Pygmy Sunfishes (Elassomatidae) and Sunfishes (Centrarchidae) in the Ohio River Drainage

Thomas P. Simon and Robert Wallus

The state of the art for identification of early life forms in the families Elassomatidae and Centrarchidae has not yet reached the level where a diagnostic key to all species can be developed. Incomplete and missing descriptions of the early life history and development of some species result in an incomplete record for the sunfishes occurring in the Ohio River drainage. In this volume, we have summarized the information available and made an effort to present it in a most usable format.

PRELIMINARY INTRODUCTION TO PHYLOGENY OF THE FAMILIES CENTRARCHIDAE AND ELASSOMATIDAE

The Elassomatidae were historically thought to be related to Cichlidae (Jordan 1877) or to the sunfishes, family Centrarchidae (Greenwood et al., 1966; Robins et al., 1991; Nelson, 1994). Recognition of a close relationship between *Elassoma* and Centrarchidae dates to Boulenger (1895). Jordan and Evermann (1896b) recognized Elassomatidae as a distinct family, but for many years it had been considered part of the Centrarchidae. Greenwood et al. (1966) included *Elassoma* within Centrarchidae but provided no morphological evidence for the monophyly of this group. Morphological differences in the acoustico-lateralis system caused Branson and Moore (1962) to remove *Elassoma* from Centrarchidae, but they still considered it closely related. Johnson (1984) and Johnson and Patterson (1993) used morphological characters to prove that elassomatids are not closely related to the Centrarchidae. Johnson (1984) discussed the possibility that elassomatids are related to Mugilidae. Johnson and Patterson (1993) proposed a new position for the Elassomatidae moving them from the Percomorpha to a newly established acanthomorph taxon. On the basis of chromosomal variability, Roberts (1964) also viewed Centrarchidae as an independent and unrelated lineage relative to Elassomatidae. An analysis of elassosomatid relationships based on mitochondrial DNA by Jones and Quattro (1999) supports a monophyletic Centrarchidae; however, the consensus is that the elassomatids merit recognition as a distinct family.

Most of the systematic studies of species of Centrarchidae used traditional morphological characters. Bailey (1938) recognized two subfamilies (Centrarchinae and Lepominae) and three unresolved tribes in each subfamily that were included in Centrarchinae: Centrarchini, Archoplitini, and Ambloplitini; and in Lepominae: Lepomini, Enneacanthini, and Micropterini. The same phylogeny was used by Smith and Bailey (1961) in an elevation of the evolution of predorsal bones. Branson and Moore (1962) studied a variety of characters derived from the acoustico-lateralis system and proposed relationships of genera and species. Branson and Moore (1962) did not support Bailey's classification but represent his subfamilies and some tribes as paraphyletic. Branson and Moore (1962) hypothesize that *Lepomis* and *Micropterus* are sister genera, the two also sister to *Chaenobryttus*. Moving to the most recent common ancestor, the base is sister to *Enneacanthus*, then *Ambloplites*, then *Centrarchus* plus *Pomoxis*, then *Acantharchus*, and finally *Archoplites* as the basal sister group.

Morphological and genetic studies were conducted by Roberts (1964), Hester (1970), Avise and Smith (1977), Avise et al. (1977), Mok (1981), and Near et al. (2005). Roberts (1964) studied the karyology of the centrarchids and indicated that the pattern of evolutionary relationships could not be deduced; however, some synapomorphies were observed. Hester (1970) evaluated patterns in hybridization frequency as an indicator of genealogical relations. Avise and Smith (1977) examined protein variation among species and using phenetic data analysis, resolved relationships among some species of *Lepomis*. The results of their study showed that *Micropterus* was within *Lepomis* and this group was sister to *Pomoxis*, *Enneacanthus*, *Ambloplites*, and *Centrarchus*, which form a monophyletic group sister to the *Lepomis* group. *Acantharchus* was a sister group to *Archoplites* plus all remaining members of the family. Near et al. (2005) found *Acantharchus* to be basal to all centrarchids with two nodes, one supporting a *Centrarchus*, *Enneacanthus*, *Pomoxis*, *Archoplites*, and *Ambloplites* clade, while a second node supported the *Micropterus* and *Lepomis* sister clade.

Mabee (1995) indicated that two patterns were observed in the ontogeny of melanistic pigment development in centrarchids. In the first, an initial group of lateral bars appears simultaneously, and bars are added anteriorly and posteriorly throughout development until the adult complement is reached. It should be noted that lateral bars, as defined by Mabee (1995), include both the vertical bars that are deeper than wide that are typical in the *Lepomis* species, but also mid-lateral blotches of pigment that are apparent on some of the black basses. Mabee (1995) suggested that the most parsimonious interpretation as the primitive condition for centrarchids is the presence of the pattern among *Micropterus*, *Lepomis*, *Enneacanthus gloriosus* and *E. obeus*, *Acantharchus*, *Pomoxis*, and possibly *Archoplites*. The other pattern of primary bar ontogeny involves a very specific temporal sequence of bar appearance in which a single bar appears first, followed by another bar until the adult complement is present. This pattern evolved three times within centrarchids among *E. chaetodon*, *Centrarchus*, and *Ambloplites*. This is the single most parsimonious optimization for this character. *Acantharchus*, *L. cyanellus*, and all *Micropterus* possess 1–14 bars. This may be considered the primitive centrarchid condition. Primary bar number is changed eight times within centrarchids and may be achieved in 13 equally parsimonious ways. Mabee (1995) indicated that the node that included *L. cyanellus* and *Enneacanthus* required an optimization and independent reduction in bar number from the primitive centrarchid condition (11–14) to 7–8 bars in *L. gulosus* and from 11–14 bars to

9–10 bars in all other *Lepomis* and *Enneacanthus*. In the lineage leading to *L. symmetricus*, *L. macrochirus*, and *Enneacanthus*, the character of 9–10 primary bars is reduced to 7–8, while in *L. symmetricus* 7–8 bars increases to 9–10. In the descendents of the second node, primary bar number is reduced from 11–14 to 4 in the ancestor of *Ambloplites*, *Pomoxis*, *Centrarchus*, and *Archoplites*. Bar number is increased to 6 in *Pomoxis* and to 5 in *Archoplites*. Light centers form in the primary bars of *M. coosae* and *M. dolomieu*, *L. gulosus*, female *L. humilis*, *L. microlophus*, *L. gibbosus*, *L. punctatus*, *L. symmetricus*, and *L. macrochirus*. In all other *Lepomis*, the primary bars remain solidly pigmented. Primary bars develop light centers in all descendents of node B with the exception of *Acantharchus*.

In many species it was observed that the development of light centers in primary bars occurred at the same time as the development of secondary bars. In *Lepomis punctatus*, *L. symmetricus*, *L. macrochirus*, and female *L. humilis*, primary bars do not develop light centers and secondary bars are absent. Primary bars do not develop light centers in *L. cyanellus*, *E. chaetodon*, or *M. punctulatus*, *M. salmoides*, and *M. treculi* but secondary bars are present. The development of secondary bars is an apomorphic trait among the most recent common ancestor of centrarchids. Secondary bars are lost in the lineage leading to *L. humilis*, *L. auritus*, *L. megalotis*, *L. marginatus*, *L. microlophus*, *L. gibbosus*, *L. punctatus*, *L. symmetricus*, *L. macrochirus*, and *Enneacanthus*. The reappearance of secondary bars is a shared derived condition of *L. microlophus* and *L. gibbosus* and an autapomorphy of *E. chaetodon* and of *L. humilis*. Secondary bars are independently lost in *Acantharchus* (Mabee, 1995).

A pigmented blotch is present posteriorly in the dorsal fin of *Lepomis macrochirus*, *L. cyanellus*, *L. symmetricus*, and *Centrarchus macropterus*. Mabee (1995) suggested that dorsal fin pigmentation evolved independently in the most recent common ancestor of *L. macrochirus* + *L. symmetricus*, in *L. cyanellus*, and in *C. macropteus*. In *C. macropterus* the dorsal fin spot develops as a dorsal extension of the third primary bar. Although a dorsal-fin spot is present in *L. macrochirus*, *L. cyanellus*, and *L. symmetricus*, it does not arise as an extension of a primary bar. In these taxa, the spot develops independently in the fin, dorsal to the fin base musculature, and separated from the body pigmentation by a non-pigmented interradial membrane.

ONTOGENY

The current state of the art for identification of early life forms of centrarchids may be aided by patterns observed in generic, subgeneric, and species groups. In *Lepomis*, we use discussions of subgeneric diagnosis to group species that are similar in developmental characters. Other information, such as restricted distributions, may aid in removing allopatric or sympatric species from consideration in taxonomically similar groups. As examples, members of the *Elassoma*, *Micropterus coosae*, *Centrarchus macropterus*, *L. symmetricus*, and *L. miniatus* are localized in their distribution. These groupings and considerations, in most instances, will be discussed in the taxonomic diagnosis sections of each species account.

The use of ontogenetic data sets to resolve relationships within the Elassomatidae and Centrarchidae provides solutions to several phylogenetic issues surrounding the problematic relationships of sunfishes. Three publications including Anjard (1974), Hogue et al. (1976), and Hardy (1978) provided descriptive and comparative information for the recognition of sunfish genera. These regional studies did not link consistent patterns across geographic areas, thus they identified important character states but did not describe consistencies or variation. Gut shape and length, air bladder position and size, myomere counts, size at ontogenetic benchmarks, and pigmentation patterns were important attributes useful for generic comparisons; however, Conner (1979) found sufficient variation in a larger amount of material to warrant concerns.

NOTES ON CHARACTERS USEFUL IN DISTINGUISHING ELASSOMATID AND CENTRARCHID GENERA

Size at Ontogenetic Benchmarks

Differences in various ontogenetic benchmarks may be diagnostic for genera of the elassomatids and centrarchids. Table 4 summarizes the size at hatching, size at yolk absorption, size at development of various fin ray elements, including the size that the full adult complement is formed, and the size that the gut first coils for all elassomatids and centrarchids that occur in the Ohio River drainage. A comparison between genera and species of the various ontogenetic events summarized may allow for genus or even species discrimination. For example, when considering these benchmarks, variation between *Elassoma* and the centrarchids is apparent. Only *Pomoxis* hatch at similar or smaller sizes; however, *Pomoxis* species have greater postanal myomere counts and are larger when fin rays first develop; *Elassoma* has a complete adult complement of fin rays before pomoxids form their first. Individuals of *Elassoma*, *Pomoxis*, and *Centrarchus* absorb yolk at lengths <4.0 mm TL. Both *Pomoxis* and *Centrarchus* have complete yolk absorption by

Table 4

Morphometric and meristic characters and developmental benchmarks of young sunfish, black basses, and crappies present in the Ohio River drainage.

| Species | Myomeres | | | Length as %TL | | | | Length (mm TL) Character Formed | | | | | | | |
	Preanal	Postanal	Total	ED	HL	PreAL	GBD	Hatching Length	Yolk-Sac Absorbed	First Ray Forms D1/D2	A	C	Adult Fin Rays Form	Air Bladder Position	Gut Coil
Elassoma alabamae	9–11	16–18	23–29	7–9	19–28	42–44	23–24	2.8–3.3	3.4–3.6	5–6	5–6	5–6	10–11	Anterior	—
Elassoma zonatum	11–15	14–17	27–31	8–9	19–25	41–47	20–23	3.0–3.8	3.9–4.0	5–6	5–6	3–4	8–9	Anterior	4.0
Ambloplites rupestris	10–14	16–19	30–32	10–11	24–30	45–49	24–29	4.7–5.6	7.4–8.6	7–9	7–9	7	17	Anterior	<8.6
Centrarchus macropterus	9–12	20–25	29–35	9–12	14–29	38–42	14–26	3.4–4.9	5.0	10–11	8–11	8	15	Posterior	4.9
Lepomis auritus	9–16	15–17	24–32	8–10	16–27	46–48	20–27	4.0–5.0	7.0–10	8	8	7	14–15	Anterior	7.0–7.7
Lepomis cyanellus	11–15	13–17	26–30	7–9	20–26	43–47	23–27	3.5–3.7	4.7–5.8	7–8	7–8	5–7	13–16	Anterior	5.4
Lepomis gibbosus	10–13	17–21	27–34	8–9	13–18	37–45	—	2.4–3.5	4.5–5.5	>5.4	>5.4	>5.4	14	Anterior	5.4
Lepomis gulosus	11–15	14–18	25–30	7–9	17–28	43–45	21–24	2.3–3.1	4.9–5.3	7–8	7–8	5	14–16	Anterior	<5.5
Lepomis humilis	11–15	14–17	27–31	5–9	18–28	43–47	14–23	4.0–5.3	7.6–7.9	8–9	8–9	7–8	9–10	Anterior	9.6
Lepomis macrochirus	11–15	15–18	27–32	7–9	17–26	42–47	11–24	2.2–3.7	5–6	4.5	4.5	4.5	12–14	Anterior	7.7
Lepomis marginatus	—	—	—	—	—	—	—	—	—	—	—	—	—	—	—
Lepomis megalotis	13–15	13–18	26–32	9–10	22–28	44–46	22–28	5.0–5.2	7.3–7.6	8	7–8	6–7	13–14	Anterior	7.2
Lepomis microlophus	12–16	14–17	28–31	7–10	19–27	45–47	19–26	4.8–5.1	5.1	8	8	6	11	Anterior	5.0
Lepomis miniatus	11–13	15–18	28–31	6–14	19–26	42–46	16–34	4.0	<7.0	8	8	8	10	Anterior	6.6
Lepomis symmetricus	10–12	16–18	27–29	8–10	19–30	32–48	17–37	4.5–5.5	6.8	7	7	5–6	12	Anterior	>6.6–<7.9
Micropterus coosae	15–18	15–18	31–35	9–12	24–30	50–54	20–25	<8.5	9.1–9.5	10	10	8–9	>13	Anterior	9.2
Micropterus dolomieu	12–16	17–19	31–34	6–12	15–31	46–53	22–38	4.5–5.5	9–10	9	9	8	16	Anterior	9.5
Micropterus punctulatus	14–17	15–20	32–35	7–12	24–32	47–53	21–27	<6	>10.2	7–8	7–8	7	18–19	Anterior	10.2
Micropterus salmoides	14–19	13–17	31–33	10–11	22–31	50–52	20–23	2.7–4.3	8–10	8–10	8–9	7–8	15–16	Anterior	8
Pomoxis annularis	10–14	17–23	28–33	8–16	—	—	—	1.2–2.6	4.0–5.0	9–11	9–11	7–8(11)	>20	Posterior	4.0
Pomoxis nigromaculatus	10–14	18–23	29–35	—	—	—	—	2.0–2.3	3.7–5.0	9–11	9–11	7–8	>20	Posterior	3.0

5 mm TL as do three sunfish (i.e., *Lepomis cyanellus*, *L. gibbosus*, and *L. gulosus*).

Preanal Lengths

Preanal lengths are variable among centrarchid genera even within a given developmental phase of a particular taxa (Table 4). Yolk-sac larvae of *Elassoma, Centrarchus,* and *Pomoxis* usually have preanal lengths less than 47% of TL, whereas the mode for genera *Lepomis* and *Micropterus* includes proportions that are greater than 47%. Overlap between centrarchid genera occurs in the ranges from 38 to 42% of TL, with the exception of *Micropterus* (Conner, 1979). As ontogenetic changes occur with growth, reliable identifications are more accurate at preanal lengths above 48 or below 47% of TL. Members of the Elassomatidae (preanal lengths 41–47% TL) are contained within the range exhibited by *Lepomis* and *Ambloplites*; however, differences in preanal lengths greater than 49% of TL are diagnostic for *Micropterus*.

Gut Morphology

Anjard (1974) and Hardy (1978) indicate that *Micropterus* is distinguishable from *Lepomis* and *Pomoxis* by the massive, thick coiled gut. Morphology of the *Elassoma* gut is similar to that of *Micropterus* but *Elassoma* are much smaller than bass when gut coiling begins (Table 4). Sunfish that are < 4 mm TL with a coiled gut can only be *Pomoxis* or *Elassoma*, while specimens < 5 mm TL would only include *Centrarchus, Elassoma,* and *Pomoxis*. *Lepomis* shows extreme variation in the length that coiling first occurs and also the gut shape among post yolk-sac larvae. Conner (1979) suggests that these characters might be of diagnostic value in distinguishing *Lepomis* subgenera, but may be problematic in distinguishing *Pomoxis* from *Lepomis* species that have short, anteriorly coiled guts.

Air Bladder Position

Perhaps the most reliable character for distinguishing post yolk-sac larvae of centrarchid genera will prove to be the position of the air bladder relative to the position of the vent and other parts of the gut. Genera with massively coiled guts, i.e., *Micropterus, Elassoma,* and *Ambloplites,* have an air bladder confined to the area above and anterior to the gut coils (Anjard, 1974; Hogue et al., 1976; Hardy, 1978). In yolk-sac and post yolk-sac larvae of the remaining sunfishes the air bladder encroaches to some extent into the space behind the section where coiling exists or is developing. Many reports indicate that the air bladder of *Pomoxis* and *Centrarchus* extends posteriorly to the anus or beyond, while in the genus *Lepomis* the air bladder terminates well in front of the anus. However, Conner (1979) suggests that in many post

yolk-sac larval *Pomoxis* the air bladder fails to attain or approach the anus.

Myomere Counts

The use of myomere counts has been considered an important discriminating character for separating the genera of centrarchids (Hardy, 1978; Conner, 1979); however, overlaps in myomere counts between the genera often confound the precision of identification. Hogue et al. (1976) relied heavily on myomere counts for distinguishing centrarchids in the Tennessee River drainage. Conner (1979) demonstrated that the couplets in a key that was developed by Hogue et al. (1976) would have misidentified *Micropterus* and *Pomoxis* from the Coastal Plain of LA. Conner (1979) documents differences in myomere counts between *Lepomis* from the Tennessee Valley (Hogue et al., 1976) and specimens he examined from LA, showing that 5–66% of the LA specimens would be misidentified using Tennessee River criteria. Postanal myomere counts can be used as a diagnostic character when >20. Only *Centrarchus* and *Pomoxis* possess postanal counts that high. Within the *Elassoma*, postanal myomere counts overlap, but can possibly distinguish *E. alabamae* (16–18) from *E. zonatum* (14–17).

Pigmentation

Pigmentation is an often used character for distinguishing centrarchid genera. Anjard (1974) and Hardy (1978) cite the presence of a supra-anal melanophore in yolk-sac and post yolk-sac larvae of *Lepomis*. Conner (1979) indicated that wild-caught larvae from southern LA did not exhibit this pigment character, possibly because of the turbid conditions of many southern LA rivers. The presence of the supra-anal melanophores is limited to select sunfish groups and Conner suggests that the character is not consistently observed.

As previously noted in systematic discussions (page 23), four sunfishes, *Centrarchus macropterus, Lepomis macrochirus, L. cyanellus,* and *L. symmetricus,* develop a pigmented blotch posteriorly in the soft dorsal fin (Plates 1 and 2). Also, discussed earlier is the development of primary and secondary bars of melanistic pigment laterally on the centrarchids. The numbers of primary bars and the presence or absence of secondary bars may aid in identifying young elassomatids or centrarchids (Table 5).

TAXONOMY

When using this volume to identify a young fish, the practitioner should first determine the genus to which that specimen belongs. This can be accomplished by evaluating the generic information included in the key at the end of this chapter, by referring to the information summarized in

Table 5

Meristic characters and pigmentation patterns of juvenile sunfish, black basses, and crappies in the Ohio River drainage.

Species	Fin Ray Counts						Lateral Line Scale Counts	P Bars	Lateral[a] S' Bars	Pigmentation Stripe	Breast	D2 Ocellus	Spot
	P1	P2	A	D1	D2	C							
Elassoma alabamae	14–19	—	I–III/5–8	II–IV	8–13	10–13	27–32	7–8	—	Absent	Absent	Absent	Absent
Elassoma zonatum	14–18	—	III–IV/4–7	III–V	8–13	10–14	31–39	7–8	—	Absent	Absent	Absent	Absent
Ambloplites rupestris	12–14	1/5	IV–VII/9–11	X–XII	10–11	—	37–51	4	Present	Absent	Absent	Absent	Absent
Centrarchus macropterus	11–14	1/6	VII–VIII/14–17	XI–XIII	12–14	20–25	36–42	4	Present	Absent	Absent	**Present**	**Present**
Lepomis auritus	13–15	1/5	III–IV/8–10	IX–XI	10–12	—	39–54	12	Absent	Absent	Present	Absent	Absent
Lepomis cyanellus	12–15	1/5	III/8–11	IX–XII	10–12	—	40–53	11–14	Present	Present	Present	Absent	**Present**
Lepomis gibbosus	11–14	1/5	II–IV/8–12	IX–XII	10–13	—	34–47	9–10	Present	Present	Present	Absent	Absent
Lepomis gulosus	12–14	1/5	III/8–10	IX–XI	9–11	—	35–48	7–8	Present	Present	Absent	Absent	Absent
Lepomis humilis	14–15	1/4–5	III/9	IX–XI	10–11	—	32–42	9–10	Absent	Absent	Present	Absent	Absent
Lepomis macrochirus	12–15	1/5	III/10–12	IX–XI	10–12	—	38–50	7–8	Absent	Absent	Absent	Absent	**Present**
Lepomis marginatus	11–13	1/5	III/9–10	IX–XI	10–12	—	34–44	9–10	Absent	Absent	Absent	Absent	Absent
Lepomis megalotis	13–15	1/5	III/8–12	IX–XII	10–12	—	33–46	9–12	Absent	Absent	Present	Absent	Absent
Lepomis microlophus	11–16	1/5	III/9–11	IX–XI	10–12	17	34–47	9–10	Present	Present	Absent	Absent	Absent
Lepomis miniatus	12–15	1/5	III/9–10	X–XI	10–12	—	33–42	9–10	Absent	Absent	Absent	Absent	Absent
Lepomis symmetricus	11–13	1/4–5	II–IV/9–11	IX–XI	9–12	17–18	30–38	9–10	Absent	Absent	Absent	Absent	**Present**
Micropterus coosae	14–17	—	III/9–11	IX–XI	11–14	—	64–72	11–12	Present	Present	Present	Absent	Absent
Micropterus dolomieu	13–18	1/5	II–III/9–12	IX–XI	12–15	—	68–81	11–14	Present	Present	Absent	Absent	Absent
Micropterus punctulatus	14–17	—	III/9–11	IX–XI	11–13	—	58–71	11–14	Present	Present	Present	Absent	Absent
Micropterus salmoides	13–15	1/5	III/10–12	IX–XI	12–14	17	55–73	10–12	Present	Present	Present	Absent	Absent
Pomoxis annularis	13–16	1/5	IV–VII/16–19	V–VII	13–16	—	34–45	6–7	Absent	Absent	Absent	Absent	Absent
Pomoxis nigromaculatus	13–15	1/5	V–VII/16–19	VI–IX	14–18	—	33–44	6–7	Absent	Absent	Absent	Absent	Absent

[a] P Bars = primary bars; S bars = secondary bars; stripe = midlateral stripe.

Tables 4 and 5, and by reviewing the information included at the beginning of each genus section. Diagnostic characters that may be used to distinguish between yolk-sac larval and post yolk-sac larval specimens are summarized in Table 4. Table 5 provides a summary of pigmentary characters and adult meristic data that may aid in the identification of juveniles. Six genera of centrarchids and elassomatids are recognized in this treatment including *Elassoma* Jordan, *Ambloplites* Rafinesque, *Centrarchus* Cuvier, *Lepomis* Rafinesque, *Micropterus* Lacepede, and *Pomoxis* Rafinesque.

Once the genus of a specimen is determined, the following possibilities are available to the user. If the genus is *Elassoma*, only two species occur in the Ohio River drainage (spring pygmy sunfish *Elassoma alabamae* and banded pygmy sunfish *E. zonatum*) and taxonomy is not an issue for either adult or larval forms, due to their restricted and allopatric distributions. Despite increased knowledge, we are still unable to identify, with certainty, early life forms of all members of the family Centrarchidae. The "Key to Genera" at the end of this section will allow species identifications for the young of the monotypic genera *Ambloplites* (rock bass *A. rupestris*) and *Centrarchus* (flier *C. macropterus*). Provisional keys to species are presented in genus introductions for the young of *Micropterus* and *Pomoxis*. The taxonomy for young *Micropterus* of the Ohio River drainage (redeye bass *M. coosae*, smallmouth bass *M. dolomieu*, spotted bass *M. punctulatus*, and largemouth bass *M. salmoides*) is provisional due to gaps in the developmental information for redeye bass and spotted bass. Identification of *Pomoxis* larvae is tentative because of reports of regional differences in character states. Also, adult and larval taxonomy in this genus is clouded by the tendency of its two species, white crappie *P. annularis* and black crappie *P. nigromaculatus*, to hybridize. No key is attempted for the young of *Lepomis* because of gaps in our knowledge of the early development of several species. However, descriptive information of the early life forms of *Lepomis* species is provided in subgeneric discussions (in the genus introduction), individual species accounts, developmental and morphometric and meristic information tabularized for all species (Tables 4 and 5), and color photographs of juveniles (Plates 1 and 2). Using these sources of information, the user may be able to construct a suite of characters that will aid in identifying larvae or juveniles of sympatric species present in the study area.

Provisional Key to Genera of Young Centrarchids and Elassomatids Present in the Ohio River Drainage

Generic Key to Yolk-Sac Larvae

1a. Gut massively coiled; air bladder confined to area above and anterior to gut coils.........................2
1b. Gut uncoiled or, if coiled, air bladder encroaches on space posterior to gut coils............................5

2a. Larvae very small – newly hatched at 2.8–3.8 mm TL; most yolk absorbed by 4.0 mm TL; profuse pigment develops over the head and body...*Elassoma*
2b. Yolk-sac larvae larger (>5.0 mm TL); sparsely pigmented or with melanophores concentrated on dorsum (mainly on head) and/or scattered profusely laterally on the body.......3

3a. Preanal lengths >50% TL; preanal myomeres >14 ....*Micropterus*
3b. Preanal lengths 45–49% TL; preanal myomeres 10–14 4

4a. Large stellate melanophores present, scattered over most of body by 6.5 mm TL; yolk absorption complete between 7.4 and 8.6 mm TL... *Ambloplites*
4b. Pigment initially visible as bands developing along junction of yolk sac and body (6–7 mm TL); at 7.3 mm TL, pigment is visible on dorsal surface of head and nape and scattered on dorsum of yolk; pigmentation becomes progressive more profuse over yolk and head and expands along dorsum and laterally onto sides of yolk, head, and body; by 7.7 mm TL, scattered pigment covers dorsal and lateral aspects of body; yolk and oil completely absorbed between 9 and 10 mm TL ....*Micropterus dolomieu*

5a. Postanal myomeres 14–18 .... *Lepomis*

5b. Postanal myomeres ≥19 .... *Pomoxis* or *Centrarchus*
(*Note*: see diagnostic discussion in *Centrarchus macropterus* species account)

Generic Key to Post Yolk-Sac Larvae

1a. Gut massively coiled; air bladder confined to area above and anterior to gut coils 2

1b. Gut uncoiled or, if coiled, air bladder, encroaches on space posterior to gut coils 6

2a. Post yolk-sac larvae small, TL ranges from about 4.0 mm to 8–11 mm; caudal fin rays visible by 5–6 mm TL; complete complement of adult fin rays formed by 8–11 mm; profusely pigmented with melanophores all over head and body .... *Elassoma*

2b. Larvae larger (TL range >7.4 mm to >13 mm); caudal fin rays first visible at lengths ≥7 mm TL; adult compliment of fin rays visible at lengths >13 mm TL (usually 16–19 mm TL); pigment concentrated dorsally on head and may or may not be widely scattered laterally on the body 3

3a. Preanal length >50% TL; lacking dense pigment or large stellate melanophores scattered over entire body .... *Micropterus*

3b. Preanal length <50% TL; pigmentation may be scattered over entire body or may consist of dark dorsal surfaces of occiput, head, and snout, a dark opercular band posterior to the eye, and a developing lateral stripe 4

4a. Pigmentation consists of dark dorsal surfaces of occiput, head, and snout, a dark opercular band posterior to the eye, and a developing mid-lateral stripe .... *Micropterus* (spotted bass)

4b. Pigmentation either densely distributed dorsally and laterally on body or with large stellate melanophores widely scattered over entire body 5

5a. Post yolk-sac larvae with widely scattered, large stellate (greatly expanded) melanophores over entire body; depth of caudal peduncle >9% TL; greatest depth usually >26% TL .... *Ambloplites*

5b. Post yolk-sac larvae with densely distributed pigmentation dorsally and laterally on body; caudal peduncle depth <9% TL; greatest depth <24% TL .... *Micropterus dolomieu*

6a. Air bladder terminates well in advance of anus or, if approaching anus, specimens <8.0 mm TL; postanal myomeres usually <19 7

6b. Air bladder extends to or beyond level of anus; postanal myomeres usually ≥19 8

7a. Postanal myomeres 13–18 .... *Lepomis*

7b. Postanal myomeres >18 .... *Lepomis gibbosus*

8a. Air bladder extends even with level of anus; for specimens <12 mm TL, eye diameter is subequal to or only slightly greater than snout length (less than 1.5 times snout length); specimens >12 mm TL have <8 dorsal spines .... *Pomoxis*

8b. Air bladder extends posteriorly beyond level of anus; for specimens <12 mm TL, eye diameter is conspicuously greater than snout length (greater than 1.7 times snout length); specimens >12 mm TL have ≥10 dorsal spines .... *Centrarchus*

Generic Key to Juveniles

1a. Caudal fin emarginated or truncate, comprised of 17 or more principal rays 2

1b. Caudal fin rounded, comprised of 16 or fewer principal rays .... *Elassoma*

2a. Anal fin spines ≥5 3

2b. Anal fin spines ≤4 5

3a. Soft dorsal rays 10–11; anal fin rays 9–11 ....*Ambloplites*

3b. Soft dorsal rays ≥12; anal fin rays ≥14...4

4a. Dorsal fin with 11–13 spines; a large ocellus present in posterior portion of soft dorsal fin ....*Centrarchus*

4b. Dorsal fin with 5–9 spines; soft dorsal fin without ocellus.......................................*Pomoxis*

5a. Anal spines 4...6

5b. Anal spines 2–3...8

6a. Dorsal spines 5–9 (usually <9); anal fin rays 16–19...*Pomoxis*

6b. Dorsal spines 9–12; anal fin rays 9–12...7

7a. Body pigmentation heavy and uniformly mottled; 4 dorso-lateral bars of primary melanistic pigment develop on body..*Ambloplites*

7b. Body pigmentation variable with many (7–14 depending on species) primary bars of pigment developing dorso-laterally on body ....*Lepomis*

8a. Dorsal fins nearly separate; body elongate and robust in specimens less than 150 mm; scales in lateral series ≥53...*Micropterus*

8b. Dorsal fins continuous; body deep and somewhat laterally compressed (occasionally less so in *L. cyanellus*); scales in lateral series usually 52 or fewer...............................*Lepomis*

Sources: Anjard (1974), Hogue et al. (1976), Conner (1979), Buynak and Mohr (1978), Mabee (1995), Etnier and Starnes (2003)

LITERATURE CITED

Anjard, C.A. 1974.
Avise, J.C. and M.H. Smith. 1977.
Avise, J.C. et al. 1977.
Bailey, R.M. 1938.
Boulenger, G.A. 1895.
Boschung, H.T., Jr. and R.L. Mayden. 2004.
Branson, B.A. and G.A. Moore. 1962.
Buynak, G.L. and H.W. Mohr, Jr. 1978.
Conner, J.V. 1979.
Etnier, D.A. and W.C. Starnes. 2003.
Greenwood, P.H. et al. 1966.
Hardy, J.D., Jr. 1978.
Hester, F.E. 1970.

Hogue, J.J., Jr. et al. 1976. (in Appendix)
Johnson, G.D. 1984.
Johnson, G.D. and C. Patterson. 1993.
Jones, W.J. and J.M. Quattro. 1999.
Jordan, D.S. 1877.
Jordan, D.S. and B.W. Evermann. 1896b.
Mabee, P.M. 1995.
Mok, H.K. 1981.
Near, T.J. et al. 2005.
Nelson, J.S. 1994.
Roberts, F.L. 1964.
Robins, C.R. et al. 1991.
Smith, C.L. and R.M. Bailey. 1961.

FAMILY ELASSOMATIDAE
Pygmy Sunfishes

Maurice F. Mettee

There are six described (Nelson et al., 2004) and either one or two undescribed species of pygmy sunfishes restricted to lowland streams, swamps, and marshes along the southern Atlantic and Gulf coasts. These are some of the smallest freshwater fishes (<40 mm TL) in North America. Boulenger (1895) suggested they were dwarf sunfishes, but Eaton (1953) recommended the term pygmy, since this condition is normal and genetically controlled, while the dwarf condition may or may not be genetically controlled. Pygmy sunfishes establish and defend territories and exhibit elaborate spawning behaviors and the males become brightly colored during their February to March spawning season. These characteristics and their ability to survive and readily reproduce in captivity have made these discrete and diminutive fishes a favorite among aquarists and fish breeders.

Pygmy sunfishes have an elongated, compressed body, a small oblique mouth with numerous conical teeth, and head length of approximately one-third standard length. Additional adult characteristics for members of this family include: 23–36 rows of cycloid scales, but no lateral line; dorsal fin with 3–5 spines and 8–13 soft rays; anal fin with 3 spines and 4–7 soft rays; pectoral fins with 13–17 soft rays; caudal rounded with 12 branched rays; gill membranes broadly connected across the isthmus (Mayden, 1993; Gilbert, 2004).

TAXONOMY AND SYSTEMATICS OF FAMILY ELASSOMATIDAE

Controversy has surrounded the taxonomic status of this group since Jordan (1877) described *Elassoma zonatum* and suggested its affinity to the Cichlidae. Hay (1883) and Jordan and Gilbert (1883) included *Elassoma* in the Elassomatidae. Boulenger (1895) assigned it and *Kuhlia* to the Centrarchidae based on similarities in vertebral structure. Jordan and Evermann (1896a) placed *Kuhlia* in the family Kuhliidae and modified the family name to Elassomidae, a format followed in Jordan et al. (1930). The use of Elassemidae by Carr (1936) possibly resulted from a printing error.

Genetic (Eaton, 1953, 1956; Roberts, 1964), morphological (Branson and Moore, 1962; Moore, 1962; Moore and Sisk, 1963); and behavorial (Mettee, 1974) studies supported *Elassoma* retention in the Elassomatidae. Hubbs and Allen (1943) used the name Elassomidae. Breder and Rosen (1966), Greenwood et al. (1966), Smith-Vaniz (1968), Eddy (1969), and Pflieger (1975b) retained *Elassoma* in Centrarchidae. Johnson (1984) and Johnson and Patterson (1993) removed *Elassoma* from the Percomorpha and placed it in the Smegmamorpha which included the silversides, sticklebacks, seahorses, and mullets. Jones and Quattro (1999) suggested Elassomatidae was a sister group to the Centrarchidae. Chang (1988) proposed that Elassomatidae and Centrarchidae formed a monophyletic group. Roe et al. (2002) recognized *Elassoma* as a sister group to the Moronidae, but they indicated it could also belong to a larger clade that included the

Centrarchidae and other perciform taxa. *Elassoma* was placed in the Centrarchidae by Robins et al. (1991) and in the Elassomatidae by Nelson et al. (2004). Elassomatidae has been used in recent publications by Conner (1979), Walsh and Burr (1984), Etnier and Starnes (1993), Mettee et al. (1996), Ross (2001), and Boschung and Mayden (2004).

Taxonomy of Ohio River drainage *Elassoma* representatives (*E. alabamae* and *E. zonatum*) is normally not an issue for adults or larval forms due to their allopatric distributions. Excellent keys to adult identities are provided by Boschung and Mayden (2004) and Mettee et al. (1996). Adult meristic characters for these two species are presented in Table 6.

Table 6

Meristic data for *Elassoma alabamae* and *E. zonatum* collected in Alabama (modified from Mayden, 1993).

	Elassoma alabamae (n = 70)			*Elassoma zonatum* (n = 76)		
	Range	Mean	SD	Range	Mean	SD
Lateral scale rows	27–32	28.9	1.2	31–39	34.6	1.73
Transverse scale rows	10–13	11.5	0.73	12–15	13.6	0.72
Caudal peduncle scale rows	15–20	17.4	1.31	20–25	22.9	1.24
Dorsal fin spines	2–4	2.9	0.41	3–5	4.4	0.54
Dorsal fin rays	8–13	10.6	0.84	8–13	10.1	0.88
Anal fin spines	1–3	2.9	0.37	3–4	3.0	0.16
Anal fin rays	5–8	6.5	0.65	4–7	5.4	0.66
Pectoral fin rays	14–19	16.5	1.02	14–18	15.7	0.87
Caudal fin rays	10–13	12.0	0.62	10–14	12.3	0.84

LITERATURE CITED

Boschung, H.T., Jr. and R.L. Mayden. 2004.
Boulenger, G.A. 1895.
Branson, B.A. and G.A. Moore. 1962.
Breder, C.M. and D.E. Rosen. 1966.
Carr, A.F. 1936.
Chang, C.M. 1988.
Conner, J.V. 1979.
Eaton, T.H. 1953.
Eaton, T.H. 1956.
Eddy, S. 1969.
Etnier, D.A. and W.C. Starnes. 1993.
Gilbert, C.R. 2004.
Greenwood, P.H. et al. 1966.
Hay, O.P. 1883.
Hubbs, C.L. and E.R. Allen. 1943.
Johnson, G.D. 1984.
Johnson, G.D. and C. Patterson. 1993.
Jones, W.J. and J.M. Quattro. 1999.

Jordan, D.S. 1877.
Jordan, D.S. and B.W. Evermann. 1896a.
Jordan, D.S. and C.H. Gilbert. 1883.
Jordan, D.S. et al. 1930.
Mayden, R.L. 1993.
Mettee, M.F. 1974.
Mettee, M.F. et al. 1996.
Moore, G.A. 1962
Moore, G.A. and M.E. Sisk. 1963.
Nelson, J.S. et al. 2004.
Pflieger, W.L. 1975b.
Roberts, F.L. 1964.
Robins, C.R. et al. 1991.
Roe, K.J. et al. 2002.
Ross, S.T. 2001.
Smith-Vaniz, W.F. 1968.
Walsh, S.J. and B.M. Burr. 1984.

GENUS
Elassoma Jordan

Maurice F. Mettee

Pygmy sunfishes are secretive little fishes that occur in lowland environments from the Brazos River in TX to northern FL along the Gulf Coast and from southern FL to the Cape Fear drainage in NC along the Atlantic Coast. Beautiful spawning colors and an elaborate spawning behavior have made the elassomids favorites among aquarists for many years (Clark, 1937; Fowler, 1936; Thompson, 1941; Bertoldt, 1949; Shortt, 1956) and resulted in their export to aquarium markets in Europe and Asia (Mayer, 1934; Nachstedt and Tusche, 1954).

Two species, the spring pygmy sunfish *Elassoma alabamae* and the banded pygmy sunfish *E. zonatum*, occur in the Ohio River drainage. Four additional species occur in Atlantic and Gulf coastal drainages. The Everglades pygmy sunfish *E. evergladei* (Jordan, 1884) is distributed from SC into FL and westward to the Mobile Basin in southern AL. The Okefenokee pygmy sunfish was originally described as *E. evergladei orlandicum* (Lönnberg, 1894) and subsequently as *E. okefenokee* (Böhlke, 1956). The former name was relegated to *nomen oblitum* due to lack of use (Gilbert, 2004). *Elassoma okefenokee* includes one described and one undescribed form. *Elassoma okefenokee* is distributed from the lower Altamaha River in GA southward into lakes near Orlando, FL. The undescribed form extends from the Choctawhatchee River eastward to the Waccasassa River. Both occur in the Suwannee River, but their distributions are allopatric (Gilbert, 2004). *Elassoma boehlkei* is restricted to the Waccamaw and Santee drainages in NC and SC and *E. okatie* occurs in the lower Edisto, New, and Savannah Rivers in SC (Rohde and Arndt, 1987). A *zonatum*-type specimen collected in a rotenone sample in TN has not been described (Wayne Starnes, 1998 personal communication).

Preferred habitats of *Elassoma* species include floating and submerged aquatic vegetation, such as *Ceratophyllum* and *Myriophyllum*, in freshwater springs, swamps, oxbow lakes, and low gradient streams (Mettee et al., 1996). With the exception of *E. alabamae*, pygmy sunfishes usually live solitary lives and only congregate to spawn, usually from February through April. After spawning is completed, individuals separate and return to a solitary existence. Direct observation of spawning behavior and daily activities is difficult. Fortunately, most *Elassoma* species readily spawn in aquaria under favorable photo period and water temperature conditions and, for this reason, considerable information has been compiled concerning the life history of these small, secretive creatures. Most of the information on habitat and spawning activities used in this report is based on publications by Barney and Anson (1920), Echelle and Echelle (2002), Mettee (1974), Mettee and Scharpf (1998), Poyser (1919), and Walsh and Burr (1984).

LITERATURE CITED

Barney, R.L. and B.J. Anson. 1920.
Bertoldt, W. 1949.
Böhlke, J.E. 1956.
Clark, E.W. 1937.

Echelle, A.F. and A.A. Echelle. 2002.
Fowler, H.W. 1936.
Gilbert, C.R. 2004.
Jordan, D.S. 1877.

Jordan, D.S. 1884.
Lönnberg, E. 1894.
Mayer, F. 1934.
Mettee, M.F. 1974.
Mettee, M.F. and C. Scharpf. 1998.
Mettee, M.F. et al. 1996.
Nachstedt, J. and H. Tusche. 1954.
Poyser, W.A. 1919.
Rohde, F.C. and R.G. Arndt. 1987.
Shortt, L.R. 1956.
Thompson, H.E. 1941.
Walsh, S.J. and B.M. Burr. 1984.

SPRING PYGMY SUNFISH

Elassoma alabamae Mayden

Maurice F. Mettee

Elassoma, Greek: "small-body"; *alabamae*, an Alabama endemic species.

RANGE

Spring pygmy sunfish is an AL endemic species represented by small sustaining populations in two small Tennessee River tributaries in northern AL.[1-5] The species was discovered in Cave Spring (Lauderdale County) in 1937, but that population was extirpated by completion of Pickwick Dam and the filling of Pickwick Reservoir. Another population, subsequently discovered in Pryor Branch (Limestone County), was eliminated in 1940 by runoff from heavy agriculture in the surrounding watershed, aquatic plant removal in contributing springs, and game fish stocking.[2,6,7] When intensive sampling failed to find any additional populations, scientists speculated the species had been lost to extinction. However in 1973, D.A. Etnier and students from the University of Tennessee discovered an unknown population in Moss Spring, a headwater tributary to Beaverdam Creek (Limestone County, AL).[2] Subsequent sampling found isolated populations in several other spring-fed tributaries to the system. Specimens collected in Moss Spring were successfully re-introduced into Pryor Branch in 1984 and again in 1987.[8]

An endangered species proposal for spring pygmy sunfish was withdrawn after the Pryor Branch population was re-established.[8,9] However, continued survival of the species is threatened by herbicide/pesticide contamination and increased sedimentation from heavy agriculture and encroaching urban development. Conservation listings include endangered,[10-14] Priority 1 (Highest Conservation Concern),[9,15] and a recommendation for state and federal protection.[5] Efforts to find other spring systems with similar water quality and habitat conditions have been unsuccessful. Monitoring of extant populations will be necessary to ensure future survival.

HABITAT AND MOVEMENT

This species usually remains in springs and spring runs throughout its life and individuals usually do not segregate or leave the immediate area after spawning. Most individuals spend their entire lives in dense, sometimes deep, suspended vegetation along the margins of spring runs. Margin habitats are slightly warmer than cooler runs.[2]

Although normally an inhabitant of heavily vegetated margins of springs and spring runs,[2] limited numbers of spring pygmy sunfish have been collected from dense aquatic vegetation along the margins of the lower reaches of a larger creek (Beaverdam Creek; Peggy Shute, TVA, personal communication, 1994).

Comparative water-quality measurements from Moss Spring and Pryor Spring Branch, at the time spring pygmy sunfish were re-introduced from the former into the latter, were as follows: water temperature 16°C vs. 14.5°C; specific conductance (μmhos/cm) 55 at 16°C vs. 142 at 14.5°C; pH units 5.7 vs. 6.3; and alkalinity (as HCO_3 mg/L) 20 vs. 66.[8]

No capture/recapture data are available, but research indicates the species is possibly mobile, and could utilize different spring and swampy microhabitats at different times of the year.[6,16]

DISTRIBUTION AND OCCURRENCE IN THE OHIO RIVER SYSTEM

Distribution in the Ohio River system is limited to isolated populations in several spring-fed tributaries to Beaverdam Creek, Limestone County, AL and a successfully re-introduced population in Pryor Branch, Limestone County, AL.[2,8]

SPAWNING

Location
Spawning usually occurs in dense submerged aquatic vegetation (*Ceratophyllum*, *Myriophyllum*) along the stream margin.[2,12,16]

Opinions differ concerning the precise spawning location of *Elassoma* species. Some researchers suggest these fishes construct a nest on the bottom like the centrarchids,[17,18,20,21] while others observed spawning in *Ceratophyllum* and *Myriophyllum* suspended above the bottom. [2,7,22,23] When spawning occurred in coarse-leave vegetation such as *Elodea*, most of the eggs fell to the bottom and on these occasions, the male cleared plant parts and other debris from around them, a behavior that could have led some workers to incorrectly suggest that pygmy sunfishes construct a "centrarchid-like" nest. Eggs falling to the bottom in the marshy, silt-laden environments inhabited by *Elassoma* species would probably not survive due to predation or because of smothering by the thick layer of bottom silt found in these areas.

Season

Spring pygmy sunfish generally spawn in February and early March.[2,4,7] The species is believed to be an annual species since most breeding adults die days to weeks after spawning.[2,12,16]

Temperature

Males establish territories in submerged aquatic vegetation along shallow stream margins and females visibly gravid appear when water temperatures reach 18–20°C. Active spawning begins at 21–22°C and continues for several days, depending on weather and flow conditions.[2,4,7]

Colorful males and slightly gravid females were collected for spawning observations in aquaria from Beaverdam Creek; water temperatures ranged from 16–17°C. Individuals were transported to the laboratory and allowed to acclimate in insulated ice chests for 10–12 h. Spawning occurred within hours after two or three males and several females were placed in aquaria maintained at 21°C.[2,4,7]

Fecundity

Six spawns observed in aquaria produced 50–65 eggs per spawn.[2]

Sexual Maturity

Most individuals reach sexual maturity several months after hatching and spawn the following spring. Most females die soon after spawning. Some males live from days to weeks longer.[1,2,5,12] Very few individuals live to age 2, even under optimum aquarium conditions. Considered young adults at about 16 mm SL.[2]

Spawning Act

Prior to and during the spawning season, males maintain territories that include suspended aquatic vegetation.[2] Intruding males and non-gravid females entering the territory were challenged with a Sidling Threat Display.[2,19] All fins on the defending male were fully extended, the pectoral and caudal fins vibrated rapidly, and the body became visibly darker. The defending male turned broadside to present the largest possible image and slowly moved forward to frighten the intruder away. If the intruder failed to flee, the defending male would arch his body, dart forward, and attempt to bite or throw his body against his opponent. This behavior usually produced a quick retreat.[2] When a gravid female entered the territory and seemed receptive, the male began a Wiggle Waggle Display,[2,19] which included continuous flexing of all his fins, especially the dorsal and anal fins, while swimming up and down in an undulating pattern toward the aquatic vegetation (Figure 3). This behavior was repeated several times

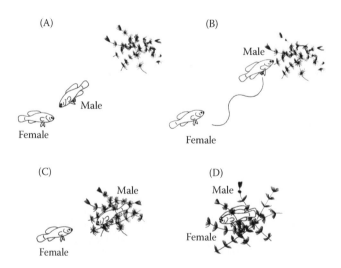

Figure 3 Reproductive behavior of elassomid fishes: (A) male approaching female; (B) wiggle-waggle dance of male; (C) female approaching spawning site; (D) the spawning act. (A–D reprinted from Figure 2, reference 2, with author's permission.)

until the female entered the vegetation and selected the spawning site. If she did not select a spawning site quickly, the male usually chased her out of his territory.[2]

Leading up to spawning, the male becomes more intensely colored and his body and fins begin to quiver as the female approaches the aquatic vegetation. When the female stops at a potential spawning site, the male begins to nudge her abdomen and genital area on one side and then the other. His body and fin coloration increases in intensity. He aligns himself vertically beside the female, they both vibrate slightly, and eggs and milt are extruded into the fine-leaved vegetation. Fertilization is immediate. Most of the semi-adhesive eggs attach to the fine-leaved vegetation in small clusters. One to several may fall to the bottom, depending on clutch size and vegetation density.[2] After spawning, the male chases the female away to avoid egg cannibalization. Males may spawn several times with different receptive females; females usually spawn once or twice.[2,4,7]

EGGS

Description
Unfertilized eggs removed from gravid females were spherical to indented with visible attachment disks, 2.0–2.1 mm in diameter, and yellowish in color. Watert-hardened eggs were 2.3–2.3 mm in diameter and had an irregularly shaped yolk, a large oil globule (diameter 0.5 mm), and varying numbers of small oil droplets.[2]

Fertilized eggs 30 min old were circular and 2.8–3.3 mm in diameter except for the area of a conspicuous attachment disc. No perivitelline space visible.[2]

Incubation
74–85 h at an average temperature of 23.5°C.[2]

Development
Composite illustrations of embryonic development for elassomid fishes are presented in Figures 4–7.

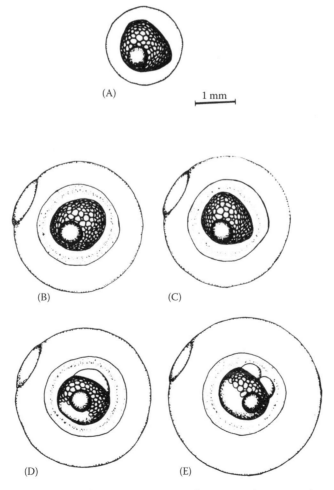

(A)

1 mm

(B)   (C)

(D)   (E)

Figure 4 Development of young pygmy sunfishes. A–E. Composite illustrations depicting embryonic development of elassomid fishes: (A) unfertilized egg; (B and C) fertilized eggs; (D) 1-celled embryo; (E) 2-celled embryo. (Reprinted from Figure 3, reference 2, with author's permission.)

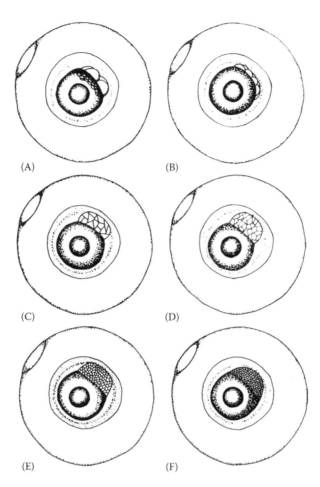

(A)   (B)

(C)   (D)

(E)   (F)

Figure 5 Development of young pygmy sunfishes. A–F. Composite illustrations depicting embryonic development of elassomid fishes: (A) 4-celled embryo; (B) 8-celled embryo; (C) 16-celled embryo; (D) 32 to 64-celled embryo; (E) early high blastula; (F) late high blastula. (Reprinted from Figure 3, reference 2, with author's permission.)

Illustrations are based on observations of eggs from multiple spawnings of four species of *Elassoma* including spring pygmy sunfish *E. alabamae*.[2]

Embryonic egg development for spring pygmy sunfish is described in detail below. Development times and descriptions come from fertilized eggs harvested from six spawns in aquaria.[2]

- 30 min — egg is circular and 2.8–3.3 mm in diameter except for area of conspicuous attachment disk; cytoplasm forms prominent blastodisk on side of yolk; yolk turns translucent, apparently due to water hardening; zona pellucida visible inside faint chorion; no evident perivitelline space.
- 50 min — chorion is slowly increasing in size; 2 equal-size blastomeres visible; zona pellucida 1.7–1.8 mm in diameter.
- 90 min — second division has occurred at right angles to first, forming 4 blastomeres.
- 130 min — 8–16 blastomeres present, some beginning to divide.

- 180 min— 32–64 smaller blastomeres present; area of cell division increasing in size.
- 4½ h — 128+ cells are present; yolk decreasing in size; oil globule still prominent.
- 8 h — early blastula; large number of very small cells visible.
- 9¼ h — late blastula; blastoderm begins to flatten and spread over yolk.
- 11–12 h — thick ridge developing in blastoderm covers >75% of yolk.
- 14–15 h — neurula; embryonic shield present as longitudinal intrusion into yolk.
- 22–24 h — embryo visible with optic placodes and 10–12 somites developing in posterior end of body; oil globule is on opposite side of yolk than developing embryo.
- 27–30 h — TL of embryo is approximately 1.6 mm; 14–16 somites are visible, some developing into myomeres toward the posterior end of the body; brain development evident in slight bulge on top of head;

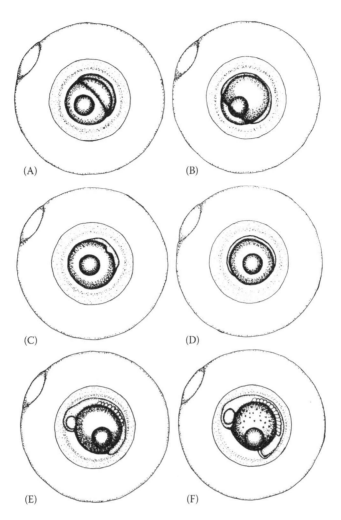

Figure 6 Development of young pygmy sunfishes. A–F. Composite illustrations depicting embryonic development of elassomid fishes: (A) early gastrula; (B) late gastrula; (C) neurula, end view; (D) neurula, lateral view; (E) early larval stage; (F) 12–14 somite stage. (Reprinted from Figure 3, reference 2, with author's permission.)

pericardial area developing as yolk disappears under head.

34–35 h — TL of embryo is approximately 2.0 mm; 20–22 myomeres; body becoming compressed; undeveloped caudal fin beginning to separate from yolk; urostyle visible in posterior end of body and finfold; primitive heart developing; head bulge slightly enlarged; scattered melanophores beginning to develop on sides and ventral surface.

42–45 h — TL of embryo is approximately 2.5 mm; notochord extends through 22–25 somites and myomeres; heart pumping at 88–90 beats per minute; colorless blood cells are visible moving in major vessels; increased melanophores laterally and ventrally; pupil and chorioid fissure evident in colorless eye; caudal finfold and posterior end of body separating from ventral surface of body; caudal finfold becoming turned to one side of body by infrequent wiggling inside the egg.

52–54 h — TL of embryo is approximately 2.7–2.8 mm; 29–31 myomeres evident; melanophore development spreading over body and head; yolk smaller and oblong shaped; oil globule oriented to posterior end of yolk; heart rate 140–144 beats per minute; circulatory system evident except for tiny vessels in head; large brain hemispheres and entire notochord visible; posterior end of body mostly detached from body near oil droplet; finfold extends from dorsal area above anterior end of yolk, around tail and along ventral surface to posterior margin of yolk; posterior end of body continuing to become oriented to one

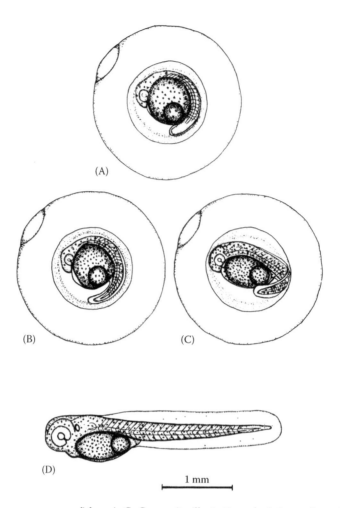

Figure 7 Development of young pygmy sunfishes. A–C. Composite illustrations depicting embryonic development of elassomid fishes: (A) 16–18 somite stage; (B) 20–24 somite stage; (C) prehatch larva; (D) newly hatched larva. (A–D reprinted from Figure 3, reference 2, with author's permission.)

side; tiny pectoral buds evident on either side of body when larvae viewed dorsally; larval movement increasing inside the egg.

68–70 h — TL of embryo is approximately 3.2–3.4 mm; larva fully formed, with numerous melanophores; smaller elongated yolk with smaller posterior oil globule; heart fully functional; blood has slight reddish tint; pupil and eye remain colorless but appear functional; brain area more enlarged with developing spinal column; increased myomere number difficult to count due to constant larval movement; finfold fully developed around end of body; pectoral fin buds slightly larger.

74–85 h (hatching) — constant movement inside the egg; tail used as lever to rupture egg wall and free posterior end of body; after a short rest, the larva shakes itself free; due to enlarged yolk and no functional fins, larva spends most of first 2 or 3 days lying on its side; short bursts of movement observed but due to poor eyesight, individuals frequently collide with side of aquarium.

YOLK-SAC LARVAE

See Figure 8.

Size Range
TL at hatching is 2.8–3.3 mm (mean = 3.1 mm, n = 15). Yolk is absorbed in 2–3 days at 3.4–3.6 mm.[2]

Myomeres [2]
Preanal 5–7; postanal 16–17; total 21–24.[2]

Morphology
2.8–3.3 mm TL. Snout rounded; oil globule at posterior end of yolk; yolk shrinking in length and diameter; urostyle straight, extending into clear posterior finfold.[2]

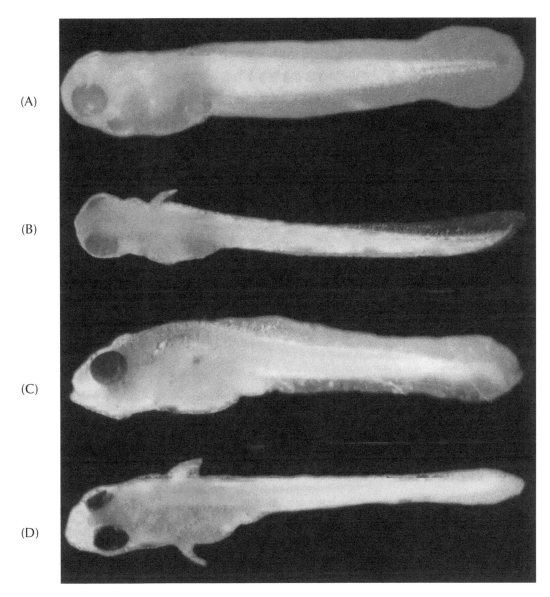

Figure 8 Development of young spring pygmy sunfish. A–B. Yolk-sac larva, 3.2 mm TL: (A) lateral view; (B) dorsal view. C–D. Post yolk-sac larva, 4.2 mm TL: (C) lateral view; (D) dorsal view. (A–D are original photographs. Specimens are spawned and cultured by author.)

3.4–3.8 mm TL. Overall body length shrinks as yolk sac and oil globule are absorbed. Mouth is inferior and becomes functional with a well-developed lower jaw. Elongated body shows slight downward bend where yolk sac is almost completely absorbed. Pericardial and abdominal areas become visible. Gill filaments and arches are visible under incompletely formed operculum. Cranial lobes over left and right sides of brain enlarge slightly. A slight lateral crease develops in nape area between head and body.[2]

Morphometry
See Table 7.

2.8–3.3 mm TL. As percent of TL: HL 21; PreAL 39; GD 21–25.[2]

3.4–3.8 mm TL. As percent of TL: HL 18–21; PreAL 38–42; GD 21–24.[2]

Fin Development
2.8–3.3 mm TL. Median finfold arises dorsally at 4th or 5th myomere, continues posteriorly around the straight urostyle, and forward along the ventral midline to the anus. Fin ray promordia lacking in finfold. Pectoral fin buds evident but lack rays. Pelvic fin buds absent.[2]

3.4–3.8 mm TL. Median finfold decreasing in height in caudal peduncle area; caudal fin ray primordia beginning to form ventrally on posterior end of urostyle; dorsal and anal fin ray primordia absent; small pectoral fin primordia beginning to form near base of fin.[2]

Table 7

Morphometric data for aquarium-raised
Elassoma alabamae expressed as average lengths
and depths (mm) related to age.[2]

Days after Hatching	N	TL	SL	HL	SNL	ED	PreAL	GD
1	15	3.09	3.0	0.6	0.1	0.2	1.3	0.7
2	5	3.3	3.1	0.6	0.1	0.3	1.3	0.7
3	4	3.7	3.5	0.8	0.1	0.4	1.5	0.8
6	3	3.4	3.2	0.6	0.1	0.2	1.3	0.6
8	4	3.6	3.4	0.7	0.1	0.3	1.4	0.7
11	2	3.6	3.4	0.8	0.1	0.3	1.5	0.7
18	5	4.4	4.2	1.0	0.2	0.4	1.9	0.9
24	1	5.5	4.8	1.4	0.3	0.6	2.5	1.1
27	2	6.0	5.1	1.5	0.3	0.6	2.7	1.2
32	3	6.9	5.7	1.7	0.4	0.6	3.1	1.3
56	1	10.7	8.8	2.9	0.7	1.0	4.8	2.4
70	1	11.6	9.2	3.5	1.0	1.1	5.0	2.6
80	3	12.3	10.0	3.4	0.8	1.1	5.9	2.8
100	1	12.8	10.3	3.8	0.8	1.1	5.8	3.1
128	5	13.4	10.9	3.6	0.9	1.2	6.2	3.1
135	4	13.9	11.3	3.9	1.0	1.3	6.5	3.1
184	2	15.1	12.5	4.1	1.0	1.2	7.2	3.5
200	1	16.1	13.4	4.0	1.1	1.4	7.8	4.1
227	5	16.9	13.7	4.7	1.1	1.6	7.5	4.3
240	1	17.8	14.5	4.8	1.3	1.6	8.5	4.5
250	2	17.9	14.8	5.3	1.3	1.6	8.0	4.4
288	1	19.8	16.3	5.3	1.2	1.6	9.7	4.8
325	5	19.7	16.4	5.5	1.3	1.7	8.7	4.6

Pigmentation

2.8–3.3 mm TL. Eyes lightly pigmented; melanophores most evident on dorsal, lateral, and ventral surfaces of head; melanophores widely scattered and inconspicuous on yolk sac and ventral surface of body.[2]

3.4–3.8 mm TL. Exterior and interior surfaces of eyes lightly pigmented to black, with lighter colored pupil; stellate melanophores present on top of head and in nape area. Pigmentation not well developed along sides of body, but distinct concentrations of melanophores are present on and along either side of the ventral midline from the isthmus to the anal area.[2]

POST YOLK-SAC LARVAE

See Figures 8 and 9.

Size Range

3.4–3.6 mm to 10–11 mm TL.[2]

Myomeres

5.5 mm TL. Preanal 6–7; postanal 17–18; total 23–26.[2]

5.9–6.5 mm TL. Preanal 6–7; postanal 16–18; total 22–25.[2]

6.9–7.2 mm TL. Preanal 6–7; postanal 16–18; total 22–25.[2]

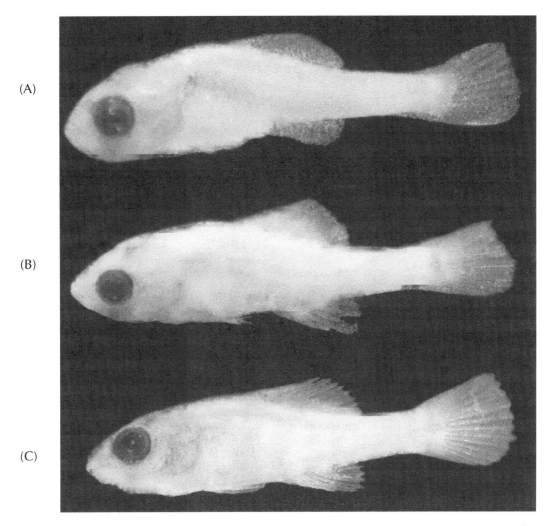

Figure 9 Development of young spring pygmy sunfish. A–B. Post yolk-sac larvae: (A) 6.6 mm TL; (B) 8.8 mm TL. (C) Juvenile, 10.4 mm TL. (A–C are original photographs of specimens spawned and cultured by the author.)

Morphology
5.5 mm TL. Eyes and mouth fully functional.[2]

5.9–6.5 mm TL. Mouth large and upturned with pronounced lower jaw; gill filaments and branchiostegal rays visible; nostrils on dorsal surface of snout.[2]

6.9–7.2 mm TL. Air bladder formed; branchiostegal rays and several opercular bones visible; body becoming elongated from increased growth.[2]

8.0–9.0 mm TL. Eyes fully developed; optic lobes well developed and visible behind eyes.[2]

Morphometry
See Table 7.

4.2–5.5 mm TL. As percent TL: HL 21–26; PreAL 32–35; GD 19–20.[2]

5.9–6.5 mm TL. As percent TL: HL 23–25; PreAL 35–37; GD 19–20.[2]

6.9–7.2 mm TL. As percent TL: HL 25–28; PreAL 35–40; GD 18–20.[2]

8.0–9.0 mm TL. As percent SL: HL 33; PreAL55; GD 27.[2]

Fin Development
5.5 mm TL. Shortened upturned urostyle visible in caudal fin; finfold between dorsal and caudal fin and anal and caudal fin further diminished; 6–7 dorsal fin primordia; 3–4 anal fin primordia; 2–3 pectoral fin primordial; 13 caudal fin primordia visible; no pelvic fin buds.[2]

5.9–6.5 mm TL. Finfold replaced by developing median fins; dorsal fin with 2–3 spines and

8–9 unbranched rays; anal fin with 1 spine and 4–5 unbranched rays; 14 unbranched caudal rays; 6–8 unbranched pectoral rays; small pelvic fin buds present; hypural plate formed.[2]

6.9–7.2 mm TL. 7–8 central caudal rays branched; 2–3 dorsal spines and 8–9 unbranched rays; 1–2 anal spines and 5–6 rays; 3–4 pectoral fin primordia; pelvic fin buds present but primordia lacking.[2]

8.0–9.0 mm TL. Pectoral fin with 13–15 rays; 11–13 caudal fin rays; dorsal fin with 2–3 spines and 9–10 rays; 1–2 anal fin spines and 4–5 rays; 1 pelvic spine and 3–4 rays.[2]

Pigmentation
5.5 mm TL. Melanophores are scattered on head and anterior half of body; limited numbers of smaller melanophores are present posteriorly on body.[2]

5.9–6.5 mm TL. Large stellate melanophores present on head; smaller ones on nape and back; a few scattered on sides and abdomen.[2]

6.9–7.2 mm TL. A dense concentration of melanophores present on top of head; pigmentation increasing on body; median brown stripe in anal fin; none in other fins.[2]

8.0–9.0 mm TL. Seven to eight distinct vertical bands are present along sides of body separated by narrow clear areas that extend from the dorsum to the midline. The dorsal ends of three bars extend into a dark band along dorsal fin base. Ventral ends of two bars extend into anal fin base. A longitudinal stripe is present from the operculum posteriorly to the caudal peduncle, passing through the vertical bars. A basicaudal spot joins with most posterior vertical bar. Melanophores separated by clear areas extend onto bases of upper, middle, and lower caudal fin rays. Stellate melanophores are scattered on the head (concentrated over each brain lobe), operculum, and pectoral fin base. Melanophores are variously scattered along some rays in all fins.[2]

JUVENILES

See Figure 9.

Size Range
Phase begins at 10–11 mm TL;[2] adults by about 16 mm TL.[2]

Myomeres
10.0–11.0 mm TL. Preanal 7–8; postanal 16–17; total 23–25.[2]

Morphology
10.0–11.0 mm TL. Branchiostegal rays and operculum bones fully formed.[2]

Larger juveniles. See Table 6. Body, except top of head, covered with scales; 28–30 lateral scales; no lateral line.[2,4]

Morphometry
See Tables 7 and 8.

10.0–11.0 mm TL. As percent SL: HL 34; PreAL 57; GD 28.[2]

14.0–24.5 mm SL. A comparison of proportional measurements for young male and female spring pygmy sunfish is presented in Table 8.

Fins
See Table 6.

10.0–11.0 mm TL. All fins completely formed with adult complement of spines and rays; 15–17 pectoral fin rays;[2,3] 11–14 caudal fin rays;[1,2] 2–3 dorsal spines and 9–11 rays;[2,3] 2–3 anal spines and 5–7 rays.[1,2]

Pigmentation
10.0–11.0 mm TL. Males distinguished by 7–8 broad, dark vertical bands separated by lighter colored, narrower, interspaces on body. Vertical bands converge at dorsal midline but anterior bands do not extend onto clean abdomen. The upper ends of three bands extend onto dorsal fin base and two extend ventrally onto anal fin base. Clean interspaces appear as light colored round spots on dorsal midline in front of dorsal fin base but are more distinct posteriorly. Dorsal, middle, and ventral spots on caudal fin base connect with basicaudal band and are separated from each other by lighter colored areas. The anterior edge of dorsal and anal fins is densely pigmented. The lateral longitudinal line is conspicuous and extends from the dorsal edge of the operculum through the eye and onto the snout. Concentrations of melanophores are visible on upper and lower jars. Female colors are more mottled brown and white with no obvious bands on their sides. Melanophores are reduced or missing on fins.[2]

TAXONOMIC DIAGNOSIS OF YOUNG SPRING PYGMY SUNFISH

Similar species in Ohio River drainages: banded pygmy sunfish and other centrarchids.

Table 8

Proportional measurements (expressed as thousands of standard length) of male and female *Elassoma alabamae* collected in Alabama. Significant (P < 0.05) differences between genders are indicated by asterisks.

	Males (n = 21)			Females (n = 20)		
	Range	Mean	SD	Range	Mean	SD
Standard length*	16.1–20.4	18.1	1.3	14.2–24.5	20.4	2.5
Preanal length*	528–574	552	15.4	528–630	583	25.4
Caudal peduncle depth*	106–147	129	10.6	99–141	114	9.7
Dorsal fin length*	364–448	405	23.9	332–393	357	17.4
Pectoral fin length*	106–148	133	12.0	74–122	105	11.9
Eye diameter	80–192	89	5.4	69–92	81	5.6
Bar width*	71–132	95	17.6	68–123	96	18.4

Source: Modified from Mayden, 1993.

Spring pygmy sunfish and banded pygmy sunfish have allopatric ranges.[2] Juveniles and adults can also be distinguished by the number of dorsal spines (3 for spring pygmy and 4–5 for banded pygmy) and the presence of light-colored narrow windows on the dorsal and anal fins of adult spring pygmy sunfish, which are lacking on the fins of banded pygmy sunfish.[5] Table 6 provides a comparison of meristic data for the two species.[1]

For taxonomic differences between the pygmy sunfishes and centrarchids see the "Key to Genera" on pages 27–29.

ECOLOGY OF EARLY LIFE PHASES

Occurrence and Distribution (Figure 10)

Eggs. The semi-adhesive eggs of spring pygmy sunfish are usually deposited in dense concentrations of suspended fine-leaved vegetation such as *Myriophyllum* or *Ceratophyllum* along the margins of springs and spring runs. These immediate areas have little or no flow, silty substrates, and are inhabited by predacious invertebrate species that would readily consume the eggs if they were not guarded by the spawning male. Most of the eggs become attached to the fine-leaved vegetation in small clusters. Eggs that fall through the spawning medium onto the silty substrate below may become encased in silt and die or may be consumed by predators.[2]

Yolk-sac larvae. Yolk-sac larval spring pygmy sunfish develop on or in the immediate spawning area. Initially, yolk-sac larvae in aquaria lay quietly on the bottom, but soon many began to exhibit short spurts of upward swimming motion followed by downward drift.[2]

Moderate to high yolk-sac larvae mortality can occur during the time period between complete yolk absorption and active feeding[24–29] if an adequate supply of small-sized prey is not available.

Post yolk-sac larvae and juveniles. Swimming skills continuously improve, but aquarium and field observations suggest that individuals spend most of their time in dense suspended vegetation.[2]

Growth

Elassoma yolk-sac larvae experienced the same critical growth period. TL increased steadily as yolk was absorbed from day 2–3, then decreased from day 4–6 with complete absorption of the yolk and development of a functional mouth.[2,7] TL increased to 3.7–4.0 mm TL (day 7–12) as feeding efficiency increased. Laboratory fish grew slower than those living in natural environments.[28] Young spring pygmy sunfish collected in Beaverdam Creek approximately 120 days after spawning were about 3–4 mm SL longer than laboratory-raised fish of the same age.[2]

Young spring pygmy sunfish cultured in an aquarium averaged 3.6 mm TL 11 days after hatching, 5.5 mm at 24 days, 10.7 mm at 56 days, 12.8 mm

Figure 10 General distribution of spring pygmy sunfish in the Ohio River drainage (shaded areas).

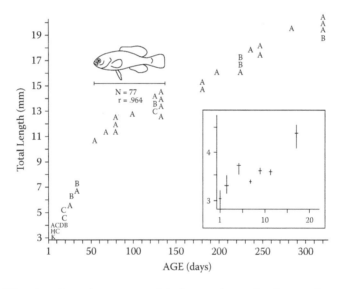

Figure 11 Scatter diagram of TL and age for spring pygmy sunfish. A = one observation, B = two observations, etc. Insert is a diagram of growth in TL the first 20 days after hatching. Vertical line = range and horizontal line = mean. (Reprinted with author's permission.)

at 100 days, and 19.8 mm at 288 days (Table 7).[2] A scatter diagram of TL related to age is presented in Figure 11.

Spring pygmy sunfish absorb most of their yolk within 5–6 days after hatching. Although not studied in nature, laboratory observations suggest that post-sac larvae decreased slightly in TL and GD during this period between yolk absorption and the onset of active feeding. If they survive this critical period, then most individuals experience consistent growth until they reach adult size.[2]

Feeding Habits

Post yolk–sac larvae spring pygmy sunfish will not survive to juvenile age unless they have a readily available supply of suitable-sized food items on which they can begin actively feeding during the final stages of yolk absorption.[2] Larval spring pygmy sunfish readily consumed brine shrimp nauplii in the laboratory.

LITERATURE CITED

1. Mayden, R.L. 1993.
2. Mettee, M.F. 1974.
3. Boschung, H.T., Jr. 1992.
4. Mettee, M.F. et al. 1996.
5. Boschung, H.T., Jr. and R.L. Mayden. 2004.
6. Jandebeur, T.E. 1979.
7. Mettee, M.F. and C. Scharpf. 1998.
8. Mettee, M.F. and J.J. Pulliam. 1986.
9. Warren, M.L. 2004.
10. Ramsey, J.S. 1976.
11. Mount, R.H., ed. 1984.
12. Mettee, M.F. and J.S. Ramsey. 1986.
13. Pierson, J.M. 1990.
14. Warren, M.L. et al. 2000.
15. O'Neil, P.E. 2004.

A 16. Darr, D.P. and G.R. Hooper. 1991.
17. Axelrod, H.P. and L.P. Schultz. 1971.
18. Axelrod, H.P. and S.R. Shaw. 1967.
19. Miller, H.C. 1964.
20. Poyser, W.A. 1919.
21. Innes, W.T. 1969.
22. Barney, R.L. and B.J. Anson. 1920.
23. Walsh, S.J. and B.M. Burr. 1984.
24. Runyun, S. 1961.
25. Lagler, K.E. et al. 1962.
26. Mansueti, A.J. 1962.
27. Mansueti, R.J. 1962.
28. Mansueti, R.J. 1964.
29. Farris, D.A. 1959.

BANDED PYGMY SUNFISH

Elassoma zonatum Jordan

Maurice F. Mettee

Elassoma, Greek: "small body"; *zona*, Greek: "band or belt."

RANGE

The banded pygmy sunfish is the most widely distributed member of the family. It occurs from Albemarle Sound in NC, south to the St. John's River in FL, westward along the Gulf coast to the Brazos River in TX, and north in the Mississippi River basin to southern IL and IN. It is seldom reported above the Fall Line in the central Mississippi and lower Ohio Valleys, except in the Wabash River in southern IL[1,25] and the Tradewater and Green Rivers in KY.[1,26] Limited collections exist from the Coosa and Tallapoosa Rivers above the Fall Line in AL.[2]

Banded pygmy sunfish is not currently listed as endangered, threatened, or of special concern.[3,4]

HABITAT AND MOVEMENT

Banded pygmy sunfish is a Coastal Plain species[2,25–27,29] that is rarely reported above the Fall Line.[2,25,26] It primarily inhabits dense submerged vegetation (*Ceratophyllum*, *Myriophyllum*, *Zannichellia*, *Lemna*, and *Wolffia*) in swamps, oxbow lakes, and low-gradient streams.[1,5–8,32] This habitat is subject to extreme temperature variations, ranging from 5–7°C in December to 32°C in August.[1] Relatively few species are found in these microhabitats in LA[5] and IL,[9] but 25 species were cohabitants with banded pygmy sunfish in Mayfield Creek, KY.[1]

In a MO lowland hardwood wetland, adult banded pygmy sunfish in semipermanently flooded habitats were consistently day-active. However, in seasonally flooded habitats their diel activity pattern changed through time. From late February until early March they were predominantly night-active. From mid-March to early April, there was no significant difference between day and night CPUE, but from mid-April until the habitats dried in May, adult banded pygmy sunfish were primarily day-active. This shift from night to day activity generally corresponded with the spawning period, which peaks from late March through April in the study region.[31]

Banded pygmy sunfish are usually widely scattered except during the period March to April, when they congregate to spawn. Longitudinal movement patterns are unknown.

DISTRIBUTION AND OCCURRENCE IN THE OHIO RIVER SYSTEM

In IL, the banded pygmy sunfish formerly occurred in the lower Wabash River system, where it has been extirpated, probably as a result of extensive oil pollution and the drainage of natural swamps.[25]

In KY, banded pygmy sunfish are generally distributed and common in tributaries of the lower Ohio and Tennessee Rivers. It is occasionally collected and uncommon in the lower Green and Tradewater River drainages.[26] It is reported in the Ohio River drainage of TN only in the Big Sandy River system, a tributary to Kentucky Reservoir.[27] In AL and MS, this species is widely distributed in Coastal Plain streams.[2,28] It has been recorded at a few locations above the Fall Line in the Coosa and Tallapoosa Rivers in AL,[2] but it is not reported from the Tennessee River drainage in northern AL[2] or MS.[28]

SPAWNING

Location
Spawning usually occurs along stream margins and in spring runs. Some authors have suggested that *Elassoma evergladei* and *E. zonatum* construct centrarchid-type nests in aquaria,[9,10] while others reported spawning occurred in submerged aquatic vegetation (*Ceratophyllum*, *Myriophyllum*) above the bottom.[1,5,8,11] The former suggestion seems less likely since silt concentrations in marshy and swampy environments could smother newly hatched eggs which would also be subject to heavy predation by abundant larval benthic invertebrates found in these environments.

Males established and defended territories in submerged vegetation that averaged 10 cm × 10 cm ×

20 cm in aquaria.[1] Dense vegetation usually prohibits direct observation of the spawning act in nature, but 145–236 eggs were harvested from submerged vegetation that included *Ceratophyllum*.[5] A few eggs fell to the bottom and were soon covered with tiny bits of plant debris and silt. Several authors have observed banded pygmy sunfish spawning on numerous occasions in *Ceratophyllum* and *Myriophyllum* in aquaria.[1,8,11,12]

Season

Ovarian growth occurs from May through August, reaching a peak in April and May.[1] Males exhibiting breeding coloration have been observed from late February until June.[1,8] Spawning occurs from March through April.[1,8,31]

Temperature

Gravid females have been collected in the wild at 12–16°C[1] and 16.7–20°C.[5] Banded pygmy sunfish have spawned in aquaria at temperatures of 21–24°C.[1,5,8,11,12] Spawning activity concluded at 27°C.[1]

Fecundity

Preserved gravid females from Mississippi contained from 96–970 ova, including immature oocytes.[5] Twelve female banded pygmy sunfish from MO contained an estimated 300 eggs per female.[15]

Kentucky females contained 350–797 oocytes (average = 537.1) and 43–255 mature and intermediate oocytes per female. An average of 37.9 eggs per spawn in Kentucky decreased with successive spawnings.[1] Significant correlations existed between total number of oocytes and SL and between total number of oocytes and adjusted body weights.[1]

Reduced clutch sizes may occur in captive populations.[1,12] The number of eggs per spawn in aquaria ranged from 20 to 68.[8]

Sexual Maturity

Individuals reach sexual maturity in less than 1 year.[1,8,5] Considered young adults at about 16 mm SL.[8]

Spawning Act

No nest is constructed by banded pygmy sunfish.[1,8,12] Males establish and defend territories in submerged vegetation aquaria.[1] Intruders are challenged and usually repelled with a Sidling Threat Display.[13] Gravid females that entered the territory are approached less aggressively. If they do not immediately retreat, the male color pattern intensifies, his body begins to quiver, and he initiates the highly animated Wiggle Waggle Display,[1,8,13,14] which involves swimming back and forth in an up

and down undulating pattern from the female to the submerged vegetation while raising and lowering his vertical fins and alternately flexing the pelvic fins. Females usually follow the male into the vegetation and select the spawning site (Figure 3).[8] The female stops in a vertical position. The intensely colored male aligns himself beside but slightly lower and behind her and begins nudging her genital area. Both fishes begin to quiver rapidly. As the female begins releasing short bursts of eggs, the male aligns himself beside her and releases milt. Most of the adhesive, demersal eggs attach to the aquatic vegetation. If eggs fall to the bottom of the aquarium, the male usually removes debris from the surrounding area and guards them, as well as those suspended in the vegetation.[8] This behavior may explain why some researchers have suggested that the banded pygmy builds a centrarchid-type nest.[9,10]

Males spawn repeatedly with several females. Clutch size gradually diminishes as females continue to spawn with the same or different partners for several days. Polyandry has been observed in aquaria.[1,11] Males will consume eggs produced by other spawns but generally not their own.

Spawning activity is usually greatest during morning hours but continues until late afternoon.[1]

EGGS

Description

Mature oocytes taken from preserved females were 0.79 mm in diameter.[1] Unfertilized water-hardened eggs were pale yellowish, 2.6–2.7 mm in diameter[8]; yolk was large (0.7–0.8 mm in diameter), granular in appearance with an indentation on one side. One large spherical oil globule (0.1–0.2 mm diameter) is present in the yolk along with several tiny droplets in some individuals.

Fertilized, water-hardened eggs, 45 min after fertilization, are spherical, 3.0–3.4 mm in diameter, with one large or two smaller attachment disks. Yolk sac is mostly filled with granular yolk that contains one large oil globule and several smaller oil droplets.[8]

Incubation

100–110 h at 21°C[2]; 97–116 h at 21 ± 1°C.[1]

Development

Composite illustrations of embryonic development for elassomid fishes are presented in Figures 4–7. Illustrations are based on observations of eggs from multiple spawnings of four species of *Elassoma* including banded pygmy sunfish *E. zonatum*.[8]

Table 9

Egg development related to age (hours) reported for aquarium-raised *Elassoma zonatum*.

Stage	Time (h) Average for 24 Clutches [2]	Time (h) Ranges from 9 Clutches [1]
One-cell egg	0.75	0.5–0.8
Two-cell egg	1.3	1.3–1.5
Four-cell egg	2.0	2.0–2.3
Eight-cell egg	2.9	2.8–3.1
16-cell egg	4.3	4.2–4.6
32- to 64-cell egg	5.8	5.8–7.3
Early high blastula	9.3	7.8–9.3
Late high blastula	11.5	—
Early gastrula	13	12.5–13.0
Late gastrula	16	13.5–17.0
Neurula	19	19–22
Early larva = tail bud	29	24–30
8–10 somites	—	30–32
12–14 somites	35	—
12–16 somites	—	36–40
16–18 somites	43	—
18–22 somites	—	42–49
20–24 somites	53	—
24–27 somites	—	52–70
Pre-hatch larva	72	70–96
Hatch stage	100–110	97–116

Embryonic egg development for banded pygmy sunfish is described in detail below and summarized in Table 9. Development times and descriptions come from fertilized eggs harvested from 24 spawns in aquaria.[8]

45 min — egg is spherical, 3.0–3.4 mm in diameter, with one large or two smaller attachment disks; yolk sac is mostly filled with granular yolk that contains one large and several smaller oil droplets.

80 min — blastodisc divides to form two equal size blastomeres.

2 h — two blastomeres divide into four.

2 h and 55 min — unsynchronized cell division forms 8, then 16 blastomeres.

3 h — 16 cells randomly divide to form 32, then 64 cells.

5¾ h — early blastula stage; multicell blastodisc on side of yolk; yolk beginning to loose its granular appearance.

9¼ h — late blastula stage; high, prominent blastodisc beginning to slide over yolk surface.

13 h — gastrula; thin blastoderm covering around 75% of yolk.

19 h — neurula; embryonic shield developing as longitudinal v-shaped furrow on yolk surface.

24–29 h — early embryo; 3–4 somites and optic placodes visible; oil globule still near anterior end of yolk.

32–35 h — 11–13 somites; posterior somites developing into myomeres; oil globule located in middle of yolk; bulge developing on top of head over mesencephalon and telencephalon regions; pupil developing in optic capsules; primordial heart developing but no heartbeat; posterior end of finfold visible around tail, not yet separating from body; pointed end of notochord visible at posterior end of body; estimated larval length 1.7–1.8 mm.

43 h — 22–24 somites; finfold oriented on one side of body; chromatophores more abundant on body behind yolk; choroid fissure in eye; heart beat around 82 beats per minute.

53 h — 24–25 myomeres; eyes, brain, otic capsules, brain stem, and circulatory system visible; heart beats 97–98 per minute; blood essentially colorless; most of finfold separated from abdominal area; chromatophores increasing in number and size.

60 h — 24–25 myomeres; finfold separated and beginning to rotate to one side; oil globule still in middle of yolk; anal area developing.

72 h — pre-hatch larva; pupil turning dark; pink blood cells flowing through circulatory system; heart beat 134–136 per minute; 26–27 myomeres (estimate); larger chromatophores present over the entire body, with greatest concentrations on top of head and abdomen; eyes colorless, fully developed, with choroid fissure; yolk is teardrop-shaped with smaller oil globule now near posterior end; tail completely rotated to side; no fin ray primordial; pectoral fin bud visible on either side of body posterior to head; no pelvic buds.

100–110 h — hatching larva exhibits vigorous movement inside the egg; tail protrudes first; alternate head shaking frees body; larvae rest on bottom for several seconds before swimming in short forward bursts; body tadpole shaped except for ventrolateral bulge caused by yolk sac; round, 0.2–0.3-mm-diameter oil globule in middle of yolk; mouth not developed; pectoral fin consists of fan-shaped membrane; fin ray primordia lacking; eyes black; chromatophores on abdomen and head, less abundant on dorsum; larger heart pumping reddish blood cells throughout most of the body; yolk shrinking in length and diameter; posterior oil globule smaller; digestive track visible along dorsal surface of yolk, ending in anus; spinal column with tiny dorsal and ventral processes; slight indentation between head and abdomen is apparent; 7–8 preanal and 16–17 postanal myomeres visible.

YOLK-SAC LARVAE

See Figure 12.

Size Range
Hatch at 3.0–3.8 mm TL (mean = 3.2 mm, n = 23).[8] Yolk absorption complete by 3.9–4.0 mm TL on some individuals,*,[8] but remnants of yolk still visible at 5.1 mm TL on some.[8]

Myomeres
Preanal 7–8; postanal 17–19.[8]

Morphology
3.0–3.4 mm TL. Snout rounded; oil globule located posteriorly in yolk; urostyle straight, extending into middle of finfold; heart, intestines, and anus visible; mouth absent.[8] Heartbeats ranged from 114 to 143 (mean = 124.4)[1] or 134–136 beats per minute.[8]

3.5–3.8 mm TL. Oil globule near the middle of diminishing yolk; mouth visible, but probably not functioning; 3–4 gill arches visible when viewed dorsally; left and right cranial lobes separated from body by slight crease; operculum incomplete.[8]

Morphometry
Morphometric data for laboratory–cultured banded pygmy sunfish are presented in Table 10.[8]

3.0–3.4 mm TL. As percent TL: HL 17–21; PreAL 32–43; GD 18–26.[8]

3.5–3.8 mm TL. As percent TL: HL 17–21; PreAL 34–37; GD 17–24.[8]

Fin development
3.0–3.4 mm TL. Median finfold originates dorsally at 6th or 7th myomere and is continuous posteriorly around end of urostyle and ventrally anterior

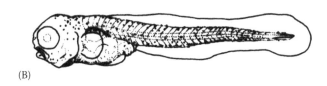

(A)

(B)

(C)

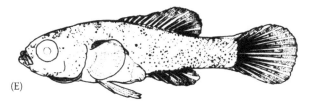

(D)

(E)

Figure 12 Development of young banded pygmy sunfish. (A) Yolk-sac larva, about 3.4 mm TL. B–D. Post yolk-sac larvae: (B) 4.0 mm TL; (C) 6.3 mm TL; (D) 8.4 mm TL. (E) Juvenile, 10.7 mm TL. (A reprinted from Figure 3, reference 8, with author's permission; B–E reprinted from Figures 2–5, reference 30, with author's permission.)

to the anus; no fin ray primordia. Pectoral fin buds are present; no pelvic fin buds.[8]

3.5–3.8 mm TL. Slight depression is apparent in median finfold dorsally and ventrally in future caudal peduncle region. One or two early caudal fin primordia are visible.[8]

Pigmentation
3.0–3.4 mm TL. Eyes lightly pigmented; most melanophores concentrated on head, nape, yolk sac, and ventral surface of head; very few melanophores on body and tail.[8]

Table 10

Morphometric data for aquarium-raised
Elassoma zonatum expressed as average lengths
and depths (mm) related to age.[8]

Days after Hatching	N	TL	SL	HL	SNL	ED	PreAL	GD
1	15	3.2	3.1	0.6	0.1	0.3	1.2	0.7
2	7	3.5	3.3	0.7	0.1	0.3	1.2	0.7
3	7	3.6	3.4	0.7	0.1	0.3	1.3	0.7
4	2	3.8	3.6	0.8	0.2	0.3	1.4	0.8
5	7	3.5	3.4	0.8	0.1	0.3	1.3	0.7
6	8	3.6	3.4	0.8	0.1	0.3	1.3	0.6
7	7	3.8	3.6	0.8	0.1	0.4	1.4	0.8
9	5	4.0	3.8	0.9	0.2	0.4	1.6	0.8
11	4	4.4	4.2	1.0	0.1	0.4	1.7	0.8
12	1	4.5	4.4	1.0	0.2	0.4	1.9	0.8
14	2	4.6	4.4	1.0	0.2	0.4	1.9	0.9
17	1	5.1	4.9	1.1	0.2	0.4	2.2	1.1
23	2	5.8	5.5	1.4	0.3	0.5	2.5	1.4
28	6	6.9	6.0	1.8	0.4	0.7	3.3	1.4
31	4	7.9	6.6	2.0	0.4	0.7	3.4	1.6
36	1	8.3	7.4	2.4	0.4	0.8	3.9	1.8
41	1	10.4	8.5	2.6	0.5	0.9	4.4	2.1
65	2	17.6	14.8	4.9	0.9	1.3	8.9	3.7
75	1	17.9	15.3	5.1	1.1	1.4	9.0	3.9
80	4	19.1	15.9	5.2	1.2	1.4	8.6	4.5
89	2	20.5	16.7	5.7	1.1	1.5	8.9	4.7
92	1	21.4	17.8	6.1	1.4	1.5	9.5	5.0
97	1	22.0	18.7	5.8	1.2	1.5	10.1	4.8
101	2	23.0	19.9	6.4	1.6	1.7	10.1	5.0

3.5–4.4 mm TL. Eyes are dark to black; choroid fissure is absent; large chromatophores are present on head and anteriorly on ventrum; yolk smaller and dark; oil globule very dark in reflected light.[8] Also reported to possess profuse pigmentation with melanophores over all of head and body.[30]

POST YOLK–SAC LARVAE

See Figure 12.

Size Range

3.9–5.1*,[8] to 8.0–8.5 mm TL in the Warrior River system, AL[8] and to 8.5–10.0 mm in the Big Sandy River, TN.*

Myomeres
Warrior River, AL:
4.0–5.1 mm TL. Preanal 7–8; postanal 16–17; total 23–25.[8]

5.7–7.3 mm TL. Preanal 8–9; postanal 16–18; total 24–27.[8]

7.7–8.3 mm TL. Preanal 8–9; postanal 18–19; total 26–28.[8]

Note: Counts reported above are lower than those below. This is due in part to different counting criteria for those above (most posterior myomere counted being the one touching the anus vs. the last one transected by an imaginary vertical line at the posterior margin of the anus) and use of polarizing light filters by the sources below.

Tallapoosa River, AL:
5.5–8.8 mm TL. Preanal 12–13; postanal 17–18.[23]

Big Sandy River, TN (Table 11):
3.9–5.6 mm TL. Preanal 11–13; postanal 16–17; total 27–30.*

Table 11

Myomere count ranges (average and standard deviation in parentheses) at selected size intervals for young banded pygmy sunfish from TN.*

Size Range (mm TL)	N	Myomeres			
		Predorsal Fin	Preanal	Postanal	Total
3.9–5.6	10	—	11–13 (11.9 ± 0.70)	16–17 (16.6 ± 0.49)	27–30 (28.5 ± 0.92)
5.7–5.8	3	—	12–13 (12.3 ± 0.47)	16–17 (16.3 ± 0.47)	28–29 (28.7 ± 0.47)
5.83–9.4	43	6–9 (7.7 ± 0.76)	12–16 (14.1 ± 0.80)	14–17 (15.6 ± 0.65)	28–31 (29.7 ± 0.63)
9.41–10.5	11	6–9 (7.6 ± 0.98)	14–15 (14.6 ± 0.48)	15–16 (15.5 ± 0.49)	29–31 (30.2 ± 0.71)

5.7–5.8 mm TL. Preanal 12–12; postanal 16–17; total 28–29.*

5.8–9.4 mm TL. Predorsal 6–9; preanal 12–16; postanal 14–17; total 28–31.*

Morphology

4.0–5.1 mm TL. Yolk and oil globule absent; liver visible on anterior abdominal wall; mouth and digestive tract functional.[8]

5.7–7.3 mm TL. Choroid fissure has disappeared; gill arches and rakers visible; jaws becoming more oblique. Flexion occurs, developing hypural plate visible.[8]

7.7–8.3 mm TL. Hypural plate fully formed.[8]

Morphometry

See Tables 10 and 12.

4.0–5.1 mm TL. As percent TL: HL 22–23; PreAL 38–43; GD 18–22.[8]

5.7–7.3 mm TL. As percent TL: HL 23–26; PreAL 44–49; GD 20–24.[8]

7.7–8.3 mm TL. As percent TL: HL 25–27; PreAL 42–49; GD 21–24.[8]

Fin Development

4.0–5.1 mm TL. Urostyle extending into finfold; no fin ray primordia visible in finfold; pelvic buds are visible at 11–17 days[8] or 23 days.[1]

5.7–7.3 mm TL. Urostyle flexes; hypural plate visible at caudal fin base; finfold begins to diminish between the dorsal and caudal fins and the anal and caudal fins; three spines and 8–9 soft rays in the dorsal fin; 2–3 spines and 5–6 soft rays in the anal fin; 16–18 rays in the caudal fin at 23–25 days; fin ray primordia visible in pelvic fin buds.[8]

7.7–8.3 mm. TL. Most of the finfold disappears from 31 to 36 days[8] or 40 days[1] after hatching. Dorsal fin with 3–4 spines and 9–10 rays; anal fin with 2–3 spines and 5–6 rays; caudal fin has 17–19 rays; pectoral fin with 6–7 rays.[1,8]

Pigmentation

4.0–5.1 mm TL. Eyes black with clear pupils; choroid fissure faintly evident at ventral margin of the eye; large melanophores are present on top of head, mid-ventrally on abdomen, and isthmus; rows of small melanophores present along dorsal and ventral finfold base.[8]

5.7–7.3 mm TL. Large numbers of melanophores concentrated dorsally on head; a few melanophores are scattered along the sides of the body.[8] Also, pigmentation was in the form of scattered melanophores covering the entire body.[23,30]

7.7–8.3 mm TL. Dense concentrations of melanophores on top of head and on isthmus;[8] small melanophores widely scattered on body.[8,23,30] Seven to eight vertical bands of pigment form along the sides of the body. A dark band of pigment is present on mandible and from corner of mouth through the eye and on to the opercule. Two light-colored basicaudal spots developing near the caudal fin base.[8]

Table 12

Morphometric data for young banded pygmy sunfish from TN* grouped by selected intervals of total length. Characters are expressed as percent total length (TL) and head length (HL) with a single standard deviation. Range of values for each character is in parentheses.

	TL Intervals (mm)						
Length Range N Characters	3.90–4.90 4 Mean ± SD (Range)	5.00–5.99 11 Mean ± SD (Range)	6.00–6.99 16 Mean ± SD (Range)	7.00–7.99 13 Mean ± SD (Range)	8.00–8.99 8 Mean ± SD (Range)	9.00–9.99 10 Mean ± SD (Range)	10.00–10.99 5 Mean ± SD (Range)
Length as % TL							
Snout	1.66 ± 0.28 (0.05–0.10)	3.60 ± 0.47 (0.15–0.25)	3.94 ± 0.45 (0.18–0.30)	4.41 ± 0.41 (0.27–0.39)	4.70 ± 0.24 (0.36–0.44)	4.11 ± 0.20 (0.35–0.43)	4.08 ± 0.22 (0.39–0.45)
Eye diameter	8.76 ± 0.46 (0.36–0.40)	9.03 ± 0.34 (0.46–0.56)	9.1 ± 0.28 (0.53–0.61)	8.78 ± 0.16 (0.63–0.71)	8.49 ± 0.14 (0.68–0.74)	8.67 ± 0.42 (0.73–0.91)	8.15 ± 0.21 (0.82–0.85)
Head	19.0 ± 0.82 (0.71–0.97)	22.2 ± 1.24 (1.04–1.42)	23.8 ± 1.01 (1.30–1.65)	25.0 ± 1.58 (1.65–2.15)	25.1 ± 0.53 (1.97–2.27)	23.4 ± 0.75 (1.97–2.35)	23.0 ± 0.76 (2.25–2.42)
Predorsal	—	—	34.2 ± 0.0 (2.25–2.25)	33.7 ± 0.58 (2.40–2.75)	34.8 ± 1.26 (2.63–3.18)	32.7 ± 4.05 (2.60–3.60)	35.1 ± 1.20 (3.40–3.75)
Prepelvic	—	31.9 ± 0.83 (1.78–1.93)	32.2 ± 0.97 (1.86–2.18)	29.8 ± 1.12 (2.10–2.50)	28.6 ± 1.18 (2.30–2.48)	27.4 ± 0.73 (2.40–2.70)	27.8 ± 0.51 (2.75–3.00)
Preanal	41.4 ± 0.97 (1.56–2.04)	44.9 ± 0.88 (2.33–2.66)	46.6 ± 1.28 (2.70–3.13)	46.0 ± 0.72 (3.38–3.68)	46.4 ± 0.47 (3.65–4.03)	45.9 ± 0.91 (4.10–4.45)	46.0 ± 0.72 (4.60–4.75)
Postanal	58.6 ± 0.97 (2.25–2.85)	55.1 ± 0.88 (2.76–3.27)	53.4 ± 1.28 (3.20–3.56)	54.0 ± 0.72 (3.80–4.42)	53.6 ± 0.47 (4.24–4.63)	54.1 ± 0.91 (4.75–5.42)	54.0 ± 0.72 (5.35–5.75)
Standard	94.5 ± 0.76 (3.66–4.60)	95.0 ± 1.31 (4.96–5.60)	93.1 ± 2.87 (5.56–6.13)	86.9 ± 2.60 (6.28–6.93)	82.1 ± 0.47 (6.61–7.08)	81.9 ± 1.71 (7.20–8.15)	79.9 ± 0.53 (8.00–8.30)
Air bladder	9.40 ± 1.49 (0.31–0.58)	10.5 ± 1.37 (0.46–0.80)	11.1 ± 1.06 (0.59–0.80)	11.7 ± 0.95 (0.71–0.98)	13.10 ± 0.0 (1.05–1.05)	14.4 ± 0.90 (1.25–1.55)	—
Eye to air bladder	9.57 ± 0.91 (0.32–0.52)	10.7 ± 0.66 (0.50–0.70)	11.1 ± 0.72 (0.61–0.85)	11.1 ± 0.97 (0.75–1.08)	12.1 ± 1.24 (0.80–1.20)	12.8 ± 0.44 (1.15–1.30)	12.9 ± 0.46 (1.25–1.40)
Fin length as % TL							
Pectoral	—	12.0 ± 1.35 (0.60–0.80)	11.6 ± 0.68 (0.62–0.80)	11.0 ± 0.64 (0.76–0.98)	10.8 ± 0.53 (0.85–0.95)	12.0 ± 1.70 (0.88–1.35)	10.6 ± 0.94 (0.92–1.20)
Pelvic	—	2.20 ± 0.24 (0.11–0.15)	3.05 ± 0.77 (0.11–0.30)	5.0 ± 0.54 (0.30–0.46)	6.27 ± 1.28 (0.36–0.73)	6.6 ± 3.84 (0.23–1.25)	12.6 ± 1.68 (1.02–1.60)
Body depth as % TL							
Head at P1	20.1 ± 0.75 (0.81–0.97)	21.9 ± 2.80 (1.05–1.80)	22.8 ± 1.17 (1.30–1.53)	22.2 ± 0.58 (1.53–1.80)	22.1 ± 0.83 (1.70–2.00)	21.0 ± 1.29 (1.80–2.30)	22.7 ± 0.20 (2.25–2.42)
Head at eyes	—	19.4 ± 0.0 (1.15–1.15)	19.7 ± 1.18 (1.13–1.35)	19.6 ± 0.47 (1.40–1.63)	19.3 ± 0.32 (1.53–1.65)	18.5 ± 1.35 (1.53–2.00)	19.5 ± 0.25 (1.95–2.05)
Preanal	15.3 ± 0.68 (0.56–0.74)	15.8 ± 0.71 (0.80–0.95)	17.0 ± 1.24 (0.88–1.25)	17.1 ± 1.11 (1.10–1.43)	18.3 ± 0.37 (1.48–1.60)	19.2 ± 1.10 (1.60–2.08)	20.2 ± 0.35 (1.98–2.15)
Caudal peduncle	—	—	7.45 ± 0.38 (0.45–0.51)	7.79 ± 0.25 (0.55–0.66)	9.43 ± 1.25 (0.64–0.94)	10.3 ± 1.53 (0.75–1.19)	11.7 ± 0.24 (1.13–1.21)
Body width as % HL							
Head	—	63.0 ± 2.18 (0.77–0.89)	64.9 ± 3.99 (0.85–1.08)	58.6 ± 2.24 (1.04–1.24)	60.1 ± 3.12 (1.15–1.41)	67.2 ± 3.16 (1.32–1.61)	69.8 ± 3.30 (1.55–1.75)

Table 13

Proportional measurements (expressed as thousands of standard length) for male and female *Elassoma zonatum* collected in AL. Significant (P < 0.05) differences between genders are indicated by asterisks.

	Males (n = 10)			Females (n = 10)		
	Range	Mean	SD	Range	Mean	SD
Standard length*	22.4–32.1	28.1	3.1	22.8–28.8	25.5	1.9
Preanal length*	601–627	612	8.9	635–667	654	9.9
Caudal peduncle depth*	145–169	155	8.5	124–156	136	9.8
Dorsal fin length*	487–546	522	19.1	447–502	469	22.7
Pectoral fin length*	184–212	197	7.6	158–184	169	8.6
Eye diameter	65–84	77	6.0	69–88	79	5.8
Bar width*	31–51	40	7.1	35–77	56	12.9

Source: Modified from Mayden, 1993.

JUVENILES

See Figure 12.

Size Range
8.0–8.5 to about 16 mm TL.[8]

Myomeres
See note in Post Yolk-sac section on page 52.

Warrior River, AL:
Preanal 8–9; postanal 18–20; total 26–29.[8]

Tallapoosa River, AL:
9.0–12.2 mm TL. Preanal 12–13; postanal 17–18.[23]

Big Sandy River, TN (Table 11):
9.4–10.5 mm TL. Predorsal 6–9; preanal 14–15; postanal 15–16; total 29–31.*

Morphology
Juveniles resemble small adults. Branchiostegal rays 5; gill rakers 6–8; vertebrae 24–25.[27] Scales cover the body; lateral scales number 31–36. Lateral line absent.[2,8] See Table 6.

Morphometry
See Tables 10, 12, and 13.

8.1–10.4 mm TL. As percent TL: HD 25–29; PreAL 42–47; GD 20–27.[8]

22.4–32.1 mm SL. A comparison of proportional measurements for young male and female banded pygmy sunfish is presented in Table 13.

Fins
See Table 6.

Dorsal fin has 4–5 spines[8,22,27,33] and 9–11 rays;[8,22,27] anal fin has 3 spines[8,27] and 5–6 rays;[8,22,27,33] pectoral fins with 13–14 rays[8] or 14–17;[22,27] pelvic fins with 1 spine and 4 rays.[8] Caudal fin rounded with fewer than 16 principal rays.[30] Most fin rays are branched.[2]

Pigmentation
Nine to 11 distinct vertical bars are evident along sides of body. Distinct pigment spots are present dorsally, ventrally, and at the middle of the caudal fin base. Two to three bands of melanophores are present in the dorsal fin and one band of melanophores is visible in the anal fin.[8]

TAXONOMIC DIAGNOSIS OF YOUNG BANDED PYGMY SUNFISH

Similar species in Ohio River drainage: spring pygmy sunfish and other centrarchids.

Spring pygmy sunfish and banded pygmy sunfish have allopatric ranges.[8] Juveniles and adults can also be distinguished by the number of dorsal spines (3 for spring pygmy and 4–5 for banded pygmy) and the presence of light-colored narrow windows on the dorsal and anal fins of adult spring pygmy sunfish, which are lacking on the fins of banded pygmy sunfish.[22] Table 6 provides a comparison of meristic data for the two species.[8]

For taxonomic differences between the pygmy sun-fishes and centrarchids see the "Key to Genera" on pages 27–29.

ECOLOGY OF EARLY LIFE PHASES

Occurrence and Distribution (Figure 13)

Eggs. Semi-adhesive eggs of this species are usually deposited in dense concentrations of suspended fine-leaved vegetation such as *Myriophyllum* or *Ceratophyllum* along the margins of springs and spring runs. These immediate areas have little or no flow, silty substrates, and are inhabited by predacious invertebrate species that would readily consume the eggs if they were not guarded by the spawning male. Eggs that drift through the spawning medium and fall onto the silty bottom may become encased in silt and die or are consumed by predators.[8]

Average egg survival in aquaria ranged from 55 to 65%.[1,8]

Yolk-sac larvae. Yolk-sac larvae of banded pygmy sunfish develop on or in the immediate spawning area. Yolk-sac larvae in aquaria initially lay quietly on the bottom, but soon many individuals began to exhibit short spurts of upward swimming motion followed by downward drift.[8]

Fungus,[11] and an absence of proper food when the mouth and digestive tract become functional,[1] may be primary causes of larval mortality. Several authors have suggested that mortality of yolk-sac larval fish can be moderate to high if a proper food supply is unavailable between the period of complete yolk absorption and active feeding,[17–21] which seems to apply to banded pygmy sunfish.[1,8,11] Survival of banded pygmy sunfish yolk-sac larvae increased when live brine shrimp nauplii were introduced into nursery aquaria 3–4 days after hatching.[8] Mortality still ranged from 45 to 55%.[8]

Post yolk-sac larvae and juveniles. Larvae and juvenile banded pygmy sunfish are usually reported to remain in dense aquatic vegetation.[1,5,8,12] However, both life phases were collected in surface drift samples from the lower Tallapoosa River, AL, in late April.[23]

Larval banded pygmy sunfish were collected in low numbers in early spring with light traps and ichthyoplankton nets from channel, tupelo, and oak habitats in a hardwood wetland of the Cache River, AR.[24] Diel activity patterns in a lowland hardwood wetland in MO were reported for larvae, juveniles, and adults based on collections in activity traps. Larvae and juveniles were primarily active at night.[31]

Growth

Three growth periods have been proposed for yolk-sac larvae: rapid growth after hatching, slower growth as yolk size diminishes, and slight body shrinkage between yolk absorption and the onset of active mouth function.[16] In a nursery area in Mayfield Creek, KY, on June 4, 1980, 96.7% of all banded pygmy sunfish collected were young-of-the-year.[1]

Figure 13   General distribution of banded pygmy sunfish in the Ohio River drainage (shaded areas).

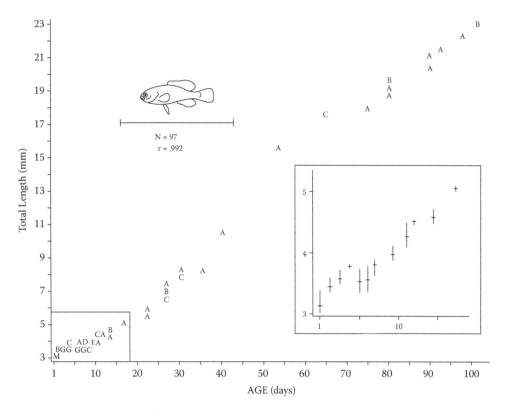

Figure 14 Scatter diagram of TL and age for banded pygmy sunfish. A = one observation, B = two observations, etc. Insert is a diagram of growth in TL the first 15 days after hatching. Vertical line = range, horizontal line = mean.

Most banded pygmy sunfish live 1 year and die days to weeks after spawning.[11] Ninety-four percent of 555 specimens collected in KY were up to 1 year of age, 4.7% were 1–2 years old, and 0.4% were older than 2 years.[1] An age-3 individual was 34 mm TL.[11]

Average TL of banded pygmy sunfish cultured in aquaria was 3.6 mm at 3 days, 4.0 mm at 9 days, 5.1 mm at 17 days, 6.9 mm at 28 days, 10.4 mm at 41 days, and 17.6 mm at 65 days (Table 10).[8]

A scatter diagram relating age to TL for banded pygmy sunfish is presented in Figure 14.

Feeding Habits
Post yolk-sac larvae banded pygmy sunfish will not survive to juvenile age unless they have a readily available supply of suitably sized food items on which they can begin actively feeding during the final stages of yolk absorption.[8]

At Reelfoot Lake, TN, 100 banded pygmy sunfish averaging 15 mm in length were collected from June 26 to August 17, 1941. These fish seemed to consume any small organisms such as *Microvelia*, *Plea*, snails, mosquito larvae, and chironomid larvae, but preferred crustaceans. Crustaceans constituted 93% by volume of their diet and included cladocerans (35%), copepods (30%), ostracods (10%), and amphipods (18%).[32]

Banded pygmy sunfish larvae were successfully cultured on brine shrimp nauplei in the laboratory.[8]

LITERATURE CITED

1. Walsh, S.J. and B.M. Burr. 1984.
2. Mettee, M.F., P.E. O'Neil, and J.M. Pierson. 1996.
3. Warren, M.L. et al. 2000.
4. O'Neil, P.E. 2004.
5. Barney, R.L. and B.J. Anson. 1920.
6. Branson, B.A. and G.A. Moore. 1962.
7. Gunning, G.E. and W.M. Lewis. 1955.
8. Mettee, M.F. 1974.
9. Axelrod, H.P. and L.P. Schultz. 1971.
10. Axelrod, H.P. and S.R. Shaw. 1967.
11. Poyser, W.A. 1919.
12. Taber, C. 1964.
13. Miller, H.C. 1964.
14. Echelle, A.F. and A.A. Echelle. 2002.
15. Pflieger, W.L. 1975b.
16. Farris, D.A. 1959.

17. Runyun, S. 1961.
18. Lagler, K.E., J.E. Bardack, and R.R. Miller. 1962.
19. Mansueti, A.J. 1962.
20. Mansueti, R.J. 1962.
21. Mansueti, R.J. 1964.
22. Boschung, H.T., Jr. and R.L. Mayden. 2004.
23. Scheidegger, K.J. 1990.
24. Killgore, K.J. and J.A. Baker. 1996.
25. Smith, P.W. 1979.
26. Burr, B.M. and M.L. Warren, Jr. 1986.
27. Etnier, D.A. and W.C. Starnes. 1993.
28. Ross, S.T. 2001.

29. Pyron, M. and C.M. Taylor. 1993.
30. Conner, J.V. 1979.
31. Stewart, E.M. and T.R. Finger. 1985.
32. Rice, L.A. 1942.
33. Mayden, R.L. 1993.

* Original morphometric data and myomere counts come from specimens collected by TVA biologists from the Big Sandy River, TN. These specimens are uncatalogued and curated at the Division of Fishes, Aquatic Research Center, Indiana Biological Survey, Bloomington, IN.

FAMILY CENTRARCHIDAE
Sunfishes

Robert Wallus

The family Centrarchidae is the second largest freshwater fish family indigenous to North America, with 31 living species representing 8 genera (Nelson et al., 2004); 19 species occur in the Ohio River drainage. Members of the family include sunfishes, black basses, crappies, and several lesser known species. This family is one of the most important groups of North American fishes, highly esteemed for sporting and food qualities. The black basses, genus *Micropterus*, are well known as sport fishes, but most centrarchid genera contain important game fishes (Jenkins and Burkhead, 1994).

Centrarchids are small- to medium-sized percoid fishes. Many species have bright yellow, orange, or red colors on the breast and belly. These color patterns, usually associated with breeding males of the genus *Lepomis*, doubtless led to the common name "sunfishes." The smallest species in the family are in *Enneacanthus*, the banded sunfish genus (not present in the Ohio River drainage); adults range from about 30 to 70 mm SL. The largemouth bass *Micropterus salmoides* is the largest centrarchid, reaching sizes in excess of 900 mm TL (about 3 ft) and 10 kg (22 lb) (Etnier and Starnes, 1993; Jenkins and Burkhead, 1994).

Except when very young, centrarchids are territorial or occur in small aggregations, often associated with submerged structure such as stumps, trees, rocks, vegetation, or overhanging banks. The crappies (*Pomoxis*), however, are often found in large aggregations, especially just prior to spawning. Centrarchids usually begin to spawn in the spring and the spawning season is often extended, lasting well into the summer or early fall. All species of centrarchids that occur in the Ohio River drainage are nest builders, and the male typically guards the nest after spawning. Nests are shallow depressions in the substrate usually created by the male with sweeping actions of his caudal fin. The typical spawning ritual includes circling activity about the nest by the male and female, punctuated by short encounters during which the female lies on her side extruding eggs that are fertilized by the male. Two or more females may spawn within the nest of a single male. After spawning is accomplished, the female either leaves the nest voluntarily or is chased away by the male. The male then guards the incubating eggs, departing the nest only to defend his territory or to begin courting another female (Breder, 1936; Etnier and Starnes, 1993; Jenkins and Burkhead, 1994).

Because of overlaps in spawning habitat and behavior, sunfishes, especially members of the genus *Lepomis*, frequently hybridize. The green sunfish *L. cyanellus* is an especially promiscuous and indiscriminate spawner (Etnier and Starnes, 1993). Hybridization usually occurs when spawning habitat is limited or disturbed, when one species greatly outnumbers another, or with introduced species (Hubbs, 1955). This frequency of hybridization should be kept in mind when contemplating any specimen that is particularly difficult to identify through the use of regional keys and illustrations (Etnier and Starnes, 1993).

The basic morphology of centrarchids is much like that of many other percoid families. They have a moderately elongate body that is moderately to very compressed; a medium or large head; a small to large mouth that is terminal or subterminal, with the upper jaw protrusile; villiform teeth are present in both jaws, on the vomer, and on the palatine bones and tongue of some species; eyes are medium or large; lateral line usually complete (except in *Lepomis symmetricus*); scales usually ctenoid, present over entire body, as well as on the breast, cheeks, and opercles. The spines (5–13) and rays of the dorsal fin are broadly joined by a membrane; caudal fin either rounded, emarginated, or slightly forked, with 17 principal rays; anal fin usually with 3 or more spines; pelvic fins in thoracic position, usually with 1 spine and 5 soft rays; pectoral fins are set high on the body. Branchiostegal rays number 5–7; psuedobranchae are small and concealed; branchiostegal membranes are separate, not connected to the isthmus (Etnier and Starnes, 1993; Jenkins and Burkhead, 1994; Boschung and Mayden, 2004).

Distinguishing characteristics for larval forms in the family Centrarchidae include a large oval yolk sac at hatching; position of the oil globule variable, but usually situated posterior in the yolk sac; the vent near midbody; the body usually robust with a large head; air bladder distinct; a short gut that coils with growth; a continuous spinous and soft dorsal fin; and total myomere counts (range: 26–31) usually less than Percidae (range: 33–52) and more than Moronidae (range: 18–22).

LITERATURE CITED

Boschung, H.T., Jr. and R.L. Mayden. 2004.
Breder, C.M. 1936.
Etnier, D.A. and W.C. Starnes. 1993.

Hubbs, C.L. 1955.
Jenkins, R.E. and N.M. Burkhead. 1994.
Nelson, J.S. et al. 2004.

GENUS
Ambloplites Rafinesque

Thomas P. Simon IV

Ambloplites is an eastern North American genus containing four species: *Ambloplites constellatus* Cashner and Suttkus (Ozark Uplands of northern AR and southern MO); *A. cavifrons* Cope (rivers of Albermarle and Pamlico sounds in VA and NC); *A. ariommus* Viosca (Gulf coastal drainages from the Apalachicola River west to Lake Pontchartrain, eastern tributaries to the Mississippi River in southern MS, and west of the Mississippi River in the St. Francis, White, and Ouachita Rivers); and *A. rupestris* (Rafinesque) rock bass (Mississippi, Great Lakes–St. Lawrence, and southern Hudson Bay basins), the only species native to the Ohio River drainage (Hubbs and Lagler, 1958; Scott and Crossman, 1973; Etnier and Starnes, 1993; Boschung and Mayden, 2004)

Characteristics of adult *Ambloplites* include an oblong body that is moderately elevated and compressed; a projecting lower jaw; well-developed supramaxilla; an opercle with two flat projections; preopercle serrate at an angle; teeth on tongue, vomer, and palatines. The dorsal fin is continuous with a shallow gap between spines and rays. Dorsal fin spines 10–11; anal fin spines 5–7. Members of this genus have no bright red, orange, blue, or green colors (Boschung and Mayden, 2004).

The rock bass was described in 1817 from lakes of New York, Vermont, and Canada (Rafinesque, 1817). This species prefers cool, clear, silt-free streams and lakes (Smith, 1979; Becker, 1983). It is often associated with pondweeds and water willow (Adams and Hankinson, 1928; Hubbs and Lagler, 1958; Trautman, 1981), while in deeper waters it occurs among stumps, logs, and roots (Jordan and Evermann, 1923; Adams and Hankinson, 1928). Spawning occurs in springtime (Forbes and Richardson, 1909; Trautman, 1981; Etnier and Starnes, 1993; Mettee et al., 2001) in shallow water (Langlois, 1954; Finger, 1982; Lester et al., 1996). Nests are usually constructed by male rock bass (Breder, 1936; Langlois, 1954; Breder and Rosen, 1966) but rock bass may spawn in the nests of other species (Carbine, 1939). The male guards and cares for the incubating eggs (Breder, 1936; Carbine, 1939; Gross and Nowell, 1980; Noltie and Keenleyside, 1987), but soon leaves the fry after they swim up (Breder, 1936; Gross and Nowell, 1980; Noltie and Keenleyside, 1987).

TAXONOMY AND SYSTEMATICS OF GENUS *AMBLOPLITES*

Ambloplites has either a basal position in centrarchid phylogeny (Branson and Moore, 1962), or is moderately advanced (Smith and Bailey, 1961; Mabee, 1988; Sachdev, 1993). Smith and Bailey (1961) considered *Ambloplites* sister to *Acantharchus* of the Atlantic Slope. Branson and Moore (1962) related *Ambloplites* to *Lepomis* and *Enneacanthus*. Mok (1981) included *Ambloplites* in an unresolved trichotomy with *Acantharchus* and a clade inclusive of *Pomoxis*, *Centrarchus*, and *Acantharchus* based on olfactory data (Eaton, 1956). Based on kidney morphology, the genus is in an unresolved polytomy with *Enneacanthus*, *Lepomis*, *Acantharchus*, *Ambloplites*, and a clade inclusive of *Pomoxis* and *Centrarchus*. Chang (1988) considers *Ambloplites* a sister clade inclusive of *Archoplites*, *Centrarchus*, and *Pomoxis*. A phylogenetic study by Mabee

(1993) and Wainwright and Lauder (1992) found the genus to be a sister clade inclusive of *Pomoxis, Archoplites,* and *Centrarchus.* Roe et al. (2002) consistently resolved *Ambloplites* as the sister group to a clade inclusive of *Pomoxis* and *Archoplites.*

Adult rock bass have five or more anal fin spines; *Micropterus* and *Lepomis* have three. Rock bass have 11 or fewer soft rays in the anal fin compared to 14 or more for *Pomoxis* and *Centrarchus* (Etnier and Starnes, 1993).

For characters used to distinguish rock bass larvae and juveniles from the young of elassomatids and other centrarchids see the "Key to Genera" on pages 27–29, Tables 4 and 5, and the "Taxonomic Diagnosis" section in the species account that follows.

LITERATURE CITED

Adams, C.C. and T.L. Hankinson. 1928.
Becker, G.C. 1983.
Boschung, H.T., Jr. and R.L. Mayden. 2004.
Branson, B.A. and G.A. Moore. 1962.
Breder, C.M., Jr. 1936.
Breder, C.M., Jr. and D.E. Rosen. 1966.
Carbine, W.F. 1939.
Chang, C.H. 1988.
Eaton, T.H., Jr. 1956.
Etnier, D.A. and W.C. Starnes. 1993.
Finger, T.R. 1982.
Forbes, S.A. and R.E. Richardson. 1909.
Gross, M.R. and W.A. Nowell. 1980.
Hubbs, C.L. and K.F. Lagler. 1958.
Jordan, D.S. and B.W. Evermann. 1923.

Langlois, T.H. 1954.
Lester, N.P. et al. 1996.
Mabee, P.M. 1988.
Mabee, P.M. 1993.
Mettee, M.F. et al. 2001.
Mok, H.K. 1981.
Noltie, D.B. and M.H.A. Keenleyside. 1987.
Rafinesque, C.S. 1817.
Roe, K.J. et al. 2002.
Sachdev, S.C. 1993.
Scott, W.B. and E.J. Crossman. 1973.
Smith, C.L. and R.M. Bailey. 1961.
Smith, P.W. 1979.
Trautman, M.B. 1981.
Wainwright, P.C. and G.V. Lauder. 1992.

ROCK BASS

Ambloplites rupestris (Rafinesque)

Thomas P. Simon IV and John V. Conner

Ambloplites, Greek: *amblys*, "blunt" and *hoplites*, "heavily armed foot soldier"; *rupestris*, Latin: *rupestris*, "living among rocks."

RANGE

The rock bass is native to the Mississippi, Great Lakes–St. Lawrence, and southern Hudson Bay basins; recorded from southern Canada from QU to southeastern SK; south through the Dakotas, NE, southern KS, northeast OK, central AR, LA,[30] and TX;[17] east to FL panhandle; northward, generally west of Appalachians to VT and QU.[15,30] It occurs in all of the Great Lakes[30] and is common to abundant throughout the Lake Michigan drainage.[48] It occurs along the pebbly beach of Lake Michigan and in some of the harbors in IL.[52] Widely introduced on the Atlantic Slope[17] in WA,[10] CO, and WY.[30] Introduced in Patuxent River and Rock Creek in 1894;[28] now in all major tributaries of Chesapeake Bay[3] and in NJ.[26]

HABITAT AND MOVEMENT

Rock bass is found in relatively shallow waters of lakes,[1,2,105,106,116] ponds,[23] clear gravelly rivers,[25,67,69,94,100,133,140] creeks,[2] and clear rocky streams[9,69,92] of moderate gradients.[4,100,123,138] Preferred substrates are bottoms of rock,[2,9,18,30,67,127,128] gravel, or boulders[4,67] often in association with aquatic vegetation[2,7,17,31,67,69,94,100,108,128,133,140] such as bulrushes,[34] *Potomogeton*,[25] and water willow.[4] Rock bass is often found in eddies behind large boulders;[67,69,94,100,133,140] also in deep holes,[25] around stumps, logs, and roots, and in shady places under high banks and projecting rocks.[2,25] It is intolerant of silt and turbidity and is most abundant in cold, well-oxygenated water;[52,54] least abundant in muddy lakes.[2]

Although still common in many streams, rock bass is less widely distributed than formerly because of siltation and general deterioration of water quality.[52] It was encountered in clear water (60% frequency), slightly turbid water (27% frequency), and turbid water (13% frequency) of varying depths, over substrates of sand (26% frequency), gravel (20%), mud (13%), rubble (12%), boulders (12%), silt (9%), bedrock (4%), clay (2%), detritus (1%), hardpan (1%), and marl (1%). It was found in lakes and reservoirs, and in streams of the following widths: 1.0–3.0 m (6% frequency), 3.1–6.0 m (15%), 6.1–12.0 m (9%), 12.1–24.0 m (31%), 24.1–50.0 m (24%), more than 50 m (16%)[54]

Maximum reported water depth for rock bass is 21.3 m.[11] Maximum temperature is unknown, but it is intolerant of temperatures greater than 26°C.[113,115,137] The maximum water temperature recommended for aquarium stock is 20°C.[35,69] Maximum recorded salinity is 0.03 ppt,[37] but it has been recorded "off Worton Point, MD," where salinities are probably somewhat higher.[3] Rock bass is intolerant to acid mine drainage and occurs in waters of variable acidity.[95]

The rock bass is a schooling species.[23] It tends to show little movement[12,79,93,132,134] and is apparently non–migratory, occupying a very limited area, and unlikely to range through more than a mile of stream in the course of a 2-year period.[32,94,118,126] It hibernates in summer haunts under leaves and debris or among roots and water willows.[4,67]

The rock bass is a "chameleon" and can undergo rapid and dramatic color changes to match its surroundings.[68]

DISTRIBUTION AND OCCURRENCE IN THE OHIO RIVER SYSTEM

Rock bass is reported as occasional in the upper two–thirds of the Ohio River and rare in the lower third. In a 1984 report, the only trend in distribution was an increase in the number of records from upstream of ORM 80, after 1970.[47] An analysis of data from collections made during the 13-year period 1973–1985 from five localities in the upper and middle Ohio River established that rock bass was significantly more abundant in the upper Ohio River than in the middle.[148]

This species is occasional in IN and occurs in the Wabash River drainage.[70] It is generally distributed in northern and central IL, but is extremely sporadic in the western and southern parts of the state.[52,66] In WV and VA, rock bass is indigenous only to the Tennessee and Big Sandy drainages, and has been introduced into the New River drainage.[69,141,142] In KY, rock bass is generally distributed throughout the eastern two-thirds of the state;[42] found in the Green, Kentucky, and Licking River drainages.[42,129–131] In TN, it is common throughout the Tennessee and Cumberland River drainages.[57]

It is limited to the Tennessee River drainage in AL, but may enter the Tombigbee River system through the Tennessee–Tombigbee Waterway.[71,73]

SPAWNING

Location

Rock bass move into areas of very shallow water to spawn,[22,106,124] such as along the shores of lakes and streams,[20,46] in the lower reaches of rivers entering lakes,[22] also in swampy bays,[2] and on gravel shoals.[30] Spawning is reported in water from a few centimeters deep to more than 1.0 m deep.[5,22,112]

Male rock bass build nests over bottoms of sand,[40,41,45] gravel,[2,27,40,41,45,112] marl[5,34] or exposed roots.[20] The nests are often close together[30] and often near cover such as aquatic vegetation,[2,20,27,38,45] sticks, stakes, rocks, or other structures.[2,27,100] The nest is a circular depression[2,20,22,112] varying in diameter from 200 mm[2,22] to 1.1 m,[20] and with maximum depth (rim to bottom) of about 51 mm.[34]

In some cases, no definite nest is built.[34] Rock bass may spawn in nests of other species.[20]

Season

On May 9, 1986, rock bass eggs were found on rocks in Coal Creek, a tributary of the Clinch River, Anderson County, TN.* Rock bass eggs were collected on May 5, 1987 from Walden Creek, Sevier County, TN.*

Additional reports of spawning season follow.

Location	Season
Canada[30,74]	16 days[74] in June
Lake Erie[1]	May and June
WI[54]	Late May to June
MN[38]	May to June
MI[20]	Late May to mid-July
MO[41]	April to June
VA[69]	April to June
IN[2,27]	Mid-May to mid-June
IA[44]	May to early June
IL[46,52,66]	June
TX[19]	Begins in March
AL[71]	March and April
NC[21]	Spring

Temperature

Nesting activity is triggered at 12.8–15.6°C.[41] Spawning occurs at 15.6–21.1°C.[30,49] Also reported that spawning is initiated when water warms to 20.5°–21°C and may continue to 26°C.[54,55,69,111] Rock bass eggs were collected in TN from water 13.3°C.*

Fecundity

Reports of fecundity for rock bass range from 2,000 to 11,000 eggs,[9,30,59,99] depending on the size of the fish.[59] Also reported that a single nest may contain 344–1,758 eggs,[20] with about 400–500 eggs released by a single spawning female.[80–82,120]

A 173-mm, 136-g female rock bass captured 7 June from Blake Creek (Waupaca County), WI had ovaries 14.2% of her body weight. She held an estimated 6,300 mature eggs that were orange and 1.7–1.9 mm in diameter, and a small number of immature (white-yellow) eggs, 0.7–1.3 mm in diameter.[54]

Sexual Maturity

There are reports of rock bass maturing at age 2, although these fish would not have spawned until end of their third full year of life.[69,121] In WI, males and females were mature at 89 mm TL.[29] In ON, the size at first breeding was 120 mm TL, weight at first breeding 33 g, and age at first spawning 4 years.[74]

Spawning Act

Rock bass is a solitary spawner, the male usually constructing its nest near vegetation,[2,20,27,38,42,45] a boulder,[52] or other structure.[2,27,100] The male fans out sand and debris from the bottom to create a depression of 20–25 cm diameter.[22] The reproductive behavior is quite similar to that of other sunfishes, but spawning may begin a bit earlier.[52]

Details of the spawning behavior of rock bass show that immediately preceding and during the spawning season, adult females congregate in pools.[5,54–58] A female approaches a nest only when she is ready to deposit her eggs. She is driven into the nest by the male, who guards her carefully until the eggs have been deposited. During spawning and fertilization, the female reclines on her side and the male remains upright. The two fish engage in a peculiar rocking motion in a head-to-tail position. Only a few eggs at a time are extruded, and at each deposit, milt is extruded by the male.[99] This behavior may continue for an hour or more, after which the female leaves the nest and does not return.[34,36]

The fertilized eggs become attached to aquatic vegetation and roots,[34,36] rocks,* or other substrate at the bottom of the nest. The male remains with the eggs and guards and cares for them. He takes up a position above the nest, and every now and then fans the eggs with his fins.[5,20,56,58,80,81,100,102] Large males are more successful in guarding nest territory and defending nests post-spawning.[104] A few days after the eggs have hatched, the fry gradually rise out of the nest. The male soon leaves them to fend for themselves.[5,56,80,81]

In an aquarium, a female with a distinctly blunt and red ovipositor attempted to mate with a male that was fanning a nest with eggs.[5,56] The female quivered considerably, and the two fish performed the peculiar rocking motion in a head-to-tail position.[5,54–58] Throughout the pairing, this male continued, with little interruption, to fan the eggs he was guarding. During the fanning period, males in the aquarium drove off large crayfish (*Cambarus*) whenever the crayfish got within a few centimeters of the eggs. Such struggles continued for an hour until the crayfish eventually withdrew.[5,54–58]

EGGS

Description
Fertilized eggs are demersal,[50] adhesive,[2,30,50] about the size of small shot.[27] Egg diameter is 2.0[49,74] to 2.1 mm.[*,49] There is a single oil globule, 0.76 mm in diameter.[49]

Incubation
Eggs incubated at 20.5–21.0°C hatched in 3–4 days.[5,30]

Development
Embryology and ontogenetic development of rock bass is described from Lake Erie.[147]

YOLK-SAC LARVAE

See Figure 15.

Size Range
Rock bass hatch between 4.7 and 5.6 mm;[49,50] or at 6.4 mm TL (Canada).[74] Yolk is absorbed between 7.4 and 8.6 mm TL.[49,50]

Myomeres
See Table 14.

Preanal 12–14; postanal 17–19; total 30–32; [50,103] or preanal 10–13 and postanal 16–18.[43]

Morphology
4.7– 5.6 mm TL. Yolk sac ovoid;[50] oil globule is located in the middle[49] or posterior portion[50] of the yolk sac; head straight, gut not apparent (newly hatched).[49]

6.4 mm TL. Notochord upturned slightly.

6.7 mm TL. Anus faintly visible, maxilla and dentary developing.[49]

6.8 mm TL. Air bladder beginning to inflate, mouth open.[50]

Morphometry
See Table 15.

As % TL: PreAL 45–46; HL 16–21;[49,50] GD 30; ED 10.[50]

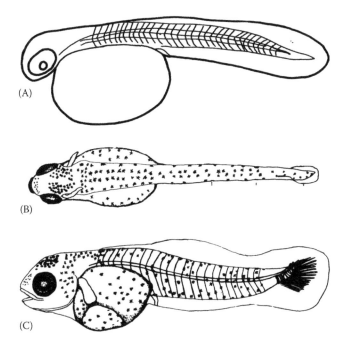

Figure 15 Development of young rock bass. (A) Yolk-sac larva, 5.6 mm TL. B–C. Yolk–sac larva, 6.7 mm TL: (B) dorsal view; (C) lateral view. (A reprinted from Figure 1, reference 50, with publisher's permission; B–C reprinted from Figure 1, reference 49.)

Table 14

Myomere count ranges (average and standard deviation
in parentheses) at selected size intervals
for young rock bass from TN.*

Size Range (mm TL)	N	Predorsal fin	Myomeres		
			Preanal	Postanal	Total
4.5–8.5	71	—	10–14 (12.6 ± 0.92)	15–18 (16.8 ± 0.81)	26–31 (29.4 ± 1.39)
9.1–11.0	5	5–8 (6.0 ± 1.09)	14–16 (15.0 ± 0.63)	15–16 (15.2 ± 0.40)	29–31 (30.2 ± 0.74)
11.6–12.7	3	7–8 (7.67 ± 0.48)	13–15 (14.3 ± 0.97)	16	29–31 (30.3 ± 0.97)

Fin Development

5.6–6.4 mm TL. Pectoral buds evident.[49,50]

6.7 mm TL. Epural and hypural elements evident.[49]

6.9–7.4 mm TL. Caudal fin rays visible.[49,50]

Pigmentation

4.7–5.6 mm TL. Newly hatched and unpigmented.[49,50]

6.4–6.8 mm TL. Eyes pigmented;[49,50] large melanophores on top of head, air bladder, lateral and ventral aspects of body, and yolk sac (6.7–6.8 mm);[49,50] a few dorsal melanophores present on each side of median finfold (6.8 mm).[50]

7.0 mm TL. Melanophores scattered over entire body.[43]

POST YOLK-SAC LARVAE

See Figures 16 and 17.

Size Range

Yolk is absorbed between 7.4 and 8.6 mm TL.[49,50] Transition to juvenile stage occurs at <17 mm TL.[53] A single specimen 10.5 mm TL[1] is described.

Myomeres

See Table 14.

Myomeres 13 + 18;[1] or preanal 12 or 13;[1,50,103] postanal 16–17[50,103] or 18;[1] total 28–31.[1,50,103]

Morphology

10.5 mm TL. Body oblong, compressed; caudal peduncle stout; mouth terminal, oblique; maxilla extends past anterior margin of pupil.[1,19,53]

Morphometry

See Table 15.

7.5–10.0 mm TL. As % TL: PreAL 41–46; HL 23 to 27;[49,50] GD 27; ED 12.[50]

10.5 mm TL. As % TL: PreAL 47.6; HL 29.5; GD 30.3. As % HL: ED 35.5.[1]

10.0–17.0 mm TL. As % TL: PreAL length 41–51; HL 28–31;[1,49,50] GD 29–30; ED 11.[1,50]

Fin Development

7.4–8.6 mm TL. Rays developing in dorsal and anal fins;[16,50] pectoral fin rays visible (8.6 mm).[50]

9.0 mm TL. Pelvic buds evident; dorsal fin with nine short spines and eight or more soft rays.[49]

11–13.5 mm TL. Pelvic fin rays visible;[49,50,53] 6 or 7 pectoral fin rays evident (13.0 mm).[49,53]

12.9–14.0 mm TL. Caudal fin forked; dorsal fin with 10–11 spines and 9–11 soft rays.[49]

Table 15

Morphometric data for young rock bass from TN grouped by selected intervals of total length. Characters are expressed as percent total length (TL) and head length (HL) with a single standard deviation. Range of values for each character is in parentheses.

	TL Intervals (mm)			
Length Range	4.55–8.48	9.10–11.02	11.62–12.65	18.20–20.55
N	71	5	3	5
Characters	Mean ± SD (Range)	Mean ± SD (Range)	Mean ± SD (Range)	Mean ± SD (Range)
Length as % TL				
Snout	4.03 ± 0.67 (0.20 – 0.45)	4.27 ± 0.39 (0.35 – 0.53)	5.37 ± 0.68 (0.55 – 0.71)	6.10 ± 0.30 (1.12 – 1.25)
Eye diameter	10.40 ± 1.32 (0.44 – 0.98)	11.2 ± 0.25 (1.00 – 1.20)	11.4 ± 0.35 (1.30 – 1.46)	10.9 ± 0.13 (2.00 – 2.22)
Head	23.8 ± 4.38 (0.88 – 2.35)	27.8 ± 1.37 (2.33 – 3.05)	28.5 ± 0.55 (3.25 – 3.70)	29.5 ± 0.60 (5.50 – 5.92)
Predorsal	33.4 ± 1.19 (2.45 – 2.90)	32.4 ± 1.64 (2.90 – 3.90)	33.4 ± 0.84 (4.00 – 4.20)	32.6 ± 0.66 (6.00 – 6.50)
Prepelvic	27.8 ± 1.20 (2.05 – 2.50)	28.1 ± 0.75 (2.45 – 3.10)	28.3 ± 1.40 (3.25 – 3.80)	27.4 ± 0.75 (5.20 – 5.50)
Preanal	46.5 ± 2.61 (2.48 – 4.08)	48.6 ± 1.65 (4.25 – 5.3)	47.2 ± 0.83 (5.60 – 5.95)	45.4 ± 0.85 (8.30 – 9.35)
Postanal	53.5 ± 2.61 (1.90 – 4.68)	51.4 ± 1.65 (4.60 – 5.72)	52.8 ± 0.83 (6.02 – 6.70)	54.6 ± 0.85 (9.90 – 11.20)
Standard	90.5 ± 6.03 (4.43 – 7.21)	81.6 ± 1.19 (7.50 – 9.00)	80.7 ± 1.23 (9.35 – 10.15)	79.2 ± 0.79 (14.5 – 16.35)
Air bladder	11.7 ± 1.47 (0.65 – 1.30)	11.9 ± 0.85 (1.05 – 1.25)	12.0 ± 1.68 (1.25 – 1.70)	7.35 ± 0.95 (1.20 – 1.50)
Eye to air bladder	11.4 ± 0.91 (0.75 – 1.08)	12.3 ± 1.27 (0.90 – 1.40)	9.97 ± 0.23 (1.15 – 1.30)	9.61 ± 0.71 (1.70 – 2.00)
Fin Length as % TL				
Pectoral	15.4 ± 3.60 (0.48 – 1.88)	19.0 ± 3.10 (1.40 – 2.60)	19.9 ± 0.24 (2.30 – 2.50)	21.0 ± 0.74 (3.80 – 4.40)
Pelvic	2.61 ± 0.48 (0.15 – 0.30)	5.8 ± 1.75 (0.30 – 0.80)	10.8 ± 2.86 (0.95 – 1.85)	15.0 ± 0.32 (2.75 – 3.05)
Body Depth as % TL				
Head at P1	23.6 ± 1.74 (1.60 – 2.41)	28.8 ± 2.21 (2.43 – 3.50)	28.1 ± 1.01 (3.30 – 3.60)	27.7 ± 1.29 (5.05 – 5.73)
Head at eyes	24.0 ± 1.40 (1.35 – 2.11)	26.9 ± 1.79 (2.30 – 3.30)	24.1 ± 0.42 (2.85 – 3.05)	23.8 ± 0.53 (4.40 – 4.85)
Preanal	20.7 ± 2.60 (1.20 – 2.08)	26.2 ± 2.01 (2.10 – 3.05)	22.8 ± 4.23 (2.00 – 3.40)	26.4 ± 1.00 (4.90 – 5.50)
Caudal peduncle	8.24 ± 0.57 (0.58 – 0.80)	9.50 ± 0.91 (0.80 – 1.10)	10.1 ± 0.39 (1.18 – 1.32)	10.6 ± 0.69 (1.90 – 2.30)
Body Width as % HL				
Head	42.3 ± 5.44 (0.70 – 1.10)	53.0 ± 9.41 (1.00 – 1.90)	47.8 ± 2.62 (1.55 – 1.85)	48.7 ± 5.43 (2.40 – 3.20)

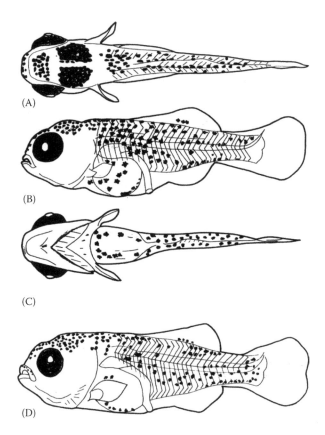

Figure 16 Development of young rock bass. A–C. Post yolk–sac larva, 8.6 mm TL: (A) dorsal view; (B) lateral view; (C) ventral view. (D) Post yolk–sac larva, 9.1 mm TL. (A–D reprinted from Figure 1, reference 50, with publisher's permission.)

Pigmentation

7.4–8.6 mm TL. Dorsally, there are dense patches of pigment on the head and behind eyes and a less dense patch on the snout between eyes.[49,50] Melanophores are present dorsally on the posterior dorsal portion of air bladder. Some pigment is visible on caudal, anal, and dorsal fin rays.[49]

10.3 mm TL. Pigment has increased on the dorsal, lateral, and ventral surfaces of the body.[50]

10.5 mm TL. Body and head have scattered large stellate melanophores, especially on tip of both jaws, over snout, and on top of head. A heavy dorsal and ventral series of pigment is present along the margins of body and around fins. The abdomen is lighter than rest of body, but has a few small chromatophores. Chromatophores are developed on dorsal, caudal, anal, and pectoral fins, usually near their bases. There are small chromatophores on the dorsal surface of air bladder.[1]

13.5–16.5 mm TL. Large melanophores are present on the gut and saddle-like bands of pigment appear on the lateral surface of body. Otherwise pigment patterns similar to those described previously.[49,50,53]

JUVENILES

See Figures 18 and 19.

Size Range

Adult complement of fin rays is formed and all finfolds are absorbed at 17 mm TL.[53] Sexual maturity is reached on some at 89 mm TL.[1]

Morphology

Early juveniles. Body oblong,[14] rather deep, and compressed;[13] mouth terminal,[30] large, directed obliquely upward;[13] maxillary reaches to middle of pupil;[27] scales ctenoid;[13] caudal fin emarginate.[14]

30.6 mm TL. Eye large, maxilla reaching middle or posterior edge of pupil.[50]

36.0 mm TL. Squamation complete.[49]

Larger juveniles. Scales above the lateral line 11[13]–16;[16] lateral line scales 37–51;[30] scales on cheek 6–8.[9] Gill rakers on lower limb 9–12.[7] Branchiostegals 5–6;[9] vertebrae 29–30[30] or 32.[27,51]

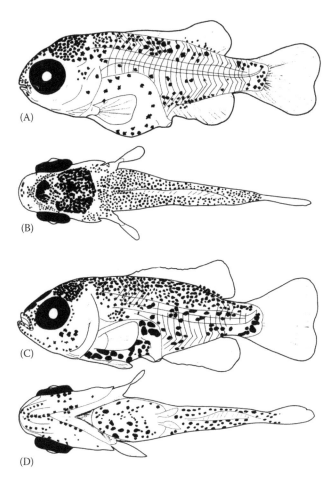

Figure 17   Development of young rock bass. (A) Post yolk–sac larva, 10.3 mm TL. B–D. Post yolk–sac larva, 13.5 mm TL: (B) dorsal view; (C) lateral view; (D) ventral view. (A–D reprinted from Figure 1, reference 50, with publisher's permission.)

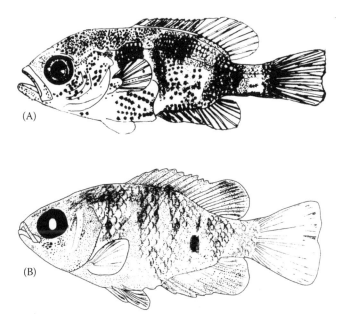

Figure 18   Development of young rock bass. A–B. Juveniles: (A) 16.4 mm TL; (B) 30.6 mm TL. (A reprinted from Figure 3, reference 49; B reprinted from Figure 1, reference 50, with publisher's permission.)

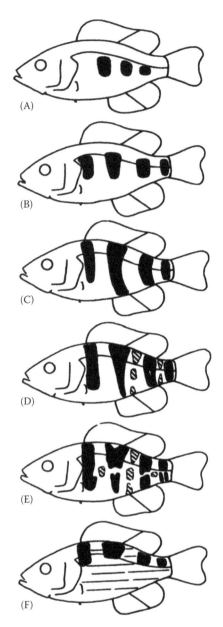

(A)

(B)

(C)

(D)

(E)

(F)

Figure 19 Development of young rock bass — schematic drawings showing ontogeny of melanistic pigment patterns. (A) 13.3 mm SL; (B) 12.4 mm SL; (C) 14.4 mm SL; (D) 15.4 mm SL; (E) 20.0 mm SL; (F) 36.7 mm SL. (A–F reprinted from Figure 8, reference 143, with publisher's permission.)

Morphometry

See Table 15.

17.0–40.0 mm TL. As % TL: PreAL 42–43; HL 31–33;[48,50] GD 37; ED 11.[50]

Early juveniles. As percent TL: GD 31.8–36.7; CPD 12–14. As percent HL: ED 26.7–40.9; maxillary 33.3–45.5.[30] Head 2.75 times in length.[21] Base of anal fin about one-half length of dorsal base.[17,50] The body of the juvenile was relatively deep and laterally compressed; the greatest depth (37% TL)

occurred at the origin of the dorsal fin. The eyes were large and located high, well in front of the center of the head. Head length was about 31% TL. The maxillary reached the middle or posterior edge of the pupil.[50]

Fins

17.0–22.0 mm TL. Full complement of adult fin rays is present (17 mm[43,53] or 22 mm[49]).

30.6 mm TL. Base of dorsal fin twice as long as anal fin base.[50]

Larger juveniles. Dorsal fin with 10[13]–12[9,50,69] spines and 10–11[34,69] soft rays; anal fin has 5–7 spines and 9–11 soft rays;[13,50,69] pectoral fin rays 12–14;[34,69] pelvic fin spines I, rays 5;[5,30] caudal fin with 17 principal rays.[51]

Pigmentation

17.0–40.0 mm TL. Several vertical bands,[43,53] blotches of melanophores,[143] or indistinct "saddles" appear laterally.[50] Each scale below the lateral line becomes marked with a black spot and the opercular flap has an indistinct blackish spot. The dorsal and anal fin spines become more pigmented than the interconnecting membranes.[50]

51.0 mm TL. Sides with general black marbling; saddle marks of adult inconspicuous by day, well developed at night. Young with irregular bars and blotches of black[24,25] or brown.[27]

Juvenile rock bass share a general pigmentation ontogeny with other *Ambloplites*.[143] Based on the size of pigment blotches on the smallest juvenile specimen examined, the second primary bar of melanistic pigment is the first to appear (Figure 19A). Four primary bars are formed. The three posterior bars extend from the dorsal to the ventral midline. The first primary bar extends from the dorsal midline to the pectoral fin base (Figure 19B and 19C). The temporal sequence of secondary bar development is variable: secondary bars may appear initially between primary bars two and three, three and four, or in both positions simultaneously (Figure 19D). In *Ambloplites*, similar to *Centrarchus* and *Archoplites*, when the secondary bars appear, the adjacent areas of the primary bars become light centered and break up dorsoventrally (Figure 19E). Primary and secondary bars eventually become difficult to distinguish, and a longitudinal striping pattern is present in adults (Figure 19F). Dorsal portions of the primary bars are retained in adults. Differences among *Ambloplites* species in longitudinal striping and other aspects of juvenile and adult patterns are described.[144–146] The temporally ordered primary bar ontogeny is present in all *Ambloplites* species.[143]

Larger juveniles. Light olive,[27] dark slate, or olive green with bronze or coppery reflections above; back crossed with 4–7 dark broad saddles;[4] upper sides olivacious[9] or brassy olivacious;[15] sides below lateral line lighter;[4] sides irregularly mottled with black, or with dorsal saddles extending to lateral line; a dark spot or bar on each lateral scale, these forming interrupted black lateral stripes;[9,13,15,27] venter white, brassy or brassy white,[4] white dusted with brassy,[31] or bluish white with dark dots; top of head dark green; sides of head brassy with dark bar downward from eye; opercular spot black,[9,27] gold rimmed in males;[13] eye described as red[9,24] or golden overlaid with crimson.[23] Dorsal, anal, and caudal fins delicate greenish,[13] olivacious yellow,[27] or brownish yellow,[13] with dark mottlling[9] and, usually, black margins;[4] vermiculations of soft dorsal fin often forming ocelli; pelvic and pectoral fins transparent, olive,[4] or pale lemon.[27] In clear water young rock bass are boldly marked with black and bronze; in turbid water often yellow bronze and lacking back markings.[4]

TAXONOMIC DIAGNOSIS OF YOUNG ROCK BASS

Similar species: other centrarchids, especially smallmouth bass.

See "Key to Genera" (pages 27–29) for characters used to distinguish young rock bass from the young of other genera of centrarchids.

Adult rock bass are somewhat similar to the warmouth but differ from that species and all *Lepomis* species in having more anal spines, a character useful for distinguishing juveniles.

Larvae and juveniles. Rock bass yolk-sac larvae can be distinguished from those of some *Lepomis* by their larger size at hatching. By 6.4 mm TL, yolk-sac larval rock bass differ from other centrarchids, with the exception of smallmouth bass, in having melanophores scattered over their entire bodies (Figure 15). Once pigmentation has formed, larval and juvenile rock bass larvae should be confused only with smallmouth bass, because body pigmentation of the two species is similar. However, the scattered melanophores on the rock bass are large and widely spaced, while melanophores on smallmouth bass are smaller and the arrangement is much denser and gives a "dusky appearance." In addition, juvenile rock bass usually have 5–7 anal spines, while smallmouth bass have only 3.[50] Young rock bass superficially resemble the flier, but juveniles of that fish have a large spot or ocellus posteriorly placed in the soft dorsal fin and lack the large melanophores scattered over the body.*

ECOLOGY OF EARLY LIFE PHASES

Occurrence and Distribution (Figure 20)
Eggs. Attached to substrate at bottom of nest;[34,36] guarded by male parent during incubation.[56,57,80,81,100,102,110]

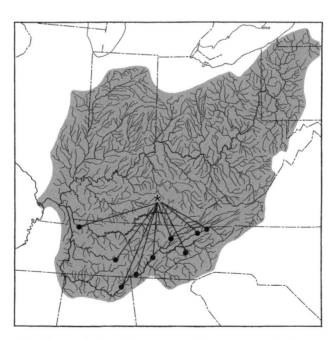

Figure 20   General distribution of rock bass in the Ohio River system and areas where early life history information has been collected (circles). Asterisk indicates TVA collection localities.

Yolk-sac larvae. Yolk-sac larvae have been observed among plants in bottom of the nest.[34,35,109] In MI, the number of larvae per nest ranged from 344 to 1,756 with an average of 796 for each of nine nests counted.[20] Yolk-sac larvae first appeared during early to mid-June.[119]

Post yolk-sac larvae. Post yolk–sac larvae rise from the nest a few days after hatching.[2,22]

Rock bass larvae were first collected in mid-July during night sampling in KY.[125] During the period 1976–1983, post yolk-sac larval rock bass 8–17 mm TL occurred in TVA ichthyoplankton collections in the Tennessee River drainage (Figure 20). Larvae were collected from mid-May to early August, usually in water 1–11 m deep. Numbers were always low, never exceeding three fish per sample, usually only one fish per sample.*

Juveniles. One-month-old rock bass have been observed along shore in grass.[36,107] Young of unspecified age are reported from shallow water, over algae-covered stones along shore, in dense aquatic vegetation, and in lower courses of steams.[1,2,6,8,18,30,33,109,121,135,152] In lakes, young rock bass move from littoral to limnetic zones as development proceeds.[46,109] Laboratory habitat preference studies indicated that rock bass juveniles preferred a heavily vegetated substrate.[75]

During the period 1976–1981, juvenile rock bass 19–49 mm TL occurred in TVA ichthyoplankton collections in the Tennessee River drainage (Figure 20). Juveniles were collected from early June to late August, usually at night in water 1–9 m deep. Numbers were always low, never exceeding one fish per sample.*

Early Growth

Growth of young rock bass varies with age (Table 16),[55] latitude, and water characteristics.[97,98] General growth curves for rock bass populations show that males grow more rapidly than females but the differential growth in the sexes is not distinctly apparent before the fourth year.[122] Most growth occurs before late July and early August.[60,61,136] Good growth was positively correlated with high temperatures, especially in June and September, as well as heavy precipitation in June.[61,91]

In ON, rock bass averaged 35 mm TL in August, 45 mm in September, and 46 mm in October.[74] In Muscote Bay (Bay of Quinte), ON, age-0 rock bass grew to 37 mm TL by late August and were 46 mm by early November.[152] In OH, rock bass YOY were 20–51 mm in October and 28–89 mm at about age 1.[56] The average summer growth rate from upper Niagara River, NY, was 0.51 mm/day.[75]

In TN, rock bass generally average about 45 mm TL the first year, 80 mm the second, 115 the third, 160 the fourth, and 195 the fifth.[149] In TN streams, young rock bass ranged from 33 to 46 mm TL (mean = 38 mm) in September.[76–78] In Norris Reservoir, TN, an age-2 rock bass measured 127 mm TL and seven age-3 rock bass averaged 178 mm TL.[39]

The calculated TL of rock bass collected from the tailwater of Nolin River Lake, KY, during 1983 was 43 mm at age 1, 86 mm at age 2, 130 mm at age 3, and 165 mm at age 4.[150]

Rock bass yolk-sac larvae weigh 0.004 g at hatching.[74] Reports of growth in weight for rock bass are extremely variable among ages (Table 16). In ON,

TABLE 16

Mean of means, central 50% mean of means, and range of mean of means of TL (mm) and weight (g) for young rock bass related to age summarized by Carlander (1977) from empirical data from numerous reports and various locations.[55]

Age	TL (mm)				Weight (g)		
	N	Mean of Means	Central 50%	Range	N	Mean of Means	Range
1	683+	78	54–93	28–241	442+		6–227
2	1604+	122	108–137	43–217	668+	43	8–142
3	1939+	152	132–168	84–279	996	71	11–207
4	1879+	172	155–193	105–279	1078	102	26–244
5	1166+	186	168–206	111–279	706	143	54–255
6	621+	204	183–228	135–315	58	162	83–365

Table 17

Length–weight relationships of young rock bass collected in AL during the period 1949–1964.[110]

TL (mm)	No. of Fish	Average Empirical Weight (g)
25	2	0.9
76	1	4.5
102	3	16.8
127	4	38.6
152	10	86.3
178	5	127.1
203	5	190.7

Source: Constructed from data presented in unnumbered table on page 44, reference 110.

age-0 rock bass averaged 1.7 g in September and 1.9 g in October, which is about 6% of adult weight.[74]

The average weight of 6-inch (152 mm TL) and 8-inch (203–204 mm TL) rock bass from AL (Table 17)[110] was heavier than the mean of means of comparable-sized fish from other areas (Table 16).[55]

The length–weight regression equation for age-0 individuals from Lake Opinicon, ON, is log W = −4.87 + 3.08 log length (r^2 = 0.61, F = 68.94).[74,148] The species attains a length of more than 300 mm (12 inches), but most individuals are less than 150 mm (6 inches).[52] Maximum length is about 346 mm.[10] The species lives 5 or more years and occasionally produces very large, old individuals.[52]

Feeding Habits

Rock bass is classified as an invertivore and carnivore.[88,89] Food items vary and insects, crustaceans, and small fishes make up the bulk of the diet of the adult.[52,72] Populations from IL fed on cladocerans, *Cyclops*, corixid, chironomid, and neuropteran larvae and some land insects.[62,86,87] In northeastern WI lakes, rock bass feed on insects (including dipteran, ephermerid and caddisfly larvae, dragonfly nymphs, and ants) followed by crayfish and fish.[63] In other WI lakes, rock bass fed mostly on crayfish (64% by volume), cladocerans (12%), insect larvae (11.8%), mites (4.2%), amphipods (2.9%), insect pupae (0.6%), ostracods (0.2%), and sand (4.2%).[64] In Lake Mendota, WI, they consumed crayfish (50%), cladocerans (14%), plant materials (10%), amphipods (5.8%), insect adults (5%), algae (3%), insect larvae (2%), and sand (11.2%).[64] In 24 rock bass stomachs from Lake Geneva, WI, 4 *Micropterus* sp., 8 *Notropis volucellus*, and 14 crayfish were found.[65]

Snails and other aquatic insects, including mayflies and stoneflies, may be important food for young rock bass.[83,96] As they grow larger, rock bass feed heavily on crayfish and fish.[83] Most feeding occurs during the day; especially in late evening.[82,101] Rock bass are known to feed at night and in some areas may be mostly nocturnal.[84,85,101] Rock bass YOY of all sizes fed on amphipods, copepods, and several other aquatic invertebrates.[75,90,114,117,135,139] In MN, rock bass 23–35 mm SL consumed algae and entomostraca; those 80–135 mm consumed mostly chironomids.[114]

Competitive interactions with smallmouth bass (*Micropterus dolomieu*) demonstrated that Levin's average niche breadth or dietary diversity value for smallmouth bass was 2.51, and for rock bass, 3.78. Levin's dietary overlap values were highest early in the season, 0.71–0.88, and decreased to 0.05–0.17 as the season progressed. Ecological segregation was by both food and habitat partitioning.[75]

LITERATURE CITED

1. Fish, M.P. 1932.
2. Adams, C.C. and T.L. Hankinson. 1928.
3. Musick, J.A. 1972.
4. Trautman, M.B. 1957.
5. Breder, C.M., Jr. 1936.
6. Reed, H.D. and A.H. Wright, 1909.
7. Bailey, R.M. et al. 1954.
8. Bailey, J.R. and J.A. Oliver. 1939.
9. Beckman, W.C. 1952.
10. Slipp, J.W. 1943.
11. Cady, E.R. 1945.
12. Werner, R.G. 1967.
13. Sterba, G. 1967.
14. Moore, G.A. 1962.
15. Eddy, S. 1969.
16. Whitworth, W.R. et al. 1968.
17. Hubbs, C.L. and K.F. Lagler. 1958.
18. Webster, D.A. 1942.
19. Leary, J.L. 1912.
20. Carbine, W. F. 1939.
21. Smith, H.M. 1907.
22. Langlois, T.H. 1954.
23. Bean, T.H. 1903.
24. Truitt, R.V. et al. 1929.
25. Jordan, D.S. and B.W. Evermann. 1923.
26. Fowler, H.W. 1920.
27. Evermann, B.W. and H.W. Clark. 1920.
28. Smith, H.M. and B.A. Bean. 1899.

29. Hile, R. 1941.
30. Scott, W.B. and E.J. Crossman. 1973.
31. Reighard, J.E. 1915.
32. Scott, D.C. 1949.
33. Brice, J.J. 1898.
34. Hankinson, T.L. 1908.
35. Meinken, H. Not dated.
36. Bade, E. 1931.
37. Martin, F.D. 1968.
38. Eddy, S. and J.C. Underhill. 1974.
39. Eschmeyer, R.W. 1948.
40. Cross, F.B. 1967.
41. Pflieger, W.L. 1975b.
42. Burr, B.M. and M.L. Warren, Jr. 1986.
A 43. Hogue, J.J., Jr. et al. 1976.
44. Harlan, J.R. and E.B. Speaker. 1969.
45. Cahn, A.R. 1927.
46. Forbes, S.A. and R.E. Richardson. 1909.
47. Pearson, W.D. and L.A. Krumholz. 1984.
48. Becker, G.C. 1976.
49. Powles, P.M. et al. 1980.
50. Buynak, G.L. and H.W. Mohr, Jr. 1979.
51. Bailey, R.M. 1938.
52. Smith, P.W. 1979.
53. Tin, H.T. 1982.
54. Becker, G.C. 1983.
55. Carlander, K.D. 1977.
56. Trautman, M.B. 1981.
57. Etnier, D.A. and W.C. Starnes. 1993.
58. Breder, C.M., Jr. and D.E. Rosen. 1966.
A 59. Vessel, M.F. and S. Eddy. 1941.
A 60. Snow, H. 1969.
61. Hile, R. 1942.
62. Forbes, S.A. 1880.
63. Couey, F.M. 1935.
64. Pearse, A.S. 1921.
65. Nelson, M.N. and A.D. Hasler. 1942.
66. Forbes, S.A. and R.E. Richardson. 1920.
67. Miller, R.J. and H.W. Robison. 2004.
68. Pflieger, W.L. 1997.
69. Jenkins, R.E. and N.M. Burkhead. 1994.
70. Gerking, S.D. 1945.
71. Mettee, M.F. et al. 1996.
72. Probst, W.E. et al. 1984.
73. Boschung, H.T., Jr. and R.L. Mayden. 2004.
74. Keast, A. and J. Eadie. 1984.
75. George, E.L. and W.F. Hadley. 1979.
76. Speir, H.J. 1969.
77. Cathey, H.J. 1973.
78. Gwinner, J.R. 1973.
79. Funk, J.L. 1955.
80. Gross, M.R. and W.A. Nowell. 1980.
81. Noltie, D.B. and M.H.A. Keenleyside. 1987.
82. Ross, S.T. 2001.
83. Elrod, J.H. et al. 1981.
84. Keast, A. and L. Welch. 1968.
85. Helfman, G.S. 1981.
86. Angermeier, P.L. 1985.
87. Angermeier, P.L. 1982.
88. Goldstein, R.M. and T.P. Simon. 1999.
89. Webb, P.W. 1984.
90. Keast, A. 1980.
91. Ryan, P.M. and H.H. Harvey. 1977.
92. Grossman, G.D. et al. 1995.
93. Storr, J.F. et al. 1983.
94. Pajak, P. and R.J. Neves. 1987.
95. Eaton, J.G. et al. 1992.
96. Roell, M.J. and D.J. Orth. 1993.
97. Putnam, J.H. et al. 1995.
98. Hile, R. 1931.
99. Ulrey, L. et al. 1938.
100. Scott, D.C. 1949.
101. Johnson, J.H. and D.S. Dropkin. 1993.
102. Sabat, A.M. 1994a.
103. Jayne, B.C. and G.V. Lauder. 1994.
104. Sabat, A.M. 1994b.
105. Hoffman, G.C. et al. 1990.
106. Lester, N.P. et al. 1996.
107. French, J.R.P. III. 1988.
108. Weaver, M.J. et al. 1997.
109. Faber, D.J. 1967.
A 110. Swingle, W.E. 1965.
111. Christie, W.J. and H.A. Regier. 1973.
112. Tyus, H.M. 1969.
113. Reutter, J.M. and C.E. Herdendorf. 1976.
114. Nurnberger, P.K. 1930.
115. Stauffer, J.R., Jr. et al. 1976.
116. Hall, D.J. and E.E. Werner. 1977.
117. Johnson, J.H. 1983.
118. Rodeheffer, I.A. 1940.
119. Engelbrecht, R.S. et al. 1984.
120. Noltie, D.B. 1986.
121. Holland, L.E. 1986.
122. Redmon, W.L. and L.A. Krumholz. 1978.
123. Kuehne, R.A. 1962.
124. Finger, T.R. 1982.
125. Kindschi, G.A. et al. 1979.
126. Gerking, S.D. 1953.
127. Matthews, W.J. 1986.
128. Hoyt, R.D. et al. 1979.
129. Weddle, G.K. 1986.
A 130. Jones, A.R. 1973.
A 131. Jones, A.R. 1970.
132. Gerber, G.P. and J.M. Haynes. 1988.
133. Lobb, M.D. III and D.J. Orth. 1991.
134. Gerber, G.P. 1987.
135. Holland, L.E. and M.L. Huston. 1985.
136. Jackson, S.W., Jr. 1957.
137. Minns, C.K. et al. 1978.
138. Lotrich, V.A. 1973.
139. Keast, A. 1985.
140. Smith, P.W. et al. 1971.
141. Hocutt, C.H. et al. 1978.
142. Hambrick, P.S. et al. 1973.
143. Mabee, P.M. 1995.
144. Cashner, R.C. 1974.
145. Cashner, R.C. and R.D. Suttkus. 1997.
146. Cashner, R.C. and R.E. Jenkins. 1982.
147. Ernest, J.R. 1960.
148. Van Hassel, J.H. et al. 1988.
149. Carlander, K.D. 1982.
A 150. Axon, J.R. 1984.

* Original information presented on reproductive biology of rock bass comes from field observations and data from the following locations in the Tennessee River drainage: a tributary of Walden Creek (Little Pigeon River drainage), in Sevier County, TN; Coal Creek, a tributary of the Clinch River, Anderson County, TN; and Big Creek, a tributary of the Holston River, Hawkins County, TN. Early life ecology information comes from TVA ichthyoplankton collections made in the Tennessee Valley during the period 1976–1983. Original descriptive information and data come from a developmental series of young rock bass cultured by TVA biologists from eggs and larvae collected from the tributary to Walden Creek. Specimens (lots TV2607, TV3002, TV3015, and TV3091) are curated at the Division of Fishes, Aquatic Research Center, Indiana Biological Survey, Bloomington, IN.

GENUS
Centrarchus Cuvier

Thomas P. Simon

The genus *Centrarchus* is monotypic and closely related to *Pomoxis* based on morphological (Bailey, 1938; Smith and Bailey, 1961; Branson and Moore, 1962), anatomical (Mok, 1981; Mabee, 1988), and genetic (Wainright and Lauder, 1992) evidence. *Centrarchus macropterus*, flier, ranges in lowlands from MD to FL, along the Gulf slope and into the Mississippi River drainage. The species occurs in sluggish channels, backwaters, and pools in streams, and ponds, swamps, and lakes. It occurs in low pH water and has been found at salinity levels as high as 7% (Musick, 1972).

The flier is a deep-bodied, laterally compressed sunfish, with the anal fin nearly as long and large as the dorsal fin. The dorsal fin origin is distinctly in front of anal fin origin. The head is small, snout pointed and short, mouth of moderate size, and the upper jaw reaches almost to middle of eye. Rear margins of the preopercle are serrate. Two patches of teeth are present on the tongue. Spinous dorsal and soft dorsal fins are broadly connected without a notch between. Dorsal fin has 11–13 spines and 12–15 soft rays; anal fin has 7 or 8 spines. The species rarely exceeds 127–153 mm TL. The maximum length and weight reported are 203 mm and 454 g (Robison and Buchanan, 1988; Boschung and Mayden, 2004).

TAXONOMY AND SYSTEMATICS OF GENUS *CENTRARCHUS*

There are two separate hypotheses concerning the systematic relationships of *Centrarchus*, which are variously grouped with genera *Pomoxis* or *Archoplites* (Bailey, 1938; Smith and Bailey, 1961; Branson and Moore, 1962; Mabee, 1988) or grouped with *Ambloplites* and *Enneacanthus* (Avise et al., 1977). Wainwright and Lauder (1992) and Mabee (1993) hypothesized that *Centrarchus* is the sister group to *Archoplites* and together form a sister clade to *Pomoxis*. *Archoplites* is restricted to rivers west of the Continental Divide. Branson and Moore (1962) hypothesized a close relationship of *Centrarchus* with *Pomoxis* on the basis of the acoustico–lateralis system and select meristics. Mok (1981) considers *Centrarchus* to be sister to *Pomoxis* based on kidney morphology. Eaton (1956) used olfactory variation data to place *Centrarchus* in an unresolved trichotomy with *Pomoxis* and *Archoplites*. Chang (1988) considers *Centrarchus* and *Pomoxis* to be sister groups. Roe et al. (2002) found this genus to be sister to *Ambloplites* plus *Pomoxis* plus *Archoplites* or this clade and *Enneacanthus*.

Numbers of anal and dorsal fin elements easily distinguish adult flier from elassomatids and other centrarchids (Table 5). Flier have 7–9 anal fin spines, *Elassoma* have 1–4, and *Micropterus* and *Lepomis* have 3. Rock bass have 11 or fewer soft anal fin rays compared to 14–17 for flier, and flier have more dorsal fin spines (usually 12–13) than the crappies (5–9) (Etnier and Starnes, 1993; Boschung and Mayden, 2004).

For characters used to distinguish flier larvae and juveniles from the young of elassomatids and other centrarchids see the "Key to Genera" on pages 27–29, Tables 4 and 5, and the "Taxonomic Diagnosis" section in the species account that follows.

LITERATURE CITED

Avise, J.C. et al. 1977.
Bailey, R.M. 1938.
Boschung, H.T. and R.L. Mayden. 2004.
Branson, B.A. and G.A. Moore. 1962.
Chang, C.H. 1988.
Eaton, T.H., Jr. 1956.
Etnier, D.A. and W.C. Starnes. 1993.
Mabee, P.M. 1988.

Mabee, P.M. 1993.
Mok, H.K. 1981.
Musick, J.A. 1972.
Robison, H.W. and T.M. Buchanan. 1988.
Roe, K.J. et al. 2002.
Smith, C.L. and R.M. Bailey. 1961.
Wainright, P.C. and G.V. Lauder. 1992.

FLIER

Centrarchus macropterus (Lacepede)

Thomas P. Simon

Centrarchus: Greek, *"kentron,"* meaning spine, and "archos," meaning anus; referring to the development of the anal spines; *macropterus*: Greek, meaning long fin.

RANGE

The flier occurs from VA along the Atlantic Slope to north central FL, and west on the Gulf Slope to Trinity River, TX. It is present northward into the Mississippi River Valley to southern IL and IN.[1-6] It was possibly more common throughout its range prior to the channelization of swamps and lowland habitats.[8]

HABITAT AND MOVEMENT

Flier is common in the clear waters of swamps, backwaters, sloughs, mill ponds, and low gradient streams usually with an abundance of rooted aquatic vegetation.[1-7] The species is usually found around sunken logs and stumps and frequents sluggish channels and pools in streams and lakes.[9] It is particularly abundant in oxbow lakes.[4]

Adult flier are often adapted to low-pH waters of the Coastal Plain and it has been reported in upper mesohaline zone with salinity as high as 7 ppt.[10] Flier do poorly when exposed to neutral to alkaline water.[4]

The flier is classified as a sedentary species, showing little seasonal or spawning migration.[22] Tagging studies showed appreciable movement only during the spring in Duke Swamp, NC,[37,38] and Okefenoke Swamp, GA.[39] Diel activity shows a behavioral ontogenetic niche shift with flier juveniles more active during the day,[22] while adults are more active at night.[17]

DISTRIBUTION AND OCCURRENCE IN THE OHIO RIVER SYSTEM

Flier is sporadic to occasional in distribution,[40] and seldom reported from the Ohio River.[15] Reports of flier from the river during the period 1957–1980 were few and all came from downstream of ORM 845. These occurrences probably represented strays from lentic floodplain waters.[15]

The flier occurs in the lower Wabash River of IN and IL,[11,12] and in tributaries such as the Patoka River, IN.[13] In KY, the flier is abundant and ubiquitous on the coastal plain; but absent from tributaries along the Ohio River.[14] It is common to abundant, but primarily restricted to Coastal Plain drainages at the western end of the state in TN. It is absent from east TN[3] and from the Tennessee River drainage of MS[4] and AL.[2,5] The range of flier has probably shrunk considerably within recent decades, perhaps assisted by human-induced habitat alterations.[8]

SPAWNING

Location

The flier does not appear to be specific with regard to the habitat in which it will spawn.[17] Gravid females were collected from a variety of habitats ranging from debris-filled ditches with soft bottoms, to streams with gravel bottoms, to isolated borrow pits filled with heavy stands of aquatic vegetation.[17] Water conditions in these habitats ranged from clear to very turbid and from stagnant to free flowing.[17]

Season

Flier spawn during the spring;[4,6] or spawning may occur as early as February and as late as May.[1] The reproductive period extends over a 2-week period in southeastern MO in April.[8,17] In MS, spawning is during late spring or early summer.[20] Fingerling fliers were found in May and June in Reelfoot Lake, TN, which indicates that spawning had occurred in April.[21] Nesting occurs from March to May in VA.[9] Western TN populations still had a considerable number of eggs, though could have been partially spent, during mid-May.[3] AL spawning occurs as early as mid-February, but usually from March until May.[5]

Table 18

Flier fecundity related to TL and age during April spawning in southeastern MO.[17]

TL (mm)	Age	Mature Eggs		
		Right Ovary Lobe	Left Ovary Lobe	Total
70	I	2,112	2,300	4,412
75	I	3,086	2,242	5,328
86	I	3,310	3,901	7,211
90	I	4,668	3,919	8,587
93	I	4,019	4,764	8,783
95	I	4,672	4,822	9,494
98	I	4,422	5,045	9,467
100	I	4,834	5,086	9,920
104	II	4,518	5,411	9,929
106	II	4,239	5,484	9,723
108	II	5,462	5,601	11,063
110	II	5,904	5,618	11,522
112	II	6,087	6,572	12,659
115	II	6,401	6,670	13,071
117	II	5,661	7,321	12,982
118	III	6,124	7,504	13,628
123	III	7,832	8,442	16,274
125	III	8,451	10,028	18,479
127	III	10,808	11,124	21,932
147	IV	10,517	11,846	22,363
152	V	12,581	13,281	25,862
192	VI	16,171	18,714	34,885
205	VI	23,908	24,346	48,254

Temperature

An absolute critical temperature for reproduction does not occur; rather it is the duration and fluctuation of temperature that influence gonadal development and subsequent time of spawning.[18] Temperatures reported during nesting ranged from 18.0 to 20.0°C (mean = 19.0°C).[19] In MO, fliers were ripe in mid-April at temperatures of 18.3–19.4°C and spawning occurred until early May at a temperature of 23.3°C.[17] AL populations spawn when water temperatures reach 17°C;[4] surface water temperatures in AL ponds at the time of hatching of larvae were 15.6–18.3°C. [19]

Fecundity

Number of ova per female generally increases with TL and age (Table 18), and with successive broods during a season.[17] In MS, a female (120 mm TL) collected mid-June contained nearly 5,600 nearly mature ova.[20] In AL, mean fecundity for age-1 females was 2,150 eggs.[5] In another report, brood size ranged from 1,900 to 37,500 ova.[4,9] Mean number of total ova was estimated as 14,449 in southeastern MO, [17] and ranged from 4,412 to 48,254 (Table 18). Differences between calculated and actual egg numbers ranged from 0.2 to 3.2% with an overall error of 1.6%.[17] The estimated number of eggs per ovary is described as log E = 1.9078 + 2.0981 log L, where E = estimate number of eggs, and L = TL of individual in mm. The left ovary lobe was consistently larger (80%, N = 23) compared to the right ovary lobe. The number of ova per gram of body weight was 925.[17]

Sexual Maturity

Minimum size at first reproduction is greater than 40 mm TL.[17] In VA, females have matured when as small as 70–75 mm TL;[9] by age 1 in AR[6] and MO at 70 mm TL.[8,17] In MO, all males >75 mm were sexually mature, while females >70 mm all had mature ovaries and appeared ready to spawn.[17] In NC, sexually maturity is reached at 2 years.[7]

Spawning Act

The flier is a guarding, nest spawning, lithophil.[34] Courtship and spawning are similar to that of other sunfishes.[16,18,19] Nesting is usually in close colonies.[5,7, 17] Males reach ripe reproductive condition slightly in advance of females and remain in spawning condition longer than females.[17] The first male in ripe condition was collected in early April and the last male was collected in early May. Females in ripe condition were only found in mid-April. Females collected either before or after the mid-April spawning date were either in pre-spawn condition or spent.[17]

Flier are known to hybridize with white crappie.[31]

EGGS

Description

Ripe eggs are about 0.3 mm in diameter,[17] are golden yellow, and adhesive.[5]

Incubation

No information.

Development

No information.

YOLK-SAC LARVAE

See Figure 21.

Size Range

3.4 mm (newly hatched) to 4.9–5.0 mm TL.*,[27]

Myomeres

From NC: predorsal 7–8;* preanal 11.2 ± 0.54; post-anal 22.6 ± 0.81; total 33.8 ± 0.83.[27]

Morphology

3.4–4.9 mm TL. Body laterally compressed; yolk sac moderate, oval (ca. 20.6% SL); yolk with a single, posterior oil globule; head not deflected over the yolk sac; eyes spherical; mouth a simple stomodeum (3.4–3.9 mm); functional jaws developed by 4.9 mm.*

Morphometry

See Table 19.

Fin Development

See Table 20.

4.0 mm TL. Pectoral fin bud evident.*

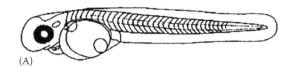

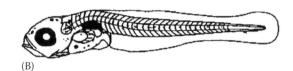

Figure 21 Development of young flier from NC. A–B. Yolk-sac larvae: (A) 3.9 mm TL; (B) 4.9 mm TL. C–E. Yolk-sac larva, 6.1 mm TL: (C) dorsal view; (D) lateral view; (E) ventral view. (A–B reprinted from Figure 2, reference 27, with publisher's permission; C–E reprinted from Figure 32, reference 24, with permission from Progress Energy Service Company, LLC.)

Pigmentation

3.4–3.9 mm TL. Eye pigmented; yolk unpigmented to pale yellow; melanophores present on the developing air bladder, but absent on the rest of the body.[27]

4.0–4.1 mm TL. Light pigmentation noted on caudal region and around the urostyle.[27]

4.5 mm TL. Moderate pigmentation present on the gut and head regions and melanophores present along the midbody line and ventral surface.*,[27]

POST YOLK–SAC LARVAE

See Figures 22, 23, and 24.

Size Range

4.9–5.0 mm*,[27] to 15.2–15.5 mm TL.*,[25,27]

Myomeres

Predorsal 6–7; preanal 9–11; postanal 20–25; total 30–34.*,[11,23]

Table 19

Morphometric data expressed as percentage of TL for young flier from North Carolina.[*, 27]

Length Range (mm TL)	N	Length (% TL)							Body Depth (% TL)		
		ED	HL	PreAL	PosAL	GL	IL	SL	GBD	BDA	CPD
3.4–4.0	16	12.0 ± 1.6	14.3 ± 1.0	40.0 ± 0.9	60.0 ± 0.9	—	—	99.6 ± 1.0	19.4 ± 0.4	8.0 ± 0.2	2.8 ± 0.1
4.1–5.0	59	12.4 ± 1.7	20.0 ± 1.1	34.3 ± 0.8	65.7 ± 0.8	—	—	94.2 ± 0.8	14.3 ± 0.3	8.6 ± 0.1	2.8 ± 0
5.1–6.0	6	12.1 ± 0.9	20.6 ± 0.9	35.3 ± 0.5	64.7 ± 0.5	23.0 ± 0.01	2 ± 0	97.1 ± 0.4	16.9 ± 0.4	8.8 ± 0.1	3.2 ± 0
6.1	1	12.2	20.6	35.2	64.8	22.0	2	96.4	16.1	8.8	3.1
7.1–8.0	2	11.0	22.9 ± 0.2	37.1 ± 0.3	62.9 ± 0.3	22.0	3	88.6 ± 0.5	20.0 ± 0.2	12.0 ± 0.1	5.7 ± 0.1
8.4–8.7	2	11.1 ± 0.2	25.1 ± 0.2	39.7 ± 0.4	60.3 ± 0.4	23.0	3	91.9 ± 0.3	20.0 ± 0.2	11.8 ± 0.2	6.2 ± 0.1
9.1–10.0	5	11.1 ± 0.5	27.3 ± 0.4	40.9 ± 0.3	59.1 ± 0.3	21.0 ± 0.01	3 ± 0	88.5 ± 0.2	21.1 ± 0.1	15.2 ± 0.2	6.6 ± 0.1
10.1–11.0	3	10.1 ± 0.5	27.9 ± 0.4	41.8 ± 0.5	58.2 ± 0.5	18.0 ± 0.01	3 ± 0	85.1 ± 0.3	23.6 ± 0.2	17.9 ± 0.1	8.9 ± 0.1
11.3–11.5	2	10.4 ± 0.2	28.6 ± 0.2	39.4 ± 0.4	60.6 ± 0.4	18.0	3 ± 0	83.6 ± 0.3	23.2 ± 0.2	17.9 ± 0.1	8.7 ± 0.1
12.2	1	10.2	28.6	42.9	57.1	18.0	3.0	85.7	25.7	20.0	9.4
15.2	1	8.9	23.4	37.5	62.5	15.6	3.1	80.2	22.9	20.3	8.8

Morphology

5.0 mm TL. Yolk sac absorbed; air bladder forming anterior to intestinal loop.*

8.7 mm TL. Air bladder extends posterior to anus.*

9.3 mm TL. Branchiostegal rays form.*

Morphometry

See Table 19.

Fin Development

See Table 20.

6.0 mm TL. Incipient caudal fin rays evident.[27]

7.5–8.0 mm TL. Anal fin anlage first observed.[27]

8.0–8.5 mm TL. Dorsal fin anlage present; incipient anal fin rays form by 8.5 mm.[27]

8.7 mm TL. Incipient dorsal fin rays form.[27]

10.0–10.6 mm TL. Dorsal and anal spines develop.[27]

10.7–11.5 mm TL. Pelvic fin buds form; spinous and soft dorsal finfold partially differentiated.[*,25]

12.2 mm TL. Both dorsal and anal fins developed;[27] dorsal finfold completely absorbed;* caudal fin completely formed.[27]

Pigmentation

5.0–6.1 mm TL. Melanophores over air bladder form a dark concentration dorsally and posteriorly. Dorsally, melanophores form over cerebellum and continue posteriorly onto nape and to anterior dorsal finfold origin. Ventral pigment is present as a series of stellate melanophores from the breast to posterior area of gut.*

7.8–9.3 mm TL. Melanophores are present on the dorso-posterior portion of opercle. Melanophores are present dorsally over air bladder. At later sizes within range, melanophores appear along mandible, extending to mid-orbit. Ventrally, melanophores form within branchiostegal rays, on lower lip, and scattered on breast. Random melanophores are visible ventrally on postanal myomeres between anus and middle of caudal peduncle.*

9.7–10.7 mm TL. Dorsally, melanophores outline edge of maxillae and mandible; cerebellum covered with punctate melanophores; and a few melanophores on anterior nape. Laterally, melanophores are present horizontally along mid-opercle. Air bladder is darkened dorsally. Ventrally, pigment is scattered on gut; melanophores outline mandible, and edges of branchiostegal rays; a line of punctate melanophores forms from isthmus to anus. A concentration of scattered melanophores forms over future anal fin lepidotrichia.*

12.2–15.2 mm TL. Pigmentation is similar to that described for previous length interval with the following changes. Dorsally, melanophores are present on otic capsule; additional melanophores extend posteriorly along dorsum to soft dorsal fin insertion. Laterally, there is a chevron cluster of melanophores mid-opercle. Internally, melanophores are scattered dorsally over gut and air bladder. A midlateral stripe of single melanophores forms from prepectoral to anus and from anus to hypural plate; pigment is scattered along myosepta

Table 20

Meristic counts and size (mm TL) at the apparent onset of development for *Centrarchus macropterus**

Attribute/Event	*Centrarchus macropterus*
Branchiostegal Rays	7[3]
Dorsal Fin Spines/rays	Xi–xiii/12–14[2–6,8,9]
First spines formed	10.0–10.6*
Adult complement formed	11.5*–12.2[27]
First soft rays formed	9.7*
Adult complement formed	10.6*
Pectoral Fin Rays	11–14*[,3,8,9]
First rays formed	9.7*
Adult complement formed	10.6*
Pelvic Fin Spines/Rays	I/6
First rays formed	>12.2*
Adult complement formed	15.2*
Anal Fin Spines/Rays	VII–VIII/14–17[2–6,8,9]
First rays formed	10.0–10.6[27]/8.4–8.7*
Adult complement formed	12.2[27]/10.6*
Caudal Fin Rays	iii–iv, 10–13+10–12, iv–v*
First rays formed	7.9*
Adult complement formed	11.5*–12.2[27]
Lateral Line Scales	36–42[2–6,8,9]
Myomeres/Vertebrae	29–35/31[25–27]
Predorsal myomeres	7–8*
Preanal myomeres	9–12[25–27]
Postanal myomeres	20–25[25–27]

in hypaxial portion of postanal myomeres. Ventrally, melanophores are scattered on chin, along edge of isthmus, and on pectoral girdle; scatted melanophores form on breast anterior to pelvic fin bud; concentration of melanophores forms from anus to anal fin insertion.*

JUVENILES

See Plate 1A following page 26.

Size Range
15.2–15.5 mm* to 70–75 mm TL, when smallest females are considered mature.[8,9,17]

Myomeres
Preanal 9–11; postanal 22.[11,23]

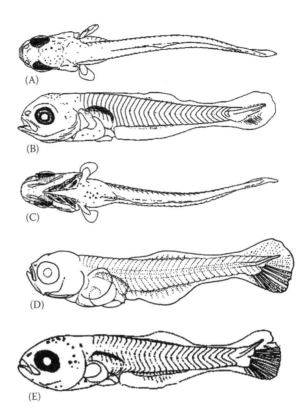

Figure 22 Development of young flier. A–C. Post yolk-sac larva from NC, 7.8 mm TL: (A) dorsal view; (B) lateral view; (C) ventral view. D–E. Post yolk-sac larvae: (D) from southern LA, 8.4 mm TL; (E) from NC, 8.7 mm TL. (A–C from Figure 33, reference 24, with permission of Progress Energy Service Company, LLC; D reprinted from Figure 1, reference 25, with author's permission; E reprinted from Figure 2, reference 27, with publisher's permission.)

Morphology

15.8–17 mm TL. Scale formation begins at 15.8 mm (Figure 25), which is a slightly smaller size than that reported for other centrarchids.[26] Scales first appear on the midlateral region of the body along the median myosepta and directly below the union of the spinous and soft rayed portions of the dorsal fin. A single row of scales develops both anteriorly and posteriorly, but progression is more rapid toward the tail. This single row of scales extended from slightly behind the operculum to the end of the caudal peduncle when the fish were 17 mm.*[,17]

18.0–23.0 mm TL. At lengths >18 mm, scales continued to develop anteriorly, and they developed more rapidly below the lateral line toward the breast than they did above the lateral line toward the nape. Only the nape was without scales when the fish attained 21 mm and the body, except the nape, was completely scaled at 23 mm (Figure 25).*[,17]

30.0–33.0 mm TL. Scales on the top of the head develop slowly, especially in the suborbital region. A 30-mm individual was the first to be completely scaled, and the largest fish not completely scaled

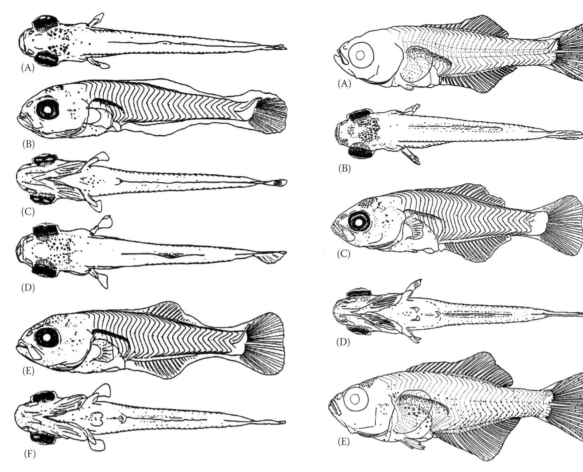

Figure 23 Development of young flier from NC. A–C. Post yolk-sac larva, 9.3 mm TL: (A) dorsal view; (B) lateral view; (C) ventral view. D–F. Post yolk-sac larva, 10.7 mm TL: (D) dorsal view; (E) lateral view; (F) ventral view. (A–F reprinted from Figures 34 and 35, reference 24, with permission of Progress Energy Service Company, LLC.)

Figure 24 Development of young flier. (A) Post yolk-sac larva from southern LA, 11.5 mm TL. B–D. Post yolk-sac larva from NC, 12.2 mm TL: (B) dorsal view; (C) lateral view; (D) ventral view. (E) Late post yolk-sac larva from southern LA, 15.2 mm TL. (A and E reprinted from Figure 1, reference 25, with author's permission; B–D reprinted from Figure 36, reference 24, with permission of Progress Energy Service Company, LLC.)

was 33 mm.[*,17] Fish were completely scaled at an average length of 31 mm.

Larger juveniles. Body deep and slab-sided with a moderately large mouth, the upper jaw reaching nearly to middle of eye;[8] eye large.[9] Margin of preopercle is serrate.[8] Snout slightly upturned; mouth oblique, supraterminal[9] or superior, tip of lower jaw a continuation of dorsal profile.[13] Ventral profile deepest just anterior to the dorsal fin origin.[*] Gill membranes separate, free from the isthmus.[*] Body and top of head scaled.[13] Lateral line scales (36)37–41(42), scales above lateral line 6–7; scales below lateral line (11)13–14.[9] Gill rakers slender, 30–35, and the most numerous of all centrarchids.[5,15]

Morphometry
See Table 19.

Fins
Spinous dorsal and soft dorsal broadly connected, without a notch between them.[8] Origin of dorsal fin distinctly anterior to anal fin origin;[6] dorsal spines 11–13, soft rays 12–15.[2–6,8,9,12] Anal fin usually as long and as large as the dorsal fin with 7–9 anal spines and 14–16 soft rays.[2–6,8,9,12] Caudal fin emarginate;[*] principal rays 20–24.[*] Pelvic fins small,[9,23] inserted midway between the anal origin and a point ventral to the pectoral insertion;[3] rays 6.[13] Pectoral fins long, pointed;[9] rays (11)13–14.[9]

Pigmentation
12.2–>54.0 mm SL. Primary bars of melanistic pigment appear on flier as an invariant temporal sequence. The third primary bar is the first to appear and forms as a midlateral patch of melanophores beneath the middle of the soft dorsal fin (Figure 26A). As the fourth, second, and first primary bars successively appear, dorsoventral elongation proceeds until all of the bars except the first extend at their fullest development from the dorsal to ventral midlines (Figure 26D). The dorsal fin spot, a distinctive feature of juvenile flier, develops

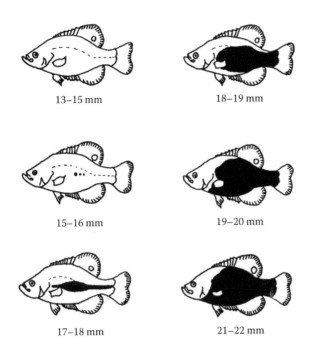

Figure 25 The origin and pattern of scale development in the flier.[27]

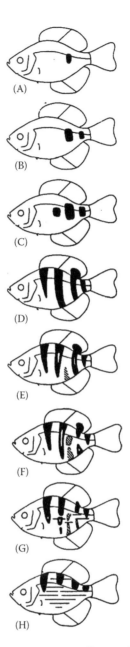

Figure 26 Development of young flier — schematic drawings showing the ontogeny of melanistic pigment patterns. (A) 12.2–12.5 mm SL; (B) 13.2–13.5 mm SL; (C) 13 mm SL; (D) 12.2–28.1 mm SL; (E) 31.8 mm SL; (F) 30.1 mm SL, 25.7–30.1 mm SL; (G) 31.7 mm SL; and (H) 54.0 mm SL and larger. (Reprinted from Figure 9, reference 41, with publisher's permission.)

as a dorsal extension of the third primary bar (Figure 26D). Other sunfish (green sunfish, bluegill, and bantam sunfish) have a dorsal fin spot, but in these species the spot does not arise as an extension of a primary bar. In these species the spot develops independently in the fin, dorsal to the fin base musculature and separated from body pigmentation by an unpigmented interradial membrane. At its fullest development, the spot, uniquely in the flier, is surrounded by a bright orange halo. The primary bar dorsal fin spot may be interpreted as a non-terminal addition relative to the developmental sequence of addition of the primary bars. Secondary bar formation in *Centrarchus* is consistently initiated ventral to and between the second and third primary bars (Figure 26E). The adjacent primary bars become light centered and narrow, eventually breaking up dorsoventrally (Figures 26F and 26G). Although the dorsal portions of the primary bars may be retained, longitudinal striping dominates the adult pattern (Figure 26H).[41]

<75 mm TL. Young and juveniles with an ocellus in second dorsal fin. Young possess 4–5 dark vertical bars.[9]

TAXONOMIC DIAGNOSIS OF YOUNG FLIER

Similar species: other centrarchids, especially *Pomoxis*.[27]

The dorsal fin spot or ocellus, a distinctive feature of juvenile flier, develops as a dorsal extension of

the third primary bar of melanistic pigment (Plate 1A, Figure 26D). Although a dorsal fin spot is present in *Lepomis symmetricus*, *L. macrochirus*, and *L. cyanellus*, it does not arise as an extension of a primary bar. In these taxa, the spot develops independently in the fin, dorsal to the fin base musculature and separated from body pigmentation by an unpigmented interradial membrane.[41]

Fliers are limited in the Ohio River drainage to southeastern IL, southwest IN, western KY, and northwest TN (Figure 27). Specimens that could possibly be flier larvae collected from outside of these areas are likely to be crappie. Conner (1979) was unable to confidently distinguish between crappie and flier prior to the completion of caudal fin ray development, though he felt that a pronounced difference in eye size might extend down through protolarval and mesolarval phases. He further stated that as a matter of practical consideration, it should be noted that flier larvae may seldom, if ever, be encountered through conventional ichthyoplankton sampling procedures.[25]

After caudal fin development is completed, for flier <12 mm TL, the eye diameter (ED) is conspicuously greater than snout length (>1.7 times SnL); for crappies <12 mm with complete caudal fins, ED is subequal to or only slightly greater than SnL (<1.5 times SnL). Post yolk-sac larvae and juvenile flier ≥12 mm TL have 10 or more,[25] usually 12–13,[5] dorsal fin spines compared to 8 or fewer for crappie.[5,25]

Flier vs. black crappie in NC. No characteristics were found that differentiated flier from black crappie at lengths <5.0 mm.[27] At all lengths, myomere counts overlapped in both species. Eye diameter (ED) and body depth (BD) were usually greater and intestine length (IL) shorter in flier larvae compared to black crappie larvae of similar lengths. However, none of these characteristics alone could be used to reliably distinguish between the two species at any length. For lengths greater than 5.0 mm, the ratios of TL to eye diameter (ED), TL to body depth (BD), and intestine length (IL) to gut length (GL) provided reliable ways of distinguishing the two species. Means of TL to ED for flier larvae ranged from 10.1 to 12.1 and generally

decreased with size. Ratio means for crappie ranged from 14.0 to 16.7 and also showed a general decrease with size. The ratio of TL to BD showed overlap between the two species from 5.1 to 7.0 mm. After 7.0 mm, BD in flier larvae was greater than that of black crappie as reflected in IL/BD means. Intestine length-to-gut length ratio was the most reliable characteristic for distinguishing between these two species. Intestine length in black crappie was substantially greater than that of fliers. Ratios (IL/GL) for black crappie ranged from 0.36 to 0.43 and generally decreased with increasing length. For flier larvae, the IL/GL ratios were from 0.19 to 0.23 and also decreased as length increased. Intestine length was usually greater than 35% of the gut length in black crappie larvae, while for flier larvae, it was usually less than 25%.[27]

ECOLOGY OF EARLY LIFE PHASES

Occurrence and Distribution (Figure 27)

Eggs. Eggs incubate in nests constructed on a variety of substrates.[17]

Yolk-sac larvae. Yolk-sac larvae remained on the substrate during the first 24 h.[27]

Larvae and juveniles. Swim–up occurred 1 day after the larvae hatched.[27] Post yolk-sac larvae nursery areas varied from debris-filled ditches with soft bottoms, to streams with gravel bottoms, to isolated borrow pits filled with heavy strands of

Figure 27   General distribution of flier in the Ohio River drainage (shaded area).

Table 21

Age and calculated lengths (mm TL) for male and female flier from southeast Missouri.[17]

| | Age Group | N | TL at Capture | Length at End of Year | | | | | | |
				1	2	3	4	5	6	7
I	Female	179	83.6	54.8						
	Male	133	83.7	55.2						
II	Female	64	102.3	57.1	94.2					
	Male	41	102.2	56.2	94.2					
III	Female	22	120.2	61.4	94.4	116.2				
	Male	15	130.9	61.4	95.9	122.6				
IV	Female	17	158.2	64.6	110.5	132.9	149.8			
	Male	9	154.5	59.4	100.0	125.6	140.2			
V	Female	9	184.6	64.8	117.1	143.3	160.4	177.4		
	Male	8	180.6	64.0	108.4	142.3	162.2	177.0		
VI	Female	8	203.3	70.0	119.2	159.0	176.4	186.1	193.0	
VII	Female	1	210.0	71.8	119.0	153.0	174.3	186.2	192.2	199.1

aquatic vegetation. Water conditions included clear to turbid and from stagnant to free-flowing.[17] Due to a high degree of parental care afforded by the flier's mode of reproduction and its tolerance of poor water quality conditions, the reproductive success is usually high.[13]

Early Growth

In two VA lakes , fliers at age 4 were 155 mm and 168 mm TL.[28,29] Females live longer and grow larger than males (Table 21).[8,17] In MO, equal numbers of both sexes are born, but for all other age classes there is a dominance of females and males older than age 5 were not found (Table 21).[17] In MO, mean TL for age-1 flier was 55 mm; for age 2, 97.5 mm; for age 3, 130 mm; for age 4, 155 mm; for age 5, 177.5 mm; for age 6, 190 mm; and for age 7, 195 mm.[8,17] The largest specimen in MD was 250 mm TL;[30] in VA, 245 mm TL.[9] The length–weight relationship is log W = –4.913 + 3.077 log L.[17] Condition factors show no significant difference between males and females nor between fish of different age groups.[36]

Flier YOY increased in length from 16.4 mm in early May to 65.2 mm in late November. The average daily gain in length varied from 0.733 to 0.037 mm (mean = 0.225).[17] The daily mean weight gain was 0.018 g/day from May to November.[17] Larval flier grew 0.3384 mm daily for 26 days in a NC laboratory. The largest young-of-the-year was 79 mm TL and weighed 7.64 g.[17] Length–weight relationships of young fliers, collected in AL, are presented in Table 22.

Table 22

Length and weight relationships of young flier collected in AL during the period 1949–1964.

TL (mm)	No. of Fish	Average Empirical Weight (g)
51	1	4.5
76	3	12.3
102	8	25.4
127	10	49.9
152	1	68.1

Source: Constructed from data presented in unnumbered table on page 27, reference 42.

Feeding Habits

The flier is a drift invertivore that feeds in the water column.[35] Food varies with size of fish. Fish less than 25 mm TL fed exclusively on copepods. Individuals <175 mm continue to eat small crustaceans, but aquatic insects increase in importance. Fliers >175 mm ate mostly aquatic insects, while fish and crustaceans each comprised 7% of the diet. Large flier ate young bluegills.[8,17] In IL, diet consists mostly of insects, crustaceans, and undigested filamentous algae.[32] In MS, flier fed heavily on aquatic insects, especially water boatmen (Hemiptera, Corixidae), which are often abundant.[4] In

VA, flier are reported to feed on small crustaceans, larval dipterans, and mayflies.[33] In western TN, flier stomachs contained terrestrial insects, indicating a primarily surface-oriented feeding behavior. Young flier consumed midge larvae and other benthic organisms.[3]

LITERATURE CITED

1. Lee, D.S. et al. 1980.
2. Mettee, M.F. et al. 2001.
3. Etnier, D.A. and W.C. Starnes. 1993.
4. Ross, S.T. 2001.
5. Boschung, H.T., Jr. and R.L. Mayden. 2004.
6. Robison, H.W. and T. Buchanan. 1988.
A 7. North Carolina Wildlife Resource Commission. 1983.
8. Pflieger, W.L. 1997.
9. Jenkins, R.E. and N.M. Burkhead. 1994.
10. Musick, J.A. 1972.
11. Gerking, S.D. 1945.
12. Smith, P.W. 1979.
13. Simon, T.P. et al. 2005.
14. Burr, B.M. and M.L. Warren, Jr. 1986.
15. Pearson, W.D and L.A. Krumholz. 1984.
16. Heinrich, O. 1921.
17. Conley, J.M. 1966.
18. Breder, C.M. 1936.
19. Breder, C.M., Jr. and D.E. Rosen. 1966.
20. Hildebrand, S.F. and I.L. Towers. 1928.
21. Baker, C.L. 1939.
22. Stewart, E.M. and T.R. Finger. 1985.
23. Swingle, H.S. 1956.
A 24. McGowan, E.G. 1988.
25. Conner, J.V. 1979.
26. Conley, J.M. and A. Witt, Jr. 1966.
27. Swing, J.M. and E.G. McGowan. 1987.
28. Rosebery, D.A. and R.R. Bowers. 1952.
A 29. Smith, P. and J. Kaufman. 1982.
A 30. Enamait, E.C. and R.M. Davis. 1982.
31. Burr, B.M. 1974.
32. Gunning, G.E. and W.M. Lewis. 1955.
33. Flemer, D.A. and W.S. Woolcott. 1966.
34. Simon, T.P. 1999.
35. Goldstein, R.M. and T.P. Simon. 1999.
36. Geaghan, J.P. and M.T. Huish. 1981.
37. Whitehurst, D.K. 1976.
A 38. Huish, M.T. and G.B. Pardue. 1978.
39. Holder, D.R. 1970.
40. Gunning, G.E. and R.D. Suttkus. 1990.
41. Mabee, P.M. 1995.
A 42. Swingle, W.E. 1965.

GENUS
Lepomis Rafinesque

Robert Wallus and Thomas P. Simon

Lepomis is the largest genus of the Centrarchidae, comprising 12 species (Nelson et al., 2004) of small- to medium-sized fishes generally referred to as sunfish. Eleven of the 12 species occur in the Ohio River drainage (Etnier and Starnes, 1993; Jenkins and Burkhead, 1994). The common name "sunfish" alludes to their deep and rounded profile, to the yellow, orange, and red colors of many species, and to the habit of some species of hovering in open water, "sunning" (Jenkins and Burkhead, 1994). Several species are popular panfish, highly valued for their taste (Etnier and Starnes, 1993).

Though members of *Lepomis* generally have similar morphologies, there is considerable variation, and they can be difficult to identify even by experts. This is compounded by the fact that natural hybrids are common and juveniles of most species often exhibit a similar barred color pattern on the sides (see Plates 1 and 2 following page 26). This pattern of coloration perhaps has adaptive significance in serving to camouflage the young against a background of weeds or other cover with which they are usually associated (Etnier and Starnes, 1993; Jenkins and Burkhead, 1994).

Adult characters of *Lepomis* include an oblong, ovate, relatively deep and more or less compressed body; jaws about equal; greatly reduced or absent supramaxilla; opercle usually with a flap (ear) and without flat spinous projections; edge of opercle smooth, not serrate; teeth on vomer and palatine bones, none on tongue; fewer than 55 scales in lateral rows; 6 (rarely 5 or 7) branchiostegal rays. Dorsal fin is continuous, with a shallow gap between spines and rays; dorsal fin spines 9–12, typically 10, soft rays 10–12, typically 11. Anal fin spines 3. Bright red, orange, blue, and green colors are typical (Etnier and Starnes, 1993; Boschung and Mayden, 2004).

SYSTEMATICS OF GENUS *LEPOMIS*

Bailey (1938) considered *Chaenobryttus* (*Lepomis gulosus*) to be the basal sister group to *Lepomis*. Branson and Moore (1962) consider the genus to be distinct and closely related to *Lepomis* and *Micropterus*. Nelson et al. (2004), based on overall similarity and the high frequency of hybridization with species of *Lepomis*, placed *Chaenobryttus* within the species *Lepomis*. In a phenetic analysis, Avise and Smith (1977) and Avise et al. (1977) placed *L. gulosus* in *Lepomis* and consider *L. macrochirus* its closest sister species. Mok (1981) included *Chaenobryttus gulosus* in *Lepomis* based on synapomorphies in kidney variation. Wainwright and Lauder (1992) and Mabee (1993) place *C. gulosus* as the sister group to a clade inclusive of species of *Lepomis* and *Enneacanthus*. Merriner (1971b), Miller and Robison (1973), Clay (1975), and Mayden et al. (1992) retain *Chaenobryttus* as a recognized genus.

Wainwright and Lauder (1992) recognize *Lepomis* as a monophyletic group, excluding *Chaenobryttus*, as a sister group to *Enneacanthus*. Mabee (1993) considers *Lepomis*, excluding *Chaenobryttus*, as a paraphyletic group. Mabee's study suggests that *L. symmetricus* and *L. macrochirus* are more closely related to *Enneacanthus* than to any other *Lepomis*, and *L. cyanellus* forms the sister group to the remaining species of *Lepomis* plus the *Lepomis–Enneacanthus* clade. Mok (1981) provides evidence based on kidney morphology to support the monophyly of *Lepomis* (inclusive of *Chaenobryttus*). Chang (1988) considers *Lepomis* to be one of the basal members of the family, sister to all genera except *Elassoma* and *Micropterus*. Avise and Smith (1977) found *Lepomis* to be either monophyletic (including *gulosus*) or paraphyletic with respect to the genus *Micropterus*. Roe et al. (2002) found extensive support for *gulosus* to be included within *Lepomis*, a classification that we follow here.

INTRODUCTION TO SUBGENERA OF GENUS *LEPOMIS*

The following subgeneric classification for *Lepomis* is based on the findings of Bailey (1938), Branson and Moore (1962), Avise and Smith (1977), Wainwright and Lauder (1992), Mabee (1993), and Near et al. (2005), and original data used to describe adult and ontogenetic classification based on ontogenetic patterns (Conner, 1979). Information contained below each subgenus includes species found within the Ohio River drainage, summaries of phylogeny, summaries of adult and ontogenetic characters that are diagnostic for the subgenus, and comparisons of early development to that of members of other similar subgenera or species groups.

Lepomis Rafinesque

Included species: *Lepomis auritus*

Bailey (1938) placed *Lepomis auritus* in the monotypic subgenus *Lepomis*. He and Branson and Moore (1962) consider this species to be closely related to *L. megalotis* and *L. marginatus*. Avise and Smith (1977) suggest a closer relationship with *L. punctatus* using a phenetic analysis, while a Wagner analysis aligns the species with *L. gibbosus*. Wainwright and Lauder (1992) place *L. auritus* as the sister group to a clade inclusive of *L. megalotis*, *L. marginatus*, *L. punctatus*, *L. macrochirus*, *L. gibbosus*, and *L. microlophus*. Using ontogenetic relationships, Mabee (1993) found relationships similar to those of Wainwright and Lauder, except that *L. macrochirus* was not included in the clade.

Ontogenetic. Hypothesized to be similar to the *Icthelis* subgenus (*Lepomis megalotis* and *L. marginatus*) by Conner (1979). Descriptions of *L. auritus* show a strong resemblance to *L. megalotis* (Taber, 1969; Anjard, 1974; Buynak and Mohr, 1978; Hardy, 1978). Diagnostic information provided in Table 4 supports Conner's (1979) hypothesis that *L. megalotis* and *L. auritus* are similar.

Allotis Hubbs

Included species: *Lepomis humilis*

Bailey (1938) placed *Lepomis humilis* in the monotypic subgenus *Allotis* Hubbs. In other studies a relationship was observed between this species and *L. macrochirus* (Branson and Moore, 1962). Avise and Smith (1977) determined that *L. humilis* was either the sister group to *L. macrochirus* plus *L. gulosus* or

the basal sister group to all other *Lepomis* (including *L. gulosus*) and *Micropterus*. Wainwright and Lauder (1992) consider *L. humilis* to be the sister group to a clade including *L. megalotis*, *L. marginatus*, *L. punctatus*, *L. macrochirus*, *L. gibbosus*, and *L. microlophus*. Based on ontogenetic analysis, Mabee (1993) found similar relationships except that *L. auritus* was part of the latter group, but *L. macrochirus* was not.

Ontogenetic. Most similar to subgenus *Helioperca* (*Lepomis macrochirus*). Early yolk-sac stages of orangespotted sunfish are separable from *L. macrochirus* by the greater preanal length (≥45% TL), number of preanal myomeres (mode = 14), and smaller eye diameters (<5.5%TL); preanal length of bluegill <45% TL, preanal myomeres (mode = 12–13), and eye diameter >5.5% TL. Post yolk-sac larval orangespotted sunfish have retarded coiling of the foregut (>9.6 mm), smaller eye diameters (<5.5 vs. >5.5% for bluegill), and greater preanal lengths (≥45 TL vs. <45% for bluegill) (Hutton, 1982). Early juvenile orangespotted sunfish have fewer anal fin rays than bluegill (9–10 vs. 10–12), and possess disjunct preoperculomandibular canals, inflated supraorbital canals, and longer caudal peduncles with melanophores in the lower vertical intermuscular septum compared to conjunct preoperculomandibular canals, supraorbital canals not enlarged, and shorter caudal peduncles with melanophores not extending up into the vertical intermuscular septum for bluegill (Hutton, 1982).

Apomotis Rafinesque

Included species: *Lepomis cyanellus*

Bailey (1938) placed *Lepomis cyanellus* in the monotypic subgenus *Apomotis* Rafinesque. Branson and Moore (1962) and Smith and Lundberg (1972) consider it the most pleisomorphic member of the genus. Branson and Moore (1962) consider this species to be most closely related to *L. symmetricus*. Phylogenetic hypotheses presented by Wainwright and Lauder (1992) and Mabee (1993) concur with the pleisomorphic placement of *L. cyanellus* with respect to other species of *Lepomis* (exluding *L. gulosus*). In Mabee's (1993) analysis, some *Lepomis* species are considered more closely related to *Enneacanthus* than to other species of *Lepomis*. Wainwright and Lauder (1992) recognize a monophyletic *Lepomis* with *L. cyanellus* as the most pleisomorphic member of the genus.

Ontogenetic. Most similar to subgenus *Bryttus* (*Lepomis miniatus*) in Ohio River drainage and *L. punctatus*. The *Apomotis* have relatively short preanal lengths (modally <45% TL) and also have deeper heads and more extensive and prominent pigment, particularly in the head and trunk regions. Post yolk-sac larvae possess profuse pigmentation, especially well-developed head pigmentation. Head pigmentation includes abundant pigmentation in the cheek and postorbital areas. Additionally, green sunfish have a characteristic mid-lateral streak of melanophores that often simulates a regularly spaced series of dashes. Green sunfish are characteristically smaller at some important morphological benchmarks compared to members of other subgenera (Table 4).

Bryttus Valenciennes

Included species: *Lepomis miniatus* (in the Ohio River drainage) and *L. punctatus*

Bailey (1938) placed *Lepomis miniatus* in the subgenus *Bryttus* Valenciennes. Warren (1992) raised *L. miniatus* from synonomy with *L. punctatus*. Branson and Moore (1962) hypothesized a close relationship between *L. gibbosus* and *L. microlophus*, but the phenetic analysis by Wainright and Lauder (1992) suggest that the *L. punctatus–L. miniatus* complex is part of a clade with *L. macrochirus*, *L. microlophus*, and *L. gibbosus*.

Ontogenetic. Limited information exists for this group and a diagnosis cannot be included.

Chaenobryttus

Included species: *Lepomis gulosus*

Smith and Lundberg (1972) recognized *Chaenobryttus* as a genus distinct from *Lepomis* and consider their fossil species *C. serratus* to be sister to *L. gulosus*. DNA studies suggest that *L. gulosus* populations west of the Mobile basin constitute a separate lineage from those to the east (Bermingham and Avise, 1986). A study examining variation in *L. gulosus* has not been conducted.

Ontogenetic. Limited information is available for *Chaenobryttus*. Conner (1979) indicated diagnostic characters may include well-developed head pigmentation, especially in the cheek and postorbital areas; a characteristic mid-lateral streak of melanophores that often simulates a regularly spaced series of dashes; and a tendency toward smaller size at comparable stages than other subgenera. Among *Lepomis* subgenera, only *Chaenobryttus* and *Helioperca* (*L. macrochirus*) possess 7–8 primary pigment bars compared to 9 or more for members of all other subgenera (Table 5).

Eupomotis Gill and Jordan

Included species: *Lepomis microlophus* and *L. gibbosus*

Bailey (1938) recognized two subspecies of *Lepomis microlophus* including one that has not been described. The distribution of these two subspecies has not been well defined. Page and Burr (1991) consider the eastern subspecies to naturally occur in FL, GA, and southern AL, while the western subspecies occupies the remainder of the species range. Numerous introductions have limited the ability to understand the distributions of these subspecies. Avise and Smith (1977) place *L. microlophus* in a group with *L. marginatus* and *L. megalotis*, while Lauder (1983a,b), Wainwright and Lauder (1992), and Mabee (1993) consider *L. gulosus* to have a sister group relationship with *L. gibbosus*.

Ontogenetic. The *Eupomotis* subgenus is intermediate between *Helioperca* (*Lepomis macrochirus*) and *Apomotis* (*L. cyanellus*) subgenera (Conner, 1979). The subgenus is most similar to *Helioperca* in meristic data and pigmentation, but in morphometric characters and gut and air bladder architecture *Eupomotis* resembles *Apomotis*. Yolk-sac and early post yolk-sac larval *Eupomotis* can be distinguished from *Apomotis* by the virtual absence of pigment in the head region. *Eupomotis* differs from *Helioperca* in having a relatively thickened (usually coiled) foregut and a more anteriorly placed air bladder. At comparable stages of development, *Eupomotis* tend to be the smaller in size; have larger eyes; and have deeper, more robust heads than *Helioperca* (Table 4). Our Ohio River specimens did not always support this hypothesis since the *Eupomotis* hatch at similar or larger sizes than *Helioperca*. *Eupomotis* with complete caudal fins are difficult to distinguish due to the similar sizes of other subgenera with complete caudal fin development. Late post yolk-sac larvae and juvenile specimens of *Eupomotis* are less strongly pigmented than both the *Apomotis* and *Icthelis* (*L. megalotis* and *L. marginatus*). *Eupomotis* lacks pigment concentrations in the ventral intermuscular septum of the caudal peduncle, while post yolk-sac larvae show retarded development of pigmentation in the interradial membranes of the soft anal and dorsal fins. Once the majority of rays are ossified in the soft anal and dorsal fins of *Eupomotis*, these fins will be virtually immaculate until very late in the post yolk-sac stage. When pigment does develop, the melanophores tend to be very tiny and distributed along the rays giving a dusky appearance to the fins rather than a speckled or banded appearance.

Helioperca Jordan

Included species: *Lepomis macrochirus*

Bailey (1938) placed *Lepomis macrochirus* in the monotypic subgenus *Helioperca* Jordan. Its closest relative is hypothesized to be *L. humilis* (Branson and Moore, 1962) or *L. gulosus* (Avise and Smith, 1977) or *L. symmetricus* (Mabee, 1993). Mabee (1993) found that the *L. macrochirus–L. symmetricus* clade is more closely related to *Enneacanthus* than to other species of *Lepomis*. Wainwright and Lauder (1992) place *L. macrochirus* in a clade with the *L. punctatus–L. miniatus* complex, *L. microlophus*, and *L. gibbosus*.

Ontogenetic. Helioperca is characterized by retarded thickenings and coiling of the foregut (Conner, 1979). Most *Lepomis macrochirus* have an essentially uncoiled gut until very late in the post yolk-sac larval phase. Yolk-sac larvae and early post-yolk sac larvae tend to have proportionally smaller eyes, greater preanal lengths, more preanal myomeres, and small, more posteriorly placed air bladders than other *Lepomis. Helioperca* are larger at comparable stages of development than either *Apomotis* (*L. cyanellus*) or *Icthelis* (*L. megalotis* and *L. marginatus*). As late post yolk-sac larvae and juveniles, *Helioperca* tend to become less distinctive than either *Apomotis* or *Icthelis,* particularly with respect to meristic and morphometric data. *Helioperca* >9.5 mm TL are best recognized by differences in pigmentation when compared to *Apomotis* or *Icthelis.*

Icthelis Rafinesque

Included species: *Lepomis megalotis* and *L. marginatus*

Bailey (1938) placed *Lepomis megalotis* and *L. marginatus* in the subgenus *Icthelis* (Rafinesque). Fowler (1945) considered *L. marginatus* a subspecies of *L. megalotis.* A morphological study examining geographic variation in *L. megalotis* (Barlow, 1980) indicated that some populations of *L. marginatus* were more similar within group than between other populations of *L. megalotis.* Barlow (1980) found that the *L. marginatus* grouping is most similar to the subspecies of *L. megalotis* from the Great Lakes area. Jennings and Philipp (1992) conducted a genetic study that found distinct populations of *L. marginatus* that differed from all other *L. megalotis,* including the Great lakes populations. Bauer (1980) indicated that *L. marginatus* may be polytypic, but no subspecies have been described. Branson and Moore (1962), Avise and Smith (1977), and Mabee (1993) suggest a close relationship between these two species. The phylogenetic study by Wainwright and Lauder (1992) proposes *L. marginatus* as the sister group to a clade including *L. punctatus, L. macrochirus, L. gibbosus,* and *L. microlophus.* Five subspecies of *L. megalotis* are recognized across the species range (Bailey, 1938). Branson and Moore (1962) noted that *L. megalotis* has a number of races, some that have yet to be described. Barlow (1980) recognized at least seven races among the five subspecies.

Ontogenetic. Our current knowledge of the *Icthelis* subgenus lacks diagnostic information. Conner (1979) suggests that free-swimming yolk-sac larval stages for this subgenus are more nest-bound as yolk-sac larvae or at least are less prone to venture into pelagic areas than are other *Lepomis.* Fin rays in the caudal fin of *Icthelis* first appear between 5.8 and 6.3 mm TL. Other characteristics of this subgenus include a robust body shape and pronounced thickening and coiling of the foregut.

Lethogrammus
Included species: *Lepomis symmetricus*

Bailey (1938) placed *Lepomis symmetricus* in the monotypic subgenus *Lethogrammus* and hypothesized the closest sister taxa to be *L. cyanellus.* Branson and Moore (1962) also placed *L. symmetricus* as sister to *L. cyanellus,* while Mabee (1987) placed *symmetricus* sister to *L. macrochirus* based on ontogenetic characters. Bantam sunfish is the only species of *Lepomis* with the adult characteristic of an incomplete lateral line.

Ontogenetic. Early yolk-sac stages are separable from all other *Lepomis* by possessing a moderate-sized head and in having postanal pigmentation present (limited to ventral postanal myosepta).* Hatching occurs between 4.5 and 5.5 mm TL with pectoral fin buds present. Yolk is absorbed by 6.8 mm TL. Post yolk-sac larvae have scattered melanophores laterally on the body with concentrated melanophores over the cranial hemisphere, and abundant ventrolateral and postanal pigmentation. Post yolk-sac larvae and juvenile *L. symmetricus* are separable from all other *Lepomis* except *L. macrochirus* and *L. cyanellus* by the presence of a black spot that forms posteriorly in the soft dorsal fin (Tables 4 and 5).

TAXONOMY OF GENUS *LEPOMIS*

Identification of adult *Lepomis* species is accommodated by numerous regional works (Clay, 1975; Smith, 1979; Etnier and Starnes, 1993; Jenkins and Burkhead, 1994; Pflieger, 1997; Ross, 2001; Boschung and Mayden, 2004) and will not be addressed here.

Despite increased knowledge, we are still unable to identify with certainty early life forms of all members of the genus *Lepomis.* Identification techniques using cellulose acetate protein electrophoresis have been developed that proved useful in identifying larval bluegill, pumpkinseed, and green sunfish in four southwest MI lakes (Rettig, 1998), but we are unaware of any broad application of this technique. Currently, no developmental information is available for dollar sunfish *L. marginatus* and additional description is needed for developmental phases of young pumpkinseed *L. gibbosus,* redbreast sunfish *L. auritus,* redear sunfish *L. microlophus,* redspotted sunfish *L. miniatus,* and bantam sunfish *L. symmentricus.*

Though we are not able to present a "Key" to the identification of young *Lepomis* from the Ohio River drainage, we have compiled much information within this work that may help the user identify larval or juvenile *Lepomis* specimens, or, at a minimum, place them into taxonomic groupings.

Discussions or tools that can be used to that end include the following.

Subgeneric Descriptions

These discussions, presented above, summarize ontogenetic characters for members of each subgenus and, if possible, compare their early development to that of members of other similar subgenera or species groups.

Tables 4 and 5

These tables, presenting morphometric and meristic data, developmental benchmarks, and pigmentation characters of all centrarchids (including 11 *Lepomis* species) present in the Ohio River drainage, allow the user easy comparison of important character states among species.

Color Pattern Ontogeny of *Lepomis*

Primary bars of melanistic pigment appear laterally on all lepomid juveniles and secondary bars appear on some. A series of pattern transformations consisting of variation in the number of primary bars, the presence or absence of secondary bars, and presence or absence and formation of light centers in the primary bars characterize different sexes and species of *Lepomis* (Mabee, 1995). Orangespotted sunfish, redbreast sunfish, longear sunfish, dollar sunfish, redear sunfish, pumpkinseed, bantam sunfish, and probably redspotted sunfish (based on this condition in spotted sunfish) develop 9–10 primary bars of pigment. Green sunfish develop 11–14 primary bars and bluegill and warmouth have 7–8 bars. Light centers form in the primary bars of warmouth, redear sunfish, pumpkinseed, bantam sunfish, bluegill, and female orangespotted sunfish. Secondary bars form in warmouth, green sunfish, redear sunfish, and pumpkinseed. Schematic illustrations and discussions of these characters taken from Mabee (1995) are presented in each species account. Also of use will be Plate 1 and Plate 2, which provide color photographs that will allow comparison of coloration and pigmentation patterns among juveniles of all species of *Lepomis* present in the Ohio River drainage.

Distribution Maps

Several of the sunfishes have restricted distribution and maps in each species account may help reduce the list of choices.

"Taxonomic Diagnosis" Section of Each Species Account

Within these sections are summaries of reports from the literature concerning the taxonomy of early life forms of selected groupings of sympatric *Lepomis* species. These reports alone may provide the user with all the taxonomic information needed.

A review of all this information may allow the user to construct a suite of characters that will aid in identification of larvae or juveniles present in the study area of concern, or, if identification to species proves impossible, allow the creation of taxonomic groupings.

LITERATURE CITED

Anjard, C.A. 1974.
Avise, J.C. and M.H. Smith. 1977.
Avise, J.C. et al. 1977.
Bailey, R.M. 1938.
Barlow, J.R., Jr. 1980.
Bauer, B.H. 1980.
Bermingham, E. and J.C. Avise. 1986.
Boschung, H.T., Jr. and R.L. Mayden. 2004.
Branson, B.A. and G.A. Moore. 1962.
Buynak, G.L. and H.W. Mohr, Jr. 1978.
Chang, C.H. 1988.
Clay, W.M. 1975.
Conner, J.V. 1979.
Etnier, D.A. and W.C. Starnes. 1993.
Fowler, H.W. 1945.
Hardy, J.D., Jr. 1978.
Hutton, G.D. 1982.
Jenkins, R.E. and N.M. Burkhead. 1994.
Jennings, M.J. and D.P. Philipp. 1992.
Lauder, G.V. 1983a.

Lauder, G.V. 1983b.
Mabee, P.M. 1987.
Mabee, P.M. 1993.
Mabee, P.M. 1995.
Mayden, R.L. et al. 1992.
Merriner, J.V. 1971b.
Miller, R.J. and H.W. Robison. 1973.
Mok, H.K. 1981.
Near, T.J. et al. 2005.
Nelson, J.S. et al. 2004.
Page, L.M. and B.M. Burr. 1991.
Pflieger, W.L. 1997.
Rettig, J.E. 1998.
Roe, K.J. et al. 2002.
Ross, S.T. 2001.
Smith, G.R. and J.G. Lundberg. 1972.
Smith, P.W. 1979.
Taber, C.A. 1969.
Wainwright, P.C. and G.V. Lauder. 1992.
Warren, M.L., Jr. 1992.

REDBREAST SUNFISH

Lepomis (Lepomis) auritus (Linnaeus)

Robert Wallus and John V. Conner

Lepomis, Greek: *lepis* "scaled" and *poma* "lid"; *auritus*, Latin: "eared," from *auris*, "ear" in reference to the long opercular flap.

RANGE

Redbreast sunfish is one of few centrarchids native to the Atlantic Coast drainages east of the Appalachian Mountains. Redbreast is believed native as far north as southern NK and present throughout the Atlantic slope from NK south to central FL and west to the Apalachicola River. It is probably native only as far west as Choctawhatchee River drainage in western FL. Mobile Bay populations are probably introduced[38] and any other populations outside this area are considered to be introduced.[40]

HABITAT AND MOVEMENT

Redbreast sunfish appears to be adaptive to a wide range of habitats[21,49] and flourishes under a wide range of ecological conditions from small headwater streams to sluggish coastal plain rivers.[44] In NC, redbreast sunfish inhabits waters reaching elevations of 3,500 feet[6] and is sometimes found in brackish water.[30,43]

Redbreast sunfish tends to be more of a river species than most other *Lepomis*.[14,38,42] It is usually found inshore,[22,29,35,49] in pools[22,46,49] and backwaters[46,49] of warm,[22,46] usually clear but occasionally turbid[46] rivers,[22,35,42,44,46] streams,[21,35,39,44,46,49] and creeks[35,42,46] of low or moderate gradient.[46,49] Redbreasts inhabit small to medium-size upland and lowland streams where they usually occur in pools or backwater areas.[39] In NY, it's a fish of standing waters and the slower parts of streams.[43] In streams with rapids, it occurs in slower, deeper areas of rock and gravel.[49]

Redbreast sunfish also are found in sloughs,[12,35] reservoirs,[39,46] lakes,[21,35,44,49] and ponds.[13,35,39,46] In piedmont SC, they are an abundant member of fish communities in beaver ponds.[14] In lakes they occur most abundantly on rocky shoals or in deeper, quieter water with sand or mud bottoms and emergent vegetation.[49]

This species is reported from a wide variety of bottom types[39] including rock,[21,35,49] gravel,[35,49] sand,[12,35,42,49] sand and gravel,[35] organic debris,[35] or mud.[35,49] It is frequently associated with aquatic vegetation,[12,22,35,39] often found at the margins of pools and near cover such as water willow and submerged vegetation.[22]

Redbreast sunfish generally avoids swamps[46] or slow-moving or stagnant, heavily vegetated sloughs. Sometimes found in brackish water.[43] Seems intolerant of acidic waters;[46] reported from water with a pH range between 4.8 and 8.4.[6] Reported from waters having up to 8% seawater equivalency[6,46] or a maximum recorded salinity of 0.98 ppt.[35]

Reported from elevated water temperatures to 39°C, below a power plant.[46] Acclimation temperature range is a significant variable affecting acute preferenda in fishes. Acclimation temperature range for redbreast sunfish is 8.9–28.9°C.[24]

Species richness and centrarchid abundance associated with coarse woody debris (CWD), residentially developed habitats (DEV), and undisturbed (UND) littoral habitats were evaluated in three southeastern reservoirs. Redbreast sunfish numbers and biomass were generally highest in DEV habitat and lowest in CWD and UND habitats during all seasons in all reservoirs. The elevated abundance of redbreast sunfish in residentially developed habitats may be related to their preference for rocky substrata and possibly the additional interstices created by riprapped shorelines.[1]

Historical data (1958) and more recent data (1990) of stream fish assemblage in a VA creek were evaluated to determine the effects of urbanization in the third-order watershed. The watershed had been altered by human activities including road and bridge construction, commercial and residential development, and riparian losses. Fish species diversity was significantly lower in 1990 than it had been in 1959 and abundance for all species and trophic guilds was lower. Redbreast sunfish were collected in low numbers (3–9 individuals) at all six collection stations in 1959 but only one specimen was collected in 1990.[2]

No significant differences in density or biomass were observed for redbreast sunfish after the introduction and establishment of a self-sustaining population of alewife into oligotrophic Mayo Reservoir, NC.[7]

Redbreast sunfish is typically a solitary species, although sometimes forming small groups.[35] They apparently live independent lives during summer,[49] but during winter, when water temperature is below 5°C, redbreast sunfish form dense hibernating schools on or near the bottom in deep water.[29,44] They move inshore in spring when water temperatures reach 10°C or more.[44] Amount of activity in summer shows a direct relation to amount of sunshine and to water temperature.[49]

Recoveries of 189 redbreast sunfish tagged in Little River, NC, indicated that this species is moderately mobile. Upstream movements predominated among fish recaptured within 60 days after their release in April and May. The extent of movement of males and females was about equal. About half of the tagged fish moved away from the locality of first capture. Of the 189 recaptures, 82 (43.4%) were made 0.16 km (0.1 mile) or more from the original tagging site; the rest were recovered at or near the locality of release. The maximum distance traveled was 6.1 km (by each of three fish).[3]

A study of redbreast sunfish movements in a large coastal plain stream in southwest GA suggested long-term residence in relatively small areas. Juveniles were marked and recaptures over a 17-month period primarily occurred within 25 m upstream or downstream of the original capture location. However, some individuals moved at least 100 m and distance moved was not strongly correlated with time between captures, suggesting that long distances traveled did not result from accumulated gradual movement.[50]

DISTRIBUTION AND OCCURRENCE IN THE OHIO RIVER SYSTEM

All populations of redbreast sunfish in the Ohio River drainage are the result of introductions.[38,40]

In KY, redbreast sunfish has been introduced in several localities including Little River, Cypress Creek, upper Cumberland River, and Kentucky River. A presumed hatchery escapee was recorded from Licking River.[39] Etnier and Starnes (1993) state that the species is established only in the extreme southeastern portion of the state.[40]

In VA, the redbreast sunfish has been transplanted to all drainages west of the Atlantic slope.[46] It has been introduced into the New and Tennessee River systems of NC.[45]

The first recognized collection of redbreast sunfish in TN was made in Boone Reservoir on the South Fork Holston River in 1956.[17] Since then, it has become established in the upper Tennessee River drainage downstream to the Chattanooga area. It is also present in the upper Cumberland River drainage, and known from the Big Sandy system in west TN. It appears to be increasing in abundance in east TN;[40] populations have become well established and are reproducing in many east TN reservoirs,[17,40] and most east TN streams and rivers have robust populations.[40]

Redbreast sunfish has been introduced and become established in the Tennessee River system in AL.[41,42]

SPAWNING

Location

Redbreast sunfish spawn in lakes,[29,35] streams,[35] canals,[13,35] rivers,[35] and beaver ponds.[14] Spawning is usually reported over bottoms of sand,[9,30,35] coarse sand,[18,37,44] or gravel.[18,35,37,44] Optimal habitat conditions for spawning are enhanced by low, stable flow.[18,19]

Redbreast sunfish will nest in flowing water, more so than other lepomids,[28] but in this environment the nests are often solitary,[18,28,35,42] whereas in still backwaters, ponds, and lakes, nests are often close together[30,32,35,42,46,49] and in the open.[49] Of 51 successful nests in a VA river, 17 were solitary and 34 were constructed in colonies, each colony with as many as 11 nests.[18] In the Delaware River, 24 male redbreast sunfish were observed guarding nests in a colony, where some nests were only 0.3–0.6 m apart.[44]

In streams redbreast sunfish typically spawn among plants,[28,35] near submerged logs,[9,10,13,28,35] fallen trees,[9,10] stumps,[9,10,13,35] and snags,[13,35] near shorelines with overhanging banks,[28,35] or downstream of protective rocks,[28,30,32,35,46,49] and sometimes nests are observed in areas with no cover.[37]

Nests are typically located in calm pool margins of streams in less than 1 m of water.[28,30,32,34,35,37,44,46,49] Nests are also reported at depths of 1.1[13,35] to 1.8 m[18] and in tidal water depth may vary from 0–1,525 mm.[30]

Redbreast sunfish in streams select nesting sites located in pools where cover is abundant and velocities relatively low. In North Anna River, VA, redbreast sunfish nests were located in pools containing ample cover and all nests were constructed of coarse sand to fine gravel substrata (0.5–16 mm diameter). Nest sites were observed in water from 0.42 m to 1.81 m deep (average 1.08 m). Mean water column velocities at nest sites were variable, ranging to 0.15 m/s; bottom velocities ranged to 0.04 m/s with a mean of 0.01 m/s. The mean distance from shore for nests was 12.4 m. Cover near nests was typically bedrock ledges and boulders.[18]

In 10 VA streams, mean current velocity at most nest sites was slow, averaging 0.9 cm/s, and was probably a function of shallow depth and abundance of adjacent cover. Vegetative and structural cover in nesting areas was consistently more abundant than in non-nesting areas and water depth was consistently shallower at nesting sites. Water depth averaged 64 cm at nest sites. Substrate particle size composition in the nest was significantly altered by the male's nest construction activities. Intermediate-sized (0.51–16 mm) substrate particles dominated (80%) the nest depression.[19]

Nests in Catatonk Creek, NY, had uniform, small-size stones about 13 mm or less (about ½ inch) in diameter. It was possible to distinguish at a glance between redbreast nests and those of pumpkinseeds on the same bottom type, since the latter nests had generally larger and heterogeneous-sized stones in them. Redbreast sunfish males in nature and in aquaria often and persistently carry larger stones in their mouths to the edge of the nest for deposit. The frequency of this activity for nesting redbreast explains the difference in appearance of redbreast and pumpkinseed nests in the same area.[28]

Spawning is reported from the freshwaters of upper estuaries; nests have been found just above the freshwater interface in DE, usually in colonies.[34] Nesting is also reported in slightly brackish water.[30] Richmond (1940)[30] found large groups of redbreast nests in a VA tidal river; these nests were exposed at high tide and water temperature, depth, and salinity varied drastically.[49]

Season

Generally May to August in Atlantic Coast drainages from NY to GA:

In NY, early June to early August[44]
In DE, late spring and early summer[37]
In Potomac River, June to late July, possibly into August[34]
In VA, late May into August[18,19,46]
In NC, early April to June[10,45]
In SC, May through August in beaver ponds[14]
In GA, during May and June[9]
In TN, late June*

Begins earlier in southern latitudes:

In AL, April to August[41,42]
In FL, April through October,[10–13] with a peak in late spring and summer[13]

Temperature

Several reports suggest that redbreast sunfish begin to spawn near or shortly after water temperatures reach 20°C.[9–11,13,18,19] However, nest building is reported at 16°C[29] and 17°C[28] and may begin at 21.1°C.[11,35] Spawning is reported at water temperatures as high as 29°C.[19]

In NY, water temperatures during spawning may vary from 18.3 to 26.7°C.[44]

In VA, mean daily water temperatures during the spawning season in the North Anna Riverwere 23.5–27.5°C.[18] Water temperatures during spawning reached as high as 29°C in a study of 10 other VA streams.[19]

Redbreast are reported to spawn at water temperatures of 22–25°C in SC,[14] NC,[10] GA,[9] and TN.*

Fecundity

Reported estimates of fecundity range from a few hundred to almost 10,000 eggs per female. Numbers of eggs increase with age[10,47] and size of the female.[9,10,14,47]

In NC, female redbreast sunfish 140–142 mm (age 2) had an average of 963 ova; at 150–155 mm (age 3) they had an average of 1,008 ova; at 171–184 mm (age 4), 3,563 ova; at 183–203 mm (age 5), 5,620 ova; and at 228–235 mm, 8,250.[10,47]

In SC, mature females >76 mm TL contained approximately 3,000 ova. Fecundity increased significantly with TL.[14] In Satilla River, GA, fecundity estimates ranged from 322 to 9,206 ova per female with a mean of 3,302 ova. Fecundity increased with both length and weight.[9] In FL, fecundity ranged from 942 to 9,968 ova per female.[13]

Sexual Maturity

Redbreast sunfish are sexually mature as early as age 2[10,14] and as small as 60–75 mm TL.[14,25,35]

In GA, 43% of females and 60% of males were sexually mature at 90–114 mm TL.[9] In VA, smallest nesting males were 114 mm TL and smallest females were estimated to be about 80 mm TL.[18]

In the Chattahoochee River, GA/AL, in 1972, examination of gonads indicated that female redbreast sunfish mature as small as 60 mm TL at 3–6 g; males were mature as small as 70 mm and 6 g.[25]

Spawning Act

Male redbreast sunfish defend territories and build nests but will occasionally spawn in the nests of other centrarchids,[35,49] including largemouth bass.[49] During the process of nest construction, redbreast males remain almost horizontal and make sweeping movements with the tail while moving forward, unlike other *Lepomis*, whose sweeping actions are from a more vertical posture.[28,42] A behavioral characteristic of redbreast sunfish, which is uncommon in other lepomids, is stone carrying. After a nest depression has been swept clean, certain stones are removed from the nest, leaving only stones of a uniform size, about 12–13 mm in diameter.[28]

The male redbreast sunfish sweeps out a nest that is usually a circular depression with a gravel bottom.[35] Nests are reported to be about 51 mm deep,[28,35] and from 0.25 m to about 1 m in diameter,[9,28,35,44] depending upon the size of the guarding male.[44]

Females approaching colonies of nests were actively courted by several males that swam in circular patterns above their nests. In one spawning event the female swam directly into a solitary nest with no courtship. While the male and female circled the nest, several egg-releasing events occurred that lasted about 10 s each. These events were characterized by the female turning on her side, quivering and releasing eggs while the male remained upright and presumably released sperm. The entire spawning act took 4 min, with the male leaving the nest many times to chase away other redbreast sunfish while the female remained in the center of the nest. Intruders repeatedly tried to swim between the male and female after they were side-by-side. After an intruder was chased away, the male and female resumed circling the center of the nest.[18,28] Redbreast males leave their nests more readily than do pumpkinseed or bluegill males.[28]

Spawning ended when the male chased the female from the center of the nest[18] and began using his caudal fin in a sweeping motion to mix the eggs into the gravel of the nest bottom.[18,28] A redbreast male studied from movies made in the field, immediately after spawning was completed, made incipient fanning movements, many of which turned into sweeping. He swept during the time eggs were in the nest, but when the eggs hatched, he ceased sweeping and guarded the nest from one side.[28]

Female redbreast sunfish may become aggressive toward, and often chase, other females ready to spawn or that have just spawned. Large females may be more aggressive, participate in dominance encounters, and become tyrants if they are the largest fish.[28]

Of seven sunfishes studied, redbreast was the only species observed that made no sounds during courtship. Seventeen courtships and three spawning events were observed.[27]

Male redbreast sunfish guard their nest, eggs, and fry.[44] They spend a considerable amount of time defending their nests from potential predators. In a VA stream, males made from 30 to 115 nest departures per hour (mean 58), while spending 72 to 92% (mean 83%) of the time on their nests.[18]

In the Susquehanna River, PA, two of three redbreast sunfish nests collected in 1974 and one in 1975 contained eggs of the swallowtail shiner *Notropis procne*.[32]

In TN[40] and VA,[46] redbreast sunfish hybridizes occasionally with bluegill, with which it most often occurs. Several other hybrid combinations involving redbreast are known.[40]

Jenkins and Burkhead (1994)[46] report that Bailey (1938)[48] documented natural hybridization between redbreast sunfish and pumpkinseed, green, bluegill, and redear sunfish.

EGGS

Description
Diameters of mature ova averaged 1.12 mm in SC;[14] 1.2 mm in GA.[9]

Fertilized eggs are demersal,[34] adhesive, bright yellow[32,35,36] or amber, and nearly transparent.[35,36] Average diameter is 1.8 mm[35,36]–2.1 mm (mean diameter of eggs fixed in 10% formalin).[32] Yolk has a single large and several small oil globules; diameter of largest oil globule about 0.6 mm.[35,36]

Incubation
No information.

Development
No information.

YOLK-SAC LARVAE

See Figure 28.

Size Range
Newly hatched redbreast sunfish larvae from the Susquehanna River, PA, were 4.6–5.1 mm TL;[32] in TN, newly hatched larvae were 4–5 mm TL.*

Earlier reports indicate that yolk was completely absorbed by 7.0 mm TL[34] or 7.9 mm,[32] while a post yolk-sac larva was 7.7 mm TL.[37] In TN larvae retained yolk longer; yolk was absorbed by about 8.4 mm TL on most specimens, but traces remained on some individuals until about 10 mm.*

Myomeres
See Table 23 for ranges of myomere counts in TN.

In PA:

5.0–7.6 mm TL. Preanal 12–14, mode 12; postanal 17–18, mode 18; total 29–31, mode 30.[32]

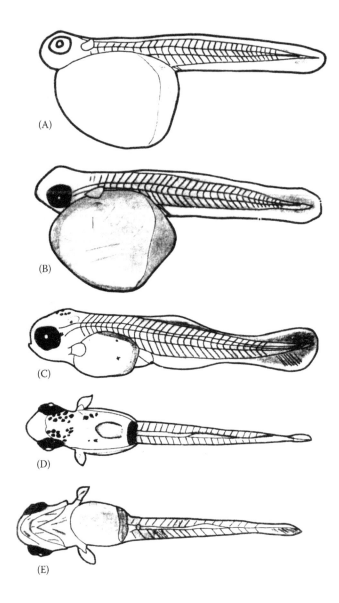

Figure 28   Development of young redbreast sunfish. A–B. Yolk-sac larvae: (A) 5.2 mm TL; (B) 6.0 mm TL. C–E. Yolk-sac larva, 7.8 mm TL: (C) lateral view; (D) dorsal view; (E) ventral view. (A–E reprinted from Figure 1, reference 32, with publisher's permission.)

Morphology

4.6–6.0 mm TL. Yolk sac large and ovoid, yolk granular; urostyle straight; mouth incomplete; a single large oil globule is present in the posterior area of the yolk sac.[32]

7.0 mm TL. Yolk absorbed and massive gut coils developing (Potomac River).[34]

7.5 mm TL. In TN, oil absent on some.*

7.8 mm TL. In PA, air bladder begins to inflate; mouths open.[32]

7.9 mm TL. In PA, yolk completely absorbed.[32]

8.4 mm TL. In TN, remnants of oil droplets are still present on some individuals; yolk is completely absorbed on most individuals (traces of yolk may still be visible until about 10 mm).*

Morphometry

See Table 24 for morphometric data from TN.

In PA:

5.0–7.6 mm TL. As percent TL: SL 97–98; PostAL 48–54; HL 14–22; ED 9–10; GD 30–33.[32]

Table 23

Myomere count ranges (average and standard deviation in parentheses) at selected size intervals for young redbreast sunfish from TN.*

Size Range (mm TL)	N	Myomeres			
		Predorsal Fin	Preanal	Postanal	Total
4.6–6.9	21	—	9–14 (11.6 ± 1.73)	15–17 (16.0 ± 0.72)	24–31 (27.6 ± 2.29)
7.0–10.0	52	5–6 (5.95 ± 0.21)	13–16 (14.8 ± 0.71)	15–17 (16.2 ± 0.41)	30–32 (31.0 ± 0.58)
10.1–13.2	35	5–7 (6.0 ± 0.41)	15–16 (15.3 ± 0.46)	16–17 (16.0 ± 0.16)	31–32 (31.3 ± 0.47)
13.3–14.0	2	6	15	16	31
14.2–17.9	12	5–7 (5.91 ± 0.52)	14–16 (15.1 ± 0.68)	16	30–32 (31.1 ± 0.68)

Fin Development

4.6–6.0 mm TL. In PA, pectoral fin buds present; median finfold present dorsally, around urostyle, and ventrally to the anus, which drops out at the posterior margin of the yolk sac.[32]

7.3 mm TL. In TN, first rays formed in caudal fin.*

7.8 mm TL. In PA, developing caudal fin rays are visible ventrally under urostyle. Median finfold reduced on caudal peduncle with future locations of dorsal and anal fins apparent; remnant of ventral finfold present between the anus and posterior margin of yolk.[32]

Pigmentation

4.6–5.2 mm TL. In PA, eyes lack pigment and no body pigment is visible.[32]

6.0 mm TL. In PA, eyes are pigmented but pigmentation is still lacking on the body.[32]

7.0 mm TL. Melanophores present on dorsal surface of brain case, a single row of melanophores along lateral line, and 20–30 small melanophores ventrally between operculum and anus (Potomac River).[34]

7.8 mm TL. In PA, two ovoid-shaped groups of melanophores are present in the occipital region. No lateral or ventral pigmentation is present.[32]

POST YOLK-SAC LARVAE

See Figures 29 and 30.

Size Range

Yolk is completely absorbed by 7.0,[34] 7.7,[37] 7.9,[32] or 8.4 mm TL.* In PA, the juvenile phase has begun by 19.0 mm.[32] In TN, the adult complement of spines and rays is present in all fins and the median finfold is absorbed on some individuals by 14.2 mm TL, but a remnant of ventral finfold anterior to anus is still present on some individuals at 14.7 mm TL.*

Myomeres

See Table 23 for ranges of myomere counts in TN.

In PA:

7.8–11.4 mm TL. Preanal 12–14, mode 12; postanal 14–18, mode 16; total 27–30, mode 28.[32]

11.8–18.5 mm TL. Preanal 12–14, mode 13; postanal 14–16, mode 15; total 27–29, mode 28.[32]

Morphology

7.7 mm TL. Robust and elongate; gut is massive and densely coiled anteriorly.[37]

7.9 mm TL. Flexion has begun and a massive coiling in the anterior portion of the gut is apparent.[32]

Table 24

Morphometric data for young redbreast sunfish from TN* grouped by selected intervals of total length. Characters are expressed as percent total length (TL) and head length (HL) with a single standard deviation. Range of values for each character is in parentheses.

	TL Intervals (mm)					
Length Range	4.61–6.93	7.00–10.05	10.08–13.15	13.35–14.00	14.18–17.90	18.00–21.30
N	21	52	35	2	12	6
Characters	Mean ± SD (Range)	Mean ± SD (Range)	Mean ± SD (Range)	Mean ± SD (Range)	Mean ± SD (Range)	Mean ± SD (Range)
Length as % TL						
Snout	3.42 ± 0.33 (0.18–0.25)	4.10 ± 0.57 (0.19–0.55)	4.87 ± 0.38 (0.41–0.73)	5.28 ± 3.82 (0.70–0.71)	5.09 ± 0.13 (0.71–0.92)	4.78 ± 0.20 (0.90–1.00)
Eye diameter	8.37 ± 0.42 (0.37–0.58)	9.01 ± 0.47 (0.57–0.98)	9.42 ± 0.31 (0.95–1.25)	9.48 ± 0.11 (1.25–1.28)	9.49 ± 0.16 (1.38–1.68)	9.16 ± 0.22 (1.70–1.90)
Head	15.90 ± 1.80 (0.68–1.35)	23.0 ± 1.61 (1.38–2.45)	25.8 ± 0.79 (2.45–3.60)	27.0 ± 0.15 (3.58–3.62)	26.4 ± 0.76 (3.73–4.60)	25.7 ± 0.57 (4.80–5.30)
Predorsal	—	31.4 ± 0.92 (2.85–3.15)	29.8 ± 0.64 (3.05–4.05)	30.1 ± 0.11 (4.00–4.03)	30.8 ± 0.61 (4.30–5.50)	32.4 ± 0.57 (6.00–6.80)
Prepelvic	—	30.2 ± 0.94 (2.75–3.00)	28.2 ± 0.84 (2.80–3.80)	29.4 ± 0.19 (3.90–3.95)	28.2 ± 0.60 (4.00–5.00)	29.7 ± 0.85 (5.60–6.00)
Preanal	47.5 ± 2.13 (2.30–3.09)	46.4 ± 0.97 (3.11–4.65)	46.6 ± 0.78 (4.55–6.20)	46.4 ± 0.38 (6.15–6.25)	46.6 ± 0.78 (6.48–8.40)	47.5 ± 1.12 (9.00–9.70)
Postanal	52.5 ± 2.13 (2.31–3.85)	53.4 ± 1.24 (3.43–5.40)	53.3 ± 1.10 (5.35–7.15)	53.6 ± 0.38 (7.10–7.20)	53.4 ± 0.78 (7.70–9.00)	52.5 ± 1.12 (9.60–11.60)
Standard	97.2 ± 1.01 (4.52–6.80)	91.5 ± 4.91 (6.81–8.55)	83.0 ± 1.55 (8.60–10.75)	82.8 ± 0.76 (10.9–11.15)	81.7 ± 0.55 (11.5–14.3)	81.9 ± 0.67 (15.3–17.4)
Air bladder	8.76 ± 0.54 (0.51–0.65)	13.7 ± 1.94 (0.68–1.65)	17.0 ± 0.99 (1.55–2.40)	—	—	—
Eye to air bladder	13.5 ± 1.17 (0.76–1.07)	12.3 ± 1.11 (0.79–1.35)	11.6 ± 0.56 (1.10–1.50)	—	—	—
Fin Length as % TL						
Pectoral	—	13.0 ± 1.23 (0.95–1.63)	15.0 ± 1.31 (1.25–2.25)	16.8 ± 0.30 (2.20–2.28)	16.0 ± 0.67 (2.20–3.05)	18.5 ± 0.69 (3.40–4.05)
Pelvic	—	2.50 ± 0.56 (0.15–0.38)	4.80 ± 0.98 (0.31–0.91)	6.63 ± 0.26 (0.85–0.92)	9.7 ± 1.57 (1.15–2.50)	14.1 ± 0.84 (2.50–3.05)
Body Depth as % TL						
Head at P1	—	20.5 ± 0.56 (1.80–2.10)	22.8 ± 1.35 (2.15–3.30)	24.5 ± 0.19 (3.25–3.30)	26.1 ± 0.96 (3.50–4.58)	27.4 ± 1.24 (4.90–5.80)
Head at eyes	—	19.1 ± 0.74 (1.70–1.95)	20.0 ± 1.25 (1.80–2.90)	22.1 ± 0.38 (2.90–3.00)	22.0 ± 0.74 (3.05–4.00)	23.5 ± 1.11 (4.30–5.00)
Preanal	—	17.5 ± 0.81 (1.45–1.85)	20.4 ± 1.57 (1.80–3.00)	22.3 ± 0.19 (2.95–3.00)	24.2 ± 0.83 (3.30–4.40)	25.3 ± 0.81 (4.65–5.30)
Caudal peduncle	—	7.97 ± 1.64 (0.07–0.88)	9.66 ± 0.48 (0.91–1.35)	10.1 ± 0.15 (1.33–1.37)	10.6 ± 0.32 (1.45–1.92)	10.5 ± 0.43 (1.93–2.20)
Body Width (as % HL)						
Head	—	—	53.9 ± 1.07 (1.90–1.95)	54.9 ± 1.02 (1.95–2.00)	67.5 ± 6.45 (2.30–3.60)	72.3 ± 2.62 (3.35–3.90)

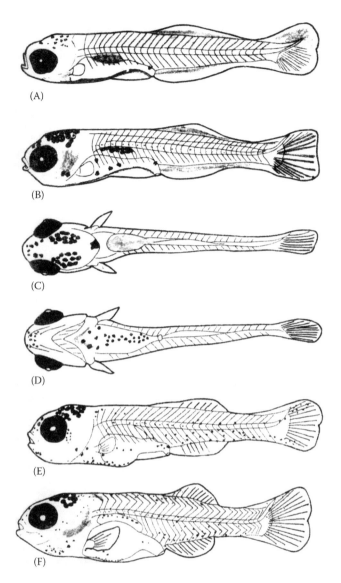

(A)

(B)

(C)

(D)

(E)

(F)

Figure 29 Development of young redbreast sunfish. (A) Post yolk-sac larva, 7.9 mm TL. B–D. Post yolk-sac larva, 8.0 mm TL: (B) lateral view; (C) dorsal view; (D) ventral view. E–F. Post yolk-sac larvae: (E) 8.1 mm TL; (F) 9.8 mm TL. (A–F reprinted from Figure 2, reference 32, with publisher's permission.)

9.8 mm TL. Flexion near completion, about 45 degrees.[32]

10.0 mm TL. Body shape and gut coiling similar to *Micropterus* spp.; mouth is small and not bass-like.[34]

≥11.0 mm TL. Deep bodied and laterally compressed; snout is short; forehead steeper than other sunfishes; gill rakers are short and stiff, usually 8 rakers on lower limbs.[37]

Morphometry

See Table 24 for morphometric data from TN.

In PA:

7.8–11.4 mm TL. As percent TL: SL 85–91; PostAL 49–53; HL 19–28; ED 9–10; GD 19–24.[32]

11.8–18.5 mm TL. As percent TL: SL 81–84; PostAL 49–50; HL 28; ED 9–10; GD 25–28.[32]

Fin Development

8.1 mm TL. In PA, insertions of soft dorsal and anal fins are nearly defined, little finfold remaining on caudal peduncle. Remnant of finfold is still present ventrally, anterior to anus. Dorsal, anal, and pectoral fin rays are visible. Caudal fin rounded with many rays developed.[32]

9.0–9.1 mm TL. In TN, first dorsal and anal fin spines forming; adult complement of caudal fin rays (17) formed.*

9.5 mm TL. In TN, adult complement of dorsal fin rays formed.*

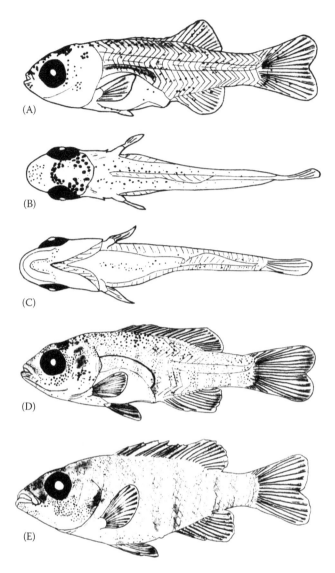

Figure 30   Development of young redbreast sunfish. A–C. Post yolk-sac larva, 11.8 mm TL: (A) lateral view; (B) dorsal view; (C) ventral view. D–E. Juveniles: (D) 19.0 mm TL; (E) 36 mm TL. (A–E reprinted from Figure 3, reference 32, with publisher's permission.)

9.8 mm TL. Little median finfold remains except for remnant anterior to anus. Anterior and posterior margins of soft dorsal and anal fin well defined.*,[32] In PA, development of dorsal, pectoral, and anal fin rays more advanced. No visible spinous dorsal development.[32]

10.2–10.4 mm TL. In TN, adult complement of anal and dorsal fin spines formed.*

11.0 mm TL. In TN, pelvic fin buds visible.*

11.8 mm TL. In PA, developing spines in dorsal fin evident.[32]

14.2 mm TL. Adult complement of pelvic fin elements (1 spine and 5 rays) formed. Remnant of ventral finfold anterior to anus is completely absorbed on some; a trace still remains at 14.7 mm on some individuals.*

Pigmentation

7.7 mm TL. A few melanophores are present on the dorsal surface of the head and along the lateral line. A large dark patch of pigment is present mid-ventrally anterior to the anus. The rest of the body is only sparsely pigmented.[37]

7.9 mm TL. A supra-anal melanophore is visible.[32]

8.0 mm TL. Lateral and ventral pigmentation is evident and the number of melanophores in the occipital region has increased. A V-shaped arrangement of several melanophores is present anterior to those in the occipital region. Ventral pigmentation consists of 20–30 small melanophores that form an elongate patch or row between the operculum and the anus. Laterally, melanophores are present on the gut and on the air bladder and a single row of melanophores is present along the lateral line.[32]

8.1 mm TL. Melanophores on lateral and ventral surfaces of the body have increased in number. Some pigmentation evident on caudal fin.[32]

10.0 mm TL. Larvae have scattered single melanophores over entire body.[34]

11.8 mm TL. Dorsal and ventral body pigmentation appears denser; melanophores on all body surfaces have increased and a few are present on the dorsal and anal fins. The ovoid patches of pigment in the occipital region and elongated patch of pigment on the venter are still present.[32]

JUVENILES

See Plate 1B following page 26 and Figure 30.

Size Range
14.2–14.7 mm TL* to as small as 60–90 mm.[9,14,18,25]

Myomeres
See Table 23.

Morphology
36 mm TL. Body is deep and laterally compressed; greatest depth is at origin of dorsal fin. Eye situated in front of head; maxillaries short, reaching only to anterior edge of the eyes. Base of dorsal fin more than twice as large as the base of the anal fin.[32]

Larger juveniles. Lateral line scales 43–49 in Canada;[49] 39–54, usually 42–46 in VA;[46] 42–51 in TN;[40] and 41–52 in AL.[41,42] Gill rakers 9–12,[40,42] length of longest rakers about four times their basal width in YOY.[40] Vertebrae 29–30;[40,49] 7 pyloric caeca.[49]

Morphometry
See Table 24 for morphometric data from TN.

In PA:

19.0 mm TL. As percent TL: HL 28; GD 30.[32]

36.0 mm TL. As percent TL: ED 9.[32]

Fins
Two dorsal fins are broadly joined and appear as one.[49] First dorsal fin has 9–11, usually 10–11 spines; second dorsal fin has 10–12 soft rays.[40,41,46,49] Anal fin has 3–4 spines, usually 3,[41,46,49] and 8–10, usually 9–10 soft rays.[40–42,46,49] There are 13–15 soft rays in

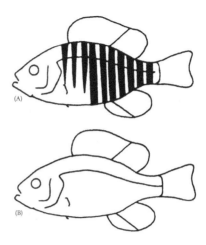

Figure 31 Development of young redbreast sunfish — schematic drawings showing melanistic pigment patterns of young redbreast sunfish. (A) Juveniles, 17–36 mm TL; primary pigment bars visible, but no secondary bars develop. (B) Larger juveniles and adults; lateral bars of pigment disappear. (A and B reprinted from Figure 14, reference 31, with publisher's permission.)

the pectoral fins[40,42,46,49] and 1 spine and 5 rays in the pelvic fins.[49]

Pigmentation
Primary pigment bars appear as an initial set of lateral patches of melanophores to which bars are added anteriorly and posteriorly in development in all *Lepomis*.[31] Secondary bars of pigment that appear between the primary bars on other *Lepomis* are not present in redbreast sunfish (Figure 31).[31] YOY redbreast up to about 25 mm TL may have about 12 vertical dark bars on sides; these bars become faint or absent in larger specimens.[40]

17.0 mm TL. Faint vertical bars may appear laterally.[34]

19.0 mm TL. Pigmentation is dense on the dorsal surface of body. Laterally, pigmentation has increased and vertical bands or bars of pigment are visible.[32]

33.0 mm TL. Opercular spot evident; about 15 narrow vertical lateral bars, each about as wide as the interspaces.[35]

Larger juveniles are quite plain, with only a trace of vertical bars in contrast to other species of *Lepomis* that have pronounced vertical bands in the young stages.[43]

TAXONOMIC DIAGNOSIS OF YOUNG REDBREAST SUNFISH

Similar species: redbreast sunfish is the only member of the subgenus *Lepomis*. Hypothesized to be similar to members of the *Icthelis* subgenus (longear and dollar sunfish).[51] Descriptions of young redbreast in the literature show a strong resemblance to young longear sunfish.[32,34,35,51,52] There are comparisons in the literature of the early development of young redbreast sunfish to bluegill, pumpkinseed, and green sunfish,[32-34,40] and longear sunfish.[40]

See *Lepomis* subgeneric discussion on page 90.

For comparisons of important morphometric and meristic data, developmental benchmarks, and pigmentary characters between redbreast sunfish and all other *Lepomis* that occur in the Ohio River drainage, see Tables 4 and 5 and Plate 1B.

Redbreast sunfish vs. green sunfish, pumpkinseed, and bluegill. From hatching to swim-up, redbreast sunfish larvae can be distinguished from green, pumpkinseed, and bluegill by their larger size. Redbreasts are 4–5 mm TL at hatching*,[32] and swim up at 6–7 mm* or 7.6–8.2 mm.[32] The others are ≤3.7 mm at hatching and <6.3 mm at swim-up.[33] The anterior coiled portion of the redbreast's gut becomes massive by about 7.0[34]–7.9 mm;[32] the others have smaller S-shaped guts. By the time redbreast sunfish larvae have developed first caudal fin rays (mesolarvae) they have numerous (20–30) small melanophores ventrally between the opercle and the vent that form an elongated patch or row. Larvae of green sunfish, pumpkinseed, and bluegill may have ventral pigmentation but not an elongated patch or row.[32] Gill rakers on the first gill arch of redbreast sunfish are shorter than those of juvenile pumpkinseed at comparable sizes.[34]

Redbreast sunfish YOY have reddish eyes and the dark vertical bars on the sides are vague or absent (Plate 1B). They are thus very similar in appearance to young green sunfish (Plate 1C), which have more and slightly longer gill rakers, and young longear sunfish (Plate 2C), which have fewer lateral-line scales and shorter gill rakers.[40]

ECOLOGY OF EARLY LIFE PHASES

Occurrence and Distribution (Figure 32)

Eggs. The demersal,[34] adhesive[32,35,36] eggs of the redbreast sunfish are deposited in the nest where they attach to gravel and rocks[32,35] and to each other, sometimes forming large clumps.[35] Redbreast sunfish use their caudal fin to evenly mix the eggs in the gravel of the nest bottom. The gravel and eggs form a stable, gelatinous matrix approximately 3 cm thick in which the eggs incubate. Males defend their nests from predators.[18]

In North Anna River, early nesting efforts were unsuccessful in 1992; 58 nests constructed in late May were destroyed by high flows on June 4. Thirty of 51 nests monitored in June and July produced free-swimming larvae. During this period flows were low and relatively stable and predation and nest abandonment by the male were suspected causes of the 21 nest failures.[18] Optimal habitat conditions for successful nesting are enhanced by low, stable water flow.[18,19]

When a guarding male was removed by angling, small sunfish in the area quickly entered the nest and ate the eggs.[44]

Larvae. In nests in holding tanks in TN, larvae began swimming up at 6–7 mm TL; about 95% of

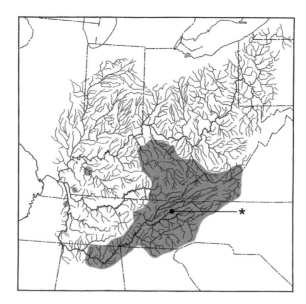

Figure 32 Distribution of introduced populations of redbreast sunfish in the Ohio River system (shaded area) and area where early life history information has been collected (circle). Asterisk indicates original TVA data.

the larvae on a nest had swum up by 7–8 mm TL.*
At 7–10 mm TL, larvae in a living stream unit were
very active and dispersed.*

Juveniles. Young redbreast sunfish remain in small
schools after leaving the nest[29,35] and throughout
the first summer.[35] They are typically associated
with aquatic vegetation,[12,35,42] often found in shallow vegetated zones in still water[42] where they
have been recorded over bottoms of mud and
rock.[35] Young redbreast sunfish also frequent
heavily vegetated tributary streams.[42]

Early Growth

Growth of young redbreast sunfish, summarized
by Carlander (1977), was slower in NY lakes than
in more southern populations in NC, WV, and GA.
Mean TL of age-1 redbreast in NY ranged from
43 to 60 mm, while the range for those southern
populations studied was 73–135 mm. At age 2,
range of average TL of redbreast in NY was 63–77,
in the south 83–163; at age 3, 68–88 mm in NY and
140–196 mm in the south.[47] Redbreast appear to
have similar growth rates throughout the southeastern United States.[5]

In VA, growth of redbreast sunfish is about the
same in streams as in reservoirs[46] and growth rates
are similar to those summarized for NC and the
Potomac River in MD.[46,47] In VA, the range of
means of length at age 3 from 11 stream and reservoir populations, plus a sample combined from
several rivers, was 119–173 mm TL, with an
unweighted mean of 140.[46] In NC coastal streams
average TL of captured redbreast at age 0 ranged
from 54 to 81 mm; at age 1, 81–117 mm; at age 2,
121–166 mm; at age 3, 152–202 mm; and at age 4,
178–226 mm (Table 25).[15]

In GA, average calculated total lengths at age 1–4
were 59, 90, 125, and 153 mm, respectively. The
length–weight for all redbreast sunfish examined
was: log W = –5.2810 + 3.2368 log L (r = 0.9759).[9]

Growth was slower in SC where averaged calculated total lengths at age 1–4 were 41, 75, 102, and
128 mm, respectively. Growth of males was faster
than females.[14]

Data from redbreast sunfish collected in 1972 from
the Chattahoochee River watershed, GA/AL, also
indicated that young males grew faster than
females. Average calculated TL (mm) related to age
is reported as follows:[25]

	Age			
	1	2	3	4
Average	44.4	78.7	120.4	147.7
Males	46.7	86.4	133.2	157.8
Females	42.9	74.7	112.1	140.1

There is a great deal of variation in growth related
to age for young redbreast sunfish among FL
streams. Calculated total lengths of redbreast at
age 1 from 13 streams ranged from 28 to 93 mm;
72–135 mm at age 2 (12 streams); 108–139 mm at
age 3 (7 streams); and 118–152 mm at age 4 in 2
streams (Table 26).[5]

Redbreast sunfish populations in a GA lake influenced by thermal discharges showed no increased
growth over populations from other thermally
uninfluenced areas.[8]

In GA, male redbreast sunfish weighed less per
given length than female specimens up to 150 mm.[9]
Length–weight relationships of young redbreast

Table 25

Mean TL (mm) at age and sample size of redbreast sunfish
collected by electrofishing from coastal NC streams.

Stream (year)	Mean Calculated TL and (Sample Size) at Age:				
	0	1	2	3	4
Black River (1994–97)	61 (33)	81 (214)	133 (79)	190 (49)	220 (16)
Cape Fear River (1988)	54 (2)	88 (12)	121 (5)	160 (3)	—
Contentnea Creek (1993)	—	117 (54)	166 (43)	202 (28)	226 (3)
Lumber River (1994–97)	60 (35)	85 (237)	133 (243)	181 (179)	218 (95)
South River (1971)	—	—	140 (a)	152 (a)	178 (a)
Trent River (1986)	81 (7)	99 (10)	145 (42)	160 (27)	199 (3)

[a] Sample size unknown.

Source: Constructed from data presented in Table 2 (in part), reference 15.

Table 26

Mean length of redbreast sunfish from FL for age classes 1–4, back calculated from examination of otolith annuli.

	Age			
	1	2	3	4
Stream samples (N)	13	12	7	2
Mean of means (mm TL)	61	104	115	119
Standard deviation	11	20	8	4
Minimum TL	28	72	108	118
Maximum TL	93	135	139	152

Source: Constructed from data presented in Table 4 (in part), reference 5.

Table 27

Length–weight relationships of young redbreast sunfish from AL. Values represent data collected during the period 1949–1964.

Approximate Average TL (mm)	N	Average Empirical Weight (g)
25	31	0.9
51	19	2.5
76	34	13.6
102	48	20.0
127	45	44.0
152	34	77.2
178	46	131.7
203	10	163.4

Source: Constructed from data presented in unnumbered table on page 51, reference 26.

sunfish, collected in AL during the period 1949–1964, are presented in Table 27.

Feeding Habits
Redbreast sunfish are insectivores.[4,9,13,16,20]

In the Juniata River, PA, in October 1989, a 24-h diel food habit study was conducted on the fish community at 4-h intervals. The major prey items (expressed as % dry weight) consumed by redbreast sunfish 26–189 mm TL at the 4-h intervals were:

> 0400 h — ephemeropteran nymphs, mainly baetids (38%) and heptageniids (21%); trichopterans, mainly hydropsychids (16%); chironomids (14%)

> 0800 h — epheremopteran nymphs, mainly potamanthid nymphs (33%) and baetids (12%); hydropsychids (16%); chironomids (14%);

> 1200 h — epheremopteran nymphs, mainly potamanthids (28%) and ephemerids (14%); hydropsychids (24%); chironomids (15%)

> 1600 h — trichopterans (41%); chiromomids (15%); ephemeropteran nymphs (40%)

> 2000 h — ephemeropterans, mainly heptageniids (30%); chironomids (21%)

> 2400 h — hydropsychids (31%) and ephemeropterans, mainly baetids (16%) and potamanthids (13%)

Over the 24-h period, ephemeropterans (57.5%), mainly baetids (25%) and heptageniids (17%), were the dominant food followed by hydropsychids (20%) and chironomids (15%).[20]

In July 1990, another 24-h diel study was conducted in the Juniata River, PA. In this study, ephemeropteran nymphs were again the major prey amounting to 70.5% dry weight of food consumed by redbreast sunfish 49–174 mm TL and were the dominant prey for each 4-h interval of the 24-h study. Ephemerids were clearly the dominant ephemeropteran consumed by redbreast. Trichopterans, mainly hydropsychids, were the second ranking aquatic invertebrate food. Chironomids and miscellaneous aquatic and terrestrial insects made up the remainder of the diet.[16]

In two piedmont lakes in NC, the percent frequency of food items in age-0 redbreast sunfish (13.5–36 mm SL) stomachs was as follows: copepods (34–59%), cladocerans (18–30%), ostracods (5–8%), rotifers (1–2%), insect larvae (15–20%) and fish (1–7%)[23]

In SC beaver ponds redbreast sunfish <75 mm TL consumed cladocerans and copepods.[14]

In the Satilla River, GA, redbreast sunfish <75 mm consumed small insect larvae such as dipterans, organic matter, detritus, and vegetation. Redbreast 86–125 mm consumed most forms of insects and larvae with the exception of larger forms such as naiads of the order Odonata. Small crustaceans such as freshwater shrimp were ingested, as well as young crayfish, organic debris, detritus, and vegetation. Fish >126 mm ingested all sizes of insects, and insect larvae and larger terrestrial insects, larger crustaceans such as crayfish, and small minnows were more frequently consumed by these larger redbreast.[9]

In small AL streams, redbreast sunfish were primarily insectivores relying heavily on mayflies, chironomids, caddisflies, and adult and larval aquatic beetles. Fish 50–100 mm TL consumed proportionally greater quantities of aquatic insects by number and volume, while fish ≥101 mm TL consumed greater quantities of terrestrial insects and other food. Crustaceans, including crayfish, were regularly consumed.[4]

In FL streams the redbreast sunfish is primarily an insectivore, feeding upon benthic larval and nymphyal stages, as well as emergent aquatic insects from the surface.[13]

LITERATURE CITED

1. Barwick, D.H. 2004.
2. Weaver, L.A. and G.C. Garman. 1994.
3. Hudson, R.G. and F.E. Hester. 1976.
4. Cooner, R.W. and D.R. Bayne. 1982.
5. Mantini, L. et al. 1992.
6. Shannon, E.H. 1967.
7. Crutchfield, J.U., Jr. et al. 2003.
8. O'Rear, R.S. 1970.
9. Sandow, J.T., Jr. et al. 1975.
10. Davis, J.R. 1972.
11. Bass, D.G., Jr. and V.G. Hitt. 1973.
12. Hellier, T.R., Jr. 1967.
13. Bass, D.G., Jr. and V.G. Hitt. 1975.
14. Levine, D.S. et al. 1986.
15. Ashley, K.W. and R.T. Rachels. 1998.
16. Johnson, J.H. and D.S. Dropkin. 1995.
17. Fitz, R.B. 1966.
18. Lukas, J.A. and D.J. Orth. 1993a.
19. Helfrich, L.A. et al. 1991.
20. Johnson, J.H. and D.S. Dropkin. 1993.
21. Houston, J. 1990.
22. Lobb, M.D. III and D.J. Orth. 1991.
23. Lemly, A.D. and J.F. Dimmick. 1982.
24. Mathur, D. et al. 1981.
25. Hiranvat, S. 1975.
A 26. Swingle, W.E. 1965.
27. Gerald, J.W. 1971.
28. Miller, H.C. 1964.
29. Breder, C.M., Jr. and R.F. Nigrelli. 1935.
30. Richmond, N. 1940.
31. Mabee, P.M. 1995.
32. Buynak, G.L. and H.W. Mohr, Jr. 1978.
33. Taubert, B.D. 1977.
34. Anjard, C.A. 1974.
35. Hardy, J.D., Jr. 1978.
36. Breder, C.M., Jr. and A.C. Redmond. 1929
37. Wang, J.C.S. and R.J. Kernehan. 1979.
38. Lee, D.S. et al. 1980.
39. Burr, B.M. and M.L. Warren, Jr. 1986.
40. Etnier, D.A. and W.C. Starnes. 1993.
41. Mettee, M.F. et al. 2001.
42. Boschung, H.T., Jr. and R.L. Mayden. 2004.
43. Smith, C.L. 1985.
44. Raney, E.C. 1965.
45. Menhinick, E.F. 1991.
46. Jenkins, R.E. and N.M. Burkhead. 1994.
47. Carlander, K.D. 1977.
48. Bailey, R.M. 1938.
49. Scott, W.B. and E.J. Crossman. 1973.
50. Freeman, M.C. 1993.
51. Conner, J.V. 1979.
52. Taber, C.A. 1969.

* Original comments concerning early development, reproductive biology, and early life ecology come from redbreast sunfish collected from the Holston River in east TN on 6-29-1978. Brood stock were captured with electrofishing gear and transported to the TVA laboratory in Norris, TN, where they were placed in cattle tanks (about 0.9 m deep and about 2.4 m wide) supplied with gravel substrate. These fish subsequently nested and spawned. Some larvae were collected soon after hatching and placed into living streams for culture. Developmental descriptions come from a series of larvae drawn from this cohort. This developmental series is catalogued TV812, DS–18, and curated at the Division of Fishes, Aquatic Research Center, Indiana Biological Survey, Bloomington, IN.

GREEN SUNFISH

Lepomis (Apomotis) cyanellus Rafinesque

Robert Wallus and John V. Conner

Lepomis, Greek: *lepis*, "scale" and *poma*, "lid"; *cyanellus*, Greek: "kyaneos," blue.

RANGE

The green sunfish is native to mid-America from the southern Great Lakes, throughout the Mississippi River Basin, south to the Gulf States, except FL.[53,69,70] It has been widely introduced elsewhere in the United States.[53,69,70] and Germany.[53] It is now present in practically all of the contiguous United States.[70]

HABITAT AND MOVEMENT

Green sunfish occupies most aquatic habitats,[53,54,60,62,65,67,71] avoiding only fast-flowing shoals and riffles.[54,60,62] It is reported from small rivers[57] and rivers of low gradient,[72] small streams,[61,65,71,72] sluggish creeks,[54,57,60,62] ditches,[61,71] lakes,[22,57,61,71,72] impoundments,[57] ponds,[22,57,61,65,71,72] pools in streams[22,57] and large rivers,[66] canals and oxbows,[57] and springs.[5] It shows no noticeable preference for bottom substrate,[71] and is reported from over rock, gravel,[57] sand, silt, mud,[57,74] clay, and marl.[57] It is often associated with aquatic vegetation[57,61,71] and brushpiles.[57,71]

In WI, green sunfish is encountered in clear to turbid water at depths generally less than 1.5 m over substrates of gravel (23% frequency), sand (17%), silt (17%), rubble (12%), mud (12%), clay (8%), boulders (8%), detritus (1%), and marl (1%). It occurred in low-gradient streams of the following widths: 1.0–3.0 m (18%), 3.1–6.0 m (22%), 6.1–12.0 m (22%), 12.1–24.0 (28%), 24.1–50.0 m (8%), and >50 m (2%).[72]

In MS, green sunfish habitat is typically in areas of slow current (5.7 cm/s) and averaged 56 cm deep, with fairly fine (silt, mud, sand) substrate.[74]

Largest populations of green sunfish occur in habitats where there is little competition from other sunfish species.[62] It is an ecologically labile species, generally tolerant of adverse environmental conditions.[57,69] It tolerates drought,[69] extremes in turbidity,[54,60,62,71,72] temperature, dissolved oxygen, and flow, making it a hardy and pioneering species in small, intermittent water bodies. It reaches its greatest abundance in small, sluggish streams and ponds where few other sunfish occur.[31,54,60,62,65] Green sunfish can tolerate more turbidity and silt than any other sunfish[62,72] except the orangespotted.[72]

The optimal thermal regime for green sunfish is cool–moderate; it is considered a warm-water species and has a maximum temperature tolerance of 34.0[43,72] or 36°C.[72]

Laboratory reports indicate that adult green sunfish have an upper avoidance temperature of 33°C, a lower avoidance temperature of 23°C, and a final preferendum of 30.6°C. For small fish the temperatures were: upper, 30.3; lower, 26.5; and preferendum 28.2°C.[11]

In OK experiments, green sunfish acclimated to 4, 10, 22, and 30°C preferred temperatures of 10.6, about 15.5, 27.0, and 26.8°C, respectively. The final temperature preferenda for green sunfish in this study was 27.3°C.[42]

Indices of hypoxia and hypothermia tolerance for MO fish assemblages were determined based on laboratory measurements of lethal dissolved oxygen concentrations and temperatures combined with field measures of relative abundances of tolerant and sensitive species. Mean hypoxia tolerance for green sunfish was 0.63 mg/l dissolved oxygen; mean hypothermia tolerance was 37.9°C.[15,16]

Preferred temperatures and upper and lower avoidance temperatures for green sunfish collected continuously during falling temperature conditions in the New and East Rivers, VA, are presented in Table 28.[24]

Green sunfish tolerated pH changes from 7.2 to 9.6, and from 8.1 to 6.0, at water temperatures of 17–19.5°C, with 4–9 ppm of oxygen. In winter they survived a dissolved oxygen level of 3.6 ppm but died when the level was 1.5 ppm for 48 hours. Green sunfish exhibited a decline in feeding and growth when exposed to concentrations of ammonia >2 ppm and concentrations of cadmium ≥3 ppm. Concentrations of hydrogen sulfide in a pond in June resulted in a complete kill of green sunfish.[72]

Table 28

Preferred and upper and lower avoidance temperatures (°C) for green sunfish from the New and East Rivers, VA, at acclimation temperatures from 30°C in August to 6°C in February. [24]

Acclimation Temperature	Upper Avoidance Temperature	Preferred Temperature	Lower Avoidance Temperature
30	33	30.6	23
27	33	30.7	23
24	33	30.4	23
21	31	28.1	20
18	29	25.2	17
15	25	20.7	14
12	24	21.1	10
9	21	18.2	6
6	20	16.9	4

Green sunfish do not normally occur in brackish water;[53,67] maximum salinity reported is 0.7 ppt.[57]

With its tolerance for adverse conditions, such as drought, green sunfish has a great capacity for colonization and exploitation of new habitats. It frequently invades ponds and is considered a pest by those trying to manage balanced bass–bluegill populations.[69] It is considered undesirable in small ponds and lakes because it grows much faster than bluegill and redear sunfish, can out-compete them for food and spawning space, and rarely reaches suitable size for table fare.[68] Green sunfish become so abundant in farm ponds that stunting often becomes a problem.[64]

In two OK ponds, green sunfish experienced decreasing condition indices during 2 years following the application of an herbicide to control dense growths of submersed rooted aquatic plants. There was, however, little change in the numbers and weight of the standing crop in either pond, and rate of growth, though expected to change, showed no clear trend indicating an increase or decrease.[4]

In the oligotrophic Mayo Reservoir, NC, green sunfish density and biomass significantly declined after the introduction of alewife. Decreased abundance may have been related to predation from an increased population of chain pickerel or competitive interactions with other lepomids.[40]

Green sunfish seem mainly solitary in habit, each adult occupying a small area in which it forages. Territories are often tiny and congested; usually they are near the water's edge, under cover such as rocks or dams (in large impoundments) or exposed

roots.[64] It appears to be a species of restricted home range;[31,60,72] marked individuals have been found in the same pool a year or more after their initial capture.[60]

In a study of movement patterns of stream fishes in a Ouachita Highland stream, only 12% of marked green sunfish were recaptured outside the pool of initial collection. Of these, the majority were recaptured in the pools adjacent to the pool of initial collection. Most moved less than 100 m; the most distant recapture for a green sunfish was 453 m.[17] Green sunfish in small pools moved more often than ones in larger pools. Fish size and sex had no effect on the tendency to move.[36]

Green sunfish in small central TN streams abandoned the main channel for the floodplain during flooding.[29]

Although most likely introduced into lakes of the Piedmont province of NC, green sunfish are now quite common in rivers and even headwater streams. The effects of their presence in these watersheds were assessed with stream surveys and removal experiments. Green sunfish occurred in most first-order tributaries surveyed; when they were present, their abundance and biomass almost always exceeded that of other coexisting species. They also occurred in all second- and third-order streams surveyed, but never dominated the fish community by numbers or weight in those situations. When removed from first-order streams, most native species increased in numbers and biomass. These removal experiments indicate that green sunfish suppress native fish populations in

Piedmont headwater streams. Predation of young of other species is one likely means by which they do this.[14]

Hybrids between green sunfish and bluegill are often stocked in small ponds for put-and-take fisheries. These hybrids have proven aggressive, fast-growing, and easy to catch and, when managed properly, produce excellent fisheries. Hybrids with redear sunfish are also marketed for pond fishing.[74]

DISTRIBUTION AND OCCURRENCE IN THE OHIO RIVER SYSTEM

The Ohio River is the type locality for the green sunfish. Although usually found in smaller streams and ponds, it is regularly reported from the Ohio River, and many individuals are probably permanent residents of the river.[51] It has been found throughout the length of the river since 1800 and reports in 1984 and 1989 suggested no evident change in its distribution or abundance had occurred since 1970.[51,52] However, lockchamber surveys during the period 1957–2001 indicated the abundance of green sunfish in the Ohio River had increased significantly over time.[37] During the period 1973–2003 green sunfish were significantly more abundant in electrofishing catches in the upper Ohio River than in the middle.[23,59]

Green sunfish probably did not occur naturally in NY[22] and was not reported from Ohio drainage waters in NY by Smith (1985).[61] It is native to the Youghiogheny River system of PA, MD, and WV.[8]

Green sunfish is considered one of the most widely distributed and numerically abundant species of fish found in OH.[62] It is abundant in the Ohio River drainage of IL, and because of its ecological tolerances, may be more abundant now than in former times.[63] It is generally distributed and common throughout the state of KY.[54]

In VA, green sunfish is probably native to the Tennessee, Big Sandy, and New River drainages and is regarded as a native to the Tennessee and Big Sandy because of that status elsewhere in these drainages. However, the species is as rare and sporadic in the VA portion of these drainages and it is suspected that many or all the records are the result of introductions. Green sunfish is considered indigenous to the New River drainage.[67]

Green sunfish is found throughout the Ohio River drainages of TN where it is abundant in lowland streams of western TN but less common in upland streams to the east.[69] It is present throughout the Tennessee River drainage of AL[68,70] and MS.[74]

SPAWNING

Location

Spawning occurs in nests constructed by the male green sunfish in shallow water[22,39,57,60,64,65] near shore;[22,39,57,60] also in shallow backwater areas.[65] Nests are between 15 and 38 cm in diameter,[72,73] and are commonly observed on smooth, clean bottoms that slope gently away from the shoreline.[64] Gravel or rocky bottoms are preferred, but nests are occasionally found fanned out on water-soaked tree leaves and twigs.[60] Green sunfish may spawn on rock breakwalls or in the nests of other species.[39]

Nests may be solitary or in colonies, some less than 2 cm apart,[25] or they may just be close together, if suitable nest sites are at a premium; some think green sunfish is not a colonial spawner.[60]

In the Great Lakes, nests were constructed in sheltered nearshore areas, including lagoons, harbors, lower reaches of rivers, and quiet waters of streams; nesting areas were usually unshaded and within 1.8 m of shore and in water up to 38 cm deep.[39]

In WI ponds male green sunfish constructed nests in shallow water seldom deeper than 35 cm and small males constructed nests in water as shallow as 4 cm. Nests were usually constructed in cinder or gravelly areas and only rarely in murky marginal areas of the ponds. When muck substrate was used, the male dug a deep nest exposing the underlying marl. Nests commonly occurred in unshaded areas that received the maximum duration of sunshine. If available, areas sheltered by rocks, logs, and clumps of grass were used.[25]

Green sunfish brood stock captured by TVA and maintained in confined environments spawned naturally. Habitat included cattle tanks and a Living Stream unit with bottoms covered with rocks and gravel. Spawning in the Living Stream unit occurred in water 45 cm deep.*

Season

Spawning by green sunfish is often reported from May through August throughout its range,[2,22,25,33,39,55,56,60,65–68,72] with some reports in April*,[65] and September.[2,39,56] Spawning peaks are reported in June in the Great Lakes region[39] and in MO[60] and in May and June in KS[64] and OK (personal communication, J.F. Heitman, American Aquatics, Knoxville, TN).

For areas in or close to the Ohio River drainage, spawning is reported mid-May to August or September in IL,[2,39] late May through July and possibly into August in VA,[67] May through July in AL,[68] and April to September in MS.[56]

Temperature

Water temperatures reported for green sunfish spawning and nesting range from 15 to 28°C.[25,55,57,60,71,72] Spawning is initiated usually with rising water temperatures between 15.6 and 21°C.[57,60] Peak spawning activity occurs at temperatures of 20–28°C.[57,71]

Water temperatures ranged from 19 to 23°C leading up to natural spawning by green sunfish held in cattle tanks; brood stock held in a Living Stream unit spawned at water temperatures of 24–26°C.*

Fecundity

2,000–10,000;[57] in MS, a female 116 mm TL contained 4,900 nearly mature eggs.[56]

Sexual Maturity

Gonadal maturation normally occurs during spring, under long day lengths and elevated water temperatures. Experiments on green sunfish with regressed gonads during winter indicated that gonadal recrudescence is stimulated in both sexes only by a combination of both elevated water temperature (>15°C) and long photoperiod (15 h).[13]

Sexual maturation of green sunfish is reported from ages 1 to 3,[22,25,63,65,67,70,73,75] and occurs earlier in areas with longer growing seasons; for example, age 1 in MO,[67,73,75] IL, and IA,[73] but not until age 3 in MI.[67,73]

Small 1-year-old males spawned late in the season in WI ponds (late July or early August).[25]

In Doe Run Creek, KY, two large females were mature in age group 1; none of the age-1 males were mature. About a third of the age-2 males and 70% of the females had reached sexual maturity. All individuals three years or older were mature. The smallest mature females were two age-1 individuals, both 67 mm SL; the smallest mature male was age 2 and 76 mm SL.[32]

In some pond populations, stunting is so severe that specimens 76 mm TL are reproductively active.[64] Minimum reported lengths for mature males are 76 and 45 mm.[57,73]

Spawning Act

There is a cyclic component in the breeding periods of green sunfish populations. In WI ponds, nesting and spawning periods occurred at an average frequency of 8 or 9 days during the breeding season and never varied by more than 2 days from the average frequency. Frequency of the nesting periods seemed to be controlled primarily by rising water temperature and changes in the reproductive state of the male. The spawning period of a male green sunfish was usually 1 or 2 days but might extend over 3 or 4 successive days.[25]

A male green sunfish constructs its nest by using vigorous outward thrusts of its tail, each series of thrusts displacing some sand and gravel, gradually forming a shallow depression.[25] Males tended to construct a series of nests in the same region; more than half the nests constructed by 25 males in WI ponds were located within a meter of their previous nest.[25]

In ponds, a day or two in advance of nest establishment, green sunfish congregated near the spawning grounds. At first these aggregations were composed primarily of large males but as nests became established and spawning commenced, the area became congested with females and males of all sizes. Males that aggregated near the nests of the first males to spawn, later constructed their own nests, thus forming colonies.[25]

Male green sunfish exhibit territorial behavior while constructing or occupying a nest. In nesting colonies, where nests were often less than 2 cm apart, the males defended only the areas encompassed by the nest. Males guarding solitary nests defended an area 1–1.5 m in diameter. Fighting between male green sunfish was observed when they were constructing nests and spawning. Any fish that came near the nest was threatened, sometimes nipped, by the guarding male, including approaching females. Other males were chased away, but females often continued to approach, swimming to the side of the male.[25] Sound production by male green sunfish is reported during the active courtship of a female. A nesting male, upon sighting a female, would rush toward her and back toward his nest repeatedly while producing a series of grunt-like sounds.[35] Females were most strongly attracted to and attempted to mate with males that had already started to spawn. Male gyrations over the nest or odors released during spawning acts may play roles in attracting the female.[25]

The spawning act consisted of the male and female circling in the nest side by side then pausing momentarily to release sperm and eggs. Spawning occurs when the female reclines on her side and vibrates while the male remains in an upright position. An isolated pair might circle and spawn in a nest for considerable periods of time but in crowded colonies, the male frequently interrupts spawning to chase away intruding fish.[25]

Occasionally a male spawned simultaneously with more than one female,[25] and several females may spawn in a single nest.[60] After spawning, the male expels the female from the nest.[25]

If females were present on the spawning grounds, males usually commenced spawning within a day or two of nest construction. If no females were present, the male might occupy his nest intermittently for as long as a week.[25]

Male green sunfish guard their nests and progeny.[25,39,64] Some reports indicate that the male guards its nest for 3–5 days[25,61] until eggs hatch,[61] or for about a week after the eggs are deposited, about the time required for hatching and emergence of the fry.[60,63] However, in WI ponds, the period that a male green sunfish guarded its nest after spawning did not necessarily correspond to the time required for its young to become free-swimming. The median period of nest occupancy for the male was 4–5 days. Larvae cultured from eggs in the laboratory at 27–28°C never became free-swimming before 7 days after oviposition.[25]

Many North American minnow species spawn in association with nest building species such as the sunfishes.[18,60] Spawning by nest associates is synchronous with their hosts, which care for the associate's brood as well as their own.[19] The redfin shiner *Lythrurus umbratilis* often spawns in association with green sunfish.[18–21] Studies suggest that the green sunfish's nest is the least important factor in the attraction of redfin shiners. More important to the stimulation of associate spawning was the presence of a guarding male sunfish and the odor of sunfish milt and ovarian fluid in the water.[19] It is the parental care of the host male green sunfish and not the physical environment of the nest that most benefits the associate spawner.[20] Experiments have also shown a pronounced effect of having associate offspring in host nests, suggesting a mutualistic relationship; in the presence of predators, hosts with associates had significantly more fry than those without.[21]

Green sunfish hybridize naturally with pumpkinseed,[55,62,71,78–80] warmouth,[2,55,62,71] orangespotted,[55,62,71] bluegill,[1,2,55,62,71,78,79] longear, and redear sunfish.[1,55,62,71]

There appears to be no direct genetic isolation between green sunfish, redear sunfish, and bluegill. Laboratory propagation of the six possible P$_1$ crosses has successfully produced young that, when released into outdoor ponds, grew to sexual maturity.[1]

An eastern CT reservoir contains green sunfish, pumpkinseed, and bluegill. Hybrids between all three species are reported, but "pumpkinseed × green sunfish" hybrids were especially abundant, outnumbering both parents in 1977. These "pumpkinseed × green sunfish" hybrids appeared to be reproducing; male hybrids guarding nests with fry and male and female hybrids mating were observed.[79,81]

EGGS

Description
Fertilized eggs are yellow,[71] demersal,[46] and adhesive.[39,46,71] Diameter is 1.2–1.3 mm[46] or 1.0–1.4 mm.[45]

A single oil globule, 0.45 mm in diameter, is present, and the perivitelline space measures about 0.08 mm wide.[46]

Incubation
29 h at 27.6°C;[57] 31 h at 27.1°C;[55] 38–42 h at 24°C;[57] 35–55 h at 24–27°C;[46] in laboratory, 2 days at 25°C;[28] 50 h at 23.8°C;[39] about 3 days at 21.1°C.[45]

Development
Development at 24°C:[58]

   2 h 1- to 8-cell
   5 h morula
   7 h blastula flattened, blastocoel formed
   8 h blastoderm over 1/3 of yolk
   11 h blastoderm over 3/4 of yolk
   14 h embryonic shield visible
   15 h neural groove prominent, brain differentiating
   19 h somites forming
   21 h 12 somites, brain clearly segmented
   25 h optic cups forming
   28 h lens formed, muscular contractions in trunk
   31 h heartbeat established
   38 h pre-hatching stage, otoliths visible

YOLK-SAC LARVAE

See Figure 33.

Size Range
Green sunfish larvae hatch at 3.5–3.7 mm TL.[46] Larvae cultured in WI, contained small amounts of yolk at 5.8 mm (Figure 33B).[46] For green sunfish larvae cultured in TN (from AR brood stock), yolk was completely absorbed on some individuals by 4.7 mm TL, on most by 5.0 mm, but trace amounts of yolk remain on some individuals at 5.2 mm.*

Myomeres
See Table 29.

3.6–5.1 mm TL. Preanal 11; postanal 16–17.[46]

Morphology
3.6–3.7 mm TL (at hatching). Intestine is developed to the anus. Head is bent ventrally around the anterior margin of the yolk sac. Yolk is oval with a single posterior oil globule.[46]

4.4–4.9 mm TL. Mouth is grooved but not open; ear has formed separate chambers.[46]

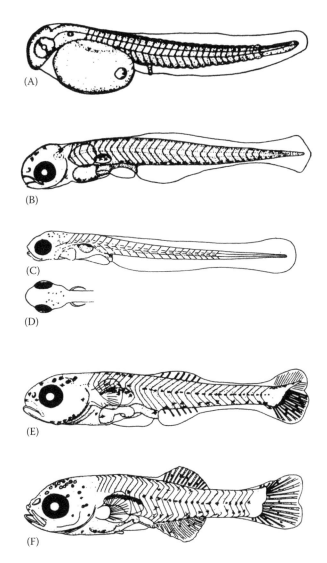

Figure 33 Development of young green sunfish. A–B. Yolk-sac larvae: (A) 3.6 mm TL; (B) 5.8 mm TL. C–D. Post yolk-sac larva, 5.4 mm TL: (C) lateral view; (D) dorsal view. E–F. Post yolk-sac larvae: (E) 8.3 mm TL; (F) 9.3 mm TL. (A–B, E–F reprinted from Figures 1 and 2, reference 46, with publisher's permission; C–D reprinted from Figure 5, reference 45, with publisher's permission.)

Morphometry
See Table 30.

Fin Development
3.6–3.7 mm TL (at hatching). Pectoral fin buds present.[46] Median finfold originates over posterior region of yolk sac and is present posteriorly along dorsum and continuous around urostyle ending ventrally at the posterior margin of the yolk sac (described from illustration).[46]

Pigmentation
3.6–3.7 mm TL (at hatching). Body is void of pigment.[46]

4.1–4.4 mm TL. Eyes are pigmented.[46]

POST YOLK-SAC LARVAE

See Figures 33 and 34.

Size Range
Phase begins between 5.0 and 6.0 mm TL.*,[46] The smallest juvenile observed was 13.4 mm TL but individuals as large as 15.7 mm had traces of median finfold remaining ventrally anterior to the anus.*

Myomeres
See Table 29.

5.6–6.3 mm TL. Preanal 11; postanal 16–17.[46]

7.8–11.4 mm TL. Preanal 13; postanal 15.[46]

Table 29

Myomere count ranges (average and standard deviation in parentheses) at selected size intervals for young green sunfish.*

Size Range (mm TL)	N	Myomeres			
		Predorsal Fin	Preanal	Postanal	Total
4.52–5.22	27	—	11–13 (11.4 ± 0.56)	15–17 (16.1 ± 0.65)	26–29 (27.6 ± 0.62)
5.23–5.84	13	—	11–13 (11.9 ± 0.61)	15–17 (16.2 ± 0.80)	27–29 (28.2 ± 0.66)
6.68–8.00	15	—	13–15 (13.6 ± 0.61)	14–16 (14.9 ± 0.61)	27–29 (28.5 ± 0.61)
8.25–13.12	51	5–6 (5.54 ± 0.49)	13–15 (14.0 ± 0.55)	13–16 (14.3 ± 0.61)	27–30 (28.3 ± 0.60)
13.37–16.90	8	5–6 (5.67 ± 0.48)	14	14	28

Morphology

5.6–5.7 mm TL. Oral velam and gill covers are functional; gill buds are present; air bladder is well developed and full; sagittae are markedly larger than other otiliths.[46]

>7.0 mm TL. Usually without a definite lump on the head.[45]

7.8–9.1 mm TL. Intestine S–shaped and has formed a loop characteristic of other *Lepomis* spp.[46]

Morphometry

See Table 30.

6.3–11.4 mm TL. Preanal length 45.3–47.0% TL.[46]

Fin Development

See Table 31.

5.3–5.6 mm TL. Caudal fin rays beginning to differentiate.[45]

5.6–6.3 mm TL. Pectoral fins fan-shaped; median finfold still present dorsally, ventrally, and around urostyle; urostyle straight.[46]

6.7 mm TL. Caudal fin rays visible.*

7.0 mm TL. Ray differentiation begins in dorsal and anal fins.[45]

7.8–8.2 mm TL. Rays are forming in dorsal and anal fins*,[46] and 13–16 principal caudal fin rays are evident; pectoral fin rays are visible and pelvic fin buds appear at about 7.9 mm* or between 8.3 and 9.1 mm.[46]

8.0–9.1 mm TL. All fin rays are formed by 8.0–8.5 mm TL[45]; or between 8.3 and 9.1 mm, pelvic fin buds appear[46] and the adult complement of soft dorsal rays (10–12), soft anal rays (9–11), and principal caudal fin rays (18) have formed.*,[46]

9.4–11.4 mm TL. On some individuals, the adult complement of fin rays and spines is present in all fins except the pelvics.[46] Remnant of preanal finfold is still evident at 10.2 mm[45] and larger (see comments next size interval).*

13.4–15.7 mm TL. Adult complement of pelvic fin elements is present. Preanal finfold may be absent on some at 13.4 mm and present on some at 15.7 mm TL.*

Pigmentation

5.0–5.1 mm TL. At 5.0 mm, heavily pigmented especially on the head, along the lateral line, on the dorsal margin of the air bladder, on either side of the gut posterior to the air bladder, on the ventral margin of the gut, and scattered ventrally, posterior to the anus;* or stellate melanophores are present over air bladder; three melanophores present near the anus, one supra-anal in position;

Table 30

Morphometric data for young green sunfish from AR grouped by selected intervals of total length. Characters are expressed as percent total length (TL) and head length (HL) with a single standard deviation. Range of values for each character is in parentheses.

Length Range N Characters	TL Intervals (mm)					
	4.52–5.22 27 Mean ± SD (Range)	5.23–5.84 13 Mean ± SD (Range)	6.68–8.00 15 Mean ± SD (Range)	8.25–13.12 51 Mean ± SD (Range)	13.37–16.90 8 Mean ± SD (Range)	18.52–22.30 5 Mean ± SD (Range)
Length as % TL						
Snout	3.29 ± 0.34 (0.13–0.20)	3.25 ± 0.39 (0.15–0.22)	3.08 ± 0.45 (0.16–0.32)	3.80 ± 0.48 (0.25–0.56)	4.57 ± 0.36 (0.55–0.82)	4.96 ± 0.43 (0.85–1.08)
Eye Diameter	7.23 ± 0.50 (0.27–0.41)	7.21 ± 0.40 (0.35–0.41)	6.88 ± 0.44 (0.41–0.60)	8.27 ± 0.76 (0.60–1.30)	9.19 ± 0.47 (1.13–1.60)	9.03 ± 0.33 (1.72–2.02)
Head	19.6 ± 0.59 (0.90–1.05)	19.6 ± 0.34 (1.01–1.11)	21.0 ± 0.81 (1.35–1.78)	23.0 ± 1.41 (1.70–3.35)	25.5 ± 1.18 (3.10–4.35)	25.7 ± 1.00 (5.0–5.7)
Predorsal	—	—	—	29.5 ± 1.17 (2.75–4.05)	31.3 ± 0.95 (4.15–5.23)	31.2 ± 0.88 (6.0–7.1)
Prepelvic	—	—	27.1 ± 3.02 (2.15–2.15)	27.1 ± 0.86 (2.18–3.65)	28.9 ± 0.86 (3.65–4.95)	28.9 ± 1.01 (5.5–6.7)
Preanal	42.7 ± 1.08 (2.05–2.25)	42.8 ± 0.70 (2.2–2.57)	46.5 ± 1.02 (3.11–3.78)	45.9 ± 0.82 (3.85–6.0)	45.7 ± 0.50 (6.15–7.75)	45.8 ± 1.17 (8.7–10.4)
Postanal	57.3 ± 1.08 (2.47–3.09)	57.2 ± 0.70 (2.93–3.27)	53.5 ± 1.02 (3.57–4.37)	54.1 ± 0.82 (4.37–7.13)	54.3 ± 0.50 (7.22–9.28)	54.2 ± 1.17 (9.82–11.9)
Standard	97.0 ± 0.82 (4.39–5.1)	96.8 ± 0.88 (5.0–5.65)	94.8 ± 2.45 (6.45–7.52)	84.3 ± 2.95 (7.35–10.89)	79.6 ± 0.74 (10.5–13.55)	77.8 ± 0.74 (14.3–17.65)
Air bladder	8.39 ± 0.62 (0.30–0.50)	8.40 ± 0.33 (0.41–0.50)	11.0 ± 0.94 (0.65–1.0)	13.7 ± 1.97 (0.90–2.25)	—	—
Eye to air bladder	14.8 ± 1.30 (0.65–1.05)	14.5 ± 0.54 (0.73–0.88)	16.6 ± 0.85 (1.05–1.40)	14.2 ± 1.72 (1.10–1.70)	—	—
Fin length as % TL						
Pectoral	—	—	10.7 ± 2.61 (0.85–0.85)	13.0 ± 2.19 (0.90–2.40)	18.0 ± 1.40 (2.10–3.20)	19.0 ± 1.81 (3.60–3.97)
Pelvic	—	—	1.26 ± 0 (0.1–0.1)	5.10 ± 3.03 (0.15–1.50)	12.2 ± 1.56 (1.25–2.40)	16.0 ± 1.19 (2.95–3.50)
Body depth as % TL						
Head at P1	—	—	—	22.8 ± 1.21 (2.30–3.13)	24.8 ± 0.94 (3.28–4.20)	26.9 ± 0.93 (5.0–6.2)
Head at eyes	—	—	—	20.5 ± 0.95 (2.10–2.90)	21.2 ± 0.57 (2.80–3.50)	20.6 ± 0.56 (4.0–4.65)
Preanal	—	—	—	21.3 ± 1.01 (2.17–3.05)	22.4 ± 1.00 (2.80–4.0)	24.9 ± 0.78 (4.55–5.70)
Caudal Peduncle	—	—	—	10.2 ± 0.53 (1.08–1.50)	1.11 ± 1.26 (1.40–2.20)	10.8 ± 0.34 (2.10–2.45)
Body width as % HL						
Head	—	—	—	—	53.4 ± 2.61 (1.85–2.40)	63.8 ± 5.23 (2.95–4.15)

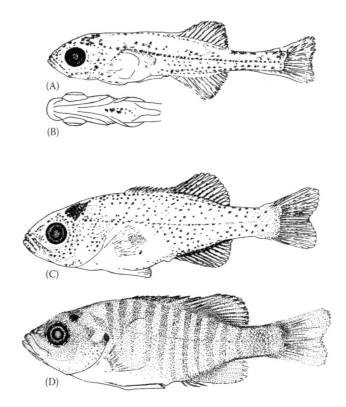

Figure 34 Development of young green sunfish. A–B. Post yolk-sac larva, 10.2 mm TL: (A) lateral view; (B) ventral view. C–D. Juveniles: (C) 15.1 mm TL; (D) 23.7 mm TL. (A–D reprinted from Figure 5, reference 45, with publisher's permission.)

Table 31

Fin development for young green sunfish from AR.*

Fins/Elements	TL (mm) at Which:	
	Fin Rays Appear	Adult Complement of Rays Present
Dorsal fin		
Spines	8.8	12.2
Soft rays	8.4	8.9
Anal fin		
Spines	8.8	9.8
Soft rays	8.3	8.8
Caudal fin rays	6.7	8.8
Pectoral fin rays	7.9	11.5–12.5
Pelvic fin	7.9	—
Pelvic fin rays	—	≥13.4

two contracted melanophores are present on the dorsal surface of the head.[46]

5.5 mm TL. Profuse spotting with small, concise chromatophores on top of head.[45]

6.0–6.3 mm TL. Stellate melanophores are present dorsally on the head and body, ventrally on the yolk sac, on dorsal surface of the air bladder, and in a row from immediately anterior to the anus to past the last myomere.[46]

>7.0 mm TL. No specific arrangement of chromatophores on the isthmus.[45]

7.8–9.4 mm TL. Melanophores increasing, present on cheeks and caudal fin membranes and a row of pigment is present along the lateral line. Between 8.6 and 9.4 mm, melanophores develop on the lips and dorsal and anal fin membranes.[46]

10.2 mm TL. Midlateral row of pigment distinct; there is an increased amount of pigment on the head and dorsum and ventrolaterally between anus and caudal fin; breast with an elongate, midventral pigment patch between the pectoral fins (description based on figure).[45]

JUVENILES

See Plate 1C following page 26 and Figure 34.

Size Range

Phase begins between 13.4 and 15.7 mm TL* and sexual maturity in the Ohio Valley may be reached by 67–76 mm SL[32] or at smaller sizes in stunted populations.[57,73]

Myomeres
See Table 29.

Morphology
Larger juveniles. Lateral line scale counts for green sunfish range from 40 to 53.[60,65,67–72,74] There are usually 11–14 gill rakers;[69,70] length of longest rakers more than four times[70] or about six times basal width.[69] Vertebrae 28–29; 6–7 pyloric caeca.[71]

Morphometry
See Table 30.

Fins
Green sunfish have two dorsal fins that are broadly joined and appear as one.[60,71,72] The first dorsal fin has 9–12,[60,64,65,67–69,71,72] usually 10,[60,67–69,71] spines; the second dorsal fin has 10–12 soft rays,[67–69,71,72,74] usually 10–11.[67,69,71,74] Anal fins have 3 spines[60,65,67–69,71,72,74] and 8–11,[67–72,74] usually 9–10 soft rays.[67,69,71,72,74] Pelvic fins are thoracic, with 1 spine and 5 soft rays,[71,72] and pectoral fin rays number 12–15,[65,67,69–71,74] usually 13–15.[65,67,69,70,74]

Pigmentation
15.1 mm TL. Entire body, except breast, is covered with widely spaced chromatophores; pigment is dense on surface of head; pigment is present on lips; pigment is visible along soft rays of dorsal and anal fins (description based on figure).[45]

Note: Primary bars of pigment appear as an initial set of lateral patches of melanophores to which bars are added anteriorly and posteriorly in development in all lepomids. Green sunfish develop 11–14 primary bars (Figure 35) with secondary bars variably present between the primary bars, sometimes as continuous lines of pigment, sometimes broken (Figure 35D). By about 32 mm SL these dark bars of pigment have disappeared from the sides of green sunfish.[77]

23.4 mm SL. Spot on soft dorsal fin is visible.[77]

23.7 mm TL. A dark patch of pigment on head and a dark spot on tip of opercle; 11–12 vertical bars of pigment are present laterally along body. There is a background of numerous small chromatophores over most of body and head, except on the breast and venter anterior to anus which are lacking in pigmentation; median fins with scattered pigmentation (description based on figure).[45]

42.8 mm SL. Melanistic pigment bars are no longer visible laterally (Figure 35E).[77]

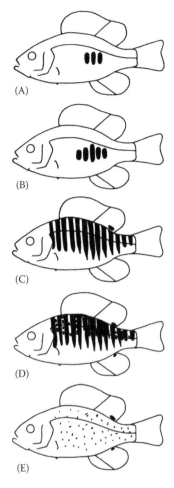

Figure 35 Development of young green sunfish. Schematic drawings showing melanistic pigment patterns of young green sunfish. (A and B) Two different specimens, both 12.0 mm SL, showing initial formation of melanistic pigment bars. (C) 24.7 mm SL; juvenile showing advanced development of primary pigment bars and appearance of dorsal fin spot. (D) 23.4 mm SL; juvenile showing secondary pigment between primary bars. (E) 42.8 mm SL; lateral pigment bars no longer visible. (A–E reprinted from Figure 15, reference 77, with publisher's permission.)

Larger juveniles. Young green sunfish have dorsal and anal fins sprinkled with brownish red and the dorsal fin has a dark blotch of pigment, posteriorly. Light bands may be present on the head of some individuals.[57] YOY green sunfish are uniformly gray on the sides and lack lateral banding.[69]

TAXONOMIC DIAGNOSIS OF YOUNG GREEN SUNFISH

Similar species: Green sunfish is the only member of the subgenus *Apomotis* and is most similar to members of the subgenus *Bryttus* (redspotted sunfish in the Ohio River drainage and spotted sunfish). Also hypothesized to be more similar to the bantam sunfish and warmouth than to other

Lepomis.[76] There are comparisons in the literature of the early development of young green sunfish to bantam sunfish and warmouth,[76] longear sunfish,[82] and bluegill and redear sunfish.[45]

See *Apomotis* subgeneric discussion on page 90.

For comparisons of important morphometric and meristic data, developmental benchmarks, and pigmentary characters between green sunfish and all other *Lepomis* that occur in the Ohio River drainage, see Tables 4 and 5 and Plate 1C.

Green sunfish is more similar to the bantam sunfish and warmouth than to other *Lepomis*. Once the yolk is mostly or entirely exhausted, these three species are characterized by relatively short preanal lengths (modally under 45% TL) and tend to have the lowest modal preanal myomere counts of all the *Lepomis*. Green sunfish types without a complete complement of caudal fin rays tend to have proportionally deeper heads and more extensive and prominent pigment, particularly in the head and trunk regions than other lepomids. Green sunfish-type larvae with more or less complete caudal fins have generally more profuse pigmentation than other lepomids, especially well-developed head pigmentation (particularly in the cheek and postorbital areas); a characteristic midlateral streak of melanophores which often simulates a regularly spaced series of dashes; a tendency toward smaller size at comparable stages of development with other lepomids.[76]

Green sunfish vs. longear sunfish. After rays appear in the caudal fin, longear have an elongate patch of small chromatophores along the venter similar to that of redbreast. Green sunfish may also have pigmentation of this pattern, but it is less dense.[82] Green sunfish become juveniles by 11.7 mm TL[46] or 13.4 to >15.7 mm TL.* Longear sunfish become juveniles by 14.2 mm TL.[82]

Green vs. bluegill and redear (from CA). First caudal fin rays are formed at 5.3–5.6 mm in green sunfish, at 6.2–6.4 mm for redear sunfish, and 7.0–7.2 mm for bluegill. Green sunfish are free-swimming by 4.2–4.7 mm, redear sunfish at 5.1–5.8 mm, and bluegill at 5.0–5.5 mm. All fin rays are formed at 8.0–8.5 mm for green sunfish, at 9.4–9.8 mm for redear sunfish, and at 8.9–9.4 mm for bluegill.[45]

Green sunfish are sympatric with all other *Lepomis*, none of which has as slender a body shape or as large a jaw — except warmouth which differs in having black bars on the cheeks and opercles. Bluegill with long pectoral fins and bantam sunfish with incomplete lateral line are the only other *Lepomis* that have a similar dark spot in the posterior portion of the dorsal fin. YOY green sunfish (Plate 1C) are usually dark gray and lacking vertical banding — most similar to redbreast sunfish

(Plate 1B), dollar sunfish (Plate 2B), and longear sunfish (Plate 2C), which have much shorter gill rakers.[69]

ECOLOGY OF EARLY LIFE PHASES

Occurrence and Distribution (Figure 36)

Eggs. Eggs are deposited in shallow nests excavated in sand, stones, gravel, or marl; also in mud or muck, where excavation reaches firm bottom.[25,39] The adhesive eggs incubate on rootlets, gravel, or lumps of clay in the bottom of the nest.[39] They have been found attached to willow, bulrush, and sedge roots, sometimes at the very edge of the water,[57] and may be deposited on rock breakwalls or in abandoned nests of other species.[39] Males guard the nest and eggs until they hatch.[60,61]

In streams, spawning success and egg survival are probably higher in pools with bass because of reduced harassment or egg predation by smaller fishes.[30]

Yolk-sac larvae. Yolk-sac larvae are guarded in the nest by the male parent.[25,60,61,63] Males remain with the nest for 3–5 days[25,61] or up to a week, until the fry have absorbed most of the yolk and become free-swimming.[60,63] It is also reported that the period that male green sunfish guarded their nests after spawning did not necessarily correspond to the time required for their young to become free-swimming.[25]

Just before swimming up to fill their air bladder with air, larvae in aquaria were attached to aquarium glass or artificial plants by the dorsal portion of the head.[46]

Post yolk-sac larvae. Larval green sunfish absorb most of their yolk and become free-swimming by 4.2–4.7 mm TL,*[45] or at 6.0–6.3 mm TL.[46]

Larval green sunfish were collected from the Middle Fork Drake's Creek, KY, in late June and early July 1982, with drift nets, light traps, and seines. They occurred in collections made along vegetated shorelines and over rocky habitat of gravel and algae-covered rocks. Most were collected with light traps from the rocky habitats.[26] Larval green sunfish are reported from the limnetic zone of lakes.[44]

Juveniles. In the Great Lakes region, YOY green sunfish are often found in the lower reaches of rivers, harbors, and marshy lagoons, often associated with vegetation over mud substrates in water 30–214 cm deep.[39]

In a 2-year study of microhabitat selection of YOY sunfishes in an OK lake, green sunfish (<80 mm

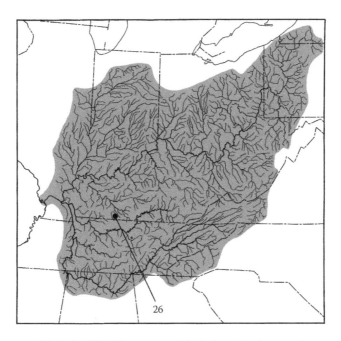

Figure 36 Distribution of green sunfish in the Ohio River system (shaded area) and areas where early life history information has been collected (circle). Number indicates appropriate reference.

TL) were collected in low numbers with an unbaited minnow trap. In 1981 they were not collected in depths >150 cm and in 1980 they were found only in the shallowest water (<50 mm deep). Nearly all green sunfish collected both years came from over fine substrates containing muskgrass or spikerush.[34] Another report indicates young green sunfish frequent sunken brush heaps and are often found in beds of aquatic vegetation.[62] In Lake Oahe, SD, YOY green sunfish were rarely collected in shoreline seining.[10]

Juvenile green sunfish were collected in large numbers from a floodplain pool of the Kankakee River, IL, after floodwaters receded, suggesting that floodplain habitats are important nursery areas for the species.[3]

In the laboratory, young green sunfish <74 mm had an upper avoidance temperature of 30°C, a lower avoidance temperature of 24°C, and a final preferendum of 27.3°C.[11]

Early Growth

Carlander (1977) summarized numerous reports of the life history and growth of green sunfish. Generally, growth is slower in northern populations than in southern[73] and males grow faster than females, but the differences may be slight.[32,73]

Growth rates of green sunfish vary widely between populations (Table 32). There is a tendency toward overpopulation and stunting, especially in closed systems such as shallow ponds.[69]

In MI, average SL of green sunfish age 1+ was 43 mm, age 2+ 63 mm, age 3+ 86 mm, and age 4+ 102 mm.[71]

Average calculated growth of young green sunfish in OK varied among years.[48,83] Growth was similar in lakes and reservoirs, slower in streams, and greatest in ponds (Table 33).[73] Average calculated growth in Canton Reservoir, OK, during the period 1964–1968 was 66 mm TL and 5 g at age 1, 104 mm and 18 g at age 2.[47] This growth in TL for green sunfish was considerably less than the state average for the preceding period 1952–1963,[48] but very similar to reports for the period 1972–1973 (Table 33).[83]

In Clear Creek, IL, average calculated TL for green sunfish in 1951 was 46 mm at age 1, 89 mm at age 2, 130 mm at age 3, and 160 mm at age 4.[38] Similar growth is reported from TN and MO.[69]

Average calculated SL for young green sunfish from Doe Run Creek, KY, during the period 1959–1961 was 35 mm at age 1, 63 mm at age 2, 90 mm at age 3, and 110 mm at age 4. Little difference was noted in the growth rates of males and females.[32]

The central 50% range in age and TL of combined reported observations for green sunfish from numerous locations summarized by Carlander (1977) is as follows: 61–95 mm at age 1, 97–135 mm at age 2, 118–165 mm at age 3, and 137–180 mm at age 4.[73]

In OK, age-0 green sunfish averaged 69 mm in June, 76 mm in July, and 86 mm in August.[73] In Lake of the Ozarks, MO, TL of young green sunfish ranged from 30 to 69 mm in August, 41–79 mm in September, 51–76 mm in October, and 36–86 mm in November and December.[75] In OH, YOY in October were 20–64 mm TL; at about 1 year, 25–81

Table 32

Age and growth calculations for young green sunfish from several locations.

Location	Mean Calculated TL (mm) at Age:			
	1	2	3	4
12 KS streams, 1995–1996[27]	35	62	87	103
Little Wall Lake, IA, 1944–1949[12]	36	82	115	195
Salt River, MO[7]	43	81	119	150
Lake of the Ozarks, MO[75]	51	84	109	135
Clear Creek, IL, 1951[38]	46	89	130	160
Barren Lake, KY, 1983[49]	46	81	140	157
Marion Cty. Lake, KY, 1983[49]	66	112	137	173
Nolin River Lake, KY, 1983[49]	36	74	104	—

Table 33

Mean calculated TL (mm) related to age of young green sunfish from OK waters for the period 1952–1963[48, 73] and 1972–1973.[83]

Waterbodies	No. of Fish	Calculated TL (mm) at Each Annulus			
		1	2	3	4
20 reservoirs[73]	476	99	163	178	203
9 large lakes[73]	128	91	142	170	201
33 small lakes[73]	245	97	150	183	190
38 ponds[73]	362	102	160	201	229
10 streams[73]	445	81	132	168	185
Statewide average (1952–1963)[48]	1,656	97	152	185	201
Statewide average (1972–1973)[83]	—	67	106	135	147

mm.[62] In Clear Creek, IL, 36 YOY green sunfish captured in August averaged 68 mm SL and 10 g.[69]

Average calculated TL and weight of young green sunfish in OK waters was 14 g and 97 mm at age 1, 64 g and 152 mm at age 2, 118 g and 185 mm at age 3, and 151 g and 201 mm at age 4.[48] These calculated growth rates in OK appear similar to length/weight relationships of young green sunfish observed in IN (Table 34) and AL (Table 35).

The following age and observed growth in length and weight of green sunfish are reported from Lake Chautauqua, IL: at age 2 (n = 7) green sunfish averaged 107 mm TL and 23 g, at age 3 (n = 25) they averaged 140 mm and 68 g, and at age 4 (n = 15) they averaged 165 mm and 114 g.[9]

In a laboratory at 21.1 C, green sunfish grew to 5.4 mm at 9 days, 10.2 mm at 30 days, 15.1 mm at 37 days, and 23.7 mm at 57 days.[45]

Feeding Habits

Young green sunfish are insectivorous, relying heavily on midge larvae and mayfly nymphs, but as they increase in size, crayfish and small fishes dominate the diet.[70]

In an OK lake, diets of YOY green sunfish <80 mm TL were variable over a 2-year study. One year they fed mostly on ostracods (57.5% abundance) and chydorids (28.3% abundance) and also ate smaller amounts of chironomids, copepods, and other items. The previous year they fed only on chironomids (40%) and items other than ostracods, chydorids, and copepods.[34]

In Bull Shoals Reservoir, AR/MO, in 1964–1965, green sunfish <48 mm TL consumed mostly aquatic insects (46% of total volume), fish eggs (24%), Entomostraca (14%), and terrestrial insects (12%). Young green sunfish 51–99 mm TL also consumed mostly

Table 34

Length and weight data for young green sunfish
collected from Long Lake, Porter County, IN,
in October 1999.*

TL (mm) Range	N	Mean TL	Weight (g) Range	Mean Weight (g)
34–38	4	36.0	0.7–0.9	0.9
41–49	12	45.8	1.2–2.1	1.7
52–58	4	53.8	2.3–3.7	2.8
61–68	4	65.3	4.0–5.3	4.7
71–79	4	75.3	6.4–9.5	7.7
81–82	3	81.3	8.3–9.3	8.8
93–97	2	95.0	13.3–15.9	14.6
100–104	3	101.7	15.2–17.0	15.8
125	1	125.0	36.1	36.1
137	1	137.0	24.4	24.4

Table 35

Length–weight relationships of young
green sunfish from AL. Values represent
data collected during the period 1949–1964.

Approximate Average TL (mm)	N	Average Empirical Weight (g)
25	5,097	1.4
51	4,428	2.6
76	2,757	7.7
102	1,400	17.7
127	462	37.2
152	111	68.1
178	24	99.9
203	5	149.8

Source: Constructed from data presented in unnumbered
table on page 34, reference 50.

aquatic insects (59% by volume); Entomostraca
(12%) and fish (10%) were next in importance followed by terrestrial insects (8%), Malacostraca
(5%), Mollusca (3%), and other items that
accounted for 1% or less of total volume. Fish
102–201 mm TL fed mostly on Malacostraca (63%)
and terrestrial insects (13%).[41]

Stomach contents of ten green sunfish (22–89 mm,
mean 51.5 mm SL) from the White River, IN, were
examined. Important foods were chironomid larvae
(39.5% volume, 80% frequency), Trichoptera larvae
(24% volume, 40% frequency), and Ephemerida
naiads (14% volume, 20% frequency). Chironomid
pupae, Plecoptera naiads, corixids, and unidentified fish were also consumed.[6]

In aquaria, larval swim-up and commencement of
feeding began 5 days after hatching, and fry fed
successfully on newly hatched San Francisco brine
shrimp nauplii.[28]

LITERATURE CITED

1. Childers, W.F. and G.W. Bennett. 1961.
2. Childers, W.F. 1967.
3. Kwak, T.J. 1988.
4. Houser, A. 1963.
5. Armstrong, J.G. III. 1969.
6. Whitaker, J.O., Jr. 1974.
7. Purkett, C.A., Jr. 1958.
8. Hendricks, M.L. et al. 1983.
9. Starrett, W.C. and A.W. Fritz. 1965.
10. June, F.C. 1976.
11. Coutant, C.C. 1977.
12. Sprugel, G., Jr. 1955.
13. Kaya, C.M. and A.D. Hasler. 1972.
14. Lemly, A.D. 1985.
15. Smale, M.A. and C.F. Rabeni. 1995a.
16. Smale, M.A. and C.F. Rabeni. 1995b.
17. Smithson, E.B. and C.E. Johnston. 1999.
18. Shute, P. and J.R. Shute. 1991.

19. Hunter, J.R. and A.D. Hasler. 1965.
20. Johnston, C.E. 1994a.
21. Johnston, C.E. 1994b.
22. Raney, E.C. 1965.
23. Van Hassel, J.H. et al. 1988.
24. Cherry, D.S. et al. 1975.
25. Hunter, J.R. 1963.
26. Floyd, K.B. et al. 1984.
27. Delp, J.G. et al. 2000.
28. Smith, W.E. 1975.
29. Barclay, L.E. 1985.
30. Harvey, B.C. 1991.
31. Gerking, S.D. 1953.
32. Redmon, W.L. and L.A. Krumholz. 1978.
33. Miller, H.C. 1964.
34. Layzer, J.B. and M.D. Clady. 1991.
35. Gerald, J.W. 1971.
36. Johnston, C.E. and E.B. Smithson. 2000.
37. Thomas, J.A. et al. 2004.
38. Lewis, W.M. and D. Elder. 1953.
39. Goodyear, C.D. et al. 1982.
40. Crutchfield, J.U., Jr. et al. 2003.
41. Applegate, R.L. et al. 1967.
42. Jones, T.C. and W.H. Irwin. 1965.
43. Wehrly, K.E. et al. 2003.
44. Rettig, J.E. 1998.
45. Meyer, F.A. 1970.
46. Taubert, B.D. 1977.
A 47. Lewis, S.A. et al. 1971.
A 48. Houser, A. and M.G. Bross. 1963.
A 49. Axon, J.R. 1984.
A 50. Swingle, W.E. 1965.
51. Pearson, W.D. and L.A. Krumholz. 1984.
52. Pearson, W.D. and B.J. Pearson. 1989.
53. Lee, D.S. et al. 1980.
54. Burr, B.M. and M.L. Warren, Jr. 1986.
55. Tin, H.T. 1982.
56. Cook, F.A. 1959.
57. Hardy, J.D., Jr. 1978.
58. Champion, M.J. and G.S. Whitt. 1976.
A 59. EA Engineering, Science, and Technology. 2004.
60. Pflieger, W.L. 1975b.
61. Smith, C.L. 1985.
62. Trautman, M.B. 1981.
63. Smith, P.W. 1979.
64. Cross, F.B. 1967
65. Robison, H.W. and T.M. Buchanan. 1988.
66. Holland-Bartels, L.E. et al. 1990.
67. Jenkins, R.E. and N.M. Burkhead. 1994.
68. Mettee, M.F. et al. 2001.
69. Etnier, D.A. and W.C. Starnes. 1993.
70. Boschung, H.T., Jr. and R.L. Mayden. 2004.
71. Scott, W.B. and E.J. Crossman. 1973.
72. Becker, G.C. 1983.
73. Carlander, K.D. 1977.
74. Ross, S.T. 2001.
75. Hoffman, J.M. 1955.
76. Conner, J.V. 1979.
77. Mabee, P.M. 1995.
78. Etnier, D.A. 1968.
79. Dawley, R.M. 1987.
80. Hubbs, C.L. and L.C. Hubbs. 1931.
81. Jacobs, R.P. 1979.
82. Yeager, B.L. 1981.
A 83. Mense, J.B. 1976.

* Original reproductive information comes from successful spawnings of green sunfish at TVA laboratories. Original morphometric and meristic data are from a series of laboratory-cultured larvae produced from a successful spawn by brood stock obtained from a fish hatchery in Stuttgart, AR (Arkansas Experimental Station) in 1979. This developmental series is catalogued TV2352, DS–33 and curated at the Division of Fishes, Aquatic Research Center, Indiana Biological Survey, Bloomington, IN. Original length–weight data from IN come from green sunfish collected in Long Lake, Porter County on 10/12/1999 by T.P. Simon.

PUMPKINSEED

Lepomis (Eupomotis) gibbosus (Linnaeus)

Robert Wallus

Lepomis, Greek: *lepis* "scaled" and *poma* "lid"; *gibbosus*, Greek: meaning wide margin or "gibbous," formed like a full moon, referring to body shape.

RANGE

Widely distributed in southern Canada from QU to ON and western ON to eastern MB through parts of ND and SD to southwestern MO; due east from MO to Appalachian Mountains then north to OH and western PA; south, east of the Appalachians, on Atlantic slope to GA. The pumpkinseed is easily established in new waters[97] and has been widely introduced into many other areas of North America and Europe.[1,73,82,83]

HABITAT AND MOVEMENT

Pumpkinseed prefer cool to moderately warm,[1,82,83,97] quiet,[1,76,83] clear[1,73,76,83,97] to slightly turbid[97] waters with aquatic vegetation[1,6,73,83] and some organic debris.[1,83]

They are often found in ponds,[1,73,76,82,84,97] small weedy lakes,[1,73,76,82,97] shallow bays of large lakes,[73,76] impoundments,[73,97] creeks,[73] slow-moving streams,[76,97] pools in streams,[1,5,73,76,82,97] rivers,[1,73,84,97] and spillpools.[73] The establishment of impoundments on creeks and rivers creates favorable habitat for the species.[97] Occupies sand or soft-bottomed embayments along the Ohio River in KY and upstream.[5]

Pumpkinseed often occur in shallow, sheltered situations;[83] typically in or adjacent to submerged aquatic vegetation[5,73,76] or brush.[5,76] In Lake Erie, they are generally restricted to weedy areas[1,71] whether along the lake shore or in adjoining streams and ponds.[71] In Lake Opinicon, ON, pumpkinseed occur in most inshore habitats with greatest densities in weedy and sandy shallow areas.[36] Surveys of 22 lakes in southern MI- indicated the presence of pumpkinseed in all lakes surveyed with increased abundance in lakes with greater vegetation density and/or more organic sediments. Pumpkinseeds of all sizes generally were concentrated in deeper areas with heavy vegetation cover.[63] In MN lakes, significant correlations were detected between occurrences of emergent and floating-leaf plant species and relative biomass and mean size of pumpkinseed.[103]

In WI, pumpkinseed occurred in low-gradient streams of the following widths: 1–3 m (13% frequency), 3.1–6 m (9%), 6.1–12 m (9%), 12.1–24 m (36%), 24.1–50 m (20%), and more than 50 m (11%).[97] They are found over a variety of bottom substrates[5,76] including sand, gravel, rock,[73] and mud or muck covered with organic debris.[5,73] In WI, pumpkinseed are found at depths of less than 1.5 m, over substrates of sand (28% frequency), gravel (19%), mud (18%), silt (13%), boulders (8%), rubble (7%), detritus (4%), clay (2%), hardpan (1%), and bedrock (1%).[97]

No significant habitat utilization was observed for pumpkinseed in four habitats (heavy macrophytes, light macrophytes, sand, and open water) studied in Cedar Lake.[26] Fish use of artificial structures with two interstice sizes (40 and 350 mm) placed at two depths (3.0 and 4.5 m) in an OH reservoir showed that pumpkinseed preferred the denser structure at the 4.5-m depth.[35]

Pumpkinseed forage in vegetated littoral areas where snails are found.[52] They are usually seen in large numbers[76,83] often at, or near, the surface of areas exposed to the sun.[76] Another report indicates that adult pumpkinseed are found in deeper water than their young and are rarely observed in aggregations of more than three or four.[98]

They are also reported to winter in deep water in schools.[34,72]

Pumpkinseed usually are found in cool to moderately warm water.[82,97] They generally live in cooler water than most other *Lepomis* and tend to inhabit denser vegetation than bluegill.[83] The pumpkinseed does fairly well in some cool Adirondack streams where it is the only native sunfish; it occurs along with brook trout in many situations.[1]

Reported temperature preferenda range from 26 to 33°C.[16,29] In laboratory situations, pumpkinseed, when free to choose, preferred water temperatures 31.1–31.7°C with a final preferred temperature of 31.5°C. They become inactive in temperatures below 10°C, but do not form hibernating schools.[1]

Pumpkinseed density and biomass significantly declined after the introduction and establishment of a self-sustaining population of alewife in oligotrophic Mayo Reservoir, NC. This decreased abundance may have been related to an increased chain pickerel population or competitive interactions with other lepomids.[104]

Pumpkinseed are tolerant of darkly stained waters of pH as low as 4.1. They sometimes occupy slightly brackish waters[90] and were commonly observed at a salinity of 18.2% in Chesapeake and Delaware Bays;[10] reported as maximum recorded salinity.[73,90]

There is evidence that pumpkinseed is a homing species. In an IA study, 64% of displaced pumpkinseeds homed. When displaced into a new territory, they initially spent time moving in circles as if searching for familiar landmarks.[99] In another report, many displaced pumpkinseed returned to their original home range within 24 h, and most fish tagged and released in the general area of capture showed little tendency to move.[100] Marked pumpkinseed moved little in a MI lake, usually traveling less than 152 m (500 ft).[28] In Cedar Lake, IL, home range sizes for four pumpkinseeds ranged from 0.23 to 1.12 ha.[25]

Pumpkinseed are active by day. At dusk they move toward the bottom, where they rest. Resting areas are usually in interstices of rocky cliff areas or near fallen logs.[97]

In the Great Lakes, pumpkinseed move short distances inshore and enter creeks to spawn.[7]

DISTRIBUTION AND OCCURRENCE IN THE OHIO RIVER SYSTEM

Based on lockchamber surveys during the period 1957–2001, the abundance of pumpkinseed in the Ohio River did not significantly change after 1957.[2] The pumpkinseed is fairly common in the upper third of the Ohio River,[4,6,11,20] less common in the middle third,[4,6,11] and present only as strays in the lower third.[4,6] The construction of high-lift dams in the 1960s and 1970s in the upper half of the river created many embayments and apparently provided habitats favorable to the pumpkinseed as far downstream as Markland Dam (ORM 531.5).[4] Records of pumpkinseed from farther down the river are considered to be strays from upstream populations and not the result of stockings.[4,5] The extreme downstream record is from ORM 754.[4]

Electrofishing surveys, as part of the Ohio River Ecological Research Program, were conducted near three electric generating stations on the upper Ohio River during the period 1981–2003. At the Cardinal Plant (ORM 76.7), pumpkinseed were collected every year from 1981 to 1995 and were generally common in samples during the 1980s. They have been found in clearly reduced numbers in recent years and were collected in only 1 of 7 years during the period 1996–2001 and 2003. At the Kyger Creek Plant (ORM 260), pumpkinseed were collected in low numbers and only occurred in samples 6 of the 17 study years between 1981 and 2003. Pumpkinseed were rarely collected (4 of 23 years) and numbers were also low in samples at the Tanners Creek Plant (ORM 494).[70]

Pumpkinseed is native to the Youghiogheny River system of PA, MD, and WV.[8]

Found throughout NY including the Allegheny River system.[92] It is not native to the Ohio River system of IL[89] or IN.*

Pumpkinseed originally occurred only in a few localities in the Ohio River drainage of central and southern OH. Scarcity of records for pumpkinseed prior to 1900 suggests that it was probably absent throughout the southern half of Ohio before the building of canals. Pumpkinseed introduced into flowing waters of southern OH between 1920 and 1950 failed to establish themselves, but those introduced into weedy impoundments flourished. Introductions were continued in more recent years. Releases in newer and larger impoundments resulted in a few permanent populations but with few exceptions the species was of rare or accidental occurrence in the flowing waters of southern OH.[96]

In KY, the pumpkinseed is known only from the Licking River, Rowan County; Greenbo Lake, Greenup County; and the Ohio River from Jefferson County upstream.[5] A record from the Ohio River, Daviess County[4] is unsubstantiated.[5]

In VA, the presence of pumpkinseed in the New and Tennessee River drainages is undoubtedly due to introductions. Middle and South Fork Holston populations also likely came from introductions. It is generally uncommon or rare in the Blue Ridge and Valley and Ridge Ecoregions of VA.[90]

It has also been introduced into the New and Tennessee River drainages of NC. Menhinick (1991) recorded reservoir collections of pumpkinseed from the New, Little Tennessee, and French Broad River drainages of western NC as well as stream captures from the Toe and New River drainages.[93]

In TN, pumpkinseed were apparently successfully introduced into South Holston Reservoir,[88] and they have moved downstream into Boone Reservoir. Early reports of pumpkinseed in Reelfoot Lake are considered in error and probably represent a hybrid combination representing other species of *Lepomis*.[87]

SPAWNING

Location

Pumpkinseed have been reported to spawn in ponds, lakes, creeks, and slow-moving streams.[73,76] In the Great Lakes region, spawning occurs in quiet near-shore areas including bays,[7,97] harbors, marshes, lagoons, backwaters, and creek mouths; also running waters of tributaries.[7] In the Thames River, southwestern ON, pumpkinseed males nested only in backwater areas, where water flow was minimal, the substrate was a deep layer of detritus and mud, and aquatic plant growth was heavy.[12]

Pumpkinseed nests in ponds and quiet waters may endure several seasons. The first reproductive behavior observed in nature is the appearance of males in shallow water of traditional nesting areas. The males gradually space themselves over old nests of the previous year, if these are present, or perhaps over nests previously used by other centrarchids earlier in the season. Therefore a pumpkinseed male does not necessarily build a completely new nest each breeding season.[101]

Nests are found over almost any kind of bottom.[101] Various authors refer to gravel, small flat rocks mixed with gravel, sandy gravel, rubble, sand, clay, muck, marl, and mud, or various combinations of most of the above;[12,50,73,76,80,90,97] sometimes in areas of exposed roots, or with overlay of sticks, leaves, or shells.[12,73] Often reported over sand and gravel.[1,80,90,97] Substrate types can be clay to sand, gravel, or rock, as the male sweeps only deep enough to expose a clean, hard bottom.[76]

Nests are sometimes constructed in areas with no plants;[66,73,90] sometimes in dense growths of algae,[73] or among other vegetation,[12,72,73,76,84] particularly emergent plants;[66,73] sometimes near solid objects such as rocks or logs on the bottom.[1,73] In a NY pond, the majority of nests were constructed within the emergent plant zone. Areas of bottom open to penetration by the sun's rays were preferred as nesting sites, although shaded areas were present. Undoubtedly, that which influences nesting areas in many habitats is shallow, sun-warmed water rather than a specific bottom type.[101] Nests that were constructed in areas of dense vegetation had runways apparently made by the guarding male that led into deeper water.[66]

Pumpkinseeds spawn in fresh and tidal freshwater.[72] They are apparently better adapted to spawning in tidal fresh and oligohaline waters than are other sunfishes in DE. In DE, they spawn in nontidal waters of lakes, large ponds, and small ponds; numerous nests also found in tidal spillpools.[80]

Nests are generally built in shallow water[72,76,80,84,90] with reported depths ranging from 76 mm to 2.1 m.[7,50,66,72,73,76,84,97] Smaller nests (approximately 20 cm in diameter) were located in shallow water of the littoral zone while larger nests (approximately 35 cm in diameter) were constructed in deeper waters.[80] Pumpkinseeds nested nearer to shore and in shallower water than bluegills.[34] Shallower nests were practically always built near a solid object such as a large boulder or submerged stump, but larger nests in deeper water were not, though suitable submerged objects were available.[101]

Some reports indicate that nests are usually in colonies[72] or large numbers of nests are normally found in the same area;[1,76,101] as many as 10–15 nests may be seen together in a small area.[97] Conflicting reports indicate that pumpkinseed nests were scattered, not colonial,[34] and that nests are typically found in small groups (2–3), sometimes close together (76–102 mm apart), rarely solitary.[50,73] In a NY pond, conditions of nest crowding did not exist; the closest proximity of one nest to another was 0.15 m (6 inches) and, in the two examples of nests this close together, they were separated by a heavy growth of vegetation.[66]

In Deep Lake, MI, nests were occasionally found within a bluegill colony.[50]

In ON ponds, shade and cover from trees overhanging the water did not appear to influence nest-site selection by pumpkinseeds and nest sites did not appear to be influenced by the proximity of aquatic vegetation or by type of substrate.[34]

Under controlled conditions, pumpkinseed spawned naturally in spawning boxes supplied with pea gravel substrate placed in circular wading pools.[24]

Season

Pumpkinseed generally spawn from May through August, often with peak spawning reported in June (Table 36).

Temperature

Pumpkinseed nesting begins at water temperatures of about 13–17°C[34,97] or 15–18°C[101] and spawning is reported at water temperatures of 12.8–30°C.[7,80] In Canada, spawning occurs between 20 and 27.8°C.[76] In Muscote Bay (Bay of Quinte), ON, first spawning is probably at 19–20°C.[64] In the Great Lakes, spawning begins at 12.8°C and occurs at water temperatures as high as 28.9°C.[7] In WI, pumpkinseed spawn at 19.4°C;[97] in MD, at 16–27°C, with optimum spawning at 21–24°C.[72] Spawning at 22–25°C is reported for pumpkinseed in Greece.[47]

Fecundity

Egg number increases with size of pumpkinseed female.[76] The numbers of eggs present in the ovaries of IN pumpkinseed 2–5 years of age, with

Table 36

Reports of spawning season for pumpkinseed from various locations.

Location	Spawning Season
Thames River, ON[12]	Mid-May through July
Lake Opinicon, ON[33]	June 12–30
Great Lakes[7,27]	May to August
WI[97]	Early May to August
MI[50]	Late May to mid-July
Mid-Atlantic Bight[73]	May into August
NY[1]	Late May to early July, peak in June
NY pond[66]	Mid-June to mid-July
NH[87]	June to August
MD[72]	May through August; peak in June
DE estuaries[80]	May through August; peak in June
IL[84]	May to June
TN[87]	Probably begins earlier than June and lasts until August
VA[90]	Begins early May, perhaps in April; may extend to August
Europe[73]	May through August
Greece[47]	Mid-June through July

average total lengths of 61–92 mm, are presented in Table 37. The average number of eggs per gram of fish for the 24 specimens represented in the table was 72.[37]

In WI, pumpkinseed 112–141 mm TL in late May had 4,100–7,000 eggs; in late June, females 119–135 mm TL had 2,400–6,700 eggs. Also in late June two females from northern WI, 122 and 126 mm TL, held 5,460 and 5,850 eggs.[97]

In Greece, pumpkinseed averaged 7,169 eggs with a range of 1,122–12,293.[47]

In ON, surviving females inhabiting two shallow ponds that experienced major winterkills were more fecund relative to body weight and their gonadosomatic index was almost double that of females from an adjacent non-winterkill lake.[42]

In a MA pond in late May 1972, gonadal recrudescence in both male and female pumpkinseeds occurred, as indicated by an increase in the gonosomatic index and the appearance of spermatocytes in the testes and active vitellogenesis in the ovaries. During this time, the pond temperature rose above 12.5°C, while day length was about 15 h.[44]

Sexual Maturity

Sexual maturity of pumpkinseed is usually achieved by age 2 in Canada,[76] but pumpkinseeds are reported mature as early as age 1.[90,91] In NY, faster growing males mature in the second year of life; all are mature in the third or fourth year.[1]

In 16 pumpkinseed populations in east-central ON, sexual maturity of males and females ranged from 2.7 to 4.6 years; size at maturity for males was 81–125 mm, for females 82–131 mm.[59] Length at first breeding in ON is also reported at 125 mm TL, weight 36 g, age 4.[33] Sexually mature pumpkinseed as small as 60 mm TL have been observed in stunted populations.[73]

In ON, pumpkinseed inhabiting two shallow ponds that experienced major winterkills matured 1–2 years earlier and at a smaller size (difference >20 mm in length) than conspecific females living in an adjacent, non-winterkill lake. Advance maturity following winterkill may have been the result of abundant food supply to survivors or release from social factors that prevented maturation of small pumpkinseed.[42]

Pumpkinseed mature at age 1+ in Greece.[47]

Table 37

Numbers of eggs found in the ovaries of pumpkinseed from IN related to age, length, and weight.

Age	No. of Specimens	Average TL (mm)	Average wt (g)	No. of Eggs		
				Average	Maximum	Minimum
2	8	61	15.3	1,034	1,684	600
3	11	77	20.4	1,491	2,366	1,108
4	3	92	31.0	2,422	2,740	2,580
5	2	92	33.5	2,436	2,923	1,950

Source: Constructed from data presented in unnumbered table on page 75, reference 37.

Spawning Act

The male pumpkinseed builds a nest[1,7,50,71,80,84,97] and then guards the nest, eggs, and young.[1,7,50,71,84] The male defends a territory and excavates a nest that consists of a circular, shallow depression created with sweeps of his tail.[80,84,97] Objects too large or heavy to be moved by this method are pulled away with the mouth.[75] Nests are usually circular[73,74,101] or oval[66,73] depressions with rims of silt and other bottom material.[101]

The size of the nest is usually proportional to the size of the fish that builds it[80] (usually about twice the length of the male[76,101] or more than twice the length[66]), but reports of nest diameter are highly variable; in one series of 36 nests, dimensions varied from 406–914 × 305–762 mm (average 643 × 523 mm).[66] Other reports of nest size include: 31–38 cm in diameter;[97] 102–406 mm in diameter;[76] and about 0.3 m in diameter.[84]

Captive pumpkinseed males, which subsequently built nests and spawned, first indicated that they were reproductively motivated by descending from mid-level water and gradually assuming a bottom territory.[101]

A female's approach to a nesting male may be slow and circuitous, with many retreats to a hiding place before she finally enters the nest and stays.[101] When a male approaches a female to induce her to enter the nest, he elevates and puffs out his gill covers and erects his ear flaps, so that there is a brilliant display of color for the female. This behavior was also observed by a male when threatening or attacking other intruding males.[84]

There is considerable display and swimming in circular pattern during courtship and mating.[76,85,101] Mutual butting and nipping serve to stimulate spawning. Eggs are laid while the pair swims in a circle over the nest with the male upright and the female at a 45° angle, so that their ventral surfaces are together. Small numbers of eggs and small quantities of sperm are emitted at irregular intervals.[76,85]

A typical successful pumpkinseed spawning event lasts 25–45 min. During this time the male leaves the nest frequently, usually in threat posture. He may swim away and back; he may rush or chase other fish; or he may threaten a neighboring male. The female remains on the nest while the male is gone. The male also shows aggressiveness by interrupting the spawning rhythm to turn and thrust at the female with a nudge in the side or display with spread opercles. If the female does not leave the nest after these threats, circling is resumed.[101]

Males may spawn more than once in the same season, in the same nest, with same or different females.[76,85] Males usually spawn with one female at a time but more than one female may contribute to the eggs in a nest.[80] On occasion male pumpkinseed will spawn with two females in immediate succession.[66] Several thousands of eggs may be deposited in a nest by one or several females.[1] Male pumpkinseed sometimes operate over two nests intermittently[34,85] and may raise more than one brood in a season. Females may spawn more than once in a season.[80]

In an aquarium in which three females and one male were present, a spawning pair was joined by a second female, and the three fish attempted, perhaps with success, to spawn together. The male was upright between the two females, which inclined on either side of him, all facing the same direction.[85,97]

The moment the female leaves the nest the male pumpkinseed begins fanning the eggs.[76,101] He guards his nest fearlessly, driving off intruders,[71] while maintaining his progeny. After hatching occurs, the male guards the young for a period of up to 11 days, returning them to the nest in his mouth if they stray. After this time the young leave the nest and the male may begin to clean the nest in preparation for a second spawning.[76]

Nesting male pumpkinseed defended their territories with typical sunfish aggressive behavior, consisting of spreading the opercle, charging, biting, chasing, and, rarely, mouth-fighting.[34,101] Nesting males defend two territory boundaries concurrently, depending on whether an intruder is approaching (the defense perimeter) or withdrawing (the attack perimeter). The defense perimeter remains constant over the breeding cycle while the attack perimeter varies.[32]

In a NY pond, when warding off would be nest intruders, guarding male pumpkinseed seemed to follow three lines of attack: (1) an eclipse charge around the nest without attempted contact with the intruder; (2) a straight line attack and return without bodily contact; and (3) a straight line attack to the body of the intruder, usually striking the latter just behind the operculum or on the caudal peduncle. They return to the nest's center after such a charge is made in a half-circle. In this study, males were observed driving 300–460 mm TL chain pickerel off their nests.[66]

In ponds containing only bluegill and pumpkinseed, approaching fish of either species were often attacked by nesting pumpkinseed when they were about 1 m from the nest. Males showed aggression initially toward any approaching female. Subsequent behavior of the female appeared to determine the outcome of these encounters. A mature female pumpkinseed threatened by a nesting male either fled or settled closer to the bottom and withstood his attack; in the latter case, the male usually began courting her. A mature female approached by an aggressive heterospecific male invariably fled. In these ponds, females were never observed trying to enter nests occupied by heterospecific males, and interspecific courtship and spawning were not seen.[34]

Sudden thunder showers during daylight caused guarding males to abandon their nests. As soon as such showers were over the males returned to their nests. Similar showers during darkness had no effect on guarding males.[66] In ON ponds, duration of nest care by the parent was dependent upon water temperature;[34] some pumpkinseeds deserted nests containing eggs when water temperature fell suddenly; others stayed on nests without eggs for over 3 weeks with no sudden temperature drop.[34] The mean number of consecutive days spent guarding a nest was 18.5 (sd 12.8).[34]

In a NY pond, guarding males seemed generally to tolerate the presence of numerous golden shiners and other small fry who hovered about the nest.[66] Golden shiners spawned in about one-third of the pumpkinseed nests present. Results of this study indicated that golden shiners benefited from spawning in pumpkinseed nests through the protection of their young by the male pumpkinseed.[55]

The likelihood and timing of reproduction are both age- and size-dependent in some populations. Small individuals that delay seasonal maturation and spawn late in the summer probably contribute little to the population due to the restricted growth and reduced overwinter survival of their progeny.[53]

Pumpkinseed sometimes spawn in nests of other species.[73]

Pumpkinseed is known to hybridize with warmouth[1,76] and bluegill[1,56,76,81] and redbreast,[1,56,76] green,[1,17–19,51,56,57,76] orangespotted,[76] and longear sunfishes.[1,12,76]

Hybrid sunfishes are often fertile and recross with other hybrids as well as with one or both parents. Hybridization occurs between pumpkinseed and bluegill to such an extent that in some eastern ON lakes it is virtually impossible to designate true parental types.[76]

Though pumpkinseed reportedly hybridize with bluegills in nature, reproductive isolation appeared complete between the two species in two ON ponds when they were stocked at the same time in equal numbers under conditions that eliminated competition for suitable spawning areas.[34]

EGGS

Description
Mature eggs are 1.0 mm in diameter.[33,71,97]

Fertilized, water-hardened eggs are adhesive and demersal;[68,71–73,76,80,97,101] transparent or whitish[73] or pale amber;[76] 0.8–1.2 mm in diameter;[47,72,76,80] and spherical[72,101] with some irregularities.[72] Egg capsule is thick, laminated.[74,80] A single large, yellow oil globule 0.37–0.40 mm in diameter,[74] or about 30% of yolk diameter,[72] is present sometimes with several smaller droplets.[72–74,80] Perivitelline space is narrow, 0.05 mm.[74]

Incubation
Reports of incubation times and temperatures vary: 10 days or less at ambient;[7] about 2 days (47.5 h) at temperatures of 19.0–24.9°C;[74,76] 2–3 days at 18–20°C;[73] about 3 days at 22°C[80] or 3 days at about 28°C;[7,76] 50–60 h at 25.5–29.4°C.[101]

Development
See Figures 37 and 38.

At 19.0–24.9°C:

- 0.25 h — blastodisc developing as a thin layer
- 0.75 h — blastodisc flattened over yolk
- 1.00 h — blastodisc peaked up over yolk
- 1.25 h — 2-cell stage
- 1.75 h — 4-cell stage
- 2.17 h — 8-cell stage
- 2.67 h — 16-cell stage
- 4.17 h — early morula
- 5.67 h — late morula
- 8.50 h — blastoderm spreading down over yolk
- 10.25 h — early gastrula
- 12.25 h — blastoderm to equator of egg
- 13.50 h — blastoderm below equator, yolk slightly constricted
- 15.25 h — yolk at maximum constriction
- 16.75 h — yolk plug stage
- 18.67 h — yolk plug greatly reduced
- 19.25 h — blastopore closed
- 21.42 h — anlagen of head differentiated
- 22.00 h — 3 somites
- 22.75 h — 10 somites, eye developing
- 24.42 h — 11 somites
- 28.75 h — 13 somites, Kupffer's vesicle developing
- 29.00 h — 18 somites, tail free, optic vesicle closed
- 36.25 h — 29 somites, lens complete
- 36.50 h — 30 somites, otoliths evident, weak body movements, heartbeats
- 39.42 h — blood vessels well-developed over yolk
- Ca. 47.50 h — hatching[74]

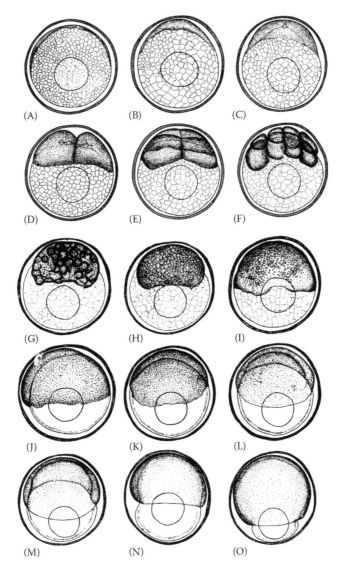

Figure 37 Development of young pumpkinseed — egg development related to age (time from fertilization): (A) 0.25 h; (B) 0.75 h; (C) 1 h; (D) 1.25 h; (E) 1.75 h; (F) 2.67 h; (G) 4.17 h; (H) 5.67 h; (I) 8.5 h; (J) 10.25 h; (K) 11.25 h; (L) 12.25 h; (M) 13.5 h; (N) 15.25 h; (O) 16.75 h. (A–F reprinted from Figure 10, G–O from Figures 11 and 12, reference 74.)

YOLK-SAC LARVAE

See Figure 39.

Size Range

Pumpkinseeds hatch between 2.4 and 3.5 mm TL;[68,72,80] yolk is completely absorbed by 4.5 mm[72] or at lengths >5.2 mm.[74]

Myomeres

4.0–5.0 mm TL. Preanal 10–13; postanal 17–21; total 27–34.[77]

Morphology

2.4–3.1 mm TL. At hatching (2.4–2.9 mm), body is elongate and tadpole-like.[80] At hatching (2.6–3.1 mm), yolk mass is egg-shaped; oil globule at or slightly posterior to midpoint of yolk. Urostyle is straight. Gut is uncoiled. Mouth not formed.[72,74] At 3.1 mm, otocysts are 0.22 mm behind eye.[74]

3.0–4.0 mm TL. Oil globule is located closer to the posterior margin of the yolk sac.[80] At 3.8 mm otocysts are 0.15 mm behind eye. At 3.9 mm, the nostrils are evident.[74] Gut is relatively thin and uncoiled.[80]

4.5–5.2 mm TL. At 4.5 mm, gills and gill arches are developing.[74] Yolk absorbed at 4.5 mm[72] or yolk still evident in some specimens at 5.2 mm.[74] At 5.2 mm, oil globule dorsoventrally compressed, lower jaw movable, and air bladder filled with air.[74] Forward part of gut coiled at 4.5–5.0 mm.[72]

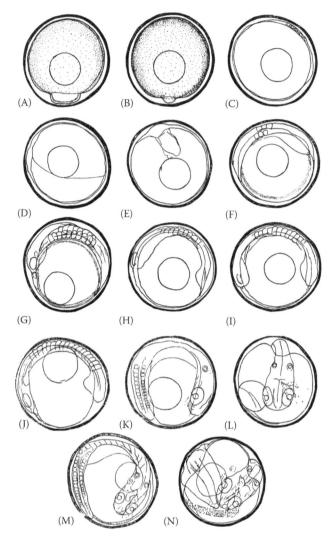

Figure 38 Development of young pumpkinseed — egg development related to age (time after fertilization): (A) 17.92 h; (B) 18.67 h; (C) 19.25 h; (D) 21.42 h; (E and F) 22.0 h; (G) 22.75 h; (H) 24.42 h; (I) 28.75 h; (J) 29.0 h; (K and L) 36.5 h; (M) 39.42 h; (N) 43.5 h. (A–I reprinted from Figures 12 and 13, J–N from Figure 14, reference 74.)

Morphometry

2.0–5.0 mm TL. As percent TL: preanal 37–45; HL 13–18; ED 8–9.[79]

4.0–5.0 mm TL. As percent TL: PreAL 40.[77]

Fin Development

2.4–2.9 mm TL. Dorsal finfold is narrower than ventral finfold.[80]

3.1 mm TL. Median finfold originates dorsally near posterior margin of yolk and is present along dorsum, around tip of urostyle and ventrally to anus, then from anus to posterior margin of yolk sac (based on illustration).[72] Pectoral buds evident.[74]

4.5 mm TL. Pectoral fins well-formed,[72,74] but without rays, oriented diagonally rather than horizontally as in earlier stages.

Pigmentation

2.6–3.1 mm TL. At hatching completely clear except for canary yellow oil globule.[74]

3.9–4.0 mm TL. Eye faintly pigmented.[72,74]

4.6 mm TL. Eye completely black; a few single melanophores ventrally between anus and tail; yellow pigment is present on yolk and posterior part of gut.[74]

5.2 mm TL. Yellow pigment on head; melanophores ventrally between anus and tail greatly increased; a few melanophores are present around tip of notochord and above notochord about halfway between anus and tail; dorsal surface of air bladder densely pigmented.[74]

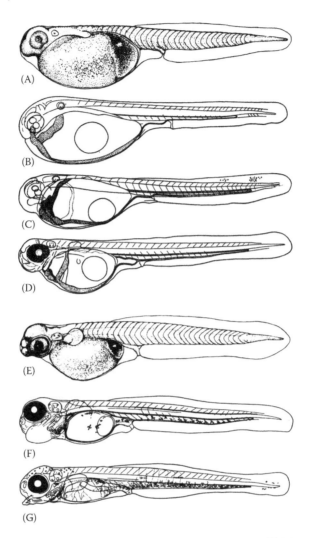

Figure 39 Development of young pumpkinseed. A–G. Yolk-sac larvae: (A) 3.1 mm TL; (B) 3.2 mm TL; (C) 3.8 mm TL; (D) 3.9 mm TL; (E) 4.0 mm TL; (F) 4.5 mm TL; (G) 5.2 mm TL. (A and E reprinted from page 185, reference 72, with editor's permission; B–D and F–G reprinted from Figures 17–19 and 21–22, reference 74.)

POST YOLK-SAC LARVAE

See Figure 40.

Size Range
Yolk generally absorbed by 4.5 mm TL[72] or may still be present at lengths >5.2 mm;[74] considered juvenile at lengths >14 mm, in MD.[72]

Myomeres
5.36 mm TL. About 12 preanal and 19 postanal (derived from illustration).[72]

Morphology
4.7–5.6 mm TL. Body is elongate. The oil globule has been absorbed by this size and the larvae move horizontally in aquaria.[80]

5.0 mm TL. Mouth well-formed; head and stomach area robust, with body tapering posteriorly (based on photograph).[78]

5.36 mm TL. Air bladder visible, mouth well-formed, operculum developing, urostyle straight (based on illustration).[72]

≥8.0 mm TL. Deep-bodied, laterally compressed, and morphologically similar to adults.[80]

Morphometry
5.2–5.4 mm TL. As percent TL: PreAL 37–42; HL 17; ED 7–10.[79]

11.0 mm TL. As percent TL: PreAL 44–45; HL 26; ED 9.[79]

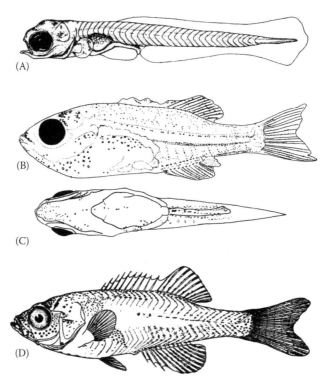

Figure 40   Development of young pumpkinseed. (A) Post yolk-sac larva, 5.36 mm TL. B–C. Post yolk-sac larva, 15.0 mm TL: (B) lateral view; (C) ventral view. (D) Juvenile, 18.5 mm TL. (A reprinted from page 185, reference 72, with editor's permission; B–C reprinted from Figures 4 and 5, reference 69; D reprinted from Figure 120, reference 71.)

Fin Development

5.36 mm TL. Median finfold originates dorsally near posterior margin of air bladder and is present along dorsum, around urostyle and ventrally to gut, then anteriorly on ventrum to an area below middle of the air bladder. Pectoral fins paddle-shaped (based on illustration).[72]

Pigmentation

Post yolk-sac larvae have a prominent melano-phore just above the anus on each side and have few or no melanophores on the breast.[72]

4.7–5.6 mm TL. Melanophores are present on the dorsal surfaces of the head, intestine, and air bladder, on the bases of the pectoral fins, and in the thoracic region. Two rows of melanophores are present on the ventral margin of the caudal pedun-cle from the anus to the tail.[80]

5.0 mm TL. Eyes black; melanophores present on head between eyes (based on photograph).[78]

5.36 mm TL. Eye pigmented. There appear to be stellate chromatophores on dorsal surface of head, pigment overlying air bladder and dorsal surface of gut, and light pigment along median myosepta at mid-body from posterior margin of air bladder to 4th or 5th postanal myomere (based on illustration).[72]

10.0–15.0 mm TL. No melanophores or only small circular ones on breast. Very little belly pigment at 15 mm, and an indefinite row of melanophores around the anal fin.[69]

JUVENILES

See Plate 1D following page 26 and Figure 40.

Size Range

>14 mm TL[72] to 81–82 mm TL;[59] as small as 60 mm TL in stunted populations.[73]

Myomeres

18.5 mm TL. Preanal 10; postanal 18.[71]

Morphology

13–19 mm TL. 12 precaudal vertebra; 18 caudal ver-tebra; and 8–9 ribs.[69]

18.5 mm TL. Body ovate, very compressed; snout short and depressed over eye, interorbital space flat; mouth small, oblique, with maxilla just reaching forward margin of the eye. Origin of spinous por-tion of dorsal fin considerably anterior to origin of anal fin.[71]

Larger Juveniles. Gill rakers short and stubby;[72,76] 8–12 on lower limb,[72,73,76,87] length of longest raker about 3–4 times basal width in YOY.[87] Branchiostegal rays 6–7.[73,76]

Pyloric caeca 7–8.[73,76] Total vertebrae 28–29;[73,87] caudal vertebrae 18.[73]

Lateral line scales 34–47;[72,73,87,90,94,97] 38–43 in WI;[97] 39–44 in MO;[94] 35–43 in TN.[87] In VA, lateral line scales 36–47, usually 37–44; scales above lateral line 6–8, usually 6–7; scales below lateral line 12–15, usually 14–15.[90]

Morphometry

18.5 mm TL. As percent TL: SL 78.4; PreAL 41.1; HL 24.4; SnL 6.0; ED 8.1; GD before vent 23.8; GD behind vent 20.0; PreDFL 31.4; PreAFL 43.2.[71]

Fins

Larger juveniles. Two dorsal fins are broadly joined and appear as one.[76,97] First dorsal fin has 9–12 (usually 10–11) spines;[72,73,76,81,87,90,94,97] second dorsal fin has 10–13 (usually 10–12) soft rays.[72,73,76,81,87,90,97] Anal fin has 2–4 (usually 3) spines[72,73,76,90,97] and 8–12 soft rays.[72,73,76,87,90,97] Pectoral fins have 11–14 soft rays.[69,73,76,87,90] Pelvic fins are thoracic[76,97] and have 1 spine and 5 soft rays.[73,76,97]

Pigmentation

> 14.0 mm TL. Usually no melanophores on venter between operculum and anus.[72]

16.0 mm TL. In a Delaware specimen, 10 faint vertical bars of pigment were visible on body between head and caudal peduncle extending slightly below lateral line. No pigment in fins.[80]

18.5 mm TL. No vertical bars of pigment on this Lake Erie specimen. Body is evenly covered with small black chromatophores that follow the margins of the myomeres. Markings are heavier around the jaws, on top of head, and on the cheeks. A single row of melanophores is present along median myosepta from above the origin of the anal fin posteriorly to the caudal fin. Pigment is present around the bases of the dorsal and anal fins. Only the belly is colorless. All fins have some pigment.[71]

25.0 mm TL. Lateral bars of pigment are visible with secondary "spots" of pigment between.[107]

Larger juveniles. Primary bars of pigment appear as an initial set of lateral patches of melanophores to which bars are added anteriorly and posteriorly in development of all *Lepomis.* Nine to 10 primary bars are visible on young pumpkinseeds by 30 mm SL (Figure 41A). Light centers form in these primary bars of pigment and secondary pigment is visible between the primary bars by 35 mm SL

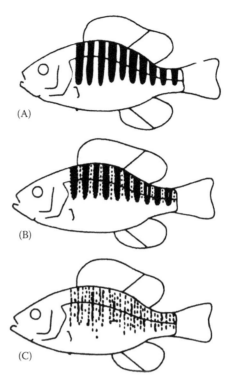

Figure 41 Development of young pumpkinseed — schematic drawings showing melanistic pigment patterns of young pumpkinseed. (A) 30.0 mm SL; primary pigment bars present. (B) 35.0 mm SL; light centers in some of the primary bars and secondary bars present between primary bars. (C) 92.0 mm SL; margin of primary bars retained, pattern dominated by thin, broken lines of pigment. (A–C reprinted from Figure 14, reference 95, with publisher's permission.)

(Figure 41B). By 92 mm SL, the margins of the primary bars are retained and the pigmentation pattern is dominated by thin broken vertical lines of pigment (Figure 41C).[95]

Young are also described as like females except that much of the bright coloring is replaced with green to olive background color[76] or silvery to light olive;[96] and more bluish than adults with pale olive-brown dorsum, indistinct vertical bands, brownish emerald and blue markings on sides of head, pearly reflections in iris, pale translucent fins, and coppery white peritoneum.[73] Yearlings and 2-year-olds are yellowish brown on back and sides, with prominent vertical bars on sides, and small orange spot on posterior edge of opercular flap.[34]

TAXONOMIC DIAGNOSIS OF YOUNG PUMPKINSEED

Similar species: pumpkinseed and redear sunfish are the only members of the subgenus *Eupomotis*, which is considered ontogenetically intermediate between *Apomotis* (green sunfish) and *Helioperca* (bluegill). The subgenus is most similar to *Helioperca* in meristic data and pigmentation, but in morphometric characters and gut and air bladder architecture *Eupomotis* resembles *Apomotis* (see Eupomotis subgeneric discussion on page 91).[108] There are comparisons in the literature of the early development of young pumpkinseed to bluegill[107] and to bluegill and redear sunfish.[69,87]

For comparisons of important morphometric and meristic data, developmental benchmarks, and pigmentary characters between pumpkinseed and all other *Lepomis* that occur in the Ohio River drainage, see Tables 4 and 5 and Plate 1D.

Pumpkinseed vs. bluegill and redear.

10–15 mm TL:

    Pumpkinseed: no melanophores or only small circular ones on breast; caudal vertebrae 18

    Bluegill: large diffuse melanophores in line on breast; caudal vertebrae 17

    Redear sunfish: melanophores in a line on belly and breast; caudal vertebrae 18[69]

15–25 mm TL:

    Pumpkinseed: 11 pectoral fin rays; caudal vertebrae 18

    Bluegill: 12 pectoral fin rays; 17 caudal vertebrae

    Redear sunfish: 12 pectoral rays; 18 caudal vertebrae[69]

25–30 mm TL — Pectoral fin rays as above

    Pumpkinseed: short gill rakers; bluegill and redear have long gill rakers[69]

Bluegill have 7–8 primary bars of pigment along their sides, but lack secondary pigment between[95,107]

Pumpkinseed[95,107] and redear sunfish[95] have 9–10 primary bars of pigment and also develop secondary pigment[95,107]

Pumpkinseed YOY (Plate 1D) are similar to bluegill (Plate 2A) and redear sunfish (Plate 2D). The redear are rarely sympatric with pumpkinseed, have more slender bodies, and as adults have longer pectoral fins, reaching to or beyond the dorsal fin base. Bluegill never has red on the opercular lobe, but red is visible on pumpkinseeds as small as about 35 mm SL.[87]

ECOLOGY OF EARLY LIFE PHASES

Occurrence and Distribution (Figure 42)

Eggs. Demersal, adhesive eggs attach to soil particles, rocks, gravel, roots, sticks, and aquatic vegetation in the bottom of the nest, where they incubate.[7,71,73,75,76]

Yolk-sac larvae. At hatching yolk-sac larvae remain motionless in nest for 2 days lying on their sides on the bottom, then begin to creep about the nest and eventually start to swim toward the surface, fall back, etc.[74] At 5–6 days they rise to the surface[73] and leave the nest.[101] Also reported attached to

Figure 42 General distribution, including areas of introduction, of pumpkinseed in the Ohio River system (shaded area).

plants.[73] Fry leave nest soon after hatching[7,97] or remain on the nest 3–4 days.[67]

In MN, fluctuations of dissolved oxygen (DO) concentrations and water temperatures in natural spawning sites were measured during the period of embryo and yolk-sac larval stages of pumpkinseed. Measured 1 cm from the bottom in nesting areas, DO concentrations ranged from 2.4 to 18.2 mg/liter and water temperatures from 15.0 to 27.5°C, with average daily fluctuations of 4.4 mg/liter and 3.3°C.[13]

In a study of the determinants of nesting success in pumpkinseed in two ON lakes, the most significant climatic variable influencing nesting success was wind-generated turbulence. Different risks of nest predation were evident in the two lakes. In one lake, piscivores were rare, planktivore–benthivore nest predators were abundant, nest-specific behavioral interactions were numerous, and nesting success was low. In the other lake, large piscivores were abundant, planktivore–benthivore numbers were low, and nesting success was high.[48]

Larvae. Larvae just out of the nest are positively phototropic. They do not school, but in aquaria tend to assemble under light. They apparently have a tendency to sink because they always swim on a slant upward; when not moving or when swimming horizontally, they drift gradually downward. They swim slowly using no fast or sudden movements. On the fourth day out of the nest they are less phototropic. Moving the light brings a less complete and rapid congregation in a new location. They swim more horizontally, but still near the surface. By the 6th day, there is no phototropic response, and they swim about the entire tank even near the bottom.[101]

Ichthyoplankton were collected weekly from Chemung Lake, ON, during the period May 13 to June 25, 1982, with Miller high-speed samplers and light traps. Pumpkinseed larvae appeared in samples taken with both gears on May 30 and were present in weekly samples collected through June 25. Peak numbers were taken with both gears on June 11 and June 18; densities were low in the Miller sampler collections (40 larvae on June 11 and 31 larvae on June 18) but more larvae were collected in the light traps (210 on June 11 and 540 on June 18). Mean lengths of specimens captured were similar for both gears except on June 25 when larvae collected in the light traps were significantly smaller (7.5 mm TL) than those from the Miller sampler (8.7 mm TL). Average lengths for larvae collected from May 30 to June 18 were 6.2–6.5 from the Miller sampler and 5.7–6.9 from the light traps. Yolk-sac larvae and post yolk-sac larvae were collected with both gears.[39]

In another Chemung Lake report, larval pumpkinseeds were collected with light traps from macrophyte-dense regions from June 3 to July 23 at water temperatures of 16–25°C. Captured yolk-sac larvae ranged from 4.7 to 7.0 mm TL and post yolk-sac larvae were 5.7–14 mm.[49]

Pumpkinseed larvae first occurred in ichthyoplankton samples from Muscote Bay, ON, on June 1 and were present in samples until late August.[64]

Pumpkinseed larvae 4–12 mm TL were collected with light traps in two littoral sites on Lac Heney, QU, during the summers of 1979 and 1980. Traps were fished in water 60–100 cm deep near beds of *Scirpus validus*. Pumpkinseed larvae appeared in samples after the middle of June when surface water temperatures were 15°C and were present in samples until mid-August.[61]

Larval pumpkinseed collected in drift samples from the upper Rhone River, France, were more abundant in near-bottom samples during the day and more abundant in surface samples at night. Changes in density corresponded to changes in light intensity.[41]

Mortality of young pumpkinseed is greatest during the transition from endogenous to exogenous nutrition. Mortality prior to yolk absorption is low, but after the yolk is absorbed, pumpkinseeds suffer higher mortality rates. Mortality and growth are directly related to prey density.[45]

Post yolk-sac larvae. In the Great Lakes, young pumpkinseed are found near spawning grounds, especially in thick weed beds over sand, in bays, harbors, and lower reaches of rivers; some are found on open shoals. Found in water depths to 1.8 m; some may assume a pelagic existence.[7] Larval pumpkinseed are found in the limnetic zone of some southwest MI lakes.[21]

In Ottawa ponds and lakes, after yolk absorption at about 5 mm TL, pumpkinseed larvae depart their nests and disperse into open, deeper limnetic waters just beyond littoral beds of aquatic vegetation.[67,68] Gradually individuals begin to associate together and develop schooling behavior. Larval pumpkinseed remain offshore in deep limnetic regions for a period of 2–4 weeks, depending on water temperature, hydrographic features of the lake, and probably other factors.[67]

Juveniles. Inshore open waters were occupied by young pumpkinseed, with weed beds forming the habitat for those over a month in age.[36] Observations made by scuba divers disclosed that juvenile pumpkinseed travel in abundant and loose schools in shallow water in areas of emergent water plants. The young swim near the surface.[98] Another study found pumpkinseed less than about 75 mm in shallow water near vegetation or in clear open water over sandy bottoms; young in schools were reported.[73] In Lake Opinicon, ON, YOY pumpkinseeds (20–28 mm SL) were first observed inshore in early August, and were found close to the bottom.[107]

In Ottawa ponds and lakes, juvenile pumpkinseed, 20–35 mm TL, return to shallow weedy regions where they live in schools. At about 50–75 mm, they live individually or in loose associations with other pumpkinseed in shallow weedy areas.[67]

In Lake Oneida, NY, almost five times as many young pumpkinseeds were taken in trawls at night than in the day,[102] but nocturnal and diurnal distribution of pumpkinseed juveniles 60–160 mm TL changed little in a shallow, sandy bay of Lake Opinicon.[36]

In laboratory, the preferred temperature of young pumpkinseed weighing between 1.2 and 7.7 g was 28.5°C.[15]

The findings of laboratory experiments with two size classes (20–24 mm and 30–37 mm TL) of age-0 pumpkinseed from a central ON pond implied that in the north temperate zone, age-0 pumpkinseed up to a certain size are unlikely to survive a given winter, regardless of their lipid reserves; pumpkinseed progeny produced late in the season are unlikely to survive in locations with long winters.[60]

Early Growth

Regional differences in the growth of pumpkinseed were not very consistent based on data tabulated from numerous reports and various locations. Ranges in TL related to age were wide for even the central 50% of mean of means (Table 38).[91] Growth

in WI was slow compared to average regional growth. Size ranges (TL) of pumpkinseed from five locations in WI were 43–54 mm at age 1, 51–87 mm at age 2, 64–137 mm at age 3, and 79–170 mm at age 4.[97] Mean calculated total lengths of pumpkinseeds from Clear Lake, IA, for the period 1947–1954[38] were within Carlander's (1977) central 50% mean of means at age 1 (57 mm) and age 2 (107 mm), and barely outside the range at age 3 (138 mm) (Table 38).

Growth is moderately fast. Reports of growth related to age for pumpkinseed populations in New Brunswick, ON, and MI (Table 39) were difficult to compare due to the different lengths used, but Smith (1952)[86] (cited in reference 76) indicated that growth in a New Brunswick population was somewhat faster to age 4 than that for particular habitats in MI; after that, growth in MI was faster.[76]

Southern populations could be expected to grow faster than northern populations but probably have shorter average life spans.[87] In VA, pumpkinseed show average to strong growth compared to regional expectations.[90,91] Range of means of TL of age-3 pumpkinseed in VA was 102–168 mm.[90]

In the Ottawa River, Canada, in 1978, pumpkinseed young attained a mean TL of 44.3 mm at the end of September.[9]

In Muscote Bay, ON, 1981–1982, young pumpkinseed averaged 5.1 mm TL on June 1, 6.0 mm on July 2, 8.0 mm on July 21, and 37.1–48.5 mm in mid-October.[64] In 16 populations in four watersheds of east-central ON, length at age 2 for male pumpkinseed ranged from 65 to 85 mm and for females 62–85 mm.[59]

In WI, YOY pumpkinseed were 28–46 mm TL in August and 39–57 mm in September.[97] In Linwood Lake, MN, young pumpkinseed were 42–44 mm

Table 38

Central 50% and range of mean of means of TL (mm) for young pumpkinseed related to age as summarized by Carlander (1977) from numerous reports from various locations.[91]

Age (years)	No. of Fish	Mean TL	
		Central 50%	Range
1	2,496+	56–81	18–139
2	1,749+	91–112	43–190
3	2,324+	109–136	65–223
4	1,521+	116–155	66–286

Table 39

Age–length relationship for northern populations of young pumpkinseed presented in Scott and Crossman (1973).[76]

Age (years)	New Brunswick Average SL (mm)	Ontario Lakes FL (mm)	Michigan Pond Average TL (mm)
1	74	89	74
2	89	107–152	104
3	105	107–183	124
4	111	122–208	145

Table 40

Average observed lengths and weights related to age of young pumpkinseed from Lake Chautaugua, IL, during the period 1950–1959.

Age (years)	No. of Fish	Average TL (mm)	Average wt (g)
2	2	104	27
3	21	142	73
4	25	165	118

Source: Constructed from data presented in Table 12 (in part), reference 3.

TL in September and October.[23] In OH, YOY in October were 20–81 mm TL; around 1 year 30–89 mm.[96]

In Lake Opinicon, ON, young pumpkinseed averaged 26 mm SL (range 20–28 mm) when the first appeared inshore in early August. By the end of September the mean SL was 33 mm (range 24–40 mm).[107] During the period October 10–16, they ranged from 34–50 mm TL (average 40.2 mm) and weighed 0.2–2.9 g (average 1.2 g).[33]

Yearling pumpkinseed in Lake St. Clair were 57–66 mm TL and 3.5–6.3 g on June 20, 1979, 86–92 mm and 13.9–16.3 g on September 5, and 90–93 mm and 16.0–17.3 g on October 30.[40]

Average observed growth in length and weight related to age of young pumpkinseed from Lake Chautauqua, IL, during the period 1950–1959 was comparatively good for the species in Midwestern lakes (Table 40).[3] Length–weight relationships of pumpkinseeds from Clear Lake, IA,[38] presented in Table 41, appear comparable to the growth in IL.

Growth of larval pumpkinseed is directly related to prey densities. In a laboratory, significant growth occurred only at prey densities higher than 0.25 plankters ml^{-1}. The minimum prey density supporting 10% survival of pumpkinseed larvae was estimated at 0.16 plankters ml^{-1}.[45]

Growth of small pumpkinseed (<40 mm SL) and bluegill (<55 mm SL) collected between 1978 and 1985 from 9 MI lakes exhibited similar responses to environmental factors. Large bluegill (>55 mm SL) and large pumpkinseed (>50 mm SL) responded differently. These breaks in growth patterns coincide with the sizes at which each species exhibits an ontogenetic shift in diet.[46]

In small productive bodies of water with large populations, stunting takes place[76,97] and maximum length may not exceed 102–127 mm.[76]

Table 41

Length–weight relationships of young pumpkinseed from Clear Lake, IA, for the period 1947–1952.

Average SL (mm)	No. of Specimens	Weight (g) Mean	Weight (g) Range
54.3	6	6.6	4–9
62.8	6	10.3	8–14
76.1	12	19.3	14–25
86.1	13	23.0	15–30
95.1	35	36.2	26–50
104.5	20	49.6	39–65
116.0	19	73.9	39–91
124.6	21	88.5	60–103
134.5	2	107.0	106–108

Source: Constructed from data presented in Table 8 (in part), reference 38.

Feeding Habits

Young and juvenile pumpkinseed fed on microcrustaceans and small aquatic insects and progressed to larger prey with growth.[58,62,68,90] In many populations pumpkinseed exhibit a strong ontogenetic diet shift. Small juvenile pumpkinseed feed primarily on soft-bodied littoral invertebrates while larger pumpkinseed feed extensively on snails.[43]

Pumpkinseed have molariform pharyngeal teeth and, like the redear sunfish, they feed extensively on snails and small clams in some environments.[58,68,90] In small MI lakes, pumpkinseed ≤75 mm SL foraged predominantly in vegetation and most of their average seasonal diet was vegetation-dwelling prey (non-gastropods). Fish >75 mm SL also foraged primarily in vegetation but fed mainly on vegetation-dwelling gastropods.[58]

In the littoral zone of a small bay in the Ottawa River, Canada, age-0 pumpkinseed fed diurnally and showed strong positive, curvilinear relationships between prey weight and fish weight. The principal seasonal foods of age-0 pumpkinseed were chironomid larvae and daphnids. Gastropods were important prey for pumpkinseed >35 mm TL.[9]

In Lake Opinicon, ON, the diet of young pumpkinseed was made up largely of chironomid larvae, mollusks, isopods, lesser quantities of amphipods, trichopteran larvae, and ephemeropteran nymphs. Cladocera accounted for a significant number and volume of prey organisms only in age-0 fish, while gastropods and plant material were prominent food items in older and larger pumpkinseed. Chironomid larvae and isopods were dominant foods throughout life and amphipods, epheremopteran nymphs, and tricopteran larvae were taken consistently but in smaller quantities. Prey items with mean weights ranging from 0.5 to 1.0 mg and lengths of 1.6–10 mm predominated in the diet of pumpkinseeds, and there was only a minor increase in average prey size with growth. The author of this study classified pumpkinseed as a feeding generalist with no great channeling by age-classes toward different resources.[30]

Lepomis larvae 5–6 mm TL collected June 15–17, 1970, from the inshore open waters of Lake Opinicon, ON, and identified as pumpkinseed larvae based on reproductive timing, began exogenous feeding on nauplii (0.1–0.2 mm long) and *Cyclops* (0.1–0.7 mm long). Larvae 6–9 mm TL primarily fed on *Cyclops*, *Bosmina* (0.1–0.4 mm), *Diaphanosoma* (0.3–0.4 mm), and *Ceriodaphnia* (0.3–0.7 mm). *Lepomis* larvae 4.5–20 mm TL, believed to be a mix of pumpkinseed and bluegill, collected July 17–19, consumed nauplii, *Cyclops*, *Bosmina*, *Sida*, and *Diaphanosoma*. Size of prey generally increased with fish growth. Small *Cyclops* (0.1–0.2 mm) and nauplii (0.1–0.2 mm) were consumed by fish 4.5–10 mm. *Lepomis* larvae 10–20 mm fed on larger *Cyclops* (0.2–0.7 mm) and the other taxa listed above. Large *Lepomis* larvae (15–20 mm TL) collected in August consumed larger *Cyclops* (0.2–0.7 mm), large *Bosmina* (0.2–0.4 mm), *Sida* (0.1–0.4 mm), *Chydorus* (0.1–0.3 mm), and *Diaphanosoma* (0.3–0.4 mm).[31] Age-3 pumpkinseed (95–115 mm TL; n = 73) from Lake Opinicon consumed the following food by volume: mollusks (33%); chironomid larvae (29%); ephemeropteran nymphs (10%); isopods (8%); anisopterans (6%); amphipods (4%); trichopterans larvae (4%); chironomid pupae (3%); cladocerans (2%); and coleopteran larvae (1%).[36]

Yearling pumpkinseed from a shallow heavily vegetated embayment of Lake St. Clair fed mainly on amphipods, chironomids, Lepidoptera, and gastropods. *Hyalella azteca* (shorter than 5.5 mm), composing 88% of seasonal total number of amphipods ingested by pumpkinseeds, were eaten extensively throughout the season (June–October 1979). Most chironomids and lepidopterans were consumed in July and September and most of the gastropods (mainly *Physa* and *Gyraulus*) in August and September. Numbers of prey and gut volumes increased with increasing plant canopy until the plant canopy reached a seasonal maximum in mid-summer.[40]

In WI, 24 pumpkinseed, 50–169 mm TL, consumed crustaceans, rotifers, snails, clams, flatworms, aquatic insect larvae, and terrestrial insects. In Green Lake, in late summer, pumpkinseed 146 mm long had eaten insect larvae (59.5% by volume), clams (6.1%), snails (26.5%), and leeches (7.5%). In Lake Mendota, pumpkinseed 118 mm long had consumed insect larvae (29.3% by volume), insect pupae (0.5%), insect adults (4.2%), amphipods (3%), ostracods (29.5%), snails (11%), leeches (5.9%), plant material (5.5%), and sand (11.1%).[97]

In a PA river in a 24–h study in October, pumpkinseed 58–147 mm TL fed primarily on ephemeropterans (53.7% dry weight), principally baetids (20.0%), ephemerids (14.3%), and chironomids (26.4%). Terrestrial invertebrates (9.7%), hydropsychids (1.3%), and miscellaneous aquatic organisms (7.7%) were also consumed.[105]

Food habits for the period March–October 1979 were reported for young pumpkinseed from two piedmont lakes in NC. The diet of age-0 pumpkinseed 13.5–36.0 mm SL in both lakes was dominated by copepods and cladocerans. The percent frequency of food items found in the stomachs from Salem Lake, NC, was as follows: copepods 64%; cladocerans 20%; insect larvae 16%. In the other lake (Lake 158), pumpkinseed ate copepods (57%), cladocerans (31%), ostracods (5%), rotifers (2%), insect larvae (4%), and fish (1%).[106]

Chironomid and dipteran larvae, terrestrial insects, and benthic prey (such as burrowing mayflies) may comprise the staples of pumpkinseed diet in TN reservoirs, as snails are scarce.[87]

During the period 1987–1989, substrate-dwelling microcrustaceans provided a diverse and abundant food resource for young fish inhabiting the littoral zone of Lake Itasca, MN. Also, free-swimming microcrustaceans occurred in the littoral zone water column but were two orders of magnitude less abundant. Larval and early juvenile pumpkinseed in these waters selectively preyed upon the free-swimming animals, especially *Bosmina*. Among the predominant prey of either prey type, young pumpkinseeds showed no preference or a preference for smaller individuals, suggesting that young fish may gain a greater benefit from ingesting smaller prey (more quickly digested) than predicted by optimal foraging models.[14]

Feeding of young pumpkinseed (3–7 cm TL, age 0 and age 1) from a creek in central ON was studied from September 21 to December 4, 1978. Volumetric analysis of stomach content showed that young pumpkinseed ceased feeding when fall temperatures reached 2°C. A decline in feeding was first noticed in mid-September, immediately after a drop from 18 to 16°C. Subsequent decrease in feeding closely followed the declining temperature regime throughout the fall and early winter, with perhaps a few irregularities or inconsistencies. This decline was not associated with reduced availability of prey or seasonal or diel foraging changes by the fish themselves. Feeding activity in the fall peaked once during daylight hours; this maximum volume of food was taken at dusk. After

October 25, food of the young pumpkinseed was composed mainly of *Asellus*, *Chironomus* pupae, *Simulium* larvae, and "other" (unidentified) species, with a proportional reduction in *Chironomus* larvae from October 9 to early December.[22]

Pumpkinseed and redear sunfish are phylogenetic sister species that are specialized for molluscivory. The greater crushing strength of redear allows them to undergo an ontogenetic niche shift from a diet of predominantly soft-bodied invertebrates to a diet of snails at a smaller size than co-occurring pumpkinseed. In MI lakes, by 55 mm SL, the diet of redear was dominated by gastropods (≥80%); the diet of pumpkinseed this size was <20% gastropods and was instead dominated by dipteran larvae.[65]

A laboratory study of the influence of habitat structure and prey type on foraging behavior of pumpkinseeds 75–80 mm TL showed that foraging behavior is relatively sensitive to temporally and spatially separated prey and habitat types.[54]

LITERATURE CITED

1. Raney, E.C. 1965.
2. Thomas, J.A. et al. 2004.
3. Starrett, W.C. and A.W. Fritz. 1965.
4. Pearson, W.D. and L.A. Krumholz. 1984.
5. Burr, B.M. and M.L. Warren, Jr. 1986.
6. Pearson, W.D. and B.J. Pearson. 1989.
7. Goodyear, C.D. et al. 1982.
8. Hendricks, M.L. et al. 1983.
9. Hanson, J.M. and S.U. Qadri. 1984.
10. Schwartz, F.J. 1964.
11. Reash, R.J. and J.H. Van Hassel. 1988.
12. Keenleyside, M.H.A. 1978.
A 13. Peterka, J.J. and J.S. Kent. 1976.
14. Hatch, J.T. et al. 1990.
15. Muller, R. and F.E.J. Fry. 1976.
16. Coutant, C.C. 1977.
17. Childers, W.F. and G.W. Bennett. 1961
18. Hubbs, C.L. 1920.
19. Hubbs, C.L. 1955.
20. Van Hassel, J.H. et al. 1988.
21. Rettig, J.E. 1998.
22. Reid, N. and P.M. Powles. 1982.
23. Lux, F.E. 1960.
24. Buynak, G.L. and H.W. Mohr, Jr. 1981.
25. Fish, P.A. and J. Savitz. 1983.
26. Savitz, J. et al. 1983.
27. Chubb, S.L. 1985.
28. Rodeheffer, I.A. 1940.
29. Cincotta, D.A. and J.R. Stauffer, Jr. 1984.
30. Keast, A. 1978a.
31. Keast, A. 1980.
32. Colgan, P.W. et al. 1981
33. Keast, A. and J. Eadie. 1984.
34. Clark, F.W. and M.H.A. Keenleyside. 1967.
35. Walters, D.A. et al. 1991.
36. Keast, A. 1978b.
37. Ulrey, L. et al. 1938.
38. DiConstanzo, C.J. 1957.
39. Gregory, R.S. and P.M. Powles. 1988.
40. French, J.R.P. III. 1988.
41. Copp, G.H. and B. Cellot. 1988.
42. Fox, M.G. and A. Keast. 1991.
43. Osenberg, C.W. et al. 1992.
44. Burns, J.R. 1976.
45. Hart, T.F., Jr. and R.G. Werner. 1987.
46. Osenberg, C.W. et al. 1988.
47. Neophitou, C. and A.J. Giapis. 1994.
48. Popiel, S.A. et al. 1996.
49. Gregory, R.S. and P.M. Powles. 1985.
50. Carbine, W.F. 1939.
51. Hubbs, C.L. and L.C. Hubbs. 1931.
52. Wainwright, P.C. 1996.
53. Danylchuk, A.J. and M.G. Fox. 1994.
54. Kieffer, J.D. and P.W. Colgan. 1993.
55. Shao, B. 1997.
56. Etnier, D.A. 1968.
57. Dawley, R.M. 1987.
58. Mittelbach, G.G. 1984.
59. Fox, M.G. et al. 1997.
60. Bernard, G. and M.G. Fox. 1997.
61. Faber, D.J. 1982.
62. Nurnberger, P.K. 1930.
63. Laughlin, D.R. and E.E. Werner. 1980.
64. Leslie, J.K. and J.E. Moore. 1985.
65. Huckins, C.J.F. 1997.
66. Ingram, W.M. and E.P. Odum. 1941.
67. Faber, D.J. 1984b.
68. Faber, D.J. 1985.
69. Werner, R.G. 1966.
A 70. EA Engineering, Science, and Technology. 2004.
71. Fish, M.P. 1932.
72. Anjard, C.A. 1974.
73. Hardy, J.D., Jr. 1978.
74. Balon, E. 1959.
75. Adams, C.C. and T.L. Hankinson. 1928.
76. Scott, W.B. and E.J. Crossman. 1973.
77. Taubert, B.D. 1977.
78. May, E.B. and C.R. Gasaway. 1967.
79. Tin, H.T. 1982.
80. Wang, J.C.S. and R.J. Kernehan. 1979.
81. Eddy, S. and T. Surber. 1947.
82. Hubbs, C.L. and K.F. Lagler. 1964.
83. Lee, D.S. et al. 1980.
84. Forbes, S.A. and R.E. Richardson. 1920
85. Breder, C.M., Jr. 1936.
86. Smith, M.W. 1952.
87. Etnier, D.A. and W.C. Starnes. 1993.
88. Fitz, R.B. 1979.
89. Smith, P.W. 1979.
90. Jenkins, R.E. and N.M. Burkhead. 1994.
91. Carlander, K.D. 1977.
92. Smith, C.L. 1985.

93. Menhinick, E.F. 1991.
94. Pflieger, W.L. 1975b.
95. Mabee, P.M. 1995.
96. Trautman, M.B. 1981.
97. Becker, G.C. 1983.
98. Emery, A.R. 1973.
99. Kudrna, J.J. 1965.
100. Reed, R.J. 1971.
101. Miller, H.C. 1964.
102. Forney, J.L. 1974.
103. Radomski, P. and T.J. Goeman. 2001.
104. Crutchfield, J.U., Jr. et al. 2003.
105. Johnson, J.H. and D.S. Dropkin. 1993.
106. Lemly, A.D. and J.F. Dimmick. 1982.
107. Brown, J.A. and P.W. Colgan. 1981.
108. Conner, J.V. 1979.

WARMOUTH

Lepomis (Chaenobryttus) gulosus (Cuvier)

Robert Wallus and John V. Conner

Lepomis, Greek: *lepis*, "scale" and *poma*, "lid"; *gulosus*, Latin: "large mouthed."

RANGE

The northern range limit of warmouth appears to be a line from southeastern MN and central WI to MD including only the western portion of PA. The range extends south to FL and the Gulf Coast to TX and the Rio Grande River and west to NM and KS. Introduced elsewhere in the United States from west of the Rockies to portions of the Atlantic slope.[1,6]

Warmouth populations are depleted in several states on the periphery of its range, especially where natural wetlands are drained and siltation has resulted in destruction of aquatic vegetation.[49]

HABITAT AND MOVEMENT

Warmouth is reported from a variety of aquatic habitats including ponds, lakes, reservoirs, and occasionally streams of all sizes.[1] The usual habitat is sluggish streams,[6,11,30,31,36,37,40,41,49] swamps,[6,30,31,34,36,49] ponds,[1,6,31,36,37,41,43,49] and lakes[1,6,30,31,36,37,41] or impoundments,[1,34,49] where it is often associated with submerged aquatic vegetation.[49] Warmouth does well in lowland habitats and is often reported from oxbow lakes,[26,30,34,37,43] sloughs,[26,30,43] wetlands,[26] overflow pools,[26,30] marshes,[37] borrow pits,[30,43] seasonal floodplains,[30,43] bayous,[34] and weedy ditches.[30] In upland areas it is found in sluggish streams and quiet pools and does well in impoundments.[34] This species is a pool inhabitant[31,34,36] in streams[1,36,41] and rivers.[34,36,41]

Warmouth is a sedentary and secretive species usually seeking cover in aquatic vegetation,[11,26,30,31,34,36,37,40,49] or around other shelter[26] such as stumps,[31,34,40,49] tree roots,[31,40,49] logs,[34,49] rocks,[49] brush,[40] and cypress knees.[34] It is usually associated with mucky[37] or soft bottom substrates[11,26,40] including mud,[6,26,31,34] silt,[6] gravel and sand,[26] and organic debris or detritus.[6,26,31,34,37] Generally avoids intense light.[49]

Young and moderate-sized warmouth are usually associated with hiding places and are often concentrated in weedy and stump-filled waters. Brush and roots attract this species and in areas lacking vegetation old tree stumps constitute a typical hiding place.[40]

Apparently small warmouth do not leave protected hiding places in shallow water even during cold weather. Larger warmouth (>127 mm TL) spend more time in deep, open water than in shallow water and exhibit no tendency to group together during winter months.[11,40] However, in Red Haw Hill Reservoir, IA, where warmouths were fairly abundant, they were seldom taken by fishermen. During summer months, warmouth were found in shallow water among weeds along shore, where angling was difficult.[3]

Over 3,600 ha of submersed aquatic vegetation were eliminated in 1 year after the stocking of 270,000 grass carp in Lake Conroe, TX, in 1981–1982. The removal of the vegetation from the coves resulted in the rapid, nearly complete, elimination of the warmouth.[62]

Warmouth may have a greater tolerance for turbidity and conditions associated with turbidity than most other sunfishes.[8,40,41] It is reported more frequently and in greater abundance in muddy or turbid waters, usually characteristic of lowland lakes, backwater areas, and sluggish streams, than in less turbid waters.[40,41] Other reports indicate that the warmouth is usually associated with clear, quiet water.[26,30,34] A report from OH suggested that warmouth was less tolerant to turbidity and siltation than the green sunfish.[37] Another study indicated warmouth preferred clear water but would tolerate moderate levels of turbidity.[34] A tolerance for turbidity may give the warmouth certain competitive advantages over other sunfishes.[40]

Warmouth is undoubtedly a warm-water species, but information on temperature tolerances was unavailable.[6] Its range extension may be limited by seasonal water temperatures.[6]

Turbid waters, organic silt deposits, and dense vegetation, usually regarded as typical features of warmouth habitats, are associated with high oxygen demands and, at times, low concentrations of dissolved oxygen.[40]

Warmouth inhabit highly acidic sections of the Dismal Swamp interior in VA.[36]

Warmouth is considered by some a strictly freshwater species, usually occupying salinities <1.5%, yet the species has been reported in salinities as high as 17.4%.[36] It is reported from brackish coastal streams,[45,51] sometimes only as strays,[45] but also as a numerically dominant species in the lower reaches of coastal streams,[8] with occasional reports from brackish water up to 4.1 ppt.[1,8] Mettee et al. (2001) found warmouth in every river system of AL, but found them in greatest abundance in the Mobile Delta during late summer when lower river discharge and greater tidal intrusion elevated salinity values to 1–15 ppt.[35]

Warmouth was found in only 2 of 67 springs surveyed in the southern bend of the Tennessee River drainage.[16]

Except when spawning, warmouth is a solitary species,[6] though aggregations may occur around desirable cover, such as riprapping along a shoreline or dam. Little social structure has been detected among such groups. Nesting colonies that have been reported are probably due to restricted spawning habitat rather than to a gregarious nature of the species.[40]

Warmouth have shown no tendency to become dominant at the expense of synoptic centrarchids or other kinds of fishes.[6,40] In 17 ponds in central IL stocked with 11 different fish combinations including warmouth with largemouth bass, smallmouth bass, and several other sunfishes, warmouths tended to establish small broods each year without restricting the reproduction or growth of companion species.[40] Availability of suitable forage species and predation by larger fishes such as largemouth bass occupying the same habitat would limit warmouth populations.[6]

In two OK ponds, warmouth experienced decreasing condition indices during 2 years following application of an herbicide that gave excellent control of dense growths of submersed rooted aquatic plants.[14]

DISTRIBUTION AND OCCURRENCE IN THE OHIO RIVER SYSTEM

Warmouth has been reported from throughout the Ohio River since 1800[32,33] but is most frequently encountered in the lower two-thirds of the river. Although normally an inhabitant of smaller bodies of water, it is encountered enough in the Ohio River to be considered a permanent resident. Collections have recorded many more warmouth in the middle third of the river since 1970 than before 1970 but this may be due to the use of more efficient collecting gear after 1970 (i.e., electrofishing gear).[32,58] Based on lockchamber surveys during the period 1957–2001, there has been no significant change in the abundance of warmouth in the Ohio River.[10]

Warmouth is absent in much of the Ohio River drainage of WV and PA[1] and was not reported by Hendricks et al. (1983) from the Youghiogheny River system of PA, MD, and WV.[12] It has been introduced into NY, but surviving populations are not present in Ohio River drainage waters.[38]

Warmouth is present throughout the Ohio River drainage of IL, but its principal habitats have been decimated due to the drainage of natural marshes, lakes, and ponds and the siltation that has destroyed aquatic vegetation in most parts of the state.[31] It was rarely collected from the Wabash River, IN.[44]

For reasons unknown, warmouth was absent in southern OH before 1925, or if present, only as strays. More recently, introductions have probably established small, local populations throughout the state.[37]

In VA, Jenkins and Burkhead (1994) regard all warmouth records from Ohio River drainage waters to be based on non-indigenous populations.[36] Though not widely distributed in western NC, the species is reported from the Hiwassee, Little Tennessee, French Broad, and New River drainages.[39]

In KY, warmouth is generally distributed and most common from the lower Green River westward; occasional in the lower Cumberland, Salt, Kentucky, and Licking Rivers. Records from the middle and upper Cumberland (above the falls), Barren, and upper Green river drainages are probably the result of introductions.[26]

The warmouth is abundant in the lowland streams of western TN and occurs less commonly in the fewer sluggish habitats of the Ohio River drainage that are afforded in eastern portions of the state, including reservoirs and large rivers. It is fairly abundant in Tennessee River reservoirs upstream as far as Watts Bar.[51] In AL, it is present in the lowland tributaries and reservoirs of the Tennessee River.[35]

SPAWNING

Location

Bottom substrate and cover apparently influence the selection of nesting sites.[40] The species is reported to nest over bottoms of mud, silt, silt covered with sticks and leaves, sand, rubble, detritus, shale, and leaf mold, apparently preferring soft substrates when firm substrates are available.[46]

However, warmouths are not so consistent in nesting on a particular kind of bottom as they are in selecting a spot near a stump, rock, root, brush, dense aquatic vegetation, or similar cover.[11,40,46]

In Venard Lake, IL, the following bottom substrates were available: loose silt, silt, rubble covered with a thin layer of silt, sand with loose silt, and clean sand. No warmouth nests were found on clean sand, habitat often selected by bluegill and pumpkinseed, and nests seen on silt were always associated with tree roots or mats of submerged vegetation. Venard Lake warmouths used all of the bottom types except sand, but showed preference for rubble lightly covered with silt and detritus. In another study area, Park Pond, warmouths nested among weed masses, stumps, roots, and brush.[40]

Warmouth build nests in a wide range of water depths, usually 5–152 cm deep;[40,57] consequently, nest locations vary in their distance from shore. In IL studies, nests were observed from 15 to 152 cm deep, with most covered with 61–76 cm of water. Nests were usually found along shallow sloping shorelines and were not found along steep banks.[40]

Warmouth brood stock were collected from Turkey Creek embayment of Ft. Loudoun Reservoir, Tennessee River, TN, on 6/14/1979 and deposited in circular tanks (outside) at a TVA Fisheries Laboratory. The tanks were provided with gravel and rubble substrate; the only cover was a large rock placed in the center of the tank and two concrete blocks along the wall of the tank. The rock and blocks were 1–1.5 m apart. Water depth was about 60 cm. After about 2 weeks, three nests were observed in the tank. Each nest was built next to the cover provided, one next to the rock and one each next to the concrete blocks.*

In the Suwannee River and Okefenokee Swamp, GA, no spawning beds were observed because of the dark tannic waters, but ripe warmouth were collected from around the bases of tupelo (*Nyssa* sp.) and cypress (*Taxodium* sp.) trees. Collections of ripe fish also came from sluggish water areas of the swamp that possessed stands of water lilies (*Nymphaea* sp.) and panic grasses (*Panicum* sp.).[19]

Warmouth tend to be more solitary in nest site selection than other sunfishes.[15,40] It is probable that reports of colony formation are the results of restricted nesting habitat.[40]

Season

The spawning period for warmouth extends over several months.[40] Length of the nesting season differs among different populations in different lakes and probably varies considerably from year to year. The length of the season also varies with the size of the warmouth, large warmouths spawning over a longer period than smaller ones. Gonadal studies have shown that a warmouth may spawn several times during a season.[40]

Warmouth spawning is often reported from mid-May into August,[3,11,30,39,40] with peaks in June,[30,34,40] but the season reportedly begins as early as April in AR[34] and GA.[19,40] In the Suwannee River and Okefenokee Swamp, GA, warmouths began spawning in April, peaked in early May, and ceased in late July or August.[19]

Temperature

Nest building begins when water temperatures reach about 13–15°C.[49] Spawning has been reported at temperatures ranging from 21.1°C[40,46] to 25.5°C.[17] Spawning was observed in TN in an outside holding tank at 20–24°C.*

Fecundity

Fecundity of warmouth may be reduced by lack of suitable nesting space, population overcrowding, unfavorable weather, or other conditions or circumstances that limit spawning opportunities and result in large numbers of mature eggs being retained and resorbed in the ovaries.[40]

Fecundity is a function of the size of the female,[11,19,40,42] with 4,500–63,000 eggs reported.[11,40] In the Suwannee River and Okefenokee Swamp, GA, the number of mature or nearly mature ova per female warmouth varied from 3,029 to 22,850. Average number generally increased with fish length.[19]

The relations of fecundity to TL (mm) were developed for warmouth from a large SC backwater lake. Fecundity (F) estimates for warmouths ranged from 798 to 34,257 eggs per female and could be expressed by the relation: $\log_{10} F = -4.678 + 3.889 \log_{10} TL$, with $r^2 = 0.67$.[2]

Sexual Maturity

There is considerable variation among warmouth in the size and age at which sexual maturity is reached.[40] In ON, warmouth are expected to mature at lengths of 50 mm or less.[6] In NY, sexual maturity is reached in the third year at about 76–89 mm TL.[11]

In central IL[40] and the Suwannee River and Okefenokee Swamp, GA,[19] size seems more important than age as a determinant for sexual maturity. In Venard Lake, IL, variation is greater between fish from different populations than among fish from the same population. Both male and female warmouth matured at age 1, at lengths of 79–86 mm TL, but in Park Pond, where growth was slower, warmouth did not mature until age 2 and 89 mm.[40] In the Suwannee River, GA, warmouth become sexually mature between 102 and 152 mm TL; most are age 3 at this size interval. In the Okefenokee

Swamp, the majority of warmouth this size are age 2.[19]

Spawning Act

As in other sunfishes, male warmouth excavate the nest. Violent sweeping motions of the tail clear loosened debris away from the selected area and produce a shallow irregular concavity. As he enters the nest site, the male turns abruptly upward and gives three or four violent sweeps of the tail while balancing in an almost vertical position. The size and neatness of the nest depend to some extent on the amount of time the male spends in its construction, which may be only a few hours in the wild. Many nests in nature are rather shapeless oval depressions 10–20 cm deep from which loose silt has been cleared. In a laboratory study, a male warmouth continued to work on the nest for a week while waiting for a female in his aquarium to become ripe. A mature female alone in an aquarium during the breeding season constructed a shallow nest, but her attachment for the nest was weaker than that characteristic of males.[40]

The preliminary courtship phase for nesting warmouth is reported from laboratory observations in IL. Courtship behavior normally appeared as an aggressive threat to other males, serving to drive them away, and as a persuasive gesture to females in spawning condition. When a female, not yet ready to spawn, was placed in an aquarium with a nesting male, she was charged, nipped, and driven to the surface. Being unable to escape the male in an aquarium, she may finally be killed by his continued aggression; in the wild a female will not become exposed to unavoidable advances of a nesting male before she is ready to spawn.[40]

Only when a female is ready to deposit her eggs will she allow the male to guide her into the nest. In getting the female to the nest, the male assumes a very aggressive attitude, approaching her with widely spread opercles and open mouth. In this condition, the body of the male very quickly becomes bright yellow in color and his eyes become blood red. If the female is ready to spawn, she is easily directed toward the nest by the male, and spawning soon occurs.[40]

Upon entering the nest, both male and female begin to circle, the female nearer the center of the nest, slightly on her side and somewhat beneath the male. As they are circling, the female works her jaws three or four times and suddenly jerks her body violently, giving the male a sharp thump on his side. Each time the female jerks she extrudes about 20 eggs. The thump she delivers to the male probably stimulates a discharge of sperm, although milt was not observed coming from the genital pore. After circling the nest several times, the female interrupts the activities and leaves the nest. The male usually follows her for a short distance

but returns quickly to the nest to assume guardianship. In nature, after a few spawning turns in the nest, the female usually retires to a clump of weeds several yards away. The male remains over the nest for a few minutes before again making advances toward the female. This behavior is repeated until the female has discharged all her ripe eggs. When spawning is completed, the female swims away, and the male settles down quietly to protect and fan the eggs.[40]

In TN, warmouth were observed spawning in two different nests in an outside holding tank. In both events, the male repeatedly herded a female into the nest. In the nest they moved about close together for several seconds and then the female would leave the nest where upon the male would go after her and herd her back into the nest. This behavior was repeated several times. About a half hour after the termination of spawning, the male turned from a very dark color to a lighter color virtually matching the color of the algae-covered substrate around the nest.*

The number of females contributing to the complement of eggs in a nest may depend upon how many females are ripe and available to the male. It is probably not uncommon for more than one female to spawn in a single nest. Such polygamy probably seldom occurs in the wild once the male assumes close guardianship of the eggs because freshly laid eggs are not found in nests containing eggs in advanced stages of development. In the laboratory, however, a male, guarding yolk-sac larvae, brought a female to the nest and spawned with her. A female in an aquarium with two nesting males alternately spawned in both nests during a continuous spawning sequence that lasted for about an hour. The nests were about 25 cm apart.[40]

Male warmouth become ready to spawn earlier and remain capable of spawning later in the season than do the females. Thus, a ripening female generally encounters plenty of males ready to spawn.[40]

Gonadal studies indicate that warmouth may spawn several times during a summer. In central IL, warmouth >137 mm TL attained spawning condition earlier in the season, and spawned over a longer period than did smaller sizes. Males matured slightly earlier in the season than did females.[40]

The presence of both mature and immature ova in female warmouth suggested that females might spawn more than once in a season, but this did not occur in the Suwannee River and Okefenokee Swamp, GA. In this area, larger females (≥200 mm TL) seemed to spawn earlier than smaller females, but a lengthened spawning season seemed to be the result of individual females becoming ripe at different times during the spring and summer.[19]

The nesting warmouth male displays an aggressive threat toward other fish that approach his nest area. He assumes a belligerent attitude by swimming toward the intruder with his mouth open and his opercles spread; at the same time, his eyes become red and his body becomes light yellow in color. As he nears the intruder, he usually turns abruptly to one side or upward and, with vigorous movements of his tail, forces small pulses of water toward the intruder. He may also nip the intruder. The entire threat attitude associated with defense of the nest is very similar to the persuasive behavior employed in courtship of a female.[40]

After warmouth larvae leave the nest area, they receive no parental care. Protection afforded by dense vegetation into which they normally scatter eliminates most of the needs for parental care.[40]

Warmouth are known to hybridize, naturally and in the laboratory, with at least four other *Lepomis* spp., rock bass, largemouth bass, and black crappie.[1,6,7,36,37,40,48] In nature, hybridization with bluegill, pumpkinseed, green sunfish, and redear sunfish is reported.[15,36]

EGGS

Description
Ovaries examined during the spawning season contained both large and small ova, but no ova in an intermediate stage of development were observed. Small ova (0.45 mm average diameter) were uniformly distributed throughout the ovaries and appeared to be poorly developed. Larger ova were in one of two stages of maturation with only one stage found in a given female. More developed, large ova (0.97 mm average diameter) were observed in warmouth ≥200 mm TL. Warmouth <200 mm contained less developed large ova (0.85 mm average diameter). The less developed large ova were opaque, yellow to dark yellow, and polygonal in shape. The more mature large ova were translucent light pinkish-orange in color and globular in shape.[19]

In IL, non–water-hardened fertilized eggs ranged from 0.95 to 1.03 mm in diameter, were translucent and light amber in color, and contained a single, dark amber oil droplet 0.35 mm in diameter.[40]

Water-hardened eggs are demersal, adhesive,[40,46,47] clear amber,[40,46] and 1.0–1.1 mm in diameter.[16,40,46] Fertilized eggs stick together in small clusters, also sometimes in linear bead-like arrangements on rootlets (Carr, 1939, cited in Hardy, 1978).[46,47]

Incubation
The following incubation times and temperatures were reported by Childers (1965)[48] and cited in Hardy (1978):[46]

29.4 h at average 26.2°C
28.8–30.1 h at average 27.3°C
28.4–29.6 h at average 27.6°C
28.9 h at average 28.1°C

Incubation is also reported in 44–53 h at 25.5°C[17] and 34.5 h at 25–26.4°C.[40]

Development
Following are stages of development of 40 warmouth eggs related to time after fertilization when incubated at 25–26.4°C:[40]

3 min — a thin perivitelline space visible between chorion and egg cell
30 min — blastodisc present
43 min — first division of blastodisc
60 min — second division of blastodisc
75 min — third division of blastodisc
90 min — fourth division of blastodisc
2.25 h — blastomeres form an oval-shaped mass at one end of the yolk
2.50 h — blastoderm begins growth down over yolk mass
11.00 h — blastoderm has grown over about 2/3 yolk mass; germ ring appears
12.25 h — blastoderm covers all but a small plug of yolk, which contains oil droplet
14.25 h — first differentiation among dividing cells becomes visible
15.17 h — neural groove visible
16.50 h — primordial form of embryo defined
25.00 h — embryo movement observed
33.30 h — hatching begins
36.00 h — hatching completed[40]

YOLK-SAC LARVAE

See Figure 43.

Size Range
In IL, warmouth hatched at 2.3–3.1 mm TL,[40,47] and yolk absorption was nearly completed by the fourth day at about 5.3 mm TL.[40] Hatching and yolk absorption occurred at about the same size for warmouth cultured in TN; yolk was absent as small as 4.85 mm TL, but trace amounts persisted on some individuals up to 5.1 mm TL.*

Myomeres
See Table 42.

3.4 mm TL. From IL, myomere development at this size was incomplete anteriorly, numbering about 8 preanal and 17 postanal.[40]

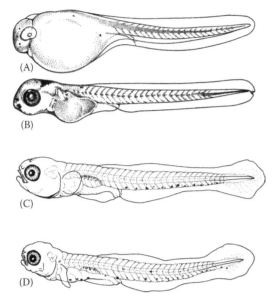

Figure 43 Development of young warmouth. A–B. Yolk-sac larvae: (A) 3.4 mm TL; (B) 4.6 mm TL. C–D. Post yolk-sac larvae; (C) 5.1 mm TL; (D) 5.5 mm TL. (A–B, delineated by E.R. Peters in Hardy 1978, from photographs presented in reference 40, Figure 14A and 14B; reprinted with publisher's permission; C–D, wild-caught in LA, identified by J.V. Conner.)

4.6 mm TL. From IL, myomeres indistinct anteriorly, about 10 preanal and 19 postanal.[40]

Note: A postanal myomere count of 19 for warmouth is higher than would be expected for a *Lepomis* species and the preanal values cited above are low compared to the following reports. We agree with Conner (1979)[54] that it is unlikely that the myomere counts reported above were made in accordance with the procedure recommended by Siefert (1969)[56] and used by the authors of this work (see definitions for "preanal myomeres" and "postanal myomeres" in the Glossary of Terms in the front of the book).

From TN:

3.3–5.1 mm TL. Preanal 11–13, postanal 14–18, and total 25–30 (Table 42).*

From LA:

5.0–5.5 mm TL. Preanal 12–13 and postanal 16–18.[54]

Morphology

3.4 mm TL. Large oval yolk mass contains one oil globule 0.3 mm in diameter. Head is deflected sharply downward in front of yolk sac, making the midbrain the most forward part of the body, the forebrain lying directly beneath. Optic capsule is faint, about 0.23 mm in diameter. Notochord straight.[40]

4.6 mm TL. Forebrain still somewhat deflected, with globular cerebellum extending high above anterior part of medulla. Optic fissure apparent; mouth indistinct; branchial elements forming.[40]

Morphometry

See Table 43.

Table 42

Myomere count ranges (average and standard deviation in parentheses) at selected size intervals for young warmouth from TN.*

Size Range (mm TL)	N	Myomeres			
		Predorsal Fin	Preanal	Postanal	Total
3.3–5.1	62	—	11–13 (12.2 ± 0.48)	14–18 (16.4 ± 0.97)	25–30 (28.5 ± 1.59)
5.2–6.1	40	—	12–13 (12.6 ± 0.48)	16–18 (16.8 ± 0.51)	28–30 (29.4 ± 0.62)
6.2–7.6	7	—	12–14 (13.1 ± 0.64)	16–18 (16.6 ± 0.73)	29–30 (29.7 ± 0.45)
8.9–13.7	9	7–9 (7.29 ± 0.70)	13–14 (13.8 ± 0.41)	15–17 (15.8 ± 0.63)	29–30 (29.6 ± 0.50)
17.9	1	7	15	15	30

Table 43

Morphometric data for young warmouth from TN* grouped by selected intervals of total length (TL). Characters are expressed as percent TL and head length (HL) with a single standard deviation. Range of values for each character is in parentheses.

	TL Intervals (mm)				
Length Range N	3.30–5.12 62	5.13–6.15 40	6.22–7.58 7	8.91–13.75 9	17.90–17.90 1
Characters	Mean ± SD (Range)	Mean ± SD (Range)	Mean ± SD (Range)	Mean ± SD (Range)	Mean ± SD (Range)
Length as % TL					
Snout	3.59 ± 0.44 (0.12–0.22)	3.41 ± 0.49 (0.12–0.24)	3.28 ± 0.48 (0.16–0.31)	4.25 ± 0.39 (0.33–0.62)	5.59 ± 2.03 (1.0–1.0)
Eye diameter	7.69 ± 0.79 (0.23–0.42)	6.64 ± 0.91 (0.24–0.43)	7.34 ± 0.53 (0.39–0.59)	8.29 ± 0.45 (0.7–1.19)	8.49 ± 2.23 (1.52–1.52)
Head	16.7 ± 2.40 (0.45–1.01)	18.2 ± 1.08 (0.89–1.12)	19.7 ± 1.79 (1.08–1.62)	24.2 ± 2.01 (1.97–3.58)	27.9 ± 1.45 (5.0–5.0)
Predorsal	—	—	—	30.2 ± 1.79 (2.6–4.0)	29.1 ± 8.92 (5.2–5.2)
Prepelvic	—	—	—	27.2 ± 0.73 (2.47–3.61)	27.1 ± 9.63 (4.85–4.85)
Preanal	42.9 ± 2.10 (1.55–2.26)	42.4 ± 1.46 (2.05–2.61)	43.8 ± 1.00 (2.63–3.36)	43.7 ± 0.72 (3.93–5.90)	45.3 ± 1.62 (8.1–8.1)
Postanal	57.0 ± 2.09 (1.73–3.10)	57.6 ± 1.46 (2.86–3.56)	56.2 ± 1.00 (3.48–4.22)	56.3 ± 0.72 (4.98–7.85)	54.7 ± 2.30 (9.8–9.8)
Standard	97.3 ± 0.87 (3.18–5.04)	97.0 ± 0.62 (4.95–5.95)	95.9 ± 2.17 (6.03–7.21)	84.0 ± 2.49 (7.75–11.10)	79.9 ± 3.56 (14.3–14.3)
Air bladder	7.31 ± 1.42 (0.16–0.51)	8.91 ± 0.48 (0.39–0.58)	10.9 ± 0.86 (0.58–0.91)	12.9 ± 1.50 (0.99–2.05)	—
Eye to air bladder	14.2 ± 1.09 (0.48–0.80)	15.0 ± 0.57 (0.74–0.97)	15.1 ± 0.81 (0.9–1.25)	14.3 ± 0.95 (1.33–1.75)	—
Fin length as % TL					
Pectoral	—	—	—	11.0 ± 1.98 (0.75–1.81)	15.1 ± 4.46 (2.7–2.7)
Pelvic	—	—	—	5.70 ± 3.82 (0.08–1.38)	13.1 ± 5.15 (2.35–2.35)
Body depth as % TL					
Head at P1	—	—	—	21.2 ± 2.04 (1.62–3.32)	23.5 ± 1.18 (4.2–4.2)
Head at eyes	—	—	—	18.4 ± 1.63 (1.40–2.65)	20.1 ± 8.14 (3.6–3.6)
Preanal	—	—	—	18.5 ± 2.85 (1.20–3.08)	22.3 ± 8.14 (4.0–4.0)
Caudal peduncle	—	—	—	9.61 ± 1.50 (0.63–1.52)	10.6 ± 5.90 (1.9–1.9)
Body width as % HL					
Head	—	—	—	57.6 ± 6.91 (0.87–2.30)	50.4 ± 1.03 (2.52–2.52)

3.4 mm TL. As percent TL: PreAL 50.0; yolk sac length 29.4; PreDFFL 29.4.[40]

4.6 mm TL. As percent TL: PreAL 43.5; HL 13.7; GD anterior to anus 15.2; GD posterior to anus 7.6. As percent HL: ED 55.6.[40]

Fin Development
3.4 mm TL. Dorsal origin of median finfold is 1.0 mm from most anterior point of body; finfold is complete around urostyle and present ventrally to posterior margin of yolk sac with a break at the anus. No visible fin ray development.[40]

4.6 mm TL. Finfold still complete from dorsal origin around urostyle and forward ventrally to posterior margin of yolk sac with a break at the anus. Caudal fin differentiation is evident above and below straight urostyle. Pectoral fin lobes are present.[40]

Pigmentation
3.4 mm TL. Eyes are unpigmented and no pigment is visible on body.[40]

POST YOLK-SAC LARVAE

See Figures 43 and 44.

Size Range
About 5.0 mm TL*,[40] to an undetermined TL >13.7* and <15.7 mm.[40]

Myomeres
From TN:

5.2–13.7 mm TL. Preanal 12–14; postanal 15–18; total 28–30 (Table 42).

From IL:

5.3–7.6 mm TL. Preanal 10; postanal 19[40] (see note in Yolk-Sac Larvae section, page 146).

Morphology
5.3 mm TL. Cranial flexures almost straightened, but cerebellum high and bulb-like; optic cavity distinct; kidney apparent through body wall; branchial arches well formed and with developing gills; mouth gape extends obliquely forward from point below middle of eye.[40]

7.6 mm TL. Urostylar flexion is evident; angle of upturn about 40°. Mouth moderately oblique.[40]

8.8 mm TL. Caudal peduncle long and narrow; anus protrudes from ventral line of body. Otic region large and clear.[40]

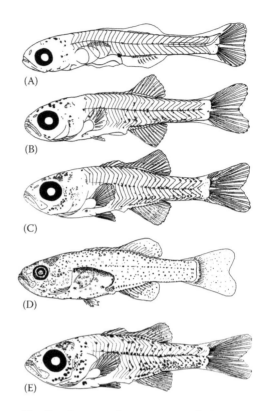

Figure 44 Development of young warmouth. A–D. Post yolk-sac larvae: (A) 8.1 mm TL; (B) 11.1 mm TL; (C) 13.3 mm TL; (D) 13.5 mm TL. (E) Early juvenile, 14.7 mm TL. (A–C and E, reprinted from Figure 18, reference 63, with permission of Progress Energy Service Company, LLC; D reprinted from Figure 16, reference 61, with senior author's permission.)

Morphometry
See Table 43.

From IL:

5.3 mm TL. As percent TL: PreAL 41.5; HL 17.0; GD anterior to anus 11.3; GD posterior to anus 6.6. As percent HL: ED 45.6.[40]

7.6 mm TL. As percent TL: PreAL 44.7; HL 19.7; GD anterior to anus 15.1; GD posterior to anus 8.6. As percent HL: ED 36.7.[40]

8.8 mm TL. As percent TL: PreAL 44.3; HL 22.7; SnL 4.5; CPD 7.4. As percent HL: ED 37.5; SnL 20.0.[40]

12.2 mm TL. As percent TL: PreAL 45.0; HL 23.3; SnL 5.8; GD 20.4; CPL 20.0; CPD 9.6. As percent HL: ED 35.7; SnL 25.0.[40]

Fin Development
5.3 mm TL. Finfold slightly narrow on caudal peduncle, but still wide around urostyle; ray differentiation evident in areas of future soft dorsal and

anal fins; distinct fin rays evident at urostyle; pectoral fin lobes well developed but with no rays.[40]

7.4–7.6 mm TL. Shapes of developing soft dorsal and anal fins forming in median finfold, little finfold remains dorsally or ventrally on caudal peduncle. Caudal fin rays (11–13) well developed on ventral side of urostyle, middle rays longest.*,[40] Rays are weak but visible in developing anal fin and soft dorsal fins. Rays are visible in pectoral fins.[40]

8.8–9.0 mm TL. Caudal fin with upturned urostyle appears homocercal; 15–17 rays. Finfold absorbed, or nearly so, on caudal peduncle; remaining finfold present only as a remnant ventrally anterior to the anus and dorsally in the area of the developing spinous dorsal fin. Soft dorsal and anal fins not complete but well formed, with 8–9 rays visible. Rays distinct in all fins present. Pelvic fins not developed.*,[40]

12.0 mm TL. Finfold remains only as a short keel ventrally anterior to anus. Pelvic fins are present with indistinct rays. Developing spinous dorsal fin low in profile with about 10 short spines.*,[40] Anal fin with 3 spines and 8–9 soft rays. Adult complement of 17 rays is present in caudal fin.*

12.1–12.6 mm TL. Adult complement of fin elements is present in caudal fin (17 rays) and dorsal fin (9–10 spines and 9–10 rays); anal fin development nearly complete with 3 spines and 8–9 soft rays. Remnant of preanal finfold persists.*

13.2–13.7 mm TL. Adult complement of fin elements developed in all fins. Small remnant of finfold is still present anterior to anus.*

Pigmentation
5.3 mm TL. A dark row of spots is visible on either side of the ventral finfold. Two large chromatophores are visible between the bases of the pectoral fins.[40]

7.6 mm TL. Pigmentation much more developed. Distinct, dark chromatophores are scattered dorsally over head. Row of spots along ventral finfold spreads as stellate chromatophores on ventral surface of the body. A series of dark dashes is present along the median myosepta. A large chromatophore is present above the anus. Six stellate chromatophores are scattered between the bases of the pectoral fins and a row of five chromatophores is present on each side across the branchiostegal rays.[40]

8.8 mm TL. Pigment along median myosepta is quite distinct. Ventral pigment spots appear larger. More chromatophores are scattered over head region.[40]

12.0 mm TL. Distribution of pigment is about the same as above, except pigment spots are more distinct. More dark pigment is present around mouth and a vertical row of pigment is present at the base of caudal fin rays.[40]

JUVENILES

See Plate 1E following page 26 and Figures 44 and 45.

Size Range
>13.7 and <15.7 mm TL*,[40] to 79–86 mm TL (in IL)[40] or by 102 mm TL in GA.[19]

Myomeres
See Table 42.

Morphology
15.7 mm TL. Body form is essentially like that of adult. Anus protrudes only slightly from abdomen.[40]

Larger juveniles. Lateral line complete, arched anteriorly.[25,34] Scales in lateral series 36–40 in WI,[42] 35–44 in KS,[25] MO,[30] AR,[34] TN,[51] MS,[50] and AL.[35,49] In VA, lateral line scales 38–48, usually 41–45.[36]

Gill rakers 9–13,[49,51] length of longest rakers in YOY about six times their basal width.[51]

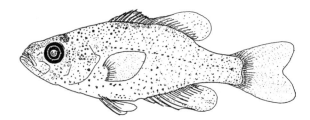

Figure 45 Young warmouth juvenile, 23.0 mm TL. (Reprinted from Figure 16, reference 61, with senior author's permission.)

Morphometry

See Table 43.

From IL:

15.7 mm TL. As percent TL: PreAL 44.6; HL 27.7; SnL 5.4; GD 23.6; CPL 18.5; CPD 9.6. As percent HL: ED 32.2; SnL 19.5.[40]

Fins

15.7 mm TL. Finfold absorbed. Pelvic and spinous dorsal fins well formed.[40]

Larger juveniles. Spinous and soft dorsal fins are broadly joined and appear as one[42] and consist of 9–11 spines[25,30,34–36,42,49–51] and 9–11 soft rays.[25,34,36,42,49–51] Anal fin has 3 spines[25,30,34–36,42,49,50] and 8–10 soft rays in WI,[42] 9–10 elsewhere.[25,34–36,49–51] Pelvic fins with 1 spine and 5 rays.[42] Pectoral fins have 12–14 rays.[25,36,49–51]

Pigmentation

15.7 mm TL. Pigmentation is much heavier than in earlier stages. More color is apparent over head and caudal peduncle. Belly is rather free of pigmentation. Many large chromatophores are scattered over dorsum of head and body. A heavy row of spots forms a circle posterior to the eye. Pigment is noticeable in soft dorsal, anal, and caudal fins.[40]

Larger juveniles. Primary bars of melanistic pigment appear as an initial set of lateral patches of melanophores to which bars are added anteriorly and posteriorly in development in all *Lepomis.* At 10–16 mm SL, seven to eight of these primary bars are apparent on warmouth (Figure 46). By 50–55 mm SL, light centers form in these primary bars and secondary bars form dorsolaterally, midlaterally, and ventrolaterally between the primary bars (Figure 46C). For large juveniles and adults, the light centers of primary bars are not as extensive, and the thick margins of the primary bars are distorted but retained. The secondary bars formed between primary bars are retained, but they often fuse with the primary bars to produce a mottled adult pattern (Figure 46D).[59]

TAXONOMIC DIAGNOSIS OF YOUNG WARMOUTH

Similar species: warmouth is the only member of the subgenus *Chaenobryttus.* Little information is available on the ontogeny of this subgenus (see *Chaenobryttus* subgeneric discussion on page 91). Conner (1979) placed warmouth larvae in the "green sunfish group," along with green and bantam sunfishes. Older larvae and juveniles in this grouping generally have more profuse pigmentation than other *Lepomis*; especially well-developed head pig-

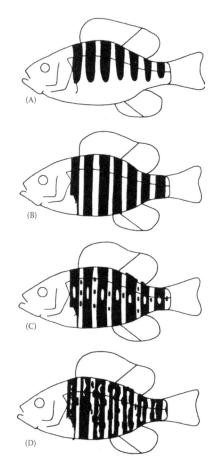

Figure 46 Development of young warmouth — schematic drawings showing melanistic pigment patterns of young warmouth. A–B. Development of primary bars of melanistic pigment: (A) pattern of development for individuals 10–16 mm SL; (B) 27 mm SL. (C) 50–55 mm SL, light centers form in primary bars and secondary bars of pigment begin to form between the primary bars. (D) 70.0 mm SL, secondary pigment fuses with primary bars to produce a mottled pattern. (A–D reprinted from Figure 12, reference 59, with publisher's permission.)

mentation (particularly in the cheek and postorbital areas); a characteristic mid-lateral streak of melanophores that often simulates a regularly spaced series of "dashes"; and a tendency toward smaller size at comparable stages than other taxa or types.[54]

There are also comparisons in the literature of the early development of young warmouth to bluegill,[49,51] bluegill and redear sunfish,[61] rockbass,[49] green sunfish, and redbreast sunfish.[55]

For comparisons of important morphometric and meristic data, developmental benchmarks, and pigmentary characters between warmouth and all other *Lepomis* that occur in the Ohio River drainage, see Tables 4 and 5 and Plate 1E.

Green sunfish larvae reportedly have more preanal myomeres and fewer postanal myomeres than warmouth.[55] Inconsistencies in myomere counting

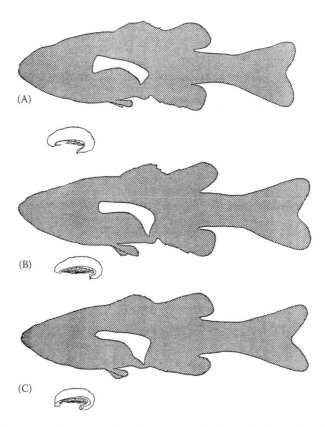

Figure 47 Shapes of air bladders and gill arches for three *Lepomis* spp. At about 14–15 mm TL: (A) redear sunfish; (B) warmouth; (C) bluegill. (A–C, reprinted from Figure 14, reference 61, with senior author's permission.)

techniques, noted in text, and original data presented in text make this a questionable character.

Warmouth vs. green sunfish. Warmouth hatch (2.3–3.1 mm TL)[47,55] and swim-up (5.3 mm TL)[55] at smaller sizes than green sunfish (hatching size 3.6–3.7 mm; swim-up size 6.0–6.3 mm).[55] Warmouths have absorbed all yolk before the appearance of head pigment; green sunfish have a great deal of yolk remaining when melanophores first appear on the top of the head.[55]

Warmouth vs. redbreast sunfish. Warmouth larvae have a single looped, S-shaped intestine; the redbreast sunfish has massive gut coils by 7 mm TL.[55]

Warmouth vs. bluegill and redear sunfish — Lake Orange, FL. At 4–5 mm TL, supra-anal pigment spot is not as prominent on warmouth as on bluegill; chromatophores on dorsum of head appear slightly later on warmouth (about 7 mm) than on redear (about 5–6 mm) and earlier than on bluegill (rarely before 8 mm TL); after about 6 mm TL, warmouth are distinguishable from bluegill and redear sunfish by mouth size (extending past the anterior edge of the orbit in warmouth, but anterior to the eye in the other two); by about 14–15 mm TL, warmouth can be distinguished from redear and

bluegill by the relative shapes of their air bladders (Figure 47). Also, by that size, gill rakers have attained adult morphologies and the redear sunfish may be distinguished from the warmouth by its short and stubby rakers (Figure 47).[61]

Warmouth (Plate 1E) and bluegill (Plate 2A) YOY have similar vertical banding laterally,[49,51] but warmouth are more darkly pigmented than other small *Lepomis*[51] and have shorter pectoral fins and larger mouths than bluegill. Warmouth young are also similar to rock bass but differ in having 3 (vs. 5 or 6) anal fin spines and dark bands of pigment radiating posteriorly from the eye.[49,51]

ECOLOGY OF EARLY LIFE PHASES

Occurrence and Distribution (Figure 48)

Eggs. Incubating eggs are readily affected by adverse weather conditions. Sudden drops in water temperature promote rapid growth of fungi that infect the eggs; often, entire nests of eggs are destroyed in the early spawning season as a result of low water temperatures and fungi.[40]

Predation is another concern. Minnows and sunfishes have been observed destroying eggs and

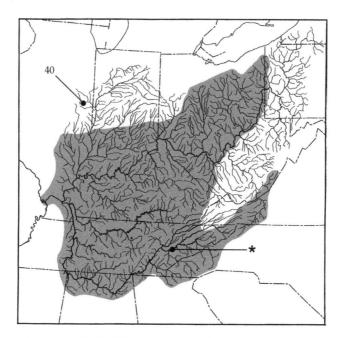

Figure 48   General distribution of warmouth in the Ohio River system (shaded area) and areas where early life history information has been collected. Numbers indicate appropriate references. An asterisk indicates TVA collection location.

larvae in unprotected warmouth nests. In laboratory aquaria, warmouth were seen robbing poorly guarded warmouth nests; they would charge into the nest snapping up eggs or larvae.[40]

Yolk-sac larvae. Temperature changes or temperature extremes, disease and predation, and dependence on the constant protection of a parent fish, which may at the same time be exposed to many dangers, result in great losses during the period of early development.[40]

Activities of warmouth larvae during their early life in the nest were limited to a few feeble movements. The following behavior of warmouth larvae kept in aquaria with water temperatures at 24–25°C is reported: immediately upon hatching the delicate yolk-sac larvae dropped down onto the sand and silt between coarse gravel on the bottom of the nest; heavy yolk restricted movement of the yolk-sac larvae and they were hard to see in the nest; at about 4.6 mm TL (1.5–2 days old), the larvae began making feeble jumps 25 mm or so above the bottom of the nest.[40]

In the laboratory, the yolk of warmouth larvae was usually exhausted within 4 days of hatching.[40]

Post yolk-sac larvae. After warmouth larvae leave the nest, they receive no parental care. In ponds and lakes, young warmouth scatter into dense vegetation, and thus it becomes impossible for the male parent to keep the young together for close care. At any rate, the male warmouth seems to lack

the drive to care for the free-swimming young; in a laboratory aquarium without vegetation in which the young could hide, warmouth males showed little interest in the fry after they left the nest.[40]

Although most yolk was absorbed by about 5.3 mm (4 days), warmouth larvae in aquaria did not begin active swimming until the end of their fifth day at about 7.6 mm TL. At this time they swam about the nest in rather compact groups; their movements were well controlled and they showed remarkable ability to avoid dip nets. School formation by post yolk-sac larvae in natural habitats was not observed. At this stage the warmouth remained either among dense submerged vegetation or in small pockets of open water surrounded by plants. Schools in the wild gradually dissolve as individuals begin independent searches for food.[40]

In Orange Lake, FL, warmouth larvae ≤8 mm TL were captured in *Hydrilla* and panic grasses, most often in panic grass habitat.[23,61] Occurrence of warmouth larvae <8 mm in collections peaked in July and August at water temperatures near 30°C.[23] The use of floating-emergent vegetation habitat increased with increasing fish size;[23] larger juvenile warmouths (>17 mm TL) predominated in the floating-emergent habitat.[61]

Larval warmouth were collected with meter and half-meter ichthyoplankton nets in early June from three habitats of the lower Mississippi River, LA. Relative abundance in samples was low, <2/100 m³. The three habitats included a tributary bayou with zero current, a river station with sluggish surface currents (5–20 cm/s), and a river station with swift surface currents (75–90 cm/s).[20]

Post yolk-sac larval warmouth, which have left the nest, are eaten in great numbers by larger fish. In Venard Lake, IL, bass about 45 mm TL fed voraciously on warmouths that were 19 mm TL. In the laboratory, warmouth juveniles 19 mm TL ate smaller warmouth larvae; a juvenile warmouth, in 5 min, ate 12 4-day-old warmouth (about 5.3 mm) larvae.[40]

Juveniles. Survival of small warmouth is closely related to the density and composition of the fish population, the time of the year, and the character of the habitat in which they are produced. Fry hatched late in the spawning season are in a population with a larger number of potential predators than are fry produced earlier. However, because the density of aquatic vegetation increases in June and July, survival in late summer broods is frequently higher than in earlier broods.[40]

Age-0 warmouth generally remain in littoral habitats.[24] Small warmouth (<127 mm TL) do not leave protected hiding places in shallow water close to the bank, even during winter, remaining in shallow vegetated areas or other dense cover for food and protection.[40] They are also found in great numbers in ripraping.[40] In IL, juvenile warmouths were never observed in large schools.[40]

In a 2-year study of the microhabitat (vegetation type, water depth, and substrate type) use of YOY of five coexisting sunfishes in Lake Rush, OK, use of water depth effectively separated warmouth from redear, longear, and green sunfish in 1980; but depth distributions of warmouth, redear, and longear were more similar the following year. The young sunfishes were collected with unbaited minnow traps; sizes ranged up to 80 mm TL. Young warmouth were caught in water depths up to 200 cm but were more frequently collected from depths >100 cm. Young bluegill, redear, longear, and green sunfish were caught proportionately more often in traps set at depths of 50 cm or less. Warmouth were more often caught over fine substrates, usually from habitats containing coontail and American lotus.[22]

Recruitment and subsequent abundance of warmouth YOY appeared unaffected following the introduction of alewife into an oligotrophic reservoir in NC.[9]

In open water areas of lakes, warmouth fry may be preyed upon heavily by small bass.[18]

Early Growth

Warmouth growth is faster in southern populations compared to northern populations, a relationship which might be expected on the basis of lengths of growing seasons.[52] Additional reports that demonstrate this relationship follow.

In WI, for combined collections of warmouth, calculated TL at each annulus was 37 mm at age 1, 74 mm at age 2, 90 mm at age 3, and 126 mm at age 4.[42]

In IA, average calculated TL at age 1 was 40.6 mm; at age 2, 91.4 mm; at age 3, 147.3 mm; at age 4, 177.8 mm.[3] The average TL (mm) of warmouth ages 2–4 captured in IA[3] compared to those captured in IN[4] and TN[5] show consistently larger individuals in TN populations, followed by IN and IA populations:

	Age		
	2	3	4
Red Haw Lake, IA	132	165	193
Muskellunge Lake, IN	140	173	183
Reelfoot Lake, TN	157	178	196

In Lake Chautauqua, IL, growth was slower with observed average TL of warmouth ages 2–4 reported at 119 mm, 137 mm, and 160 mm, respectively (Table 44).[21] Average calculated lengths of warmouth in Lake Chautauqua, IL, during the period 1950–1959 were 38 mm at age 1, 99 mm at age 2, 140 mm at age 3, and 163 mm at age 4.[21] Comparable growth was calculated for warmouth in Park Pond, IL (Table 45).[40] Early growth of warmouth from 13 water bodies in IL, during the period 1941–1953, ranged from 33 to 53 mm at age 1, 61–122 mm at age 2, 89–158 mm at age 3, and 112–180 mm at age 4.[40]

Average calculated TL for warmouth from a KY stream was about 50 mm at age 1, 79 mm at age 2,

Table 44

Average observed lengths and weights of young warmouth related to age in years at Lake Chautaugua, IL, during the period 1950–1959.

Age	No. of Fish	Average	
		TL (mm)	Weight (g)
2	9	119	36
3	65	137	59
4	74	160	100

Source: Constructed from data presented in Table 11 (in part), reference 21.

Table 45

Average calculated TL related to weight for each year of life of 1,063 warmouth collected from Park Pond, IL, during the period 1948–1949.[40]

Age	TL (mm)	Weight (g)
1	41.7	1.8
2	85.9	11.8
3	124.7	39.9
4	162.6	90.7

Source: Constructed from data presented in Table 22 (in part), reference 40.

Table 46

Age and growth determinations for young warmouth collected from three KY lakes in 1983.

| Lake | Mean TL (mm) at Age: | | | |
	1	2	3	4
Briggs Lake	51	112	155	170
Nolin River Lake	33	71	109	152
Shanty Hollow Lake	43	81	147	196

Source: Constructed from data presented in Tables 36, 61, and 70, reference 27.

Table 47

Central 50% and range of mean of means of TL (mm) for young warmouth related to age as summarized by Carlander (1977) from numerous reports from various locations.[52]

| Age (years) | No. of Fish | Mean TL | |
		Central 50%	Range
1	1,364+	74–127	51–150
2	789+	120–147	71–206
3	780+	132–168	91–198
4	542+	154–188	114–234

Table 48

A summary of calculated growth data for young warmouth from a variety of habitats in OK.[53]

| | Average TL (mm) at Age: | | | |
	1	2	3	4
Slowest growth	38	76	117	132
Fastest growth	155	203	234	246
15 reservoirs	84	132	155	185
8 large lakes	86	140	170	188
33 small lakes	84	130	157	175
8 ponds	84	130	165	188
4 streams	79	132	170	196

108 mm at age 3, and 134 mm at age 4.[13] Growth was variable in three KY lakes (Table 46), but by age 3, warmouth in the lakes grew to greater lengths than did those from the stream.

Young warmouth grew faster in the Okefenokee Swamp than in the Suwannee River, GA. Average calculated TLs of Suwannee River warmouth ages 1–4 were 52, 73, 105, and 132 mm. Okefenokee Swamp warmouth were calculated to be 54, 90, 127, and 154 mm at ages 1–4.[19]

The reports above from KY[13] and GA[19] suggest that growth of warmouth depends on habitat conditions[3,6] and reported rates of annual growth in TL are highly variable (Tables 47 and 48),[52] yet in OK, there seemed to be little difference in the average growth rates of warmouth from different types of water (Table 48).[53]

In OH, YOY in October were 20–51 mm TL; at about 1 year, 38–81 mm.[37]

In a TN laboratory, the following growth was observed for larvae that hatched on 6/30/1979 and were cultured in aquaria and fed brine shrimp:*

Date	TL (mm)
7/2	3–4
7/3	4–5
7/14	5–6
7/30	5–8
8/2	8–10
8/13	9–12
8/22	13–14
9/10	17

Table 49

Average observed lengths and weights of young warmouth from Red Haw Lake, IA.[3]

TL (mm)	N	Weight (g)
56	1	4
64	2	4
84	1	10
94	2	16
104	2	22
117	3	31
132	7	48
142	8	57
157	16	79
168	9	101
180	8	120
193	11	143
203	6	172
221	1	264

Table 50

Statewide average growth of young warmouth recorded in OK waters for the period 1952–1963.

Age (years)	Average Growth	
	TL (mm)	Weight (g)
1	84	11
2	132	48
3	160	88
4	183	135

Source: Constructed from data presented in Table 15 (in part), reference 28.

Table 51

Average empirical weight related to TL of young warmouth collected from AL waters during the period 1949–1964.

Mean TL (mm)	No. of Fish	Average Weight (g)
25	4,704	1.3
51	2,068	2.1
76	572	9.1
102	685	23.6
127	300	35.9
152	149	77.2
178	64	131.7
203	21	199.8

Source: Constructed from data presented in unnumbered table on page 75, reference 29.

A comparison of observed average lengths and weights of warmouth from IL (Table 44), IA (Table 49), OK (Table 50), and AL (Table 51) suggests comparable growth in weight related to length for young warmouth among populations. However, by 178 mm TL, warmouth in AL are heavier than warmouth from OK or IA.

The relationship of SL in mm (L) to weight in grams (W) was expressed for 866 Park Pond, IL, warmouth by the equation $\log W = -4.49867 + 3.04902 \log L$.[40]

There appears to be little difference in rate of growth between sexes of warmouth.[3,5,40] No difference in growth between sexes was observed in IA.[3] In each year-class in Park Pond, IL, young warmouth males were larger than the females, though actual differences in lengths were rather small.[40] Lengths and weights of male and female warmouth from Reelfoot Lake, TN, were either the same or only slightly different.[5]

In IL, warmouth year-classes that were largest at the end of the first year of life increased this length advantage in the second year. Warmouths that were smallest at the end of the first year showed no compensatory growth in the second year and only slight compensatory growth in the third year.[40] Fish from 84–107 mm showed wide seasonal variation in average condition probably because of their dependence on cladocerans and certain insects that varied in abundance. Larger warmouth generally had a decline in condition in early fall and again in winter. No consistent differences in condition were evident between males and females.[40]

Feeding Habits

Larval warmouth feed on small crustaceans (copepods, ostracods, cladocerans), which continue to be important for juveniles. With increasing size, warmouth include other crustaceans and aquatic insect larvae such as dragonfly nymphs and dipteran larvae in their diet.[36,40,49,50] As they increase in size, warmouth feed more on crayfish and small fish.[49] These prey items indicate a bottom-oriented feeding behavior. Terrestrial insects as prey or other evidence of surface feeding was not found in the literature.[51]

In a piedmont lake in NC, age-0 warmouths 13.5–36.0 mm SL ate primarily copepods, mostly adult *Cyclops* and *Diaptomus*. The percent frequency of food items found in the warmouth stomachs was as follows: copepods 55% (*Cyclops* 43% and *Diaptomus* 9%); cladocerans 8%; ostracods 7%; rotifers 1%; insect larvae 20%; and fishes 9%.[60]

In Park Pond and Venard Lake, IL, cladocerans, copepods, and ostracods were often consumed by warmouth <89 mm TL and were the principal foods of warmouth <43 mm, which also occasionally took small mayfly nymphs and dipteran larvae. In the pond, amphipods were an important food for small warmouth but were seldom eaten by fish >125 mm. Snails were usually found in the stomachs of warmouth 64–125 mm. In Park Pond, crayfish were consumed more often by large warmouth than by smaller warmouth. However, in Venard Lake, crayfish were eaten by more small warmouth than large ones, perhaps because large numbers of small crayfish were available to the fish there during the summer months. In both study areas, mayfly nymphs were largely consumed by small warmouth, but in Park Pond, in the summer, 30% of warmouth >127 mm ate them. In Venard Lake, where *Caenis* and *Siphlonurus* occurred in large numbers, the nymphs of *Caenis* were eaten mostly by warmouth 51–86 mm; *Siphlonurus*, which were larger, were eaten most often by larger warmouth, up to 132 mm. Caddisfly larvae were often consumed by small warmouth. Damselfly and dragonfly nymphs were also used as food by warmouth of all sizes except those <51 mm. Warmouth >127 mm more often ate other fish than did smaller ones.[40]

Warmouth from IA, ranging in length from 40 to 177 mm, consumed small fish (50–100 mm), crayfish, vegetable debris, insect larvae, leeches, dragonfly naiad, insects, and snails. Fish and crayfish were the most important items by volume.[3]

Food habits determined for warmouth from the Suwannee River and Okefenokee Swamp are reported for the following length intervals:

25–76 mm TL. Warmouth this size, from both habitats, fed mainly on insects, primarily larval odonates and dipterans. Fish were found in 14% of the Suwannee River stomachs examined, but no fish were found in the stomachs of the swamp warmouths. Crustaceans were consumed in both habitats and were second in frequency of occurrence in the swamp warmouth stomachs.[19]

77–127 mm TL. Insects were the main food items from both habitats. Odonates and coleopterans were frequently encountered in the stomachs of warmouth from both localities; trichopterans were also important in the stomachs of warmouth from the river. Crustaceans (crayfish and freshwater shrimp) were found in 31% of the river stomachs and in 45% of the swamp stomachs. Fish occurred in 17% of the river stomachs and in 9% of the swamp stomachs.[19]

128–178 mm TL. A change in the feeding habits of warmouth from the two habitats occurred at this size interval. Insects, primarily larval odonates and coleopterans, remained the principal food of swamp warmouth, while crustaceans were the main food for the river fish. Crayfish was the most common item found in the stomachs of river warmouth and fish occurrence also increased (35% of stomachs examined). Crustaceans were present in 45% of the swamp stomachs and fish were found in 27% of the stomachs from swamp warmouth.[19]

179–229 mm TL. Crayfish was the most important food item in this size range from both habitats, followed by insects, primarily odonates, and fish. The occurrence of fish in stomachs decreased, amounting to about 14% from both habitats.[19]

After yolk absorption in aquaria, without food, larval warmouths starved to death in 10–11 days at 24–25°C. Ordinarily, post yolk-sac larval warmouths began feeding by at least the 7th day of life. Stomachs of larvae collected from outdoor tanks contained flagellates, ciliates, and many bacteria. By 14 days old, warmouth larvae ate considerably larger organisms, including small mosquito larvae. Both in natural waters and in the laboratory, warmouth 19 mm long were observed feeding on smaller warmouth larvae; a juvenile warmouth, in 5 min, ate 12 4-day-old warmouth (about 5.3 mm) larvae.[40]

LITERATURE CITED

1. Lee, D.S. et al. 1980.
2. Panek, F.M. and C.R. Cofield. 1978.
3. Lewis, W.M. and T.S. English. 1949.
4. Ricker, W.E. 1945.
5. Schoffman, R.J. 1940.
6. Crossman, E.J. et al. 1996.
7. West, J.L. and F.E. Hester. 1966.
8. Carver, D.C. 1967.
9. Crutchfield, J.U., Jr. et al. 2003.
10. Thomas, J.A. et al. 2004.
11. Raney, E.C. 1965.
12. Hendricks, M.L. et al. 1983.
13. Redmon, W.L. and L.A. Krumholz. 1978.
14. Houser, A. 1963.

15. Childers, W.F. 1967.
16. Armstrong, J.G. III. 1969.
17. Merriner, J.V. 1971.
18. Illinois Natural History Survey. 1971.
19. Germann, J.F et al. 1975.
20. Gallagher, R.P. and J.V. Conner. 1983.
21. Starrett, W.C. and A.W. Fritz. 1965.
22. Layzer, J.B. and M.D. Clady. 1991.
23. Conrow, R. et al. 1990.
24. Meals, K.O. and L.E. Miranda. 1991.
25. Cross, F.B. 1967.
26. Burr, B.M and M.L. Warren, Jr. 1986.
A 27. Axon, J.R. 1984.
A 28. Houser, A. and M.G. Bross. 1963.
A 29. Swingle, W.E. 1965.
30. Pflieger, W.L. 1975b.
31. Smith, P.W. 1979.
32. Pearson, W.D. and L.A. Krumholz. 1984.
33. Pearson, W.D. and B.J. Pearson. 1989.
34. Robison, H.W. and T.M. Buchanan. 1988.
35. Mettee, M.F. et al. 2001.
36. Jenkins, R.E. and N.M. Burkhead. 1994.
37. Trautman, M.B. 1981.
38. Smith, C.L. 1985.
39. Menhinick, E.F. 1991.
40. Larimore, R.W. 1957.
41. Forbes, S.A. and R.E. Richardson. 1920.
42. Becker, G.C. 1983.
43. Baker, J.A. et al. 1991.
44. Gammon, J.R. 1998.
45. Wang, J.C.S. and R.J. Kernehan. 1979.
46. Hardy, J.D., Jr. 1978.

47. Carr, A.F., Jr. 1939.
48. Childers, W.F. 1965.
49. Boschung, H.T., Jr. and R.L. Mayden. 2004.
50. Ross, S.T. 2001.
51. Etnier, D.A. and W.C. Starnes. 1993.
52. Carlander, K.D. 1977.
A 53. Jenkins, R. et al. 1955.
54. Conner, J.V. 1979.
55. Taubert, B.D. 1977.
56. Siefert, R.E. 1969b.
57. Tin, H.T. 1982.
A 58. EA Engineering, Science, and Technology. 2004.
59. Mabee, P.M. 1995.
60. Lemly, A.D. and J.F. Dimmick. 1982.
61. Conrow, R. and A.V. Zale. 1985.
62. Bettoli, P.W. et al. 1993.
A 63. McGowan, E.G. 1984.

* Original information on the reproductive biology of warmouth comes from brood stock collected on 6/14/1979 from Ft. Loudoun Reservoir, Tennessee River, TN, that was placed in holding tanks at TVA's fisheries laboratory and allowed to spawn naturally. Original descriptive information of larval development comes from the progeny of these natural spawnings. Specimens used for developmental descriptions are in a lot catalogued as TV2353, DS–34. This material is curated at the Division of Fishes, Aquatic Research Center, Indiana Biological Survey, Bloomington, IN.

ORANGESPOTTED SUNFISH

Lepomis (Allotis) humilis (Girard)

Thomas P. Simon

Lepomis: Greek *lepis* "scale" and *poma* "lid" for the scaled operculum; *humilis*: Latin meaning low or humble.

RANGE

The orangespotted sunfish occurs throughout the Mississippi River valley from the southern Great Plains east to the Ohio River basin and south to the Gulf Slope from Pearl River, MS, west to Rio Grande River, TX, and east from OH to ND. Possibly non-native to coastal drainages east of Mississippi; introduced into Mobile Basin;[1–10,21,22] recently collected from the Coosa River, AL.[15]

HABITAT AND MOVEMENT

Orangespotted sunfish inhabits quiet streams and vegetated lakes and ponds.[1] It is found in creeks, small rivers, ponds, large lakes, and lowland lakes;[1–3] often occurs in turbid, sluggish waters[5,18,20,39] and thrives in soft-bottomed pools of streams and in reservoirs.[9] In streams, this species is associated with low or intermittent flow; less common in headwater streams.[4,41,42] It is often abundant in association with aquatic vegetation in backwater sloughs[6,22] and is sometimes found along the vegetated shores of impoundments.[4,9] Orangespotted sunfish is rarely found in pools of the lower Mississippi River, but is abundant in borrow pits and commonly collected in sloughs, oxbow lakes, and seasonal floodplain waters.[48]

Orangespotted sunfish occurs over a variety of substrates including sand, gravel, mud,[14,22] and organic debris.*,[4,5] In creeks or streams and small rivers, it inhabits pools with sand,[4,9] silt,[2–4,9] or debris[2,3] substrate, especially where there is brush cover.[4,9]

Orangespotted sunfish is very tolerant of low flows, warm water, and silty conditions, perhaps more so than any other species of sunfish.[9] Individuals are reported from salinities up to 0.74 ppt.[1] It is tolerant of wide range of pH conditions, and reported to spawn in water with pH as high as 9.3.[26]

Indices of hypoxia and hypothermia tolerance for MO fish assemblages were determined based on laboratory measurements of lethal dissolved oxygen concentrations and temperatures combined with field measures of relative abundances of tolerant and sensitive species. Mean hypoxia tolerance for orangespotted sunfish was 0.62 mg/l dissolved oxygen; mean hypothermia tolerance was 36.4°C.[50,51]

Lateral movement occurs seasonally between floodplain and channel, and floodplain waters are used by juveniles as nursery habitat.[16]

DISTRIBUTION AND OCCURRENCE IN THE OHIO RIVER SYSTEM

The orangespotted sunfish occurs throughout much of the Ohio River basin including tributaries of the Cumberland and Tennessee river drainages.[1–10] Its native range was restricted in the Mississippi drainage and it was hypothesized that this confinement was first broken in 1929 when specimens were discovered in Lake St. Mary, OH.[22] In IL, the species is found from smaller rivers, and in lakes and ponds in the bottomland.[2] It is ubiquitous in the Wabash River drainage, where it has increased its range as a result of habitat alteration (stream channelization).[23] Species expansion occurred when agricultural practices transformed clear prairie-type streams to turbid plains-type streams.[22,23] The species range has apparently shrunk in some states as a result of prairie drainage, dredging, and reservoir construction.[3,21,24,25]

In KY, orangespotted sunfish are occasional and common on the Coastal Plain, but sporadic and rare throughout the rest of the state.[27] It occurs in mainstream impoundments of the Cumberland and Tennessee Rivers.[5] In AL, the species is native to the Tennessee River drainage.[9]

SPAWNING

Location

Orangespotted sunfish often build nests close together in open near-shore water in colonies of up to 1,000 nests.[11] Nests are usually constructed over sand or gravel, but sometimes the male will clear an area of silt to a depth of 75–100 mm to reach firm substrate of gravel or hard mud.[9,11,17,21,34] Nests are constructed at depths of 46 cm,[35] 30–91 cm,[11] 10–61 cm,[21] or less than 90 cm;[36] occasionally the spawning pair may breed in water that barely covers their backs.[36] Nests are circular depressions 150–160 mm in diameter and 30–40 mm deep;[7,9,11,17] or spawning may occur without nest construction.[11]

Season

Spawning occurs during the spring and extends until late into the summer.[5,9] In AL, spawning begins in April[9] or May[6] and continues until August or September.[6,9] Breeding males and ripe females were collected from IL in early June.[2] Spawning occurs from May to August in MO[4] and OK[30,37] and from April to September in LA,[11] IA,[11] and MS.[7,12,29] Females with ripe ova were observed in August in TN[5] and in late May to August in WI.[10] Gonad examination confirms a peak spawning season in May and early June.[31,32] The extended spawning period occurs because some younger fish attain sexual maturity late in the season.[9] Younger fish nest later in the season and geographical location and weather conditions also modify season length.[11]

Temperature

Spawning begins at 18°C,[9,11] 25°C,[30] or occurs from 18°C, continuing to 24–32°C.[10]

Fecundity

Fecundity of orangespotted sunfish as small as 30 mm TL is 175 eggs, while females 105 mm TL produce up to 4,700 eggs.[9,11,17] In IN, fecundity ranges from 200–3,500 eggs depending on the size of the female.* An OK female 108 mm TL had 5,000 eggs. In WI, females had between 15,000 and 58,000 eggs per female.[10,38] Increasing fecundity is proportional to increasing female length and age.*,[11]

Sexual Maturity

Specimens born early in April were sexually mature by late fall and possessed ripe ova the following August.[9] Usually, all individuals are sexually mature by age 2.[9,11] Specimens >30 mm TL were sexually mature,[9,11] or typically 60 mm TL.[5,12]

Spawning Act

Orangespotted sunfish is classified as a nest spawning lithophilic fish that guards its young.[33] Breeding males are brilliantly colored; the belly, anal and pelvic fins are bright orange; entire margin of dorsal fin orange.[9] Courtship and spawning behaviors are typical of the genus,[9] and spawning usually occurs soon after nest completion.[6,22] The male is aggressive in guarding the nest, but generally defends a territory that reaches only to the edge of the nest.[9] He defends the nest territory from all but the largest intruders.[4,9]

In preparation for spawning, a male orangespotted sunfish fans out a depression in the substrate in shallow water using powerful sweeping motions of his tail, while almost vertical above the nest area.[14] Males actively court females by rushing out toward them and then returning rapidly to the nest, all the while producing a series of distinctive grunting noises.[13] The female when ready to spawn will change eye color from red to almost black.[14] The pair adopt a parallel position over the nest, whereupon eggs and sperm are released; usually the female will release 6–15 eggs per spawning event, although the pair may repeat spawning behavior numerous times in succession.[11,14]

Males usually spawn with a single female per event; however, eggs from several females are known from a single nest.* After spawning is completed, the male orangespotted sunfish chases the female away and then will ready himself to attract another ripe female.*,[11]

The fertilized eggs sink to the bottom and attach to stones on the bottom of the nest; the male stands guard over the nest while the eggs are developing, fanning the eggs to keep them oxygenated and free of silt.[11]

Red shiner and redfin shiner are known to quickly invade and spawn in orangespotted sunfish nests.[4] The male orangespotted sunfish then affords protection throughout the larval development of both his progeny and that of the invader species.[4]

EGGS

Description

Fertilized eggs average 1.0 mm[11] or 0.5 mm in diameter.[10] Eggs are demersal, spherical, adhesive,[11] and translucent with a narrow perivitelline space and an unsculptured and unpigmented chorion.* An oil globule is located near the surface of the yolk.*,[10] Yolk coloration ranges from yellow to pale translucent amber.*,[10,11]

Incubation

Fertilized eggs incubated in water 18–21°C hatched in 120 h.[7,9,11]

Development

No information.

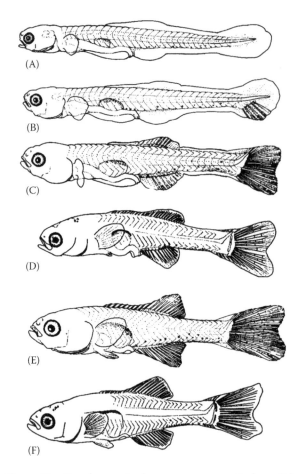

Figure 49 Development of young orangespotted sunfish. (A) Yolk-sac larva, 5.75 mm TL. B–E. Post yolk-sac larvae: (B) 6.85 mm TL; (C) 8.0 mm TL; (D) 9.6 mm TL; (E) 11.4 mm TL. (F) Early juvenile, 13.2 mm TL. (A–B, D and F reprinted from Figures 2 through 5, reference 44; C and E reprinted from Figures 4 and 5, reference 43, with author's permission.)

YOLK-SAC LARVAE

See Figure 49.

Size Range
Hatching occurs at 4.0[44] or 5.3 mm TL;[40] yolk sac absorbed by 6.85–7.6[44] or by 7.9 mm TL.[40]

Myomeres
Predorsal 5–6;* preanal 13–14 (11–15);[43,44] postanal 14–17;[43,44] and total 27–31.[43,44]

Morphology
4.0 mm TL (newly hatched). Body laterally compressed, yolk sac moderate, oval; yolk with a single oil globule; head slightly deflected over the yolk sac; eyes spherical.*,[40,43,44]

5.3–6.5 mm TL. Swim bladder distended with air; yolk sac noticeably decreased; cloaca open externally and feeding initiated.[40,43,44]

Morphometry
See Table 52.

4.0–7.6 mm TL. Swim bladder ovoid and situated posteriorly.[44]

Fin Development
4.0 mm TL (newly hatched). Pectoral fin buds developed.[40,43,44]

5.3–6.5 mm TL. Pectoral fin developed, without incipient rays; dorsal and anal finfolds continuous.*

Pigmentation
4.0–5.2 mm TL. Yolk yellow to amber otherwise larvae unpigmented.[22,33]

5.3–6.5 mm TL. Scattered chromatophores evident on the dorsal surface of the air bladder.[43,44]

6.6–7.0 mm TL. Dorsal surface of air bladder is densely pigmented. A distinct pair of supraanal melanophores is present. A row of stellate melanophores is present along the myosepta of the lower lateral areas, from myomeres 18–25. Some specimens have 1–3 melanophores on the breast region.[44]

POST YOLK-SAC LARVAE

See Figure 49.

Size Range
Beginning at lengths 6.85 mm TL[44] to >10.0 mm TL.[44]

Myomeres
Predorsal 5–6;* preanal 13–14 (11–15);[43,44] postanal 14–17;[43,44] and total 27–31.[43,44]

Morphology
6.85–7.6 mm TL. Yolk sac completely absorbed; foregut thickened, but remains straight; the hindgut forms a sigmoid curvature corresponding to deepening of the air bladder; eye round and small (<5.5% TL).[44]

7.9–9.3 mm TL. Mouth small, rudimentary maxilla reaches beyond the anterior margin of the eye. Head has developed various cephalic acoustico-lateralis features; preoperculomandibular canal disjunct from the lateral line and supraorbital pores distinctly enlarged.[19,40,43,44]

9.6–10.0 mm TL. Foregut coiling; air bladder deepened considerably to form a heart-shaped chamber;

Table 52

Morphometric and meristic data for young orangespotted sunfish grouped by selected intervals of total length (N = sample size).[*,43,44]

	Total Length Intervals (mm)							
	4.0–5.9 (N = 123) Mean + SE	6.0–7.9 (N = 114) Mean + SE	8.0–9.9 (N = 64) Mean + SE	10.0–13.9 (N = 84) Mean + SE	14.0–17.9 (N = 55) Mean + SE	18.0–22.9 (N = 27) Mean + SE	23.0–27.9 (N = 16) Mean + SE
Length (% of TL)							
Snout	1.8 ± 0.52	2.5 ± 0.33	4.28 ± 0.35	5.7 ± 0.40	6.3 ± 0.52	6.7 ± 0.66	7.2 ± 0.40
Eye diameter	5.3 ± 0.31	4.96 ± 0.10	5.53 ± 0.11	6.9 ± 0.48	7.7 ± 0.60	8.4 ± 0.78	8.7 ± 0.66
Head	17.6 ± 1.22	18.2 ± 1.22	22.9 ± 0.98	24.0 ± 1.37	24.7 ± 1.7	26.1 ± 2.1	28.2 ± 2.3
Preanal	46.2 ± 2.10	46.6 ± 2.08	45.6 ± 2.09	43.9 ± 2.46	42.9 ± 0.5	43.4 ± 0.9	44.2 ± 0.6
Postanal	53.8 ± 2.10	53.4 ± 2.08	54.4 ± 2.09	56.1 ± 2.46	57.1 ± 0.5	56.6 ± 0.9	55.8 ± 0.6
Standard	97.2 ± 3.56	95.6 ± 3.93	88.7 ± 6.15	82.2 ± 4.42	79.1 ± 6.5	78.4 ± 8.77	78.4 ± 2.1
Yolk sac	15.9 ± 4.88						
Fin Length (% of TL)							
Pectoral	6.5 ± 0.23	8.1 ± 0.40	12.1 ± 0.98	9.4 ± 1.3	12.8 ± 1.2	13.1 ± 1.7	13.4 ± 2.0
Caudal	2.8 ± 0.44	4.4 ± 0.51	11.3 ± 0.87	17.8 ± 0.76	20.1 ± 2.3	21.6 ± 1.3	21.6 ± 1.3
Body depth (% of TL)							
Head at eyes	7.01 ± 0.45	13.7 ± 0.43	13.7 ± 0.41	16.0 ± 0.39	17.2 ± 0.42	17.6 ± 0.7	18.1 ± 0.6
Preanal	8.1 ± 0.35	9.5 ± 0.31	12.3 ± 0.40	16.3 ± 0.43	19.1 ± 0.45	20.8 ± 0.8	23.1 ± 2.1
Caudal peduncle	4.1 ± 0.13	6.5 ± 0.28	7.8 ± 0.33	8.9 ± 0.24	9.7 ± 0.33	9.8 ± 0.3	10.3 ± 0.5
Myomere Number							
Preanal	13.5 ± 0.5	14.2 ± 0.8	14.4 ± 1.0	14.3 ± 0.9	14.4 ± 0.8	14.3 ± 0.8	14.4 ± 0.8
Postanal	16.2 ± 1.2	16.3 ± 2.1	16.4 ± 1.1	16.2 ± 1.4	16.2 ± 1.1	16.4 ± 1.3	15.8 ± 1.2
Total	29.3 ± 2.3	29.3 ± 2.4	30.2 ± 1.0	30.3 ± 0.9	29.6 ± 0.7	29.6 ± 0.8	29.6 ± 0.8

hindgut sigmoid; mouthparts complete and maxilla reaching beyond the anterior edge of eye.[40,43,44]

Morphometry
See Table 52.

Fin Development
6.85–7.6 mm TL. All finfolds present; fin rays formed in anal and caudal fins; no pelvic bud formed; notochord flexion occurred.[40,43,44]

7.9–9.3 mm TL. Fin rays formed in spinous and soft dorsal, anal, and caudal fins.[40,43,44]

9.6–10.0 mm TL. Fin rays formed in all paired and median fins with the exception of the pelvic fin; pelvic fin bud formed; dorsal and anal finfold partially absorbed.[40,43,44]

Pigmentation
6.85–7.6 mm TL. Sparsely pigmented but postanal pigment is consistently present on myosepta between myomeres 18–27.[40,43,44]

7.9–9.3 mm TL. Melanophores are scattered from nape to dorsal fin origin and concentrated over dorsum of air bladder. A series of melanophores forms mid-ventrally from anal fin insertion to caudal peduncle base.[40,43,44]

9.6–10.0 mm TL. Pigmentation is present on dorsal wall of the air bladder and in the postanal myosepta. Melanophores are present at the base of the hypural plate on some individuals. A series of melanophores is evident in the vertical intermuscular septum of the lower caudal peduncle.[40,43,44]

JUVENILES

See Plate 1F following page 26 and Figures 49 and 50.

Size Range
Juvenile stages 10.7 mm TL[43,44] to 34–60 mm TL.[5,40,43,44]

Myomeres
Preanal 13–14; postanal 15–17; total 29–31.[43,44]

Morphology
10.7–17.0 mm TL. Subtle flattening of the profile of the infraorbital area.[40,43,44]

>30–50 mm TL. Mouth large in size, maxilla reaching to or just beyond anterior margin of eye; dorsal section of air bladder extended forward, almost reaching the wall of the branchial chamber; opercle stiff to its posterior margin. Lateral scale rows

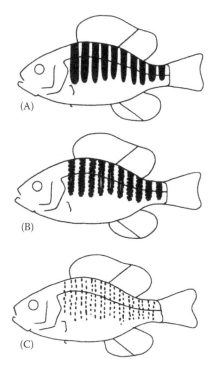

Figure 50 Development of young orangespotted sunfish — schematic drawings showing ontogeny of melanistic pigment patterns. (A) Juvenile male and females 16–21 mm SL; (B) juvenile female 23.8 mm SL; (C) adult female 43.9 mm SL. (A–C reprinted from Figure 13, reference 45, with publisher's permission.)

32–42;[2,5,9,10] gill rakers 10–15, length 5 times greatest width; palatine teeth present.[2,5,9,10]

Morphometry
See Table 52.

Fin Development
>34–60 mm TL. Pectoral fin short and rounded, its tip far short of nares when laid forward across check.[2,5,9,10,22]

Larger juveniles. Adult complement of fin spines and rays: spinous dorsal X (IX–XI); soft dorsal rays 10–11; pectoral fin rays 14–15, usually 13–14; anal spines and rays, III and 9; pelvic fin spines and rays I and 4–5.*,[2,5,9,10,22]

Pigmentation
See Figure 50.

10.7–17 mm TL. Melanophores are evident in myosepta posterior to the anal fin, in the vertical intermuscular septum of the lower caudal peduncle, and at the base of the caudal fin rays. Postanal myosepta have more than 25 melanophores between the vent and the isthmus.[22]

>34–60 mm TL. Eye is red with a black vertical band running through the center. Iridescent blue covers the lower margins of the operculum and the

area beneath the eye. Body color is tan with a white venter; numerous brown to orange red spots randomly distributed on entire body; spinous dorsal and pectoral fins clear and without pigment. Pelvic and proximal portion of anal fin orange; distal margins of soft dorsal, anal, and caudal fins with orange red pigment.*

Orangespotted sunfish develop 9–10 primary bars of melanistic pigment. Light centers form in the primary bars of females (Figure 50A and 50B), but secondary bars do not form. As the centers of the primary bars become increasingly lighter, only the anterior and posterior margins of each bar remain visible. The anterior and posterior edges of the bars become broken as pigment is lost at approximately every other scale (Figure 50C). Through the loss of pigment in many scales in the margins of the light centered primary bars, and the intensification of the pigment in other scales, the vertical orientation of the bars is lost. Solid primary bars are retained as the adult condition of male orangespotted sunfish (Figure 50A).[45]

TAXONOMIC DIAGNOSIS OF YOUNG ORANGESPOTTED SUNFISH

Similar species: other *Lepomis* species, most similar to *L. macrochirus*.[5,43,44]

See *Allotis* subgeneric discussion on page 90.

For comparisons of important morphometric and meristic data, developmental benchmarks, and pigmentary characters between orangespotted sunfish and all other *Lepomis* that occur in the Ohio River drainage, see Tables 4 and 5 and Plate 1F.

Larvae. Early yolk-sac stages of orangespotted sunfish are separable from bluegill by a greater preanal length (≥45% TL compared to <45% TL for bluegill), number of preanal myomeres (mode = 14 compared to mode = 12–13 for bluegill), and smaller eye diameters (≤5.5% TL compared to >5.5% TL for bluegill). Coiling of the gut in post yolk-sac larval orangespotted sunfish did not occur until 9.6 mm TL; coiling of the foregut of bluegill was obvious by 7.7 mm TL. Eye diameters of orangespotted post yolk-sac larvae were still smaller (5–7% compared to 7–9% for bluegill). Early juveniles had fewer anal fin rays (9–10, mode = 9 compared to 10–11, mode = 10 for bluegill). Preoperculomandibular canals of orangespotted are disjunct from the lateral line, conjunct with lateral line of bluegill. Supraorbital canals are enlarged on juvenile orangespotted sunfish but not enlarged on bluegill. For orangespotted, melanophores are in the vertical intermuscular septum of lower caudal peduncle (i.e., "imbedded" in lateral aspect); for bluegill, melanophores of the lower caudal peduncle are superficial (do not extend up into the vertical intermuscular septum).*,[44]

Juveniles. Juvenile orangespotted sunfish may be confused with bluegill, but differ by possessing 9–10 dark vertical bands on the side compared to 7–8 for bluegill.[3,17] Orangespotted sunfish YOY (Plate 1F) have the entire caudal fin pigmented on rays and membranes. In other *Lepomis* (except green sunfish and warmouth over 30 mm SL) the caudal fin of young is pale throughout or is pigmented only distally. Young green sunfish (Plate 1C) and warmouth (Plate 1E) have a darkly pigmented breast (pale in orangespotted).[5]

ECOLOGY OF EARLY LIFE PHASES

Occurrence and Distribution (Figure 51)

Eggs. Eggs are buried in the substrate, which ranges from sand, gravel, or other fine material that the male has cleaned prior to reproduction.[9] The eggs remain in the substrate for 120 h prior to hatching.[9]

Larvae. Larval orangespotted sunfish were collected with two ichthyoplankton sampling gears in early June from the lower Mississippi River. They were present in samples collected from three stations representing swift surface currents (75–90 cm/s), sluggish surface currents (5–20 cm/s), and zero current.[53]

Juveniles. Juvenile stages occur until 34–60 mm TL. Juvenile habitat includes the floodplain swamps and ditches.[16] They are tolerant of a wide variety of conditions including low dissolved oxygen and high temperatures.[16]

Early Growth

Individual growth during the first year of life reaches 30 mm TL and growth increments are about 10–12 mm during the next 3–4 years.[5,11] Average growth of young in MO is 25 mm TL at the end of the first summer and 35, 45, and 57.5 mm TL in succeeding years.[4] Growth in TN averaged 50, 60, 75, and 85 mm TL for ages 1–4.[5]

During the period 1972–1973, calculated growth of age-1 orangespotted sunfish in 23 OK reservoirs ranged from 34 to 57 mm TL; TL of age-2 fish from 16 reservoirs ranged from 55 to 82 mm; and TL of age-3 fish from five reservoirs ranged from 79 to 106 mm TL.[46] These growth rates were comparable to statewide average growth during the period 1952–1963 (Table 53).[49]

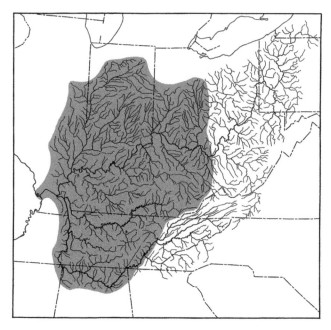

Figure 51   General distribution of orangespotted sunfish in the Ohio River system (shaded area).

Table 53

Statewide average calculated growth of young orangespotted sunfish in OK waters for the period 1952–1963.

Age in Years	Average Growth	
	TL (mm)	Weight (g)
1	53	2
2	81	10
3	99	18

Source: Constructed from data presented in Table 17 (in part), reference 49.

Table 54

Statewide length–weight relationships for orangespotted sunfish in OK.

Length (mm TL)	Weight (g)	
	1972–1973[48]	1952–1963[49]
20	0.1	0.0
25	0.2	0.1
30	0.4	0.3
35	0.7	0.5
40	1.0	0.9
45	1.5	1.3
50	2.0	1.8
55	2.7	2.5
60	3.5	3.3
65	4.4	4.4
70	5.5	5.6
75	6.7	7.0
80	8.1	8.7
85	9.7	10.6
90	11.4	12.8
95	13.4	15.2
100	15.5	18.0
105	17.9	21.1
110	20.5	24.6
115	23.3	28.5
120	26.4	32.8
125	29.7	37.4

Length–weight relationships of young orangespotted sunfish from OK were variable but similar among years (Table 54). During the period 1972–1973, weight was slightly greater related to length for fish up to about 60 mm TL when compared to the period 1952–1963. After that, the trend reversed with weight related to length of fish >70 mm TL becoming increasingly greater with growth in length for fish collected during the 1952–1963 period compared to the 1972–1973 period.[46,49] In AL, orangespotted sunfish 25 mm TL (Table 55)[47] were heavier than those from OK at the same length (Table 54), but at total lengths >50 mm orangespotted sunfish were heavier in OK.[46,47,49]

Table 55

Length and weight relationships of young orangespotted sunfish collected in AL during the period 1949–1964.

TL (mm)	No. of Fish	Average Empirical Weight (g)
25	3,500	1.2
51	3,833	1.7
76	1,303	5.0
102	125	13.2
127	10	35.0
152	6	50.0

Source: Constructed from data presented in unnumbered table on page 45, reference 47.

Feeding Habits

Orangespotted sunfish are invertivores[28] or considered insectivores.[9] This species eats drift organisms from the water column or is a benthic and surface feeder.[28] It consumes cladocerans, copepods, midge larvae, and occasionally small fish;[9,28] also reported to consume microcrustaceans and aquatic insect larvae.[28] In TN, stomachs contained chironomid larvae and terrestrial insects.[5]

In the White River, IN, 10 orangespotted sunfish stomachs were examined for food content. The most abundant food was chironomid larvae which comprised 54.5% of the volume and was eaten by 7 of the 10 fish examined. The number of chironomids per fish ranged from 4 to 98 and averaged 24.9. Other foods included *Plumatella*, Corixidae, damselfly naiads, Trochoptera larvae, unidentified invertebrates, and chironomid pupae.[52]

LITERATURE CITED

1. Lee, D.S. et al. 1980.
2. Forbes, S.A. and R.E. Richardson. 1920.
3. Smith, P.W. 1979.
4. Pflieger, W.L. 1975b.
5. Etnier, D.A. and W.C. Starnes. 1993.
6. Mettee, M.F. et al. 1996.
7. Ross, S.T. 2001.
8. Miller, R.J. and H.W. Robison. 2004.
9. Boschung, H.T., Jr. and R.L. Mayden. 2004.
10. Becker, G.C. 1983.
11. Barney, R.L. and B.J. Anson. 1923.
12. Cook, F.A. 1959.
13. Gerald, J.W. 1971.
14. Miller, H.C. 1964.
15. Sizemore, D.R. and W.M. Howell. 1990.
16. Kwak, T.J. 1988.
17. Noltie, D.B. 1989.
18. Moore, G.A. 1962.
19. Curd, M.R. 1959.
20. Branson, B.A. and G.A. Moore. 1962.
21. Cross, F.B. 1967.
22. Trautman, M.B. 1981.
23. Gerking, S.D. 1945.
24. Larimore, R.W. and P.W. Smith. 1963.
25. Smith, P.W. 1968.
26. Wiebe, A.H. 1931.
27. Burr, B.M. and M.L. Warren, Jr. 1986.
28. Goldstein, R.M. and T.P. Simon. 1999.
29. Hildebrand, S.F. and I.L. Towers. 1928.
30. Cross, F.B. 1950.
31. Stegman, J.L. 1958.
32. Stegman, J.L. 1959.
33. Simon, T.P. 1999.
34. Buss, D.G. 1974.
35. Richardson, R.E. 1913.
36. McClane, A.J. 1978.
37. Ward, H.C. 1953.
38. Eddy, S. and T. Surber. 1943.
39. Witt, L.A. 1970.
40. Auer, N.A. 1982.
41. Harrel, R.C. et al. 1967.
42. Medford, D.W. and B.A. Simco. 1971.
43. Conner, J.V. 1979.
44. Hutton, G.D. 1982.
45. Mabee, P.M. 1995.
A 46. Mense, J.B. 1976.
A 47. Swingle, W.E. 1965.
48. Baker, J.A. et al. 1991.
A 49. Houser, A. and M.G. Bross. 1963.
50. Smale, M.A. and C.F. Rabeni. 1995a.
51. Smale, M.A. and C.F. Rabeni. 1995b.
52. Whitaker, J.O., Jr. 1974.
53. Gallagher, R.P. and J.V. Conner. 1983.

* Original information on early life ecology and reproductive biology comes from data and specimens collected from Interior River Lowland Ecoregion of IN during the period 1992–1993. Specimens and observations were from a variety of sites on tributaries of the Ohio River. Measurements come from Louisiana specimens listed by Conner (1979) as *Lepomis* "B"–type specimens.

BLUEGILL

Lepomis (Helioperca) macrochirus Rafinesque

Robert Wallus and John V. Conner

Lepomis, Greek: lepis, "scale" and poma, "lid," in reference to scaly operculum; *macrochirus*, Greek: "large hand," probably in reference to body shape.

RANGE

Originally restricted to eastern and central North America where it ranged from VA to FL, west to TX and northern Mexico, and north to western MN to western NY. Widely transplanted elsewhere in North America.[1] Introductions have greatly extended the range in North America, Europe, and South Africa.[1,137] Three subspecies are recognized, but widespread introductions have resulted in extensive mixing of these gene pools.[1]

HABITAT AND MOVEMENT

Bluegill occur in a wide variety of habitats and may be expected in any water body, avoiding only extremely polluted or ephemeral aquatic habitats and high-gradient headwater streams with little pool development.[140]

Bluegill inhabit streams and lakes of all sizes.[170,176] They are most abundant in lakes and ponds[1,4,137,140,170,172,174,176,177,192,197] and are prevalent in reservoirs.[140,172,177,192]

Reported to prefer non-flowing water,[178] they are often found in oxbow lakes,[140] swamps,[172,174] and backwater sloughs.[177] This species is commonly found in pools[4,197] of streams of various sizes,[172,174,192] including tidal creeks,[176] and in the heavily vegetated, slowly flowing areas of small rivers and large creeks.[1,4,137,176]

Bluegill is commonly found in springs;[170,172] it was the second most common centrarchid captured in springs of the southern bend of the Tennessee River.[70]

Bluegill frequent shallow waters,[1,170,176,178] near-shore[176] often in areas of moderate to abundant vegetation[1,4,78,79,96,140,170,174,176,177,179,192,197] or other cover. Submerged wood,[170] rocks,[140,170] stumps,[78,79,140] or logs[140] are favored habitats where the bottom substrates may be of mixed sand,[78,79,140,192] gravel,[78,79,140,192] muck,[192] or detritus.[140] This species thrives best in warm,[1,96,176,177,179] clear,[140,174,177–179] quiet waters.[140,177,179]

It is intolerant of continuous high turbidity and siltation.[179]

Bluegill occupy most inshore habitats in Lake Opinicon, ON, including weedy shallows and weedy deep water, mixed bottom shallows, sandy shallows, gravel beds, rocky shelves, submerged stumps, and open water inshore.[78,79]

In WI, bluegill are found in quiet to moderately swift waters in streams of the following widths: 1–3 m (12% frequency); 3.1–6 m (7%); 6.1–12 m (20%); 12.1–24 m (32%); 24.1–50 m (22%); more than 50 m (6%).[177] They are encountered most frequently in clear water (occasionally in slightly turbid to turbid water) at varying depths, over substrates of sand (29% frequency), gravel (20%), mud (17%), silt (11%), rubble (8%), boulders (7%), clay (4%), detritus (2%), hardpan (1%), marl (1%), and bedrock (trace).[177]

Fluctuation in the abundance of bluegill in Clear Lake, IA, during the period 1947–1954, appeared to be closely associated with changes in the density of aquatic vegetation.[19] In MN lakes, significant correlations were detected between occurrence of emergent and floating-leaf plant species and relative biomass and mean size of bluegill.[139] Heavy macrophyte habitat was ranked the highest for bluegill in an IL lake, but its utilization was not significantly different than that of light macrophytes, sand, and transition areas among macrophytes, sand, and open water.[40]

When an infestation of hydrilla exceeded 80% coverage of Orange Lake, FL, in 1977, numbers of harvestable bluegill were negatively correlated with hydrilla coverage.[49] Many large bluegill were present in two MI Lakes that had macrophyte coverage of 41–83% of the surface. Studies on these lakes indicated that weedy lakes need not be dominated by small, stunted bluegills but are capable of producing large bluegills if fishing harvest is restricted and a favorable food chain is present.[166]

In clear Ozark lakes, underwater behavioral observations (26 h) using SCUBA were used to determine microhabitat use and time budgets of adult and juvenile sunfishes in the vicinity of weed beds

(*Justacia americana*). Adult bluegill spent 78% of their time in open water.[27]

Bluegill suffered a decline in condition index in a 2.7-acre pond in OK following aquatic vegetation control using the herbicide Silex.[68] After 3,600 ha of submerged aquatic vegetation in Lake Conroe, TX, were eliminated with grass carp, bluegill biomass declined significantly, but the species still remained abundant after the vegetation removal.[33]

The use of artificial structures with two interstice sizes (49 and 350 mm) placed at two depths (3.0 and 4.5 m) in an OH reservoir was evaluated. Bluegill preferred 40-mm interstices at 3.0 m suggesting that, to attract bluegill, fisheries managers should use dense materials in structures placed in shallow water.[20]

Bluegill will tolerate quite warm water.[96] In the laboratory, it preferred 32.2°C when all temperature ranges were available.[4] Coutant (1977) summarized reports of several authors concerning temperature preferenda for bluegill. In laboratory studies of adult bluegill, the upper avoidance temperature was 35°C, the lower 26°C with final preferenda between 27.4 (winter) and 32°C. In WI, bluegills 100–193 mm preferred water 29.6–32.6°C in the daytime and 27.2–29.0°C at night.[26] Preferred and upper and lower avoidance temperatures for bluegill from the New and East Rivers, VA, at acclimation temperatures ranging from 30°C in August to 6°C in February are presented in Table 56.[62]

The ultimate incipient lethal temperature in one study was 35.5°C, but a few fish survived to 41.5°C.[145] Upper lethal temperatures have also been reported at 33.8–35.5°C or even 37.0°C for a few days when acclimated to 30°C. Temperature extremes of 33.8 and 2.7°C caused mortality.[100]

Bluegill are reported from less brackish areas of coastal estuaries,[171,186] but salinity appeared to be a major factor limiting distribution of the species in a LA wildlife refuge.[134] In MS there seems to be no difference between coastal and inland populations of bluegill in terms of their salinity preference or tolerance.[186,187] Bluegill is a native of coastal rivers of MS, where it may occur in salinities of up to 10 ppt.[186] The species is reported as rarely found in salinities of 12.0 ppt,[23] and about 10–12% sea water is reported to be the maximum concentration in which bluegill could successfully reproduce.[135]

The lowest observed oxygen tensions at which all bluegills survived 24 h in summer were 3.4 ppm at 23°C or for 48 h in winter were 3.6 ppm, but the highest tensions that killed all bluegill were 3.1 ppm at 15°C in summer and 0.8 ppm in winter.[146] Other dissolved oxygen concentrations relative to temperature that bluegill survived are presented in Table 57.

In the Finger Lakes backwater complex, Pool 5, on the upper Mississippi River, the effects of dissolved oxygen (DO), water temperature, and current velocity on winter habitat selection by bluegill were examined. Areas with water temperatures <1°C and current velocity >1 cm/s were avoided. When DO was above 2 mg/l, bluegills selected areas with water temperature >1°C and undetectable current. As DO concentrations fell below 2 mg/l, bluegill moved to areas with higher DO, despite entering areas with water temperatures of 1°C or lower and with current velocities of 1 cm/s.[167]

During summer stratification of AL farm ponds, bluegill moved into shallow waters. They were unable to live for extended periods at water depths where the dissolved oxygen concentration was 0.3 ppm or less accompanied by a carbon dioxide concentration of 4.4 ppm or more. This condition was considered the critical combined dissolved

Table 56

Preferred and upper and lower avoidance temperatures (°C) for bluegill from the New and East Rivers, VA, at acclimation temperatures from 30°C in August to 6°C in February.[62]

Acclimation Temperature	Upper Avoidance Temperature	Preferred Temperature	Lower Avoidance Temperature
30	35	31.7	26
27	34	31.4	24
24	33	31.2	23
21	33	30.1	20
18	31	29.2	17
15	26	25.9	14
12	25	23.9	12
9	23	19.6	7
6	22	18.7	5

Table 57

Minimum dissolved oxygen (DO) concentration at which bluegill survived.[137,142]

Temperature (°C)	DO Exposure (ppm)	
	Immediate	Acclimated
25	0.75	0.70
30	1.00	0.80
35	1.23	0.90

oxygen/carbon dioxide concentration for bluegill and the depth that these concentrations usually occurred was designated as the critical depth. The critical depth for bluegill was normally 5 feet in 2-acre ponds and 7 feet in a 22-acre pond.[90]

In a 22-month exposure of bluegill to copper (Cu) in soft water, adult bluegill survival was reduced, growth was retarded, and spawning was inhibited at 162 µg/l. Survival, growth, and reproduction were unaffected at 77 µg/l and below. The maximum acceptable toxicant concentration for bluegill exposed to copper in water with a hardness of 45 mg/l (as $CaCO_3$) and a pH range of 7–8 lies between 21 and 40 µg/l.[109]

Bluegill has been an important species in pond management.[137] American warm-water pond-fish culture is concerned principally with the production of sport fishing, which is produced by the establishment and management of balanced fish populations. The simplest such combination and the one most widely used throughout the United States includes only two species — bluegill and largemouth bass.[38]

A study of 38 OK farm ponds indicated that in ponds stocked at an approximate 1:1 ratio of bluegill or redear sunfish–to–largemouth bass and in quantities of somewhat less than 75 fish of each species per acre, the bass–bluegill combination is capable of providing more satisfactory fish populations and fishery.[37]

A study of 13 small SD impoundments indicated that the relationships between bluegills and largemouth bass in small northern impoundments are similar to those in small impoundments in both midwestern and southeastern areas of the United States. The proportional stock density (PSD) of bluegill in these small impoundments ranged from 15–82 and was correlated with largemouth bass catch per unit effort (CPUE); as the density of largemouth bass increased, the percentage of bluegill ≥15 cm increased. The PSD for bluegill was inversely correlated with PSD and mean length of largemouth bass as well as mean relative weight. Mean back-calculated lengths of bluegill at age 5 and age 6 were positively correlated with largemouth CPUE.[50]

Abundant small largemouth bass in two AL impoundments limited recruitment of harvestable-sized bluegills. Following treatments of marginal poisoning with rotenone and electrofishing, largemouth bass numbers were reduced, resulting in the annual numbers of bluegill harvested more than tripling.[51]

To better understand differences among populations of bluegill in approximately 2,600 MN lakes, relationships between bluegill recruitment, growth, population size structure, and associated factors were analyzed. Bluegill year-class strength, growth, and population size structure were more strongly related to each other than to predator and lake characteristics, temperature, or season. Growth of bluegill at age 6 was positively associated with population size structure and inversely related to year-class strength, suggesting density-dependent growth effects for adult bluegill. Growth of bluegill at age 3 was inversely related to Secchi depth, so early growth and productivity may be linked. Bluegill population size structure was positively associated with length at age 5 and was inversely associated with mean year-class strength. Bluegill year-class strength was negatively associated with population size structure, length at ages 4 and 5, and lake area. Bluegill growth was positively related to relative abundance of yellow perch, walleye, black bullhead, and brown bullhead and negatively related to that of largemouth bass and yellow bullhead.[169]

Quality bluegill populations in the NE sandhills are influenced by predators, prey, and the environment. Growth, condition, and size structure were not density-dependent for bluegills in 30 shallow (<4 m maximum depth) NE sandhill lakes, but bluegill abundance, size structure, and condition were positively related to yellow perch abundance, size structure, and condition. Bluegill quality tended to increase with increased emergent vegetation. Submergent vegetation coverage ranged from 5 to 97% of lake surface area but was not related to bluegill quality. The mean relative weight of larger bluegills (200–250 mm) was positively associated with high *Daphnia* and *Cyclops* abundance.[163]

DeVries and Stein (1990) reviewed literature to determine if the manipulation of forage fish populations (gizzard and threadfin shad) to enhance sport fisheries is a successful practice. It was previously assumed that shad introductions would negatively affect competitors, such as bluegill. DeVries and Stein's author-generated conclusions and interpretations of 12 studies of shad introductions into waters inhabited by bluegills showed negative effects in 3 studies, mixed negative and neutral effects in 3 studies, neutral effects in 4 studies, mixed neutral and positive effects in 1 study, and a positive effect in 1 study. In 5 studies of situations where shad were removed from waters inhabited by bluegill, the effects were negative in 2 studies, neutral in 2 studies, and positive in 1 study.[67]

In IL, bluegill–gizzard shad interactions across 10 reservoirs each with and without gizzard shad were examined to determine direct and indirect effects of gizzard shad on bluegill size structure. The presence of gizzard shad was significantly related to bluegill growth and size structure. Both male and female bluegill were smaller in reservoirs containing gizzard shad than in those without gizzard shad. Bluegill density also differed in lakes with gizzard shad compared to those without gizzard shad. Absolute and relative numbers of large bluegill were higher in lakes without gizzard shad. Conversely, overall bluegill density was higher in reservoirs with gizzard shad than in those without gizzard shad because of higher numbers of bluegill <150 mm in the gizzard shad reservoirs. These higher densities of bluegill may have resulted from decreased largemouth bass predation due to gizzard shad availability as alternative prey.[147]

In a MO study of small lakes and ponds, the proportion of lakes containing gizzard shad decreased with increasing macrophyte coverage and abundance of age-0 gizzard shad declined in ponds with increasing macrophyte coverage. Bluegill populations in small lakes containing large adults (>203 mm TL) increased with macrophyte coverage for lakes with gizzard shad but not for lakes without gizzard shad. Although the mechanisms were not clear, the authors suggested that competitive interactions between gizzard shad and adult bluegills are density dependent and lessen with increasing macrophyte coverage because of decreasing gizzard shad abundance. In ponds, the abundance of age-0 bluegills was unrelated to abundance of gizzard shad or macrophyte coverage.[152]

Bluegill in West Point Reservoir, MS, in 1980–1981 had poor population structure and relatively low condition. Results of fish food inventory and gut content analysis revealed a shortage of preferred food for bluegills of all sizes. The food shortage was attributed to competition with abundant threadfin and gizzard shads and to the effects of an annual 3-m water level fluctuation, which reduced standing stocks of benthic invertebrates.[150]

No significant differences in density or biomass were observed for bluegill after the introduction of alewife into oligotrophic Mayo Reservoir, NC, during the period 1992–1993. Alewife established a self-sustaining population and increased from <1% of total fish biomass in 1993 to 31% in 2000.[133]

Bluegill generally travel in schools[4] or in small loose aggregations of 10–20 fish.[96]

Bluegill exhibited several types of seasonal movements in MI lakes including extensive along-shore migrations through the littoral, less extensive onshore–offshore movements, and vertical shifts in the water column.[35] Bluegill were active along the shore of Buckeye Lake, OH, between March and August. The most active period was in May, June, and July corresponding to the spawning period. Toward the end of the spawning period, in late July and August, when temperature reached 30.6°C, the activity of the fish reached a low ebb.[117]

In lakes the bluegill may spend winter near the bottom but is found in all depths during spring.[4] Groups of bluegills retreat to deeper water in winter,[176,192] where they congregate in colonies and continue feeding. Winter aggregations break up at 10°C, with males moving inshore before females.[176,192] In summer they utilize small territories and move little; an individual can be observed in the same place for hours.[176] In the summer during midday hours, bluegills will often be found lying in the shade of overhanging brush or in the deeper waters of shallow lakes. In the early morning and evening, they are usually found in shallow water, where they feed.[96] In the hottest periods of summer the largest individuals may move down as deep as 6 m.[176]

Bluegill seek warm water in winter as shown by their concentration downstream of power plants,[96] by their ascending warmer, shallow, sluggish streams early in the spring, and by some aggregation in power plant discharges in the summer.[100] Maximum temperature tolerated in the Tennessee Valley exceeded 37°C.[100]

The relative motility of bluegill in an AL reservoir was determined based on tissue polychlorinated biphenyl residues. Bluegill 150–207 mm TL (n = 45) moved an average of 16.8 km (0–45.6 km) from their location of capture.[164]

In small central TN streams, bluegill abandon the main channel for the floodplain during flooding.[58]

Homing ability is present in bluegill. It is suggested that they make use of the sun in orientation.[76] In Cedar Lake, IL, bluegill established home ranges of from 0.15 to 0.75 ha. Of nine bluegill tagged, four established home ranges in locations other than where they were captured and released. During the course of the study, one bluegill abandoned a home range and established another.[30] In LA stream, the home range of bluegill was estimated to be approximately 38 linear meters of stream.[34]

DISTRIBUTION AND OCCURRENCE IN THE OHIO RIVER SYSTEM

The Ohio River is the type locality of the bluegill. It is distributed throughout the length of the river,[2,3,8] and is the most common centrarchid collected.[2,178] It may be most abundant in the middle third of the river, responding to the creation of embayment habitats in much the same manner as the pumpkinseed.[2]

During the period 1981–2003, bluegill were abundant to common most years in electrofishing surveys of the Ohio River conducted near three electric generation stations: Cardinal (ORM 77), Kyger Creek (ORM 260), and Tanners Creek (ORM 494). At the two upper localities they were occasionally ranked among the top three species in terms of abundance.[190] Based on lockchamber surveys during the period 1957–2001, bluegill had increased in abundance in the Ohio River.[9]

The bluegill is native to the Allegheny River system of NY[4] and to the Youghiogheny River system of PA, MD, and WV.[22] In IL, bluegill is statewide and common.[174] In IN, bluegill is the most abundant sunfish and has been extensively stocked throughout the state.[197] It is generally distributed and abundant throughout KY.[140]

In VA, the native range of bluegill encompasses the Tennessee and Big Sandy drainages and it has been introduced into all other drainages.[185] It is present throughout the Tennessee and Ohio River drainages of NC.[180]

The bluegill is common throughout TN except in small Blue Ridge streams,[170] and it is found throughout the Tennessee River system of AL[172,189] and MS.[186]

SPAWNING

Location

In the Great Lakes, the bluegill spawns in quiet near-shore areas of lakes or stream pools, including wetlands, marshes, bays, coves, harbors, lagoons, and creek mouths.[175]

Nests are often constructed in shallow water[96,117,130] close to shore,[56,117] commonly over sand[13,28,56,76,96,99,114,117,130,175] or gravel[13,28,56,76,96,99,114,117,130,131,175] bottoms. Nests are also reported in substrates of muck,[13,99] marl,[13,175] clay,[175] mud,[114,117,175] detritus,[13,114,117,175] rootlets,[175] or whatever substrate is available,[28,96] or combinations of the above.[13]

The male bluegill excavates his nest down to firm bottom or rootlets,[175] and may build his nest in protective beds of vegetation.[56,175] Bluegill nests are reported on rock breakwalls.[175] Bluegills spawn over nests of other centrarchids,[13,175] such as pumpkinseeds and rockbass.[13]

In a VA lake, nest preparation by male bluegill exposed coarse gravel (8–32 mm diameter) and pebbles (32–64 mm diameter) in nest substrate and removed particles smaller than 2 mm.[131] In an OH farm pond overpopulated with bluegills and with a moderate largemouth bass population, availability of nesting substrate was an important variable affecting spawning success of bluegill in the pond. Nests with fine gravel and sand substrates produced the most fry; those with mud and debris bottoms produced fewer fry.[114]

In three northern WI lakes, spawning occurred on muck, sand, and gravel, which represented all bottom types available. Bottom type was not considered to be a limiting factor in spawning because sufficient areas of all bottom types were available.[99]

The same appears true in ponds and lakes in OH, where nests are reported from along the shore over sand and small pebbles in water 0.3–0.9 m deep, in vegetation at the bottom in water 0.9–1.5 m deep, along a shallow shoreline over masses of dead leaves, sticks, and mud, and in mud alongside a dock in water 0.9–1.5 m deep.[117]

Bottom gradients of 1–13%, 7:1 slope, and general statements of flat or gradual slope are reported for bluegill spawning sites.[115] In a WI lake, bluegill built colonies in areas with mean (range) bottom gradient of 6% (1.1–13.2%), depth of 89 cm (65–119 cm), and distance from shore of 19.1 m (1.5–86.9 m). Spawning substrate size ranged from fine (0.5 mm) to cobble (64–256 mm) and included sand, silt, sticks, gravel, and cobble. The greatest distance between individual nests in a colony was 80 cm, but most nests were closer together. About half the colonies (56%) were greater than 5 m from cover. Certain areas of the lake were used annually.[115]

In the Great Lakes, nests are constructed in water up to 4.6 m deep, usually in water <1.5 m.[175] In NY ponds, bluegill nests are usually in groups in water from 0.3 to 0.9 m deep.[76]

Bluegill generally nest in deeper water and farther from shore than pumpkinseed.[28]

In a CO pond, reproductively motivated male bluegill moved to shallow water. Water depth ranged from about 10 cm for smaller males to about 100 cm for larger males. Selected nesting areas were usually free of plants or other structure and usually had a sand or gravel bottom. Nests were rarely constructed in mucky margins of the pond and were most commonly constructed in unshaded areas that received a maximum of sunshine.[130] In ON ponds, bluegill nests were located under or close to shade if it was available. No consistent relationship was found for bluegill between nest location and aquatic vegetation, or between nest location and substrate material.[28]

In captivity, bluegill spawned in wood spawning boxes filled with pea gravel placed in circular wading pools 0.5 × 2.4 m.[10]

Season

Most reports indicate that bluegill have a protracted spawning season that may last from May through August in northern portions of the range, with peaks usually reported in June or possibly July, in more northern populations (Table 58).

In WI, the length of the spawning season varied considerably from year to year, ranging from 31 to 112 days. A longer spawning season produced a greater number of individual spawning periods. Water temperature fluctuations provided a stimulus for repeated spawning in the study lakes and thus had an important influence on length of the

Table 58

Reports of season and water temperatures for bluegill spawning.

| Location | Spawning Season | Spawning Temperature (°C) | |
		Begins	Range
Canada	Late spring to early and mid-summer, peak probably in July[176]	—	—
ON	Late June to late July (26 days)[21]	17–23[28]	—
WI	Late May to mid-August, [96, 99, 115, 177] peak in June[96]	19–21[96, 99]	19–27[99, 115, 177]
Upper Mississippi River	May to early July[182]	—	—
MI	June to September[13,137]	—	—
Great Lakes	Early May to mid-August, probably peaks in July[175]	—	15.6–30.6[175]
CO	May and June[130]	—	—
NY (ponds)	May and June until early August[76]	21[76]	—
NY	May to June or July,[181] late June and July covers most spawning[4]	—	—
DE estuaries	May through August, peak in June[194]	—	18.5–25[194]
Potomac River estuary	May to August, peak in June[195]	—	16–27; optimum 21–24
NJ	Mid-July[4]	—	24.4[4]
KS	April to September, mostly May or June[183]	—	>20
MO	Late May to August, usually peaks in June[179]	—	—
Northern IN	Late May into August[86]	—	—
IL	Early May to September[137, 161]	After rising temps exceed 20[161]	—
OH	—	—	17.2–26.7[114]
VA	May to August or September[185]	—	—
NC	May to October[180]	—	—
TN	Late spring well into summer[170]	—	—
AL (ponds)	April through September[154]	—	—
AL	April into September[172]		
MS	—	21[149]	21–32[186]
AR	April through August, peaks in June and early July[188]	—	>20[188]
FL	February to October[137]	—	—
TX	March to September[137]	—	—
OK	Early may to late September[191]	—	—

spawning season, but no significant relationship between water temperature and duration of the spawning season was observed.[99]

The season begins earlier in some southern states, with reports of spawning in April in KS,[183] AR,[188] and AL[154,172] and as early as March in TX[137] and February in FL[137] (Table 58). Male bluegills may spawn almost throughout the growing season.[137]

Temperature

Bluegill are reported to spawn at temperatures from 15.6 to 32°C (Table 58), and have been observed on nests at 35°C.[116]

Low temperatures may cause a failure to spawn, especially in ponds. In northern populations spawning may be delayed up to 4–5 weeks by low temperatures, and some summers the temperature may never reach suitable levels.[43,137] Late spawning by bluegill in OH in 1950 was associated with a cold spring and summer.[117]

In WI, temperatures below 21°C postponed or interrupted spawning and slowed development of eggs and larvae.[99] Several reports indicate that spawning begins soon after rising water temperatures exceed 20°C.[76,149,161] However, another WI report indicated that spawning begins at 19–21°C[96,99] and in an ON pond, bluegill nests were first observed when the mean daily surface water temperature was 17°C.[28] Spawning is reported as low as 15.6°C in the Great Lakes,[175] 16°C in the Potomac River estuary,[195] 17.2°C in an OH farm pond,[114] and 18.5°C in DE estuaries.[194]

An early report suggested that bluegill require a water temperature of 26.7°C or higher to spawn.[4] In an IL pond, successful reproduction did not begin until turbidity declined in early July, even though surface water temperatures had reached 27°C by early June.[101] In an OH farm pond, in all cases water temperature was rising when spawning began and was at least 26.7°C, but in one instance in mid-May, spawning occurred when water temperature was 17.2°C.[114]

Reproduction of age-1 bluegill in temperature-controlled outdoor channels was investigated at four temperature regimes: ambient Tennessee River temperature and nominal elevations of 2, 4, and 6°C above ambient. The only consistent temperature effects were early inception of spawning and corresponding larger numbers of nests in the warmer channels. No relationship was apparent between temperature and (1) the number of eggs per spawn, (2) time between spawns per female, (3) hatching success, or (4) biomass of YOY recovered.[95]

Fecundity

Reports of bluegill fecundity range from 2,300 to 80,000 eggs[137] and numbers of eggs increase with age and size of the female.[18,96,99,117]

In WI, a female weighing about 4 ounces produces about 12,000 eggs. Larger females may carry as many as 38,000 eggs. A female 114 mm TL contained 2,540 eggs.[96] In another WI report, the number of eggs per female increased with fish size and varied from 6,500 to 23,500.[99] Also in WI, estimates of fecundity for females 122–144 mm on May 25 were 2,900–8,000 (average 4,800); on June 29 fecundity was estimated at 1,900–4,600 (average 2,900).[177]

In NY, a 127-mm TL female may produce as many as 6,000 eggs and a 229-mm female may produce almost 50,000 eggs.[4]

The following fecundity information was reported for bluegills from an IN lake: age-2 females (n = 7) averaging 130 mm TL and 51 g had 2,360–5,066 eggs (mean = 3,820); age-3 females (n = 9) averaging 148 mm TL and 71 g had 6,518–13,137 eggs (mean = 9,264); age-4 females (n = 2) averaging 187 mm TL and 198 g had 16,220–22,119 eggs (mean = 19,169).[18]

In OH, fecundity varies from 2,540 eggs in females 102–114 mm TL to 64,000 eggs in a female 235 mm TL.[117]

In Lake Robinson, a large blackwater lake in SC, fecundity estimates for bluegill ranged from 571 to 27,027 eggs per female and were related to TL by the expression $\log_{10} F = -2.337 + 2.839 \log_{10} TL$ (mm) with $r^2 = 0.59$ (where F = fecundity and TL = total length). Bluegill fecundity was slightly below average compared with measures of fecundity from other habitats. Diameters of mature eggs (0.5–0.79 mm) were smaller than the mean diameter (1.09 mm) reported by Carlander (1977). The authors suggest that the low productivity typical of blackwater lakes may be related to the low fecundity and smaller egg size.[75]

Sexual Maturity

Non-experimental studies suggested that the differences in age at maturity among populations of bluegill were not genetically based, but rather were a phenotypic response to the presence of predators. Subsequent laboratory study also suggested that predators can influence prey population dynamics by altering age at maturity. Bluegill placed in the same water with predators showed significantly increased use of refugia and decreased growth. Bluegill from a natural population with a 30–40 year history of high predation levels were placed in the same water with predators in the laboratory. Significantly fewer bluegill from this test were mature at age 1 and their gonadal/somatic index was significantly lower compared to bluegill placed in aquaria where they could only see predators and compared to controls. Bluegill from a pond population that contained only bluegill placed in similar laboratory treatments did not respond to the presence of a predator in a significant manner.[128]

In Canada, sexual maturity is probably attained by age 2 or 3 for males and 3 or 4 for females.[176] Bluegills stocked in NY ponds have been known to spawn in their second summer and all reproduced at least once in their third summer.[4] Maturity is reached at age 2 in MO and WI.[137] In AL, some bluegill may reproduce their first summer, but most mature at age 1; however, if they are not at least 14 g at age 1, they will not spawn.[189]

In Lake Opinicon, ON, length at first breeding was 125 mm TL; weight 35 g; age 4 years.[21] In WI, minimum size for sexual maturity is about 114 mm TL.[96] Smallest mature males are reported at 58 mm and females 89 mm.[137]

In an OH farm pond overpopulated with bluegill, sexual maturity was related to size rather than to age in both male and female bluegills. Though males 112 mm TL were sexually mature, none <125 mm successfully maintained a nest. Some females were mature at 90 mm.[114]

Spawning Act

Bluegill nests are often observed in groups or colonies,[4,13,28,47,76,97,115,117] but single nesting is reported.[13,117,130] Nests per colony usually range from a few to 100,[13,97,115] but densely packed colonies of up to 500 are reported.[97,115] In two ON ponds, the mean distances to the nearest conspecific nesting neighbor were 50 ± 19 and 32 ± 12 cm.[28]

Nests sometimes are associated with other sunfish species.[117]

A nest is constructed and guarded by a single male.[18] Construction of the nest by a male bluegill is variously described: sweeps out a nest in the substrate with side-to-side movements of the caudal fin;[47,76] a shallow depression is fanned out with violent swishes of the body;[96] the male excavates the nest with pushing actions of the head and flipping the tail.[117]

Bluegill nests are quite similar to the nests of other sunfishes[117] and are described as shallow[86,96] saucer-like,[86] bowl-like,[117] or nearly circular depressions,[13,76] with well-formed rims.[76] Nest depressions vary with the size of the male.[76] Reported diameters range from 0.2 to 0.6 m.[13,76,96,117] Depths of 51–76 mm,[117] 51–152 mm,[96] 0.3–1.5 m,[13] and 0.1–3.1 m[115] are reported. Nests built on muck bottoms were considerably larger and not always circular in outline.[13]

Bluegill nests in ponds and quiet waters may endure several seasons. Therefore, bluegill males do not necessarily build a completely new nest each breeding season; experienced males often located themselves over old nests.[76,130] The male bluegill's behavior, once he is ready to nest, is probably not influenced so much by the existence of an old nest as it is by the proximity of neighboring males.[76] When occupying an old nest, the male would clear away debris from the bottom by tail sweeping the nest.[130]

Once the nest is constructed, sweeping activity may continue if there are other bluegill and nests in close proximity. If nests are far apart or separated by luxuriant plant growth, males hover over nests faithfully but engage in few other activities such as sweeping. Males over their nests sometimes repeatedly circle the rim of the nest with fins in display position. The sight of neighboring fish spawning stimulates males without partners to circle their nest rims faster and more persistently.[76]

The nesting territory is protected by the male, and although there may be many nests almost touching, individual nesting territory is vigorously protected.[96] Males competed aggressively for positions within colonies, and the ability to maintain residency on a nest was related to size. Males defended their nests against intruding adult males an average of 3 times per 10-min observation period. Nest defense most frequently consisted of swimming at the intruder occasionally making contact, and in some cases prolonged fighting was observed.[97]

Within nesting areas where the nests were often less than 10 cm apart rim to rim, the territory of the male encompassed the area within the nest plus a distance of 5–10 cm beyond the rim. In one area of the pond, there was only one nesting bluegill. The nearest nest was 10 m away and the male defended an area 2–2.5 m from his nest. Several times other bluegill would try to establish nests in this area but were chased away by the resident. The diameter of this isolated nest was three times the length of the resident instead of the usual two times and the nest was deeper (8.5 cm) in the center compared to the usual nest (3–8 cm).[130]

All male bluegill regardless of size were chased away from the territory of a nesting male. Aggressive behaviors included chasing, nipping, swiping, displaying, and thrusting. Usually there was no physical contact with an intruder. Successful defenses result in the male vigorously tail-sweeping for 1–2 days and completing his bowl-shaped nest. If a male cannot successfully defend his territory, he will move to another area and reestablish a nesting site.[130]

Once males were established on nests, females arrived in large numbers[47,76,97,130] at the colony and remained in the water column until spawning was initiated.[76,97] Females were attracted to areas of spawning activity and formed aggregations above the spawning nests.[76,97] The males were circling the rims of their nests and continuing to tail-sweep. To initiate spawning, a female leaves the water column and approaches a nesting male while displaying a coloration pattern (dark background, dark eyes, and pronounced vertical barring) which is known to reduce aggression in the nesting males of other sunfish.[76,97] Courtship begins when a gravid female appears in the nest area of a male.[130] A nesting male, upon sighting a female, would rush toward her and back toward his nest repeatedly while producing a series of grunt-like sounds.[5]

Territory guarding male bluegill show aggression initially toward any approaching female. Subsequent behavior of the female appeared to determine the outcome of these encounters. A female bluegill threatened by a male either fled or settled closer to the bottom and withstood the attack; in the latter case, the male usually began courting her.[28] The male chases and encircles the female and guides her to his nest. As the male and female approach the nest they splash about, the male nudging the female at the vent with his head.[117] After entering the nest, the female may swim away or remain. In the latter case, courtship is completed.[130] Once the female is in the nest, the male begins to swim in tight circles, turning, with the female following. Every few seconds as the pair turns, the female turns on her side, presses her genital pore against that of the male, quivers, and releases eggs that the male fertilizes.[76,97,117]

The bluegill is polygamous; more than one female apparently lays eggs in a given nest[18,179] and a female may deposit her eggs in more than one nest.[179] Some male bluegills occupy more than one nest; some occupy nests built by conspecific males and some occasionally occupy nests built by pumpkinseed males.[28] There may be more than one male and female spawning in a nest at the same time. A second female was observed slipping into the nest and aligning herself with a spawning pair of bluegills. Two males and four females were also observed spawning at the same time over a nest.[130] As many as about 224,000 fry have been reported from a single bluegill nest.[18]

Within a colony, spawning is highly synchronous, with the spawning period for an entire colony cycle occurring in a 1- to 2-day period.[76,97] Spawning activity in the colony lasts for several hours and then the females leave.[47] The males remain to care for the eggs, guarding them against predators and fanning them,[47,86] a behavior that apparently keeps silt from settling on the eggs and also aerates the eggs.[117] As soon as a spawning bout has finished numbers of large and small sunfish will dart in to devour the eggs. The male bluegill dashes about, seemingly in all directions at once, until he has driven the intruders away from his nest. He then stands guard above the nest, keeping any new intruders away.[117] The male guards and fans the eggs for 2–5 days until they hatch. After hatching, the fry are guarded (but not fanned) for an additional 3–6 days[47,175] or the male guards the nest until the eggs hatch, but does not guard the fry once they leave the nest.[179]

In ON ponds, male bluegill guarded their nests an average of 8.7 days. Bluegill nest occupancy was determined primarily by presence of eggs or fry in the nest. Some bluegills stayed on nests containing eggs through periods of fluctuating water temperature; those without eggs usually deserted their nests within days of occupancy.[28]

Nesting male bluegill defended territories with typical sunfish aggressive behavior, consisting of spreading the opercula, charging, biting, chasing, and, rarely, mouth fighting. In ON ponds, bluegill defended smaller territories than did pumpkinseed. Approaching fish of either species were often attacked by nesting pumpkinseed when they were about 1 m from the nest. Bluegill seemed to defend only the nest itself, seldom attacking an intruding fish until it was within a few cm of the nest.[28]

There are advantages to nesting in colonies, especially when there is synchrony in spawning and subsequent protection of developing eggs and fry afforded by the tightly clustered group of adults. Another advantage to colonial nesting is safety in numbers.[189] Mobbing, the assemblage of individuals around a potentially dangerous predator, was instigated with colonially nesting bluegills and then also observed as a natural occurrence in a NY lake. In an attempt to initiate mobbing, a large snapping turtle was released into several actively spawning colonies of bluegill. In each of five trials, the response of the bluegills was immediate and unlike any behavior previously reported. As the turtle moved through a colony, dozens of nesting males, gravid females, and males without nests rapidly approached and followed the turtle across numerous territories until it left the colony area. Subsequent to these trials, the author observed a natural occurrence of bluegill mobbing, again with a snapping turtle. A similar trial with a painted turtle, which is incapable of preying on adult bluegill, did not elicit mobbing.[118]

Bluegill may have 1–3 brood cycles per season, and breed for 1–3 years.[47]

In species where male reproductive success is dependent on male competition and aggression, alternative reproductive patterns, thought to represent a reduction in male reproductive effort, sometimes occur. Female mimicry in bluegill is an example of an obligate alternative male strategy. Female mimics are small, sexually mature males that mimic the details of female behavior, and gain access to functional females attracted to the nests of large, aggressive territorial males.[97]

After bluegill successfully reproduced for the first time in IA farm ponds, they generally reproduced successfully each year, except where prevented by winterkill or other cause such as low water temperatures, high bass populations, or human activities.[43] In an IL pond high turbidity delayed successful reproduction.[101]

In the laboratory, bluegill were prompted to spawn out of season by the manipulation of water temperature and photoperiod and by the manipulation of photoperiod alone.[52]

Bluegills were reproductively isolated from pumpkinseeds in two ON ponds. Mature females of either species approached by an aggressive male invariably fled. Females were never observed trying to enter nests occupied by heterospecific males, and interspecific courtship and spawning were never seen. If incompatible reproductive behavior prevented interspecific breeding in the two ponds, at least two separate mechanisms may have been involved: nesting males may have courted conspecific females but rejected or ignored heterospecific females, and gravid females may have distinguished between the two types of males.[28] Other reports indicate that bluegills hybridize in nature with pumpkinseed[12,48,69,176] as well as with seven other lepomids including: warmouth[69,176] and green,[12,48,69,176] redear, orangespotted, longear, redbreast, and spotted sunfishes.[69,176]

In many cases, hybrids are fertile and able to mate with both parents and with hybrids. In Canada, the hybrid between bluegill and pumpkinseed occurs virtually everywhere these two species live. The degree of hybridization is so great and the range of characters so confusing as to make it almost impossible to separate hybrids from backcrosses with parental species, or even to identify pure parental forms.[176]

There appears to be no direct genetic isolation between bluegill, redear, and green sunfish. In the laboratory, the six possible crosses of these three species all produced large numbers of F_1 hybrid embryos that developed normally, hatched, and became free-swimming fry. The fry were released into outdoor ponds, where they grew to sexual maturity.[25]

In the laboratory, intergeneric hybridization is reported between bluegill and *Micropterus*, *Pomoxis*, and *Chaenobryttus*.[74]

Golden shiners were observed spawning in bluegill nests in a NC pond. Fifty-three percent of the larvae that hatched from the eggs taken from one of the nests were golden shiners.[89]

EGGS

Description
Unfertilized, water-hardened eggs averaged 1.04–1.1 mm in diameter.[21,71,116]

Fertilized eggs are amber in color,[91] demersal,[91,116,117,192] adhesive,[91,16,117,192] spherical,[91,116] and 1.2–1.4 mm in diameter.[91,92,117] Yolk is granular in appearance[91,117] and contains a very large (approximately 30% of yolk diameter[91]) oil globule[91,117] and occasionally a small supplemental oil droplet.[91] The chorion is very thick[91] and the perivitelline space is about 0.045 mm wide.[117]

Incubation
In bioassay experiments bluegill eggs hatched at temperatures from 18 to 36°C during two incubation tests; maximum hatching occurred at 22.2–23.9°C.[83]

In northern WI, incubation of bluegill eggs takes 2–3 days if water temperature is above 21°C and 4–5 days if below.[99] In the Great Lakes, 2–4 days at 19.4–24.4°C.[175]

Other reports include:

85 h at 18.5 ± 1.0°C[193]
71 h at 22.6°C[137,177]
31.5–62 h at 22.2–23.3°C[117]

32–62 h at ambient[86]
30–34 h at 22–30°C[91]
41.33 h at 24.5 ± 1.0°C[193]
32.5 h at 27.3°C[137,177]
34 h at 26.9°C[137,177]
29.17 h at 28.5 ± 1.0°C[193]

Development
In OH, eggs incubated at 22.2–23.3°C did not develop at the same rate even though fertilization of all occurred within a 12-min period. Some of the eggs had reached early blastula stage while others were still in the two- or four-cell stages.[117] At 22.2–23.3°C, general development related to age (time after fertilization) was as follows (Figures 52 and 53):[117]

0.58 h — 2–cell stage
0.92 h — 4–cell stage
1.28 h — 8–cell stage
1.33 h — 12- to 16-cell stage
1.67 h — early blastula
3.00 h — many-celled blastula
4.00 h — blastoderm cells have covered 1/2 the yolk
6.00 h — early embryo
17.00 h — 10–12 somites
21.00 h — 20–22 somites
25.00 h — 25–27 somites
26.50 h — 28 somites
29.00 h — 32 somites
31.67 h — 33–35 somites
31.83 h — hatching

Differences in rate of development were also seen when incubation temperature ranged from 22 to 30°C. Development related to age (time after fertilization) is reported as follows:[91]

0.25 h — 1-cell, blastodisc broad and deep
0.75 h — 2-cell stage
1.25 h — 4-cell stage
2.00 h — early morula; cells crowded and bunched together, almost form a peak
6.00 h — late morula; blastomeres moving down over yolk
11.50 h — early embryo covering about 2/3 of yolk circumference; somites visible; anterior end of embryo distinct
22.00 h — optic vesicle forming; tail not free; about 5 somites formed
27.50 h — eye formation evident; heart beating; somites added posteriorly; tail free
30–34 h — hatching

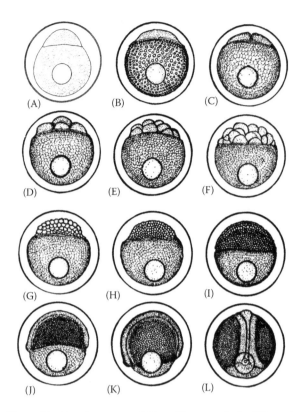

Figure 52 Development of bluegill eggs, A–L: (A) early blastula; (B) late blastula; (C) 2-cell stage; (D) 4-cell stage; (E) 8-cell stage; (F) 12- to 16-cell stage; (G) early morula; (H) late morula; (I) blastoderm over one-half of yolk; (J) early embryo; (K) embryo, lateral view, 9 h; (L) embryo, dorsal view, 11 h. (A reprinted from Figure 1A, reference 91, with permission of Chesapeake Biological Laboratory; B–L reprinted from Figures 1–11, reference 117, with publisher's permission.)

YOLK-SAC LARVAE

See Figure 54.

Size Range

In most reports, bluegill newly hatched individuals range from 2 to 3.7 mm TL;[15,91,96,117,193] hatching at 4.5 mm TL in ON is also reported.[21] Yolk is completely absorbed at 5–6 mm TL.[86]

Myomeres

See Table 59.

2.8 mm TL. Total myomeres 33–35.[15]

2.9–4.5 mm TL. Total myomeres 23–27 (counts appear low; methodology not described).[91]

≤6 mm TL. There was statistically significant geographic and temporal variation in numbers of myomeres in protolarvae bluegills[107] (≤6.0 mm TL[110]) collected throughout the 1978 spawning season

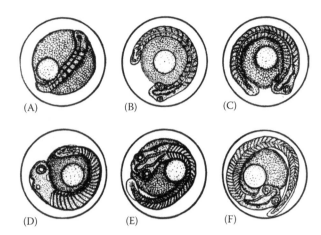

Figure 53 Development of bluegill eggs, A–F: (A) 10–12 somites; (B) tail-free stage, 20–22 somites; (C) 25–27 somites; (D) 28 somites; (E) 32 somites; (F) 33–35 somites, just before hatching. (A–F reprinted from Figures 12–17, reference 117, with publisher's permission.)

from the Mississippi River at Eudora, AR, and St. Francisville, LA. Mean number of total myomeres was higher at St. Francisville than 400 river kilometers north at Eudora. The mode of total myomeres through time at St. Francisville was 29 compared to 27–28 at Eudora (Figure 55).[107]

Morphology

From OH:

2.79 mm TL. Recently hatched; auricle and ventricle of heart are well developed, blood is red.[117]

3.51 mm TL. Gut visible from the yolk sac to the vent; heart visible in front of yolk sac; gill arches clearly evident in front of yolk sac and above the heart; yolk sac large with a large oil globule; brain differentiated; air bladder developing.[117]

3.96 mm TL. Air bladder visibly larger; yolk sac smaller.[117]

4.4–4.8 mm TL. Body elongate and transparent; oil globule located posteriorly in yolk sac.[194] At 4.54 mm, auditory vesicles and otoliths are visible; mouth is more developed, lower jaw moveable; heart beat is rapid.[117] At 4.75 mm, yolk and oil almost completely absorbed; opercle is present; gut is visible all the way to the pharynx.[117]

From MD:

2.9 mm TL. Recently hatched; otocyst formed, two otoliths present; head bent down over yolk.[91]

3.0 mm TL. Air bladder forming; three otoliths in some individuals.[91]

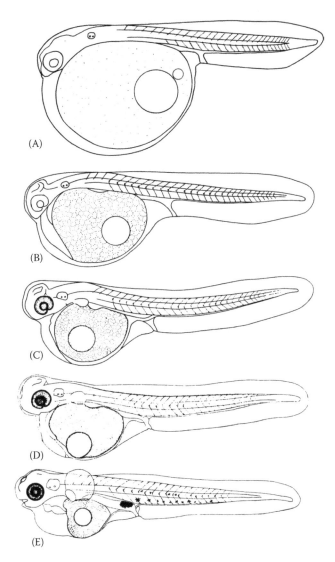

Figure 54 Development of young bluegill. A–E. Yolk-sac larvae: (A) 2.9 mm TL; (B) 3.6 mm TL; (C) 3.9 mm TL; (D) 4.0 mm TL; (E) 4.5 mm TL. (A–E reprinted from Figures 2, 4, 5, 6, and 8, reference 91, with permission of Chesapeake Biological Laboratory.)

4.5 mm TL. Choroid fissure closed; mouth well developed, probably not yet functional on some individuals but open and functional on others; heart sigmoid.[91]

From IN:

3.26–3.72 mm TL. Two days old. The alimentary canal is apparent from yolk sac to anus. Optic vesicles and lens of eye are present. The auricle of the heart is visible in front of the yolk sac; heart is beating and the blood is red. Yolk is lemon-shaped with one oil globule.[116]

3.48–4.40 mm TL. Three days old. Yolk is smaller but still lemon-shaped. Anus clearly open.[116]

4.27–5.54 mm TL. Four days old. The alimentary canal clearly open from above the yolk sac to the anus; jaws present but mouth not open; blood visible flowing through the gill arches.[116]

4.64–4.82 mm TL. Five days old. Mouth open and jaws weakly functional; air bladder evident.[116]

4.58–5.21 mm TL. Six days old. Alimentary canal clearly open from the pharynx to the mouth; peristalsis evident.[116]

4.93–5.74 mm TL. Seven days old. Larvae are free-swimming; yolk smaller; oil globule now in the anterior region of the yolk.[116]

4.99–5.81 mm TL. Eight days old. Yolk is almost gone. Larvae have large heads and a narrow trunk.[116]

Table 59

Myomere count ranges (average and standard deviation in parentheses) at selected size intervals for young bluegill from TN.*

Size Range (mm TL)	N	Myomeres			
		Predorsal Fin	Preanal	Postanal	Total
4.31–5.89	138	4–8 (6.5 ± 0.85)	11–14 (12.4 ± 0.61)	14–18 (16.2 ± 0.70)	26–30 (28.6 ± 0.82)
5.91–6.71	66	6–10 (7.65 ± 0.80)	12–14 (13.1 ± 0.58)	15–18 (15.9 ± 0.68)	28–30 (29.0 ± 0.52)
6.72–8.78	86	6–12 (9.47 ± 1.11)	13–16 (14.4 ± 0.64)	13–16 (14.7 ± 0.63)	28–30 (29.0 ± 0.54)
8.79–10.03	65	9–11 (9.89 ± 0.58)	13–16 (14.5 ± 0.55)	13–15 (14.4 ± 0.58)	28–30 (28.9 ± 0.49)
10.13–11.92	47	9–12 (10.0 ± 0.61)	13–15 (14.3 ± 0.50)	14–16 (14.7 ± 0.50)	28–30 (29.0 ± 0.41)
11.97–15.93	45	9–11 (10.0 ± 0.72)	13–16 (14.3 ± 0.55)	14–15 (14.7 ± 0.45)	28–30 (29.0 ± 0.51)
16.0–21.0	13	10–11 (10.3 ± 0.45)	14–15 (14.3 ± 0.45)	14–15 (14.6 ± 0.49)	28–30 (28.9 ± 0.63)

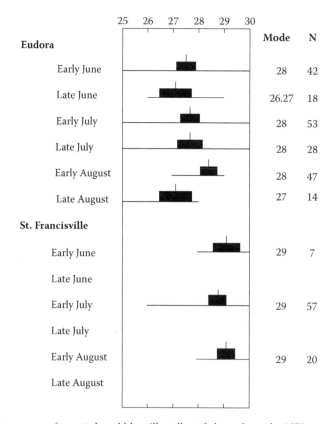

Figure 55 Variation in total myomeres for protolarval bluegills collected throughout the 1978 spawning season in the Mississippi River at Eudora, AR, and St. Francisville, LA. Horizontal line is the range; vertical line is the mean; black bar is plus or minus two standard errors of the mean. Information is printed, in part, from Figure 1, reference 107.

4.92–5.84 mm TL. Nine days old. Yolk is absorbed except for a tiny oil droplet. Air bladder appears to be composed of many vacuoles. Skull appears to be chondrifying.[116]

5.41–5.80 mm TL. Ten days old. Tiny oil droplet still present.[116]

Morphometry

See Table 60 for morphometric data collected from TN specimens.

See Table 61 for selected measurements of yolk-sac larvae from southeastern LA. (*Note:* Additional morphometric data for LA bluegills can be found in reference 84.)

From OH:

2.79 mm TL. As percent TL: PreAL 54.8; PostAL 45.2; depth before the vent 22.6; HL 22.6; ED 4.7; YS length 32.3.[117]

3.51 mm TL. As percent TL: PreAL 48.1; ED 6.3; YS length 28.2.[117]

3.96 mm TL. As percent TL: PreAL 46.5; PostAL 53.3; ED 6.8; YS length 19.2; oil globule width 6.8.[117]

4.54 mm TL. As percent TL: PreAL 43.6; PostAL 56.4; ED 7.9; YS length 15.9; oil globule width 4.5.[117]

Fin Development

From OH:

2.79 mm TL. Recently hatched; a continuous median finfold extends along the back around the tail and on the underside of the body to the region of the vent, and forward from the vent to the yolk sac.[117]

3.51 mm TL. Tiny pectoral fins visible above the yolk sac.[117]

4.54 mm TL. Pectoral fin growth is evident; caudal fin differentiation begins. On some individuals, ray development begins in both dorsal and anal fins.[117]

4.75 mm TL. Mesenchyme cells present in both dorsal and anal fins.[117]

From MD:

3.9 mm TL. Pectoral fin forming.[91]

From IN:

3.26–3.72 mm TL. Two days old. Tiny pectoral fins apparent above the yolk sac. A continuous median finfold starts just above the posterior end of the yolk and extends around the end of the urostyle forward to the anal region.[116]

4.27–5.54 mm TL. Four days old. Faint rays appear in pectoral and caudal fins. Caudal fin is homocercal but the continuous finfold is still present.[116]

4.64–4.82 mm TL. Five days old. Pectoral fins beat rapidly.[116]

4.58–4.82 mm TL. Six days old. Pectoral fins fan-like and move rapidly; finfold reduced except in the tail region; mesenchyme cells present in dorsal and anal fins.[116]

4.92–5.84 mm TL. Nine days old. Pelvic fins are present.[116]

5.41–5.80 mm TL. Ten days old. Finfold reduced; caudal fin with more rays.[116]

Pigmentation

From MD:

2.9–3.6 mm TL. No pigmentation.[91]

3.9 mm TL. Eye lightly pigmented.[91]

4.0 mm TL. Heavy eye pigment.[91]

4.5 mm TL. A row of melanophores is present on the ventrum posterior to the anus. A mid-lateral series of melanophores is present on some individuals as well as a prominent supra-anal melanophore.[91]

From IN:

4.27 mm TL. Eye pigmented.[116]

4.58–4.82 mm TL. Six days old. A pigment cell is located on each side of the body just dorsal to the anus.[116]

4.99–5.81 mm TL. Eight days old. Pigment cells are present in pairs lateral and dorsal to the anus, along the abdomen, and posterior to the liver.[116]

4.92–5.84 mm TL. Nine days old. Air bladder black.[116]

5.41–5.80 mm TL. Ten days old. Lens of the eye appears orange.[116]

From OH:

3.51 mm TL. The only pigmentation evident is in the eyes.[117]

3.96 mm TL. In life, blood is deep red.[117]

4.54 mm TL. Eyes deeply pigmented; very little pigment on body, if present consists of a few cells dorsally and ventrally.[117]

4.75 mm TL. More pigment present dorsally and ventrally.[117]

Table 60

Morphometric data for young bluegill from TN* grouped by selected intervals of total length. Characters are expressed as percent total length (TL) and head length (HL) with a single standard deviation. Range of values for each character is in parentheses.

	TL Intervals (mm)						
Length Range N	4.31–5.89 138	5.91–6.71 66	6.72–8.78 86	8.79–10.03 65	10.13–11.92 47	11.97–15.93 45	16.00–21.10 13
Characters	Mean ± SD (Range)	Mean ± SD (Range)	Mean ± SD (Range)	Mean ± SD (Range)	Mean ± SD (Range)	Mean ± SD (Range)	Mean ± SD (Range)
Length as % TL							
Snout	3.05 ± 0.38 (0.10–0.23)	3.07 ± 0.36 (0.15–0.25)	3.88 ± 0.52 (0.18–0.43)	4.79 ± 0.47 (0.32–0.57)	5.29 ± 0.64 (0.42–0.75)	5.21 ± 0.41 (0.52–0.95)	5.39 ± 0.34 (0.82–1.25)
Eye diameter	6.77 ± 0.36 (0.31–0.41)	6.76 ± 0.42 (0.36–0.51)	7.26 ± 0.38 (0.42–0.68)	8.26 ± 0.62 (0.66–0.92)	9.02 ± 0.48 (0.79–1.12)	9.28 ± 0.27 (1.08–1.45)	9.21 ± 0.16 (1.49–1.98)
Head	17.2 ± 0.88 (0.61–1.10)	18.6 ± 1.07 (1.01–1.35)	21.2 ± 1.49 (1.00–2.05)	23.7 ± 1.01 (1.95–2.60)	25.0 ± 1.64 (1.63–3.10)	25.3 ± 0.69 (2.89–4.10)	25.7 ± 0.37 (4.10–5.40)
Predorsal	31.2 ± 1.97 (1.50–2.00)	34.0 ± 2.34 (1.73–2.50)	33.6 ± 1.75 (2.08–3.00)	30.2 ± 1.07 (2.60–3.10)	30.5 ± 1.03 (3.00–3.80)	31.3 ± 0.83 (3.60–4.95)	30.9 ± 0.52 (4.90–6.50)
Prepelvic	—	—	—	28.3 ± 1.29 (2.45–3.00)	28.2 ± 0.98 (2.80–3.45)	28.4 ± 1.00 (3.20–4.55)	28.8 ± 0.78 (4.25–6.20)
Preanal	42.6 ± 1.09 (1.90–2.60)	44.2 ± 1.56 (2.48–3.20)	46.7 ± 0.94 (3.01–4.18)	46.4 ± 1.37 (4.09–4.93)	45.5 ± 1.35 (4.50–5.50)	44.6 ± 0.79 (5.20–7.25)	45.3 ± 1.05 (7.10–9.85)
Postanal	57.4 ± 1.08 (2.38–3.41)	55.6 ± 2.26 (2.68–3.71)	53.3 ± 0.95 (3.61–4.60)	53.6 ± 1.37 (4.54–5.58)	54.5 ± 1.35 (5.25–6.68)	55.4 ± 0.79 (6.59–8.75)	54.7 ± 1.05 (8.85–11.38)
Standard	96.7 ± 1.23 (4.13–5.75)	97.0 ± 0.65 (5.73–6.52)	94.3 ± 3.08 (6.53–8.11)	84.7 ± 1.53 (7.50–8.52)	82.3 ± 1.25 (8.45–9.81)	80.6 ± 1.09 (9.70–12.86)	79.2 ± 0.42 (12.60–16.80)
Air bladder	6.19 ± 1.39 (0.17–0.52)	8.36 ± 0.81 (0.38–0.65)	10.2 ± 0.84 (0.55–1.01)	11.6 ± 1.41 (0.85–1.58)	14.9 ± 1.98 (1.00–2.02)	18.0 ± 1.61 (1.95–3.20)	19.4 ± 5.31 (3.10–3.10)
Eye to air bladder	16.7 ± 1.08 (0.65–1.08)	16.6 ± 0.68 (0.85–1.15)	16.5 ± 0.94 (0.95–1.65)	15.6 ± 1.85 (1.12–1.82)	12.8 ± 1.93 (1.13–1.95)	11.3 ± 0.58 (1.27–1.85)	10.5 ± 0.12 (1.68–1.70)
Fin Length as % TL							
Pectoral	—	—	12.9 ± 0.76 (0.85–1.25)	14.0 ± 1.38 (1.05–1.72)	16.4 ± 1.69 (1.40–2.20)	17.0 ± 1.32 (1.85–2.95)	19.8 ± 1.60 (2.82–4.80)
Pelvic	—	—	—	3.6 ± 1.08 (0.12–0.55)	6.98 ± 1.77 (0.31–1.28)	10.0 ± 1.49 (0.90–1.97)	13.9 ± 0.87 (2.05–3.20)
Body Depth as % TL							
Head at P1	11.2 ± 0.98 (0.50–0.87)	12.8 ± 1.33 (0.56–1.03)	14.9 ± 1.23 (0.85–1.52)	17.5 ± 1.29 (1.30–2.00)	20.2 ± 1.29 (1.82–2.74)	22.2 ± 1.45 (2.37–3.87)	24.4 ± 1.38 (3.70–5.51)
Head at eyes	11.4 ± 0.64 (0.53–0.75)	12.3 ± 0.90 (0.65–0.97)	14.2 ± 0.83 (0.81–1.43)	16.1 ± 0.75 (1.28–1.72)	18.1 ± 1.05 (1.70–2.30)	19.2 ± 0.91 (2.10–3.20)	19.8 ± 0.70 (3.00–4.35)
Preanal	10.9 ± 0.85 (0.46–0.85)	11.9 ± 1.03 (0.55–0.95)	13.9 ± 0.97 (0.81–1.42)	15.4 ± 1.32 (1.18–1.83)	17.5 ± 1.03 (1.50–2.23)	20.0 ± 1.21 (2.13–3.45)	22.6 ± 1.34 (3.40–5.15)
Caudal peduncle	—	—	6.69 ± 0.49 (0.41–0.68)	7.94 ± 0.61 (0.62–0.92)	8.82 ± 0.48 (0.80–1.15)	9.44 ± 0.35 (1.01–1.52)	10.0 ± 0.45 (1.55–2.25)
Body Width as % HL							
Head	33.7 ± 4.16 (0.24–0.46)	37.9 ± 5.90 (0.26–0.80)	41.6 ± 4.88 (0.45–0.95)	46.4 ± 6.00 (0.80–1.42)	48.2 ± 6.49 (0.94–1.73)	51.6 ± 5.25 (1.30–2.50)	58.5 ± 5.97 (2.10–3.60)

Table 61

Means of selected measurements of larval and early juvenile bluegill from southeastern LA; all measurements are in mm.

TL Range	N	Preanal Length	Eye Diameter	Depth behind Vent	Depth to Base of Eye
4.00–4.99	9	2.05	0.29	0.35	0.46
5.00–5.99	41	2.40	0.31	0.48	0.51
6.00–6.99	82	2.79	0.37	0.63	0.60
7.00–7.99	47	3.23	0.44	0.79	0.67
8.00–8.99	47	3.74	0.54	0.96	0.80
9.00–9.99	39	4.20	0.68	1.32	0.97
10.00–10.99	33	4.55	0.77	1.61	1.09
11.00–11.99	35	5.07	0.88	1.91	1.22
12.00–12.99	21	5.36	1.00	2.29	1.42
13.00–13.99	10	5.84	1.09	2.51	1.53
14.00–14.99	7	6.35	1.25	2.77	1.72

Source: Constructed from data presented in Table 2 (in part), reference 84.

POST YOLK-SAC LARVAE

See Figures 56 and 57.

Size Range
Yolk and oil are completely absorbed between 5 and 6 mm TL.[116,117] All fin development is completed and finfold absorbed by 14.5 mm TL in OH,[117] by 12–12.6 mm TL in TN,* and by about 13.0 mm TL in Japan.[193]

Myomeres
See Table 59.

4.0–7.6 mm TL. Mode 12 preanal and 13 postanal.[84]

Morphology
5.04 mm TL. In OH, yolk and oil completely absorbed.[117]

5.40–5.63 mm TL. In IN, yolk and oil absorbed.[116]

7.6 mm TL. Jaws functional; foregut coils are evident.[84]

<13 mm TL. About 200 young bluegills less than 13 mm from Burgess Falls Lake, an impoundment of Falling Water River, TN, were examined and all found to be scaleless.[94]

Morphometry
See Table 60 for morphometric data collected from TN specimens.

See Table 61 for selected measurements of post yolk-sac larvae from southeastern LA. (*Note:* Additional morphometric data for LA bluegills can be found in reference 84.)

4.0–7.6 mm TL. ED > 5.5% TL; PreAL < 45% TL.[84]

5.04 mm TL. In OH as percent TL: PreAL 41.9; PostAL 57.9; ED 6.3.[117]

14.5 mm TL. As percent TL: PreAL 44.8.[15]

Fin Development
5.04 mm TL. No evidence of pelvic fins.[117]

6.0–8.0 mm TL. Caudal fin rays begin forming by 6.6–7.0 mm* or by 7.0–7.2 mm.[92]

Between 7.0 and 8.0 mm, ray differentiation begins in pectoral fins, first soft rays form in dorsal and anal fins, and the adult complement of caudal fin rays develop.*

8.0–9.0 mm TL. Adult complement of fin elements develop in median fins and by 9.0 mm, rays are visible in the pectoral fins.*

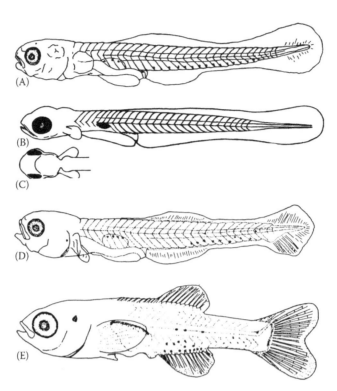

Figure 56   Development of young bluegill. (A) Post yolk-sac larva, 5.4 mm TL. B–C. Post yolk-sac larva, 5.5 mm TL: (B) lateral view; (C) dorsal view. D–E. Post yolk-sac larvae: (D) 6.8 mm TL; (E) 10.1 mm TL. (A and D–E reprinted from Figures 2, 3, and 4, reference 84; B and C reprinted from Figure 3, reference 92, with publisher's permission.)

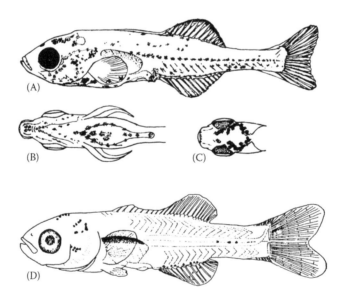

Figure 57   Development of young bluegill. A–C. Post yolk-sac larva, 10.3 mm TL: (A) lateral view; (B) ventral view; (C) dorsal view. (D) Early juvenile, 13.3 mm TL. (A–C reprinted from Figure 3, reference 92, with publisher's permission; D reprinted from Figure 5, reference 84.)

8.9–9.4 mm TL. All fin rays have formed.[92]

9.0–10.0 mm TL. Ray differentiation begins in pelvic fins; rays visible on some by 10.0 mm; median finfold absent on caudal peduncle on some individuals but traces of finfold still present on some.*

10.0–11.0 mm TL. Last remnants of finfold on caudal peduncle and ventrally anterior to the anus disappear.*

12.0–12.6 mm TL. Adult complement of pelvic fin rays visible.*

Pigmentation

5.40–5.63 mm TL. No pigmentation on top of head.[92] On each side of the body about 11 pigment cells are present between the ventral somites from the anus to the tail; 2–3 pigment cells are lateral and dorsal to the anus[116] and 3–4 are located ventrally on the abdomen between the heart and the anus.[84,116] There is a large pigment cell on each side of the body just posterior to the pectoral fins.[116] Scattered chromatophores are evident on the dorsal surface of the air bladder. [84]

7.0 mm TL. A row of chromatophores is developing along the posterior margin of the head.[92]

8.0 mm TL. Chromatophores have developed along the sides of the head.[92]

9.0–11.1 mm TL. Chromatophores cover the dorsal surface of the head. Ventral pigment is heavy on the isthmus and stomach and sometimes appears egg-shaped with a line of pigment down the center; less distinct patterns are present on some individuals.[92]

JUVENILES

See Plate 2A following page 26 and Figure 57.

Size Range

All fin development is completed and finfold absorbed by 12.0–12.6 mm TL in TN* and by 14.5 mm TL in OH.[117] Smallest mature males are reported at 58 mm and females 89 mm.[137] Sizes when males are capable of successfully reproducing may be larger and dependent upon their ability to defend their nest.[114]

Myomeres

See Table 59.

Morphology

13.0–14.0 mm TL. Scales first appear along the midline immediately anterior to the caudal fin (Figure 58).[94]

14.0–15.0 mm TL. In TN, scales develop anteriorly in a single row along the midline.[94] At 14.5 mm TL, in OH, scales have not yet developed.[117]

15.0–16.0 mm TL. Dorsal and ventral rows of scales form next to the row along the midline.[94]

16.0–25.0 mm TL. Progression of scale development continues anteriorly, more so ventrally than dorsally. The last area of the body to become scaled is the dorsal area of the nape. The first individuals observed to be 100% scaled were 24–25 mm TL (Figure 58).[94]

≥25.0 mm TL. All bluegills 25 mm and longer were completely scaled.[94]

Larger juveniles. Lateral line scales 38–50,[185,189] usually 38–45;[170,172,176,177,179,181,184,186,188] gill rakers 12–16;[170,176,189] vertebrae 28–30;[170,176,181] In Canada: 12 gill rakers; usually 6 branchiostegal rays; usually 6 pyloric caeca.[176]

Morphometry

See Table 60 for morphometric data collected from TN specimens.

See Table 61 for selected measurements of early juvenile bluegills from southeastern LA. (*Note:* Additional morphometric data for LA bluegills can be found in reference 84.)

Mean maximum depths related to TL of juvenile bluegills ≥25 mm TL are presented in Table 62.

From OH:

14.5 mm TL. As percent TL: PreAL 44.8; PostAL 55.2; GD 19.0; ED 8.3.[117]

Fins

14.5 mm TL. All fins now developed; pelvic fins are small, but all elements are present.[117]

Larger juveniles. Two dorsal fins are broadly joined and appear as one.[176,177] First dorsal fin has 9–11,[170,172,176,179,186,188] usually 10 spines;[170,177,179,181,184,186,188,195] second fin with 10–12 soft rays.[170,172,176,177,181,184,186,188,195] Anal fin has 3 spines[172,176,177,181,184,186,188,195] and 10–12 soft rays.[170,172,176,177,181,184,186,188,189,195] Pectoral fins have 12–15,[170,186] usually 13–14 soft rays.[170,176,181,189] Pelvic fins have 1 spine and 5 soft rays.[176,177,181]

Pigmentation

13.5 mm TL. There is a small, faint patch of pigment on the head and a similar patch on the opercle between the eye and pectoral fin base. Pigment is developed along the mid-lateral line on posterior third of the body. Scattered chromatophores are present ventrally from the anal origin to the tail, these extending dorsally almost to the mid-lateral line. Pigment is evident on the caudal and anal fins.[191,192]

14.0–15.0 mm TL. Lateral stripe evident (Japan).[193] At 14.5 mm, pigment is present on the head, opercle, jaws, fins, and body.[117] At 15 mm, a ventral line of melanophores is interrupted and not present on the belly; melanophores on the breast are dispersed.[86]

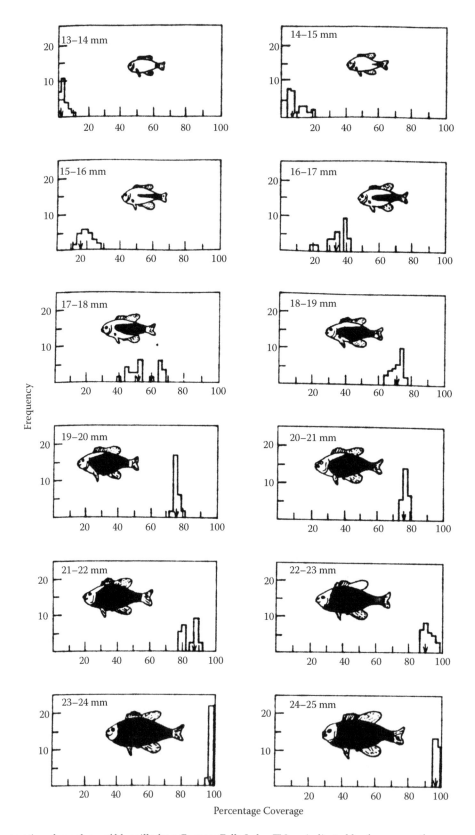

Figure 58   Squamation chronology of bluegills from Burgess Falls Lake, TN, as indicated by frequency of occurrence and diagrammatic representation of median (arrow) percentage of scale coverage. (Reprinted from Figure 2, reference 94, with publisher's permission.)

Table 62

Mean maximum body depth (mm) related to TL (mm) for young bluegill.[137]

TL	No. of Fish	Depth Mean	Depth Percentage TL
25	141	7.1	28.4
51	195	14.0	27.5
76	189	24.0	31.6
102	146	34.0	33.3
127	180	44.0	34.6
152	176	54.0	35.5

21.5–24.0 mm TL. Pigment is evident in all fins except the ventrals. About 11 narrow bands of pigment are visible on the body between the head and the end of the caudal peduncle.[191,192,198]

<50 mm TL. A few large melanophores are present on the breast.[178]

32.0–59.7 mm SL. Primary bars appear as an initial set of lateral patches of melanophores to which bars are added anteriorly and posteriorly in development of all *Lepomis*. Seven to eight primary bars are well developed on bluegill juveniles by 32 mm SL. Light centers develop in the primary bars on bluegill. Secondary bars of melanistic pigment do not form on juvenile bluegills (Figure 59).[93]

In life, immature specimens have light flesh-colored breasts like females, but blue-green body pigment and vertical bars are more pronounced. Young are lighter than adults, usually silvery, sometimes with a purplish sheen.[192] Yearlings and 2-year-olds are grayish blue on the back and sides, with indistinct lateral barring and a solid black opercular spot.[28,192]

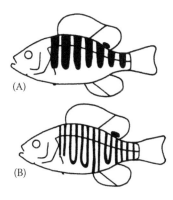

Figure 59 Development of young bluegill — schematic drawings showing the development of melanistic pigment patterns in young bluegills. (A) 32.0 mm SL; (B) 59.7 mm SL. (A and B reprinted from Figure 14, reference 93, with publisher's permission.)

TAXONOMIC DIAGNOSIS OF YOUNG BLUEGILL

Similar species: other *Lepomis* species, very similar to orangespotted sunfish. There are comparisons in the literature of the early development of young bluegill to redear sunfish and warmouth,[31] pumpkinseed and redear sunfish,[86] green sunfish,[92] and green sunfish and longear sunfish.[110]

See *Helioperca* subgeneric discussion on page 91.

For comparisons of important morphometric and meristic data, developmental benchmarks, and pigmentary characters between bluegill and all other *Lepomis* that occur in the Ohio River drainage, see Tables 4 and 5 and Plate 2A.

Bluegill vs. redear and warmouth (Orange Lake, FL). At 4–5 mm TL, the supra-anal pigment spot is more prominent on bluegill than on warmouth or redear; after about 6 mm TL, bluegill larvae can be distinguished from warmouth by mouth size (mouth extends past the anterior edge of the orbit in warmouth; mouth is anterior to the eye in bluegill); chromatophores on dorsum of the head rarely appear before 8 mm TL on bluegill larvae but are visible by about 5–6 mm TL on redear sunfish and by about 7 mm on warmouth; by about 14–15 mm TL, the relative shapes of the air bladders are different for bluegill, redear, and warmouth (Figure 47). Also by that size, gill rakers have attained adult morphologies and the rakers are short and stubby in redear sunfish and longer and slimmer for bluegill (Figure 47).[31]

Bluegill vs. Pumpkinseed. For comparisons of bluegill vs. pumpkinseed, see discussion in pumpkinseed species account (page 134).

Bluegill vs. redear sunfish.

> 7.0–7.2 mm TL — Caudal fin ray formation begins for bluegill; begins for redear at 6.2–6.4 mm.[92]
>
> 8.9–9.4 mm TL — All fin rays are formed for bluegill, not until 9.4–9.8 for redear.[92]
>
> 12–17 mm TL — Mean ratio of body depth into TL was 4.6 for bluegill and 4.1 for redear; bluegill has 17 caudal vertebra; redear has 18 (requires clearing and staining).[86]
>
> <15 mm TL — Redear has a series of melanophores running in a straight line posteriorly along its ventrum from the angle of the gill covers across the breast and belly to the anus, where they branch into two lines, one on either side of the anal fin. Bluegill also has a line of melanophores on its ventrum,

but this line is interrupted and not present on the belly. Also, the melanophores on the breast of preserved bluegill are large and diffuse; melanophores, if present, on the breast of redear are small and circular.[86]

25–30 mm TL — Bluegill have long gill rakers; redear have short gill rakers.[86]

Bluegill vs. green sunfish. See discussion in green sunfish species account (page 118).[92]

Bluegill vs. green sunfish and longear sunfish (in LA).[110] Bluegills have an essentially uncoiled foregut until most caudal fin rays are formed. Prior to completion of caudal fin development, bluegills have proportionally smaller eyes and more posteriorly placed air bladders than green sunfish and longear sunfish. Also, bluegills are markedly larger at comparable stages of development.[110]

(*Note:* Conner [1979] indicates that bluegill have greater preanal lengths and more preanal myomeres than longear and green sunfish; however, data in his tables [2 and 3] do not support this statement.)

For larger bluegill larvae, if a mid-lateral streak of "dash"-like melanophores develops at all (occasionally in bluegill), it tends to be much less prominent than in green sunfish and is usually confined to the caudal peduncle. Prior to the appearance of juvenile coloration (vertical bars narrower than the interspaces), no completely diagnostic characters are apparent for bluegills.[110]

Juveniles. Juveniles of some other *Lepomis* species have barred color patterns on their sides (Plates 1 and 2) and may closely resemble the bluegill. Useful characters for distinguishing bluegills <50 mm SL from other *Lepomis* include the lack of red or orange pigment in the eye, the pale breast and lower head, and the unpigmented basal portion of the caudal fin.[170]

ECOLOGY OF EARLY LIFE PHASES

Occurrence and Distribution (See Figure 60)

Eggs. Fertilized bluegill eggs are slightly adhesive and stick to pebbles, small rocks, sticks, and leaves that may be present in the nest.[117,175] The male parent normally remains to care for the eggs, guarding them against predators and fanning them,[47,86] a behavior that apparently keeps silt from settling on the eggs and also aerates the eggs.[117] The male guards and fans the eggs for 2–5 days until they hatch.[47,86] However, when food is scarce bluegills commonly eat their own eggs, and earliest eggs are usually eaten by the male bluegill, so that early spawns are rarely successful.[137]

In MN, fluctuations of dissolved oxygen (DO) concentrations and water temperatures in natural spawning sites of bluegill were measured during embryo and yolk-sac phases of development. DO concentrations 1 cm from the bottom of bluegill nests ranged from 2.4–18.2 mg/l and water temperatures from 15.0–27.5°C, with average daily

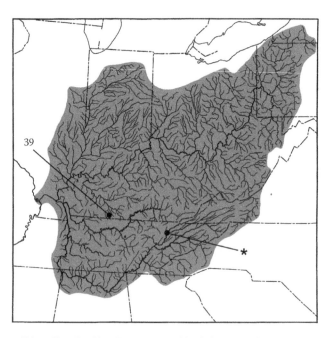

Figure 60 General distribution of bluegill in the Ohio River system (shaded area) and areas where early life history information has been collected (circles). Number indicates appropriate reference. Asterisk indicates TVA collection locality.

fluctuations of 4.4 mg/l and 3.3°C.[14] Bluegill nesting may be arrested by unseasonably cold water temperatures.[96]

The application of herbicides (Aquathol–K or 2,4–D) caused nest abandonment by guarding male bluegills. Abandonment occurred at herbicide concentrations commonly used to control vegetation (4 ppm), but well below concentrations lethal to sunfishes (≥125 ppm). When abandonment occurred, time away from the nest averaged 5–6 min, and in nearly all the trials where abandonment occurred, congeners intruded upon the nests consuming eggs and fry.[36]

The sensitivity of bluegill embryos and larvae to low pH in soft (12 or 18 mg/l $CaCO_3$) and hard (165 or 197 mg/l $CaCO_3$) water was compared in 5-day laboratory toxicity tests. Embryo and larval bluegills were exposed to pH levels ranging from 3.8 to 7.0 in soft water and 3.8 to 8.0 in hard water. An on-site toxicity test, using lake water (3.4 mg/l $CaCO_3$) adjusted to pH levels ranging from 3.5 to 7.3, was conducted to compare laboratory and field results. At low pH, hatching was reduced, the hatching period prolonged, and the incidence of partial hatching increased. Increased water hardness mitigated acid toxicity, enhanced larval survival, and promoted hatchability. Hatching rates were decreased over those of the controls by 76% in soft water and by 23% in hard water at pH 4.0, and hatching was negligible at pH 3.8. Partial hatching averaged 43% at pH 4.4. Increasing acidity resulted in increased embryo-larval mortality, averaging >62% at pH levels <4.6; mortality was 100% at pH levels <4.4 in soft water and 3.8 in hard water. Bluegill larvae were more sensitive than eggs.[105]

Yolk-sac larvae. After hatching, bluegill fry are guarded (but not fanned) by the male parent for an additional 3–6 days[47,175] or the male guards the nest until the eggs hatch, but does not guard the fry once they leave the nest.[179]

Newly hatched yolk-sac larvae are weak swimmers[86] and are not capable of much action.[117] Handicapped by a large amount of yolk and the absence of fins, they can only move short distances at the bottom of the nest. Initial movement consists of circling on their sides, pivoting around their large yolk sacs. Some may remain quiet, lying on their sides for some time. Eventually they swim by an undulating movement of the body and are quite active for short periods followed by short resting periods and then movement again.[117] Absorption of the yolk occurs rapidly, and by the eighth day after fertilization the alimentary tract has developed and the yolk-sac larvae are able to begin exogenous feeding. Two days later, at 5–6 mm, the yolk has been completely absorbed and the free-swimming bluegills leave the nest.[86]

In northern WI, yolk-sac larvae remain on the nest 4–10 days after hatching. They leave the nest when the yolk is almost absorbed and disperse over 1–3 days, depending on water temperature. The larvae average 5 mm TL at this time. There was a significant positive relationship between dispersal of bluegill larvae and the date of first dispersal.[99]

In a VA lake, nest preparation by male bluegill exposed coarse gravel (8–32 mm in diameter) and pebbles (32–64 mm in diameter) and removed particles smaller than 2 mm. Particles larger than 8 mm provided suitable interstitial space to accommodate bluegill larvae. Survival of larvae was directly correlated with the proportion of coarse substrate in the nest. Mortality of larval bluegill from predation was measured in 56 nests guarded by males and 21 nests from which the male guardian was removed. Mortality was significantly greater in unguarded nests (median = 68%) than in guarded nests (median = 14%). Bluegill 30–120 mm TL were the most abundant nest predators; pumpkinseed (70–110 mm TL), largemouth bass (40–50 mm TL), and whitefin shiners (50–60 mm TL) were also nest predators.[131]

Larvae. In Ridge Lake, IL, bluegill had protracted spawning seasons during the period 1987–1989 and produced several peaks in larval abundance each year (Figure 61). In each year, peaks in abundance of newly hatched larvae occurred during May, June, and July, and there was a general decline in larval abundance throughout the summer.[161]

Post yolk-sac larvae. As soon as the yolk is absorbed and the fins have developed, bluegill larvae rise from the nest and remain in schools hiding in vegetation along the shore, no longer guarded by the male.[117] During the period immediately following parental care, young bluegills are especially vulnerable to predation, but they minimize their availability to predators by abandoning the nests synchronously, with most fry in an entire colony leaving within a period of a few hours.[189]

In IN and MI lakes, ecology studies have shown that bluegill larvae leave the littoral zone shortly after leaving the nest and move to the limnetic region,[80,86,113] arriving there when they are 10–12 mm long, spending their time in the epilimnion, most commonly in the upper 2 m of water. Distribution of the fry in the limnetic region was patchy, suggesting that they were aggregated, possibly in schools.[86,113] In northern WI lakes, bluegill larvae also demonstrated an intralacustrine migration pattern. They were collected in the limnetic area of the lakes about 2 days after dispersal from the nest was complete. They remained in the limnetic zone for 30–40 days and were 15–20 mm TL when they returned to the littoral zone. The migrational behavior of bluegill fry may be an adaptation to

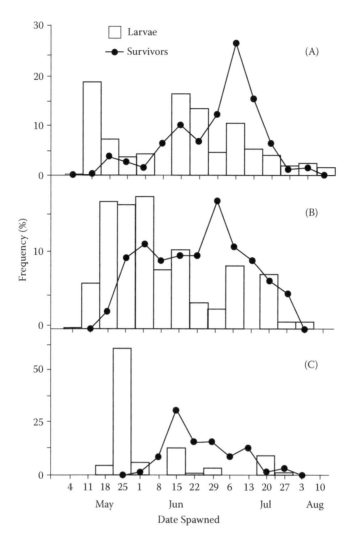

Figure 61 Comparison of spawning date distribution of newly hatched larval bluegills (5 and 6 mm TL) and juvenile bluegills in Ridge Lake, IL, that survived to fall during (A) 1987, (B) 1988, and (C) 1989. Larvae were sampled weekly at night from late April through August; juveniles were sampled annually in mid-September in shoreline rotenone surveys. (Reprinted from Figure 1, reference 161, with publisher's permission.)

move to an environment with fewer predators and a greater food supply.[99] In the Great Lakes, larval bluegill <12.7 mm TL moved into the limnetic zone and remained there for 6–7 weeks. When about 20–25 mm TL they returned to the littoral zone and remained in dense vegetation in marshes, bays, or lower reaches of lower rivers for the rest of the summer.[175]

In the Buncombe Creek arm of Lake Texoma, OK, larvae and YOY bluegill from <5.5 mm to 48 mm TL were collected by shoreline seining and open-water trawling. With the trawls, 18.1% of the young bluegills were taken in daytime collections (120 samples); 81.9% of the young bluegills came from night collections (175 samples). Bluegill larvae were active and widespread in the lake at an early age. Larvae ≤5.5 mm were often found in surface collections from open water. In the daytime, these smaller bluegill larvae were concentrated near the

surface. At night they showed a more even distribution but were still more abundant in surface samples than at mid-water or near the bottom. Larvae 6–12 mm were infrequent from deepwater stations and were more concentrated in the coves. They were collected primarily from the bottom in the daytime, but many were found in mid-water near the shore. At night they were still abundant on the bottom but were more evenly distributed through the water column. Larvae >12.5 mm were essentially absent from open water areas and rarely caught in surface collections. Bottom distribution of these larger larvae was similar to that of the 6–12 mm fish with an even higher percentage of their number taken in coves. Larvae >12.5 mm stayed close to the bottom both day and night. Larvae and YOY bluegills 5–48 mm were collected by shoreline seining. They were most commonly collected in quiet, silt-bottom areas and appeared to be in loose aggregations near the shore.[191]

In two AL ponds in which successful bluegill spawning occurred from April through September 1994, larval bluegill density was negatively correlated with zooplankton density in both ponds, suggesting that larvae reduced zooplankton abundance via predation. Cohorts of larvae whose densities were low tended to have faster growth and higher survival than cohorts with high larval densities. Relative survival of larval bluegills to the juvenile stage was negatively correlated with limnetic larval fish density, but no relations were detected between larval survival and juvenile recruitment across cohorts, suggesting that juvenile recruitment was set at larval sizes larger than those examined in this study (5–13 mm TL). The authors suggested that growth and mortality of larval bluegills are influenced by larval abundance and intraspecific competition for zooplankton when resources are limiting, and that recruitment of juveniles did not appear to be set during the larval stage of the bluegill's life history.[154]

In Orange Lake, FL, peak occurrence of larval bluegill in larval fish collections was in July and August when surface water temperatures were 24–27°C. Bluegills <8 mm TL were captured from April through September.[31] Smallest bluegill were captured in open water habitat and in hydrilla and panic grass habitat; most bluegill <8 mm were found in panic grass habitat.[31,72] Few bluegill >12 mm TL were captured in open water and few <20 mm were captured in floating-emergent vegetation.[31]

In a KY creek, larval bluegill were collected with drift nets, light traps, and seines from mid-June to mid-July. Most of the larvae were collected in low-flow areas with light traps. Bluegill larvae were captured in most habitats sampled, but were most numerous from areas with algae-covered rocks or gravel substrate and from near root wads in undercut banks.[39]

Bluegill larvae 5.1–12.6 mm TL were collected from the Tallapoosa River, AL, in 1988 and 1989 with seines and larval pushnets. Most were captured with seines in shallow (0.67–0.77 average depth) shoreline (1–5.5 m from shore) areas with slow current (average current 0.61–1.31 cm/s).[60]

Bluegill larvae were collected in early June from the lower Mississippi River with meter and half-meter ichthyoplankton nets. Three stations were sampled to include a wide range of surface current velocities. Bluegill were captured from a station with swift current (75–90 cm/s) and from a bayou station that registered zero current, but no bluegill larvae were captured from a station with sluggish current (5–20 cm/s).[85]

Larval bluegill survival after 90 days' exposure to copper (Cu) was adversely affected at 40–162 µg Cu/l. At 40 and 77 µg Cu/l, survival was significantly lower than those of the controls. No larvae survived at 162 µg Cu/l.[109]

In small hypereutrophic reservoirs (<100 ha; >100 µg total phosphorus/l), larval gizzard and threadfin shad reach high densities in the limnetic zone, virtually eliminate zooplankton, and perhaps compromise success of other planktivorous larvae, such as bluegill. To examine how the relative timing of appearance of shad and bluegills influences their relative success, laboratory experiments were conducted and densities of bluegill larvae and zooplankton were quantified during spring through summer in 3 OH reservoirs across 8 years (1987–1994). These experiments suggest that bluegill success should vary among reservoirs and years as a function of their appearance relative to gizzard shad. In reservoirs, zooplankton availability and bluegill abundances were consistently low during years when gizzard shad dominated reservoir fish assemblages.[162]

Juveniles. Young bluegill are gregarious and frequent weedbeds or other areas of heavy cover.[170] In OH, early juveniles 14.5 mm TL were quite active and traveled in schools.[117]

In WI, young bluegill will often be seen in the hottest and shallowest water in midsummer.[96]

In a northern IN lake, bluegill fry returned from the limnetic region to the littoral zone at lengths of 20–25 mm.[86,113] In northern WI lakes, juvenile bluegills 15–20 mm moved from the limnetic zone back into littoral habitats.[99] This migrational behavior was also reported from an IL pond. Juvenile bluegills 14–20 mm TL were generally most abundant at the 1–m depth at offshore locations, and bluegills 20.5–26.5 mm were generally most abundant at shoreline locations. However, small bluegill were abundant at two shore locations, indicating that some bluegill larvae did not migrate to the limnetic zone or that they returned from the limnetic waters to the shoreline at a smaller size than given in other reports.[101]

Young bluegill 35–50 mm SL are restricted to vegetated habitats by predation pressure. Vegetation provides refuge by hindering predator foraging success. Laboratory experiments suggest that juvenile bluegill are capable of perceiving and selecting plots of vegetation that offer safety from predation.[159] In Lake Mendota, WI, age-0 bluegill were more abundant where Eurasian water milfoil was abundant and patchy; yearling and older bluegill were more abundant in areas of dense and species-rich vegetation.[46]

In MO ponds with macrophyte coverage ranging from 0 to 100%, age-0 gizzard shad abundance declined with increasing macrophyte coverage; however, age-0 bluegill abundance was unrelated to macrophyte coverage.[152] In clear Ozark lakes, underwater behavioral observations (26 h) using SCUBA were analyzed to determine microhabitat use and time budgets of adult and juvenile sunfishes in the vicinity of weed beds (*Justacia americana*). Juvenile bluegill spent 92% of their time in the vegetation at mid-depths.[27]

In Orange Lake, FL, most bluegill 20–40 mm were captured in floating-emergent vegetation,[72] but few <20 mm came from this habitat.[31] In Reelfoot Lake, TN, juvenile bluegill 16–36 mm in length were abundant in shallow water and heavy vegetation.[59]

In a MI lake, bluegill <80 mm TL were confined throughout the year to vegetation, apparently to avoid predation; larger fish moved up in the water column. These smaller bluegill were observed in schools and exhibited differences in vertical distribution of size classes. Bluegill averaging 25 mm TL occurred in schools averaging 64 individuals. Eleven such schools ranged in vertical distribution from near the bottom to an average upper boundary 56 cm above the bottom. Bluegill averaging 50 mm TL were observed in seven slightly smaller schools (average 46 individuals) with average lower and upper boundaries of 79 and 146 cm above the bottom. Six schools of bluegill >75 mm TL averaging 47 individuals all occurred between 150 and 300 cm above the bottom. Bluegill 50 mm and shorter were clearly confined to the vegetation; those 75 mm and longer had moved into the 0.5- to 1.0-m stratum above the vegetation.[35]

Juvenile bluegill (20–90 mm TL) were collected with minnow traps in stands of aquatic vegetation from Bluff Lake, MS, during July–October 1990. Traps were fished in stands of water lily, water shield, and pondweed. For water lily, water shield, and pondweed, average stem density within macrophyte stands was about 19, 209, and 867 stems/m^2, respectively. Pondweed exhibited significantly higher surface coverage than the other two species. Bluegill fry were attracted to the relatively dense, underwater leaf/stem complexities and relatively small interstices afforded by the pondweed. Catch per unit effort (CPUE) for numbers of bluegill (fish/trap-night) and CPUE for weight of bluegill (g/trap-night) were significantly greater for samples from pondweed than for those from water lily or water shield.[148]

Bluegill in a small MI lake over-wintered in two localized regions of tall perennial vegetation in the deep littoral apparently in response to predation and food availability. YOY bluegill migrated to the limnetic zone in early summer and returned to the littoral in August. This movement behavior was also observed in an IN lake.[35]

In a 0.6-ha pond in central IL, juvenile bluegill were captured with plexiglass traps from May 27 to September 22, 1982. They were collected in all three habitats sampled which included shoreline (0.5 m deep in littoral zone within 1 m of shore; traps set at the surface), near-shore (1.5 m deep, 1–3 m outside the littoral zone; traps set at the surface and 1-m depths), and mid-pond (water depth 3 m, about 30 m from shore; traps set at the surface and 1- and 2-m depths). Though occupying all habitats sampled, they became increasingly restricted to shoreline habitat as they increased in size. Small bluegill (14–20 mm TL) were abundant in all three habitats. Medium fry (20.5–26.5 mm TL) were abundant in near-shore and shoreline habitats but not in mid-pond habitat. Bluegill ≥27 mm TL were chiefly shoreline residents. All sizes avoided surface waters at near-shore and mid-pond locations and, in early and mid-summer, the 2-m depth at the mid-pond location. The catch of bluegill fry at the 2-m depth increased as dissolved oxygen concentrations gradually increased in late August and September.[42]

In Lake Rush, OK, seines captured more YOY bluegill <30 mm TL than minnow traps, but larger YOY were more abundant in traps. Unbaited minnow traps were used in 1980 and 1981 to collect YOY of five species of sunfishes to determine their diets and use of microhabitats (type of vegetation, water depth, and substrate type). In 1980, bluegill up to 80 mm TL were caught proportionately more often in traps set at depths of 50 cm or less, but their distribution was bimodal with high abundances in the shallowest habitats and at depths between 151 and 200 cm. In 1981 most bluegill were captured at depths between 151 and 200 cm but they were present at all depths. In 1980 most bluegill were collected in habitats containing coontail and water milfoil, but many were also caught from areas where muskgrass and quillwort grew. With the disappearance of water milfoil in 1981, bluegill greatly increased their use of coontail stands. During the 2 years bluegill were more often found over fine substrates (72–76%) than coarse substrates (24–28%).[73]

In four flood-control reservoirs in NE, trap nets with 13-mm mesh caught significantly more bluegill <80 mm TL than did 16-mm mesh nets. The 16-mm mesh trap nets caught significantly more bluegill ≥80 mm TL.[165]

Three juvenile bluegill 17.2–31.6 mm TL were captured with seines from the Cahaba River, AL, 1 m from shore in water 0.47–0.63 m deep, in current that ranged from 0.0 to 1.74 cm/s.[60]

In the New River, WV, young bluegill ≤99 mm TL and 100–199 mm were captured by electrofishing from backwater, side channel, pool margin, lateral pool, and submerged vegetation habitats. The smaller bluegill were also captured around water willow. Neither size class of bluegill occurred in mid-pool habitat.[6]

In Lake Opinicon, ON, bluegill fry born earlier in the spawning season had strikingly higher survivorship to age 1 than fry from middle- or late-season spawning bouts. Young bluegill produced during the first trimester increased 231% in representation among yearlings. This increased survivorship was probably due to increased over-winter survivorship and possibly also size selective predation. The authors felt that their results highlighted the importance of the temporal dimension in understanding the nature of fish recruitment.[112]

No differences were found in the date of reappearance (August 4) at three sites in the littoral zone of Lake Opinicon, ON, of YOY bluegill and pumpkinseed >24 mm TL, but spatial separation was evident between the two species when they first appeared inshore. Pumpkinseed YOY were found close to the substrate while bluegill occurred in the upper stratum (>1 m of the water column). However, an observed shift in bluegill distribution was evident for the period August 5–18 and the period August 21 to September 30. Bluegill occurred significantly more often in the upper stratum (>1 m) in the period August 5–18 than after.[122]

In three northern WI lakes there was a significant positive relationship between date of first fry dispersal and fall standing crop of fingerlings in weight and numbers.[99]

Early studies of centrarchid fishes suggested that recruitment was higher for fish hatched early in the spawning season. Similar recruitment patterns were observed for bluegill, but only in piscivore–free waters at the northern extent of their range. The role of predation and first-winter mortality in governing bluegill recruitment was investigated in Ridge Lake, IL. During the period 1987–1989, spawning date distributions for larval bluegill collected in ichthyoplankton tows were compared with distributions of juveniles in shoreline rotenone samples from the fall and following spring. For juvenile bluegill sampled in fall, the range of dates on which they were spawned was similar to that observed for newly hatched larvae (Figure 61). The period of greatest larval abundance occurred early in the spawning season in all years, but bluegill from later spawns showed the highest contribution to fall survivors. Lack of recruitment of early-spawned bluegill was not explained by estimates of mortality of limnetic larvae 5–11 mm TL. Age-0 bluegill consumption by largemouth bass indicated that predation was an important source of mortality for early-spawned bluegill. Estimates of bluegill abundance and size structure in fall and spring showed that losses over the first winter were high (75–88%), but unlike studies from northern populations, there was little evidence of size-specific mortality.[161]

Degray Lake, AR, biomass of bluegill increased from May to August and decreased in September, probably as a result of predation. Because of higher summer water temperatures, bass would more likely be found offshore in limnetic waters where bluegill were available for forage.[132]

In a study of the effectiveness of a drop net, a pop net, and an electrofishing frame for collecting quantitative samples of juvenile fishes in vegetation habitats of a backwater lake in the upper Mississippi River, juvenile bluegill dominated the catch of all three gear types.[87]

Laboratory experiments showed that with a relatively minor increase in turbidity an apparent reduction in open-water predation risk for juvenile bluegill from largemouth bass existed, suggesting that size-specific habitat use by bluegill in turbid systems may not be as simply defined as in clearwater lakes.[155]

Coutant (1977) summarized reports of several authors concerning temperature preferenda for young bluegill. In laboratory studies, upper avoidance temperature for young bluegill (size not provided) ranged from 32.1–33.1°C, lower avoidance temperature ranged from 28.5–29.3°C, and final preferenda ranged from 30.2–31.5°C. For bluegill 45–110 mm, the upper avoidance temperature was 33.0°C, lower avoidance was 26°C, and preferred temperature was 32.3°C. Bluegill 120–155 mm preferred 30.5°C water. In WI, young bluegill 53–99 mm preferred water 28.8–31.2°C in the daytime and 27–29°C at night.[26]

In the laboratory, the upper 96-h thermal limit (TL_{50}'s) were 27.5 and 37.3°C for juvenile bluegills acclimated at 12.1 and 32.9°C, respectively. The lower 96-h TL_{50}'s were 3.2 and 15.3°C for the same acclimation temperatures. TL_{50} temperatures for juvenile bluegill increase with increasing temperature of acclimation.[83]

Indices of hypoxia and hypothermia tolerance for MO fish assemblages were determined based on laboratory measurements of lethal dissolved oxygen concentrations and temperatures combined with field measures of relative abundances of tolerant and sensitive species. Mean hypoxia tolerance for bluegill was 0.66 mg/l dissolved oxygen; mean hyperthermia tolerance was 37.9°C.[65,66]

Young bluegill 30–80 mm TL were the most abundant species found drifting in the tailwaters downstream of Lock and Dam 13 and Lock and Dam 14 on the main channel of the Mississippi River during late winter (February 1984 and March 1985). During winter the main channel of the Mississippi River can fall to 0°C for extended periods. Warmer alternative habitat exists in off-channel areas. Most of the fish caught were live or recently dead, and their presence in the drift of the main channel strongly suggests that their likelihood for survival was less than those in less hostile microhabitats in the pool.[126]

In OH hatchery ponds, preferences of juvenile bluegill 35–95 mm TL for artificial structure interstices of 40, 150, and 350 mm were determined both before and after largemouth bass were added to the study ponds. In the absence of predators, these yearling bluegill preferred small- and medium-interstitial structures over large interstitial regardless of major differences in water clarity among ponds. In the presence of largemouth bass, bluegill preferred small over medium and large interstitial structure in all ponds.[158] In another study in OH hatchery ponds, the preferences of juvenile bluegill (50–100 mm TL) for interstitial sizes (40, 150, and 350 mm) within artificial structures and shaded vs. unshaded structures in the absence or presence of largemouth bass were examined. The juvenile bluegill consistently preferred 40-mm interstitial structure regardless of shading or presence of largemouth bass.[168]

Nocturnal habitat selection by young bluegill under the risk of predation was examined in the laboratory. This study quantified the diel use of artificial macrophytes and open water when the open water was empty (control), contained food, or contained both a caged predator and food. Juvenile bluegill 62–77 mm TL spent significantly more time in macrophytes in the predator and food treatment than in the control, followed by the food-only treatment. Bluegill this size spent significantly more time in macrophytes during the day than at night in all treatments. Larger bluegill (102–130 mm) showed no difference in habitat use among treatments but spent significantly less time in macrophytes at night than during the day. The results of these experiments suggest the potential for a diel littoral–pelagic habitat change by juvenile bluegill that would have important implications for the role of bluegills in lake food webs, including the possibility of nutrient translocation that could generate alternate stable states in lakes.[160]

In laboratory studies, juvenile bluegill 75–125 mm TL, when given a choice of open water and either structure or low light intensity, or both, most often occupied areas that had very low light intensity and vertically oriented structure. The choice of vertically or horizontally oriented structure was influenced by the size of the interstitial spacing between the structural components. When interstitial spacing of horizontally oriented structure exceeded the body depth of the fish, horizontal and vertical structures were used more evenly.[54]

Eleven parasite taxa and six endoparasitic taxa were observed in 705 juvenile (25–65 mm TL) bluegill collected from a floodplain pond of the lower Mississippi River from March 19 through December 16, 1986. A trematode, *Allacanthochasmus* sp., was the most prevalent parasite. Parasite loads increased quickly as spring progressed, and nearly all the juvenile bluegill over 25 mm long supported a parasite fauna. Mean endoparasite intensities in pre-winter bluegill were 14.0–35.6 times higher than those exhibited by similar-sized bluegills in the spring. *Allacanthochasmus* sp. accounted for 92% of the endoparasite load, and mean intensities were 17.2–47.0 times higher than those of similar-sized post-winter bluegills. There was no relationship between endoparasite load and bluegill condition for any length group studied; however, there was little evidence of parasite mortality at low temperatures and the authors suggest that if the pre- and post-winter intensity data are representative, over-wintering mortality of parasitized bluegill may be substantial.[45,156]

Bluegill juveniles ≤70 mm TL in a NC lake were susceptible to infestation by metacercariae of *Uvulifer ambloptitis*, the black spot parasite of centrarchids. Prevalence and density of infection were greatest in bluegills 31–50 mm TL. Data indicated that heavily infected hosts (>50 metacercariae per fish) were most abundant during September, and were eliminated from the population by January. This removal appears to be selective and may involve effects of the parasite on host swimming and/or feeding ability.[151]

Early Growth

Growth of bluegill varies considerably from one body of water to another.[179] Most growth is completed in spring and summer in some populations,[137] but growth may continue into September (OH).[117]

Growth is often good in new impoundments (Table 63)[137,179] with slower growth characteristic of later years. Rapid growth in the first year has been associated with influx of nutrients and zooplankton, rising water levels, and low population density.[137] In MO, growth is most rapid in new ponds

Table 63

Calculated TL (mm) by age (years) reported for young bluegill in OK.[81,82,137,141]

	Mean Calculated TL at Age:			
	1	2	3	4
Statewide average 1963[81]	81	127	152	175
Statewide average 1972–1973[82]	64	107	132	150
Reservoirs	81	132	160	183
Large lakes	79	117	140	160
Small lakes	81	127	155	178
Ponds	84	127	150	170
Streams	71	117	152	185
10 new waters	97	147	173	201
59 clear waters	81	127	155	178
24 turbid waters	76	114	142	165

Table 64

Central 50% and range of mean of means of TL (mm) for young bluegill related to age summarized by Carlander (1977) from empirical data from numerous reports from various locations.[137]

Age in Years (months)	No. of Fish	Mean TL Central 50%	Mean TL Range
0 (June)	337+	39–64	8–76
(July)	764+	22–56	10–129
(August)	1,590+	28–38	10–74
(September)	716+	38–64	15–118
(October)	1,067+	38–80	5–190
(Nov–Dec)	684+	29–67	19–127
0	669+	45–81	27–147
1	15,341+	75–119	13–203
2	18,172+	107–149	54–234
3	22,399+	132–170	73–288
4	13,784+	149–190	76–272

Table 65

Growth rate of larval and juvenile bluegill raised in IN hatchery ponds.[113,137]

Age (days after hatching)	No. of Fish	Mean TL (mm)	TL Range (mm)
16–19	32	13.2	12–14
20	76	13.8	13–15
21	77	14.5	13–15
22	149	15.1	13–16
23	134	15.8	12–18
24–26	19	16.6	15–19
27–29	9	18.7	18–20
30–39	24	22.1	20–26
40	11	24.2	21–27
43–46	8	25.8	23–29

Table 66

Average TL (mm) of young bluegill by age, collected in 1939 from six localities in Norris Reservoir, TN. The number of specimens measured is in parentheses.

Location	Age 1	Age 2	Age 3	Age 4
Lost Creek	—	102 (9)	119 (58)	—
Norris Dock	97 (4)	102 (3)	112 (21)	125 (4)
Sequoyah	58 (9)	71 (33)	81 (38)	—
Big Ridge Dam	—	89 (9)	109 (7)	—
Cove Creek	—	102 (8)	130 (14)	—
Big Creek	—	—	140 (10)	—
Simple Average	79	94	114	125

Source: Constructed from data presented in Table 5, reference 44.

and reservoirs and slowest in turbid or overpopulated ponds. Stunting commonly occurs in the latter situation.[179]

Considerable variation in the first year's growth can be expected because of the prolonged spawning season (Table 64). First-year growth has been correlated with fertility and growth of bluegills has been shown to increase with an increase in phytoplankton in ponds.[137] Growth is generally better in clear than in turbid waters (Table 63).[137]

Bluegill were 20–24 mm SL (mean = 21 mm) when they first reappeared in the littoral zone of Lake Opinicon, ON, on August 4, 1980. They averaged 27 mm with a range of 24–35 mm by the end of September[122] and in another study were 22–39 mm TL (average 30.9 mm) and 0.2–0.8 g (average 0.5 g) by October 10–16.[112]

In a MN lake, age-0 bluegill collected with seines averaged 36–39 mm TL in September and October.[32] In OH, YOY bluegill were 18–81 mm in October and 25–100 mm at about 1 year.[178] In IN hatchery ponds, 20 days after fertilization, young bluegill averaged about 14 mm TL; 43–46 days after hatching, average TL was about 26 mm (Table 65).[113,137] In Reelfoot Lake, TN, in 1941, young bluegill grew about 10 mm in length between June 26 and August 17.[59]

Young bluegill grow very slowly in Norris Reservoir, TN, a tributary reservoir used for power, flood control, and as an aid to navigation in the Tennessee River. In connection with flood control, the reser-

voir is drawn down seasonally as much as 15–18 m limiting or preventing growth of aquatic vegetation and greatly influencing the number of bottom organisms in shoreline areas when the lake is full. In 1939, 3 years after completion of Norris Dam, the rate of growth varied throughout the reservoir and was invariably poor (Table 66). This slow growth was attributed to a lack of insect life in the reservoir due to fluctuations in water levels, suggesting that local conditions are probably as important as the length of the growing season in determining rate of growth.[44]

Table 67

Regional means of calculated TL (mm) of young bluegill at each annulus summarized by Carlander (1977) from numerous reports from various locations.[137]

Region	Mean Calculated TL at Age:			
	1	2	3	4
CA, OR	57	110	148	165
Northeast (unweighted)	72	112	141	171
MI, WI, MN, SD	45	86	115	145
OH, IN, IL, IA	51	98	134	158
TN, KY	63	108	138	158
MO	48	89	122	154
AR, OK	75	122	150	170
Southeastern U.S.	70	110	142	169
TX, GA	49	85	114	144
Mean of regional means	53	95	128	153

Table 68

Age and growth determinations for young bluegill collected from six KY lakes and two KY lake tailwaters (TW) in 1983.

Lake	Mean TL (mm) at Age:			
	1	2	3	4
Barren River Lake	61	112	142	168
Briggs Lake	86	150	185	198
Green River Lake	48	107	130	—
Marion County Lake	71	142	188	208
Spurlington Lake	61	127	165	183
Dewey Lake	61	94	122	—
Rough River Lake (TW)	43–81	81–119	119	—
Nolin River Lake (TW)	36	71	99	109

Source: Constructed from data presented in Tables 29, 34, 45, 48, 74, and 80, reference 196.

Table 69

Mean length of bluegill from FL for age classes 1–4 back-calculated from examination of otolith annuli. The number of lakes in which fish were examined, the mean of means, the corresponding standard error of the mean, and minimum and maximum length at age are listed.

	Age (years)			
	1	2	3	4
Lake samples (N)	58	58	58	48
Mean of means (mm TL)	61	125	165	193
Standard deviation	17	21	23	22
Minimum TL (mm)	30	83	108	130
Maximum TL (mm)	115	180	211	234

Source: Constructed from data presented in Table 4 (in part), reference 136.

Table 70

Average total lengths (TL) and weights (g) of bluegill for each summer of life from Reelfoot Lake, TN, in 1947 and 1950.

Year	Age in Summers	N	Average TL (mm)	Average Weight (g)
1947	2	10	163	101
	3	65	179	132
	4	86	201	167
	5	34	216	221
1950	2	5	137	79
	3	22	164	112
	4	73	184	157
	5	64	194	182

Source: Constructed from data presented in Table 1, reference 111.

Growth of bluegill is variable among regions (Table 67). Both slow and rapid growth are reported for bluegill from most regions, indicating that population and edaphic conditions may have a greater effect on average growth rates than do growing season or latitude. Review of tabulated data does indicate more rapid growth in the southern part of the range than in the north[137] (Tables 67, 68, and 69).

Within regions, there is great variation in the growth of bluegill among populations and years. Statewide average growth of bluegill in OK waters was consistently greater for ages 1 through 4 in 1963[81] than in 1972–1973[82] (Table 63). In Reelfoot Lake, TN, the average rate of growth in length and weight of bluegill was greater in 1947 than in 1950 (Table 70). In KY waters, growth is generally greater in lakes than in tailwater populations

(Table 68) and was highly variable among lakes.[196] Growth in a KY stream was also slower than growth in lakes. Average calculated SL of bluegill from Doe Run, KY (1959–1961), at the end of each year of life was 35 mm at age 1, 63 mm at age 2, 86 mm at age 3, 106 mm at age 4, and 124 mm at age 5.[63]

During the period 1947–1954, bluegill grew more rapidly in Clear Lake, IA, than in most midwest waters. Mean calculated TL (mm) at ages 1–4 was 61, 108, 142, and 158.[19] Growth of bluegills in IA farm ponds having balanced populations was generally good, TL (mm) at successive annuli being 43, 104, 155, and 178. Growth in ponds not displaying balanced populations was noticeably slower.[43]

Average back-calculated TL for bluegill from seven KS streams sampled during 1995 and 1996 was 39 mm (32–45) at age 1, 63 mm (55–71) at age 2, 89 mm (79–102) at age 3, and 112 mm (103–131) at age 4. Growth was consistently lower in KS streams when compared to streams in MO and OK.[57] In the Black River, MO, 1947, average TL for each age group was 48 mm at age 1, 117 mm at age 2, 152 mm at age 3, 175 mm at age 4.[104]

In Clearwater Lake, a reservoir formed in 1948 on the Black River, MO, for the period 1948–1951, average TL for each age group by year classes for bluegill was 64 mm TL at age 1, 107 mm at age 2, 142 mm at age 3, 168 mm at age 4. There was a small, consistent decline in the rate of growth of the bluegill population during the period. Male bluegill outgrew females in each age group during the period 1949–1952:[104]

| | Average TL (mm) at Age: | | | |
	1	2	3	4
Males	79	117	**152**	179
Females	74	104	**137**	150

Carlander (1977) reported that males grew more rapidly than females in some populations, but differences were usually small. In some lakes no sexual difference in growth was detected, and in a few populations females grew larger than males.[137]

In Lake Wappapello, MO, during the period 1948–1951, average calculated TL at the end of each year of life was 43 mm at age 1, 76 mm at age 2, 112 mm at age 3, and 142 mm at age 4.[53]

Growth related to age for bluegill from OK waters summarized from empirical data from numerous reports from various locations is presented in Table 64 and calculated TL by age for bluegill from OK is presented in Table 63.

In OH, average calculated TL at each annulus was 40 mm at age 1, 73 mm at age 2, 105 mm at age 3,

Table 71

Estimated TL (mm) at each annulus for young bluegill from TVA reservoirs during the 1970s.[41]

| Reservoir | N | \multicolumn{4}{c}{Estimated TL (mm) at Each Annulus} |
		1	2	3	4
Chickamauga	128	43	93	136	158
Pickwick	421	45	100	145	161
Guntersville	1,361	41	100	142	164
Kentucky	380	40	102	149	169
Barkley	216	47	102	146	165

131 mm at age 4.[117] The range of average lengths at age 3 from 69 VA populations, mostly in impoundments, was 96–180 mm TL.[185]

Studies conducted in Reelfoot Lake, TN, indicating over 140 mm growth in the first year appear to be in error as a result of sampling deficiency for smaller specimens caused by using "inch and a half mesh traps." Most studies in TN indicate growth increments of about 40 mm in the second and third years and 12–25 mm the fourth and fifth years.[170]

Growth of young bluegill was very similar in five TVA reservoirs during the 1970s (Table 71).[41] Growth was faster in the first year for bluegills from Cordell Hull Reservoir, Cumberland River, TN, similar in the second year, and slower thereafter. Mean calculated TL related to age for 270 bluegills collected from March 1979 through February 1980 was 72 mm at age 1, 107 mm at age 2, 130 mm at age 3, 142 mm at age 4.[88]

In FL lakes, annulus formation for bluegill occurred between March and July. Mean calculated lengths at ages 1–4 for bluegill from numerous FL lakes were 61, 125, 165, and 193 mm TL, respectively (Table 69). There is a great deal of variation in length at age among lakes with ages 1–4 ranging from 30 to 115, 83 to 180, 108 to 211, and 130 to 234, respectively.[136] Many factors that probably contribute to this variation include both density-dependent and density-independent factors.[137] The mean lengths at ages 1–4 for bluegill from FL are larger than the mean of regional means reported for most northern lakes in the United States.[136,137]

Since very little variation existed in the size of young bluegill taken each day from hatchery ponds in IN (Table 65), it was assumed that a single spawning had occurred and that the mean length of the daily catch would give a reliable estimate of the growth of the fry. These data showed that the rate of growth was about 0.4 mm/day.[113] Growth of bluegill 12–25 mm was 0.1 mm/day in IN ponds

Table 72

Central 50% and range of mean of means of weight (g) related to TL (mm) for young bluegill summarized by Carlander (1977) from empirical data from numerous reports from various locations.[137]

TL Range	No. of Fish	Mean Weight Central 50%	Mean Weight Range
15–50	53,845+	0.5–2.2	0.2–4.0
51–75	36,678+	2.5–5.7	0.7–10.0
76–101	24,322+	10–17	1–36
102–126	18,206+	25–32	9–100
127–151	15,112+	46–59	6–116

Table 73

Statewide average growth of young bluegill recorded in OK waters for the period 1952–1963.

Age in Years	Average Growth TL (mm)	Average Growth Weight (g)
1	81	8
2	127	35
3	152	67
4	175	107

Source: Constructed from data presented in Table 7 (in part), reference 81.

Table 74

Statewide length–weight relationships for bluegill in OK.

Length (mm)	Weight (g) 1963 [81]	Weight (g) 1972–1973 [82]
20	<0.1	0.1
25	0.1	0.2
30	0.2	0.4
35	0.4	0.6
40	0.6	1.0
45	1.0	1.4
50	1.4	2.0
55	2.0	2.7
60	2.7	3.5
65	3.5	4.5
70	4.6	5.7
75	5.8	7.0
80	7.2	8.5
85	8.9	10.3
90	10.9	12.2
95	13.1	14.4
100	15.6	16.9

Source: Constructed from data presented in Table N, page 101 (in part), reference 82.

with poor growth and 0.6 mm/day in ponds with good growth.[143]

Growth of bluegill fry during the first 14 days after dispersal from the nests into the limnetic zone was similar in three northern WI lakes, ranging from 0.4 to 0.7 mm/day. Fastest growth generally occurred between July and August with growth slowing down by September.[99] In a MN lake, the daily increase in mean TL varied from 0.2 to 0.5 mm/day for bluegills 11–39 mm.[32] In Ridge Lake, IL, during the period 1987–1989, average absolute growth of 5–11 mm TL bluegill larvae across all months and years was 0.55 ± 0.14 mm/day. Estimates of instantaneous growth were lowest for bluegill spawned in May and highest for fish spawned later in the year.[161]

No consistent regional differences in length–weight relationships are evident. The average weights from AL, MO, and AR were often low and those from CA and OR were often high but probably do not indicate regional differences.[137] Length–weight relationships summarized by Carlander (1977) from empirical data from numerous reports from various locations are presented in Table 72. Relationships were similar in OK (Tables 73 and 74), though variable among years (Table 74), and in AL (Table 75).

In a series of IL lakes, no association was evident between growth rate and lake size or depth, transparency of the water, abundance of vegetation, or abundance of predators or competing fish.[144] In 1,146 MN lakes, secchi depth and maximum depth were negatively correlated with bluegill length at ages 1–6. Total alkalinity, percent of littoral area, and mean maximum July air temperature were positively correlated with bluegill length at those ages. Sixteen to 33% of the variation in growth of bluegill during their first 5 years was explained by secchi depth, maximum depth, and total alkalinity.[153]

Growth and condition of bluegill from acidic lakes (pH 5.1–6.0) and circumneutral lakes (pH 6.7–7.5) in northern WI were compared. Although mean

Table 75

Length–weight relationships of young bluegill from AL collected during the period 1949–1964.

Approximate Average TL (mm)	N	Average Empirical Weight (g)
25	52,376	0.5
51	34,314	2.1
76	21,516	7.3
102	13,103	18.6
127	8,259	36.8
152	5,020	63.6
178	1,810	104.4
203	324	158.9

Source: Constructed from data presented in unnumbered table on page 13, reference 173.

condition factors and mean back-calculated TL at ages 1–4 varied significantly among lakes, the differences were not related to pH in the lake. The ranks of mean condition factors and back-calculated lengths at ages 2, 3, and 4 were negatively correlated with relative density of bluegill among the lakes.[121] In three northern WI lakes, growth of age-0 bluegill was inversely related to density in the fall.[99]

In a moderately fertile WV reservoir, no significant positive or negative effects of liming were observed on the growth of YOY bluegill.[106]

Temperature was found not to be the controlling factor of bluegill growth in the discharge area of a power generation plant in GA. Bluegill from the thermally influenced area generally did not show increased growth over other areas in the reservoir.[103] In four heated reservoirs in TX, size attained at an annulus and K-factors were found to be not significantly different between bluegill collected in the discharge areas of power plants and other parts of the reservoir.[102]

In a laboratory study, juvenile bluegill under fixed daily rations had similar growth rates when held at 28 and 30°C and growth at these temperatures was greater than at 34°C. Growth differences of bluegill held at 34°C and the other temperatures increased with larger daily rations.[120]

Juvenile bluegill weighing 1.8–8.0 g were individually marked and fed to excess during a 30-day constant temperature test. Day length was 16 h and fish were tested at temperature intervals of 2° from 20–36°C. The highest specific growth rate occurred at 30°C; however, growth rates of only the groups

held at 20 and 36°C were statistically different from the rate for the group held at 30°C.[98]

In Lake Shelbyville, IL, the interactions among larval bluegill, gizzard shad, and their zooplankton prey were examined. Bluegill growth was highest mid to late summer and rates of growth of bluegills and shad were not correlated with fish density. Bluegill growth was correlated with abundance of prey, which might become limiting when larval fish densities are high.[64]

Bluegill are protracted spawners, producing fry over a several-week breeding season. Studies on Lake Opinicon, ON, relating the body size and survival of yearlings to the day that they were born, demonstrated that this time dimension has important implications for recruitment. The production of bluegill was estimated throughout the 1993 spawning season. In 1994, yearlings were collected and otolith daily growth rings used to determine their 1993 date of birth. Yearling body size decreased with date of birth in 1993. There was a progressive decline in body size across eight spawning bouts, resulting in a 50% difference in the size at age 1 of early- and late-born fry. This difference is apparently due to the longer growing season available to the early-born fry.[112]

A study that compared growth and reproduction of bluegill in a reservoir where no fishing was allowed to reservoirs that are fished suggested that differences in bluegill growth and reproduction were largely attributable to differences in abundance and size structure of natural predators (e.g., largemouth bass) in SC reservoirs.[119]

For YOY bluegill from the littoral zone of six large (2–3 ha) AL ponds, the back-calculated growth rate for the first 10 days after swim-up, when they were limnetic, was similar to the growth rate of larvae from the same cohort, collected in the limnetic zone. Size-selective mortality for young bluegill ≤25 mm TL was not important. However, a negative relationship between the density of bluegill >25 mm TL in the littoral zone and the density of YOY largemouth bass in the littoral zone suggests increasing importance of size in influencing bluegill survival as fish grow.[127]

Feeding Habits

Young and juvenile bluegill consume small crustaceans and aquatic insects and sometimes eat plants.[4] Bluegill shift their diets from littoral prey to open-water prey between 55 and 85 mm SL.[11,16]

In Lake Opinicon, ON, the bluegill is a generalist feeder and the diets of bluegill year-classes are substantially the same. Consumption of chironomid larvae and pupae, trichopteran larvae, and amphipods varied little with age. Cladocerans, the most important single food, accounted for 75% of prey items consumed and 70% of the total volume of prey utilized by age-0 bluegill. Cladocerans showed a decline in importance with age but was compensated for in year 2 fish by increased trichopteran larvae consumption and the utilization of several different food types that temporarily became abundant.[77]

Lepomis larvae 4.5–20 mm TL, believed to be a mix of pumpkinseed and bluegill, collected July 17–19 from Lake Opinicon, ON, consumed nauplii, *Cyclops*, *Bosmina*, *Sida*, and *Diaphanosoma*. Size of prey generally increased with fish growth. Small *Cyclops* (0.1–0.2 mm) and nauplii (0.1–0.2 mm) were consumed by fish 4.5–10 mm. *Lepomis* larvae 10–20 mm fed on larger *Cyclops* (0.2–0.7 mm) and the other taxa listed above. Large *Lepomis* larvae (15–20 mm TL) collected in August consumed larger *Cyclops* (0.2–0.7 mm), large *Bosmina* (0.2–0.4mm), *Sida* (0.1–0.4 mm), *Chydorus* (0.1–0.3 mm), and *Diaphanosoma* (0.3–0.4 mm).[29]

In three northern WI lakes, feeding habits of young bluegill changed as their behavior patterns changed. When they first left the nest at about 5 mm TL and migrated to the limnetic areas of the lakes, food selection for bluegill larvae was limited to zooplankton ≤0.25 mm, which were predominantly copepod nauplii and cladocerans. As the larvae reached 8–10 mm they were able to select organisms up to 0.5 mm in size, predominantly copepod nauplii, *Bosmina* spp., and early instar stages of *Daphnia* spp. From 11–15 mm, the larval and early juvenile bluegills continued to feed on limnetic zooplankton but were able to take larger sizes of *Bosmina* and *Daphnia* as well as other species of limnetic zooplankton. Once the juveniles reached 15–20 mm and started to move back into the shoreline areas of the study lakes, their diets became much more diverse and littoral organisms dominated. Among young bluegills 11–19 mm, cladocerans, principally *Daphnia* spp., *Bosmina* spp., Chydorinae, and *Ceriodaphnia* spp., were the dominant foods in all three study lakes; copepods and insects were also consumed in good numbers. A critical survival period exists for bluegill larvae at the time they first leave the nest and change from an endogenous to exogenous source of energy. If the right size and amounts of food are not available during the littoral to limnetic migration, the larvae could starve.[99]

In MN, food of bluegill at the initiation of feeding consisted of *Polyarthra* and copepod nauplii. At 7.0 mm, rotifers and cyclopoid copepods appeared in the diet, and at 8.0 mm, cladocerans became the dominant food and remained so for larger fish.[125]

In two AL ponds in 1994, successful spawning of bluegill occurred from April through September (15 cohorts). Most larval bluegill >7 mm TL had consumed prey. Some trends were evident within and among early, mid-season, and late-spawned bluegill relative to the percent biomass of prey taxa in diets. In pond S-3, rotifers never exceeded 64% of the diet biomass of early-spawned bluegill. *Daphnia* were consumed by all sizes of early-spawned bluegills whereas no *Daphnia* were consumed by mid- or late-spawned larvae. Copepod nauplii were an important diet component for early-spawned 6- and 7-mm bluegill larvae but were less important for larger larvae. Rotifer biomass in larval bluegill diets increased in mid-season–spawned and late-spawned bluegill. In mid-spawned fish, rotifers contributed more than 95% of the diet of 6–8 mm larvae. Rotifers remained important for larvae up to 10 mm but became less important in longer larvae. Ostracods, *Ceriodaphnia*, cyclopoids, and *Chaoborus* were important foods for larval bluegill >9 mm. Diets of late-spawned fish were similar to those of mid-spawned fish. In the other pond (S-8), diets of the bluegill larvae were less diverse in terms of the number of taxa consumed. Early-spawned larvae up to 10 mm TL primarily consumed rotifers and *Bosmina*. Calanoid copepods and ostracods were important in the diets of larvae 10–11 mm but rotifers dominated the diets of these larvae. For mid-spawned fish, rotifers contributed more than 97% of the diet of 5–9 mm larvae. For late-spawned fish, rotifers contributed more than 89% of the total prey biomass for all sizes of larvae.[154]

In a northern IN lake stomach contents of young bluegill in the limnetic zone contained planktonic crustaceans with a dominance of copepods and *Bosmina*. Upon returning to the littoral zone at about 20–25 mm, the food habits changed and insect larvae became more important in the diet.[86]

In three piedmont lakes in NC, age-0 bluegill ≤12.5 mm SL consumed primarily copepods (90–95% frequency) followed by rotifers (4–7%) and cladocerans (1–2%). The food of age-0 bluegill 13.5–36.0 mm SL was also dominated by copepods (66–71%), followed by cladocerans (11–18%), insect larvae (9–16%), ostracods (1–3%), and rotifers (1–2%).[7]

In Reelfoot Lake, TN, young bluegill 16–25 mm in length showed a preference for crustaceans including cladocerans (75% by volume); copepods (10%); amphipods (2%); ostracods (3%); chironomid larvae (5%); and miscellaneous insect larvae and adults (5%) were also consumed. Slightly larger bluegills (25–36 mm) preferred insect larvae and adults including chironomid larvae (45%), miscellaneous dipteran larvae (10%), and miscellaneous insects (10%). Crustacea made up 30% of the food including mostly larger amphipods (25%) and the smaller forms copepods, cladocerans, and ostracods (5%). These data suggest that food preferences change to a slightly larger form after bluegills exceed 25 mm in length.[59]

In Bull Shoals Reservoir, AR, bluegill ≤48 mm TL (N = 4) consumed mostly aquatic insects (60% of total volume); Entomostraca (31%), Acari (8%), and Malacostraca (2%) were also consumed. Stomach contents of bluegills 51–99 mm TL (N = 69) included terrestrial insects (26%), aquatic insects (26%), bryozoans (25%), filamentous algae (13%), Entomostraca (4%), fish eggs (3%), Acari (1%), and traces of Malacostraca, mollusca, and detritus. Bluegill 102–201 mm TL consumed terrestrial insects (23%), filamentous algae (23%), fish eggs (19%), bryozoans (16%), aquatic insects (8%), fish (6%), Malacostraca (3%), detritus (1%), and traces of Acari, Entomostraca, and mollusca.[138]

Grouped data on feeding habits of 63 bluegill 90–110 mm for June, 1969–1971, in Lake Opinicon showed percent volume of food taken as follows: cladocerans 28%; chironomid larvae 24%; trichopteran larvae 14%; dipteran larvae and pupae 5%; isopods, zygopterans, and anisopterans each 4%; dipteran adults, amphipods, mollusks, and plant material each 3%; ostracods 2%; chironomid pupae, ephemeropterans, and *Hydracarina* each 1%.[78] In another report, age-2 bluegill 60–85 mm TL differed in their diet from bluegill 130–170 mm (ages 3 and 4) only in changes in the relative volumes of different items taken, the elimination of cladocerans from the diet, and the occurrence of fish fry in the larger fish.[79]

The principal food items of bluegill 88–191 mm TL from Cordell Hull Reservoir, TN, were dipterans, primarily chironomid larvae. Cladocerans and copepods were frequently found in the diet of smaller bluegill. Chironomids were consumed most heavily from November through February and filamentous algae were also consumed in large quantities. Terrestrial and aquatic insects made up a large portion of the diet from July through October.[88]

The stomachs of 59 bluegill 17–132 mm SL from the White River, IN, were examined. The fish were captured from June through January (28 in summer, 29 in fall, and 2 in winter). Food habits data were examined by season and size groupings, but were quite similar. The principal food was chironomid larvae (72.8% by volume). Of the 59 stomachs examined, 54 contained chironomids and many of them only this food. The top four foods all were aquatic insect larvae (chironomid larvae, chironomid pupae [8.7%], ephemerid naiads [6.9%], and trichopteran larvae [2.7%]), collectively accounting for 91.1% of the volume of food in stomachs.[61]

In Lake Shelbyville, IL, the interactions among larval bluegill, gizzard shad, and their zooplankton prey were examined. Diet overlap was substantial; gizzard shad and bluegills fed selectively on smaller prey items in June, switching to larger cladocerans and copepods by July. Results of this study suggested that growth and survival of planktivorous larval gizzard shad and growth of larval bluegill are affected by availability of zooplankton prey, which may become limiting when larval fish densities are high.[64]

In a series of *in situ* enclosure experiments, turbidity from suspended sediments reduced consumption by bluegill (about 12.5 mm TL) of crustacean zooplankton, primarily cyclopoid copepods and copepod nauplii. This reduction occurred when light intensity in parts of enclosures fell below a threshold, estimated at <450 lx.[55]

A study in the summers of 1987–1989 in Lake Itasca, MN, found that substrate-dwelling microcrustaceans provided a diverse and abundant food resource for young fish inhabiting the littoral zone. Free-swimming microcrustaceans occurred in the littoral zone water column but were at least two orders of magnitude less abundant. Still larval and early juvenile bluegill selectively preyed upon the free-swimming animals, especially *Bosmina*. Among their predominant prey, bluegill showed no size preference or a preference for smaller sizes, which suggests that young fish may gain a greater benefit from ingesting smaller prey (more quickly digested) than predicted by optimal foraging models.[24]

In two FL lakes, the greater similarity of young bluegill (≥75 mm TL) diets to epiphytic than to benthic macroinvertebrate assemblages suggested that bluegills obtained most of their macroinvertebrate prey in vegetation and little from the benthos. Linear selectivity indices indicated positive selection for chironomids, caddisflies, mayflies, odonates, hemipterans, and *Palaemonetes* spp. Except for *Palaemonetes* spp., these taxa were more abundant among epiphytic than among benthic assemblages.[156]

In MI lakes small bluegill juveniles compete with small pumpkinseed juveniles for soft-bodied littoral invertebrates; larger pumpkinseed feed extensively on snails and are not competitors.[17] Bluegill and pumpkinseed >75 mm SL exhibited distinct diet and habitat separation. Bluegill foraged primarily on open-water zooplankton (*Daphnia*), while pumpkinseeds specialized on vegetation-dwelling gastropods. In contrast, the diets and habitat use by smaller bluegill and pumpkinseed (≤75 mm SL) were quite similar. The smaller fish of both species foraged predominantly in the vegetation and 76–91% of their average seasonal diet was vegetation-dwelling non-gastropod prey. The increase in shared resources by the smaller fish was due to two factors: (1) piscivorous fish restrict small size-classes to the vegetation where they are less vulnerable; (2) small pumpkinseeds were unable to feed effectively on snails, the adult resource in the vegetation.[11]

In a macrophyte bed of a backwater lake in the upper Mississippi River, the diets of age-0 bluegill and adults were similar and changed seasonally, probably in response to changes in life stages of macroinvertebrates (i.e., emergence of winged adults). Diets and diel patterns of age-0 bluegills suggested that they were feeding in the vegetated, littoral zone and predation by age-0 largemouth bass appeared to influence their use of this foraging habitat. In summer, when most age-0 bluegill were vulnerable to predation by age-0 largemouth bass, bluegill abundance was strongly correlated with vegetation biomass. In October and November, piscivory by age-0 largemouth bass was limited by gape and the relationship between the abundance of age-0 bluegill and vegetation biomass was weakened due to the reduced risk of predation.[129]

Studies in the laboratory and an OH lake indicated that vegetation did not influence growth or diet of bluegill beyond relatively low densities owing to the interaction between capture probabilities and macroinvertebrate densities.[108]

In a laboratory study, the effects of variation in prey abundance on foraging site selection by juvenile bluegill (35–50 mm SL), in the presence and absence of a predator was examined. Groups of bluegill chose between two plots of artificial vegetation differing in plant stem density (i.e., 100 and 500 stems/m²) with two prey distribution treatments. In the first treatment, the 500 stem plot had sufficient prey to satiate all group members; the

100 stem plot had one-fifth this amount. In the second treatment, prey distribution was reversed. With no predator present bluegill preferred to forage, and achieved a higher foraging rate, in the vegetation plot with the greatest number of prey. The presence of a predator had no significant effect on foraging rate when the most rewarding plot of vegetation also offered a refuge from predation (i.e., 500 stems/m² plot). In contrast, when the 100-stem plot contained the greatest number of prey, bluegill reacted to the presence of a predator by increasing time spent in the 500-stem plot while simultaneously increasing their foraging rate when in the 100-stem plot. Therefore, juvenile bluegills altered their foraging behavior to reduce time spent exposed to predation.[123]

In a FL lake the number of bluegill 40–150 mm TL present in an area was significantly related to plant density, but not to the presence of a predator (largemouth bass). Fewer bluegills were observed at low plant densities than at medium or high plant densities, but there was no significant difference in bluegill abundance between medium and high plant density sites. The proportion of bluegill feeding in an area was not related to plant density or predator presence. The largest bluegill shoals were observed in areas of moderate plant density and the smallest in dense vegetation, but most bluegill traveled alone or in pairs.[124]

LITERATURE CITED

1. Lee, D.S. et al. 1980.
2. Pearson, W.D. and L.A. Krumholz. 1984.
3. Pearson, W.D. and B.J. Pearson. 1989.
4. Raney, E.C. 1965.
5. Gerald, J.W. 1971.
6. Lobb, M.D. III and D.J. Orth. 1991.
7. Lemly, A.D. and J.F. Dimmick. 1982.
8. Van Hassel, J.H. et al. 1988.
9. Thomas, J.A. et al. 2004.
10. Buynak, G.L. and H.W. Mohr, Jr. 1981.
11. Mittelbach, G.G. 1984.
12. Etnier, D.A. 1968.
13. Carbine, W.F. 1939.
A 14. Peterka, J.J. and J.S. Kent. 1976.
15. Taubert, B.D. 1977.
16. Osenberg, C.W. et al. 1988.
17. Osenberg, C.W. et al. 1992.
18. Ulrey, L. et al. 1938.
19. DiConstanzo, C.J. 1957.
20. Walters, D.A. et al. 1991.
21. Keast, A. and J. Eadie. 1984.
22. Hendricks, M.L. et al. 1983.
23. Schwartz, F.J. 1964.
24. Hatch, J.T. et al. 1990.
25. Childers, W.F. and G.W. Bennett. 1961.
26. Coutant, C.C. 1977.
27. Dibble, E.D. 1991.
28. Clark, F.W. and M.H.A. Keenleyside. 1967.
29. Keast, A. 1980.
30. Fish, P.A. and J. Savitz. 1983.
31. Conrow, R. and A.V. Zale. 1985.
32. Lux, F.E. 1960.
33. Bettoli, P.W. et al. 1993.
34. Gunning, G.E. and C.R. Shoop. 1963.
35. Hall, D.J. and E.E. Werner. 1977.
36. Clark, P.W. 1990.
37. Gasaway, C.R. 1968.
38. Swingle, H.S. 1956.
39. Floyd, K.B. et al. 1984.
40. Savitz, J. et al. 1983.
A 41. Tennessee Valley Authority. 1978.
42. Dimond, W.F. and T.W. Stork. 1985.
43. Moorman, R.B. 1957.
44. Eschmeyer, R.W. 1940.
45. Fischer, S.A. and W.E. Kelso. 1988.
46. Weaver, M.J. et al. 1997.
47. Coleman, R.M. and R.U. Fischer. 1991.
48. Dawley, R.M. 1987.
49. Colle, D.E. et al. 1987.
50. Guy, C.S. and D.W. Willis. 1990.
51. McHugh, J.J. 1990.
52. Mischke, C.C. and J.E. Morris. 1997.

53. Patriarche, M.H. 1953.
54. Johnson, S.L. 1993.
55. Miner, J.G. and R.A. Stein. 1993.
56. Mraz, D. and E.L. Cooper. 1957.
57. Delp, J.G. et al. 2000.
58. Barclay, L.A. 1985.
59. Rice, L.A. 1942.
60. Scheidegger, K.J. 1990.
61. Whitaker, J.O., Jr. 1974.
62. Cherry, D.S. et al. 1975.
63. Redmon, W.L. and L.A. Krumholz. 1978.
64. Welker, M.T. et al. 1994.
65. Smale, M.A. and C.F. Rabeni. 1995a.
66. Smale, M.A. and C.F. Rabeni. 1995b.
67. DeVries, D.R. and R.A. Stein. 1990.
68. Houser, A. 1963.
70. Armstrong, J.G. III. 1969.
71. Merriner, J.V. 1971.
72. Conrow, R. et al. 1990.
73. Layzer, J.B. and M.D. Cady. 1991.
74. West, J.L. and F.E. Hester. 1966.
75. Panek, F.M. and C.R. Cofield. 1978.
76. Miller, H.C. 1964.
77. Keast, A. 1978a.
78. Keast, A. 1978b.
79. Keast, A. 1970.
80. Rettig, J.E. 1998.
A 81. Houser, A. and M.G. Bross. 1963.
A 82. Mense, J.B. 1976.
A 83. Banner, A. and J.A. Van Arman. 1973.
84. Hutton, G.D. 1982
85. Gallagher, R.P. and J.V. Conner. 1983.
86. Werner, R.G. 1966.
87. Dewey, M.R. 1992.
88. Martinez, G.J. 1980.
89. DeMont, D.J. 1982.
90. Byrd, I.B. 1952.
A 91. Carver, D.M. 1976.
92. Meyer, F.A. 1970.
93. Mabee, P.M. 1995.
94. Hudson, W.F. and F.J. Bulow. 1984.
95. Wrenn, W.B. and K.L. Grannemann. 1980.
96. Snow, H. et al. 1966.
97. Dominey, W.J. 1981.
98. Lemke, A.E. 1977.
99. Beard, T.D. 1982.
A 100. Wojtalik, T.A. 1970.
101. Dimond, W.F. et al. 1985.
102. Serns, S.L. 1972.
103. O'Rear, R.S. 1970.
104. Lane, C.E., Jr. 1954.
105. Moynan, K.M. 1989.
106. Coahran, D.A. 1990.
107. Bosley, T.R. and J.V. Conner. 1984.
108. Savino, J.F. et al. 1992.
109. Benoit, D.A. 1975.
110. Conner, J.V. 1979.
111. Schoffman, R.J. 1952.
112. Cargnelli, L.M. and M.R. Gross. 1996.
113. Werner, R.G. 1969.
114. Stevenson, F. et al. 1969.
115. Hatleli, D.C. 1996.
116. Toetz, D.W. 1965.
117. Morgan, G.D. 1951b.
118. Dominey, W.J. 1983.
119. Belk, M.C. and L.S. Hales, Jr. 1993.
120. Beitinger, T.L. and J.J. Magnuson. 1979.
121. Wiener, J.G. and W.R. Hanneman. 1982.
122. Brown, J.A. and P.W. Colgan. 1982.
123. Gotceitas, V. and P.W. Colgan. 1990.
124. Butler, M.J. IV. 1988.
125. Siefert, R.E. 1972.
126. Bodensteiner, L.R. and W.M. Lewis. 1994.
127. Putman, J.H. II and D.R. DeVries. 1993.
128. Belk, M.C. 1994.
129. Dewey, M.R. et al. 1997.
130. Avila, V.L. 1976.
131. Bain, M.B. and L.A. Helfrich. 1983.
132. Dewey, M.R. et al. 1981.
133. Crutchfield, J.U., Jr. et al. 2003.
134. Carver, D.C. 1967.
135. Tebo, L.B. and E.G. McCoy. 1964.
136. Mantini, L. et al. 1992.
137. Carlander, K.D. 1977.
138. Applegate, R.L. et al. 1967.
139. Radomski, P. and T.J. Goeman. 2001.
140. Burr, B.M. and M.L. Warren, Jr. 1986.
A 141. Jenkins, R. et al. 1955.
142. Moss, D.D. and D.C. Scott. 1961.
143. Krumholz, L.A. 1949.
144. Richardson, R.E. 1942.
145. Hickman, G.D. and M.R. Dewey. 1973.
146. Moore, W.G. 1942.
147. Aday, D.D. et al. 2003.
148. Duffy, J.J. and D.C. Jackson. 1994.
149. Miranda, L.E. and R.J. Muncy. 1987a.
150. Ali, A.B. and D.R. Bayne. 1985.
151. Lemly, A.D. 1980.
152. Michaletz, P.H. and J.L. Bonneau. 2005.
153. Tomcko, C.M. and R.B. Pierce. 2001.
154. Partridge, D.G. and D.R. DeVries. 1999.
155. Miner, J.G. and R.A. Stein. 1996.
156. Fischer, S.A. and W.E. Kelso. 1990.
157. Schramm, H.L., Jr. and K.J. Jirka. 1989.
158. Johnson, D.L. et al. 1988.
159. Gotceitas, V. and P. Colgan. 1987.
160. Shoup, D.E. et al. 2003.
161. Santucci, V.J., Jr. and D.H. Wahl. 2003.
162. Garvey, J.E. and R.A. Stein. 1998.
163. Paukert, C.P. et al. 2002.
164. Bayne, D.R. et al. 2002.
165. Jackson, J.J. and D.L. Bauer. 2000.
166. Schneider, J.C. 1999.
167. Knights, B.C. et al. 1995.
168. Lynch, W.E., Jr. and D.L. Johnson. 1989.
169. Tomcko, C.M. and R.B. Pierce. 2005.
170. Etnier, D.A. and W.C. Starnes. 1993.
171. Desselle, W.J. et al. 1978.
172. Mettee, M.F. et al. 2001.
A 173. Swingle, W.E. 1965.
174. Smith, P.W. 1979.
175. Goodyear, C.D. et al. 1982.

176. Scott, W.B and E.J. Crossman. 1973.
177. Becker, G.C. 1983.
178. Trautman, M.B. 1981.
179. Pflieger, W.L. 1975b.
180. Menhinick, E.F. 1991.
181. Smith, C.L. 1985.
182. Holland-Bartels, L.E. et al. 1990.
183. Cross, F.B. 1967.
184. Clay, W.M. 1975.
185. Jenkins, R.E. and N.M. Burkhead. 1994.
186. Ross, S.T. 2001.
187. Peterson, M.S. et al. 1993.
188. Robison, H.W. and T.M. Buchanan. 1988.
189. Boschung, H.T., Jr. and R.L. Mayden. 2004.
A 190. EA Engineering, Science, and Technology. 2004.
191. Taber, C.A. 1969.
192. Hardy, J.D., Jr. 1978.
193. Nakamura, N. et al. 1971.
194. Wang, J.C.S. and R.J. Kernehan. 1979.
195. Anjard, C.A. 1974.
A 196. Axon, J.R. 1984.
197. Gerking. S.D. 1945.
198. Brown, J.A. and P.W. Colgan. 1981.

* Original comments concerning early development, reproductive biology, and early life ecology come from bluegill collected from a rock quarry on Norris Reservoir, TN, June 6, 1978. Eggs and larvae were collected with a slurp gun from a nest guarded by a male bluegill and then transported to the TVA laboratory in Norris, TN, where they were cultured in aquaria and holding nets in a pond. Developmental descriptions come from a series of larvae drawn from this cohort. This developmental series is catalogued TV909, DS–21, and curated at the Division of Fishes, Aquatic Research Center, Indiana Biological Survey, Bloomington, IN.

DOLLAR SUNFISH

Lepomis (Icthelis) marginatus (Holbrook)

Robert Wallus

Lepomis, Greek: *lepis*, "scale" and *poma*, "lid," in reference to scaly operculum; *marginatus*, Latin: "bordered, enclosed with a border," in reference to light-margined gill cover.

RANGE

Dollar sunfish are distributed in Atlantic and Gulf Coastal drainages, mostly below the Fall Line,[2] from NC to TX and north through the central Mississippi basin to KY and AR.[1]

Due to its great similarity in appearance to younger specimens of longear sunfish, the distribution of the dollar sunfish has not been well understood in certain portions of its range.[2]

HABITAT AND MOVEMENT

Dollar sunfish is a common to abundant inhabitant of good-quality,[2,7] lowland habitats such as spring-fed wetlands and sloughs,[7] unaltered sluggish streams,[1,2,4] vegetated swamps,[1,2,12] and natural lakes.[2] It is often found in slow-moving, small to large streams,[4,7,12] floodplain pools[4] and oxbow lakes,[4,10] ponds,[4] and vegetated areas of large reservoirs.[4,6,10–12] The dollar sunfish is usually found over substrates of sand or clay overlain with silt and organic debris, and is often associated with submerged aquatic vegetation, hydrophytes, and overhanging vegetation along undercut banks.[7]

In MS, dollar sunfish are commonly collected in oxbow lakes characterized by tannin-stained water and mud substrate.[10]

Although dollar sunfish can reproduce in reservoir habitats, they are unable to permanently coexist with large centrarchids such as largemouth bass, bluegill, and redear sunfish that are common in such areas.[10,11]

In Lake Conroe, TX, catches of dollar sunfish declined to nearly zero in cove habitats, after submersed aquatic vegetation was removed by grass carp, and seine catches indicated that this littoral species may have been extirpated from the main basin and embayments of the reservoir.[6]

In NC, dollar sunfish inhabit deeper water in winter and move to shallow water in May to spawn.[3]

DISTRIBUTION AND OCCURRENCE IN THE OHIO RIVER SYSTEM

In the Ohio River system, distribution of the dollar sunfish is restricted to the Clarks River in western KY, where it is uncommon,[7] and tributaries to the lower Tennessee River in TN,[2] MS,[10] and AL.[12,13]

In TN, dollar sunfish are restricted to the Coastal Plain and streams tributary to the lower Tennessee River such as the Big Sandy River. Channelization has altered suitable habitats for the dollar sunfish in western TN. While still common locally, it may have formerly been much more abundant in the state.[2]

In AL, the dollar sunfish is distributed primarily below the Fall Line; scattered records occur in the Tennessee River drainage.[12]

SPAWNING

Location
Dollar sunfish nests are typically located in shallow water 10–50 cm deep and less than 2 m from the shoreline.[5] Nests are sometimes found in high-density colonies.[3]

Season
In NC, May into August;[3] in SC, May through August, with peak spawning from mid-May to the end of June;[5] in AL, April into August;[12] in FL, April to September.[9]

Temperature
No information.

Fecundity
No information.

Sexual Maturity
In NC, dollar sunfish attain sexual maturity at age 2 and 60 mm TL.[3]

Spawning Act
The male dollar sunfish constructs his nest,[3,4] usually on hard sand substrates.[3] Males create their nests by fanning away silt and fine organic matter until a small, clean depression is made (mean nest diameter is about 30 cm).[5]

Once nest construction is completed, males display for mates and defend their nests against other nesting males. Defensive behavior consists of chasing away fishes that are near or on the nest. Males also display to intruding males, but then chase them away.[5] Males engage in much display and fighting, and smaller males are usually not successful in maintaining a nest.[2,3]

Male dollar sunfish exhibit parental care and defend eggs and larvae against predators.[4,5] Spawning occurs repeatedly over a season with males sometimes simultaneously guarding both eggs and broods of 150–200 larvae.[2,3]

Nesting dollar sunfish face a trade-off between guarding their current brood from predation and the need to avoid their own predation. In a SC study, when males were threatened with model aerial predators (kingfishers and herons), they retreated from the nest regardless of the reproductive stage of the nest (nests empty or nests with eggs and/or larvae). However, they returned more quickly to nests when offspring were present.[4]

EGGS

Description
No information.

Incubation
No information.

Development
No information.

YOLK-SAC LARVAE

Size Range
No information.

Myomeres
No information.

Morphology
No information.

Morphometry
No information.

Fin Development
No information.

Pigmentation
No information.

POST YOLK-SAC LARVAE

Size Range
No information.

Myomeres
No information.

Morphology
No information.

Morphometry
No information.

Fin Development
No information.

Pigmentation
No information.

JUVENILES

See Plate 2B following page 26.

Size Range
Size at transformation to juvenile is unknown. Sexual maturity is reached at 60 mm TL in NC.[2,3]

Morphology
Large juveniles. Lateral line scales 34–44;[2,8,10,12,13] cheek scales 3–4.[2,8,12,13] Gill rakers 9–10; for YOY, length of longest rakers is equal to about twice the basal width.[2,13]

Morphometry
No information.

Fins
Large juveniles. Dorsal fin has 9–11,[2,8,10,12] usually 10[2,8,10] spines and 10–12 soft rays.[2,8,10,12] Anal fin has 3 spines[8,10,12] and 9–10 soft rays.[2,8,10,12,13] Pectoral fins have 11–13,[2,8,10,13] usually 12[2,10,13] rays.

Figure 62   Development of young dollar sunfish — schematic drawings showing melanistic pigment patterns of young dollar sunfish. (A) 18 mm SL, primary melanistic pigment bars present; pattern is characteristic for dollar sunfish 12–35 mm SL. (B) 67.9 mm SL; lateral pigment bars no longer visible. (A–B reprinted from Figure 14, reference 14, with publisher's permission.)

Pigmentation

In the development of all *Lepomis*, primary bars of pigment appear as an initial set of lateral patches of melanophores, to which bars are added anteriorly and posteriorly. Nine to 10 primary bars develop on juvenile dollar sunfish between 12 and 35 mm SL, but by 68 mm SL no primary or secondary bars of pigment are present (Figure 62).[14]

Immature dollar sunfish have a light, blue-green background color dorsally and laterally and the sides are flecked with iridescent blue spots. The belly is white or pale orange and the fins are lightly pigmented.[10]

TAXONOMIC DIAGNOSIS OF YOUNG DOLLAR SUNFISH

Similar species: dollar sunfish is in the subgenus *Icthelis* along with longear sunfish (see *Icthelis* subgeneric discussion on page 92). There is a great similarity in appearance of adult dollar sunfish with young longear sunfish[2] and dollar sunfish larvae are probably very similar to longear larvae.[15]

The larvae of dollar sunfish have not been described, so comparison to larval development of longear sunfish is impossible. Juvenile dollar sunfish (Plate 2B) are very similar to young longear sunfish (Plate 2C).[2] Larger dollar sunfish juveniles differ in having 3–4 scale rows on the cheeks vs. 5–7 for the longear and 11–13 (usually 12) pectoral fin rays vs. 13–15 for the longear.[8] Also, dollar sunfish tend to occur in swamps and sluggish water rather than flowing streams as does the longear.[8,10]

ECOLOGY OF EARLY LIFE PHASES

Occurrence and Distribution (Figure 63)

Eggs. Males exhibit parental care by constructing nests and guarding their eggs.[2–5]

Yolk-sac larvae. Guarded in the nest by the male parent.[2–5]

Conner (1979) suggested that young dollar sunfish were probably very similar to young longear sunfish. Longear "type" larvae were rare in Conner's

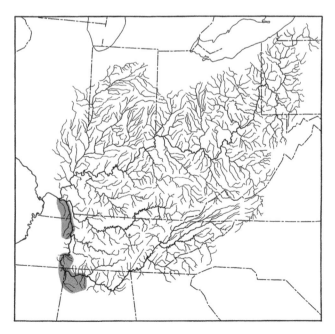

Figure 63   General distribution of dollar sunfish in the Ohio River system (shaded area).

collections and the smallest individuals had only remnants of yolk remaining, leading him to conclude that even as free-swimming individuals, these larvae were much more nest-bound or at least much less prone to venture into pelagic areas than other *Lepomis*.[15]

Post yolk-sac larvae. No information.

Juveniles. No information.

Early Growth
Young dollar sunfish attain a length of about 10 mm TL in 1 month.[2,3]

Size classes within a series of dollar sunfish collected from western TN in August averaged 57 mm TL for age 1, 75 mm for age 2, 83 mm for age 3, and 95 mm for age 4; no YOY were present in the collection.[2]

Feeding Habits
Adult dollar sunfish reportedly have benthic and surface-oriented feeding behavior, feeding principally on midge larvae and microcrustaceans, as well as on filamentous algae, detritus, and various aquatic and terrestrial insects.[2,3,9,10,13]

The authors could find no information on the feeding habits of young dollar sunfish.

LITERATURE CITED

1. Lee, D.S. et al. 1980.
2. Etnier, D.A. and W.C. Starnes. 1993.
3. Lee, D.S. and B.M. Burr. 1985.
4. Winkelman, D.L. 1996.
5. Winkelman, D.L. 1994.
6. Bettoli, P.W. et al. 1993.
7. Burr, B.M. and M.L. Warren, Jr. 1986.
8. Robison, H.W. and T.M. Buchanan. 1988.
9. McLane, W.M. 1955.
10. Ross, S.T. 2001.
11. Paller, M.H. et al. 1992.
12. Mettee, M.F. et al. 2001.
13. Boschung, H.T., Jr. and R.L. Mayden. 2004.
14. Mabee, P.M. 1995.
15. Conner, J.V. 1979.

LONGEAR SUNFISH

Lepomis (Icthelis) megalotis (Rafinesque)

Robert Wallus and John V. Conner

Lepomis, Greek: *lepis*, "scale" and *poma*, "lid"; *megalotis*, Greek: "great ear," in reference to the prominent opercular flap.

RANGE

Longear sunfish is restricted to the freshwaters of east-central North America. This species occurs west of the Appalachians from southern QU south to the Gulf of Mexico in AL and western FL. Range extends west through TX and Rio Grande tributaries in northeast Mexico, north through eastern parts of the states from OK to southern ON.[1]

HABITAT AND MOVEMENT

Longear sunfish occurs in relatively unpolluted lacustrine and riverine habitats of all sizes[64,65] from upland regions to lowland swamps and oxbows on the coastal plain.[64] It is most commonly reported from free-flowing streams[25] of permanent[64,65,72] or semi-permanent flow[72] where it is found in sluggish[18,23] or moderate-flow[73] habitats. It typically inhabits small streams and upland parts of rivers[1,72] and is generally absent from lowland sections of large streams.[1] Longear sunfish also thrives in littoral habitats[24] of reservoirs and natural lakes often occurring in numbers along the shorelines.[1,64,65,72] It does well in ponds.[72]

Longear sunfish occur in a variety of habitats,[64,72,73] but are often found along the margins of moderately flowing streams where they occupy quiet areas[26,64] in cool,[25] clear,[23,25,65] permanent waters over gravel,[25,64] rocky,[65,72] or sandy[64,65] substrates. The longear is often associated with cover such as vegetation,[23,65,69,73] undercut banks,[73] logs,[73] and brush.[73]

This species is found in streams of all sizes but is more abundant in creeks than in large rivers.[65] In a KY creek system, longear sunfish were limited in distribution to the main channel (fourth order) and larger tributaries.[13,22] No longear sunfish were found in second-order streams and they were rarely collected in third-order streams.[13] In other studies in the Paducah area of KY, longear sunfish were abundant in third-order tributary streams to the Ohio River (J.F. Heitman, personal communications, American Aquatics, Knoxville, TN).

Longear sunfish is an abundant species in tributary streams in the Tennessee Valley 6–45 m wide char-acterized by stream flows of less than 1,000 cfs and frequent riffles of gravel and rubble and sometimes exposed limestone or sandstone bedrock.[12]

It avoids strong currents,[26,65] occurring most commonly in pools,[26,41,65,72] inlets, and overflow waters adjacent to the stream channel.[65] However, in two areas of the Pearl River, LA and MS, longear sunfish was by far the most abundant centrarchid and its abundance was attributed to its tolerance of the current along the shore of the river.[6]

Though often associated with aquatic vegetation, this is not an essential requirement.[65] In clear Ozark lakes, underwater behavioral observations (26 h) using SCUBA were used to determine microhabitat use and time budgets of sunfish and juvenile bass in the vicinity of weed beds (*Justacia americana*). Adult longear sunfish spent 66% of their time in open water habitats.[89] Longear sunfish did not associate with aquatic plants in MI lakes,[33] but showed a distinct preference for densely vegetated areas of large NY streams.[69] In Lake Conroe, TX, longear sunfish was the only centrarchid species to respond favorably to vegetation removal. Average size decreased, but density increased following vegetation removal. Despite increasing nearly an order in magnitude between 1980 and 1986, longear biomass did not differ significantly after vegetation removal.[51]

Longear sunfish is reportedly intolerant of turbid, silted waters[23,53] and appears to be sensitive to large amounts of silt in NY.[18] Abundance of longear sunfish decreased in central OH during the period 1920–1950. The decrease was undoubtedly associated with increased turbidity and siltation.[70] In VA, longear tolerate substantial siltation.[68]

Longear sunfish is intolerant of salinity[53] or often frequents brackish waters of coastal streams, where salinities range from 2–23 ppt.[75,84]

Indices of hypoxia and hyperthermia tolerance for MO fish assemblages were determined based on laboratory measurements of lethal dissolved oxygen concentrations and temperatures combined with field measures of relative abundances of tolerant and sensitive species. Mean hypoxia tolerance for longear sunfish was 0.68 mg/l dissolved oxygen; mean hyperthermia tolerance was 37.8°C.[4,5]

In the laboratory, longear sunfish preferred low light compared to high light intensity and preferred submerged cover compared to no cover. When given a choice, they preferred to occupy low light intensity conditions without cover rather than submerged cover under higher light intensity. This suggests that low light intensity, such as is found at greater pool depths, may be the preferred refuge from predation for longear.[34]

Longear show little movement in streams;[26,78,79] one report indicated that when movement did occur, there was more movement downstream than upstream.[78,79] The estimate of 70 linear feet (about 21 m) as the approximate home range of longear sunfish in a LA creek[14] was considerably smaller than estimates of 100–200 feet (30–60 m) previously reported in IN.[26] In another LA study, average length of home range in streams was 42 m and some longear seemed to desert their home ranges during winter months.[31] Twenty of 24 marked longear sunfish in one LA creek were recaptured in their respective home ranges after 1 year had elapsed.[44]

Of 364 marked longear sunfish in an AR creek, 231 (65%) were recaptured. Of those recaptured, 86% were recaptured in the pools of initial collection. The majority of those recaptured outside their home pools were found in pools adjacent to the pool of initial collection. Most moved less than 100 m; the most distant recapture was 506 m. Of 28 longear recaptured at least twice, 17 (61%) exhibited no movement, 5 (18%) exhibited one-way movement, and 6 (21%) exhibited complex movement.[3] Longear were more likely to move from small pools than larger ones.[37]

Longear sunfish showed a temporal pattern of use in an Ozark spring branch characterized by a fairly constant temperature of 13.5 ± 1°C. The lowest use was during the warmwater period of April–October and the greatest use was during the November–March coldwater period. During the coldwater period, biomass of longear in the spring branch was significantly greater than in similar habitats in the receiving stream.[2]

In small TN streams with floodplains, longear sunfish abandoned the main channel for the floodplain during flooding.[28] In reservoirs, longear sunfish maintain a greater degree of residency at normal reservoir levels than during periods of high water.[42]

DISTRIBUTION AND OCCURRENCE IN THE OHIO RIVER SYSTEM

The longear sunfish is the second most abundant centrarchid in the Ohio River.[63] It has been reported from throughout the river since 1800[62] and is commonly encountered in the lower[63] and middle third[11,16,67] of the river; and is less common in the upper river.[11,16,67] Based on lockchamber surveys (1957–2001) it has increased in abundance in the Ohio River since 1957,[50] and the number of reports of this species in the middle third of the river, near the embayments, increased dramatically after 1970.[63]

During the period 1981–2003, longear sunfish were collected in Ohio River Ecological Research Program electrofishing surveys near three electric generation stations: Cardinal (ORM 76.7), Kyger Creek (ORM 260), and Tanners Creek (ORM 494). They were collected more consistently and in greater numbers at ORM 494 than at the two upstream stations.[67]

Longear sunfish is not naturally present in the Ohio River drainage of NY.[69] It occurs throughout OH and is numerous and often abundant in the streams of southern OH.[70] The longear is widely distributed and abundant in the Wabash River drainage in IL and IN,[76,77] where it is much more common in the small and medium-sized tributaries than in the river.[76] It is generally distributed and abundant throughout KY and is the most common sunfish in KY streams and rivers.[64] In VA, longear sunfish is native to the Tennessee and Big Sandy river drainages and has apparently been introduced into the New River drainage.[68]

Longear sunfish is widely distributed throughout the Ohio River drainages of TN except in the highest portions of the Blue Ridge. It is much less common in the upper Tennessee drainage where it is broadly sympatric with, and possibly being supplanted by, the ecologically similar redbreast sunfish.[73] In 1991, there had been no recent records in the Tennessee drainages of NC; it is probably extirpated in that state.[66] In addition to stream populations, the longear sunfish is abundant in some Cumberland River reservoirs.[73] It occurs throughout the Tennessee River drainage of northern AL[71,75] and MS.[78]

SPAWNING

Location

Male longear sunfish build nests that may be more or less isolated from each other or crowded together in small colonies.[18,84] Nests are usually built in shallow freshwater and are found in both calm water,[48,84,85] with little or no current,[21,35] and running water;[84] inshore in lakes[18,48,84] and at the mouths of rivers entering lakes or reservoirs;[71,84] in creeks and streams;[8,39,84] over bottoms of rock,[48,84,85] rubble,*[82] sand,*[8,19,21,35,48,71,82,84,85] gravel,*[8,19,21,35,39,71,84,90] or marl.[84] Nesting is also reported in ponds.[39]

Table 76

A summary of reproductive information for longear sunfish collected by TVA biologists in the Tennessee Valley.

Locality	Date	Progeny	Source	Habitat	Substrate	Water Temperature	Water Depth
Powell River, TN	5/28/87	Eggs	Nest	Shallow water near river shoreline	Gravel	27.2°C	Shallow
Norris Reservoir, Clinch River, TN	6/2/78	Eggs and larvae	Nests	Rock quarry contiguous with reservoir	Gravel	21.5°C	90–180 cm
Pickwick Reservoir, Tennessee River, AL	6/4/75	Eggs	Nest	Shoreline, backwater side of island	Sand	26°C	46 cm
Little Bear Creek, Tennessee River tributary, AL	6/8/76	Eggs	Nest	Gravel bar downstream of a large pool	Gravel	21.7°C	20 cm
Wilson Reservoir, Tennessee River, AL	6/16/75	Eggs	Nest	Shoreline	Gravel	26.5°C	180 cm

In IL, spawning occurred in broad, shaded, shallow parts of large creeks with gravel bottoms in areas under 0.6 m deep with moderate current.[39] In headwater streams of the Little Miami River, OH, colonies of longear nests were observed on the bottom of long, slow-moving pools with sand and pebble bottom deposits.[8] In the Thames River, ON, some longear sunfish nests were observed in backwater habitats, but most were in the main river, where flow was variable, the substrate mainly sand and gravel, and vegetation sparse.[19] In another report, groups of nests were observed in open areas, but solitary nests were often located near instream structure.[21,35] In the Tennessee Valley, nests are often observed along the gravelly margins of streams.[*,73]

Spawning colonies in two Ozark reservoirs were generally found in brush-free areas having a gradually sloping gravel substrate.[42] Water depth in nesting areas ranged from 0.2 to 3.4 m with an average of 1.5 m.[42] Other reports of depth include 30 cm,[18] 1–2 m,[*,82] 12–50 cm,[21,35] and 20–60 cm.[44] A nest observed in Lake of the Ozarks, MO, was 1.2 m from shore in water 0.25 m deep.[90]

In the Tennessee Valley, longear nests have been observed along the shallow shorelines of reservoirs, in the backwater sides of islands, and in a rock quarry contiguous with a reservoir (Table 76).[*]

Season

Longear sunfish in northern populations of WI[80] and MI[84] begin spawning in June and continue into August. Spawning is reported from May to August throughout the rest of the range.[18,23,42,44,48,52,54,65,85,90]

Reports of spawning times for populations in or near the Ohio River drainage include: late May to July in IL;[39] May and June in IN;[26] late May to early

July in VA;[68] May into August in AL;[71,75] and late May to mid-June in TN (Table 76).[*,82]

Temperature

Reports of spawning temperatures for longear sunfish range from 20 to 30.6°C,[42,44,52,68,85] with optimum spawning occurring at temperatures of 23–25°C.[85]

In MD, spawning begins at 20°C.[85] In AR, spawning began when temperatures at nest depths reached 21–23°C.[42] Spawning has been reported in water 25°C in VA,[68] 23.9–30.6°C in KS,[52] and 29°C in LA.[44] In the Tennessee Valley, nests with progeny have been observed at 21.5–27.2°C (Table 76).[*]

Fecundity

Reports for fecundity of longear sunfish are variable. One report indicated that mature females 2–4 years of age contained 2,360–22,119 eggs.[74] In another report, the number of ova stripped from 5 female longear, 83–98 mm TL and age 2 and 3, averaged 414 and ranged from 177 to 717. The total number of developed ova from these fish averaged 548 and ranged from 236 to 940.[54]

In Beaver and Bull Shoals Reservoirs, AR, estimates of the number of spawned ova (May to July) for longear sunfish <101 mm ranged from 1,417 to 3,600; for those 101–129 mm, the estimate was 3,440–4,136; and for larger females from Beaver Reservoir, 4,213.[54]

In WI, on August 1, an age-2 female 75 mm and 9.5 g held an estimated 745 eggs; an age-3 female 93 mm and 19 g had an estimated 1,620 ripe eggs.[80]

In AR, 12 nests in varying stages of development contained 52–2,836 eggs and/or larvae.[42]

Sexual Maturity

Longear sunfish attain sexual maturity usually between ages 2 and 4.[18,25,70,74,75]

In NY, sexual maturity is attained in the third summer at about 76 mm TL.[18] In VA maturity is reached in about 2 years;[68] 2–3 years in AL.[75]

In KY, 95% of females were sexually mature at age 2, but only 9% of males. All age-3 and age-4 females were mature. Sixty-four percent of age-3 males were mature and 96% of age-4 males. The smallest mature female was age 1 and 59 mm SL, the only mature individual in the age class. The smallest mature male was age 2 and 79 mm SL, the only mature male found in the age class.[25]

Other reports of sexual maturity include males at about 100 mm and females at about 75 mm.[14] In stunted populations sexual maturity may be reached by 60 mm.[84] The subspecies *Lepomis megalotis peltastes* matures as small as 53 mm.[70]

In culture tanks, longear sunfish males 100–120 mm and females 70–90 mm spawned 22 weeks after hatching.[29]

Spawning Act

Although a few male longear sunfish nest solitarily,[20] most territorial males build their nests in dense aggregations or colonies;[20,30,65] often nests are so close together that their rims almost touch.[65] Spatial clumping of nests can result from habitat limitation or social interactions, and social interactions that favor nesting colonies can be responses to natural selection (predation) or sexual selection.[21] Field studies have indicated that large areas of unused habitat suitable for nesting are often present, suggesting that longear nesting aggregations are not formed as a result of limited preferred habitat. Males appear to be socially attracted during breeding,[40,65] and social nesting presumably is a successful reproductive strategy for at least a portion of the population.[40]

Nesting activity occurs in synchronous bouts that begin with nest construction lasting about 2 days, followed by spawning, which takes place over 1–2 additional days.[21,35]

At the onset of occupying a territory and building a nest, the male longear sunfish swims into shallow water and remains over an area, circling the area and chasing away other fish, tail wagging, circling back out to deeper water, and then back to the shallow area.[44] The male creates a circular[44] or slightly ovoid[90] depression about 0.5 m in diameter[42,90] by vigorous actions of his tail. He assumes a position at a 45° angle with the bottom and moves his whole caudal peduncle area vigorously from side to side, thus hollowing out the depression in the gravel.

Once built the male guards the nest aggressively, chasing off other fish including other male and female longear. Approaching male longear will turn and flee when chased by the guarding male; female longear ready to spawn will remain in the territory and be courted by the male. He displays violently, dashing to the surface of the water and back to the bottom of the nest, turning on his side and displaying his bright orange ventral surface to the female. She then enters the nest and circles with the male. The process of spawning consists of the male and female circling the nest, the female always between the male and the center of the nest. Both circle in an upright position and every 50–60 s the female turns over on her side at a 20–30° angle with the bottom and brings her vent close to the male's vent. This spawning posture lasts from 2–20 s while both fish continue circling slowly and quiver rapidly while releasing gonadal products. If another fish comes near the nest while the male is spawning, he immediately leaves the female to chase the intruder away and then returns to continue circling with the female who remains in the nest until his return. After spawning is completed, the male chases the female away from the nest[38,44,90] and returns to fan the eggs and guard the nest.[18,21,23,30,38,42,44,90] Initially, the male gently fans the nest, presumably mixing the eggs and sperm and cleansing the nest of excess milt and silt. This is followed by a period of violent fanning activity performed in a vertical tail-stand position that may serve to drive the eggs into the gravel interstices for added protection.[73,90]

Territorial aggressiveness of the male increases after eggs and young are present in the nest.[44] Parent male longear protect eggs and developing young until they are free-swimming and leave the nest.[38] In addition to other nest predators, nest-guarding male longear sunfish have to deal with conspecific disruptive intrusion behavior of neighboring territorial males, smaller non-territorial males, and female longear. The main function of nest intrusion by females is eating eggs; by males it is apparently the fertilization of eggs in another male's nest.[30]

Male longear sunfish have been observed guarding their nests against all types of potential predators including large suckers. However, when approached by a sizable largemouth bass, the male fled and hid, then returned to the nest after the bass went away. This behavior emphasizes the male longear's ability to distinguish between potential predators on itself and other large but non-piscivorous species.[90]

Male aggressiveness decreases after the young have departed the nest. Males have been observed guarding a nest 15–16 days after the young had left, occasionally chasing fish away and leaving periodically to spend more and more time in an adjacent pool.[44]

Sound production by male longear sunfish is reported during active courtship of females. A nesting male, upon sighting a female, would rush toward her and back toward his nest repeatedly while producing a series of grunt-like sounds. In subsequent study, males were induced to court and call to dead females manipulated on a string. Whether or not females produce sounds was not determined. However, the female may have produced a single grunt as she approached the male, perhaps to attract his attention.[45]

Among group nesters, females spawned preferentially with males nesting early within a spawning period and occupying a central nest. However, males that spawned later in the breeding season obtained significantly more larvae than those breeding earlier. Male size and nest diameter were negatively correlated with spawning success.[20] There was no difference in reproductive success between solitary and social nesters,[20,35] but aggregations may nonetheless result from sexual selection: males unlikely to attract females may nest around more attractive males to steal fertilizations from them.[20] In two Ozark reservoir studies, females frequently spawned in more than one nest and often more than one female was observed spawning with the same male.[42]

Solitary nesters tend to be large, successfully mating males that experience lower rates of fertilization stealing (cuckoldry) compared to colonial males. Solitary nesters may favor sites containing structure because it physically blocks access to the nest by both cuckolding neighbors and predators.[21]

Previously used nests are frequently the site of subsequent nesting activity. The same individual often reuses his nest in multiple bouts, but construction of more than one nest in a year by the same individual is reported. In a field study, tagged longear males built an average of 2.8 nests (range = 1–5) in a year, including reused nests. Longear sunfish also use nests constructed by other centrarchids including smallmouth bass, largemouth bass, and rock bass, and interspecific aggressive encounters around nest sites are reported.[21]

Nesting activity can be limited by flow regimes.[21,35] In an IL stream, the number of nests constructed tended to be lower during years with more variable flow.[35]

In a KY creek, rosefin shiners *Lythrurus fasciolaris* were commensal spawners in the nests of longear sunfish. The spawning behavior of both species was observed and in at least one instance a male longear sunfish and a male rosefin shiner were seen guarding the same nest. The odors of longear sunfish milt and ova apparently attract rosefin shiners to the longears' nests.[88]

Hybridization is reported between longear sunfish and pumpkinseed,[19,68,70] green, bluegill,[68,70] redear, and orangespotted sunfishes.[70]

EGGS

Description

Fertilized longear sunfish eggs are spherical,[85] demersal, and adhesive.[39,44,74,82,85] Live fertilized eggs are amber to light yellow[74,82] with a single oil globule.[82] Egg diameter is reported at 1.0 mm in northern populations,[74,85] but in TN, fertilized eggs ranged from 1.65 to 1.84 mm and averaged 1.71 mm in diameter.[82]

Incubation

In aquaria, 2 days at 25°C;[29] 3 days at 25°C is also reported;[44,86] 3–5 days at ambient temperatures.[74]

Development

No information.

YOLK-SAC LARVAE

See Figure 64.

Size Range

Longear sunfish from TN hatch at 5.0–5.2 mm TL and absorb their yolk by 7.3–7.6 mm TL.[82]

Myomeres

6.0–7.6 mm TL. Preanal 13–15; postanal 15–18; total 29–32.*,[82]

Morphology

5.0–5.2 mm TL (newly hatched). Head is decurved over the yolk sac; no stomodeum present. Otic vesicles are visible but otoliths are not. Urostyle is straight. Yolk sac is large and ovoid with a single oil globule located posteriorly and dorsally near the forming gut.[82]

5.9 mm TL. Mouth is open, head is in line with the body axis, and otiliths are forming.[82]

6.5 mm TL. The yolk sac is ovoid.[82]

7.3–7.6 mm TL. Last remnant of yolk is absorbed. Digestive tract is functional prior to complete yolk absorption. At 7.4 mm, lower jaw is developed to

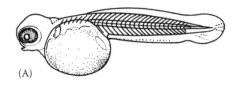

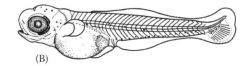

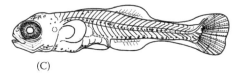

Figure 64 Development of young longear sunfish. A–B. Yolk-sac larvae: (A) 5.5 mm TL; (B) 7.0 mm TL. (C) Late yolk-sac larva, 7.2 mm TL. (A–C reprinted from Figures 1–3, reference 82, with author's permission.)

tip of snout. Opercle covers gill cavity and is developed to base of pectoral fins. Gill rakers appear as blunt knobs on the arches.[82]

Morphometry
See Tables 77 and 78.

Fin Development
See Table 79.

5.0–5.2 mm TL (at hatching). A median finfold originates at the 8th preanal myomere and is present along the dorsum and around the urostyle forward ventrally to the yolk sac. Pectoral fins are broad-based flaps positioned dorsally and anteriorly on the yolk sac.[82]

6.5–7.1 mm TL. Hypural complex forms. At 6.5 mm, the caudal fin has opaque areas dorsal and ventral to the slightly flexed urostyle. Caudal fin ray differentiation begins by 6.8 mm. Local widenings of the finfold indicate positions of the future dorsal and anal fins and fin ray definition has begun in some individuals by 6.9 mm.[82]

7.2–7.6 mm TL. At 7.2 mm, the caudal fin is truncate and slightly notched with the urostyle in the upper lobe and rays are apparent. Soft ray formation begins in developing dorsal and anal fins. Pectoral fins are still moderately broad-based. Pelvic

Table 77

Means (in millimeters) or model numbers of morphometric and meristic characters examined for longear sunfish from TN.

Size Class	No. of Specimens	Total Length	Urostyle Length	Preanal Length	Postanal Length	Head Length	Modal No. Preanal Myomeres	Modal No. Postanal Myomeres
5.0–5.2	2	5.19	5.03	2.39	2.80	0.75	14	18
5.5–6.49	5	5.82	5.54	2.56	3.26	0.87	14	18
6.5–7.49	5	7.14	6.71	3.20	3.94	1.60	14	16
7.5–8.49	4	7.84	7.00	3.48	4.36	1.80	14	16
8.5–9.49	5	9.02	7.71	4.06	4.86	2.30	14	16
9.5–10.49	5	9.95	8.37	4.51	5.44	2.50	13	16
10.5–11.49	5	10.88	8.93	5.02	5.86	2.91	13	15
11.5–12.49	5	12.06	9.89	5.47	6.46	3.36	13	14
12.5–13.49	5	13.06	10.62	5.82	7.23	3.65	13	14
13.5–14.49	5	14.11	11.65	6.59	7.46	4.22	13	13
14.5–15.49	5	15.23	12.26	6.72	8.45	4.29	13	13
15.5–16.49	5	15.71	12.67	7.04	8.74	4.33	—	—
16.5–17.49	5	17.09	13.70	7.65	9.44	4.54	—	—
17.5–18.49	5	18.05	14.18	7.81	10.25	4.99	—	—
18.5–19.49	5	18.94	15.01	8.35	10.59	5.09	—	—
19.5–20.49	5	19.84	16.03	8.77	10.85	5.31	—	—

Source: Reprinted from Table 1, reference 82, with author's permission.

Table 78

Morphometric data for young longear sunfish from TN* grouped by selected intervals of total length (TL). Characters are expressed as percent TL and head length (HL) with a single standard deviation. Range of values for each character is in parentheses.

	TL Intervals (mm)					
Length Range N Characters	6.01–7.61 42 Mean ± SD (Range)	7.70–8.08 13 Mean ± SD (Range)	8.15–12.90 70 Mean ± SD (Range)	12.96–16.95 30 Mean ± SD (Range)	17.10–19.23 17 Mean ± SD (Range)	20.95–30.55 8 Mean ± SD (Range)
Length as % TL						
Snout	3.72 ± 0.41 (0.21–0.31)	3.79 ± 0.26 (0.25–0.32)	4.40 ± 0.37 (0.32–0.63)	5.07 ± 0.24 (0.61–0.88)	5.43 ± 0.25 (0.85–1.05)	5.26 ± 0.60 (1.08–1.6)
Eye diameter	8.64 ± 0.39 (0.49–0.68)	8.59 ± 0.33 (0.63–0.72)	9.08 ± 0.34 (0.68–1.18)	9.44 ± 0.58 (1.22–1.63)	9.56 ± 0.36 (1.5–1.9)	9.19 ± 0.38 (1.98–2.7)
Head	21.6 ± 0.71 (1.35–1.7)	21.8 ± 0.81 (1.63–1.9)	24.0 ± 1.02 (1.75–3.25)	26.1 ± 0.72 (3.2–4.45)	27.6 ± 1.09 (4.3–5.45)	27.4 ± 1.03 (5.75–8.15)
Predorsal	—	—	29.7 ± 1.20 (2.6–3.9)	31.0 ± 0.85 (3.9–5.25)	30.7 ± 0.90 (5.0–6.0)	31.3 ± 0.87 (6.35–9.55)
Prepelvic	—	—	27.8 ± 1.11 (2.4–3.55)	29.6 ± 7.20 (3.5–8.85)	28.5 ± 1.06 (4.5–5.8)	29.8 ± 0.61 (6.0–9.1)
Preanal	45.2 ± 1.12 (2.71–3.45)	44.4 ± 0.65 (3.43–3.6)	45.7 ± 0.82 (3.63–5.9)	45.9 ± 1.15 (5.9–7.8)	46 ± 1.00 (7.65–8.9)	46.2 ± 0.67 (9.7–13.65)
Postanal	54.8 ± 1.12 (3.3–4.27)	55.7 ± 0.81 (4.25–4.57)	54.1 ± 1.34 (4.35–7.03)	54.1 ± 1.15 (7.01–9.25)	53.4 ± 3.00 (7.89–10.33)	53.8 ± 0.67 (11.2–16.9)
Standard	92.7 ± 2.30 (5.71–7.05)	87.5 ± 1.71 (6.6–7.15)	83.0 ± 1.51 (6.85–10.6)	80.3 ± 0.85 (10.3–13.5)	79.1 ± 0.74 (13.6–15.1)	79.0 ± 0.91 (16.5–23.75)
Air bladder	9.88 ± 1.33 (0.54–0.92)	11.5 ± 0.57 (0.83–0.98)	14.5 ± 1.84 (0.85–2.15)	—	—	—
Eye to air bladder	14.7 ± 1.16 (0.75–1.15)	13.6 ± 0.55 (1.0–1.15)	12.7 ± 0.87 (1.05–1.55)	12.0 ± 0.41 (1.45–1.81)	—	—
Fin length as % TL						
Pectoral	—	—	14 ± 1.25 (0.95–2.2)	17.2 ± 1.14 (2.0–3.3)	18.0 ± 0.92 (3.1–3.6)	18.1 ± 1.26 (3.5–5.4)
Pelvic	—	—	4.9 ± 2.03 (0.11–1.1)	11.5 ± 1.61 (1.12–2.3)	15 ± 0.83 (2.4–3.15)	16.1 ± 1.64 (2.85–5.0)
Body depth as % TL						
Head at P1	—	—	21.7 ± 1.26 (1.8–3.0)	24.2 ± 0.94 (2.95–4.25)	25.4 ± 0.73 (4.3–5.05)	27.8 ± 1.51 (5.2–8.9)
Head at eyes	—	—	19.1 ± 1.02 (1.6–2.55)	20.8 ± 0.81 (2.65–3.63)	21.0 ± 0.64 (3.5–4.15)	20.5 ± 0.83 (4.2–6.0)
Preanal	—	—	20.2 ± 1.46 (1.6–2.85)	23.1 ± 0.82 (2.88–4.15)	24.4 ± 0.76 (4.15–4.75)	25.9 ± 1.17 (4.83–8.2)
Caudal peduncle	—	—	9.92 ± 0.70 (0.75–1.35)	10.6 ± 0.39 (1.25–1.79)	11.1 ± 0.41 (1.82–2.15)	10.7 ± 0.28 (2.17–3.25)
Body width as % HL						
Head	—	—	50.4 ± 3.97 (1.35–1.7)	51.4 ± 5.94 (1.4–2.4)	51.7 ± 4.11 (2.2–3.1)	58.6 ± 5.66 (2.85–5.05)

Table 79

Fin development for young longear sunfish.

Fins	TL (mm) at Which:	
	Fin Rays Appear	Adult Complement of Rays Present
Dorsal		
Spines	8.1*	10.6*
Soft rays	7.6,[82] 7.9*	11.8,[82] 8.45*
Anal		
Spines	7.9*	9.4*
Soft rays	7.2,[82] 7.9*	11.8,[82] 9.0*
Caudal	6.6–7.2,[82] 6.0*	8.8,[82] 7.2–7.5*
Pectoral	8.8,[82] 8.2*	12.4[82]
Pelvic	11.7,[82] 8.2*	14.2[82]

fins not yet present. At 7.6 mm, segmentation of caudal fin rays is apparent.[82]

Pigmentation
5.0–5.2 mm TL (at hatching). Except for light pigment around the eye lenses, larvae are devoid of pigment.[82]

7.2 mm TL. Dorsum of the yolk sac, the cleithrum, the cheek posterior to the eye, and the gill arches are pigmented.[82]

7.4 mm TL. The top of the head has 4–5 large chromatophores and thereafter becomes heavily pigmented. Pigment is present at the posterior margin of the head. The remaining yolk and visceral area have large randomly spaced chromatophores. A midlateral line of pigment is present from the incipient air bladder to the caudal peduncle. Ventrally, a double row of melanophores is visible posteriorly from the anus along the caudal peduncle. Two to three large chromatophores are located on the dorsum of the anus. The ventral portion of the caudal fin has 2–3 proximal melanophores.[82]

POST YOLK-SAC LARVAE

See Figure 65.

Size Range
Phase begins at about 7.6 mm TL; phase ends with completion of pelvic development between 13.6 and 14.2 mm TL.[82]

Myomeres
Preanal 13–15; postanal 13–17; total 26–31.*,[82]

Morphology
7.6–7.7 mm TL. Well-formed recurved teeth are present on the upper and lower jaws. The air bladder begins to fill. Body depth increases, giving the already robust larvae a deeper appearance. Gut is thick and muscular with a left-hand bend when viewed from the ventral perspective. Larvae at this size are at the "swim-up" stage.[82]

8.0 mm TL. Choroid fissure apparent in the eye.[82]

14.0 mm TL. Gill rakers moderately long, length about two times the width.[82]

Morphometry
See Tables 77 and 78.

Fin Development
See Table 79.

7.8 mm TL. Eight dorsal fin soft rays and five anal fin soft rays are faintly discernable on some individuals; 16 caudal fin rays apparent on some. Urostylar flexion is at 45° or more. Dorsal and ventral finfolds still narrowly connected to the caudal fin.[82]

8.0 mm TL. Caudal fin indented. Opaqueness near the base of the pectoral fins indicates early differentiation of fin rays.[82]

8.5–8.8 mm TL. Pelvic fins become apparent as crescents even with the posterior edge of the pectoral fins by 8.5 mm and as flaps by 8.8 mm. At 8.8 mm, the finfolds are completely absorbed between the dorsal and anal fins and the caudal fin. Anal spines are apparent and differentiation of dorsal spines has begun. Six pectoral rays are visible on some individuals, 3–4 of the longest soft dorsal rays are segmented, and 10 dorsal and 10 anal fin soft rays are visible. The caudal fin has attained the adult complement (17–19) of rays.[82]

9.7–9.9 mm TL. Five dorsal spines apparent. Anal fin rays segmented on some individuals. Remnant finfold still present ventrally just anterior to the anus.[82]

11.5–12.2 mm TL. The third anal spine becomes easily discernable from the soft rays. By 11.8 mm, dorsal and anal fins have attained the adult complement of soft rays and spinous dorsal fin has attained the adult complement of spines.[82]

12.4 mm TL. Pectoral fin ray development complete.[82]

13.6 mm TL. Finfold completely absorbed.[82]

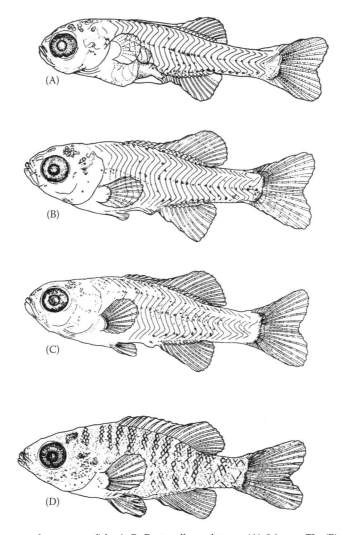

Figure 65 Development of young longear sunfish. A–B. Post yolk-sac larvae: (A) 9.1 mm TL; (B) 11.9 mm TL. C–D. Juveniles: (C) 14.2 mm TL; (D) 21.0 mm TL. (A–D reprinted from Figures 4–7, reference 82, with author's permission.)

Pigmentation

7.6–7.8 mm TL. Dorsally, the head posterior to the eye is darkly pigmented, a double row of melanophores extends from the head to the middle of the soft dorsal fin, and 2–3 melanophores are present posteriorly. Distinct ventral pigmentation develops, varying from a triangular patch to a very elongated concentration of melanophores. Internal gut wall is pigmented.[82]

8.0–8.8 mm TL. A few melanophores are present along the ventral edge of the opercle. Branchiostegal rays and snout become pigmented. Internal pigment is present along vertebrae to middle of air bladder. A well-defined line of melanophores appears along the posterior edge of the hypural complex. By 8.2 mm, pigmentation has spread laterally on air bladder and ventral pigmentation appears more as an elongate patch. Otoliths pig-

mented by 8.6 mm. By 8.8 mm, dorsally, a double row of melanophores extends to base of caudal fin.[82]

9.9 mm TL. A spot is apparent dorsally at the base of the caudal fin. The posterior ventral lobes of the brain have a circular "stitched" pigment pattern.[82]

10.1–11.1 mm TL. Ten to 12 melanophores appear in the gular region. A forked patch of dark pigment develops on the opercle, posterior to and even with the ventral edge of the orbit; this pattern remains apparent into the juvenile phase.[82]

11.7–12.0 mm TL. Ventrally, a "V" pattern of melanophores is present in the branchiostegal region. The premaxilla is pigmented. Dorsal, anal, and caudal fins are pigmented to their edges, but there is no pigmentation apparent on the pectoral and pelvic fins. Scattered pigment is present above and below the midlateral line and increases dorsally between the double row.[82]

JUVENILES

See Plate 2C following page 26 and Figures 65 and 66.

Size Range
Between 13.6 and 14.2 mm TL[82] to about 79 mm SL.[25]

Other reports of sexual maturity include males at about 100 mm and females at about 75 mm,[14] but in stunted populations 60 mm.[84] The northern subspecies *Lepomis megalotis peltastes* matures as small as 53 mm.[70]

Morphology
15.7 mm TL. Squamation begins along the midlateral line of the caudal peduncle and then rapidly extends forward.[82]

17.5 mm TL. A wide based opercular flap is present.[82]

17.8–18.3 mm TL. Body begins to assume the typical adult deep-bodied outline. The body depth from the nape to the posterior margin of the soft dorsal fin rapidly increases. By 17.9 mm, scale formation extends forward to the cleithrum along the midline; dorsum and ventrum of the caudal peduncle are fully scaled.[82]

20.9 mm TL. No scales have developed on the breast, belly between the pelvic fins, or anteriorly on the dorsum from about the middle of the spinous dorsal fin.[82]

21.3 mm TL. Scales forming on the nape and belly.[82]

22.3 mm TL. Opercle and nape are completely scaled.[82]

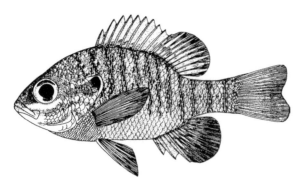

Figure 66 Juvenile longear sunfish, 61.5 mm TL. (Reprinted from Figure 137c, reference 70, with publisher's permission.)

31.0 mm TL. Gill rakers are relatively shorter and wider in relation to the growing arch but are still not of the typical adult form. The opercular flap has not yet elongated to the adult form.[82]

Larger juveniles. Numbers of lateral line scales range from 33–38 in northern populations of Canada[74] and MI.[80] The range is wider to the south: 35–41 in MO;[65] 38–44 (36–46) in VA;[68] 36–43 in AR;[72] 33–45 in TN and AL;[71,73,75] 36–40 in MS.[78]

Longear have 9–11 gill rakers; for YOY, length of longest rakers equals to about 2 times their basal width.[73,75]

Vertebrae 28–30.[73,74]

Morphometry
See Tables 77 and 78.

Fins
14.2 mm TL. Pelvic fins completely develop, marking the beginning of the juvenile phase.[82]

Larger juveniles. Dorsal fins of longear sunfish consist of a spinous fin broadly connected to a soft dorsal fin; numbers of dorsal spines range from 9–12, usually 10–12;[65,68,71–74,78,80] dorsal soft rays are usually 10–12.[68,71–75,78,80] Anal fins have 3 spines[65,68,71,72,74,78,80] and 8–12, usually 9–11, soft rays.[68,71–74,78,80] Canadian populations of longear have 10 to 13, usually 12, pectoral fin rays.[74] Counts are higher for southern populations with ranges reported from 13 to 15, usually 13–14.[68,71–73,75,78] Pelvic fins have 1 spine and 5 soft rays.[74,80]

Pigmentation
15.0 mm TL. A patch of small melanophores is visible just posterior to the eye. Midlateral pigment is heaviest posterior to the air bladder. Ventrally, a double row of melanophores has formed along the base of the anal fin and diffuse pigment is present postanally.[82]

17.3–18.3 mm TL. Primary bars of pigment appear as an initial set of lateral patches of melanophores to which bars are added anteriorly and posteriorly in the development of all *Lepomis*.[81] Nine to 12 lateral bands of pigment appear, developing dorsal to the median myosepta, the first located between the anus and the insertion of the anal fin. These bands begin to fuse dorsally below the dorsal fin. Most bands are wider than the spaces between them. The spinous dorsal fin is pigmented interradially and along the spines; pigment is present between the rays of the soft dorsal. Anal and caudal fins have a "stitched" pattern of pigmentation along the rays. A few chromatophores are present between the anal spines and the first and second soft rays.[82]

19.8 mm TL. A conspicuous dark spot is present on the opercular flap.[82]

31.2 mm TL. Lateral pigmentation has increased. Body is heavily pigmented except on the breast.[82]

Larger juveniles. Young longear sunfish have reddish eyes[73] and are olivaceous with chainlike bars on back and side, yellowish or whitish below.[75] They are similar to adult females except that oranges and blues are lacking. Bands[70] or dusky vertical bars[73,81] are usually prominent on the sides of young as are spots on the fins of larger young; smaller juveniles have transparent fins.[70]

TAXONOMIC DIAGNOSIS OF YOUNG LONGEAR SUNFISH

Similar species: other *Lepomis* species, especially redbreast sunfish.

Longear sunfish is in the subgenus *Icthelis* along with the dollar sunfish (see *Icthelis* subgeneric discussion on page 92). There is a great similarity in appearance of adult dollar sunfish with young longear sunfish[73] and dollar sunfish larvae are probably very similar to longear larvae.[83] The larvae of dollar sunfish have not been described, so comparison to larval development of longear sunfish is not possible.

For comparisons of important morphometric and meristic data, developmental benchmarks, and pigmentary characters between longear sunfish and all other *Lepomis* that occur in the Ohio River drainage, see Tables 4 and 5 and Plate 2C.

Longear sunfish vs. redbreast sunfish, green sunfish, warmouth, bluegill, and redear sunfish.[82] The species most closely resembling the longear sunfish in larval development is the redbreast sunfish.[82] Longear sunfish in TN hatch at 5.0–5.2 mm TL compared to reports for redbreast sunfish of 4.6–5.1 mm in PA[92] and 4.3–4.8 in TN.[82] Hatching sizes for the other species listed are also smaller than those reported for the longear sunfish (Table 4).[82] Swim-up size of longear sunfish (7.3–7.6 mm TL) is smaller than that reported for redbreast sunfish (7.6–8.2 mm),[82,92] but larger than reports for other *Lepomis* species.[82] However, it should be noted that swim-up size may vary substantially within a species.[93]

From the late mesolarval phase to the early juvenile period, longear sunfish larvae are less robust than redbreast sunfish larvae. Differences in head, body, and peduncle depth were readily visible in metalarvae and early juveniles. Cultured specimens of both species showed this characteristic difference. The gut of mesolarval and metalarval longear sunfish was not as massive as that of the redbreast sunfish but was thick and muscular.[82]

Mesolarval and metalarval longear sunfish have an elongate patch of small melanophores along the venter similar to that of redbreast sunfish. This character superficially distinguishes these two species from other mesolarval and metalarval lepomids. Green and redear sunfish have pigmentation of this pattern but it is less dense.[82]

Longear sunfish become juveniles at about 14.2 mm TL,[82] a much shorter length than that reported for redbreast sunfish (19 mm)[92] and warmouth (15.7 mm)[94] and a greater length than reported for green sunfish (11.7 mm)[93] or redear sunfish (9.8 mm).[95]

Longear sunfish YOY (Plate 2C) can be distinguished from the very similar young of green sunfish (Plate 1C) and redbreast sunfishes (Plate 1B) on the basis of lateral-line scale counts (Table 5) and their shorter gill rakers.[73]

ECOLOGY OF EARLY LIFE PHASES

Occurrence and Distribution (Figure 67)

Eggs. Eggs after deposition in the nest are found attached to stones and roots in the bottom of the nest.[39] Eggs may also be in the gravel interstices at the nest bottom, having been pushed there by fanning action of the guarding male.[73,90] Eggs of longear sunfish are guarded by the male parent in the nest.[18,21,42]

Brood loss may be attributed to biotic factors including predation, fungus, or desertion of the nest by the guarding male.[21] However, variation in flow regime modifies the importance of biotic effects on the survival.[35] Destruction of nests by flooding is as important a source of brood mortality as biotic factors.[21] Survival of eggs and larvae in the nests of longear sunfish was monitored in an IL stream for 4 years. During 2 years characterized by low, relatively stable flow, nest failures were attributed to biotic interactions. During 2 years with more variable flow, most brood losses occurred during floods. Floods led to nest desertion and loss of offspring regardless of nest location.[35]

A great nuisance to a nest-guarding longear male is other small, marauding males that overrun the nest and devour the eggs.[75] In streams, spawning success and egg survival are probably higher in pools with bass because of reduced harassment or egg predation by smaller fishes.[27]

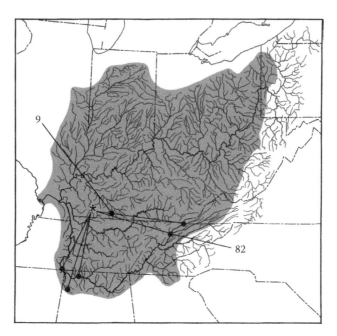

Figure 67   General distribution (native and introduced) of longear sunfish in the Ohio River system (shaded area) and areas where early life history information has been collected (circles). Number indicates appropriate reference. Asterisk indicates TVA collections.

Yolk-sac larvae. After hatching, young longear remain on the bottom of the nest hidden among rocks. At first they have large yolk sacs and remain motionless on their sides. After about 2 days they are able to propel themselves with quick flexing actions along the bottom of the nest. After about 7 days they have lost most of their yolk and become free-swimming.[44]

While in the nest, yolk-sac larval longear sunfish are guarded by the male parent. Males guard offspring until they are free-swimming and leave the nest,[38,42] which may be up to 9 days depending on water temperature.[21,35] Young longear probably become free-swimming before yolk is completely absorbed.[83]

In LA, "longear-type" larvae (thought to be either longear or dollar sunfish) were extremely rare in collections, suggesting that, even as "free-swimming" individuals, they may be much more nest-bound or at least much less prone to venture into pelagic areas than other *Lepomis*.[83]

For the first week of life, longear sunfish larvae cultured in glass aquaria were quiescent and rarely moved, suggesting that larvae remain in or near nests for a time after hatching. After swimming commenced in an aquarium, larvae tended to move about in schools.*

Post yolk-sac larvae. In reservoirs, age-0 longear sunfish generally remain in littoral habitats.[46] In Lake Texoma, larval longear sunfish were seldom taken in areas far from the shoreline and were most abundant near rocky shorelines. Longear 10–73 mm TL were rarely taken with seines along wave-swept shorelines, but otherwise were widely distributed. Young were apparently restricted to shallow water areas near the bottom in daytime, but some smaller larvae swam upward and dispersed at night.[48]

Longear sunfish larvae were collected from a KY creek in 1982 with drift nets, light traps, and seines. Most were captured with light traps and associated with gravel substrates and algae-covered rocks; many were also collected along vegetated shorelines. They were present in samples from mid-June into July.[9]

Longear sunfish larvae were collected from the Tallapoosa and Cahaba Rivers, AL, with seines and pushed plankton nets. Larvae 6.2–8.3 mm TL were collected 1–4.5 m from shore in water 0.2–0.8 m deep with current speeds ranging to 7.14 cm/s.[15]

Longear "type" (probably longear and/or dollar sunfish) larvae were collected with plankton nets in low numbers (relative abundance <2/100 m³) from a bayou tributary of the Mississippi River that had zero current, but were not collected from areas in the river with sluggish (5–20 cm/s) or swift (75–90 cm/s) flows.[17]

Juveniles. In an OK lake, longear sunfish YOY up to 80 mm TL were collected with unbaited minnow traps. They were caught proportionately more often in traps set at depths of 50 cm or less, but over a 2-year period used decidedly different depths up to 200 cm. Longear were primarily

Table 80

Average calculated growth of young longear sunfish
from OK waters[53] with statewide averages
for the periods 1952–1963[59] and 1972–1973.[91]

Waterbodies	No. of Fish	Calculated TL (mm) at Each Annulus			
		1	2	3	4
12 reservoirs	109	76	102	124	137
5 large lakes	38	76	102	—	—
21 small lakes	247	66	102	114	130
7 ponds	19	61	104	124	—
7 streams	243	61	94	114	127
Statewide avg. (1952–1963)	—	69	102	117	132
Statewide avg. (1972–1973)	—	53	88	109	136

collected in habitats with coarse substrate; other sunfish species were most abundant in areas having fine organic substrates. Microhabitats lacking vegetation were used extensively both years. Extensive use by longear sunfish of non-vegetated habitat with coarse substrates clearly separated them from other congeners.[47]

In a MO river, YOY longear sunfish were collected with seines and activity traps usually near cover in the form of water willow, rootwads, and rubble.[49]

In clear Ozark lakes, underwater behavioral observations (26 h) using SCUBA were used to determine microhabitat use and time budgets of adult and juvenile sunfishes in the vicinity of weed beds (*Justacia americana*). Though adult longear sunfish spent 66% of their time in open water, juveniles were in open water only 25% of the time observed. They spent 75% of their time in the weeds at the bottom.[89]

Upper lethal temperatures for young longear sunfish acclimated to 25, 30, and 35°C were 35.5, 36.6, and 38.2°C, respectively.[72]

Longear sunfish YOY feed in littoral areas of reservoirs which makes them susceptible to early summer predation by largemouth bass.[56]

Early Growth
In general, growth in rivers and streams is slower than in lakes and reservoirs of the same area (Table 80).[53] Growth is variable in streams[7] (Table 81) and better growth is reported downstream than in the headwaters of streams;[53] growth is generally more rapid in southern parts of the range than in the north[53] (Table 81).

In the Milwaukee River, WI, YOY longear sunfish caught on August 1, 1963, ranged from 52 to 73 mm and averaged 65.5 mm TL.[80] Average calculated TL at annulus for these fish was comparable to growth in KS streams and slower than growth reported for southern streams (Table 81).

Young longear sunfish grow slowly in NY, reaching only about 102 mm TL by age 4.[18] In OH, YOY longear in October were 20–56 mm long and at about 1 year 30–71 mm.[70]

Calculated SL for longear sunfish from Doe Run, KY, during the period 1959–1961 was 40 mm for age 1, 64 mm for age 2, 83 mm for age 3, and 99 mm for age 4. There was little difference in the rate of growth for young male and female longear.[25]

Table 81

Reports of average calculated growth for
longear sunfish in streams.

Stream (s)	Calculated TL (mm) at Age:			
	1	2	3	4
Milwaukee River, WI[80]	29	58	76	92
KS streams[7]	25	53	73	85
OK streams[53]	61	94	114	127
Clear Creek, IL[10]	41	79	107	132
KY streams[53]	66	107	135	150
Station Camp Creek, TN[36]	43	61	81	99
New River, TN[87]	43	69	88	103

Table 82

Statewide average growth of young longear sunfish recorded in OK waters during the period 1952–1963.

	Average Growth	
Age in Years	TL (mm)	Weight (g)
1	69	7
2	102	23
3	117	36
4	132	54

Source: Constructed from data presented in Table 11 (in part), reference 59.

Table 83

Length–weight relationships of young longear sunfish from AL collected during the period 1949–1964.

Approximate Average TL (mm)	N	Average Empirical Weight (g)
25	1,292	1.4
51	4,539	2.5
76	3,210	9.1
102	2,872	20.0
127	867	39.5
152	153	59.0
178	8	99.9

Source: Constructed from data presented in unnumbered table on page 39, reference 60.

Calculated TL and weight for young longear sunfish in Canton Reservoir, OK, during the period 1962–1967 were as follows: 56 mm and 4 g at age 1; 91 mm and 18 g at age 2; 114 mm and 32 g at age 3; 130 mm and 44 g at age 4.[58] These rates of growth were below the state average for the period 1952–1963 though comparable at ages 3 and 4 (Tables 80 and 82).[59] Statewide average growth among years in OK was variable, especially during the first 3 years (Table 80).[59,91]

Growth of longear sunfish in TN appears to approximate that of OK populations.[73] Length–weight relationships of young longear from AL waters (Table 83) also appear comparable to growth in OK (Table 82).

Table 84

Age and growth determinations for young longear sunfish collected from three KY lakes in 1983.

	Mean TL (mm) at Age:			
Lake	1	2	3	4
Barren River Lake	51	107	137	150
Nolin River Lake	43	79	104	117
Shanty Hollow Lake	51	122	—	—

Source: Constructed from data presented in Tables 30, 57, and 71, reference 61.

Growth of young longear sunfish was variable among three KY lakes in 1983 (Table 84).[61]

Feeding Habits

Longear sunfish is usually classified as an insectivore, but is also reported to eat other small invertebrates.[36]

In MI lakes, longear sunfish about 60–80 mm SL fed primarily on aeschnid, libellulid, and gomphid dragonfly nymphs, *Caenis* nymphs, the amphipod *Hyallela azteca*, and the mayfly *Hexagenia limbata*. Most prey were classified as sediment-dwelling, averaging about 54% of the diet by weight. Large aeschnid dragonfly nymphs comprised the bulk of vegetation prey.[33]

In an OK lake, the most abundant item in the diet of longear sunfish YOY was ostracods, followed generally by chydorids, chironomids, and copepods.[47]

In DeGray Reservoir, AR, aquatic insects were slightly more important than terrestrial insects in the diet of all longear sunfish examined. There was little correlation between size of longear and the consumption of immature or adult aquatic insects. By contrast, the consumption of terrestrial insects was positively correlated with fish length. Crayfish were eaten in significant amounts only by fish over 76 mm and made up a maximum of 15.1% of the diet of longear 101–125 mm. Fish eggs were among the more important seasonal foods, appearing mostly April through June and in the stomachs of fish <100 mm; fish eggs composed 38% of the diet of longear <50 mm. Of plant material in longear stomachs, filamentous algae was the most important.[43]

In Bull Shoals Reservoir, AR, aquatic insects and Entomostraca were equally important and made up 86% of the diet of longear sunfish <48 mm TL. Aquatic insects (48%), fish eggs (23%), terrestrial insects (9%), and Bryozoa (9%) were the predominant foods eaten by longear 74–99 mm TL. Longear >102 mm TL showed a selectivity for terrestrial insects.[57]

In a MO river, young longear sunfish fed on benthic invertebrates throughout their first year; diets became more diverse as the young fish grew. Longear 6–25 mm TL fed primarily on chironomids (45% of total prey items) and Cladocera (37%). Longear 26–84 mm ate primarily chironomids (73% of total prey), Cladocera (10%), and other invertebrates (7%). Crayfish were a minor prey item for this size group. The diets of longear 85–163 mm TL consisted of chironomids (32%), stoneflies (16%), coleopteran (13%), other invertebrates (12%), and decopods (6%).[49]

In small AL streams, longear sunfish were primarily insectivores, relying heavily on mayflies, chironomids, caddisflies, and adult and larval aquatic beetles. Longear 50–100 mm TL consumed proportionately greater quantities of aquatic insects by number and volume, while fish ≥101 mm consumed greater quantities of terrestrial insects and other food. Crustaceans, including crayfish, were regularly consumed.[55]

In a LA creek, small longear sunfish (50–100 mm TL) were observed eating the eggs in a longear sunfish nest.[44]

For longear sunfish feeding is highly variable among individuals for time of day and months. In the Vermilion River, IL, longear fed with greatest intensity near sunrise and sunset. During their high-feeding periods, longear averaging 88 ± 17 mm TL fed in habitat that was deeper and swifter than that of their low-feeding periods. They were found more frequently over gravel substrates during high-feeding times and used areas with bedrock only during low feeding. Cover in the form of woody structure, rocks, plants, and roots was associated with equivalent proportions of longear sunfish habitat during high-feeding periods and they were found using cover in 100% of locations during low-feeding hours, with woody structure used most frequently.[32]

In aquaria, swim-up and commencement of feeding began 7 days after hatching. Larval longear accepted newly hatched San Francisco brine shrimp readily as a first food.[29] Larvae cultured in glass aquaria also did well on a diet of primarily *Artemia*.*

LITERATURE CITED

1. Lee, D.S. et al. 1980.
2. Peterson, J.T. and C.F. Rabeni. 1996.
3. Smithson, E.B. and C.E. Johnston. 1999.
4. Smale, M.A. and C.F. Rabeni. 1995a.
5. Smale, M.A. and C.F. Rabeni. 1995b.
6. Gunning, G.E. and R.D. Suttkus. 1990.
7. Delp, J.G. et al. 2000.
A 8. Brown, E.H., Jr. 1960.
9. Floyd, K.B. et al. 1984.
10. Lewis, W.M. and D. Elder. 1953.
11. Reash, R.J. and J.H. Van Hassel. 1988.
12. Ruhr, C.E. 1957.
13. Kuehne, R.A. 1962.
14. Gunning, G.E. and C.R. Shoop. 1963.
15. Scheidegger, K.J. 1990.
16. Van Hassel, J.H. et al. 1988.
17. Gallagher, R.P. and J.V. Conner. 1983.
18. Raney, E.C. 1965.
19. Keenleyside, M.H.A. 1978.
20. Dupuis, H.M.C. and M.H.A. Keenleyside. 1988.
21. Jennings, M.J. 1991.
22. Lotrich, V.A. 1973.
23. Miller, H.C. 1964.
24. Hall, D.J. and E.E. Werner. 1977.
25. Redmon, W.L. and L.A. Krumholz. 1978.
26. Gerking, S.D. 1953.
27. Harvey, B.C. 1991.
28. Barclay, L.A. 1985.
29. Smith, W.E. 1975.
30. Keenleyside, M.H.A. 1972.
31. Berra, T.M. and G.E. Gunning. 1972.
32. Kwak, T.J. et al. 1992.
33. Laughlin, D.R. and E.E. Werner. 1980.
34. Goddard, K. and A. Mathis. 1997.
35. Jennings, M.J. and D.P. Philipp. 1994.
36. Swann, D.L. et al. 1991.
37. Johnston, C.E. and E.B. Smithson. 2000.
38. Smith, R.J.F. 1969.
39. Hankinson, T.L. 1919.
40. Bietz, B.F. 1981.
41. Jackson, W.D. and G.L. Harp. 1973.
42. Boyer, R.L. and L.E. Vogele. 1981.
43. Bryant, H.E. and T.E. Moen. 1980.
44. Huck, L.L. and G.E. Gunning. 1967.
45. Gerald, J.W. 1971.
46. Meals, K.O. and L.E. Miranda. 1991.
47. Layzer, J.B. and M.D. Clady. 1991.
48. Taber, C.A. 1969.
49. Livingstone, A.C. 1987.
50. Thomas, J.A. et al. 2004.
51. Bettoli, P.W. et al. 1993.
52. Cross, F.B. 1967.
53. Carlander, K.D. 1977.
54. Boyer, R.L. 1969.
55. Cooner, R.W. and D.R. Bayne. 1982.
56. Dewey, M.R. et al. 1981.

57. Applegate, R.L. et al. 1967.
A 58. Lewis, S.A. et al. 1971.
A 59. Houser, A. and M.G. Bross. 1963.
A 60. Swingle, W.E. 1965.
A 61. Axon, J.R. 1984
62. Pearson, W.D. and B.J. Pearson. 1989.
63. Pearson, W.D. and L.A. Krumholz. 1984.
64. Burr, B.M. and M.L. Warren, Jr. 1986.
65. Pflieger, W.L. 1975b.
A 67. EA Engineering, Science, and Technology. 2004.
68. Jenkins, R.E. and N.M. Burkhead. 1994.
69. Smith, C.L. 1985.
70. Trautman, M.B. 1981.
71. Mettee, M.F. et al. 2001.
72. Robison, H.W. and T.M. Buchanan. 1988.
73. Etnier, D.A. and W.C. Starnes. 1993.
74. Scott, W.B. and E.J. Crossman. 1973.
75. Boschung, H.T., Jr. and R.L. Mayden. 2004.
76. Gammon, J.R. 1998.
77. Smith, P.W. 1979.
78. Ross, S.T. 2001.
79. Funk, J.L. 1957.
80. Becker, G.C. 1983.
81. Mabee, P.M. 1995.
82. Yeager, B.L. 1981.
83. Conner, J.V. 1979.
84. Hardy, J.D., Jr. 1978.

85. Anjard, C.A. 1974.
86. Tin, H.T. 1982.
87. Bettoli, P.W. 1979.
88. Steele, B.D. 1981.
89. Dibble, R.D. 1991.
90. Witt, A., Jr. and R.C. Marzolf. 1954.
A 91. Mense, J.B. 1976.
92. Buynak, G.L. and H.W. Mohr. 1978.
93. Taubert, B.D. 1977.
94. Larimore, R.W. 1957.
95. Meyer, F.A. 1970.

* Original information on reproductive biology and early life ecology of longear sunfish comes from collections of eggs and larvae made by TVA biologists in the Tennessee Valley (see Table 76). Original descriptive information of larval development comes from the progeny cultured from some of the above sources and from a developmental series of larvae from Beaver Reservoir, White River drainage, AR. Specimens used for developmental descriptions are in lots catalogued as TV806, TV815, TV2582, and TV2616 and curated at the Division of Fishes, Aquatic Research Center, Indiana Biological Survey, Bloomington, IN.

REDEAR SUNFISH

Lepomis (Eupomotis) microlophus (Guenther)

Robert Wallus and John V. Conner

Lepomis, Greek: *lepis*, "scale" and *poma*, "lid"; *microlophus*, Greek: "small nape or crest."

RANGE

The redear sunfish is native to penisular FL, lower Atlantic slope and Gulf slope drainages west to the Rio Grande River in TX, and northward in the Mississippi Valley to southern IL and IN. It has been introduced widely into areas outside its natural range.[26,27,43,45,46,53]

HABITAT AND MOVEMENT

The redear sunfish is an inhabitant of sluggish open waters[26,40] such as lakes and reservoirs,[26,35,40,41,44] ponds,[40,44,53] swamps,[41,44,53] lagoons,[40] bottomland lakes,[41] bayous,[35,40] and the backwaters,[26,41,44] protected bays, side channels,[44] and overflow pools[44,46] of streams including small- to medium-sized[53] and large[35,40,53] rivers. It is seldom found in flowing waters.[45]

Redear sunfish does best in warm,[27,35,44,46,49] clear[27,46,49] waters with no noticeable current[27,46,49,53] and an abundance of aquatic vegetation,[27,40,44,46,49,53] where it is often found in congregations around cover such as brush,[40] stumps, and logs.[27,40,44,49] It is often associated with substrates of rock, sand, silt, and organic debris,[40,44] and is usually found in waters 20–34°C.[40]

This species is uncommon in sloughs and borrow pits of the lower Mississippi River drainage, but is commonly collected in moderated numbers from oxbow lakes and seasonal floodplain waters.[3]

The deepest portions of an OK lake were little used by redear sunfish, while shallow shoreline waters produced high populations, possibly by virtue of a greater abundance of food.[21] In the oligotrophic Mayo Reservoir, NC, redear sunfish abundance increased after the introduction of alewife, which corresponded with increased aquatic vegetation in the reservoir littoral zone.[7]

Redear persisted in coves of Lake Conroe, TX, after the removal of aquatic vegetation by grass carp, but little recruitment occurred; after 3 years without vegetation, the redear sunfish population was composed of relatively few, large individuals.[8]

Numbers of harvestable redear sunfish were negatively correlated with hydrilla coverage in Orange Lake, FL.[9]

This species has also been reported from less brackish portions of coastal environments;[26,27,35,40,47,53] recorded from waters where salinity ranged from 5.0 to 24.4 ppt.[27,40,47,53]

Its affinity for logs, stumps, and other standing cover has led to another common name, "stumpknocker"; in some areas it is also called "bream," a catch-all common name used by anglers for the sunfishes, and "shellcracker" in reference to its food habits.[49]

Redear sunfish have been introduced into ponds and artificial lakes, usually along with largemouth bass.[41] A study in 1964 of 38 OK ponds found that bass–bluegill stocking combinations provided more satisfactory fish populations than did bass–redear combinations.[6]

Though commonly referred to as "shellcrackers" in some localities, studies suggest that the redear sunfish not be considered a panacea for biological control of the invasive zebra mussel and that it should not be introduced beyond its normal range expressly for the control of zebra mussels without careful evaluation.[24]

Redear sunfish is a southern species and those introduced into OH ponds had difficulty maintaining their numbers in many waters, apparently subject to winter kill.[45]

DISTRIBUTION AND OCCURRENCE IN THE OHIO RIVER SYSTEM

The redear sunfish has been reported from throughout the Ohio River since 1800[31] but is now taken only occasionally.[30,37] It has been reported from each 100-mile segment of the river from ORM 13 to ORM 952. Unlike other sunfishes, redear sunfish apparently did not increase in abundance after the creation of embayments in the upper half of the river.[30] Based on lockchamber surveys during the period 1957–2001, there was no significant change in the abundance of redear sunfish.[4]

The original range of redear sunfish probably did not extend up the Ohio River beyond the mouth of the Wabash River, but during the period 1955–1980 it was introduced into many farm ponds in OH, especially in the southern half of the state.[45] It is present in the Youghiogheny River system of PA, MD, and WV.[1] In IL, it is native to the southern third of the state including the Wabash River drainage and the lower Ohio River. Through stocking, the species now occurs in all parts of the state.[41]

It is sporadic and uncommon throughout KY; up to 1986, there were no records from above the Falls of the Cumberland River. Records from east of the lower Green River may have been based on introductions.[44] All VA populations of redear sunfish have been recently established by introduction.[48] In TN, the redear sunfish is widespread but not abundant in the Ohio River drainage.[26] It occurs throughout the Tennessee River drainage of northern AL.[47,53]

SPAWNING

Location
Male redear sunfish sometimes construct single nests[40] and sometimes construct their nests in colonies[15,26,35,40,46,53] of up to several hundred[40] with the rims of the nests often almost touching.[46] Nests are reported from shallow water[26,42] at depths of 45–90 cm, or as deep as 2–3 m, often in submerged vegetation.[35,40,53]

The nests are saucer-shaped depressions built on firm substrates[40] or fanned out in silt if no gravel is present.[46] Nests are sometimes only relatively indistinct depressions in vegetation.[40]

Season
Redear sunfish breed throughout the warmer months.[27] Onset of spawning varies by latitude, generally occurring in the spring when water temperatures approach 20–21°C. Spawning normally ends by mid-summer, but extends into the fall in some southern states (Table 85).[35,48]

Temperature
Spawning generally begins when surface-water temperatures reach 20–21°C (Table 85).[35,40]

Fecundity
In FL, female redear sunfish 190–242 mm TL averaged 16,000–26,000 mature eggs. In AR, the number of mature ova in fecund females 230–260 mm TL ranged from 35,500 to 64,000.[49] Other reports indicate 2,000–10,000 eggs,[35] and 15,000–30,000 mature eggs.[26,48]

Sexual Maturity
Some redear sunfish in the deep south mature at age 1; those in the middle and northern parts of the range first spawn at age 2.[48] In FL and TX, redear sunfish may spawn late in their first year of life,[35] but not until the second summer in TN,[54] IL,[35] and MI.[36]

Sexual maturity of redear sunfish may be reached in 1 year at 140 mm TL, but is generally attained at 2 years or after 2 summers.[27,46,53] In TN, redear spawned in their second summer and averaged about 109 mm TL.[54]

In FL lakes, competition may result in spawning being delayed until age 2. Maturity may be reached at total lengths of 134–147 mm if growth is rapid, but not until 188 mm, otherwise.[35]

Spawning Act
Redear sunfish males may use the production of sounds during the active courtship of females as a cue in mate recognition. The males produce popping sounds near the sides and around the head of the female during courtship. These sounds may also help orient the female toward the courting male and may stimulate the female sexually and help overcome her tendency to flee the aggressive attention of the courting male. Sound production, along with other courtship and spawning activities, may stimulate other members of the species and act to synchronize the reproductive efforts of the entire colony.[18]

Modes of communication other than acoustic are also involved in reproductive isolation in the genus *Lepomis*. Olfactory cues are probably involved in colony formation and may be important in mate recognition. Also, visual cues, such as coloration, patterning, and locomotor behavior, play important roles in courtship.[18,19]

Peaks in spawning activity may correspond to times of full or new moons.[52]

After spawning, redear sunfish males presumably defend and maintain their nest until the eggs hatch.[26] However, males reportedly abandoned their nests as a result of herbicicde application. Abandonment occurred at concentrations commonly used to control vegetation, but well below the concentrations lethal to sunfish.[23]

Redear sunfish introduced into OH ponds apparently had difficulty competing for living space and hybridized readily with bluegill, green, and pumpkinseed sunfish.[45] Redear, bluegill, and green sunfish usually nest in colonies. In IL, mixed colonies containing two and, less frequently, all three of these species were not uncommon.[15]

There appears to be no direct genetic isolation between redear sunfish, green sunfish, and bluegill.

Table 85

A summary of reports on redear sunfish spawning seasons
and associated water temperatures.

Location	Season	Associated Water Temperatures
MI[35]	July	Begins when water temperatures have exceeded 21°C for several days
IL[17, 35, 42]	May and June	20.5–24.0°C
MO[46]	May and June 2nd nesting in August (pond)	—
AR[49]	April to August	—
OH[42]	June and July	—
NC[50]	May to August	—
TN[26]	May to August	—
AL[47]	Late April to early June	—
AL[2, 35, 53]	Peaks in spring and fall, sparingly in summer	When surface water temperatures are about 24°C
FL[35]	Late February to October	Begins when water temperature approaches 21°C — may continue in waters up to 32°C
TX[35]	Early May to early July	—

Laboratory propagation of the six possible P_1 crosses has successfully produced young that, when released into outdoor ponds, grew to sexual maturity.[10]

Redear are also known to hybridize occasionally with warmouth.[15]

Redear and green sunfish reproduced naturally when redear males and green females were isolated in small ponds that contained no other fish. No F_1 hybrids were produced when green males and redear females were isoloated.[10] The same approach with male bluegill and female redear produced many hybrids that were raised to sexual maturity.[11,12]

EGGS

Description
Redear sunfish eggs are demersal and adhesive.[42] In CA, six eggs ranged from 1.3 to 1.6 mm in diameter.[13]

Incubation
49–52 h at 23.6°C; about 27–28 h at 28.7°C;[35,42] 3 days at 21.1°C;[13,42] or 6–10 days.[35]

Development
No information.

YOLK-SAC LARVAE

Size Range
Unclear. Redear sunfish are reported to hatch at 4.8–5.1 mm TL when incubated at 23.6–28.7°C,[35] but yolk is also reported absorbed at 5.1 mm TL.[40]

Myomeres
No information.

Morphology
No information.

Morphometry
No information.

Fin Development
No fin rays visible during this phase of development.*

Pigmentation
5.0 mm TL. Pigment is present dorsally on the air bladder and yolk. A few scattered chromatophores are present on the head. There is a row of pigment along the dorsal wall of the gut posterior to the air bladder. A series of widely spaced chromatophores, approximately one per myomere, is visible ventrally between anus and tail.[40,42]

POST YOLK-SAC LARVAE

See Figure 68.

Size Range
About 5.1 mm TL[40] to about 11.25 mm TL.*

Myomeres
5.9–8.4 mm TL. Preanal 12–15; postanal 14–17; total 28–30 (Table 86).*

8.8–11.1 mm TL. Preanal 14–16; postanal 14–16; total 29–31 (Table 86).*

Morphology
7.0–11.0 mm. A noticeable dorsal hump is present on the head and becomes prominent by 9.0 mm.[13]

10.0–15.0 mm TL. Body compressed.[39]

Morphometry
See Table 87.

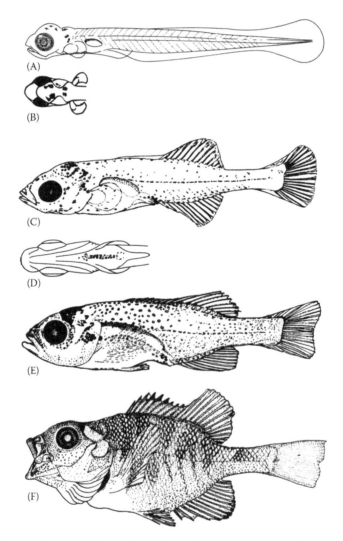

(A)

(B)

(C)

(D)

(E)

(F)

Figure 68   Development of young redear sunfish. A–B. Post yolk-sac larva, 5.5 mm TL: (A) lateral view; (B) dorsal view. C–D. Post yolk-sac larva, 10.2 mm TL: (C) lateral view; (D) ventral view. (E) Early juvenile, 14.0 mm TL. (F) Juvenile, 32.0 mm TL. (A–F reprinted from Figure 4, reference 13, with publisher's permission.)

Table 86

Myomere count ranges (average and standard deviation in parentheses) at selected size intervals for young redear sunfish.*

Size Range (mm TL)	N	Myomeres			
		Predorsal Fin	Preanal	Postanal	Total
5.9–6.3	7	—	12–14 (12.86 ± 0.69)	15–17 (15.71 ± 0.76)	28–29 (28.57 ± 0.53)
6.4–8.4	36	—	12–15 (13.19 ± 0.67)	14–17 (15.64 ± 0.72)	28–30 (28.83 ± 0.51)
8.8–11.1	19	6–7 (6.29 ± 0.47)	14–16 (14.68 ± 0.58)	14–16 (15.11 ± 0.46)	29–31 (29.79 ± 0.54)
11.3–20.3	12	6–7 (6.92 ± 0.29)	14–15 (14.92 ± 0.29)	14–15 (14.58 ± 0.51)	29–30 (29.5 ± 0.52)

Fin Development

5.5 mm TL. Pectoral fins are rayless.[40]

6.2–6.4 mm. Caudal fin ray formation begins.*,[13]

8.3 mm. Dorsal and anal fin rays begin to form.[13] No pelvic fins visible.*

9.4–9.8 mm. All fin rays formed but no dorsal fin spines are visible.[13]

10.2 mm. Median finfold is still present anterior to soft dorsal fin in area of future spinous dorsal and a remnant is visible ventrally anterior to the anus. Finfold is absorbed on caudal peduncle.[13]

11.25 mm TL. All of median finfold is absorbed and an adult complement of spines and fin rays is present in all fins.*

Pigmentation

5.1–7.0 mm. Two patches of broad chromatophores are present on dorsum of head.[13]

7.0–10.2 mm. A dark bar of chromatophores is present on the isthmus and becomes prominent at about 9 mm. At 10.2 mm (Figure 68), a dark blotch of pigment is present dorsally on the head posterior to the eyes; lateral pigment on the body consists of a few scattered chromatophores above and below a single row of chromatophores along the median myosepta, which is present from about the middle of the air bladder to the base of the caudal fin; internally, pigment is visible dorsally on the air bladder.[13]

JUVENILES

See Plate 2D following page 26 and Figure 68.

Size Range

11.25 mm TL* to 109–150 mm TL, in areas of rapid growth,[27,35,46,53,54] or to 188 mm TL, otherwise.[35]

Myomeres

11.3–20.3 mm TL. Preanal 14–15; postanal 14–15; total 29–30 (Table 86).*

Morphology

Young redear sunfish look like adult females.[45]

Lateral line scales 34–47;[26,46,48,49,51,53] gill rakers 9–11 and short, length of longest rakers are about three times their basal width (YOY).[26,53]

Mean number of precaudal vertebrae 12.03, mean number of caudal vertebrae 17.93;[39] total vertebrae 30.[42]

Morphometry

See Table 87.

12.0–17.0 mm TL. Ratio of body depth into total length is 4.1.[39]

Fins

14.0 mm TL. Dorsal spines are well formed but low in profile, about one quarter the length of the longest dorsal fin ray (Figure 68).

Table 87

Morphometric data for young redear sunfish from GA* grouped by selected intervals of total length. Characters are expressed as percent total length (TL) and head length (HL) with a single standard deviation. Range of values for each character is in parentheses.

	TL Intervals (mm)					
Length Range N Characters	5.94–6.31 7 Mean ± SD (Range)	6.36–8.40 36 Mean ± SD (Range)	8.85–11.05 19 Mean ± SD (Range)	11.25–12.85 7 Mean ± SD (Range)	13.72–16.95 4 Mean ± SD (Range)	20.25–20.25 1 Mean ± SD (Range)
Length as % TL						
Snout	2.02 ± 0.59 (0.07–0.20)	3.21 ± 0.91 (0.05–0.38)	4.88 ± 0.53 (0.35–0.58)	5.86 ± 0.34 (0.6–0.8)	6.34 ± 0.29 (0.82–1.13)	6.52 ± 2.84 (1.32–1.32)
Eye diameter	7.41 ± 0.49 (0.42–0.49)	7.76 ± 0.64 (0.45–0.71)	8.79 ± 0.42 (0.75–1.05)	9.95 ± 0.34 (1.11–1.32)	8.85 ± 0.34 (1.29–1.5)	8.99 ± 2.93 (1.82–1.82)
Head	19.2 ± 1.53 (1.05–1.38)	21.2 ± 1.78 (1.1–1.94)	24.9 ± 1.15 (2.09–2.83)	27.5 ± 0.88 (2.95–3.65)	26.5 ± 0.25 (3.6–4.47)	26.9 ± 6.56 (5.45–5.45)
Predorsal	—	36.1 ± 5.83 (3.03–3.03)	30.7 ± 2.02 (2.6–3.5)	31.9 ± 1.74 (3.3–4.35)	30.4 ± 1.27 (4.45–5.05)	30.6 ± 0.0 (6.2–6.2)
Prepelvic	—	—	27.6 ± 1.72 (2.45–3.1)	27.7 ± 1.18 (3.0–3.8)	27.3 ± 0.47 (3.75–4.7)	27.2 ± 5.08 (5.5–5.5)
Preanal	45.2 ± 2.00 (2.65–3.13)	44.8 ± 1.13 (2.83–3.85)	46.7 ± 0.91 (4.1–5.0)	46.6 ± 1.16 (5.2–6.12)	45.6 ± 0.22 (6.2–7.75)	47.2 ± 1.76 (9.55–9.55)
Postanal	56.0 ± 1.28 (3.19–3.68)	55.2 ± 1.12 (3.51–4.85)	53.3 ± 0.91 (4.65–6.05)	53.4 ± 1.16 (5.95–6.95)	54.4 ± 0.22 (7.52–9.2)	52.8 ± 1.85 (10.7–10.7)
Standard	98.4 ± 2.83 (5.75–6.63)	93.6 ± 3.45 (6.1–7.8)	83.7 ± 1.92 (7.38–9.17)	79.6 ± 0.69 (8.95–10.22)	79.4 ± 0.70 (10.9–13.55)	78.3 ± 2.34 (15.8–15.8)
Air bladder	10.9 ± 1.24 (0.61–0.80)	11.6 ± 1.56 (0.57–1.33)	15.8 ± 1.31 (1.23–1.90)	16.1 ± 0.82 (1.8–1.9)	—	—
Eye to air bladder	14.1 ± 1.10 (0.73–1.0)	13.4 ± 1.27 (0.75–1.27)	12.3 ± 0.95 (1.0–1.4)	11.4 ± 0.48 (1.22–1.5)	11.4 ± 0.40 (1.5–1.9)	11.9 ± 4.15 (2.4–2.4)
Fin length as % TL						
Pectoral	—	11.9 ± 5.40 (0.9–1.0)	14.0 ± 1.61 (1.05–1.8)	18.1 ± 1.21 (1.8–2.55)	18.0 ± 0.78 (2.4–3.25)	21.2 ± 7.19 (4.3–4.3)
Pelvic	—	—	4.8 ± 2.19 (0.16–1.15)	12.4 ± 1.14 (1.38–1.7)	12.0 ± 0.78 (1.6–2.11)	15.0 ± 4.15 (3.03–3.03)
Body depth as % TL						
Head at P1	—	18.9 ± 0.58 (1.5–1.55)	21.4 ± 2.51 (1.68–3.2)	23.3 ± 0.39 (2.63–3.03)	25.0 ± 1.00 (3.25–4.43)	25.7 ± 7.76 (5.2–5.2)
Head at eyes	—	18.4 ± 0.49 (1.45–1.53)	20.1 ± 0.62 (1.7–2.3)	21.2 ± 0.32 (2.4–2.7)	21.4 ± 0.10 (2.95–3.6)	22.2 ± 0.0 (4.5–4.5)
Preanal	—	16.1 ± 0.50 (1.23–1.4)	18.6 ± 1.26 (1.4–2.25)	21.3 ± 0.56 (2.4–2.85)	23.1 ± 0.88 (3.05–4.1)	23.2 ± 0.0 (4.7–4.7)
Caudal peduncle	—	7.79 ± 6.56 (0.6–0.65)	8.91 ± 0.55 (0.7–1.03)	9.27 ± 0.19 (1.05–1.18)	9.85 ± 0.11 (1.35–1.7)	9.88 ± 2.07 (2.0–2.0)
Body width as % HL						
Head	—	47.2 ± 4.16 (0.78–1.0)	48.0 ± 3.24 (1.0–1.33)	44.6 ± 1.65 (1.35–1.7)	48.5 ± 2.67 (1.62–2.3)	56.9 ± 8.30 (3.1–3.1)

Larger juveniles. Spinous dorsal fin is broadly connected to the soft dorsal. Reports of dorsal spines range from 9 to 11,[26,43,46–49,51] usually 10 in TN,[26] KS,[43] MO,[46] and AR[49] and 10–11 in VA.[48] Soft dorsal fin has 10–12 rays.[26,43,47–49,51,53] Anal fin has 3 spines[43,47–49,51] and 9–11 soft rays.[26,43,48,49, 51] Pectoral fins have 11–16 rays,[26,39,43,47,48,51,53] usually 13–14.[26,43,47,48,51,53] Pelvic fins have 1 spine and 5 soft rays. Caudal fin has 17 primary rays.[43]

Pigmentation

11.25–15.0 mm TL. A conspicuous dark bar of pigment is present on the breast and belly and a double row of pigment is present ventrally posterior to the anus.[39,40] At 14.0 mm (Figure 68), the body appears fairly heavily pigmented. The dark blotch of pigment persists on the head along with pigment around the eyes and a bar of pigment on the opercle posterior to the eyes. A mid-lateral row of pigment is present along with considerable scattered pigment on sides of body; anteriorly over air bladder, dorsolateral pigment consists of large, scattered chromatophores. Internal pigmentation is dark over the air bladder.[13]

19 mm TL. Vertical bars are present laterally on the body.* Primary bars of melanistic pigment appear as an initial set of lateral patches to which bars are added anteriorly and posteriorly in development in all *Lepomis* species.[25]

23.0 mm SL. Laterally, 9–10 primary bars are discernable (Figure 69A).[25]

27.0 mm SL. Laterally, secondary bars of pigment are visible between the primary bars and light centers have appeared in some of the primary bars (Figure 69B).[25]

32.0 mm TL. Head and most of body are covered with scattered pigment overlaid laterally with 6–7 vertical bars of darker pigment. Breast and branchiostegals appear to have little pigment (Figure 68).[13]

70.0–100.0 mm SL. Lateral bars of pigment are no longer noticeable on body (Figure 69C).[25]

Older juveniles. Young redear sunfish have a whitish or white-yellow breast, an opercle spot that is light gray or pale yellow, and a silvery body with 5–10 distinct vertical, lateral bands.[45]

TAXONOMIC DIAGNOSIS OF YOUNG REDEAR SUNFISH

Similar species: other *Lepomis*, especially pumpkinseed. Redear sunfish and pumpkinseed are the only members of the subgenus *Eupomotis*.

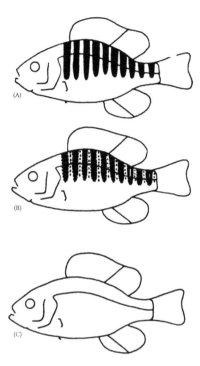

Figure 69 Development of young redear sunfish — schematic drawings showing melanistic pigment patterns of young redear sunfish. (A) 23.0-mm SL juvenile showing primary melanistic pigment bars. (B) 27-mm SL juvenile showing primary and secondary melanistic pigment bars (note that light centers have appeared in some of the primary bars). (C) 70–100 mm SL showing the absence of all melanistic pigment bars. (A–C reprinted from Figure 14, reference 25, with publisher's permission.)

Redear sunfish vs. pumpkinseed. See *Eupomotis* subgeneric discussions on page 91. Also, see discussions in pumpkinseed species account on page 134.

For comparisons of important morphometric and meristic data, developmental benchmarks, and pigmentary characters between redear sunfish and all other *Lepomis* that occur in the Ohio River drainage, see Tables 4 and 5 and Plate 2D.

Redear sunfish vs. warmouth. See discussions in warmouth species account on page 151.

Redear sunfish versus bluegill. See discussions in bluegill species account on page 185.

Redear sunfish YOY (Plate 2D) show red on the opercular lobes at about 30 mm SL but are similar to young bluegill (Plate 2A) in shape and coloration; also their sides tend to be more spotted and less vertically banded than bluegill.[26]

ECOLOGY OF EARLY LIFE PHASES

Occurrence and Distribution (See Figure 70)

Eggs. Redear sunfish eggs are spawned into nests where they are presumably defended and maintained by the male parent until they hatch.[26]

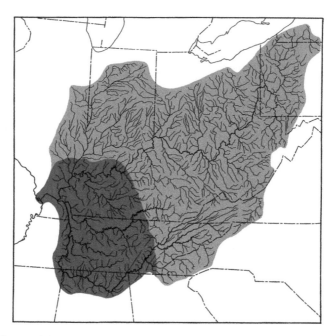

Figure 70 Distribution of redear sunfish in the Ohio River system. Dark-shaded area represents approximate native distributiozin[27] and light-shaded area the general distribution resulting from introductions.

Yolk-sac larvae. Redear sunfish yolk-sac larvae remain in the nest for about a week and are guarded by the male parent.[35]

Post yolk-sac larvae. Larvae become free-swimming about 3 days after hatching at 5.1–5.8 mm.[13] Young redear sunfish usually remain in littoral habitats after hatching.[14,39]

In Orange Lake, FL, redear sunfish larvae were collected in low numbers with towed nets in open water and around panic grasses, hydrilla, and floating emergent vegetation.[5,55] They were found in low numbers in light trap samples taken near floating emergent vegetation, but none were collected with light traps from the other habitats.[5]

Redear sunfish fry were collected from an IL pond in plexiglass fry traps from May 26 through September 22, 1982. Larvae <12 mm TL were restricted to shoreline habitats (75%) and near-shore habitats (25%). Young redear ≥12 mm TL were almost exclusively restricted to shoreline (97%); few young redear were collected in mid-water habitat.[17]

Juveniles. Juvenile redear sunfish are frequently reported from near-shore, vegetated shallow water.[40]

Inconsistent recruitment of age-1 and older fish is often reported for redear sunfish populations. Lack of winter hardiness has been suggested as a reason, but low densities of redear sunfish fry in the fall are a more likely cause for this species' inconsistent recruitment. Low densities of redear young in the

fall have been correlated with inadequate vegetative cover and high turbidity, but more often the causes remain unidentified. Another potential cause of low fall densities of young redear sunfish is their vulnerability to predation in littoral habitats.[17]

YOY redear sunfish <80 mm TL were collected with seines along the shorelines of lower Mississippi River floodplain ponds. Mean larval densities of several species of sunfishes in the ponds in May were correlated with the numbers of sunfishes subsequently seined during August and November.[22]

In an OK lake, redear sunfish and bluegill were co-dominant species in unbaited minnow traps set in waters up to 200 cm deep and in shoreline seine hauls. Minnow traps collected slightly larger redear than did seine hauls; by late August only 24% of the redear collected by seining were <30 mm TL. Minnow traps collected redear up to 80 mm TL. YOY were most abundant in the shallowest areas. Though collected in traps set up to 200 cm deep, redear were caught proportionately more often in traps set at depths of 50 cm or less. In a comparison of catches for 2 years, the distribution of redear was inversely related to depth both years, but the proportion collected at depths less than 50 cm or less was only 0.32 in 1981 compared to 0.75 in 1980. Greatest densities of redear occurred in habitats where spike rush and muskgrass grew.[38]

Redear sunfish serve the same management purpose in ponds as do bluegill, in that their young provide forage for bass while the adult redears themselves afford angling catches. Substitution of

Table 88

A summary of calculated growth data for young redear
sunfish from a variety of habitats in OK[35,56]
with statewide average growth given
for the periods 1952–1963[29] and 1972–1973.[58]

	Avg. Calculated TL (mm) at Each Annulus			
	1	2	3	4
Slowest growth	38	79	117	163
Fastest growth	185	221	251	277
16 reservoirs	99	135	178	211
10 large lakes	102	155	183	206
48 small lakes	91	150	183	208
21 ponds	91	145	178	206
2 streams	74	130	170	206
Statewide avg. (1952–1963)	94	147	180	208
Statewide avg. (1972–1973)	62	109	142	156

redears for bluegills in initial stocks usually is based on an assumption that redears reproduce successfully but less prolifically than bluegills in ponds. If that assumption proves correct (indeterminate in previous documented studies) in a particular pond, a troublesome problem of overpopulation and stunting by the forage species is avoided or postponed.[43]

Early Growth

Carlander (1977) summarized many reports of growth of redear sunfish and concluded that, in general, growth of young redear is slower in the northern part of the range (MI), moderate in IL and IN, and faster in the south.[35]

In OK, growth in streams appeared slower, at least for the first 2 years, than in ponds, lakes, and reservoirs.[35] Statewide average growth among years was variable (Table 88).[29,58]

Young redear sunfish from an IL pond averaged 39 mm TL in September.[17] In OH, YOY are 20–75 mm TL in October and age-1 redear range from 25 to 100 mm TL.[45]

In Reelfoot Lake, TN, redear sunfish grew to over 50 mm TL the first year, to about 109 mm the second year, 152 mm their third year, and 175 mm their fourth year.[54] This report suggests that, except in well-managed ponds, growth of most TN populations of redear sunfish is probably less than that reported in FL (Table 89),[33] OK (Table 90),[29] or KY (Table 91)[28] studies.[26] This observation was also

supported by reports of redear sunfish growth in two TVA reservoirs. During the 1970s, estimated TL for young redear at each annulus from Chickamauga Reservoir, Tennessee River, TN, was 51 mm at age 1, 123 mm at age 2, 173 mm at age 3, and 194 mm at age 4. Growth in Guntersville Reservoir, AL, was very similar: 52 mm at age 1, 124 mm at age 2, 169 mm at age 3, 191 mm at age 4.[57]

The range of average lengths of age-3 redear sunfish from 10 VA impoundments was 135–216 mm TL (unweighted mean 182 mm),[48] a typical growth range for the species.[35]

Calculated TLs for redear sunfish ages 1–3 from a LA pond were 30, 185, and 201 mm.[20]

In 46 FL lakes, first annulus formation for redear sunfish occurred between February and July. Length at age varied greatly among the lakes (Table 89), ranging from 37 to 130 mm TL at age 1.[33]

Length–weight relationships of young redear sunfish from AL (Table 92)[34] appear similar to those in OK (Table 90), though it appears that OK redear may be slightly heavier at comparable lengths.

In IL, male redear sunfish were slightly longer and heavier than females at age 2.[10] No difference in growth related to sex was noted in Reelfoot Lake, TN.[54]

Pope et al. (1995) applied the regression-line-percentile (RLP) technique to weight–length data for 150 redear sunfish populations to develop a new 75th-percentile standard weight (Ws) equation. They found the proposed RLP Ws equation,

Table 89

Mean length of redear sunfish from FL for age classes 1–4 back calculated from examination of otolith annuli. The number of lakes in which fish were examined, the mean of means, the corresponding standard error of the mean, and minimum and maximum length at age are listed.

	Age			
	1	2	3	4
Lake samples (N)	46	45	43	35
Mean of means (mm TL)	72	138	183	213
Standard deviation	19	24	27	33
Minimum TL (mm)	37	90	125	137
Maximum TL (mm)	130	197	242	319

Source: Constructed from Table 4, in part, reference 33.

Table 90

Statewide average growth of young redear sunfish recorded in OK waters for the period 1952–1963.

	Average Growth	
Age in Years	TL (mm)	Weight (g)
1	94	13
2	147	54
3	180	106
4	208	171

Source: Constructed from data presented in Table 9 (in part), reference 29.

Table 91

Age and growth determinations for young redear sunfish collected from three KY lakes in 1983.

	Mean TL (mm) at Age:		
Lake	1	2	3
Briggs Lake	69	160	246
Marion County Lake	61	160	—
Shanty Hollow Lake	89	—	—

Source: Constructed from data presented in Tables 35, 49, and 68, reference 28.

$\log_{10} Ws = -4.968 + 3.119 \log_{10} TL$, where Ws is the standard weight in grams and TL is the total length in millimeters, valid for redear sunfish ≥70 mm TL.[16]

Feeding Habits

Young redear sunfish feed on microcrustaceans and small chironomid larvae.[26] Principal foods of redear 16–30 mm TL were copepods, cladocera, and amphipods.[53]

In MI lakes young redear sunfish ≤40 mm SL consumed mostly zooplankton and snails, followed in descending order of proportion by dipteran larvae, insect nymphs, and amphipods. The diet of redear >40 mm SL was dominated by snails, which con-stituted about 80% of the prey identified. For smaller size-classes of sunfishes, the extent of molluscivory is constrained by crushing strength. This study showed that the greater crushing strength of redear allowed them to undergo an ontogenetic niche shift from a diet of predominantly soft-bodied invertebrates to a diet of snails at a smaller size than co-occurring pumpkinseed sunfish.[32]

In an OK lake, YOY redear sunfish consumed large numbers of prey in mid-morning and late afternoon or evening. In a 2-year study (1980–1981), ostracods were the most abundant item in the diets of YOY redear, constituting 56 and 84% of the diet. Chironomids, copepods, chydorids, and *Daphnia*

Table 92

Length–weight relationships of young redear sunfish from AL collected during the period 1949–1964.

Approximate Average TL (mm)	N	Average Empirical Weight (g)
25	55	0.5
51	1,993	2.9
76	290	6.8
102	420	22.2
127	550	34.5
152	266	63.6
178	232	99.9
203	124	145.3

Source: Constructed from data presented in unnumbered table on page 51, reference 34.

were also preyed upon. Redear seemingly fed over a greater range of habitats in 1981, but became more selective in their feeding. During the 2 years, diet and habitat sometimes overlapped for YOY bluegill and redear sunfish. Diets of these species varied somewhat at different depths indicating that diets diverged more when the species occurred in the same habitats. In 1981, bluegill and redear were distinctly separated on the basis of diet even though their use of habitat was similar.[38]

LITERATURE CITED

1. Hendricks, M.L. et al. 1983.
2. Swingle, H.S. 1956.
3. Baker, J.A. et al. 1991.
4. Thomas, J.A. et al. 2004.
5. Conrow, R. et al. 1990.
6. Gasaway, C.R. 1968.
7. Crutchfield, J.U., Jr. et al. 2003.
8. Bettoli, P.W. et al. 1993.
9. Colle, D.E. et al. 1987.
10. Childers, W.F. and G.W. Bennett. 1961.
11. Ricker, W. 1948.
12. Krumholz, L.A. 1950.
13. Meyer, F.A. 1970.
14. Meals, K.O. and L.E. Miranda. 1991.
15. Childers, W.F. 1967.
16. Pope, K.L. et al. 1995.
17. Dimond, W.F. and T.W. Storck. 1985.
18. Gerald, J.W. 1971.
19. Gerald, J.W. 1970.
20. Muncy, R.J. 1966.
21. Ghent, A.W. and B. Grinstead. 1965.
22. Sabo, M.J. and W.E. Kelso. 1991.
23. Clark, P.W. 1990.
24. French, J.R. III and M.N. Morgan. 1995.
25. Mabee, P.M. 1995.
26. Etnier, D.A. and W.C. Starnes. 1993.
27. Lee, D.S. et al. 1980.
A 28. Axon, J.R. 1984.
A 29. Houser, A. and M.G. Bross. 1963.
30. Pearson, W.D. and L.A. Krumholz. 1984.
31. Pearson, W.D. and B.J. Pearson. 1989.
32. Huckins, C.J.F. 1997.
33. Mantini, L. et al. 1992.
A 34. Swingle, W.E. 1965.
35. Carlander, K.D. 1977.
36. Cole, V.W. 1951.
A 37. EA Engineering, Science, and Technology. 2004.
38. Layzer, J.B. and M.D. Clady. 1991.
39. Werner, R.G. 1966.
40. Hardy, J.D., Jr. 1978.
41. Smith, P.W. 1979.
42. Tin, H.T. 1982.
43. Cross, F.B. 1967.
44. Burr, B.M. and M.L. Warren, Jr. 1986.
45. Trautman, M.B. 1981.
46. Pflieger, W.L. 1975b.
47. Mettee, M.F. et al. 2001.
48. Jenkins, R.E. and N.M. Burkhead. 1994.
49. Robison, H.W. and T.M. Buchanan. 1988.
50. Menhinick, E.F. 1991.
51. Ross, S.T. 2001.
52. Wilbur, R.L. 1969.
53. Boschung, H.T., Jr. and R.L. Mayden. 2004.
54. Schoffman, R.J. 1939.
55. Conrow, R. and A.V. Zale. 1985.
A 56. Jenkins, R. et al. 1955.
A 57. Tennessee Valley Authority. 1978.
A 58. Mense, J.B. 1976.

* Original morphometric and meristic data were obtained from specimens cultured in TVA laboratories from eggs and larvae produced at Cohutta National Hatchery, Cohutta, GA (1977). Specimens are catalogued TV818 and curated at the Division of Fishes, Aquatic Research Center, Indiana Biological Survey, Bloomington, IN.

REDSPOTTED SUNFISH

Lepomis (Bryttus) miniatus Jordan

Thomas P. Simon

Lepomis: Greek *lepis* "scale" and *poma* "lid" for the scaled operculum; *miniatus*: Latin *minium* meaning stained red with cinnabar, a sulphide of mercury, a bright red pigment.

RANGE

The redspotted sunfish occurs in the Mississippi River valley from IL south to the Gulf of Mexico; west to the Rio Grande in TX; and east to Mobile Basin, AL. Gulf Slope drainages from the Apalachicola drainage to the Nueces River, TX; north in the Mississippi River to central IL; introduced into Devils River (Rio Grande drainage), TX; common, locally abundant in south.[1,2,10,11,15,16]

HABITAT AND MOVEMENT

Inhabits heavily vegetated sluggish streams, swamps, sloughs, bottomland lakes, pools of creeks, and small to medium rivers, less brackish portions of coastal estuaries; common in quiet or moderately flowing waters with heavy vegetation or other cover and bottom of mud or sand.[1,2,10,11,15,16] The redspotted sunfish is a habitat generalist and is tolerant of saline conditions.[1,26] The species occurs over a variety of substrates including sand, gravel, and clay;* occurring along vegetated shorelines of oxbow lakes with bottoms of mud or sand overlain by organic sediment at depths of 0.5–1 m.[29] Individuals were collected from northwestern FL in salinity up to 11.8 ppt.[1,3] Hybrids in the lower Escambia River were found at sites with salinities from 5.0 ppt at the surface to 24.4 ppt at the bottom.[1,4]

Some lateral dispersal of redspotted sunfish is expected with individuals known from floodplain wetlands.[27] Nocturnal shift in position occurs close to shore.[30] Water level fluctuations are hypothesized to be a problem; however, no genetic bottlenecks were observed.[22]

DISTRIBUTION AND OCCURRENCE IN THE OHIO RIVER SYSTEM

The redspotted sunfish occurs in the lowermost portions of the Ohio River basin including tributaries of the Cumberland, Tennessee, and Wabash river drainages.[1,2,10] In the lower Wabash River drainage, the species is common in Posey, Vanderburgh, and Spencer counties, IN.*,[10] This species occurs in tributaries of the lower and middle Tennessee River from AL, TN, and KY.[1,2,10,11,15,16,26] In KY, the species reaches its eastern limit in the Ohio River basin in the Green River where it is threatened by mining and oil extraction.[18] It occurs in Terrapin Creek (Obion River drainage)[19] and in Crooked Creek (Cumberland River drainage), the only substantiated record for the Cumberland River.[20]

In IL, the species is known from Goose Pond, Yellowbank Slough, Beaver Pond, Black Lake, Big Lake, Long Pond, Beaverdam Pond, Loon Lake, and Bushy Slough.[28]

SPAWNING

Location

Nests most often occur in isolation and almost invariably in very shallow water near some kind of cover.[1,4–7] Nests are generally not aggregated, but occasionally two to four nests were observed in close proximity.[6] Most nests were in the main channel and principal flow.[6]

Nests are usually constructed near instream cover such as woody debris, large boulders, or usually near aquatic plants.* Nests are usually in water depths <75 mm and are shallow depressions less than 30 mm in diameter* or 30–60 mm in diameter.[6] Nests are excavated about 25–50 mm deep, and found in water depths of 0.1–0.38 m[6] or typically observed in water less than 0.6 m in depth.* Substrates are usually over sand or gravel, but sometimes nests are built over soft material that has been cleaned by the male prior to spawning.*

Season

Redspotted sunfish reportedly spawns from March to September,[5–7] but spawning may occur year-round in southern portions of the range,[4] peaking in mid-summer.[4,15] Also reported to peak from May to June[22] or April to June.[6,22]

In AL, spawning occurs from May to July;[13,31] mid-July in KY;[29] April to August in OK;[30] May in IL.[32]

Temperature

Gametogenesis in IN populations occurs at temperatures from 18 to 20°C and 14 h of daylight.* Spawning occurs at temperatures from 22 to 28°C, but redspotted sunfish do not spawn when temperatures exceed 30°C.*

Fecundity

Fecundity in IN populations ranges from 2,000 to 10,000 eggs depending on the size of the female.* Increasing fecundity is proportional to increasing female length and age.*

Sexual Maturity

Almost all specimens were sexually mature by age 2, but some were mature at age 1. In IN, redspotted sunfish >50 mm TL were sexually mature;* also reported mature at about 55 mm TL.[1,5-7]

Spawning Act

Redspotted sunfish are classified as nest-spawning lithophilic fishes that guard their young.[35] During the spawning period, the male body color is brilliant and his fins are very dark.[2] Spawning usually occurs soon after nest completion.[6,22] The male is aggressive in guarding the nest, defending the nest territory from all but the largest intruders.[1] Males usually spawn with a single female per event; however, eggs from several females are known from a single nest.*[,23] After spawning, the male is known to drive away the spawned female and then ready himself to attract the next ripe female.[1,5-7] Redspotted sunfish produce grunting sounds as part of spawning behavior.[14]

Golden shiners have been observed quickly invading redspotted sunfish nests and spawning. Subsequently, the male redspotted sunfish then afforded protection throughout both species' larval development.[1,5-7]

EGGS

Description

Fertilized eggs average 1.6 mm in diameter (range 1.4–1.8 mm)[1,5,6] and are demersal, spherical, and translucent; yolk coloration ranges between yellow to pale translucent amber*[,6] or distinctively blue;[11] perivitelline space is narrow; chorion is unsculptured and unpigmented.*[,6] An oil globule is located near the surface of the yolk.[6]

Incubation

Eggs incubated at 20°C hatched in 48 h and those incubated at 24°C hatched in 52 h.[1,5,6] Although this temperature range is counterintuitive, it is more probable that these rates are reversed in the literature.[1]

Development

Development has not been described.

YOLK-SAC LARVAE

See Figure 71.

Size Range

Hatching occurs at 4 mm TL;[5] yolk sac is absorbed by 6.6 mm TL[9] or at lengths <7.0 mm TL.[5]

Myomeres

Predorsal 3–4; preanal 11–13; postanal 15–18; total 28–31.[33,*]

Morphology

4.0 mm TL (newly hatched). Body laterally compressed, yolk sac moderate, oval; yolk with a single oil globule; head appears to be deflected over the yolk sac; eyes spherical.*

6.5–7.0 mm TL. Swim bladder distended with air; yolk sac noticeably decreased; cloaca open externally and feeding initiated.[6]

Morphometry

See Table 93.

Fin Development

3.6–4.1 mm TL (newly hatched). Pectoral fin buds not developed.[21]

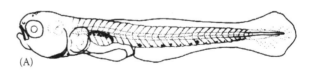

(A)

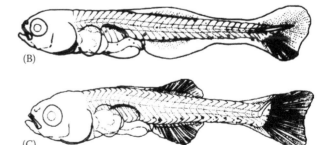

(B)

(C)

Figure 71 Development of young redspotted sunfish. (A) Yolk-sac larva, 5.0 mm TL. B–C. Post yolk-sac larvae: (B) 6.6 mm TL; (C) 7.9 mm TL. (A–C reprinted from Figures 2 through 4, reference 9.)

Table 93

Morphometric data expressed as percentage of TL for young redspotted sunfish from Louisiana.*

Total Length (mm)	N	Length (% TL)							Body Depth (%TL)		
		ED	HL	PreAL	PosAL	GL	IL	SL	GBD	BDA	CPD
5.0	1	14.5	18.7	42.3	57.7	—	—	95.9	16.3	8.1	4.1
6.6	1	6.3	20.3	43.8	56.2	7.8	14.1	95.3	15.6	8.6	4.7
7.9	1	6.8	24.0	44.8	55.2	14.4	11.2	87.2	18.4	12.8	7.2
11.0	1	7.8	26.4	42.2	57.8	20.2	9.4	82.8	20.3	14.8	9.4
39.2	1	8.8	21.6	46.4	53.6	27.2	8.8	81.6	33.6	29.6	12.0

4.5 mm TL. Pectoral fin not well developed, without incipient rays, dorsal and anal finfolds continuous.*

Pigmentation

3.6–4.1 mm TL. Larvae unpigmented;[22,33] yolk yellow to amber.*

6.0–7.0 mm TL. Half-moon–shaped patch of melanophores is present over each cranial hemisphere. There is a paucity of ventrolateral and postanal pigmentation.[22,33]

POST YOLK-SAC LARVAE

See Figures 71 and 72.

Size Range

Beginning at lengths between 6.5 and 7.0 mm TL.[5,6,9] A 10.3-mm TL juvenile is reported.[33]

Myomeres

Preanal 12–14; postanal 16–18; total 29–31.*[,9,33]

Morphology

6.6–7.0 mm TL. Yolk sac completely absorbed.*[,9,33]

Morphometry

See Table 93.

Proportionately deeper heads than *Lepomis cyanellus.*[9,33]

Fin Development

7.9 mm TL. Fin rays formed in pectoral, soft dorsal, anal, and caudal fins; median finfold partially

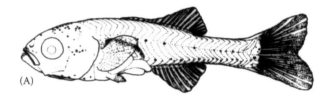

Figure 72 Development of young redspotted sunfish. (A) Post yolk-sac larva, 11.0 mm TL. (B) Juvenile, 32 mm SL. (A reprinted from Figure 5, reference 9; B, original photograph.)

absorbed dorsally on caudal peduncle, completely absorbed ventrally on caudal peduncle, and present ventrally anterior to the anus; no pelvic bud formed.*[,33]

10.3 mm TL. Fin rays formed in all paired and median fins with the exception of the pelvic fin; pelvic fin bud formed; median finfold partially absorbed ventrally anterior to the anus.*[,33]

Pigmentation

>7.0 mm TL. Postanal myosepta with more than 25 melanophores between the vent and the isthmus.[22,33]

JUVENILES

See Plate 2E following page 26 and Figure 72.

Size Range

Juvenile phase is reached at lengths >10.3 mm TL.[33]
Sexual maturity is reached at >50–55 mm TL.[*,1,5–7]

Myomeres

Preanal 13–14; postanal 15–17; total 29–31.[33]

Morphology

>30–50 mm TL. Mouth moderate in size, maxilla reaching to or just beyond anterior margin of eye; opercle stiff to its posterior margin; not fimbriate along its posterior edge. Lateral scale rows 33–42, usually 35–41 [1,2,11] or 36–39.[29] Transverse scale rows 17–23, usually 19–22.[1,2,11] Gill rakers 8–11, length 2.5–3.5 times greatest width; no palatine teeth.[1,2,11]

Morphometry

See Table 93.

Fin Development

>30–50 mm TL. Pectoral fin short and rounded, its tip far short of nares when laid forward across check.[1,2,11]

Larger juveniles. Adult complement of fin spines and rays: spinous dorsal fin with X (X–XI) spines; soft dorsal fin rays 10–12; pectoral fin rays 12–15, usually 13–14; anal spines III, and soft rays 9–10, usually 10; pelvic fin rays I 5.[*,1,2,11]

Pigmentation

See Figure 73.

>15 mm TL. Postanal myosepta with more than 25 melanophores between the vent and the isthmus.[22]

>30–50 mm TL. Eye red with a black vertical band through the center. Lower margins of operculum and area beneath the eye covered with iridescent blue. Body color is tan with a white venter. Numerous brown to orange-red spots are randomly distributed on entire body. Spinous dorsal and pectoral fins are clear and without pigment. Pelvic and proximal portion of anal fin orange; distal margins of soft dorsal, anal, and caudal fins with orange-red pigment.[*]

Nine to 10 primary melanistic bars develop laterally on redspotted sunfish (Figures 73A and 73B). Light centers form in these primary bars, but secondary bars are not formed. As the centers of the primary bars become increasingly lighter, only the anterior and posterior margins of each bar remain

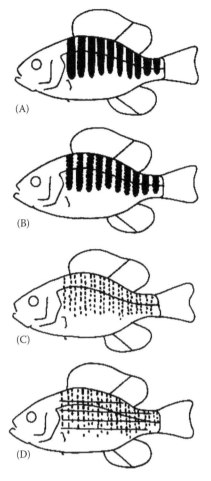

Figure 73 Development of young redspotted sunfish — schematic drawings showing ontogeny of melanistic pigment patterns. (A) 17 mm SL; (B) 19 mm SL; (C) 28 mm SL; and (D) 57 mm SL. (A–D reprinted from Figure 13, reference 36, with publisher's permission.)

visible. The anterior and posterior edges of the bars become broken as pigment is lost at approximately every other scale (Figure 73C). Through the loss of pigment in many scales in the margins of the light-centered primary bars, and the intensification of the pigment in other scales, the vertical orientation of the bars is lost and a spotted pattern characterizes adults, though longitudinal stripes are variably developed (Figures 73C and 73D).[36]

TAXONOMIC DIAGNOSIS OF YOUNG REDSPOTTED SUNFISH

Similar species: other *Lepomis* species. Redspotted sunfish is in the subgenus *Bryttus* along with the spotted sunfish that does not occur within the Ohio River drainage. Limited early life history information exists for members of this subgenus (see page 91).

Since limited information is available for this species, we are unable with certainty to distinguish its larvae from those of other *Lepomis*. Yolk-sac larvae between 3.6 and 4.1 mm TL do not have pectoral fin buds present. Post yolk-sac larvae have a pronounced half-moon patch of melanophores over each cranial hemisphere and a paucity of ventrolateral and postanal pigmentation. Post yolk-sac larvae and juveniles are separable from all other *Lepomis* by the occurrence of more than 25 melanophores between the vent and isthmus.[21]

Juvenile specimens of redspotted sunfish (Plate 2E) may be confused with *Lepomis symmetricus* (Plate 2F) but differ in having no black spot in the posterior portion of the soft dorsal fin, and possessing scattered dark spots along the sides of the body.[*,11] Redspotted sunfish YOY (Plate 2E) are dark and lacking vertical bands and are also similar to green sunfish (Plate 1C), warmouth (Plate 1E), dollar sunfish (Plate 2B), and longear sunfish (Plate 2C). They differ in tending to be deeper bodied than these species, and in having at least a few dark spots scattered along their sides.[11]

ECOLOGY OF EARLY LIFE PHASES

Occurrence and Distribution (Figure 74)

Eggs. Eggs are laid in the nest and buried in the substrate, which ranges from sand, gravel, or other fine material that the male has cleaned prior to reproduction.[*,1,5–7] The eggs remain in the substrate for 48–52 h prior to hatching and are guarded by the male parent.[*,1,5–7]

Yolk-sac larvae. Yolk-sac larvae remain in the substrate for 9 days.[1,5–6] No information is available on the abundance or survival of early life stages. Larvae 6.5–7 mm TL fed by 9 days post-hatching and left the nest on the tenth day.[6,22]

Post yolk-sac larvae. Larvae are free swimming at 7 mm TL.[1,5,6] Larvae form loose schools between days 10 and 13 and then disperse.[6,22]

Juveniles. Juvenile stages occur until 50–55 mm TL. The habitat use of juveniles is similar to adults occurring in heavily vegetated streams and swamps. They are tolerant of a wide variety of conditions including low dissolved oxygen and high temperatures.[*,1–7]

Early Growth

Average growth rates for redspotted sunfish in OK were 33, 74, 114, and 150 mm TL for ages 1–4, respectively;[1,7,8] IN populations were 30, 69, 112, and 139 mm TL for similar age classes,* while TN populations were 40 mm at age 1.[11] The relation of length and weight does not vary significantly from month to month or between sexes.[5,24] Formulas for the relationship between standard length and weight include: log Wt = 3.002 log SL − 4.32;[5] and log Wt = 3.05 log SL − 4.53.[23] The mean K is 4.82, ranging from 4.38 to 5.59, with no obvious increase in K with fish size.[24] Growth rates of 0.12 mm/day and 44 mm/year were measured for specimens between 90 and 120 mm SL.[5]

Figure 74 Distribution of redspotted sunfish in the Ohio River system (shaded area).

Feeding Habits

Redspotted sunfish are insectivores[9] or invertivores[34] and consume drift, feeding in the water column and near the surface.[34] They consume benthic midge larvae and other immature insects and to a lesser extent microcrustaceans, such as amphipods and cladocerans.[9,12,17,34] Insects were found in 100% of the stomachs including aquatic beetles, chironomids, hemipterans, mayfly nymphs, and terrestrial hymenopterans; also found in small amounts were amphipods, decapods, and snails.[25] Chironomids, mayfly nymphs, odonate nymphs, and terrestrial insects were commonly eaten; also found were amphipods, insects, cladocerans, and copepods.[9] One study found relatively large quantities of twigs and leaf fragments in the stomach, but did not quantify the volume or mass.[25]

LITERATURE CITED

1. Boschung, H.T., Jr. and R.L. Mayden. 2004.
2. Warren, M.L., Jr. 1992.
3. Kilby, J.D. 1955.
4. Bailey, R.M. et al. 1954.
5. Caldwell, R.D. et al. 1955.
6. Carr, M.H. 1946.
7. Carlander, K.D. 1977.
8. Finnell, J.C. et al. 1956.
9. McLane, W.M. 1955.
10. Gerking, S.D. 1945.
11. Etnier, D.A. and W.C. Starnes. 1993.
12. Goldstein, R.M. and T.P. Simon. 1999.
13. Mettee, M.F. et al. 1996.
14. Childers, W.F. 1967.
15. Ross, S.T. 2001.
16. Lee, D.S. et al. 1980.
17. Vanderkooy, K.E. et al. 2000.
18. Warren, M.L., Jr. and R.R. Cicerello. 1982.
19. Burr, B.M. and M.L. Warren, Jr. 1990.
20. Warren, M.L., Jr. et al. 1991.
21. Moody, D.P. 1979.
A 22. Hill, J.E. and C.E. Cichra. 2005.
23. DeWoody, J.A. et al. 2000.
24. Anderson, R.O. and R.M. Neumann. 1996.
25. Chable, A.C. 1947.
26. Ross, S.T. 2001.
27. McElroy, T.C. et al. 2003.
28. Burr, B.M. et al. 1988.
29. Burr, B.M. and R.L. Mayden. 1979.
30. Miller, R.J. and H.W. Robison. 2004.
31. Pflieger, W.L. 1997.
32. Forbes, S.A. and R.E. Richardson. 1920.
33. Conner, J.V. 1979.
34. Goldstein, R.M. and T.P. Simon. 1999.
35. Simon, T.P. 1999.
36. Mabee, P.M. 1995.

* Original information on early life ecology and reproductive biology comes from data and specimens collected from the Interior River Lowland Ecoregion of IN during the period 1992–1993. Specimens and observations were from a variety of sites from the Pigeon River, Little Pigeon River, and tributaries of the Ohio River from Posey, Vanderburgh, and Spencer counties IN. Measurements from Louisiana specimens were listed by Conner (1979) as *Lepomis* "A"-type specimens. Juvenile illustration is of a 32-mm TL specimen collected by Vertebrate Field Zoology, Samford University, March 15, 2004, in Ebenezer Swamp, Shelby County, AL.

BANTAM SUNFISH

Lepomis (Lethogrammus) symmetricus Forbes

Thomas P. Simon

Lepomis: Greek, *lepis* "scale" and *poma* "lid" for the scaled operculum; *symmetricus*: Latin, symmetrical.

RANGE

The bantam sunfish occurs below the Fall Line in Gulf Coast drainages from Jordan River, MS, west to the Colorado River, TX. In the Mississippi River basin it presently extends north to the bottomland oxbow lakes and swamps of southern IL and southwest IN. It is virtually absent east of the Mississippi River, though locally common in extreme western KY.[14,21] The species is common in LA, eastern TX, western TN, and southern AR.[1–4,6]

HABITAT AND MOVEMENT

Bantam sunfish inhabit good quality, generally clear and well-vegetated lowland habitats such as oxbow lakes and overflow swamps;[1–6] often found in lentic waters with standing timber, submerged logs and stumps, and vegetation;[6] occurs in sloughs, oxbows, ponds, backwaters, lakes, and swamps typical of the coastal plain;[6] also reported from springs.[11]

Bantam sunfish are found over bottoms of decomposed vegetation, silt, mud, and some sand;[1] greatest numbers occur over substrates of mud, detritus, and decayed plant materials.[6] During the summer they are often found at depths of 60–120 cm and during the fall and winter at depths of 15–30 cm;[6,7] depths from 300 mm to 18 m are reported.[6]

The bantam sunfish is usually found in intimate association with submerged aquatic vegetation such as coontail (*Ceratophyllum demersum*).[1,6] It occurs along vegetated shorelines dominated by spatterdock (*Nymphaea advena*), American lotus (*Nelumbo lutea*), common arrowhead (*Sagittaria latifolia*), coontail, and duckweed (*Lemna* spp., *Wolffia* spp);[6] also reported among thick growths of sedges, rushes, arrowhead, and *Polygonum* among cypress knees;[10] or among green alga–yellow pond lily–duckweed plant assemblages.[11] In Lake Conroe, TX, over 3,600 hectares of submersed aquatic vegetation was eliminated 1 year after 270,000 grass carp were stocked in 1981–1982. As a result, catches of bantam sunfish declined to nearly zero by 1986. Coupled with severe declines in density in cove habitats, seine catches indicated that this littoral species may have been extirpated from the main basin and embayments of the reservoir.[20]

DISTRIBUTION AND OCCURRENCE IN THE OHIO RIVER SYSTEM

It occurs in the lowermost portions of the Ohio River basin including the backwater ponds and sloughs of the Wabash River, White County, IL.[6] The species is found primarily in Mississippi River tributaries with an almost continuous occurrence in Clear Creek, Cache River drainage, IL.[12,13,15] It has been recently collected from single tributary of the West Fork White River (Wabash River drainage), Greene County, IN.[5] In KY, bantam sunfish is known from several localities in the extreme western portion of the state, where it is locally common; otherwise it is virtually absent from the rest of the state.[14,21] This species is uncommon in TN, persisting in a few natural lakes and overflow swamps in west TN and still common in Reelfoot Lake and Isom Lakes; not reported from the Tennessee River drainage.[1] Channelization is suspected to have destroyed many former habitats.[1,6] The bantam sunfish has been extirpated from its type locality in the Illinois River near Pekin, IL.[6]

SPAWNING

Location

In IL, the silty, dark-stained water of lakes where the species occurs prohibited observation of nests or territory size.[6] Bantam sunfish spawn in mud depressions and leaf litter substrates in AR.[8] A male bantam sunfish in an aquarium constructed a shallow nest over sand and gravel substrates about 90–120 mm in diameter.[6] It is likely that the species builds shallow depressions in the mud bottom of Wolf Lake, IL, along the shallow edges close to vegetation, where egg attachment may take place.[6]

Season

Spawning occurs from April to early June throughout the species range.[6] Ripe females in IL were found from March to May, but post-spawning females were collected in June; most females probably spawned in May.[6] Spawning in TN occurred during April and May.[1]

Temperature

Spawning occurs at temperatures from 18 to 22°C. Aquarium-held individuals exhibited pre-spawning behavior for 7 days at water temperatures varying from 24 to 28°C.[6]

Fecundity

Fecundity is proportional to increasing female length and age (Table 94).[6] Fecundity ranged from 219 to 1,600 ripe ova per female. The relationship between the number of mature ova (F) and the adjusted body weight (W) was F = –50.94 + 210.70W and between the number of mature ova and standard length (L) was log F = –2.785 + 3.383 log L.[6]

Sexual Maturity

Females were sexually mature by age 1 at lengths of 34 mm TL, while males were age 1 and >40 mm.[6]

TABLE 94

Ovarian and fecundity data for bantam sunfish from Wolf Lake, IL.[6]

SL (mm)	GSI	Age (years)	No. of Mature Ova	Ovary Weight (g)
34	7.0422	1	326	0.10
34	8.0537	1	219	0.12
36	7.3034	1	491	0.13
37	12.1212	1	368	0.20
37	8.7379	1	403	0.18
38	8.6207	1	330	0.20
39	9.9526	1	432	0.21
40	7.4074	1	417	0.18
42	10.6556	1	421	0.26
43	7.2727	1	374	0.20
45	10.2484	1	364	0.33
45	9.2814	1	378	0.31
51	30.8204	2	1,600[a]	1.39
52	11.6248	2	1,400[a]	0.88

GSI = gonadosomatic index.

[a] Estimated count.

Spawning Act

Bantam sunfish is classified as a nest-spawning lithophilic fish that guards its young.[17] The spawning act is suspected to be similar to other members of *Lepomis*.[1] No field observations have been made; however, aquarium observations found that male stimulation occurred after feeding. The male courts the female by nudging her with his snout along the posterior regions of her body and continually nipping at her caudal fin. The female did not respond to these actions, but the male continued to nip at her fins and nudged the female with his snout between the pelvic fins while chasing her. The female remained unresponsive. After 3 days of this behavior, the male began to charge the female at rapid speeds with his opercles flared out and the irises of his eyes more intense in color. As he approached the female, he turned his body to a vertical position (with his snout pointing upward) and gently swam around her in a closed circle while fanning his tail. After 7 days of constant nipping, nudging, and displaying to the uncooperative female, the male completely mutilated the female's caudal fin and the female expired on the 8th day.[6]

EGGS

Description

Mature eggs ranged from 0.6 to 0.9 mm in diameter and were demersal, spherical, and translucent. Yolk coloration is orange. A single oil globule is present.[6]

Incubation
No information.

Development
No information.

YOLK-SAC LARVAE

See Figure 75.

Size Range

Length at hatching is unknown; smallest available specimens range from 4.5 to 5.5 mm to 6.8 mm TL.*,[9]

Myomeres

Preanal 10–12; postanal 16–18; total 27–29.[9]

Morphology

4.0–5.4 mm TL. Head large; body elongate and laterally compressed; yolk sac small to moderate (19.6% TL); air bladder small (8.69% TL).*

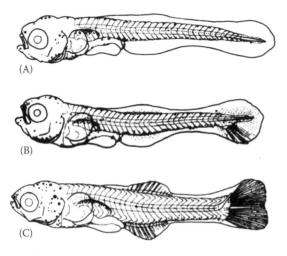

Figure 75 Development of young bantam sunfish. (A) Yolk-sac larva, 4.3 mm TL; B–C. Post yolk-sac larvae: (B) 5.3 mm TL; (C) 7.4 mm TL. (A–C reprinted from Figures 2 through 4, reference 9, with author's permission.)

5.2–5.7 mm TL. Head large; gut formed; yolk sac 10.9% TL; notochord flexion occurred.*

6.8 mm TL. Yolk sac mostly absorbed.[9]

Morphometry
See Table 95.

Fin Development
4.0–5.4 mm TL. Pectoral fin bud formed; incipient median finfolds continuous; no incipient rays formed.*

5.2–5.7 mm TL. Four incipient rays formed in caudal finfold.*

Pigmentation
4.0–5.4 mm TL. Lateral pigmentation absent; melanophores distributed dorsally over swim bladder;

ventral postanal pigmentation forming a series of melanophores along the myosepta of 14 to 15 myomeres.*

5.2–5.7 mm TL. No difference from previous length interval.*

POST YOLK-SAC LARVAE

See Figures 75 and 76.

Size Range
7 mm TL to about 12.0 mm TL.[6,9]

Myomeres
Preanal 11; postanal 17–18; total 28–29.[9]

Morphology
7 mm TL. Yolk completely absorbed.*

Morphometry
See Table 95.

Proportionately deeper heads than *Lepomis cyanellus*.[9]

Fin Development
7.4 mm TL. Fin rays formed in pectoral, soft dorsal, anal, and caudal fins; median finfold partially absorbed dorsally on caudal peduncle and nearly so ventrally, and still present ventrally anterior to the anus; no pelvic bud formed.*

10.3 mm TL. Fin rays formed in all paired and median fins with the exception of the pelvic fin; pelvic fin bud formed; ventral finfold anterior to anus partially absorbed.*

Table 95

Morphometric data expressed as percentage of TL for young bantam sunfish from Louisiana.*

Length (mm TL)	N	Length (% TL)							Body Depth (%TL)		
		ED	HL	PreAL	PosAL	GL	IL	SL	GBD	BDA	CPD
5.0	1	7.9	18.8	36.9	63.1	—	—	96.4	16.7	7.9	2.9
6.6	1	7.6	19.7	42.4	57.6	15.2	12.1	96.2	18.2	7.6	4.5
7.9	1	7.4	21.5	45.9	54.1	18.5	14.1	83.0	17.0	13.0	7.8
11.0	1	9.8	29.3	48.0	52	26.0	5.7	79.7	25.2	21.1	10.6
15.3	1	8.7	27.8	48.7	51.3	28.7	4.3	78.3	31.3	27.8	11.7
35.2	1	6.7	30.4	32.2	67.8	30.4	4.1	85.2	37.4	33.5	7.0

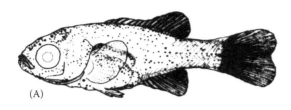

Figure 76 Development of young bantam sunfish. (A) Post yolk-sac larva, 11.0 mm TL; B–C. Juveniles: (B) early juvenile, 12.0 mm SL; (C) 30 mm SL. (A reprinted from Figure 5, reference 9, with author's permission; B–C reprinted from Figure 3, reference 6, with publisher's permission.)

Pigmentation

>7.0 mm TL. Postanal myosepta with more than 25 melanophores between the vent and the hypural plate.[9]

6.8–8.0 mm TL. Melanophores restricted to dorsum form a row from nape to soft dorsal fin insertion; pigment along postanal myosepta forms a row from the anus to caudal peduncle base.[*]

10.3–11.5 mm TL. Scattered melanophores cover the entire body. Dorsally, pigment is concentrated over cerebral hemispheres; scattered dorsally over the remainder of the body. Pigment outlines orbit and lower jaw. Laterally, pigment is concentrated over the air bladder and stomach and at base of caudal peduncle. Ventrally, pigment is concentrated at isthmus and breast anterior to the pelvic fin origin; scattered melanophores from anus to caudal peduncle form a median stripe.[*]

JUVENILES

See Plate 2F following page 26 and Figure 76.

Size Range

Juvenile stages of development range from 12 mm TL to 34–40 mm TL.[6]

Myomeres

Preanal 13–14; postanal 15–17; total 29–31.[9]

Morphology

12.0 mm TL. Nape, breast, and sides of head incompletely scaled, all other areas completely scaled.[6]

14 mm TL. Squamation patterns similar to 12 mm TL.[6]

19 mm TL. Squamation nearly complete.[6]

25 mm TL. Lateralis system developed; squamation complete.[6]

>27–40 mm TL. Mouth small in size, maxilla reaching to, or just beyond, anterior margin of eye. Lateral scale rows 30–38, usually 20 or fewer pored scales;[1] gill rakers 11–15, length 6–8 times greatest width; palatine teeth present.[1]

Morphometry

See Table 95.

Fin Development

12.0 mm TL. Median finfold completely absorbed; adult complement of fin spines and rays formed in all paired and median fins.[*]

>12–40 mm TL. Pectoral fin short and rounded, its tip far short of nares when laid forward across check.[1]

Larger juveniles. Adult complement of fin spines and rays: spinous dorsal X (IX–XI); soft dorsal fin rays 10 (9–12); pectoral fin rays 12–13 (11–13); anal spines and rays, III (II–IV) 9–11; pelvic fin rays I 4–5; principal caudal rays 17–18.[*,1–6,9]

Pigmentation

See Figure 77.

>12 mm TL. Scattered melanophores cover entire body. Small melanophores outline the scale borders on the body and some of the fin rays but were concentrated heavily on the dorsum of the cranium, on the lips, and around the eye. Soft dorsal fin ocellus was beginning to develop.[6]

14 mm TL. Many more melanophores were present in the fins and patterns were beginning to form on the body.[6]

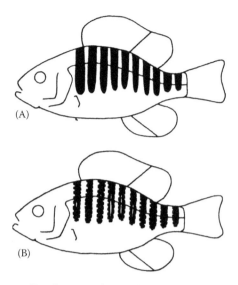

Figure 77 Development of young bantam sunfish — schematic drawings showing ontogeny of melanistic pigment patterns. (A) Juvenile 15 mm SL and (B) adult 40.0 mm SL. (A–B reprinted from figure 13, reference 19, with publisher's permission.)

19 mm TL. Vertical bars formed.[6]

25 mm TL. Overall pigmentation similar to adults.*

30–50 mm TL. Body dark olivaceous to brownish above and straw yellow below; dark spots scattered over the body with vertical bars visible on the sides. A distinct basal ocellus is present in the posterior soft dorsal fin which has a dark center surrounded by a pale area with maybe red.[1,6]

Pigmentation transformations show that 9 to 10 primary bars develop laterally in bantam sunfish. Light centers form in the primary bars (Figure 77A and 77B), but secondary bars do not form. As the centers of the primary bars become increasingly lighter, only the anterior and posterior margins of each bar remain visible. The anterior and posterior edges of the bars become broken as pigment is lost at approximately every other scale. Through the loss of pigment in many scales in the margins of the light-centered primary bars, and the intensification of the pigment in other scales, the vertical orientation of the bars is lost.[19]

TAXONOMIC DIAGNOSIS OF YOUNG BANTAM SUNFISH

Similar species: other *Lepomis* species. Bantam sunfish is in the monotypic subgenus *Lethogrammus* and hypothesized the closest sister taxa to green sunfish (see discussion on page 92).[22]

For comparisons of important morphometric and meristic data, developmental benchmarks, and pig-

mentary characters between bantam sunfish and all other *Lepomis* that occur in the Ohio River drainage, see Tables 4 and 5 and Plate 2F.

Larvae. Early yolk-sac larvae are separable from all other *Lepomis* by possessing a moderate-sized head and having postanal pigmentation present (limited to ventral postanal myosepta).* Hatching occurs between 4.5 and 5.5 mm TL. Pectoral fin bud is present at hatching. Yolk is absorbed by 6.8 mm TL. Post yolk-sac larvae have concentrated melanophores over cranial hemisphere and abundant scattered lateral, ventrolateral, and postanal pigmentation. Post yolk-sac larvae and juveniles are separable from all other *Lepomis* except bluegill and green sunfish by the occurrence of a dark spot that forms in the posterior portion of the soft dorsal fin.*,1,6,9

Juveniles. Bantam sunfish YOY (Plate 2F) are most likely confused with young specimens of redspotted sunfish (Plate 2E), dollar sunfish (Plate 2B), and possibly several other species in which the juveniles may have a barred color pattern. However, no other species of *Lepomis* has an incomplete lateral line. Flier (Plate 1A) is the only other centrarchid with a dark spot in the posterior portion of the dorsal fin in juveniles, but they have many more anal fin spines (7–8) than bantam sunfish (2–4).[1–4]

ECOLOGY OF EARLY LIFE PHASES

Occurrence and Distribution (Figure 78)

Eggs. Eggs are laid in a small nest depression over vegetation, mud, or sand substrates usually near aquatic vegetation.[1,5–7,18] The nest is constructed near cover, usually detritus, woody debris, or aquatic plants.[18] Eggs are partially buried in the substrate, which ranges from sand, gravel, or other fine material that the male has cleaned prior to reproduction.[1–6]

Yolk-sac larvae. Presumably, yolk-sac larvae remain in the substrate until the yolk is absorbed. No information is available on the abundance or survival of early life stages.[18]

Post yolk-sac larvae. Larvae are free swimming at 6.8 mm TL.[9]

Juveniles. Juvenile stages occur until 34–40 mm TL. The habitat use of juveniles is similar to adults with reports from heavily vegetated streams, oxbow lakes, and swamps. They are intolerant of varying conditions, especially low dissolved oxygen.[1,6]

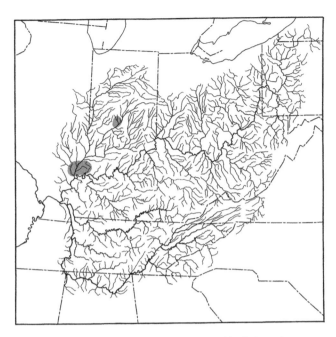

Figure 78 General distribution of bantam sunfish in the Ohio River system (shaded area).

Early Growth

Wolf Lake, IL, populations of bantam sunfish grew at a decreasing rate and reached half of the first year's mean growth in 10 weeks. The relationship between standard length (Y) and age in months (X) is $Y = 5.91 + 32.97 \log X$. Males grew at a slightly more rapid rate than females but were not significantly larger.[6] Average growth rates in IL at 13–18 months were 42.7–45.9 and at 19–24 months 47.5–49.3 mm TL.[6] In TN, bantam sunfish were 45 mm TL at age 1, 55 mm TL at age 2, and exceeded 60 mm TL at age 3.[1] Low survivorship was observed after the first year of life with less than 1.5% of the population surviving to age 3.[6]

Feeding Habits

Bantam sunfish are insectivores[1-6] or invertivores,[18] feeding on drift and feeding in the water column and near the surface on invertebrates.[18] They consume benthic midge larvae and other immature insects and to a lesser extent microcrustaceans;[1-6] or small crustaceans, midge larvae, snails, supplemented with surface feeding on hemipterans and terrestrial insects.[18] Small bantam sunfish (<21 mm TL) fed predominantly on microcrustaceans, dragonfly naiads, and chironomids, while larger individuals fed on gastropods, dragonfly naiads, and amphipods.[6] In MS, dragonfly and midge larvae were consumed.[16] Some seasonal variation in diet was observed with gastropods being consumed in the winter and spring months.[6] The largest percentages of most food items including gastropods, stratiomyids, chironomids, and some microcrustacea were eaten in months prior to spawning.[6] Aquatic Hemiptera were eaten exclusively during the summer.[6]

LITERATURE CITED

A

1. Etnier, D.A. and W.S. Starnes. 1993.
2. Lee, D.S. et al. 1980.
3. Smith, P.W. 1979.
4. Pflieger, W.L. 1975b.
5. Simon, T.P. 1994.
6. Burr, B.M. 1977.
7. Boyd, J.A. et al. 1975.
8. Robison, H.W. 1975.
9. Conner, J.V. 1979.
10. Moore, G.A. and F.B. Cross. 1950.
11. Gunning, G.E. and W.M. Lewis. 1955.
12. Burr, B.M. et al. 1996.
13. Cook, K.M. 1994.

14. Burr, B.M. et al. 1990.
15. Burr, B.M. et al. 1988.
16. Hildebrand, S.F. and I.I. Towers. 1928.
17. Simon, T.P. 1999.
18. Goldstein, R.M. and T.P. Simon. 1999.
19. Mabee, P.M. 1995.
20. Bettoli, P.W. et al. 1993.
21. Burr, B.M. and M.L. Warren, Jr. 1986.

* Original developmental information is taken from LA specimens listed as green sunfish "type" by Conner (1979).

GENUS
Micropterus Lacepede

Robert Wallus

Members of this genus are collectively called "black basses," a name derived from the dark coloration of the larvae of the smallmouth bass, the type species for the genus (Boschung and Mayden, 2004). Although called basses, members of this genus should not be confused with the temperate basses of the North American family Moronidae or any of the sea basses, family Serranidae. *Micropterus* is comprised of seven species, all native to North America east of the Continental Divide (Boschung and Mayden, 2004). This genus contains some of the best-known game fishes in the freshwaters of North America. All members are highly predaceous and function as top carnivores in their respective environments. They are highly valued as game and food fishes, and several species of *Micropterus* have been stocked in many parts of the world as sport fishes (Etnier and Starnes, 1993).

Black basses are extremely adaptive centrarchids. They have an elongate, oblong sub-compressed body; relatively small dorsal and anal fins; a relatively large, more emarginated caudal fin; a large mouth with projecting lower jaw; greater association with current, and a tendency to chase prey. Additional characters for the adults of this genus include: 3 anal spines; 6–7 (usually 6) branchiostegal rays; well-developed supramaxilla; opercle with two flat projections; teeth present on tongue, vomer, and palatines; a distinct notch is present between the spinous and soft dorsal fins; and dorsal fin spines are typically 10 (Etnier and Starnes, 1993; Jenkins and Burkhead, 1994; Boschung and Mayden, 2004).

Nest building and other reproductive behaviors of the black basses are similar to those described for *Lepomis*. Typically, the male remains over the nest, protecting the eggs and fanning them with movements of the pectoral fins. After leaving the nest, the larvae spend several weeks in a dense school guarded by the male. The guarding male is very aggressive and strikes at any strange or threatening object and is very vulnerable to anglers (Boschung and Mayden, 2004).

TAXONOMY AND SYSTEMATICS OF GENUS *MICROPTERUS*

The taxonomy of *Micropterus* was reviewed and revised by Hubbs and Bailey (1940) and Bailey and Hubbs (1949). Ramsey (1975) listed seven species in the genus contained in three "evolutionary lineages." He proposed largemouth bass *Micropterus salmoides* as the sole member of one lineage; smallmouth bass *M. dolomieu* and the redeye bass *M. coosae* constituted the second lineage; and the third lineage is represented by the widely distributed spotted bass *M. punctulatus* and three species with more restricted ranges including the shoal bass *M. cataractae* endemic to the Apalachicola drainage, the Suwannee bass *M. notius* found native to the Suwannee and Ochlockonee Rivers in FL, and the Guadalupe bass *M. treculi* of central TX. Branson and Moore (1962) felt that redeye bass was more closely related to Suwannee and the other spotted bass-like members of the genus than to smallmouth.

Kassler et al. (2002) state that phylogenetic analyses of mitochrondial DNA sequence variation indicate that the genus *Micropterus* is represented by four lineages, not three as reported by Ramsey (1975): (1) smallmouth bass and spotted bass; (2) largemouth bass, Florida bass, Suwannee bass, and Guadalupe bass; (3) shoal bass; and (4) redeye bass and Alabama spotted bass. They assert that it is likely through either natural or human-induced changes that hybridization has occurred between Alabama spotted bass and redeye bass and between spotted bass and Guadalupe bass. They contend, based on this new information, that management agencies should alter policy that results in the stocking of non-native species and that, specifically, they should terminate the stocking of Florida bass outside of FL.

Based on variation in meristic characters, allozymes, and mtDNA, Kassler et al. (2002) assert that two subspecies of *Micropterus salmoides* (*M. s. salmoides* and *M. s. floridanus*) are clearly distinct from one another and warrant elevation to species status. *Micropterus s. floridanus* should be recognized as the Florida bass *M. floridanus* and *M. s. salmoides* as the largemouth bass *M. salmoides*. Nelson et al. (2004) deferred formal recognition of this classification pending further study.

Kassler et al. (2002) point out that although the Alabama spotted bass *Micropterus punctulatus henshalli* is morphologically and genetically distinct from northern spotted bass, a thorough taxonomic assessment is still required before revision. Another spotted bass subspecies, *M. p. wichitae* (Wichita spotted bass) from OK was invalidated by Cofer (1995).

Hubbs and Bailey (1940) examined variation in *Micropterus dolomieu* and described a new subspecies, *M. d. velox*, from the Neosho River system of KS, OK, MO, and AR. Phylogenetic analyses by Stark and Echelle (1998) supported recognition of three clades of smallmouth bass from the Interior Highlands (Ozark and Ouachita uplands): (1) Neosho smallmouth bass in Ozark tributaries of the middle Arkansas River; (2) Ouachita smallmouth bass in the Little and Ouachita River drainages of the Ouachita Highlands; and (3) a clade that includes populations from the White, Black, and Missouri Rivers and other streams in the northern and eastern Ozarks. This third clade is similar to populations from the Ohio and upper Mississippi River basins, and, on the basis of allele frequency parsimony, more closely related to them than to the Neosho and Ouachita smallmouth basses.

Avise and Smith (1977) hypothesized a closer relationship between *Micropterus* and the genus *Lepomis* than to the other sunfishes, but Mabee (1987) considered *Micropterus* a primitive sister group to other centrarchid genera with living species.

Although interspecific hybridization among basses is rare in nature (Hubbs and Bailey, 1940), hybridization and genetic introgression have been reported in areas where native basses came into contact with otherwise allopatric species as a result of fish stocking (Morizot et al., 1991; Koppelman, 1994; Pierce and Van Den Avyle, 1997). Many published accounts of natural hybridization involve smallmouth bass, a species that has been stocked

extensively outside its native range (Lee et al, 1980; Morizot et al., 1991). In TX, native Guadalupe bass *Micropterus treculi* and northern largemouth bass *M. salmoides salmoides* now coexist with introduced Florida largemouth bass *M. s. floridanus* and smallmouth bass *M. dolomieu*. Interspecific hybridization is reported from three of four populations from the Blanco and San Marcos Rivers. Complex hybridization patterns are reported; at least one individual exhibited genetic markers of largemouth, smallmouth, and Guadalupe bass. The authors suggest that extensive multispecies hybridization threatens the survival of the endemic Guadalupe bass (Morizot et al., 1991).

Only four bass species are reported from the Ohio River drainage. Largemouth, smallmouth, and spotted bass are native, and redeye bass has been introduced (Parsons, 1954; Cathey, 1973; Gwinner, 1973; Pipas and Bulow, 1998). Adult basses from the Ohio River drainage are easily identified with regional keys (Smith, 1979; Etnier and Starnes, 1993; Jenkins and Burkhead, 1994; Ross, 2001; Boschung and Mayden, 2004).

Characters for distinguishing young *Micropterus* from elassomatids and other centrarchid genera are presented in the Provisional Key to Genera on pages 27–29.

The following key to the young of four *Micropterus* species that occur in the Ohio River drainage was constructed using original data and observations on the development of young basses and from assimilated data and information found in numerous literature sources (cited in species accounts that follow), and specifically from information obtained from Ramsey and Smitherman (1971), Etnier and Starnes (1993), Jenkins and Burkhead (1994), and Mabee (1995). This key is considered provisional because of gaps in our knowledge of the development of the redeye bass. Also of some concern is that most of the characters presented are descriptions of pigment pattern development. Patterns and rate of development of pigmentation may be affected by latitude, water temperature, turbidity, and locality of capture as well as other biotic and abiotic phenomena. For example, hatchery stocks of smallmouth bass larvae vs. wild-caught larvae may be lighter in overall appearance due to contraction of melanophores. This was reported for Lake Erie smallmouth bass (Fish, 1932) and also for smallmouth from a GA hatchery (original observation). In the latter occurrence, some, but not all, healthy fish from the hatchery were lighter in overall coloration due to the contraction of melanophores.

The developmental series of redeye bass used for this work was incomplete; the smallest yolk-sac larva examined was 8.5 mm TL and no specimens were available between 13 and 26 mm TL for comparison to the other species. Young redeye bass are more similar in development and pigmentation to smallmouth bass than to largemouth or spotted bass. The same barring pattern described by Mabee (1995) for smallmouth bass was used for redeye bass, but based only on preliminary observations. It should be noted, however, that in the Ohio River drainage, distinguishing between smallmouth and redeye bass is not that problematic because of the limited distribution of the latter. Also, because of the similarities of young smallmouth and redeye bass, characters presented in the key will usually take the user to couplets that contain both species, in which case locality of collection may preclude redeye as a possible identification.

Provisional Key to the Young of *Micropterus* Species Present in the Ohio River Drainage

Note: Identifications made with this key should be considered tentative and should be verified, if possible, with data and information presented in Tables 4 and 5, and with size-specific detailed information found in the species accounts that follow.

1a. ≤9.0 mm SL..2
1b. >9.0 mm SL..4

2a. <8.0 mm SL...*Micropterus* spp.
2b. 8.0–9.0 mm SL..3

3a. Small chromatophores densely scattered over most of body gives fish a dark appearance to unaided eye...Smallmouth bass
3b. Pigmentation concentrated on dorsal surface of head and melanophores may be present along dorsum and ventrum of body, and laterally on yolk sac or gut; otherwise widely scattered, if present laterally on body...*Micropterus* spp.

4a. 9.0–15.0 mm SL..5
4b. >15 mm SL..7

5a. Developing lateral stripe along median myosepta visible and easily discernable.............................
....*Micropterus* (largemouth or spotted bass)
5b. No prominent lateral stripe of pigment; pigment may be present midlaterally as a series of small separate blotches or as a series of separate dorsoventrally elongated bars or blotches; additionally, body completely or partially covered with small scattered chromatophores....................................6

6a. Small chromatophores densely scattered over much of body gives fish a dark appearance to the unaided eye ....Smallmouth bass
6b. Pigmentation less dense and may not completely cover body, fish does not appear dark to unaided eye...Redeye bass

7a. <20.0 mm SL..8
7b. ≥20.0 mm SL..11

8a. Prominent visual feature of lateral pigmentation is a dark midlateral stripe or band; caudal spot well developed (Figures 79H and 79L)..9
8b. Lateral pigmentation marked by scattered chromatophores over all or portions of body; midlateral pigmentation may consist of a series of 4–5 small separate ovoid to squarish blotches of melano- phores along the median myosepta in the abdominal area or a series of dorsoventrally elongated narrow bars of pigment; caudal spot weakly developed or absent (Figures 79A and 79D).........10

9a. Lateral blotches of pigment more confluent, extending little above the lateral stripe; lateral stripe a well-defined band (Figure 79L)...Largemouth bass
9b. Dorsal and ventral margins of lateral stripe less well defined; lateral blotches less confluent, some extending above narrow profile of lateral stripe (Figure 79H)...Spotted bass

10a. Dorsal fin with 12–15 (usually 14) soft rays; body heavily pigmented with scattered chromatophores; supralateral markings not readily visible (Figure 79D) .... Smallmouth bass

10b. Dorsal fin with 11–14 (usually 12) soft rays; supralateral markings readily visible (Figure 79A).... .... Redeye bass

11a. 20.0–40.0 mm SL 12

11b. 40.0–100 mm SL 15

12a. Lateral stripe consists of confluent blotches of primary melanistic pigment; secondary bars of melanistic pigment form dorsally; little pigment ventrolaterally (Figures 80M, 80N, 80R, 80S and Figures 79I, 79M) 13

12b. No midlateral stripe; predominant lateral pigment consists of 11–14 supralateral bars of primary melanistic pigment, taller than wide, with scattered chromatophores interspersed dorsally and laterally between; no secondary bars of melanistic pigment are formed at 25 mm SL and little secondary pigment is present at 40 mm SL (Figures 80B, 80C, 80H, 80I and Figures 79B, 79E) 14

13a. Lateral blotches more confluent, extending little above broad lateral stripe; caudal spot dark posteriorly, horizontally elongated onto anterior portion of caudal rays, twice as long as wide, length equal to or greater than eye diameter (Figure 79M) .... Largemouth bass

13b. Lateral blotches less confluent, extending well above narrow lateral stripe; caudal spot about as long as wide, not defined on anterior portion of caudal rays, length less than eye diameter (Figure 79I) .... Spotted bass

14a. Submarginal band on caudal fin becomes strongly developed (Figure 79E); dorsal fin with 12–15 (usually 14) soft rays .... Smallmouth bass

14b. Submarginal band on caudal fin usually indistinct or absent (Figure 79B); dorsal fin with 11–14 (usually 12) soft rays .... Redeye bass

15a. Submarginal band of pigment on caudal fin more or less conspicuous (Figure 79F, 79J, and 79N) 16

15b. Submarginal band of pigment on caudal fin indistinct or absent (Figure 79C) .... Redeye bass

16a. Lateral pigment marked by a bold irregular lateral stripe; caudal spot well developed (Figures 79J, 79K, 79N, and 79O) 17

16b. Lateral pigmentation consists of weakly to well-developed narrow dorsoventrally elongated bars; caudal spot weakly developed or absent (Figures 79F and 79G) .... Smallmouth bass

17a. Depigmentation of posterior margin of caudal fin equal to or slightly more at lobes than at fork (Figure 79N); suborbital stripe is usually absent or represented as blotches on the cheek; highest vertical expansions on anterior half of midlateral stripe developed in 2–4 horizontal scale rows; pyloric caeca branched near bases, 18 or more tips; tongue lacking teeth .... Largemouth bass

17b. Depigmentation of posterior margin of caudal fin much greater at lobes than at fork (Figure 79J); suborbital stripe usually well developed and continuous on cheek; highest vertical expansions on anterior half of midlateral stripe developed in 4–7 horizontal scale rows; pyloric caeca simple, 9–13 tips; tongue with narrow medial patch of small teeth (may be felt with a fine point on small fish) .... Spotted bass

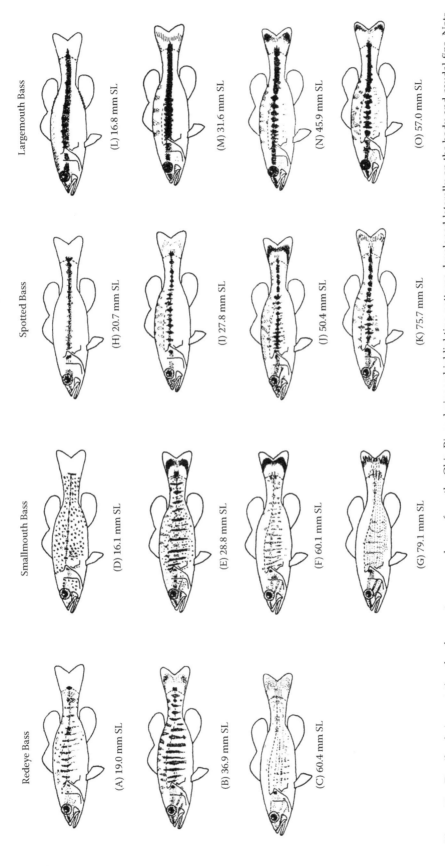

Figure 79 Details of pigmentation development on young basses from the Ohio River drainage highlighting patterns developed laterally on the body and caudal fins. Note that body proportions are constant in the drawing outlines and do not reflect proportional differences between sizes. (A–C) Redeye bass; (D–G) smallmouth bass; (H–K) spotted bass; and (L–O) largemouth bass. (A–C reprinted from Figure 5, D–G from Figure 4, H–K from Figure 2, and L–O from Figure 1 in Ramsey and Smitherman, 1971, with publisher's permission.)

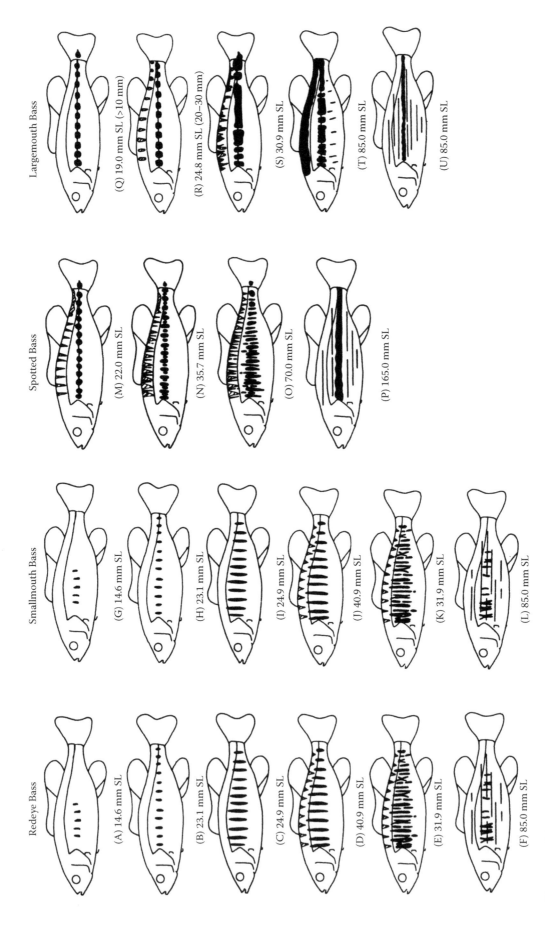

Figure 80 Schematic drawings showing ontogeny of melanistic pigment patterns for young basses in the Ohio River drainage. Note that body proportions are constant in the drawing outlines and do not reflect proportional differences between sizes. (A–F) Redeye bass; (G–L) smallmouth bass; (M–P) spotted bass; (Q–U) largemouth bass. (A–F and G–L reprinted from Figure 3, M–P from Figure 5, and Q–U from Figure 4 in Mabee, 1995, with publisher's permission.)

Largemouth Bass

(Q) 19.0 mm SL (>10 mm)
(R) 24.8 mm SL (20–30 mm)
(S) 30.9 mm SL
(T) 85.0 mm SL
(U) 85.0 mm SL

Spotted Bass

(M) 22.0 mm SL
(N) 35.7 mm SL
(O) 70.0 mm SL
(P) 165.0 mm SL

Smallmouth Bass

(G) 14.6 mm SL
(H) 23.1 mm SL
(I) 24.9 mm SL
(J) 40.9 mm SL
(K) 31.9 mm SL
(L) 85.0 mm SL

Redeye Bass

(A) 14.6 mm SL
(B) 23.1 mm SL
(C) 24.9 mm SL
(D) 40.9 mm SL
(E) 31.9 mm SL
(F) 85.0 mm SL

LITERATURE CITED

Avise, J.C. and M.H. Smith. 1977.
Bailey, R.M. and C.L. Hubbs. 1949.
Boschung, H.T., Jr. and R.L. Mayden. 2004.
Branson, B.A. and G. A. Moore. 1962.
Cathey, H.J. 1973.
Cofer, L.M. 1995.
Etnier, D.A. and W.C. Starnes. 1993.
Fish, M.P. 1932.
Gwinner, H.R. 1973.
Hubbs, C.L. and R.M. Bailey. 1940.
Jenkins, R.E. and N.M. Burkhead. 1994.
Kassler, T.W. et al. 2002.
Koppelman, J.B. 1994.

Lee, D.S. et al. 1980.
Mabee, P.M. 1987.
Mabee, P.M. 1995.
Morizot, D.C. et al. 1991.
Nelson, J.S. et al. 2004.
Parsons, J.W. 1954.
Pierce, P.C. and M.J. Van Den Avyle. 1997.
Pipas, J.C. and F.J. Bulow. 1998.
Ramsey, J.S. 1975.
Ramsey, J.S. and R.O. Smitherman. 1971.
Ross, S.T. 2001.
Smith, P.W. 1979.
Stark, W.J. and A.A. Echelle. 1998.

REDEYE BASS

Micropterus coosae Hubbs and Bailey

Robert Wallus

Micropterus, Greek: *micro*, "small" and *pteron*, "fin"; *coosae*: named for the Coosa River, AL and GA.

RANGE

Redeye bass is native to the lower Appalachian regions of AL, TN, GA, NC, SC, and FL.[19] It is endemic to the Tombigbee and Alabama Rivers of AL and GA, the Chattahoochee, Altamaha, and Savannah Rivers in GA, NC, and SC, and the Coosa River in TN.[1,3,8,19,23] It is endemic to the Mobile basin in AL, GA, TN, NC, and SC; the headwaters of the Savannah River in GA, NC, and SC; and the Chattahoochee River of the Appalachicola River system in AL and GA.[10,11,23] It is unknown from the coastal plain of any river system.[20]

This species has been successfully introduced into the Cumberland and Tennessee River systems.[3,6,9,10,12,19] Hybridization with smallmouth bass resulted from some of these introductions.[10] Also introduced in CA[4] and Puerto Rico.[8]

HABITAT AND MOVEMENT

Redeye bass habitat requirements are considered very stringent.[23] This species is found in small to large streams,[8,13,20] but only rarely in large rivers or impoundments.[13] It is often abundant in small, cool, upland headwater streams[3,18,22] with clear water and sand, rubble, or limestone bottoms.[3,22] Redeye bass is a secretive fish often found in proximity to heavy cover such as undercut banks, logs, stumps, boulders, and water willow (*Justicia*) beds or other aquatic vegetation.[11,13] It has been observed orienting to body contact with the environment; believed to be a specialization related to the preferred habitat — pools in small streams and rocky rapids in larger streams.[22]

Redeye bass is adapted to streams too cold for other warm-water fishes and too warm for trout.[3,15] It is valued as a game fish because it fills a niche in cooler streams between trout and warmer-water species such as largemouth and smallmouth bass;[11] usually found upstream of smallmouth bass, with little overlap.[3,5]

Though redeye bass are reported as poorly adapted for living in lakes or larger quiet waters,[3,11,13,17,22]

they survived in ponds built on small streams in AL and became an important part of the creel for the first 2 years. Numbers steadily declined thereafter; none were caught after 3 years.[21]

Redeye bass abundance in Allatoona Reservoir, GA,[27] and Lewis Smith Reservoir, AL,[28] gradually declined after impoundment, but in Keowee and Jocassee Reservoirs, SC, there was an increase in redeye bass several years after impoundment. Annual catches per 100 gill-net sets increased in Keowee Reservoir from 0.4 fish in 1973 (5 years after impoundment) to 13.8 in 1981. In Jocassee Reservoir, catches increased from 2.0 fish in 1975 (4 years after impoundment) to 17.4 in 1981. Growth is also rapid in these two reservoirs compared to stream populations. The increased abundance and rapid growth in Keowee and Jocassee Reservoirs suggest that redeye bass may be better suited for some lentic waters containing largemouth bass than was originally believed. Both Allatoona and Lewis Smith Reservoirs have black bass populations that include largemouth bass and spotted bass in addition to redeye bass. No spotted bass are present in Keowee and Jocassee Reservoirs, suggesting some as yet undetermined interaction between these two species that ultimately favors the spotted bass.[26]

Movement downstream out of small tributaries in the fall is reported with return in the spring.[3] In a transplanted population of redeye bass, little movement into areas vacated by other fish was noted.[5,6]

DISTRIBUTION AND OCCURRENCE IN THE OHIO RIVER DRAINAGE

Redeye bass is not native to the Ohio River drainage, but has been successfully introduced into the Cumberland and Tennessee River systems.[3,6,9,10,12,19] The earliest introduction was in 1943 into Sylco Creek, a tributary of the Ocoee River, Tennessee River drainage, Polk County, TN. Between 1953 and 1960, 16 upper Cumberland region streams of the Tennessee and Cumberland River systems were stocked with this species.[3,10,12] Several hundred miles of Cumberland River streams in middle and eastern TN and KY were stocked.[3,5–7,9]

SPAWNING

Location

Redeye bass spawn in headwater streams[5] in nests built in eddy waters at the heads of pools.[3] Nests are similar to smallmouth bass nests, usually shallow depressions in coarse (12–13 mm diameter) gravel.[3,5,6] In ponds, redeye bass spawned over artificial gravel and limestone chips and natural hard clay substrates.[1]

Season

Most spawning activity occurs in May and June with some possibly in July.[18] Redeye bass in TN spawn in late May, June, and early July in the Conasauga River drainage[3] and in May and early June in the Cumberland River drainage.[5] Spawning occurs from April to June in AL.[13] In AL ponds, initial spawning of Appalachicola River brood stock was on May 5; Alabama River bass spawned on April 14 and April 19.[1]

Temperature

Redeye bass in the Cumberland River drainage of TN spawned in water ranging from 15.6 to 21.1°C[7] with an average temperature of 18°C reported.[5,6] Spawning occurred in higher temperatures, 21.2–26.1°C, in AL ponds.[1] Spawning has also been reported at water temperatures of 17–21°C.[3,18]

Fecundity

An age-5 female redeye bass, 145 mm TL weighing 35 g, had 2,084 eggs; a female of unknown age, 205 mm TL and 98 g, had 2,334 eggs.[3]

Sexual Maturity

Under stream conditions, redeye bass become sexually mature at 3 or 4 years of age.[18]

Some age-3 females are mature at 120 mm TL and 16 g, males at age 4 and 122 mm TL and 19 g; most redeye bass >120–122 mm are mature.[3]

Spawning Act

Spawning occurs in nests or shallow depressions in coarse gravel in eddy waters at the heads of pools.[3,5,6,11] After spawning, males remain with the nest until the young have hatched and moved away from the nest, usually about a week.[11]

EGGS

Description

Mean diameter of redeye bass eggs spawned in a pond (Alabama River brood stock) was 3.5 mm.

Average diameter of eggs spawned in ponds by brood from the Appalachicola River was smaller, 2.0 mm.[1]

Incubation

Six days at 22.2°C.[1]

Development

No information.

YOLK-SAC LARVAE

See Figure 81.

Size Range

Hatching size is unknown. Yolk was completely absorbed on some wild-caught redeye bass (Falling Water River, TN) by 9.1 mm TL, but persisted on others until 9.5 mm TL.*

Myomeres

8.5–9.5 mm TL. Preanal 16–17; postanal 15–18; total 31–35.*

Morphology

8.5–9.5 mm TL. Late yolk-sac larvae, remnants of oil globule still present; at 8.5–9.0 mm, oil globule about 0.4–0.6 mm long and positioned posteriorly in abdominal cavity over gut. Digestive system is functional by 8.5 mm.*

Morphometry

See Table 96.

Fin Development

8.5–9.0 mm TL. Flexion begins; hypural elements visible by 8.9 mm; 5–10 caudal fin rays become visible. Positions of future soft dorsal and anal fins are obvious as elevations in the median finfold; pterygiophores appear at bases of soft dorsal and anal fins and ray differentiation begins.*

9.0–9.5 mm TL. Adult caudal fin shape forming; urostyle well elevated; developing hypural elements obvious; 10–15 rays visible. Fin shapes of soft dorsal and anal fins well formed; anlagen of soft rays appear; 9–10 pterygiophores evident at bases of fins; at 9.3–9.5 mm little finfold remains ventrally anterior to anus. Pelvic buds appear, barely visible at 9.2 mm TL.*

Pigmentation

8.5–9.0 mm TL. Dorsally, pigment is present over snout and pigment is heavy over brain, occiput, and posteriorly to about mid-body. At 8.5 mm, there is little dorsal or ventral pigment from about

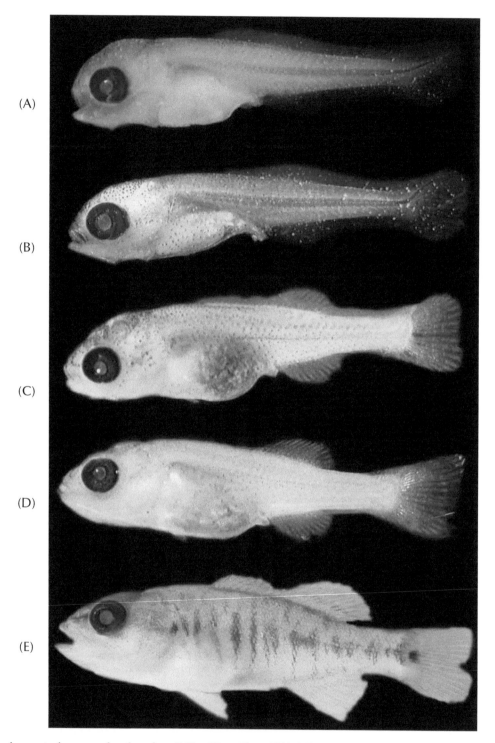

Figure 81 Development of young redeye bass from Falling Water River, TN.* A–B. Late yolk-sac larvae: (A) 8.9 mm TL; (B) 9.2 mm TL. C–D. Post yolk-sac larvae: (C) 11.2 mm TL; (D) 13.0 mm TL. (E) juvenile, 27.0 mm TL. (A–E, original photographs of young redeye bass from TN.)

the middle of the future dorsal and anal fins posteriorly to the caudal fin. Lateral pigment on the head is mostly posterior to the eye; scattered pigment is present laterally in the gut region and internally over the abdominal cavity and the gut.

Lateral pigment on the body is mostly restricted to a stripe of small chromatophores along the median myosepta (about 2–4 chromatophores per myomere) that extends posteriorly to about the middle of the future dorsal and anal fins at 8.5 mm. Some

Table 96

Morphometric data for young redeye bass from Falling Water River, TN.
Data are presented as mean percent total length (TL) and head length (HL)
for selected intervals of TL (N = sample size).

Length Range (mm TL)	TL Groupings						
	8.5–9.5	9.1–9.8	10.1–10.9	11.1–11.9	12.0–12.6	13.0–13.2	26.4–32.1
N	12	6	5	5	5	3	3
Length as Percent TL							
SL	94.8	90.6	87.6	88.8	88.3	86.8	84.8
PreAL	50.2	50.7	52.4	53.0	53.1	53.9	52.1
GD	20.3	21.3	21.7	24.9	23.2	24.2	23.7
Depth at anus	15.5	16.7	17.0	19.0	18.4	18.7	21.3
CPD	6.6	7.2	8.2	9.0	8.9	8.9	10.9
ED	10.7	11.5	11.2	11.3	11.4	11.8	9.4
HL	24.0	26.6	27.8	25.7	27.5	28.5	29.9
Diameter as Percent HL							
ED	44.5	43.1	40.4	44.0	41.4	41.5	31.4

pigment is scattered ventrally in the gular region and posteriorly to about mid-gut; little, if any, pigment is visible under the mouth and jaws. Pigment rows over the gut extend externally along the ventrum under the anterior half of the anal fin. Mid-lateral stripe becomes diffuse but remains visible and is present to about the middle of the caudal peduncle by 8.7 mm. At 8.7 mm, scattered dorso-lateral and lateral pigment has developed on the body anterior to the developing dorsal and anal fins and some scattered pigment appears on the body above and below the diffuse mid-lateral stripe. Scattered pigment may develop ventrally under jaws and scattered small chromatophores become visible ventrally and ventrolaterally on gut.*

9.0–9.5 mm TL. Pigmentation on wild-caught specimens was variable. Some individuals >9.0 mm were pigmented as described for smaller specimens above. At 9.2 mm, pigmentation is generally as described for smaller sizes and some individuals have an internal row of pigment visible laterally midway between the dorsum and the median stripe; this row of pigment runs parallel to the mid-lateral stripe to at least the middle of the caudal peduncle. Dorsal and lateral body pigment is heaviest anterior to the anus. The lateral stripe becomes more diffuse as additional pigment appears laterally on the body. Ventrally, pigment is scattered on the chin and two rows of chromatophores are visible along the base of the developing anal fin. Small chromatophores develop ventrally on the caudal peduncle.*

POST YOLK-SAC LARVAE

See Figure 81.

Size Range
Phase begins at about 9.1–9.5 mm TL. End of phase is unknown but occurs at >13.2 mm TL.*

Myomeres
9.1–9.8 mm TL. Preanal 15–17; postanal 16–17; total 32–34.*

10.1–11.9 mm TL. Preanal 15–17; postanal 15–17; total 31–33.*

12.0–13.2 mm TL. Preanal 17–18; postanal 15–16; total 32–33.*

Morphology
9.1–9.8 mm TL. Head large; opercular flap covers gill chamber.*

10.1–13.2 mm TL. Body robust, head large. Guts of wild-caught larvae often distended and full; greatest depth often measured at mid-gut.*

Morphometry
See Table 96.

Fin Development
9.1–9.8 mm TL. Dorsally, median finfold profile is low anterior to soft dorsal fin and restricted to area

of developing spines; posterior margins of dorsal and anal fins become well defined by notches in the finfold; by 9.8 mm, finfold greatly reduced posterior to the dorsal and anal fins and only a small remnant of ventral finfold is still present anterior to anus. By 9.8 mm, urostyle flexion is at about 45°, with tip confined within upper lob of caudal fin; caudal fin becomes nearly adultlike (bilobed); segmented caudal fin rays 11–16. Soft dorsal and anal fin profiles are evident; dorsal fin with 0 spines and 7–12 soft rays; anal fin with 0 spines and 7–10 soft rays. Pectoral fins are fan-like and about 0.6 mm long; no visible rays. Pelvic fins are flap-like and about 0.2 mm long; no visible rays.*

10.1–12.6 mm TL. By 10.4 mm the posterior margins of the soft dorsal and anal fins are defined and little finfold remains dorsally or ventrally on the caudal peduncle; a small remnant of ventral finfold remains anterior to the anus until about 12.6 mm when all of the median finfold is absorbed. Caudal fin is well formed, bilobed; tip of urostyle remains visible, but barely extends posterior to hypural elements; segmented caudal fin rays 15–17. Developing dorsal fin has 1–10 developing spines visible in low profile; 9–12 soft rays. Developing anal fin has 1–3 spines and 9–11 soft rays visible. Rays form in pectoral fins. Pelvic fins remain small and flap-like with no ray development discernable.*

13.0–13.2 mm TL. Segmented caudal fin rays 17; dorsal fin with 7–10 spines, still in low profile, and 12 soft rays; anal fin with 1–2 spines and 10 soft rays. Pelvic fins are still flap-like, 0.5–0.6 mm long with no visible ray development.*

Pigmentation

9.1–9.8 mm TL. Small chromatophores become scattered over much of the body laterally making the diffuse mid-lateral stripe less obvious. Otherwise, pigmentation is as described for smaller sizes.*

10.1–13.2 mm TL. Pigmentation still concentrated dorsally over brain; otherwise scattered small chromatophores are present over much of the body — laterally on gut region and along the sides of the body. Some stellate chromatophores may be present at about mid-body (under developing dorsal fin), appearing as large pigment blotches or spots which spread dorsolaterally along the body (mostly above the midline) on larger specimens. The diffuse mid-lateral stripe of concentrated small chromatophores remains discernable beneath the scattered body pigment — the stripe consists of 3–5 chromatophores per myomere and extends from about the mid-gut region posteriorly onto the caudal peduncle. Little pigment is present on the hypural plate. Scattered pigment is visible under the jaws. Internal pigment over the air bladder and

gut is still visible. The internal stripe of pigment midway between the dorsum and median myosepta is visible from near the head posteriorly to the bend in the urostyle. By 12 mm, several stellate chromatophores appear distally on urostyle.*

JUVENILES

See Figure 81.

Size Range

Size at beginning of this phase is unknown but >13.2 mm TL.* Redeye bass mature at 120–122 mm TL in the Conasauga River, TN.[3]

Morphology

Scales developed to the opercular flap by 19 mm SL.[2] Lateral scales 64–72.[17]

Morphometry

See Table 96.

Fins

Dorsal fin contains 9–11 (usually 10) spines and 11–14 (usually 12) rays;[*,2,13,17] anal fin has 3 spines and 9–11 (usually 10) rays;[*,13,17] pectoral fin rays 14–17, usually 15 or 16.[17]

Pigmentation

Note: Conversions from SL to TL are based on the formula TL = 1.21 SL, determined from measurements of 157 fish.[3]

The posterior margin of the caudal fin of juvenile redeye bass has a colorless border, especially on the inner edges of the caudal lobes.[5] The sides of the body usually have 10–12 dark blotches that do not join to form a lateral stripe.[13]

26.4–32.1 mm TL. (Pigmentation observed on wild-caught redeye bass from the Cumberland River drainage, TN.) Humeral spot present. A variable-shaped caudal spot is well defined (Figure 81E). * Predominant lateral pigment consists of 11–12 supralateral bars of pigment, taller than wide. Secondary mid-lateral bars of melanophores (as described by Mabee, 1995) forming.[*,25] Scattered chromatophores are interspaced dorsally and laterally between the bars. Head is dark dorsally and dorsolaterally. Scattered stippled pigment is visible ventrolaterally on the head, chin, and under the mouth. Belly is unpigmented. There is little or no pigment around the base of the pelvic fins and in the gular region. Spinous dorsal fin is stippled with pigment. Scattered pigment is present on the interradial membranes in middle of soft dorsal fin and pigment blotches are present proximally on the fin;

most rays are outlined with pigment. Small chromatophores outline caudal and anal fin rays. Little or no pigment is present on pectoral and pelvic fins.*

Note: Mabee (1995) reported that, based on preliminary observations, her descriptions of the ontogeny of lateral pigmentation on smallmouth bass also applied to redeye bass.[25] So, the descriptions that follow on development of primary and secondary bars of melanistic pigment on juvenile redeye bass are those presented by Mabee for smallmouth bass.

14.0–25.0 mm SL (about 17–30 mm TL). Four or five primary bars of melanistic pigment are initially visible as ovoid to squarish mid-lateral blotches of melanophores (Figure 82A). These are located over the horizontal septum in the abdominal area. Primary bars are added anteriorly and posteriorly to this initial set until the adult complement of 11–14 is reached. All primary bars are subsequently dorsoventrally elongated (Figure 82B and 82C).[25] At 19 mm SL, pigmentation of scaled young suggests that a vague, narrow lateral stripe was present before scale imbrication (Figure 83A). Lateral bars are lightly pronounced. A small, distinctly rounded or wedge-shaped caudal spot is present and usually disjunct from and darker than mid-lateral pigment (Figure 83A). No submarginal band is present on caudal fin.[2]

24.0–28.0 mm SL (about 29–34 mm TL). A lateral stripe was evident on 33-day-old individuals in this size range raised in a brood pond, but had disappeared in faster growing young at the same age from a rearing pond. Lateral bars become darker and more contrasting (some with light centers). Submarginal band on caudal fin may be defined, but usually not very dark or distinct.[2]

30.0–40.0 mm SL (about 36–48 mm TL). Secondary bars of melanistic pigment develop dorsal to the lateral line, their ventral extensions interdigitating with the dorsal tips of the primary bars (Figure 82D). The formation of the midlateral and ventral portions of the secondary bars (Figure 82E) corresponds to a lightening in the centers of the adjacent primary bars. The extension of this light area results in the distortion and bilateral splitting of the primary bars such that they cannot be distinguished from the secondary bars (Figure 82E).[25] At 36.9 mm SL (35 days), the caudal spot is still distinct and the submarginal band on the caudal fin is well defined on some but usually not very dark or distinct (Figure 83B).[2]

48.0 mm SL (about 58 mm TL). Lateral bars vague on the anterior half of the body and light-centered on posterior half of body.[2]

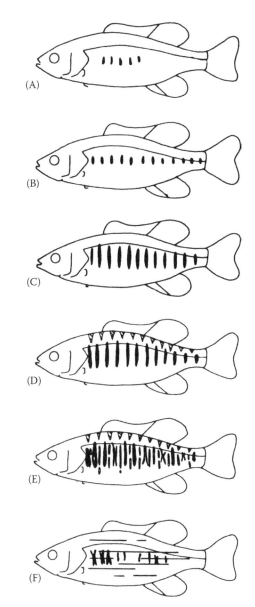

Figure 82 Development of young redeye bass — schematic drawings showing ontogeny of melanistic pigment patterns of young redeye bass. (A) 14.6 mm SL; (B) 23.1 mm SL; (C) 24.9 mm SL; (D) 40.9 mm SL; (E) 31.9 mm SL; (F) 85 mm SL. (A–F reprinted from Figure 3, reference 25, with publisher's permission.)

61.0 mm SL (64 days) (about 74 mm TL). Lateral bars almost indiscernible and lateral streaking on ventrolateral scale rows visible. Caudal spot vague or absent. Caudal band diffuse (Figure 83C).[2]

85.0–89.0 mm SL (about 103–108 mm TL). Longitudinal stripes dominate the pattern, although individuals may facultatively express bars.[25] Caudal spot absent.[2] Adult pigmentation is present on some individuals this length.[2]

Juvenile specimens usually have white iridescent color at the posterior margin of the caudal fin, and the soft dorsal, anal, and caudal fins all are washed

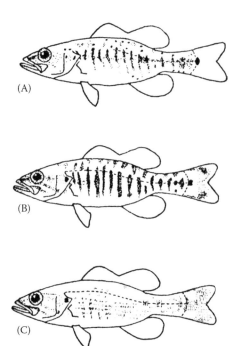

Figure 83 Development of young redeye bass — details of pigmentation for redeye bass from the Tallapoosa–Alabama River drainage. Body proportions are constant in the drawing outlines and do not reflect proportional differences between sizes. (A) 19.0 mm SL; (B) 36.9 mm SL; (C) 60.4 mm SL. (A–C reprinted from Figure 5, reference 2, with publisher's permission.)

with pale red-orange or orange in life.[2] Coloration of the smallest scaled young redeye bass from the Apalachicola River drainage is about the same as described above for Alabama River stock. Lateral bars persist vaguely into late juvenile phase and were somewhat broader and more widely spaced.[2]

TAXONOMIC DIAGNOSIS OF YOUNG REDEYE BASS

Similar species: other centrarchids, especially other *Micropterus* species; also superficially similar to elassomatids.

For comparisons of important morphometric and meristic data, developmental benchmarks, and pigmentary characters between redeye bass and all other centrarchids and elassomatids that occur in the Ohio River drainage, see Tables 4 and 5.

For characters used to distinguish between young basses and the young of other centrarchid genera and *Elassoma*, see provisional key to genera in "Taxonomic Diagnosis of Young Sunfishes, Black Basses, and Crappies" section (pages 27–29).

For identification of young basses in the Ohio River drainage, see "Provisional Key to Young of *Micropterus* Species" in genus introduction (pages 249–250).

ECOLOGY OF EARLY LIFE PHASES

Occurrence and Distribution (Figure 84)
Eggs. Eggs are laid in nests or shallow depressions in coarse gravel in eddy waters at the heads of pools[3,5,6] where they are protected by the male until hatching.[11]

Yolk-sac larvae. Swim-up fry (7–8 mm TL) were visible in the nest 5 days after hatching.[1] Male parent stays with larvae until they leave the nest, usually

Figure 84 Areas where redeye bass have been introduced into the Ohio River system (shaded areas) and areas where early life history information has been collected (circles). Numbers indicate appropriate references. Asterisk indicates TVA collection locality.

about a week.[11] Newly hatched larvae remain in schools for only a short time.[3]

Larvae. Some larvae broke away from schools 6 days after swim-up at 12–18 mm TL.[1]

Redeye bass larvae 8–12 mm TL were collected with dip nets from Falling Water River, TN, in late April. The larvae were suspended throughout the water column near an empty nest in water about 0.5 m deep. The collection was made at the lower end of a pool. Larvae 11.1–13.2 mm TL were collected in early May near the shore in shallow water and slow current.*

Juveniles. Complete school dispersal occurred by 14 days after swim-up when the young redeye bass were 16–25 mm.[1] Fry from Apalachicola River brood raised in ponds appeared over beds 10 days after spawn, but subsequently disappeared from surface and inshore areas, apparently moving to deeper waters. Fry from Alabama River brood raised in ponds inhabited shallow water at the pond's edge, especially areas where aquatic vegetation provided cover.[1]

Juvenile redeye bass occur in shallow runs and riffles over sand and gravel substrates.[13] Specimens 26.4–32.1 mm TL were collected along the shoreline of Falling Water River in shallow water with gentle current over gravel substrate.*

Early Growth

Growth of young redeye bass is greatest in the first year and gradually decreases with age (Table 97); growth is slow compared to other game fishes.[3]

Mean TL at capture of YOY redeye bass from Roaring River, TN, was 89 mm for age-1 fish, 118 mm for age 2, 149 mm for age 3, 186 mm for age 4, and 211 mm for age 5.[7]

Table 97

Average calculated TL (mm) related to age of young redeye bass from several locations.

Location	Average TL at Each Annulus			
	1	2	3	4
Spring Creek, TN (1973)[5,6]	59	93	120	160
Spring Creek, TN (1965)[6]	62	105	138	179
Roaring River, TN[7]	63	99	131	167
Halawakee Creek, AL[24]	76	131	215	301
Shoal Creek, AL[16]	61	100	129	173
Little Shoal Creek, AL[16]	51	90	124	154
Keowee Reservoir, SC[26]	125	226	270	309
Jocassee Reservoir, SC[26]	122	216	276	313
Chipola River, FL[24]	97	206	292	331

In the Conasauga River system, redeye bass grow to about 55 mm TL during their first year and an additional 25 mm or so per year for the following 3 years.[11]

Wild-caught redeye bass from Falling Water River (Cumberland River drainage), TN, were 8–10 mm TL on April 29, 11.1–13.2 mm on May 8, and 26.4–32.1 mm on June 19, 1986.*

Alabama River redeye bass spawned and hatched in AL ponds grew to 7–8 mm TL in 5 days, 16–25 mm in 23 days, and 53 mm in 99 days.[1] Growth in SL in AL ponds was also reported: 19 mm in 22 days; 37 mm in 35 days; 60 mm in 64 days; and 87–106 mm in 134 days.[2]

Young redeye bass, the result of Apalachicola and Alabama River brood spawning in ponds, were cultured in earthen rearing ponds with fathead minnows as forage and exhibited the following growth (mm TL and weight in grams): Apalachicola bass 21–24 mm stocked at 300/ac grew to 142 mm and 44.5 g in 191 days; Apalachicola bass 21–24 mm stocked at 80/ac grew to 169 mm and 49.5 g in 190 days; and Alabama bass 22–25 mm stocked at 300/ac grew to 134 mm and 19.1 g in 224 days.[1]

Minimum and maximum total lengths for second-year, pond-reared redeye bass were 152 and 245 mm. These lengths were generally much greater than those reported for redeye bass from streams (Table 97), but this growth can be accounted for, at least in part, by warmer water and a greater abundance of food.[17] The greatest growth rate reported from stream populations came from the warmer waters of the Chipola River, FL (Table 97), postulated to be the result of a long annual growing season, large stream habitat, clear water, and an abundance of productive shoals.[24] Growth of young redeye bass in SC reservoirs was faster in comparison with that of stream fish (Table 97) and was related primarily to a piscivorous diet beginning at age 2.[26]

Length–weight relationships of young redeye bass from TN and AL appear comparable (Tables 98 and 99).

The following length–weight relationships were reported for young redeye bass from the Chipola River, FL: at 191–201 mm TL, average weight was 88 g; at 203–213 mm TL, average weight was 125 g; at 267–277 mm TL, average weight was 247 g; at 279–290 mm TL, average weight was 269 g; and at 292–302 mm TL, average weight was 369 g.[24]

After 6 months in AL ponds, stocked redeye bass had grown from an average of about 21 mm TL and weight of 0.14 g to about 176 mm and 73 g; available forage consisted of bluegill fingerlings and fathead mnnows.[21]

Additional growth information is reported in Carlander (1977).[15]

Table 98

Length–weight relationships of young redeye bass from Sheed's Creek, TN (Conasauga River drainage), 1950–1952.

Average TL (mm)	Number	Average Weight (g)
81	5	6
91	5	8
102	9	11
112	11	17
127	7	21
140	9	32
150	4	42

Source: Constructed from data presented in Table 1 (in part), reference 3.

Table 99

Length–weight relationships of young redeye bass collected from AL during the period 1949–1964.

Average TL (mm)	Number	Average Empirical Weight (g)
51	54	1.8
76	603	5.0
102	235	8.2
127	12	24.1
152	18	45.4
178	5	72.3
203	4	118.0

Source: Constructed from data presented in unnumbered table on page 52, reference 14.

Feeding Habits

Zooplankton is the main diet of all young bass until they are old enough to engulf small fish or other animals.[22]

In streams young redeye bass feed primarily on terrestrial and aquatic insects.[18] In ponds they reportedly forage on fathead minnows,[1] but in one report, young redeye bass stocked in a pond containing an abundance of fathead minnows ate 85% insects and only 9% minnows.[21] Redeye bass age 2 and older from Keowee and Jocassee Reservoirs, SC, were primarily piscivorous, threadfin shad composing most of their diet.[26]

LITERATURE CITED

1. Smitherman, R.O. and J.S. Ramsey. 1972.
2. Ramsey, J.S. and R.O. Smitherman. 1971.
3. Parsons, J.W. 1954.
4. Lambert, T.R. 1980.
5. Gwinner, H.R. et al. 1975.
6. Gwinner, H.R. 1973.
7. Cathey, H.J. 1973.
8. Lee, D.S. et al. 1980.
9. Burr, B.M. and M.L. Warren, Jr. 1986.
10. Pipas, J.C. and F.J. Bulow. 1998.
11. Etnier, D.A. and W.C. Starnes. 1993.
12. Bulow, F.J. et al. 1988.
13. Mettee, M.F. et al. 2001.
A 14. Swingle, W.E. 1965.
15. Carlander, K.D. 1977.
16. Catchings, E.D. 1979.
17. Boschung, H.T., Jr. and R.L. Mayden. 2004.
18. Hurst, H. et al. 1975.
19. MacCrimmon, H.R. and W.H. Robbins. 1975.
20. Ramsey, J.S. 1975.
21. Smitherman, R.O. 1975.
22. Miller, R.J. 1975.
23. Koppelman, J.B. and G.P. Garrett. 2002.
24. Parsons, J.W. and E. Crittenden. 1959.
25. Mabee, P.M. 1995.
26. Barwick, D.H. and P.R. Moore. 1983.
27. Wood, R. et al. 1956.
28. Webb, J.F. and W.C. Reeves. 1975.

* Developmental descriptions and comments concerning early life ecology and growth of young redeye bass are from specimens wild-caught from Falling Water River, Cumberland River drainage, TN. Some wild-caught specimens cultured in aquaria in TVA's laboratory were also used for descriptive purposes. Specimens are housed at the Division of Fishes, Aquatic Research Center, Indiana Biological Survey (catalogue number TV3083), Bloomington, IN.

OTHER IMPORTANT LITERATURE

Hurst, H.N. 1969.

SMALLMOUTH BASS

Micropterus dolomieu Lacepede

Robert Wallus

Micropterus, Greek: *micro*, "small" and *pteron*, "fin"; *dolomieu*, French: a patronym for M. Dolomieu, a French mineralist for whom the mineral Dolomite was named.

RANGE

Smallmouth bass originally ranged from the Great Lakes and St. Lawrence River drainages in Canada south to the Tennessee River drainage of northern AL and GA, west to eastern OK, and north into MN.[73,75,76] Eastern distribution was restricted by the Appalachian Range. It was generally absent from the Mississippi River lowlands in southern IL, IN, and AR, but to the west in the Ozark highlands, from central MO to southwestern AR and extreme eastern OK, the smallmouth bass was abundant in all forested uplands. Its range has been greatly extended by introductions in North America and abroad.[75,76]

Smallmouth bass can be intensively cultured[94,95] and it has been introduced, and self-sustaining populations established, across the continental United States and Canada,[13,24,26,70,110] and in Hawaii, Asia, and Africa; it has been introduced to Europe and South America, but its status there is questionable.[73]

HABITAT AND MOVEMENT

Smallmouth bass is a natural inhabitant of cool, clear[47,73,77,83,139,142,146,150,164] upland streams, rivers,[47,73,77,83,139,142,146,150] reservoirs, and lakes.[47,73,77,142,146] It is found in streams of permanent flow,[150,164] often in moderate[73] to fast[139] current such as riffles, runs, and pools below riffles,[83,143,146] near shelter and cover[77] such as submerged logs, stumps, or rock outcrops.[47] Substrate preferences for smallmouth bass appear to be site specific[26] but they are generally reported from substrates of rock, rubble, and gravel.[73,77,139,142,146,150,164] They often become a dominant species in reservoirs that impound streams with the above attributes[142,146] and are prolific in many natural northern lakes.[146] In streams they are usually found in water more than 1.2 m deep.[143]

In streams, smallmouth bass seem to be attracted to dark, quiet waters and are usually found in the lee of structure at or near the edge of current; usually not found in areas of strong current.[73] The largest inland populations of smallmouth bass in

OH occurred in stream sections where about 40% of the stream consisted of riffles flowing over clean gravel, boulder, or bedrock bottoms; where pools normally had visible current and maximum depth of more than 1.2 m. There were considerable amounts of water willow, and gradients were between 0.8 and 4.8 m/km.[140]

In the Tennessee River, AL, smallmouth bass were most abundant in deep main-channel habitats with rock substrate, moderate surface currents, cover in the form of broken rocks, boulders, and ledges, and access to deeper water. Habitats most frequently occupied were riprap, rock bluff, barge wall, and turbulent shallow areas. All sizes of smallmouth bass were associated with main-channel habitats that had continuous surface currents and access to deep water.[97] In riverine waters of the New River, VA, smallmouth bass used shoreline areas more than mid-river areas.[9]

Year-class abundance in the Tennessee River, AL, fluctuated greatly and was negatively related to discharge through Wilson Dam from April to July, which corresponds to spawning and post-spawn periods.[18] In VA rivers, the recruitment success of smallmouth bass was significantly related to mean June stream flow, using a nonlinear model, with strongest year-classes produced in years with moderate flows. Data from this study suggested that stream discharge during and immediately after spawning could be critical to smallmouth bass recruitment success.[36]

In a survey of a 4th-order tributary basin (about 44–square mile drainage) of the Kentucky River, smallmouth bass were not found in 1st- or 2nd-order streams, were rare in 3rd- and 4th-order streams, and limited to the main channel and larger tributaries.[68]

Variability in density and biomass of smallmouth bass in an Ozark stream was best associated with amounts of large substrate such as boulders and cobble, undercut banks, and aquatic vegetation. Variability in condition was best explained by water temperature and density of crayfish. Variability in proportional stock density (PSD, an index of population size structure) was best explained by the presence of woody structure, aquatic vegetation, and large boulders.[35]

Population densities of smallmouth bass have declined in streams of the MO Ozark Border region since the 1940s with replacement by largemouth bass occurring in some cases. Two habitat variables, known to have been influenced by human activities, largely explain present densities of the two species in the streams of this region. Densities of smallmouth bass declined with increasing maximum summer temperature (range, 23–33°C) and percent pool area, while largemouth bass densities increased with these variables. Increases in pool quantity are accompanied by reduced food production and the loss of prey important to smallmouth bass. Changes in growth conditions for both species may proximate cause for shifts in distribution.[11] In another report from the Ozark Border region of MO, maximum summer temperature and percent pool area of streams explained most of the variability in total density, adjusted density (fish >100 mm TL), and biomass of smallmouth bass; they were negatively associated with each variable.[115]

Stream communities in which smallmouth was the predominant bass had a relatively high number of species (median 63); median proportions of associated groups of fishes were: minnows 16%, darters 17%, suckers 10%, and sunfish 19%.[80]

In lakes, smallmouth bass are reported from steep, rocky[9,40,47,143] shorelines[13,47,123,143] along submerged river and creek channels[47] or over bars or submerged humps, shoals,[40,73,77,143] reefs, or rocky ledges in water depths ≥1 m.[73,143] They are typically found over bottoms of rock, sand, or gravel and definitely avoid muddy bottoms. They are usually associated with protective cover such as riprap or large rocks,[12,70,143] boulders, talus slopes, submerged trees, roots, and logs; infrequently associated with aquatic vegetation.[143]

The best smallmouth bass lakes are over 100 acres, not less than 6–9 m deep with thermal stratification, have clear or light brown water with scanty vegetation, have large shoals of rock and gravel or sand with patches of gravel, have an ample supply of food organisms such as small fish, crayfish, and insects, and have moderate summer temperatures, neither very cold nor excessively warm.[79,164] Among 25 ON lakes (0.29–11.42 km²), smallmouth bass tended to occur in deep lakes with proportionately small littoral areas.[45]

In a central MA lake, large smallmouth bass (>406 mm) were more often observed in deep water (>8 m) and at mid-depths (4–8 m) than smaller bass in summer. During early summer, small bass (248–279 mm) and medium-sized bass (305–356 mm) used cover more frequently than large ones and small smallmouth exhibited significantly smaller summer ranges than did large individuals.[120]

In Norris Reservoir, TN, smallmouth bass is reported to be an open-water species.[2,3] Specimens collected in gill nets in March and April were about evenly distributed between the surface and 12.8 m; few fish were caught in gillnets in the summer and those caught in October ranged in depth from 2.4 to 15.8 m.[2] Abundance of smallmouth bass, based on spring and fall electrofishing catch rates in Normandy Reservoir, TN, was greatest in riprap habitats; abundance was uniform and consistently low in all non-riprap habitats.[12]

In Cave Run Lake, KY, smallmouth bass were subject to summer stress when deprived of suitably oxygenated water at preferred temperatures.[29] A distinct longitudinal distribution was observed for electrofishing catch rates of smallmouth bass from the lake; they were significantly more abundant in the oligotrophic lower lake area.[8,29] Abundance in electrofishing and angler catches declined in more fertile areas of the middle and upper lake. This difference may also have been affected by the availability of smallmouth bass habitat in the lower lake area.[8] Smallmouth bass introduced into Skiatook Lake, OK, also seemed to respond to reservoir water quality, being restricted to the clear water of the lower reservoir.[13]

The presence of smallmouth bass in small lakes (≤50 ha) in central Ontario appears to reduce abundance, alter habitat use, and extirpate many small-bodied fish species.[69]

Smallmouth bass crops in reservoirs were negatively related to those of gizzard shad, blue and flathead catfish, redbreast sunfish, crappies, longnose gar, and largemouth and spotted bass. Smallmouth crops were positively related to redhorse, spotted sucker, and rainbow trout crops. Smallmouth bass crops appear highest in older, larger, less fertile reservoirs with shorter growing seasons. Higher crops are also linked with greater annual mean water level fluctuation in more bowl-shaped reservoirs.[78]

Considered by some a habitat generalist,[118] smallmouth bass is also reported from small ponds and turbid streams[73] with warm water.[47,146] In the summer, tagged smallmouth bass were collected in warm shallow habitats of Melton Hill Reservoir, TN, and were often associated with submerged or overhead cover.[38]

In WY, altitude is a determinant of distribution; smallmouth bass become established at altitudes of less than 1,900 m above mean sea level where an agricultural growing season of at least 100 days occurs.[32]

In Lake Ontario, climate and global warming may affect year-class strength and relative abundance of smallmouth bass. During the period 1973–1996, the strongest year-classes came from El Niño years and some weak year-classes came from La Niña years. Year-class strength and July–August water temperatures were positively correlated and highly significant.[126]

Coutant (1975 and 1977) summarized reports of temperature preferenda for smallmouth bass in laboratory studies; range for final preferendum was 12–30°C depending on season. Upper and lower avoidance temperatures were 33 and 26°C.[50,90] In western Lake Erie, laboratory-tested adult smallmouth bass occupied waters 30–31°C in summer, 21–27°C in fall, 13–26°C in winter, and 18–26°C in spring.[63] Preferred temperature for smallmouth bass has also been variously estimated between 20.3 and 28°C.[143,144]

In two VA rivers preferred temperature ranged from 20.2 to 31.3°C depending upon acclimation temperature (Table 100).[167]

Summer field studies suggest that smallmouth bass prefer temperatures near 21°C in lakes and streams.[90,149] Because of this preference, in northern waters they are found near shore in summer, whereas in warmer southern waters they seek deeper water habitats.[90] Smallmouth bass generally became inactive at temperatures below 10–15°C.[149] In the Laramie River, WY, where water temperatures exceeded 16°C for 102–135 days (range for three stations) and 20°C for 16–38 days during 1992, smallmouth bass were able to reproduce and recruit, but their growth was very slow.[23]

Laboratory studies demonstrated that the presence of food and cover (presented in separate experiments) significantly affected temperature selection by smallmouth bass. When limited amounts of food were presented in water with higher than preferred temperature, they stayed longer at the high temperatures, whereas fish in this environment that were allowed to feed until satiated

Table 100

Preferred and upper and lower avoidance temperatures (°C) for smallmouth bass from the New and East Rivers, VA, at acclimation temperatures from 30 to 15°C.[167]

Acclimation Temperature	Upper Avoidance Temperature	Preferred Temperature	Lower Avoidance Temperature
30	33	31.3	26
27	31	30.1	24
24	31	29.8	22
21	30	26.5	20
18	27	22.9	17
15	26	20.2	13

retreated to the cold end of the experimental tank for most of the time. When cover was present at the warm end of the tank, smallmouth stayed in this area five times longer than when no cover was present in the tank.[117]

Carlander (1977) summarizes additional reports concerning temperature selection by smallmouth bass.[52]

Hypoxia tolerance mean for SMB is reported as 1.19 mg/L; hyperthermia tolerance mean 36.9°C.[64] Smallmouth bass are not reported from waters with a pH below 6.0.[91]

Age at quality length for smallmouth bass (280 mm), used as an index of growth rate, is positively correlated with latitude and negatively correlated with mean air temperature and degree-days exceeding 10°C; natural mortality is not correlated with these variables.[25]

In spring in OH, as water temperatures warmed to above 4°C smallmouth bass moved upstream to spawn.[140] In the Great Lakes, pre-spawning movement inshore and into bays and tributaries begins when water temperature rises above 4°C, with peak movement occurring at 12.8°C. After spawning, adults move downstream or offshore to depths of 11–13 m as water temperature approaches 25°C (usually by July).[65]

Smallmouth bass move to deeper water when water temperatures begin to fall; this migration may begin when temperatures are as high as 15.6°C, but becomes pronounced at water temperatures of 10°C and lower.[73,164] In the Shoals Reach of the Tennessee River, AL, during spring and summer, smallmouth bass were most abundant in shallow water habitats; as temperatures cooled in the fall and winter, abundance decreased in shallow habitats and increased in deeper areas.[97] Another report indicates that decreasing water temperatures in the fall may stimulate smallmouth bass activity in littoral zones, especially larger fish.[60]

Winter movements of smallmouth bass to deeper water are often accompanied by aggregation behavior. Entrance into caves, crevices, or fissures in the substrate are also reported.[77]

In the northern part of its range, smallmouth bass return to shallow water in spring about the time ice disappears.[73] In WI, smallmouth bass >200 mm TL (age 2 and older) migrated from the Embarrass River downstream 69–87 km to the Wolf River when water temperature fell below 16°C in autumn. They returned to the Embarrass River in April and May, and most individuals returned to the same 5-km reach of river.[30] Abundance of smallmouth bass in an Ozark spring branch was greatest during cold-water months (November–March) when temperatures in the spring branch exceeded the temperature of the receiving stream.[21]

Movement tendencies differ dramatically among stream-dwelling smallmouth bass populations. In summer, they typically remain in localized areas, with net movements less than 1 km. In the Otter Creek/Pecatonica River system in southwestern WI, movements were less than 200 m. Some smallmouth populations remain sedentary while others migrate more than 75 km to reach winter habitat. Migration distance is correlated with winter severity; smallmouth bass move little in streams that do not freeze, but travel more than 5 km in systems with ice. In the fall, most smallmouth left Otter Creek and entered the much larger Pecatonica River where they over-wintered; mean net movement between summer and winter habitats was 6.5 km. In the Otter/Pecatonica system smallmouth occupied slow-moving runs (maximum depth 0.9–1.8 m) with limited cover during the winter and had home ranges that averaged 299 m in length.[131]

In Center Hill Reservoir, TN, only surface temperature of the water was found to be significantly related to the rate of smallmouth bass movement. Increased movement and activity were associated with water surface temperature between 11.1 and 23.9°C. In this study, only one of five tagged fish that was displaced returned to its capture site. For the 11 fish tracked, the daily averaged distance moved varied from 15 to 1,208 m; mean distance moved per day was 345 m. Majority of movement was during daytime. Movement seemed not to be affected by noise or light.[43]

In Melton Hill Reservoir, TN, in the summer, though expected to inhabit areas with optimal temperatures except when food availability required them to move to areas of less-desired temperatures to feed, no significant changes in temperature or depth were observed during daily tracking sessions. Instead of demonstrating diel movements offshore and inshore, smallmouth bass remained in relatively warm water (>28°C) even though cooler water was available. Activity peaked in the afternoon and was minimal at night.[38]

The behavior and physiology of smallmouth bass were monitored in the discharge canal of a thermal generating station on Lake Erie during summer, winter, and the spring transition period. At lower temperatures during winter and spring, smallmouth spent most of their time in the warmest and most thermally variable areas of the discharge canal. At these times cardiac function was highly correlated with temperature. As temperatures increased in the summer this trend began to reverse with a decrease in correlation between cardiac function and temperature until a critical point was reached at 25–30°C, where most smallmouth attempted to locate thermal refugia by leaving the discharge canal.[134]

Reports indicate that smallmouth bass leave daytime habitats and move to quiet waters on the bottom where they assume positions on or beneath cover and exhibit little if any nocturnal activity.[77]

In Jacks Fork River, MO, smallmouth bass exhibited definite patterns of diel activity and habitat use that were modified by seasonal changes in water temperature. Fish remained in restricted home ranges for most of the year except in spring when all tagged fish left the home pool; 75% returned during the same season. Upstream and downstream movement was about equal, but the average distance moved upstream was greater. Intrapool movement peaked soon after sunrise and again after sunset in all seasons. Average daily movement in the pool was greater when water temperatures were highest. Movements in floods did not differ from those observed during normal flows.[104] In warmer seasons, smallmouth bass were associated with logjams and root wads during the day and increased their use of boulders by night. In cooler seasons, boulders were used almost exclusively. Boulders were the most preferred substrate and gravel the least. Smallmouth bass used intermediate depths the most and showed no daily or seasonal changes in depth preference. They preferred velocities <0.2 m/s at all times of day and in all seasons.[104]

Smallmouth bass were classified as a sedentary species in a MO study; tagged fish remained near the release point; fish of intermediate age were more mobile than younger or older fish.[1] Other reports indicate that in streams, a smallmouth bass may occupy a single pool throughout an entire season.[73,74] In an IL creek, 80% of fin-clipped smallmouth bass retaken the same year and 67% of those retaken the following year were in sections of the stream in which they had been originally marked.[74]

In both streams and reservoirs, smallmouth bass may establish home ranges of several hundred meters.[47] In large lakes, smallmouth bass often occur as several distinct populations.[73] They occupied discrete home range areas in Meredith Reservoir, TX. From July through February they occupied areas ranging from 1.32 to 43.23 ha, but moved up to 6.5 km during spring months. Minor inshore movements occurred during spring and fall, and minor offshore movements in summer and winter.[40] In a north temperate lake, smallmouth bass demonstrated significant homing tendencies.[122]

In Ontario, 15 of 18 displaced smallmouth bass returned to home ranges they held before removal; displaced fish remained at the release site for an average of 1 week and returned to their home ranges in about 4 days. The authors concluded that smallmouth bass displaced by tournament anglers will return to their home ranges and use them in a manner similar to that of undisturbed fish.[22]

Evidence of homing was obtained in an IL creek. Smallmouth bass transferred overland and released in other parts of the stream showed an ability to return to their home pools from either upstream or downstream. Twenty-three bass were involved in 31 transfers. Of the 31 transfers, 17 (55%) were followed by homing responses. Four of seven fish that were moved twice returned a second time to their home pools. One fished moved three times returned three times to its home pool.[74]

Thirty-nine smallmouth bass were radio-tagged in May and June (2000) after spawning in the Kentucky River, KY. These fish exhibited both migratory (69%) and sedentary (31%) behavior. In 2001, 15 smallmouth bass tagged in late March, prior to spawning, were mostly year-round residents with only 20% making limited movement out of the capture pool. Fifteen smallmouth bass were displaced from Elkhorn Creek, a tributary of the Kentucky River, to observe and quantify homing; 60% of these fish returned to the creek from which they were displaced.[16]

Distribution of smallmouth bass during the spawning season in a VA impoundment was classified by three patterns of movement: (1) sedentary; (2) down-lake movements; and (3) up-lake movements. In 2 years of study, 50% of the tagged fish did not leave the lake section in which they were tagged, 27% moved down-lake, and 23% moved up-lake. After the spawning season ended, all fish that did migrate and were not harvested returned to the lake section in which they were tagged. Spawning fidelity was determined from 6 smallmouth bass tracked for two consecutive years; 3 of these fish returned to the same sites during both spawning seasons.[39]

Smallmouth bass fisheries respond favorably to appropriate regulations; heavily exploited populations can substantially increase in biomass and can produce larger fish, and higher quality fishing can be provided if effective protection of bass stock is achieved.[82]

DISTRIBUTION AND OCCURRENCE IN THE OHIO RIVER SYSTEM

Smallmouth bass has been collected from throughout the Ohio River since the early 1800s.[57] It is present and common throughout the length of the river[151] but more abundant in the upper two-thirds of the river than in the lower third.[56,141] It was probably more abundant in the Ohio River before impoundment. It is the least abundant of the three black basses in the Ohio River.[141]

Smallmouth bass were commonly collected (20 or more specimens per year) near Ohio Edison Company's electric generating plant W.H. Sammis

(ORM 54) from 1981 to 1987 and thereafter (1988–1993) became abundant in collections (100 or more specimens per year).[152,153] During the period 1981–2003, smallmouth were common or abundant in collections near ORM 77 (Ohio Power Company's Cardinal Plant); they were abundant most years near this site, especially after 1985.[152–154] Though uncommon during the same period (1981–2003) from collections near ORM 260 (Ohio Valley Electric Corporation's Kyger Creek Plant), they were common in collections most years near ORM 494 (Indiana–Michigan Power Company's Tanners Creek Plant).[152–154] They were also collected in low numbers at TVA's Shawnee Steam Plant at ORM 946 (1987–1992).[152]

Based on lock chamber surveys conducted during the period 1957–2001, abundance of smallmouth bass has increased in the Ohio River since 1957. Smallmouth bass is considered pollution intolerant and these changes in abundance coincide with a marked improvement in the water quality of the Ohio River over the last 50 years.[62]

Smallmouth bass is present in the Wabash River, IN, but common only in smaller, clear, clean tributaries;[58] its distribution is sporadic in the Wabash River system in IL.[139]

It is probably absent from the lowland environments of southwestern IN.[168] It is generally distributed in upland streams throughout the eastern two-thirds of KY; occasionally in the Land-Between-the-Lakes area where it was probably more common before impoundment.[142] In VA, smallmouth bass is native to the big Sandy and Tennessee River drainages; not native to the New River.[146] It occurs throughout the Tennessee and Cumberland River drainages of TN[47] and occurs naturally in AL only in the Tennessee River system. The 20-km section of the Tennessee River downstream from Wilson Dam (often referred to as the "Shoals Reach") has a national reputation as an excellent smallmouth sport fishery.[148]

SPAWNING

Location
Smallmouth bass spawn in lakes, reservoirs, and streams and may ascend tributaries to spawn.[73] In the Great Lakes, nests are built in clear water of tributaries, river mouths, bays, harbors, shores, or shoals; spawning may also occur on harbor breakwalls.[65]

Nests are usually built in areas of good water movement,[65,124] protected from wind[158] and wave action,[65,157,158,164] near the margins of streams or lakes.[47,84,124,143] Nests are excavated in clean gravel,[47,48,65,84,97,124,143,144,158] rock,[47,48,65,97,124,143,144,158] rubble,[65,84,143] or sand.[65,123,143,144] Smallmouth bass will not spawn on heavily silted substrate.[91] Nests are usually built close to wood or rock cover[130,157,164] such as boulders, logs, stumps, fallen trees, docks, or other such structures, sometimes among rooted macrophytes.[65,124,143,144,158] Nests may be less than a meter deep in streams[97,157] and are variously reported from less than 1 m to 6 m deep in lakes and reservoirs.[47,48,65,77,97,130,143,144,150,157,158,164] Some males return to the same nest in subsequent years and over 85% return to within 150 yards of where they nested in previous years.[144]

Some smallmouth bass populations make annual spawning migrations into small streams from lakes or rivers.[82,140,143] This occurrence is not typical of stream bass and seems to be restricted to special habitats where spawning runs are necessary to find suitable spawning substrates. Smallmouth bass that migrate from lakes or large rivers into tributaries to spawn may occupy more habitat types and provide larger fisheries than sedentary populations that complete their entire life cycle in small segments of streams or lakes.[82]

In NS, smallmouth bass nests were reported from four different habitats: (1) shallow (gradient <10%) with cover consisting of logs, stumps, or boulders; (2) shallow with no cover; (3) deep (gradient ≥10%) with cover; and (4) shallow, marshy areas. Nest density was highest in shallow areas with cover and gravel–cobble mix compared with mud or silt substrate and lowest in marshy areas. Mean water depth at nest sites and mean distance from shore in favorable habitat were 130 cm and 18.7 m. In this study 76–100% of all nests observed in the four habitats were located close to boulders, logs, stumps, or other cover.[24]

In OH, smallmouth bass spawn in streams 6.1–30.5 m in average width, having gradients of 1.3–4.8 m/km. Spawning populations were low or absent in streams with gradients less than 0.6 m/km and also in streams with very high gradients.[140]

Season

In Canada, smallmouth bass usually spawn over a period of 6–10 days in late spring and early summer; most often late May to early July.[24] In Ontario lakes, spring water temperatures and the size distribution of mature males can predict both temporal and spatial variations in the timing of spawning.[135] In NS, nest building and spawning commence around the end of May and early June; egg deposition peaked around mid-June and continued until late June.[24]

In the Great Lakes spawning is reported from March to mid-August,[65] but typically occurs April–July.[48,65,123,143] In MI, the timing of spawning depends on latitude and water temperature but spawning is usually from late April to late June.[124]

Smallmouth bass spawned during the period April through May in KS[137] and April through June in AR.[77,150] Nesting in an Ozark stream in MO was generally between mid-April and early June, but timing and duration of the nesting period varied considerably from year to year.[84] Spawning occurred in May or June in IL.[139]

In the southeast, spawning occurs in April and May[47,54,72,97,138,146,149] or into early June;[106] early April to early May in AL farm ponds.[54]

Smallmouth bass may spawn several times each summer.[58] In a MO Ozarks creek, two nesting periods were reported: the first was April 26 to May 31 and the second began about June 5 and ended in the first week of July.[157] In a MA lake, two spawning periods were reported in 1 year of a 2-year study. The major spawning period both years occurred in the last week of May; the second period occurred in mid-June after a drop in water temperature.[158]

Spawning is sometimes delayed by high flows and drops in temperature.[72]

Temperature

Several reports indicate year-to-year differences in reproductive success of smallmouth bass are closely dependent on summer air temperatures in the year of hatching.[81] Temperature thresholds for spawning are variable.[145] Spawning usually occurs on rising water temperatures and has been observed over a range of 12.8–26.7°C.[48,73,97,98,106,124,146,158] Though one report suggests that spawning is probably not successful below 18.3°C,[143] there are many reports of spawning in the 15–18°C range or lower.[24,47,58,65,77,90,97,124,144,148,164]

In MO[84] and OH,[86] nesting was usually preceded by an abrupt[84] or progressive rise in water temperature over several days,[86] and began when daily minimum temperature reached 12.8°C[84,86] and daily maximum temperatures were near 18.3°C.[84]

Other reports of spawning temperatures follow. Many additional references are summarized in Carlander (1977).[52]

Spawning is generally reported at 15–18°C in northern waters.[90] In Canada, nest building and, in some areas, spawning begin over a range of temperatures from 12.8 to 20°C, but egg deposition takes place mostly at 16.1–18.3°C.[144,164] In NS, nest building and spawning commence when water temperature reaches 16–18°C.[24] In the Great Lakes region spawning usually does not begin until water temperature reaches about 16.7°C.[65] Spawning occurs at 15–18.3°C in MI.[124,164]

There are reports of spawning at 14–21°C in AR[77] and 16°C in IN.[58] Spawning has been observed at 15–22°C in VA.[106,146] In a regulated stream in VA, temperature and flow exerted dominant influences on the time of spawning.[10] In TN, spawning is apparently induced by rising water temperatures and generally occurs at 15–18°C.[47] Smallmouth bass spawn in water 16–18°C[97,148] or higher in AL;[148] 22.7–25.6°C in AL farm ponds.[54]

Fecundity

Egg number for smallmouth bass depends on the size of the female and is reported to approximate 7,000–8,000 eggs per pound of female.[52,144] Reports of fecundity range from 2,000 to 27,716.[48,52,73,81,82,144,145,147]

A 305-mm TL female smallmouth bass is reported with about 4,500 eggs.[81] In WI, females 335–414 mm TL had 4,896–5,402 eggs.[145] Estimated numbers of ova for AL females 305–521 mm TL were 2,601–27,716.[147]

A 300-mm TL female smallmouth bass weighing 661 g, with ovaries that weighed 51 g, had 5,440 mature eggs; a second specimen with ovaries that weighed 38.4 g had 3,664 mature eggs.[124]

Carlander (1977) summarizes additional information on smallmouth bass fecundity.[52]

Sexual Maturity

Most smallmouth bass mature in their 3rd or 4th year[137,145] at lengths of 229–305 mm.[137] Some reports have males maturing at ages 2–4 and females at ages 3–5.[73,148,163] Size rather than age seems to be the determining factor in sexual maturity of smallmouth bass,[81,113,145] with some reports indicating smallest sizes from 185 to 254 mm TL for males[73,106,145] and 254–305 mm TL for females.[73,81,145]

In Lake Ontario–St. Lawrence River waters, the smallest reported mature male was about 197 mm TL; smallest mature female is reported at about 254 mm TL.[73] In the Great Lakes and other large northern lakes, males mature in 3–4 years at 254–259 mm TL and females in 4–5 years at 305–325 mm TL.[81,145]

In VA, nesting males were as small as 185 mm TL and weighed as little as 74 g.[106]

Spawning Act

The male smallmouth bass builds a nest[58,157,164] with sweeping motions of the tail.[157,164] When completed, the nest is a circular, saucer-shaped depression, 51–102 mm deep and 0.3–1.8 m across[73,143,144,157] (usually about 0.6 m across)[73,143,157] and an average of 1.2 m from escape cover and 1.4 m from shore.[157] Spawning may occur as soon as the nest is constructed, but it may be delayed a week or more depending on temperature and availability of a ripe female.[73,157,164]

Females approach spawning areas from deeper water. When the male sees a female, he rushes toward her and attempts to drive her to the nest.

This behavior may not be successful at first but if she returns, he repeats the process. As time passes, the pair will remain together progressively longer and begin to approach the nest. As the pair moves slowly over the nest, the male may gently nudge the female and bite her frequently, but gently, on the opercle, cheek, and corner of the mouth.[73,124] They swim slowly about over the nest; most of the time the female swims slowly on her side in a circle. Frequently she floats motionless, partly or wholly turned on her side; at such times the male often lies beside her.[124] Immediately before spawning the pair settle down over the nest, side by side, usually facing the same direction. The female turns partly on her side,[73,124] her body quivers,[158] and she discharges eggs near the vent of the male. The male emits milt simultaneously or within seconds. At each emission, which lasts 4–10 s, 20–50 eggs are discharged. Intervals between emissions may be 22–45 s, 3 min, or 5–15 min.[73,124]

Several females may spawn in the nest of a single male, resulting in the deposition of thousands of eggs.[47,58,124,164] A male may spawn with several females successively;[124,164] reported intervals between successive spawning events range from 1 to 36 h.[164] Spawning of one male simultaneously with two females has been reported.[73,158] In South Branch Lake, MA, a male spawned with two females of comparable size; 120 min with female 1, 35 min with both females in attendance, and 50 min with female 2. Female 1 went through the body quivering–egg release phase 94 times; female 2, 103 times. Simultaneous egg-laying occurred on several occasions.[158]

A female may spawn in more than one nest.[124]

After the eggs are deposited, the male guards the nest diligently, driving away intruders.[47,58,73,84] Brood-guarding smallmouth bass exhibit a generalized, aggressive response to any fish viewed as a potential nest predator.[67] The male guards the offspring from egg deposition to fry dispersal,[47,58,73,84] often a period of a month or more.[73]

A nesting smallmouth bass male swims in place constantly day and night over and around the nest.[111,124] Alternating movements of his pectoral fins as well as vibrating movements of his caudal fin gently fan the eggs and thus keep them free from sediment.[124] These energetic costs of nest protection are exacerbated by food deprivation and can possibly result in nest abandonment because of energetic constraints.[111]

In South Branch Lake, MA, protection time afforded by the male parent ranged from 14 to 47 days after hatching. Nest fanning by the male did not measurably increase egg survival but did reduce mortality of yolk-sac larvae. Males were reported in attendance with fingerlings (mean length 28 mm) that were well beyond the dispersal size.[158]

A second nesting and spawning period may take place if adverse weather conditions (floods, drops in water temperature) cause failure of initial spawning efforts.[73]

In a regulated VA stream, large male smallmouth bass spawned earlier than smaller ones.[106,165] Large males (>305 mm TL) accounted for the highest production of free-swimming larvae and also made the most re-nesting attempts after nesting was disrupted by high flows.[106]

Introductions of spotted bass[4,66,121] and redeye bass[103] into smallmouth bass waters have lead to hybridization.

EGGS

Description

In the spring before spawning, eggs of three sizes are found in the ovaries of smallmouth bass: (1) large, opaque yellow eggs about 2.5 mm in diameter and nearly ready to be laid; (2) medium-sized, white opaque eggs 0.5–1.5 mm in diameter, probably eggs of the next season; and (3) very small, transparent colorless eggs uniformly about 0.25 mm in diameter.[124]

Unfertilized eggs taken from the ovary of a 300 mm TL female from Lake Erie measured 1.2–2.52 mm in diameter, most were about 2.2 mm. They were round, semitransparent, light amber in color, with six to many large (largest 0.9 mm in diameter), clear dark amber oil globules. These eggs were not adhesive and only loosely joined together.[123]

Fertilized water-hardened eggs are demersal[48,71,73,143] and initially adhesive.[48,71,73,124,143] They ultimately lose their adhesiveness and settle into the substrate at the bottom of the nest,[143] becoming very difficult to see.[73] They have a narrow perivitelline space, and a light yellow–colored and transparent yolk. Embedded in the yolk is an oil globule about half the diameter of the yolk with numerous smaller oil droplets about its edges.[48,124,164]

The newly fertilized water-hardened egg is variously described as grayish white, white, opaque, light amber, or pale yellow.[48,73,143,164] Reports of diameter vary: 0.9–1.7 mm,[48] 1.8–2.2 mm,[49] and 2.2–2.8 mm.[124,164] Eight fertilized eggs from an AR fish hatchery averaged 2.44 mm in diameter.*

Incubation

Temperature greatly affects the development of eggs, by influencing both the rate of embryonic growth and the percentage survival.[90]

Smallmouth bass eggs hatch in 3–6 days at normal spawning temperatures;[47,48,90,158,164] about 3 days at spawning temperatures in MO.[157] In Lake Erie, 5–15 days depending on water temperature.[123]

Other reports of incubation period for smallmouth bass eggs include 2–10 days at water temperatures of 15–25°C;[148] 94 h at 19°C;[158] 2–15 days at temperatures of 12.8–21.1°C;[65] 9–10 days at 12.2–15°C; 6 days at 15.6°C; 3–4 days at 19.4–21.7°C; 2–2.5 days at 22.2–25.6°C.[73,143]

Incubation times (days) until hatching with constant temperatures were as follows:[162]

Temperature (°C)	Days
25.0	2.17
23.9	2.25
21.7	2.92
21.1	3.25
19.4	3.75
18.3	4.08
15.6	6.25
15.0	6.96
12.8	9.83

Carlander (1977) summarizes additional information concerning incubation of smallmouth bass eggs.[52]

Development

Eggs at 24 and 48 h after fertilization and 24 h before hatching are illustrated.[124]

YOLK-SAC LARVAE

See Figures 85, 86, and 87.

Size Range

Newly hatched smallmouth bass are reported between 4.5 and 5.5 mm TL.[48,71,124,164] They also are reported to hatch at 5.6–5.9 mm TL in Canada and completely absorb their yolk by 8.7–9.9 mm TL.[144] In MI, yolk is not completely absorbed at 9.3 mm TL.[124] Yolk absorption is completed by about 10 mm TL in southern populations, but remnants of oil may still be present at 10.8 mm.*

Myomeres

5.5 mm TL. Preanal 12; postanal 19; total 31.*

6.1 mm TL. Preanal 13; postanal 19; total 32.*

6.7–6.9 mm TL. Preanal 14–15; postanal 18–19; total 32–34.*

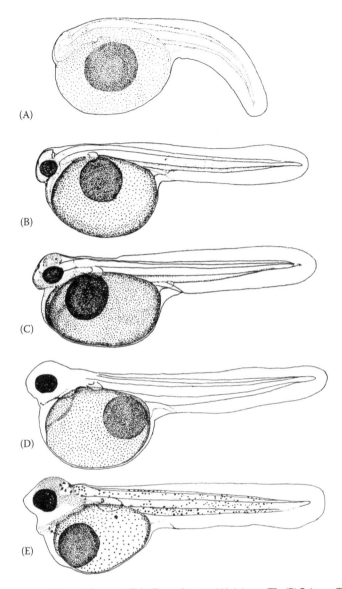

Figure 85 Development of young smallmouth bass. A–E. Yolk-sac larvae: (A) 4.6 mm TL; (B) 5.4 mm TL; (C) 6.0 mm TL; (D) 6.75 mm TL; (E) 7.5 mm TL. (Original artwork from Figures 4–6 and 8–9, reference 124, redrawn by Joan Ellis in Hardy, 1978.)

7.3–7.7 mm TL. Preanal 15–16; postanal 17–19; total 33–34.*

8.0–8.2 mm TL. Preanal 15–16; postanal 17; total 32–33.*

8.8 mm TL. Preanal 10; postanal incomplete, but at least 19.[123]

Note: On GA smallmouth bass 7.3–7.7 mm TL, pigmentation is dense enough dorsally and anteriorly on the torso to make it difficult to count the most anterior myomeres;* this could perhaps partially explain Fish's (1932)[123] lower preanal myomere counts.

Morphology

4.6–5.0 mm TL (newly hatched). Head deflected over yolk sac.[124]

5.4 mm TL. Head extends somewhat beyond yolk sac; parietal flexure is formed so that the head points downward.[124]

5.5–6.1 mm TL. Yolk sac is large and oval with a large oil globule (about 1.0 mm in diameter).* At 6.0 mm, the mouth cavity is formed but there are no jaws, and the head is directed nearly forward.[124]

6.7–6.9 mm TL. Torso curled; yolk sac is large, 2–2.25 mm long, with a large oil globule (1–1.3 mm in diameter).* Head is slightly depressed;* mouth is open; lower jaw partially formed.*,[124] The heart chamber is visible between the posterior margin of the lower jaw and the yolk sac.*

7.5 mm TL. Mouth large, open; lower jaw moves.*,[124] Yolk sac is large and oval and extends posterior to

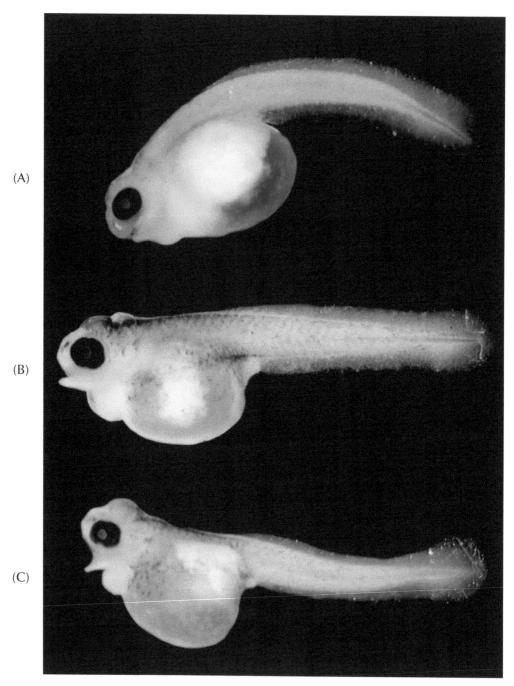

Figure 86   Development of young smallmouth bass from GA. A–C. Yolk-sac larvae: (A) 7.0 mm TL; (B) 7.5 mm TL; (C) 8.2 mm TL. (A–C, original photographs of young smallmouth bass from GA.)

anus on some individuals. A large oil globule is positioned posteriorly and dorsally in the yolk. The heart cavity and developing heart are visible under the head and anterior to the yolk. There is a notice-able notch in the dorsal profile at the juncture of the head and the occiput.*

8.0–8.3 mm TL. Yolk egg-shaped,[124] still promi-nent;* head and jaws relatively small;* snout rounded,* upper and lower jaws of nearly equal length; pericardial cavity nearly horizontal, under

and posterior to lower jaw.[124] Head has an indentation over the eye that gives a knobby-head appearance. Branchiostegal rays and operculum are developing.*

Mouth and operculum functional at 8.2 mm (onset of free-swimming in ON).[166]

8.6–8.8 mm TL. There are noticeable changes in body and head morphology from previous size range. Head and yolk regions are robust. Body has

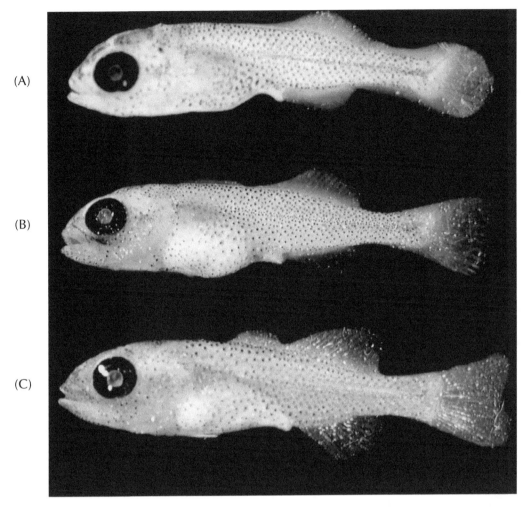

Figure 87   Development of young smallmouth bass from GA. (A) Late yolk-sac larva, 8.6 mm TL. (B) Post yolk-sac larva, 10.8 mm TL. (C) Late post yolk-sac larva, 15.0 mm TL. (A–C, original photographs of young smallmouth bass from GA.)

become bass-like, with a large head and mouth, and a compressed body posterior to vent. The eye is large and the mouth is oblique — snout pointed and adult-like; maxilla extends to middle of pupil. The operculum covers the gill chamber; intestine opens at edge of ventral finfold.*,[123] Oil is still visible — on most individuals, yolk, if present, is obscured by heavy pigmentation.*

9.0–9.7 mm TL. Reports indicate that most, but not all, yolk is absorbed by 9.0 mm[164] and that yolk is still present at 9.3 mm.[124] Pigmentation obscures yolk, but remnants of oil are still visible in all specimens examined from GA, AR, and TN.* Air bladder is developing.[124]

Morphometry
See Table 101.

8.8 mm TL. As percent TL: PreAL 45.5; GBD 20.5; ED 10.[123]

Fin Development
5.4–6.1 mm TL. Pectoral buds visible just posterior to the occiput.*,[124] The median finfold is present dorsally along the body from about the second myomere, is continuous around the urostyle and present ventrally to the posterior margin of the yolk sac.*

7.3–7.7 mm TL. Pectoral fins are triangular and shifted to 45° angle.[124] Median finfold originates dorsally at myomere 3–5 and is continuous to and around the urostyle and present ventrally to the anus. A small segment of finfold is present between the anus and the posterior margin of the yolk sac. The profile of the dorsal finfold is slightly elevated in the area of the future soft dorsal fin. The urostyle is straight; mesenchymal differentiation is visible on the ventral margin of the urostyle at the position of the future hypural complex. No other caudal fin development was noted. Developing pterygiophores are visible at the base of the future dorsal and anal fins. No other fin development noted.*

Table 101

Morphometric data for southern stocks of young smallmouth bass.* Data are presented as mean percent total length (TL) and head length (HL) for selected intervals of TL (N = sample size).

Length Range (mm TL)	TL (mm) Groupings									
N	5.5	6.1–6.9	7.3–7.7	8.0–8.8	9.2–9.8	10.2–11.7	12.0–13.6	14.1–15.9	16.3–17.8	18.0–19.7
	1	5	7	8	15	13	11	14	18	9
Length as percent TL										
SL	98.2	97.3	96.9	93.7	92.4	87.0	85.4	83.1	83.2	82.4
PreAL	47.3	46.0	46.2	49.8	50.3	51.6	51.7	52.1	52.6	51.8
GD	38.2	34.2	29.7	23.1	23.1	22.4	21.9	22.6	23.9	24.1
Depth at anus	11.8	11.6	11.8	15.6	16.7	17.5	16.7	17.8	19.3	20.3
CPD	4.5	5.3	4.8	7.0	7.7	8.0	7.9	8.7	9.3	9.9
ED	6.4	8.0	8.4	11.6	11.4	11.6	11.8	11.8	11.3	11.0
HL	14.5	13.9	14.8	25.4	25.4	27.9	28.6	29.7	30.3	30.7
Diameter as percent HL										
ED	43.8	57.6	56.8	45.9	45.1	41.4	41.3	39.9	37.4	36.0

8.0–8.2 mm TL. There is little change in the median finfold other than a slight elevation at future position of soft dorsal fin on some individuals. Flexion of the urostyle begins on some individuals.*

8.3–8.8 mm TL. Median finfold originates over posterior portion of yolk sac and continues around urostyle ventrally past the anus to the posterior margin of the yolk sac. Soft dorsal and anal fin profiles become visible as elevations in the finfold.*[,123] Caudal fin lophocercal;[123,124] urostyle is flexed to about 45°; hypural plate is well formed* and caudal fin rays develop.*[,49,124] For smallmouth bass 8.6–8.8 mm from GA and AR, 15–17 developing rays were visible in the caudal fins, but only 3–12 were segmented.* Ten to 14 pterygiophores are visible at the bases of the developing dorsal and anal fins; ray differentiation begins in these fins but no rays are formed.* Pectoral fins are rounded and without rays.*[,23,124] Pelvic buds were barely visible on an 8.7-mm individual.*

9.2–9.7 mm TL. Median fin development continues, with dorsal and anal fin rays visible in some individuals. Urostyle is flexed up to 45° on some individuals. Caudal fin is rounded, becoming bilobed, with up to 14 segmented rays. The finfold becomes deeply notched at the posterior margins of the dorsal and anal fins, but is still present, dorsally and ventrally, on the caudal peduncle and, ventrally, anterior to the anus. Pelvic buds are barely visible.*

Pigmentation

4.6–5.0 mm TL. Body transparent;[124] yolk pale yellow[124] or amber;* oil globule bright golden.[124]

5.4 mm TL. Eye pigment forming (1 day old), none elsewhere.*[,24]

6.0–6.9 mm TL. Eyes are uniformly black at 6.0 mm, becoming iridescent by 6.75 mm. A band of pigment is present along junction of yolk sac and body at 6.0 mm and, by 6.75 mm, a second band of pigment is visible on the body parallel to the first from the auditory vesicle to the posterior margin of the yolk sac.[124] Yolk is golden yellow, oil amber.*

7.3–7.7 mm TL. Smallmouth bass from GA in this size range have darkly pigmented eyes. Pigment is visible initially on the dorsal surface of the head and nape and in a band of scattered chromatophores on the dorsal aspect of the yolk (7.3 mm). Pigmentation becomes progressively more profuse over the yolk sac and head and expands along the dorsum to the urostyle and anteriorly to between the eyes. Laterally, pigmentation expands onto the sides of the yolk, head, and body and scattered pigment appears the length of the torso.*[,124] Little pigment is present ventrally on the head anterior to the middle of the eye — no pigment on lips or snout. Pigment is dense enough dorsally and anteriorly on the torso to make discernment of the most anterior myomeres difficult.*

8.0–8.2 mm TL. Dorsally, pigmentation is heavy on the head, snout, nape, and along the length of the body. Laterally, scattered pigment forms along the sides of the body and on the dorsal half of the yolk sac. Pigment is present laterally on the head to about mid-eye; there is little or no pigment ventrally on the head, pericardial cavity, or yolk sac.*

8.3–9.7 mm TL. Entire body is darkly spotted with pigment. Many large stellate chromatophores are concentrated over the head, with fewer on the sides of the head and around the jaws. The yolk sac is well covered with pigment except ventrally. The eye is dark*,[23,124] or may appear iridescent.[164] By 9.2–9.7 mm, pigment is present laterally on pterygiophores of dorsal and anal fins, and a few chromatophores are scattered proximally on the caudal fin. A 9.5-mm specimen examined from GA had no stellate chromatophores and appeared much lighter to the naked eye than other individuals with typical stellate pigmentation.*

POST YOLK-SAC LARVAE

See Figure 87.

Size Range

Yolk is completely absorbed between 8.7 and 9.9 mm TL,*,[124,144] or at 9.8–10 mm in southern populations, but remnants of oil may still be present at larger sizes.* Fin development is complete at about 19 mm TL (based on illustration in Fish, 1932),[123,144] or at about 16 mm TL in southern populations.*

Myomeres

9.5–10.0 mm TL. Preanal 10–11; postanal 19–22 (incomplete).[123]

For post yolk-sac larvae of southern smallmouth bass, an accurate count of myomeres was difficult due to heavy pigmentation.*

Morphology

9.5 mm TL. Head and mouth large, rest of body tapers to caudal region; intestine may or may not be visibly coiled at this size.*,[123]

Morphometry

See Table 101.

9.5–10.0 mm TL. As percent TL: SL 89.5–92.1; PreAL 52.5–53.7; HL 27.4; GBD 21.7–27.4; ED 10.5.[123]

Fin Development

9.5–10.0 mm TL. Median finfold is elevated in regions of future soft dorsal and anal fins; element of ventral finfold still present anterior to anus.[123] Dorsal and anal fin rays visible.[49,123] A wild-caught Lake Erie specimen 9.5 mm had many caudal fin rays and 13 dorsal and 11 anal fin pterygiophores and rays visible. Fin development was not as advanced in hatchery-reared fish at 10 mm TL.[123]

9.8 mm TL. Hypural development is nearly complete in a laboratory-cultured individual spawned by AR brood stock. The urostyle barely extends beyond the hypural plate. The caudal fin is becoming bilobed with 15–16 rays visible — 14 segmented. Insertions of dorsal and anal fins are established with no finfold left immediately posterior to either, but some finfold remains on caudal peduncle between dorsal, anal, and caudal fins. Dorsal fin has 11 visible soft rays, anal fin with 10. Spinous dorsal fin differentiation is beginning. Pelvic buds are visible.*

10.2–10.9 mm TL. No spinous dorsal fin differentiation was visible on laboratory-cultured specimens spawned by GA brood stock. Insertions of dorsal and anal fins were established on all specimens.* Some median finfold still present on the caudal peduncle near the base of the caudal fin,*,[124] and a remnant is still present, ventrally, anterior to the anus.* Median fin development proceeds:*,[124] 14–17 segmented caudal fin rays apparent; dorsal and anal spines are visible on some individuals; soft dorsal fin has 9–14 visible rays; and anal fin has 8–11 rays.* One report indicated all fin rays are formed by 10.7 mm.[49] The caudal fin is well-developed, visibly bilobed on most; urostyle still extends to near the margin of the fin. Pectoral fin ray differentiation may begin on some individuals.*

11.1–11.7 mm TL. Urostyle still visible but receding. Caudal fin is well formed with an adult complement of primary rays (17) present on most individuals. Little median finfold remains. Dorsal fin has 2–7 developing spines and 13–15 soft rays. Developing anal fin spines are visible on some, 11–13 soft rays apparent. Pectoral fin rays are visible.* Pelvic fins evident,*,[124] still bud-like and small.*

12.0–12.9 mm TL. Median finfold disappears, dorsally and ventrally, on caudal peduncle. Fin development progresses, with adult complement of soft rays present in all median fins on some individuals. Urostyle still extends beyond hypural plate. Caudal fin forked. Ray development begins in pelvic fins.*

13.0–13.6 mm TL. Adult complement of rays and spines present in median fins. Spinous dorsal fin is low and level in profile. Pelvic fins are flap-like, about 0.6–0.7 mm long.*

14.1–14.8 mm TL. Spinous dorsal fin becomes rounded in appearance, spines 2–5 are longer than the others. Pelvic fins about 1.0 mm long.*

15.1–16.0 mm TL. Urostyle no longer extends posterior to hypural plate. Development is nearly complete in pelvic and pectoral fins, the former with 5 visible rays and the latter with 15–16.* Dorsal fin spines growing in length, adult morphology of fin developing.*,[48,49] By about 16 mm, the adult complement of spines and rays is present and fully formed in all fins.*

Pigmentation

10.0 mm TL. Hatchery stock at this size were lighter in overall appearance than wild-caught larvae from Lake Erie due to the contraction of chromatophores in hatchery fish; the number and arrangement of chromatophores were identical between the two stocks.[123]

This phenomenon was observed on one 9.5-mm specimen from hatchery stock from GA, but not on all.*

10.2–15.7 mm TL. Pigmentation is as described above — entire body uniformly dark with pigment, laterally and dorsally; venter less pigmented than sides and dorsum.*,[48,49] There is little change in pigmentation pattern in this size range other than an intensification of mid-lateral pigment that becomes evident on larger specimens.* Color in life becoming bronze or green.[124]

JUVENILES

See Figure 88.

Size Range

Transformation from larvae to juvenile occurs in about 15 days.[47] Fin development is complete at about 19 mm TL[123] or at about 16 mm TL;* sexual maturity is reported at lengths of 185–197;[106,143] generally larger in northern waters, 200–300 mm TL.[73,81]

Morphology

19.0 mm TL. Mouth large and oblique, lower jaw projects, maxilla extends to middle of eye.[123]

18.0–23.0 mm FL. Scales are first evident on the caudal peduncle just anterior to the caudal fin (Figure 89).[156]

22.0–31.0 mm FL. Scales spread predominantly anterior along the middle of the body, less rapidly dorsally or ventrally (Figure 89).[156]

32.0 mm FL. Scales formed in the pectoral region[44,156] and have reached the dorsal and ventral aspects of the body along the caudal peduncle and anteriorly to about mid-body (Figure 89).[156]

Lateral line scales 68–81;[47, 138,140,144,146] gill rakers 6–8; vertebrae 31–32.[47,144]

Morphometry

16.3–19.7 mm TL. See Table 101.

19.0 mm TL. As percent TL: SL 79; PreAL 50.5; HL 28.4; GBD 22.6; ED 7.9.[123]

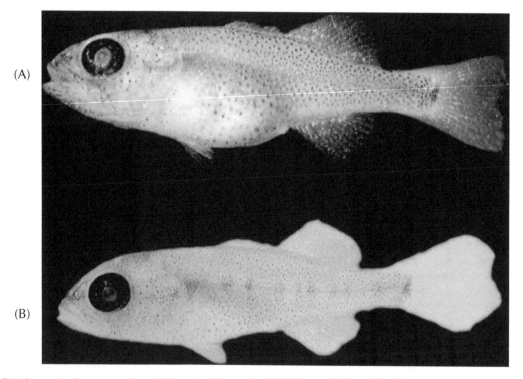

(A)

(B)

Figure 88   Development of young smallmouth bass. A–B. Juveniles: (A) 17.5 mm TL from GA; (B) 18.3 mm TL, wild-caught from TN. (A–B, original photographs.)

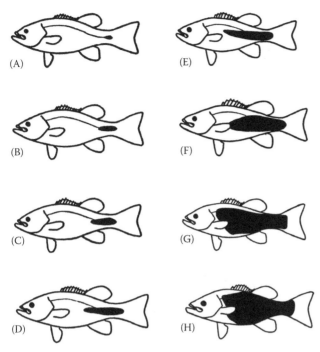

Figure 89 Scale development related to size for young small-mouth bass. (A) 21.0 mm FL; (B) 21.6 mm FL; (C) 21.9 mm FL; (D) 23.0 mm FL; (E) 24.9 mm FL; (F) 26.1 mm FL; (G) 31.1 mm FL; (H) 32.0 mm FL. (A–H reprinted from Figures 1–8, reference 156, with publisher's permission.)

Fins

16.3–19.7 mm TL. Caudal fin rays 16–17; dorsal fin spines 9–10, soft rays 13–15; anal fin spines 3, soft rays 10–12.*

Larger juveniles. Dorsal fin is generally reported to have 9–11 spines[47,123,138,140,144,146,149,150] and 12–15 soft rays;[47,123,138,140,144,146,148–150] anal fin with 3 spines,[123,138,140,144,146,149,150] rarely 2,[146] and 9–12 soft rays;[47,123,138,140,144,146,149,150] pectoral fins have 13–18 rays;[47,144,146,149] pelvic fins have 1 spine and 5 soft rays.[144]

Pigmentation

(*Note:* Conversions from SL to TL are based on the formula TL = 1.216 SL presented in Carlander, 1977.)[52]

16.3–17.8 mm TL. A row of melanophores along the posterior margin of the hypural plates becomes the foundation of a wedge-shaped blotch of pigment that develops at the center of the hypural plate. The point of the wedge is directed anteriorly. By 17 mm, this blotch of pigment is visible to the naked eye. Otherwise, body pigmentation is as before. Little, if any, pigment is visible in the fins.*

14.6–23.1 mm SL (about 18–28 mm TL). At 14.6 mm four or five primary bars of pigment are initially

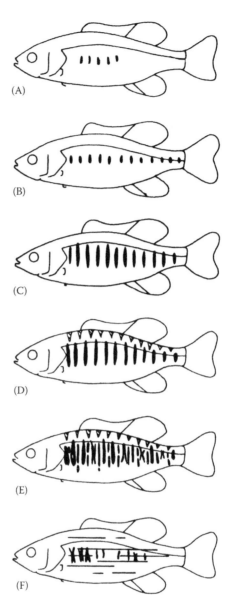

Figure 90 Schematic drawings showing ontogeny of melanistic pigment patterns of young smallmouth bass. (A) 14.6 mm SL; (B) 23.1 mm SL; (C) 24.9 mm SL; (D) 40.9 mm SL; (E) 31.9 mm SL; (F) 85 mm SL. (A–F reprinted from Figure 3, reference 51, with publisher's permission.)

visible as ovoid to squarish, midlateral blotches of melanophores located over the horizontal myoseptum in the abdominal area. Primary bars of pigment are added anteriorly and posteriorly until the adult complement of 11–14 is visible (Figures 90A and 90B).[51]

16.1 mm SL (about 20 mm TL). A faint, narrow stripe is visible along the horizontal myoseptum (in unscaled fry), but the most conspicuous feature of pigmentation is still the persistence of large, uniformly distributed melanophores over most of the body.[49,53,123] The lateral stripe disappears with scale formation. Caudal spot faint, no pigment pattern visible on caudal fin (Figure 91A).[53]

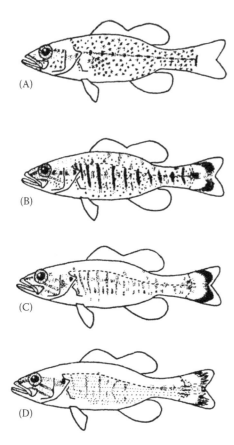

Figure 91 Details of pigmentation for smallmouth bass from AR stock, presumably from the White River drainage. Body proportions are constant in the drawing outlines and do not reflect proportional differences between sizes. (A) 16.1 mm SL; (B) 28.8 mm SL; (C) 60.1 mm SL; (D) 79.1 mm SL. (A–D reprinted from Figure 4, reference 53, with publisher's permission.)

24.9–28.8 mm SL (about 30–35 mm TL). Primary bars of pigment becoming prominent with scale imbrication[53] and appear dorsoventrally elongated (Figures 90C, 90D, and 91B).[51,53] At 28.8 mm, the caudal spot is large and fairly distinct and a submarginal band of pigment is present on the caudal fin; in life, the tail appears tricolored, opaque yellow to yellow-orange anterior to the submarginal band and iridescent white along the outer margin of the fin (Figure 91B).[53]

41.0–50.0 mm SL (about 50– 61 mm TL). Large melanophore scattered over body conspicuous until about 49 mm SL.[53] At about 41 mm SL, secondary bars of pigment develop dorsal to the lateral line, their ventral extensions interdigitating with the dorsal tips of the primary bars.[51] Midlateral and ventral portions of these secondary pigment bars develop with growth;[51] this development corresponds to a lightening in the centers of the adjacent primary bars (Figure 90).[51,53] Submarginal band of pigment prominent on caudal fin; tricolor pattern persists (Figure 91).[53]

60.0–80.0 mm SL (about 73–97 mm TL). Caudal spot has faded centrally to two faint, horizontally zigzag lines on the caudal base. These lines persist sporadically, but the caudal base eventually becomes unmarked. At about 60 mm SL, lateral bars of pigment becoming vague or absent; faint parallel streaks developing along the horizontal scale rows become the prominent pigment pattern (Figure 91).[53] The tricolor pattern of the tail persists until about 65 mm SL, then tends to become disrupted as the submarginal band of pigment expands posteriad.[53]

Very young specimens are generally boldly patterned with vertical bars and blotches on the body and have a patterned caudal fin. This pattern disappears with maturity and specimens 50–75 mm TL are much more drab.[47]

TAXONOMIC DIAGNOSIS OF YOUNG SMALLMOUTH BASS

Similar species: other centrarchids, especially rock bass and other *Micropterus* species; also superficially similar to elassomatids.

For comparisons of important morphometric and meristic data, developmental benchmarks, and pigmentary characters between smallmouth bass and all other centrarchids and elassomatids that occur in the Ohio River drainage, see Tables 4 and 5.

For characters used to distinguish between young basses and the young of other centrarchid genera and *Elassoma*, see provisional key to genera (pages 27–29).

For identification of young basses in the Ohio River drainage, see "Provisional Key to Young of *Micropterus* Species" in genus introduction (pages 249–250).

See Taxonomic Diagnosis section in rock bass species account (page 71) for discussion of differences between rock bass and smallmouth bass.

ECOLOGY OF EARLY LIFE PHASES

Occurrence and Distribution (Figure 92)

Eggs and yolk-sac larvae. Fertilized eggs settle to the bottom of the nest[48,71,73,143] where they initially[48,71,73,124,143] adhere to stones, gravel, short stems, or roots.[65,144] They ultimately lose adhesiveness and settle into the substrate at the bottom of the nest,[143] becoming very difficult to see.[73]

Newly hatched larvae are nearly transparent and also difficult to see.[73] They remain in

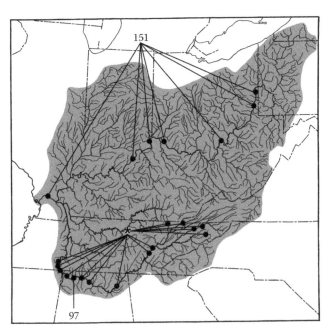

Figure 92 Present distribution of smallmouth bass in the Ohio River system encompassing native range and areas of introduction (shaded area). Areas where early life history information has been collected are indicated with circles. Numbers indicate appropriate references. Asterisk indicates TVA collection localities.

crevices[73,124,143,157,158] in the nest for several days[73,157,158] depending on water temperature, living off the yolk and growing. Larvae rise from the nest 11–12 days after hatching or 6–13 days after fertilization depending on water temperature.[52]

At 5.4 mm TL (1 day after hatching) smallmouth bass larvae are able to move along the bottom by movements of the tail. By 7.5 mm TL, pectoral fins begin vibrating and movements along the bottom become more vigorous. At 8.3 mm, the larvae can swim up from the bottom of the nest for an instant but return to it immediately. The frequency of these excursions from the bottom and their length steadily increase. By 9.3 mm and about the time the air bladder begins to develop and most of the yolk has been absorbed, they are capable of swimming up and moving about over the nest.[124] At about this time, they become increasingly conspicuous as they become progressively darker with pigmentation development.[73] For some days the fry continue to hover over the nest in a school in which the individual fish are darting here and there irregularly. The school gradually leaves the nest accompanied by the male.[124]

Smallmouth bass fry swim up from the nest 5–7 days after hatching.[47,157,158] The darkly pigmented fry begin to work their way up out of the gravel at the bottom of the nest, and in MO, 9–10 days after hatching they are swimming over the nest; they begin to disperse on days 11–12. As the season advances and water temperatures rise development and movement in and out of the nest are more rapid.[157] Larvae begin to rise off the nest

in 11 days in MA.[158] Larvae are also reported as free-swimming 6 days after hatching at about 8.7 mm TL.[49] In Lake Opinicon, ON, swim-up occurred on June 10 and the larvae averaged 8.2 mm TL; yolk was present for 2 days after free-swimming began.[166]

A successful spawn of smallmouth bass was observed in an AL pond on May 4 when water temperature was 25.6°C; swim-up fry were seen on May 7. The larvae began moving from over the bed on May 9, scattering to 6.1 m, but returning to the nest. Complete school dispersal occurred by May 16; larvae remained close to shore, staying 50–178 mm from the shoreline or submerged structure.[54]

Common hazards to smallmouth bass eggs and yolk-sac larvae include temperature fluctuation,[52,73,84,86,90] floods and water level recessions,[52,73,84,86] predation,[52,73,84,86,92] nest destruction by humans,[73] and fungus.[52,73,84,86] In lake environments, high winds and the resulting wave action often destroy nests.[90,92]

In the North Anna River, VA, in 1992, 45 of 105 smallmouth bass nests (43%) produced swim-up fry. Stepwise discriminate analysis distinguished successful nests from unsuccessful nests by: higher mean temperature (19.2 vs. 17.6°C); lower mean stream discharge (5.03 vs. 7.67 m³/s); lower distance to shore (11.7 vs. 13.2 m); and higher male aggression. Swim-up production for the 45 successful nests was positively correlated with male weight, distance to cover, nest diameter, and male aggression.[165]

In streams, high water levels during the spawning period tend to reduce spawning success.[91] Floods prior to larval dispersal from the nest can result in the loss of smallmouth bass broods.[72] Swim-up larvae (7–9 mm SL) were displaced from field nests and from laboratory flumes at water velocities of 8 mm/s. Nests in areas where velocities were 15 mm/s failed to produce young.[107]

In the North Anna River, a perennial, regulated stream in VA, in 1992, high flow (>10 m³/s) was responsible for most nest failures (85%). The temporal pattern of stream-flow fluctuation was the most important abiotic factor determining reproductive success or failure for smallmouth bass during this study. In the study period, mid-April to early June, high flows disrupted spawning five times even though males selected nest sites that had smaller increases in mean velocity with increased discharge than random locations in pool habitat.[106,165]

Low water temperatures prolong incubation and increase chances of decimation from fungus infection, siltation, or other conditions that might be operative.[86] Development of eggs and yolk-sac larvae is retarded by temperatures near or below 15.6°C.[84] For larvae, low temperature may produce deformity, stunting, or modified behavior patterns, all of which can increase their vulnerability to starvation or predation.[92] Nest desertion by the guardian male resulting from temperature drops subjects nests to oxygen deficits and predation.[92] If nests are left unguarded, eggs and larvae are quickly consumed by other fishes such as longear sunfish and bluegill.[149] In the Tennessee River drainage of AL, a maximum weekly average temperature of 26°C during the spawning season will allow survival of eggs and larvae.[148]

In four northern WI lakes, habitat characteristics affecting nesting success of smallmouth bass are extremely variable; no single habitat feature consistently predicted nest success across lakes. Measures of substrate size were the most predominant nest characteristic associated with variation in both egg survival and fry production among lakes but they were not significant in all lakes. Other features related to nest cover, nest position, and morphology explained some variation in egg survival and fry production but they were inconsistent across lakes.[129]

Catch-and-release angling for bass during the brood-guarding stage can induce premature nest abandonment by the male parent.[105,136] Captured and released bass may be less able or less willing to protect their brood.[105] Smallmouth bass males that have the largest broods and the greatest potential to contribute to annual recruitment are the most likely to be caught by anglers, indicating that angling for nesting bass during the brood-guarding period has the potential to negatively impact bass populations.[99] Significant levels of angling-induced brood loss can cause a population-level decrease in reproductive success. If that decrease causes a decrease in recruitment, the population density and/or size structure of the bass population may be affected negatively.[136] In South Branch Lake, MA, the permanent removal of males by angling or displacement resulted in empty nests the following day.[158] Also, the physiological and behavioral effects of exhaustive exercise caused by angling capture indicate a potential for catch–release angling of smallmouth bass during their spawning season to negatively affect reproductive success.[114]

Parental investment theory predicts that parents should adjust the level of care given to offspring relative to brood size and stage of brood development. A study on the effect of catch-and-release angling on parental care behavior of male smallmouth bass revealed that after catch-and-release events, angled males were less willing or less able to defend their broods than were control fish. The same study revealed that with or without angling, male smallmouth subjected to simulated brood predation were the least aggressive in defending their remaining broods.[105]

In a study of round goby predation on smallmouth bass offspring in nests, during simulated catch-and-release angling, the authors determined that if the number of surviving smallmouth bass embryos drives adult population size, managers should consider angling regulations that reduce interference with nesting males.[100]

Larvae. Smallmouth bass larvae leave the nest accompanied by the male parent and usually disperse to shallow water[65,124,143] where they seek calm marginal areas, sometimes near vegetation; they often find cover in the lee of rocks over bottoms of sand, rock, gravel, and rubble and are rarely found over mud.[143] Schools of smallmouth larvae are often observed along the shore along with the male parent who swims back and forth in a half-circle outside his school.[124]

In Lake Opinicon, ON, smallmouth bass fry remained in the littoral from the time they vacated their nests around mid-June until early October. Once the larvae became free-swimming, they remained closely associated with the nest for 2–3 days. At this time they were darkly pigmented and guarded by the male. After leaving the nest area the young spread out along a portion of the shoreline. In late June the young bass were observed in schools of 10–30 individuals dispersed along a greater area of the shoreline — the male was still present. The male deserted the young by early July and as time progressed the schools broke apart. After July 15 all young smallmouth bass were observed as single individuals along the shoreline.[166]

For the first 7–9 days after larvae leave the nest (swim-up stage, 14–18 mm TL), their access to microhabitats is affected by the behavior of the male parent. The male guards the area occupied by the brood and prevents larvae from leaving the site. Individual larvae forced to move throughout the area in response to changes in density of their siblings or the presence of predators cannot remain in one place or seek preferred microhabitats within the brood site.[72]

The male smallmouth bass protects his brood through their larval development until they reach 25–32 mm TL. They then scatter and the male leaves them.[124]

Most smallmouth bass larvae use microhabitats where rock crevices are present; large woody debris is frequently used by larvae and a high proportion of individuals use aquatic vegetation. Larvae were regularly observed within 1 m of multiple cover types. Substrate and cover used by age-0 smallmouth bass varied little, with fractured bedrock and boulder always the most frequently used substrate type; sand was also regularly used. Substrates of silt, small and large gravel, pebble, or cobble were infrequently used by age-0 smallmouth bass. All stages of age-0 smallmouth bass generally used mean velocities <20 cm/s, but individuals were observed in microhabitats with mean velocities up to 86 cm/s.[72]

In the North Anna River, VA, in 1990–1991, more than 30% of the total smallmouth bass brood area studied had a mean water velocity higher than 4 cm/s, had no woody debris or rock crevices, and had sandy substrate. After dispersal from the brood area, less than 7% of the dispersed larvae observed occupied these microhabitats; the majority occupied areas sheltered by woody cover or shelves of fractured bedrock. Large cover objects not only provided protection from predators but also created microhabitats with low water velocities throughout much of the water column. Because larvae have limited mobility in the water column when velocities exceed 4 cm/s, a larger portion of the water column would be available to them behind cover objects. Regardless of substrate or cover type, more than 27% of dispersed larvae occupied microhabitats with mean water velocities less than 4 cm/s. Differences were less distinct and less consistent between the range of depths at brood sites and the depths used by dispersed larvae.[72]

Smallmouth bass larvae in streams are subjected to high mortality from catastrophic floods, predation, starvation, and lethal temperatures.[82] Turbidity, turbulence, and velocity are all factors associated with flooding of warmwater streams and the disappearance of smallmouth bass fry. These stream flow characteristics influence fry orientation, feeding, and tolerance of temperature fluctuations, but ultimately, temperature is the major factor affecting bass fry under floodwater conditions.[128]

Smallmouth bass fry employ visual and tactile orientation in maintaining their positions in flowing water. A small amount of light allows larvae <25 mm TL to orient themselves visually; without light they cannot maintain their positions following a change in water flow (direction, velocity, turbulence). Increasing turbidity has similar effects on orientation as has decreasing light. Tactile orientation of smallmouth bass fry is influenced by changes in water velocity and turbulence. Losses of fry during floods apparently are caused by relatively rapid changes in velocity, turbulence, turbidity, and light, often with all the factors exerting simultaneous influences.[93]

Once larvae reach the free-swimming stage, dissolved oxygen concentrations above 3.0 ppm appear to be safe, in terms of mortality.[92]

In the North Anna River, VA, there is an absence of size-selective mortality among larval smallmouth bass probably because they were most susceptible to predators that were not gape-limited.[119]

At 9.3 mm TL, smallmouth bass larvae show no marked reaction to mechanical shock. Nearly a whole school may often be collected with a single sweep of the net. By 10.5–11 mm, they begin to react to mechanical shock and flee actively from nets and seines.[124]

Larval smallmouth bass 9.5–10.0 mm TL were collected from Lake Erie with a trawl at 6 m depth on July 11, 1928.[123] Smallmouth bass larvae were collected in drift nets April through July from the New River, VA; an estimated 46,900 larvae 7–18 mm TL were collected in June and July 1976.[7]

During the period 1978–1990, smallmouth bass larvae were reported in ichthyoplankton samples from several locations in the Ohio River (Table 102). They were more commonly collected from upper-river sites (ORM 54 and 77) and were not observed in 4 years of collections from the lower river at the Shawnee Plant (ORM 946).[151]

During the period 1972–1982, very few young smallmouth bass were observed in ichthyoplankton collections made by TVA in the Tennessee Valley. Twenty-nine larvae 8–19 mm TL were captured. They were present in samples from late April to early June and were usually captured at night in water less than 10 m deep; most came from water temperatures of 18–26°C.*

Juveniles. Young smallmouth bass are guarded by the male parent until 25–32 mm TL.[124]

Habitat preferences for juvenile smallmouth bass appear to be site specific.[26] In streams, fingerlings are often observed in calm marginal regions

Table 102

Presence (P) or absence (blank) of smallmouth bass larvae in ichthyoplankton collections made near several electric generating plants on the Ohio River during the period 1978–1990. A dash indicates that no collections were made at that site that year.

Location	ORM	1978	1979	1980	1981	1982	1983	1984	1985	1986	1987	1988	1989	1990
W.H. Sammis Plant	54			P		P			P	P	P	P	P	
Cardinal Plant	77	—		P		P		P		P	P			P
Kyger Creek Plant	260			P		P				P				
W.C. Beckjord Plant	453			P				—	—	—			—	
Tanners Creek Plant	494			P						P				P
Clifty Creek Plant	560	—	—	—	—	—	—	—	—	—		—	—	P
Shawnee Plant	946	—	—	—	—	—	—	—	—	—				

Source: Tables 3-1, 3-8, 3-14, 3-20, 3-22, 3-28, and 3-29 in reference 151.

containing rocks and vegetation.*[,73] In the Great Lakes, YOY smallmouth bass are found usually in water less than 6 m deep in littoral zones of streams, river mouths, bays, harbors, coves, or lake shores, usually near structure over substrates of rock, gravel, rubble, sand, or vegetation; seldom found over mud bottoms.[65] Young smallmouth bass from the Shoals Reach of the Tennessee River, AL, were most frequently collected from deep main channel habitats with rock substrate, broken rock and ledges as cover, and moderate current.[97] Age-0 smallmouth bass were collected with seines along the sides of coves adjoining the main body of Lake Texoma, TX/OK; sites farther into the coves and sites along river channels were less productive.[70]

In the North Anna River, VA, juvenile smallmouth bass moved into shallower microhabitats as the growing season progressed. Dispersed larvae were found farther from the bottom than juveniles observed 14–20 days later. In 1 year of the study, the mean distance off the bottom for dispersed larvae was 23 cm; the mean distance off the bottom for juveniles observed 2 weeks later was 14 cm. Age-0 smallmouth bass were occasionally seen at the surface, but more than 90% of all individuals occupied the bottom half of the water column.[72] In 1990, 76% of juvenile smallmouth bass observed occupied deep pools in July, but in August 58% inhabited runs or riffles. In May 1991, 84% of the juveniles observed were in deep pools, but 61% of juveniles in June and 49% of the juveniles in August occupied run and riffle areas.[72]

In the Buffalo River, AR, both age-0 (<100 mm TL) and larger smallmouth bass occupied runs and pools, but a higher proportion of age-0 fish were observed in pools. Age-0 fish were not restricted to stream edges and runs as in some eastern U.S. streams.[118]

Optimum flow for age-1 smallmouth bass in the Maquoketa River, IA, was 10 m³/s; no clear relationships were evident for ages 2–4.[34] In the upper Mississippi River, young smallmouth bass 43–116 mm SL were most frequently found in current velocities of 80–130 mm/s; this velocity range produced maximum growth in flume studies.[107]

In the upper Mississippi River, relative abundance of age-0 smallmouth bass in autumn electrofishing samples was significantly higher during years when spawning and growth period degree-days were high and discharge was low. Relative year-class abundance of age-1 and age-2 smallmouth bass increased significantly with mean back-calculated TL, which was highest during warm, low discharge years. Abundances of age-1 and age-2 year-classes were depressed during years when first summer growth was rapid and winter discharge was high.[127]

In VA rivers, electrofishing catch per unit effort of age-0 smallmouth bass in September–November is a reliable measure of year-class strength and mean June stream flow reasonably predicts the abundance of age-0 smallmouth in the fall.[36]

In Bull Shoals Reservoir, AR, the biomass of smallmouth bass 100–200 mm TL was inversely related to January–August inflow in the previous year.[46]

Behavioral carrying capacity (BCC) experiments performed at various stream-flows indicated no positive relation between available smallmouth bass habitat (weighted usable area) and BCC, and a negative relation existed between stream discharge and BCC. High BCC values recorded during low-flow experiments were the result of less upstream and downstream movement of small-mouth bass juveniles; such behavioral changes may prevent smallmouth bass populations from being limited by habitat availability during low flows.[14]

In Elkhorn Creek, KY, smallmouth bass year-class production and growth to age 1 were both highly variable and inversely related to April–July rainfall. Recruitment to the fishery as age-4 and older fish was positively correlated to densities of age-1 fish collected in the spring.[17]

A study designed to predict the effects of angling for nesting smallmouth bass found that abundance of age-0 smallmouth bass decreased as the probability of capturing a nesting male increased in both catch-and-keep and catch-and-release scenarios. Opening dates during the nesting season, when males were guarding broods, also decreased abundance of age-0 fish. It was also noted that stress resulting from handling time for catch-and-release can have a significant impact on the abundance of age-0 fish because nesting males may abandon guarding behavior.[19] Results of an Ontario study indicate that the illegal activity of preseason angling for black basses (smallmouth and large-mouth) is substantial enough in some waters to decrease fry production.[20]

Reports of preferred temperatures for young small-mouth bass range from 18 to 31°C, depending on season (laboratory studies); upper and lower avoidance temperatures were 33 and 25°C.[50,90] In western Lake Erie, laboratory tested YOY small-mouth bass occupied water with temperatures of 29–31°C in summer, 26–30°C in fall, 24–28°C in winter, and 22–28°C in spring.[63]

Lethal dissolved oxygen concentration for juvenile smallmouth bass in a river was reported at 0.98 ppm at 15.6°C and 1.32 ppm at 26.7°C; in a lake population the concentrations were 1.10 ppm at 15.6°C and 1.56 ppm at 26 7°C (Burdick et al., 1958; cited in reference 91).

In studies investigating the effects of starvation on acid tolerance of juvenile smallmouth bass, fish held under simulated winter conditions for up to 5 months experienced losses in weight, but tests failed to demonstrate any decline in the short-term tolerance of these fish for pH levels ranging from 3.0 to 4.5. Calcium and sodium concentration of the water and fish body size were all positively related to acid tolerance; a doubling in fish size (length) increased survival time at any pH by 100%.[108]

An estimated 41–55% of smallmouth bass stocked into an Ontario Lake with pH 5.9 survived; no survival was reported from two lakes with pH 4.9–5.2.[109]

Density-dependent growth during the mid- to late-juvenile period reflects a process of movement away from the natal area to locations where home ranges will be established. The authors of this study felt that this movement is a density-dependent process based on competition and possibly social foraging, observed to increase juvenile bass after their first year.[125]

Young (17–35 mm TL) smallmouth bass were first collected from the Shoals reach of the Tennessee River in May with backpack electrofishing samples from gravel bank habitat; specimens 18–35 mm TL were collected in late May from slough habitat. Subsequent samples in mid-June yielded small-mouth bass ≥24 mm TL from shallow boulder, riprap, barge wall, slough, creek, and gravel bank habitats. Young bass were collected within some type of cover, especially aquatic vegetation and woody debris in the form of roots, fallen trees, and limbs. Specimens collected from shallow boulder, riprap, and barge wall habitats were all closely associated with large rocks near shore. Young bass collected in habitats lacking rock cover were closely associated with cut banks, root wads, vegetation, and log jams. These collections of young small-mouth bass were generally in water <0.7 m deep and within 1 m of shore.[97]

During the period 1972–1982, only 14 juvenile smallmouth bass were observed in TVA ichthyo-plankton collections from numerous locations in the Tennessee Valley. Juveniles 21–45 mm TL were present in samples in May and June and captured at night from depths of 1–6 m in water temperatures of 18–26°C. Four juveniles 82–98 mm TL (age 1) were collected in daylight on April 6, 1977 in water 16°C and 3–5 m deep.*

Laboratory experiments suggest that smallmouth bass fed at higher rates in cobble habitats than in vegetation and that predation vulnerability was higher in vegetation than in cobble habitats.[102]

After smallmouth bass reach about 152 mm TL (6 inches), the angler is probably the greatest cause of mortality in many populations.[73]

Early Growth

Growth rates of young smallmouth bass vary substantially among bodies of water (Table 103).[146] Growth is generally the same for both sexes[73,81,90,148,162] and varies from one population to another depending largely on temperature,[73,81,90] food availability,[73,81] and number of bass present.[79,81]

Young smallmouth bass grow faster and to a much larger size in reservoirs, probably due to higher average temperatures and longer growing seasons, greater food availability (such as forage species like shad), and less energy expenditure.[47] One of the

Table 103

Reports of average TL (mm) related to age of young smallmouth bass from various locations.

Location	Length at Age:			
	1	2	3	4
Norris Reservoir, TN (1939)[163]	168	318	368	417
Norris Reservoir, TN (1936–45)[161]	117	259	358	412
Norris Reservoir, TN (1936–73)[96]	114	254	355	411
Spring Creek, TN (1972)[159]	49	90	118	165
Roaring River, TN (1972)[160]	95	186	267	355
Tennessee River, AL (1973–75)[37]	112	216	302	380
Tennessee River, AL (1988–89)[28]	98	179	273	367
Tennessee River, AL (1995–96)[18]	179	261	337	414
Barren River Lake, KY (1981)[155]	94	180	241	315
Barren River Lake, KY (1982)[155]	76	163	267	—
Green River Lake, KY (1981)[155]	114	226	300	401
Green River Lake, KY (1983)[155]	109	188	277	371
New River, VA[33]	107	176	236	281
New River, WV[33]	96	187	244	331
Fall Creek, NY[a]	—	140	178	216
Glover River, OK[41]	91	168	239	299
Baron Fork River, OK[41]	89	161	228	282
Big Buffalo Creek, MO[31]	78	134	183	233
Courtois Creek, MO[87]	79	150	213	272
MO average[b]	84	160	226	271
McConaughy Reservoir, NB[59]	74	169	263	327
Lake Texoma, TX/OK (introduced)[70]	103	257	341	431

[a] From Suttkus (1955), cited in reference 73.

[b] From Purkett (1958), cited in reference 31.

fastest growing populations studied was in Norris Reservoir, TN (Table 103), during the early, more productive years of the reservoir.[47,161] Growth rates of smallmouth bass in Norris Reservoir were determined on several occasions during the period 1936–1973.[96] Growth decreased after the first few years and then increased again when water levels raised, flooding new areas.[161] Regression analysis of TL vs. year indicated that smallmouth bass of all age groups grew slower as the reservoir aged.[96]

Growth was also comparatively rapid in the Tennessee River, AL, near the southeastern edge of the natural range of smallmouth bass in the United States. During the period 1995–1996 in the Tennessee River downstream of Wilson Dam, AL ("Shoals Reach," a 20-km stretch of the river that flows within its original banks and is distinctly riverine in nature), growth rates of young smallmouth bass[18] were higher than in previous studies conducted in 1988–1989[28] and 1973–1975[37] (Table 103) but were similar to those reported for fast-growing populations across the United States (Table 104), especially for ages 2–4. The comparatively rapid growth of smallmouth bass in this riverine portion of the Tennessee River is probably at least partially due to the extended growing season to which it is exposed.[37]

At a given latitude, smallmouth bass in small streams grow more slowly than those in large rivers and reservoirs.[146] In VA, river populations have faster growth rates than stream populations and size at age 4 in VA reservoirs was greater than sizes reported from rivers.[146]

Growth in TN streams varies greatly with stream size and temperatures. In 1972, calculated TL of age-1 smallmouth bass in Roaring River, TN, was nearly twice that of those from Spring Creek, a small, cool water tributary (Table 103), and thereafter exceeded twice the rate.[159,160]

Smallmouth bass in MO grew slowly in the first 4 years (Table 103) but more rapidly thereaf-

Table 104

Average growth of young smallmouth bass
reported for North American populations.

Location	TL (mm) at Age:			
	1	2	3	4
U.S. (mean)[52]	97	176	246	298
U.S. (fast-growth populations)[a]	118	258	358	411
U.S. (moderate-growth populations)[a]	94	173	249	309
Southeastern U.S. (mean)[52]	107	202	292	346
North America (mean)[73]	94	170	234	279

[a] From Anderson and Weithman (1978), cited in reference 18.

ter.[31,82,87,89] In Courtois Creek, over 83% of the smallmouth bass were caught by anglers before they completed their fifth year and most were <305 mm TL (12 inches).[88]

Implementation of a minimum-length-limit regulation on the New River, VA, resulted in a change in the growth pattern of smallmouth bass; growth of ages 1–3 increased but growth of ages 4–6 decreased.[33]

Growth of an introduced population of smallmouth bass in Lake Texoma, TX/OK, equaled or exceeded that of most other populations, including those from Tennessee reservoirs, by the second and subsequent years (Table 103).[70]

In Canada, smallmouth bass grow to 32–37 mm TL by July; 51–102 mm TL by the decline of temperatures in the fall.[144]

In Lake Michigan, it takes about 2 years for smallmouth bass to reach 200 mm TL.[61]

In another Lake Michigan study conducted in 1995–1996, approximate length ranges reported related to age of young smallmouth bass were as follows: age 1, 76–102 mm TL; age 2, 132–155 mm; age 3, 157–203; and age 4, 211–246 mm.[85]

In AR streams, SL of captured smallmouth bass related to age is reported as follows: age 0, 25–150 mm; age 1, 75–175 mm; age 2, 100–325 mm; age 3, 125–275 mm; age 4, 175–300 mm.[44] In OH, YOY in October reach 38–110 mm TL; at about 1 year 76–170 mm.[140] In IN, smallmouth bass are 50–75 mm by fall, 150–200 mm at age 2, and 225–275 at age 3.[58]

Larval smallmouth bass were dipped from their nest in the Nolichucky River, TN, on 5-7-1986 before swim-up and subsequently cultured in an aquarium at TVA's laboratory in Norris, TN. Larvae cultured at 18.9–22.2°C and fed brine shrimp daily grew steadily from 8.5–9.2 mm TL on 5-9-1986

to 12.5–13.5 mm on 5-29-1986. Growth of the series is charted below:*

Date	N	TL Range (mm)
5-9	2	8.5–9.2
5-11	6	9.5–10.0
5-13	2	9.5–10.0
5-15	2	10.5–11.0
5-17	2	10.5–11.5
5-19	6	11.0–12.5
5-21	2	11.0
5-23	4	11.0–12.0
5-25	3	11.5–12.0
5-27	5	11.0–13.0
5-29	2	12.5–13.5

YOY smallmouth bass from the Tennessee Valley grew to 8–14 mm TL by April 29, 17–19 mm by May 19, 21 mm by May 22, and 35–45 mm by June 2. Age-1 fish were 82–98 mm TL on April 6.*

In Lake Opeongo, Ontario, growth rates of juvenile age groups are higher when summers are warmer and when population abundance is reduced.[135]

There was no significant difference in the growth rate of age-1 and age-2 smallmouth bass after the introduction of threadfin shad into Dale Hollow Reservoir, TN.[6]

Weight–length data for 50 populations of smallmouth bass (N = 6,731) were used to develop the following standard weight (W_s) equation: $\log_{10} W_s$ (weight, g) = −5.329 + 3.200 $\log_{10}$ L (total length, mm). This equation is valid for fish ≥150 mm.[27]

The greatest growth rate (age 1 and 2) for smallmouth bass reported in WI is from Red Cedar River where average TL and weight by age are reported as: age 1, 100 mm and 0.01 kg; age 2, 190 mm and 0.09 kg; age 3, 274 mm and 0.29 kg; and age 4, 329 mm and 0.52 kg. Other reports of age and

Table 105

Length–weight relationships of young smallmouth bass collected in AL during the period 1949–1964.

Average Total Length (mm)	Number	Average Empirical Weight (g)
51	130	1.9
76	561	5.9
102	546	12.7
127	255	25.0
152	114	45.4
178	63	68.1
203	41	113.5
228	32	145.3
254	19	222.5
279	15	322.0
305	4	349.6

Source: Constructed from data presented in unnumbered table on page 60, reference 55.

growth from several WI localities are as follows: age 1, 61–92 mm; age 2, 135–173 mm; age 3, 185–277 mm; and age 4, 240–356.[145]

It took the average bass in Courtois Creek, MO, about 4 years to reach a weight of 224 g (about a half pound), whereas annual weight increments for older fish often exceeded 224 g.[87]

In AL ponds, smallmouth bass stocked in combination with fathead minnows grew from 17–20 mm TL to 171–177 mm TL and 55–57 g in about 200 days.[54]

Length–weight relationships for smallmouth bass collected in AL during the period 1949–1964 are presented in Table 105.[55]

Carlander (1977) summarizes many additional reports concerning length–weight relationships of smallmouth bass and concludes that, in general, there is little consistency in differences in weights at various lengths in different parts of the country.[52]

Feeding Habitats

Numerous published reports indicate that age-0 smallmouth bass are habitat and food specialists, while fewer report that they are generalists. After a review of the literature and their personal findings and observations, Pert et al. (2002) suggest that age-0 smallmouth bass may have an exceptional ability to take advantage of food and habitat resources that may be underutilized within a given system. They further hypothesize that age-0 smallmouth have evolved as flexible resource users to increase their survival in naturally variable environments by minimizing interactions with other fishes.[133]

From the time they begin feeding, the diet of smallmouth bass changes from small to large items as the fish grows. Larval smallmouth bass begin feeding on small crustaceans and continue on this diet exclusively until they are large enough to capture insect larvae, after which they progress to crayfish and fish (if available).[47,73,97,124,164] Young smallmouth will quickly assume a diet dominated by fish larvae (if available) and graduate quickly to progressively larger fish prey.[47,148] They appear to feed continuously day and night.[124]

In streams, larvae restricted to the bottom of the nest by water velocities eat small, slow-moving chironomids. By 14–18 mm TL smallmouth bass disperse from the brood area and begin consuming invertebrates.[72] Smallmouth bass up to 20 mm TL reportedly feed on midge and mayfly larvae. Larval smallmouth bass as small as 15 mm may also feed heavily on fish larvae, but in some populations the switch to piscivory begins later (40–70 mm).[148]

In MO, smallmouth bass larvae began feeding about the time they left the nest; larvae 7.3–8 mm SL ate dipterous larvae. Small dipterous larvae were the most important food for bass 7.4–9.9 mm SL with copepods next in importance; rotifers occurred in many stomachs, but were of minor importance because of their small size. Dipterans and copepods were also important food for larvae 10–20 mm SL. Fish were found in the stomach of an 11.8-mm smallmouth and were the most important item in the stomachs of larvae >15 mm; ephemeropterans were also consumed.[157]

In north-temperate lakes smallmouth bass shift from pelagic to benthic prey during their first summer.[132] In NY lakes, aquatic insects dominated diets of smallmouth bass 25–75 mm TL with zooplankton and crayfish also comprising a small percentage of the diet. Fish appeared in the diet at about 75 mm, but percent occurrence was highest (25%) at about 175 mm. Crayfish were the predominant food item for smallmouth 125–325 mm.[101] In Bear Creek, MN, age-0 smallmouth bass (<110 mm TL) fed mostly on corixids; older, larger fish were mostly piscivorous.[116]

In Lake Texoma, a shift from a diet of zooplankton and insects to one of fish and crayfish occurred at about 125 mm TL. Insects remained an important portion of the diet of larger fish (up to 500 mm TL), while fish were found in the stomachs of smallmouth bass as small as 67 mm TL.[70]

In the Tennessee River, AL, smallmouth bass larvae began feeding on entomostracans and switched to aquatic insects at 50 mm TL; they fed on fish after 150 mm TL, but in the 101- to 200-mm TL size range they fed primarily on malacostracans and aquatic insects. Smallmouth bass >200 mm fed primarily on fish with unidentified fish and clupeids comprising the bulk of fish eaten; crayfish were also eaten but infrequently.[97]

Smallmouth bass 109–515 mm TL from Norris Reservoir, TN, consumed mostly fish, including other black bass and crayfish.[5]

In the New River, VA and WV, the diet of smallmouth bass 152–228 mm TL was dominated numerically by insects (82–94% of all items found in stomachs). Ephemeropterans were most abundant; trichopterans, odonates, and terrestrial and unidentified insect parts comprised the remainder. Numerically, insects were also found to be common in the stomachs of smallmouth bass >228 mm TL, however, crayfish and fish combined to form a large part of the diet for these larger bass.[42] Another New River study also indicates that production of age-0 and age-1 smallmouth bass is supported primarily by aquatic insects and that age-2 and older fish consume mostly crayfish.[112]

Macroinvertebrates or crustaceans were the primary prey items by weight for smallmouth bass 70–174 mm TL in the Snake River (1996–1997) and crustaceans comprised the greatest percent of the diet for smallmouth 175–249 mm TL; fish were also important food items, including salmonids, which were consumed by all size classes.[15] Introduced smallmouth bass preyed on sockeye salmon smolts during out-migration from Lake Washington, WA. Juvenile salmon constituted 28% of the diet in the lake and 38% of the diet in a connected canal for smallmouth bass >150 mm TL.[10]

When juveniles move into shallow microhabitats, they frequently use locations that border high-velocity areas that probably have higher densities of larval mayflies and caddisflies, which have been identified as primary prey for juvenile smallmouth bass in late summer.[72]

LITERATURE CITED

1. Funk, J.L. 1957.
2. Cady, E.R. 1945.
3. Eschmeyer, R.W. 1944.
4. Pierce, P.C. and M.J. Van Den Avyle. 1997.
5. Raborn, S.W. et al. 2003.
6. Range, J.D. 1973.
7. Potter, W.A. et al. 1978.
8. Buynak, G.L. et al. 1989.
9. Scott, M.C. and P.L. Angermeier. 1998.
10. Fayram, A.H. and T.H. Sibley. 2000.
11. Zweifel, R.D. et al. 1999.
12. Sammons, S.M. and P.W. Bettoli. 1999.
13. Long, J.M. and W.L. Fisher. 2005.
14. Zorn, T.G. and P.W. Seelbach. 1995.
15. Naughton, G.P. et al. 2004.
16. VanArnum, C.J.G. et al. 2004.
17. Buynak, G.L. and B. Mitchell. 2002.
18. Slipke, J.W. et al. 1998.
19. Ridgway, M.S. and B.J. Shuter. 1997.
20. Philipp, D.P. et al. 1997.
21. Peterson, J.T. and C.F. Rabeni. 1996.
22. Ridgway, M.S. and B.J. Shuter. 1996.
23. Patton, T.M. and W.A. Hubert. 1996.
24. McNeill, A.J. 1995.
25. Beamesderfer, R.C.P. and J.A. North. 1995.
26. Barrett, P.J. and O.E. Maughan. 1994.
27. Kolander, T.D. et al. 1993.
28. Weathers, K.C. and M.B. Bain. 1992.
29. Buynak, G.L. et al. 1991.
30. Langhurst, R.W. and D.L. Schoenike. 1990.
31. Reed, M.S. and C.F. Rabeni. 1989.
32. Hubert, W.A. 1988.
33. Austen, D.J. and D.J. Orth. 1988.
34. Paragamian, V.L. and M.J. Wiley. 1987.
35. McClendon, D.D. and C.F. Rabeni. 1987.
36. Smith, S.M. et al. 2005.
37. Hubert, W.A. 1976a.
38. Bevelhimer, M.S. 1995.
39. Garren, D.A. et al. 2001.
40. Kraai, J.E. and C.R. Munger. 2000.
41. Balkenbush, P.E. and W.L. Fisher. 1999.
42. Austen, D.J. and D.J. Orth. 1985.
43. Peterson, D.C. and A.I. Myhr III. 1977.
44. Peek, F. 1966.
45. Weaver, M.J. et al. 1996.
46. Ploskey, G.R. et al. 1996.
47. Etnier, D.A. and W.C. Starnes. 1993.
48. Tin, H.T. 1982.
49. Meyer, F.A. 1970.
50. Coutant, C.C. 1977.
51. Mabee, P.M. 1995.
52. Carlander, K.D. 1977.
53. Ramsey, J.S. and R.O. Smitherman. 1971.
54. Smitherman, R.O. and J.S. Ramsey. 1972.
A 55. Swingle, W.E. 1965.
56. Van Hassel, J.H. et al. 1988.
57. Pearson, W.D. and B.J. Pearson. 1989.
58. Gammon, J.R. 1998.
59. McCarraher, D.B. et al. 1971.
60. Blackwell, B.G. and M.L. Brown. 2000.
61. Savitz, J. and G. Funk. 2001.
62. Thomas, J.A. et al. 2004.
63. Barans, C.A. and R.A. Tubb. 1973.
64. Smale, M.A and C.F. Rabeni. 1995a.
65. Goodyear, C.D. et al. 1982.
66. Koppelman, J.B. 1994.
67. Ongarato, R.J. and E.J. Snucins. 1993.
68. Kuehne, R.A. 1962.
69. MacRae, P.S.D. and D.A. Jackson. 2001.
70. Gilliland, E. et al. 1991.
71. Faber, D.J. 1984a.
72. Sabo, M.J. and D.J. Orth. 1994.
73. Coble, D.W. 1975.
74. Larimore, R.W. 1952.
75. Lee, D.S. et al. 1980.
76. MacCrimmon, H.R. and W.H. Robbins. 1975.

77. Miller, R.J. 1975.
78. Jenkins, R.M. 1975.
79. Carlander, K.D. 1975.
80. Funk, J.L. 1975a.
81. Latta, W.C. 1975.
82. Fajen, O.F. 1975a.
83. Funk, J.L. and W.L. Pflieger. 1975.
84. Pflieger, W.L. 1975a.
A 85. Indiana Division of Fish and Wildlife. No Date.
A 86. Brown, E.H. 1960.
87. Fajen, O.F. 1975b.
88. Fleener, G.G. 1975.
89. Funk, J.L. 1975a.
90. Coutant, C.C. 1975.
91. Bulkley, R.V. 1975.
92. Eipper, A.W. 1975.
93. Larimore, R.W. 1975.
94. Inslee, T.D. 1975.
95. Flickinger, S.A. et al. 1975.
96. Chance, C.J. et al. 1975.
97. Janssen, F.W. 1992.
98. Wrenn, W.B. 1984.
99. Suski, C.D. et al. 2004.
100. Steinhart, G.B. et al. 2004.
101. Olson, M.H. and B.P. Young. 2003.
102. Olson, M.H. et al. 2003.
103. Pipas, J.C. and F.J. Bulow. 1998.
104. Todd, B.L. and C.F. Rabeni. 1989.
105. Suski, C.D. et al. 2003.
106. Lukas, J.A. and D.J. Orth. 1995.
107. Simonson, T.D. and W.A. Swenson. 1990.
108. Shuter, B.J. and P.E. Ihssen. 1991.
109. Snucins, E.J. and B.J. Shuter. 1991.
110. Morizot, D.C. et al. 1991.
111. Hinch, S.G. and N.C. Collins. 1991.
112. Roell, M.J. and D.J. Orth. 1993.
113. Baylis, J.R. et al. 1993.
114. Kieffer, J.D. et al. 1995.
115. Sowa, S.P. and C.F. Rabeni. 1995.
116. Waters, T.F. et al. 1993.
117. Bevelhimer, M.S. 1996.
118. Walters, J.P. and J.R. Wilson. 1996.
119. Sabo, M.J. and D.J. Orth. 1996.
120. Cole, M.B. and J.R. Moring. 1997.
121. Pierce, P.C. and M.J. Van Den Avyle. 1997.
122. Hodgson, J.R. et al. 1998.
123. Fish, M.P. 1932.
124. Reighard, J. 1906.
125. Ridgway, M.S. et al. 2002.
126. Casselman, J.M. et al. 2002.
127. Swenson, W.A. et al. 2002.
128. Larimore, R.W. 2002.
129. Saunders, R. et al. 2002.
130. Bozek, M.A. et al. 2002.
131. Lyons, J. and P. Kanehl. 2002.
132. Zanden, M.J.V. and J.B. Rasmussen. 2002.
133. Pert, E.J. et al. 2002.
134. Schreer, J.F. and S.J. Cooke. 2002.
135. Shuter, B.J. and M.S. Ridgway. 2002.
136. Suski, C.D. et al. 2002.
137. Cross, F.B. 1967.
138. Mettee, M.F. et al. 2001.
139. Smith, P.W. 1979.
140. Trautman, M.B. 1981.
141. Pearson, W.D. and L.A. Krumholz. 1984.
142. Burr, B.M. and M.L. Warren, Jr. 1986.
143. Hardy, J.D., Jr. 1978.
144. Scott, W.B. and E.J. Crossman. 1973.
145. Becker, G.C. 1983.
146. Jenkins, R.E. and N.M. Burkhead. 1994.
147. Hubert, W.A. 1976b.
148. Boschung, H.T., Jr. and R.L. Mayden. 2004.
149. Ross, S.T. 2001.
150. Robison, H.W. and T.M. Buchanan. 1988.
A 151. Environmental Science and Engineering, Inc. 1992.
A 152. EA Engineering, Science, and Technology. 1994.
A 153. Environmental Science and Engineering, Inc. 1995.
A 154. EA Engineering, Science, and Technology. 2004.
A 155. Axon, J.R. 1984.
156. Everhart, W.H. 1949.
157. Pflieger, W.L. 1966.
158. Neves, R.J. 1975.
159. Gwinner, H.R. 1973.
160. Cathey, H.J. 1973.
161. Stroud, R.H. 1948.
162. Webster, D.A. 1948.
163. Eschmeyer, R.W. 1940.
164. Hubbs, C.L. and R.M. Bailey. 1938.
165. Lukas, J.A. and D.J. Orth. 1993b.
166. Brown, J.A. and P.W. Colgan. 1985.
167. Cherry, D.S. et al. 1975.
168. Gerking, S.D. 1945.

* Original information concerning early life ecology, reproduction, and early growth of smallmouth bass come from field-caught specimens, field-caught specimens subsequently cultured in TVA's laboratory in Norris, TN, and field observations made by the author at several TN locations including: Big Creek, Holston River drainage; Powell River; Tackett Creek, Cumberland River drainage; Nolichucky River; Coal Creek, Clinch River drainage. Spatio-temporal observations for larval and juvenile smallmouth bass come from ichthyoplankton data collected by TVA in the Tennessee Valley during the period 1972–1982. Original descriptive comments on early development come from developmental series cultured from larvae provided by Cohutta National Fish Hatchery, Cohutta, GA, Stuttgart National Fish Hatchery, Stuttgart, AR, and wild-caught larvae from Coal Creek, TN. All specimens are housed at the Division of Fishes, Aquatic Research Center, Indiana Biological Survey, Bloomington, IN (catalogue numbers: TV709, TV740, TV759, TV789, TV800; TV3005, TV3011, TV3081, TV3099).

OTHER IMPORTANT LITERATURE

Anderson, R.O. and A.S. Weithman. 1978.
Burdick, G.E. et al. 1958.
Doan, K.H. 1940.
Jackson, D.A. 2002.
Kerr, S.R. 1966.
Orth, D.J. and T.J. Newcomb. 2002.
Stone, U.B. et al. 1954.
Suttkus, R.D. 1955.

SPOTTED BASS

Micropterus punctulatus (Rafinesque)

Robert Wallus

Micropterus, Greek: *micro,* "small" and *pteron,* "fin"; *punctulatus,* "dotted," in reference to rows of spots on the lower sides.

RANGE

The original range of spotted bass extended from the Appalachian divide in the east to the Great Plains in the west, and from the Gulf of Mexico north to southern OH and IN.[1] Spotted bass occurs naturally in the Ohio River drainage from WV, NC, and OH through IN, IL, KY, and TN. It is also native to the central and lower Mississippi River drainage in MO, KS, OK, AR, LA, and MS and in parts of the Gulf Coast drainage of GA, FL, AL, and TX.[39]

Spotted bass has been introduced into AZ, CA, IA, NB, NM, SC, VA,[1] and MO.[3] The introduction of this species into reservoirs and streams containing smallmouth bass has resulted in hybridization.[10,90]

HABITAT AND MOVEMENT

Spotted bass is found in a variety of stream habitats[13,39,49,50,60–63,65,72,83–85] and has had varied success in reservoirs following impoundment.[39,48,60,62,63,65,85] In streams it generally inhabits permanently flowing,[13,61,62,65,84] warm, moderate-gradient,[63,83,85] small-sized to large streams,[49,62,65,83–85] that are warmer and slightly more turbid than those where smallmouth bass occur.[61,62]

Spotted bass seem to utilize environments intermediately between those preferred by smallmouth and largemouth basses.[48,82,85] Where the three species occur together, the largemouth can be found in quiet backwaters and deep pools, the smallmouth close to or in fast-moving waters, and the spotted bass along the edge of swift water and in shallower pools.[48] It is more tolerant of turbid water and silt substrate than smallmouth or largemouth bass[49] and occurs in the main channels of large rivers almost to the exclusion of other black basses.[61]

Though tolerant of turbidity,[49,61,65,85] spotted bass is also reported from clear streams in many areas including AR,[62] KS, OK,[50] VA,[85] and MO,[83] and has achieved some success in clear reservoirs in AR, OK, and elsewhere.[48,62]

Spotted bass is often abundant in streams with gravel substrate, where it spends much of its time in deep firm-bottomed pools,[62,83] but is also reported to favor the shelter of logs, stumps, and other submerged cover[60,84] over substrates of gravel and sand.[60] Spotted bass is also reported to be quite tolerant of silty or muddy substrates.[49,65] They are often found near aquatic vegetation, submerged logs, and rock or riprap walls in small to large flowing streams, rivers, and reservoirs.[84] In riverine sections of the New River, VA, spotted bass were usually near the banks in areas away from high current velocities in habitats with fine substrate, woody debris, and overhanging bank vegetation.[18]

In KS streams, habitat, such as woody debris and undercut banks, influences spotted bass density and biomass.[19] Woody debris, especially, influences populations; variability in density was best explained by the area of rootwads present and by percent pebble substrate; biomass was positively correlated with the amount of log structure and catch per effort for quality-length spotted bass was positively correlated with the area of bank roots and area of rootwads.[5]

In another KS study spotted bass use of pool habitat in a creek was significantly higher than use of riffles and runs. Cover-habitat was usually woody debris and undercut banks; clay and bedrock substrates were used in proportion to availability and large substrates less than availability. Fine substrates, often associated with logs, were used more than the proportion of availability. Spotted bass were found in the low-velocity areas of pools in flows that varied from 0 to 0.46 m/s. Depths used were similar to availability.[38]

Spotted bass populations in prairie streams of KS and OK generally are controlled by environmental variables associated with late summer conditions when there are low flows, an almost yearly occurrence.[16]

Stream fish communities in which spotted bass was the predominant bass species had a median of 60 species that were 30% cyprinids, 16% darters, 12% sunfish, and 8% suckers.[52]

Spotted bass is intolerant of ice-covered, alkaline, or highly turbid waters and distribution is limited by altitude.[1] In MO, stocked spotted bass were unable to adapt to shifting and unstable stream beds resulting from channelization and sedimentation.[3] This species is rarely found in pools in the lower Mississippi River and is uncommon in oxbows and borrow pits.[72]

In reservoirs, spotted bass prefer rocky shorelines[11,22,28,65,88] and main channel areas[11,28,88] and exhibit longitudinal trends in abundance coincident with trends in water chemistry.[37] Recruitment is reportedly not linked to reservoir hydrology.[24] In deep oligotrophic reservoirs in AL, spotted bass selected rocky substrates and steeply sloping shorelines, avoiding mud bottoms and dense, submerged vegetation.[65]

In an impoundment of the New River, VA, spotted bass were widely distributed but were most commonly found in areas over fine substrate with woody debris and bank vegetation as cover. Ecological segregation between spotted and smallmouth bass in this study appeared to be along the spatial rather than the trophic axis.[18]

In Norris Reservoir, TN, spotted bass were consistently found in deep water in close association with the bottom.[30,39–43,73] In 1943 and 1944, they were generally captured between 7 and 7.6 m deep in water temperatures between 21 and 27°C.[41,42] In April and May distribution was fairly even throughout the upper 9 m of water; during June and July, they were most abundant between 6 and 9 m,[73] but moved into deeper water in the fall (maximum depth about 21 m).[40] Very few spotted bass were netted in the upper 3 m of water after late May.[40]

A study of fish standing crops in 44 hydropower storage reservoirs, 52 hydropower mainstream reservoirs, and 77 flood control, irrigation, water supply, or recreation reservoirs indicated that spotted bass standing crops are positively related to potential prey species such as rock bass, bluegill, longear and green sunfishes, and darters, and to the cohabiting predators largemouth bass, walleye, and grass pickerel. Standing crops of redhorse, bullhead, and flathead catfish are also positively related to spotted bass crops, but those of carp, smallmouth buffalo, drum, trout, smallmouth bass, and chain pickerel are negatively related. Crappies, white bass, and spotted gar crops are not correlated with spotted bass crops.[51]

After state-mandated reductions in point- and non-point–source pollution, oligotrophication of West Point Reservoir, GA, resulted in a restructuring of the black bass population, with fewer, smaller, and less robust largemouth bass and more abundant, but smaller, spotted bass in the fishery.[25]

Spotted bass preferred water temperatures of 23.5–24.4°C in Norris Reservoir, TN.[57,71] In the Wabash River, IN, upper and lower avoidance temperatures were reported as 27 and 22°C.[7] Preferred and upper and lower avoidance temperatures of spotted bass from VA rivers at acclimation temperatures from 30°C in August to 6°C in February are presented in Table 106.[93]

Table 106

Preferred and upper and lower avoidance temperatures (°C) for spotted bass from the New and East Rivers, VA, at acclimation temperatures from 30°C in August to 6°C in February.[93]

Acclimation Temperature	Upper Avoidance Temperature	Preferred Temperature	Lower Avoidance Temperature
30	34	32.1	27
27	34	31.4	27
24	33	32.2	27
21	32	29.5	23
18	31	26.7	21
15	29	24.8	19
12	25	20.1	13
9	21	17.9	7
6	18	16.9	6

Along the Gulf Coast, spotted bass may occur in salinities up to 10–12 ppt;[65,67] growth is poor at salinities above 4 ppt.[67]

There is some indication of annual movement between large rivers and reservoirs and their tributaries; spotted bass reportedly migrate into small streams after high water in the spring to spawn and return to larger waters in the fall.[53,61] Spotted bass migrate up the Ohio River in spring. In OH, they move into larger streams by late summer and spend their winter there, returning to small streams in the spring when water temperature reaches about 10°C.[49] Similar observations are reported from a southern IL stream.[2]

Data point to the existence of both mobile and sedentary individuals in populations of spotted bass in MO streams. Nearly 92% of 24 marked fish remained in the home area; those spotted bass that did travel averaged 39 km upstream and 11 km downstream.[44]

A radio telemetry study in Otter Creek, KS, indicated that spotted bass have small home ranges; mean size home range was 3,954 m[2]. Mean movement varied from 7.2 to 18.1 m/h and differed significantly among seasons, with more movement recorded during spring and fall than in summer and winter. During all seasons, movement was typically lowest at night; peak movement in daylight occurred when water temperature was 16°C.[38] Another telemetry study indicated that total daily movement of spotted bass varied from 255 to 2,815 m using a 15-min tracking interval. Thirty-minute, 1-h, and 2-h simulations showed that diel movements determined by the 15-min tracking interval were always higher than other time interval simulations.[27]

Individual spotted bass in Norris Reservoir, TN, traveled distances as great as 51 km with ranges averaging 9 and 5 km for fish recovered within 1 year of release.[39]

In Lake Logan Martin, AL, the relationship between polychlorinate biphenyl (PCB) concentrations in fish tissue and the distance of fish at capture from the source of the pollution was used to infer the relative motility of several species of fishes; the movement of 99 adult spotted bass averaged 17.2 km and ranged from 0 to 45.6 km.[26]

DISTRIBUTION AND OCCURRENCE IN THE OHIO RIVER SYSTEM

The Ohio River is the type locality of spotted bass.[58] It occurs naturally in the drainage from WV, NC, and OH through IN, IL, KY, and TN,[39] and is commonly reported throughout the length of the Ohio River.[1,58] Based on lockchamber surveys during the period 1957–2001, it has increased in abundance in the Ohio River.[92]

Spotted bass have been reported from the upper, middle, and lower Ohio River since the early 1800s,[68] but considered common in the middle river and less common in the upper river.[70] This species was caught in significantly increasing numbers during the period 1973–1985, with catch per unit effort (CPUE) in gill nets, hoop nets, and electrofishing increasing significantly in upstream sites on the river.[69] Spotted bass were not collected in great abundance in the late 1950s and the change in catch rate may reflect improvements in water quality in the upper portion of the Ohio River.[69]

During the period 1981–2003, the numbers of spotted bass collected from the Ohio River near three electric generating stations at ORM 54 (Ohio Edison Company's W.H. Sammis Plant), ORM 77 (Ohio Power Company's Cardinal Plant), and ORM 260 (Ohio Valley Electric Corporation's Kyger Creek Plant) fluctuated from year to year, but the species was generally classified as common or abundant in these areas of the river.[79–81] From 1981 to 1999 spotted bass was recorded in low numbers from ORM 494 (Indiana-Michigan Power Company's Tanners Creek Plant), but became common in collections during the period 2000–2003.[81] It was also collected in low numbers at TVA's Shawnee Steam Plant at ORM 946 (1987–1992).[79]

Spotted bass is native to the Wabash River drainages of IL[1,83] and IN.[1] Prior to 1980 it was mostly confined to the lower Wabash River, IN, but during the 1980s and early 1990s distribution moved northward, possibly indicative of improved water quality.[59,77] In IL, spotted bass is widely distributed and common in the Wabash-Ohio drainage.[83] Tributaries of the Ohio River in southeastern IN and in the southern tier of counties in OH contain native populations. North-flowing tributaries of the Ohio River from New Martinsburg, WV, downstream all have native populations except at high altitudes.[1] In VA, spotted bass is native only to the Tennessee and Big Sandy drainages. It is regarded as introduced in the New River drainage of VA for lack of an early record.[85] It is present in tributaries of the Tennessee River in western NC.[66]

The spotted bass is most abundant throughout KY and TN and southward into AL.[1] In KY, it is generally distributed and common throughout the state except in the extreme west, where it is uncommon.[60,82] It is common throughout TN[63] and large populations are present in the mid and lower Tennessee River.[1] Spotted bass is found throughout AL,[65,84] where it is represented by two subspecies: *Micropterus punctulatus punctulatus* in all drainages except those of the Mobile Basin and *M. p. henshalli* (Alabama spotted bass) in the Mobile Basin above the Fall Line; intergrades of the two subspecies occur elsewhere in the Mobile Basin.[65] Spotted bass occurs in most river systems of MS including the Tennessee River drainage.[67]

SPAWNING

Location

Spotted bass males may construct their nests on a variety of substrates, including sand and gravel, flat rocks, and compact soil.[65] Nearby cover may be more important in nest site selection than is the nature of the substrate.[39,45,65] In Lake Fort Smith, AR, spotted bass usually used rocky or gravelly spawning substrates.[14] It is reported to spawn in the mouths of tributary streams,[84] on gravel bars, in hatchery ponds, along steep shorelines, and in coves of reservoirs at depths of 2–122 cm.[39]

During the period 1966–1971 in Bull Shoals Reservoir, AR, underwater observations were conducted weekly during the spotted bass spawning season in habitats along a steep bluff and in shallow coves. During two seasons in which water levels were low, nest density was greatest along the bluff; during two high-water seasons, the bluff was not used for spawning. Nests along steep shorelines were on solid rock ledges, large flat rocks, and patches of rubble and gravel and those in coves were on patches of broken rock, large flat rocks, and gravel, compacted soil, and exposed root hairs of flooded trees. Nest depth was correlated with water clarity along the bluff, but not in coves; depths ranged from 0.9 to 6.7 m.[11,39] Nesting sites near cover appear to be preferred.[39,45]

In a study of the effects of reservoir hydrology on basses in Normandy Reservoir, TN, the initiation of spotted bass spawning was not positively related to the first day of full pool, as was the case for largemouth bass; mean hatching dates for both species were positively related to the first day of full pool.[21]

Spotted bass collected in breeding condition in April from the Holston River, TN, and placed in large circular tanks with pea gravel substrate spawned within 3–6 days.*

Season

Beginning of spawning season varies and appears more dependent on water temperature than on calendar date. Generally, throughout their range, spotted bass start spawning about the same time as other black basses, usually during the period April through May and possibly into early June.*,[11,13,14,23,39,46,61,63,66,83–85]

In MO, nesting activity is most intense from mid-April to early June.[61] In AR, spawning began as early as April 10 and extended into late May or early June.[11,14,39] Other reports include: May and June in LA[13] and IL;[83] early April to early June in NC;[66] April and May in VA,[85] TN,*,[63] and AL;[65,84] also March through May in AL.[23]

In AL ponds, brood stock from the Apalachicola River system (*Micropterus punctulatus punctulatus*) spawned May 3–9; stock from the Talapoosa River system (*M. p. henshalli*) spawned April 5–10.[46]

Temperature

Spotted bass are reported to spawn at water temperatures from 12.8 to 25.6°C.[11,13,14,46,56,65,85]

Spotted bass in breeding condition and swim-up larvae were collected by TVA biologists from the Holston River, TN, in water 20–21°C.*

In AR, nests with fresh eggs were found in water temperatures from 12.8 to 22.8°C;[11] but active spawning usually begins when surface water temperatures reach 15°C.[14] Bennett (1965) found that spotted bass spawn at 18°C and Towery (1963 and 1964) concluded that spotted bass spawn at 17°C (cited in reference 14). Spawning in LA occurs at water temperatures of 20–21.1°C.[13]

In AL ponds, brood stock from the Apalachicola River system (*Micropterus punctulatus punctulatus*) spawned at temperatures ranging from 23.3 to 25.6°C; stock from the Talapoosa River system (*M. p. henshalli*) spawned at temperatures ranging from 20.6 to 21.1°C.[46]

Timing of spawning in Normandy Reservoir, TN, was not related to water temperature as much as reservoir hydrology.[21]

Fecundity

In Bull Shoals Reservoir, AR, fecundity of 27 females, estimated by counting the most advanced ova (diameters of ova were multimodal) from a cross section of the ovary and expanding this number by a gravimetric procedure, ranged from 3,249 to 30,586; fecundity was directly correlated to TL, body weight, and age. Mean number of eggs in 21 nests was 5,016. Resorption of residual ova began soon after spawning, and regeneration of ova was in progress by October.[11]

In Lake Fort Smith, AR, fecundity, based on mature ova, ranged from 1,727 to 9,552 with a mean of 4,304, and was significantly correlated with TL, SL, weight, age, and condition factor.[14]

In Halawakee Creek, AL, fecundity generally increased with size of female; estimates ranged from 1,150 for a female that weighed about 85 g to 26,555 for a 953-g fish.[94] In MS, 47,000 eggs are reported for an age-5 fish (Towery, 1964, cited in reference 39).

The southern subspecies of spotted bass appears to produce fewer eggs than the northern.[39]

Sexual Maturity

In Bull Shoals Reservoir, AR, some age-2 spotted bass may have spawned late in the season; not all 3-year-olds could be assumed mature, but by age 4, all females were probably mature and spawning.[11,39] Findings were similar in Lake Fort Smith, AR; no age-1 spotted bass were sexually mature, 34.2% of age-2 fish were mature, 85.5% of age 3 were mature, and 100% of age 4 were mature.[14] Age-2 spotted bass were reported mature in Norris Reservoir and averaged 292 mm TL.[89]

In MS reservoirs, both sexes are mature at the end of their second year and some may reach maturity after 1 year. Females 249 mm TL and males >241 mm TL are generally mature.[39,67]

In AL, stream-dwelling spotted bass usually do not mature until their third year;[65,94] the majority spawn at age 2 in reservoirs.[54,65] In Smith Lake, AL, 90% of all spotted bass were sexually mature at age 2; no differences were observed between sexes in minimum length at maturity. Precocious individuals were mature at 178 mm TL, but most were not mature until they reached 305 mm TL.[54]

Spawning Act

Male spotted bass prepare nests by vigorous sweeping of the substrate with the caudal fin. Nests are usually circular depressions up to 76 cm wide, depending on compactness or particle size of bottom materials.[39]

Once a female is attracted to a nest, courtship and spawning sequences begin. The male guides the female in circles about the nest, repeatedly biting at her opercle and vent. After about 20 min to an hour of such courtship, the female begins depositing her eggs. She aligns herself alongside the male in the nest, tilts away from the male, assuming an angle of about 45° from vertical, and deposits eggs. Egg deposition continues periodically, each deposition requiring 1.5–5 s. Courtship and spawning are a lengthy process, requiring up to 3.5 h for completion.[11,39] Spawning usually involves a single pair in a nest, but more than one female has occasionally been observed attempting to spawn with a single male.[11,39] If initial nests are unsuccessful, males will renest.[11]

Males are reported to defend nests during incubation and larval development,[11,61] but in AL ponds, males were never observed guarding the nest.[46] Spotted bass males in MO moved off the nest once the eggs hatched, remaining in the vicinity, but deserted the fry about the time they left the nest.[61]

EGGS

Description
No information.

Incubation
Five days at 14.4–15.6°C;[11] 2–3 days in AL at about 21°C.[46]

Development
No information.

YOLK-SAC LARVAE

See Figures 93 and 94.

Size Range
Spotted bass that spawned naturally in circular tanks in TN hatched at <6.0 mm TL. Remnants of yolk were present on some cultured specimens at 10.2 mm TL.*

Myomeres
5.9–6.9 mm TL. Preanal 14–16; postanal 17–20; total 32–34.*

7.0–7.9 mm TL. Preanal 15–17; postanal 17–19; total 33–35.*

8.0–10.2 mm TL. Preanal 15–17; postanal 15–17; total 32.*

Morphology
5.8–5.9 mm TL. Body is straight; head large and not deflected; mouth open, lower jaw forming; heart chamber visible anterior to yolk sac. Yolk sac is oval with a large oil globule (0.8–0.9 mm in diameter) placed posterior-dorsally in sac.*

6.0–6.9 mm TL. Developing heart is visible in chamber. Mouth development progresses, lower jaw extends to the middle of the eye by 6.3 mm.*

7.0–7.2 mm TL. Head large with a fully formed mouth and a rounded snout; air bladder visible.*

7.45 mm TL. Snout less rounded. Oil globule still large (0.5–0.7 mm in diameter) and positioned posterior-dorsally in yolk sac.*

7.5–7.9 mm TL. Late yolk-sac larvae; oil greatly reduced, may be present only as small droplets and only small amounts of yolk remain. Changes in head appearance are very noticeable; snout becomes pointed, jaws well developed and projecting; branchiostegal rays visible; opercle well developed, covers gill chamber. Gular musculature is developing and the digestive tract is functional.*

8.0–8.9 mm TL. Most of the oil is gone and only small amounts of yolk remain.*

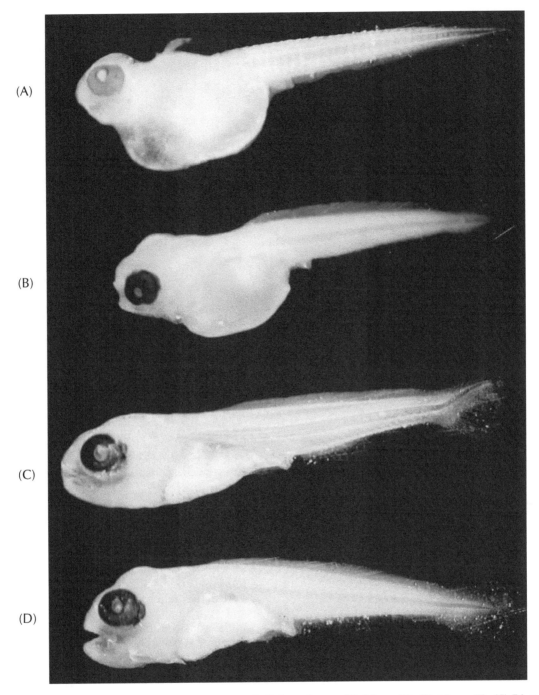

Figure 93 Development of young spotted bass from TN.* A–D. Yolk-sac larvae: (A) 6.0 mm TL; (B) 6.6 mm TL; (C) 7.2 mm TL; (D) 7.8 mm TL. (A–D, original photographs of young spotted bass from TN.)

9.0–10.2 mm TL. Late yolk-sac larvae — small amounts of yolk were still present on all specimens examined.*

Morphometry
See Table 107.

Fin Development
See Table 108.

5.8–5.9 mm TL. Median finfold is continuous from origin at about 6th myomere around urostyle to anus and past anus to posterior margin of the yolk sac. Pectoral fins are flap-like, about 0.5 mm long.*

6.9 mm TL. Basal elements of caudal fin barely visible; urostyle straight.*

7.0–7.2 mm TL. Slight elevations in the profile of the median finfold are noticeable at future locations

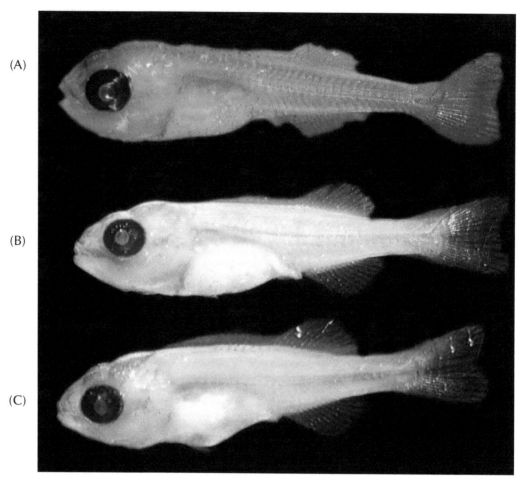

(A)

(B)

(C)

Figure 94 Development of young spotted bass from TN.* (A) Late yolk-sac larva, 8.9 mm TL. B–C. Post yolk-sac larvae: (B) 10.7 mm TL; (C) 11.8 mm TL. (A–C, original photographs.)

of soft dorsal and anal fins. Incipient caudal fin rays appear; basal elements of caudal fin developing; urostyle straight.*

7.45 mm TL. Urostyle flexion begins; 7–8 developing caudal fin rays visible. Pterygiophore differentiation is apparent at bases of developing dorsal and anal fins. Pectoral fins are still flaplike, 0.6–0.7 mm long.*

7.5–7.9 mm TL. Median finfold is still present dorsally and ventrally on caudal peduncle and anterior to the anus ventrally; little dorsal finfold remains anterior to the margin of the developing spinous dorsal fin. Urostylar flexion progresses to about 45°. Hypural plate forms and 13–15 caudal fin rays become apparent; 11–12 are segmented. Structure of developing dorsal and anal fins apparent; 9–10 pterygiophores become visible at the base of both fins and soft fin rays begin to form.*

8.0–8.9 mm TL. At 8.0 mm, the posterior margins of the dorsal and anal fins are defined by notches in the median finfold. By 8.9 mm the posterior profiles of the dorsal and anal fins are formed. Median finfold is absent anterior to the developing spinous dorsal fin, much reduced posterior to the dorsal and anal fins, and still present, ventrally, anterior to the anus. Caudal fin becomes bilobed with 14–17 segmented rays; urostyle remains visible above the hypural plate. Development proceeds in dorsal and anal fins with well-defined pterygiophores at their bases and the formation of spines and rays (Table 108). Pectoral fins are fan-shaped and about 1.0 mm long at 8.9 mm; differentiation begins, but rays are not apparent or barely visible on some specimens. Pelvic buds are visible at 8.0 mm and appear as flaps about 0.3 mm long at 8.9 mm.*

9.0–9.7 mm TL. At 9.0 mm, the median finfold is much reduced posterior to the dorsal and anal fins and only a remnant remains ventrally anterior to the anus. By 9.7 mm, the finfold is completely absorbed. Tip of urostyle extends posterior to the hypural plate. Caudal fin development nearing completion with 16–17 segmented rays. Dorsal and anal fin development progresses (Table 108). At

Table 107

Morphometric data for young spotted bass laboratory cultured by TVA from Holston River, TN, brood stock. Data are presented as mean percent total length (TL) and head length (HL) for selected intervals of TL (N = sample size).

Length Range (mm TL)	5.8–6.9	7.05–7.45	7.5–7.9	8.0–8.95	9.0–9.7	10.0–10.7	11.5–11.8	12.2–14.6	15.1–18.4	19.7–21.7	25.5–31.4	39.5–46.8
N	11	5	5	5	5	5	5	8	5	5	6	2
Length as percent TL												
SL	97.3	96.7	94.1	89.4	88.7	83.3	85.5	82.0	82.2	81.0	81.6	81.5
PreAL	47.6	49.2	50.6	50.8	52.7	52.7	52.0	51.1	51.8	50.8	50.2	52.1
GD	25.0	21.8	20.7	21.9	21.9	22.3	22.2	23.1	24.0	22.9	21.7	22.2
Depth at anus	14.0	15.7	14.0	15.6	16.1	16.3	17.8	18.6	20.0	19.7	19.5	19.7
CPD	5.6	7.6	6.8	6.9	7.7	7.9	8.8	9.1	9.9	9.5	10.1	10.4
ED	11.3	11.3	12.0	12.3	11.6	11.6	11.6	11.3	10.6	10.4	9.8	7.3
HL	23.7	23.9	27.2	28.6	28.8	29.3	29.9	29.2	31.9	28.9	28.5	27.1
Diameter as percent HL												
ED	47.6	47.4	43.9	43.1	40.3	39.7	38.8	38.6	33.1	35.8	34.2	26.9

Table 108

Appearance and numbers of median fin elements related to development and total length (TL) of young spotted bass spawned and cultured from Holston River, TN broodstock.*

TL (mm) Groupings	No. of				
	Segmented Caudal Rays	Dorsal Fin Spines	Dorsal Fin Rays	Anal Fin Spines	Anal Fin Rays
Yolk-sac larvae					
5.9–7.45	0	0	0	0	0
7.5–7.9	0–12	0	0	0	0
8.0–8.9	14–17	0–8	7–10	0–3	6–10
9.0–10.2	16–17	2–9	8–12	0–3	9–11
Post yolk-sac larvae					
10.4–11.8	17	7–9	11–12	3	9–11
12.2–14.6	17	9–10	12	3	10–11
15.1–18.4	17	9–10	12	3	9–11
Juveniles					
19.7–28.9	17	9–10	11–12	3	10
30.9–46.8	17	9–10	12	3	10

9.7 mm, developing dorsal spines are low in profile (about 0.1 mm high). Pectoral fins of some individuals with visible rays. Pelvic fins are flap-like with no visible rays.*

10.0–10.2 mm TL. Adult complement of primary caudal rays (17) present. Rays are visible in pectoral and pelvic fins.*

Pigmentation
5.8–5.9 mm TL. Eye lightly pigmented; yolk yellowish orange. No other pigment obvious.*

6.0–6.3 mm TL. Internal pigment appears on the surface of the yolk posterior to the eye.*

6.5–6.9 mm TL. Pigment is present over the brain and dorsally on the body over the yolk sac. Internal pigment is visible on dorsal and anterior surfaces of the yolk sac. By 6.9 mm, large, stellate chromatophores appear dorsally on the head over the brain; dorsal chromatophores on the body are smaller.*

7.0–7.45 mm TL. Eye is dark. Dorsally, large stellate chromatophores are present on the head over the brain; little pigment is present between the eyes and on the snout; small chromatophores are scattered on the occiput and pigment continues posteriorly on the dorsum in the form of 2–3 rows

of small chromatophores that are present to about the origin of the soft dorsal fin pterygiophores; little, if any, pigment visible on the rest of the dorsum. Laterally, chromatophores are scattered on the yolk; a patch of pigment forms posterior to the eye. By 7.4 mm, a few small chromatophores are scattered, laterally, on the torso from mid-body on to the caudal peduncle and a few are visible at midbody along the median myosepta. Internally, pigment covers the dorsum of the air bladder, yolk, and intestine and is present along the anterior portion of the yolk sac in the gular region. The underside of the head and chin has little or no pigmentation. Many scattered chromatophores are present ventrally and ventrolaterally on the yolk sac and, ventrally, small scattered chromatophores are present on the body posterior to the anus and under the developing anal fin.*

7.5–7.9 mm TL. A bar of pigment is visible posterior to the eye. Scattered pigment is present dorsolaterally on the body to the midline from about the middle of the air bladder posteriorly to the caudal peduncle. At mid-body, a scattered row of small chromatophores is present along the median myosepta. A few scattered chromatophores are present ventrolaterally from the yolk sac/gut region to the area between the developing soft dorsal and anal fins; ventrolateral pigmentation is sparse on the caudal peduncle. Internally, pigment is heavy over the air bladder and present in increased amounts in the gular region. Ventral pigmentation is little changed, with the exception of small chromatophores scattered on the chin.*

8.0–8.9 mm TL. Patterns of pigmentation are as described above with the following additional observations. Dorsal head pigment extends onto snout and lips. Scattered, small chromatophores form a band of pigment posterior to the eye. On some specimens, several stellate chromatophores are present at the base of the tail — located at the juncture of the two hypural elements. This pigment is the beginning of a future caudal spot. A few small chromatophores are scattered ventrolaterally from the anus onto the caudal peduncle and almost to the base of the caudal fin. Internally, a row of chromatophores is visible dorsolaterally (between the dorsum and the median myosepta); appears as a solid line from the occiput to about the origin of the soft dorsal fin, then continues to about the middle of the caudal peduncle as a single row of separated larger chromatophores, each positioned at the juncture of myosepta. Also, from about the anus to the base of the caudal fin, there is an internal stripe of scattered, small chromatophores just ventral to the external lateral stripe described above.*

9.0–9.7 mm TL. Pigmentation at the base of the caudal fin of some specimens extends dorsally and ventrally from the juncture of the hypural elements along the bases of 4 fin rays in each direction. The bar or band of pigment on the opercle posterior to the eye was present on all specimens examined.*

POST YOLK-SAC LARVAE

See Figures 94 and 95.

Size Range
About 10.4 to 18–19 mm TL.*

Myomeres
10.4–10.7 mm TL. Preanal 16–17; postanal 15–16; total 32.*

11.5–11.8 mm TL. Preanal 16–17; postanal 15–16; total 32.*

12.2–14.5 mm TL. Preanal 16–17; postanal 15–16; total 32.*

Morphology
10.0–10.7 mm TL. Head large; snout pointed, jaws well developed and projecting; branchiostegal rays visible; opercle well developed, covers gill chamber.*

11.5–11.8 mm TL. Head and anterior portion of body robust; greatest depth is in the gut region. Becoming bass-like in appearance.*

Morphometry
See Table 107.

Fin Development
See Table 108.

10.4–11.8 mm TL. Adult complement of spines and rays is discernible in caudal, dorsal, and anal fins of most individuals. Profile of the first seven dorsal spines remains low and about 0.3 mm high. Developing rays are visible in pectoral and pelvic fins. Urostyle still extends beyond hypural elements on some individuals.*

12.2–12.6 mm TL. The adult complement of developing spines and rays is discernible in caudal, dorsal, and anal fins. The receding urostyle is about even with the hypural plate.*

15.0–18.0 mm TL. The profile of the spinous dorsal fin becomes elevated, with anterior spines 0.7–0.8 mm long.*

18.0–19.0 mm TL. Development appears complete in all fins.*

Pigmentation
10.4–11.8 mm TL. Prominent pigment features at this size are dark dorsal surfaces of the occiput, head, and snout, a dark opercular band posterior to the eye, and a developing lateral band or stripe. External pigment is present along the median myosepta from the head to the base of the caudal fin. This stripe of external pigment consists of loosely scattered small chromatophores (3–5 per myomere anteriorly and 2–3 per myomere from mid-body onto caudal peduncle). Internally, just below the median mysepta, profuse pigment is present in a stripe from near the bend in the gut, where the anus drops out, posteriorly to the base of the caudal fin. The overall effect of this midlateral external and underlying internal pigmentation is the presence of a lateral stripe, narrow near the head and above the gut, becoming broader posteriorly on the caudal peduncle (due to the effect of the internal pigment). The external pigment becomes profuse near the base of the caudal fin, but there is an area between the forming caudal spot and the profuse pigment that has little pigment — this serves to highlight the caudal spot. A scattered double row of chromatophores is present dorsally and slightly dorsolaterally from the occiput to the spinous dorsal fin; posteriorly there is little dorsal pigment on the caudal peduncle. There is little pigment

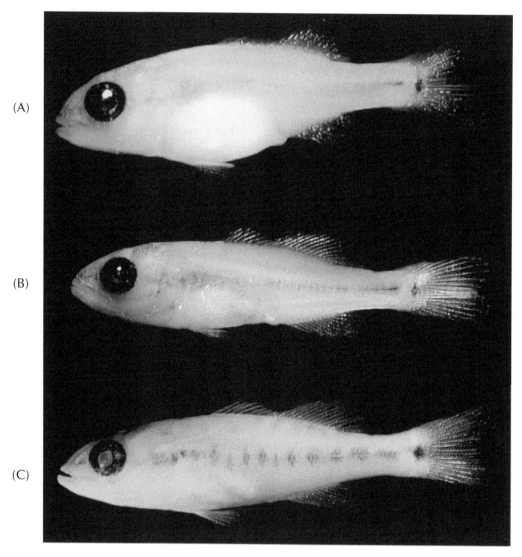

(A)

(B)

(C)

Figure 95 Development of young spotted bass from TN.* (A) Post yolk-sac larva, 16.8 mm TL. B–C. Juveniles: (B) 20.0 mm TL; (C) 31.0 mm TL. (A–C, original photographs of young spotted bass from TN.)

ventrally or ventrolaterally posterior to the anus, though a few scattered small chromatophores may be present at the base of the anal fin and ventrolaterally on the caudal peduncle.*

12.2–12.6 mm TL. The postorbital bar of pigment remains prominent. Also immediately noticeable are the two splotches of pigment at the base of the tail — one an extension of the lateral stripe, the other the caudal spot. Pigment is visible on the spinous dorsal fin of some individuals.*

14.6 mm TL. The caudal spot and lateral stripe remain distinctive. The stripe consists of scattered chromatophores on each myomere along the median myosepta, with a darker splotch of pigment near the base of the caudal fin. The caudal spot consists of three splotches that, from a distance,

appear "V"-shaped on some individuals; some of this pigment begins to outline the proximal ends of the middle two caudal rays. Pigment is now visible along the base of the dorsal and anal fins. Dorsally, the two rows of pigment that were present from the nape to the spinous dorsal fin now extend posteriorly along the base of the soft dorsal fin. A distal splotch of pigment is visible on each spine of the spinous dorsal fin.*

15.1–18.4 mm TL. Bases of the dorsal and anal fins have rows of pigment lateral to the pterygiophores and also rows at the proximal ends of the fin rays. Pigmentation on fins is variable. Middle of the dorsal rays of some individuals is outlined with pigment and pigment is present on the spines of some. The caudal spot remains distinct; pigment extends onto the proximal ends of 3–4 middle rays.*

JUVENILES

See Figure 95.

Size Range

Phase begins at 18–19 mm TL. Sexual maturity is reached as small as 178 mm TL;[54] generally between 241 and 249 mm TL.[39,67]

Morphology

20.7 mm SL. Scales not present.[55]

27.8 mm SL. Squamation has begun, but nape and belly are unscaled.[55]

Scales in lateral series 60–70 in KS,[50] AR,[62] and KY;[82] 55–77 in TN;[63] 58–71 in AL Tennessee River specimens (*Micropterus p. punctulatus*).[84] Scale rows around caudal peduncle usually 24–25 in AR;[62] 22–27 in Tennessee River specimens from AL.[84] Gill rakers 5–7;[63] vertebrae 31–33.50.[63]

Morphometry

See Table 107.

Fins

See Table 108.

Spinous dorsal fin and soft-rayed fin broadly connected;[63,65,84] dorsal spines 10 (9–11);[*,62,63,65,67,82,84] soft rays 11–13, usually 12.[*,50,62,63,65,67,82,84] Anal fin usually with 3 spines[*,50,62,65,67,82,84] and 9–11 (usually 10) soft rays.[*,50,62,63,65,67,82,84] Pectoral fin with 14–17 soft rays.[50,63,65,67,84] Caudal fin has 17 primary rays.[*]

Pigmentation

Note: Conversions from SL to TL in the following section are approximations and based on the formula TL = 1.209 SL presented in Stroud (1948) for Norris Reservoir, TN.[30]

19.7–21.7 mm TL. The lateral stripe is broader (0.2–0.4 mm deep) and remains a prominent visual feature. On larger individuals, scattered chromatophores, above and below the stripe, form 7–8 indistinct blotches. Caudal spot is still prominent, its appearance oblong (0.5–0.7 mm deep). It extends along the base of the 4–6 middle caudal rays. Pigment outlines the lower quarter of the middle caudal fin rays.[*]

24.0–31.4 mm TL. Submarginal pigment (precursor to the submarginal band) is visible in the caudal fin of TN specimens[*] but in AL, specimens 20.7 mm SL (about 25 mm TL) lack any indication of submarginal banding (Figure 96A).[55] Pigment outlines the middle portion of dorsal and anal fin rays. Distal

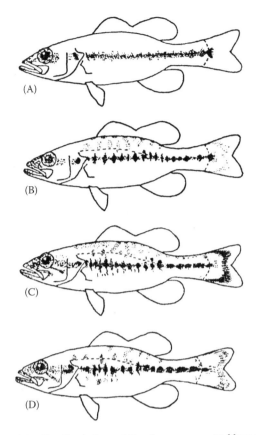

Figure 96 Details of pigmentation for young spotted bass from AL. Body proportions are constant in the drawing outlines and do not reflect proportional differences between sizes. (A) 20.7 mm SL; (B) 27.8 mm SL; (C) 50.4 mm SL; (D) 75.7 mm SL. (A–D reprinted from Figure 2, reference 55, with publisher's permission.)

halves of dorsal fin spines are heavily pigmented. Laterally, 11–15 blotches of pigment are prominent, extending above and below the underlying midlateral stripe. These blotches are the primary bars of melanistic pigment described by Mabee (1995). Secondary bands of pigment (10–12) appear dorsolaterally along the body (Figure 97A).[*,75] At 20.7 mm SL (about 25 mm TL), the caudal spot is triangular and distinct and does not extend very far onto base of caudal fin rays (Figure 96A).[55] The band of pigment is still visible on the opercle posterior to the eye. The lips are pigmented and the top of the head and the snout are still heavily pigmented.[*]

27.8 mm SL (about 34 mm TL). Lateral pigment persists as a deep, narrow stripe,[63] but is dominated by the development of bars on the scales overlaying the lateral stripe. These bars are dorsoventrally elongate on the anterior half of the body and squarish to horizontally elongate on the caudal peduncle. Caudal spot is still present and distinct. Caudal fin lacks distinct submarginal band (Figure 96B).[55]

39.5–46.8 mm TL. Between the submarginal pigment and the caudal spot, a band of pigment is

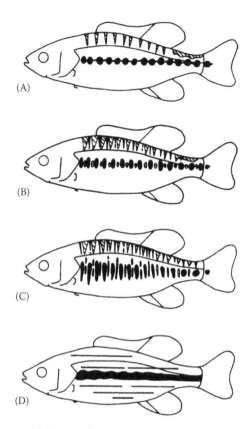

Figure 97 Schematic drawings showing ontogeny of melanistic pigment patterns of young spotted bass. (A) 20–25 mm SL; (B) 35.7 mm SL; (C) 70.0 mm SL; (D) 165 mm SL. (A–D reprinted from Figure 5, reference 75, with publisher's permission.)

present on 4–5 caudal rays. The soft dorsal fin has pigment scattered around and on its rays, but there is little or no pigment on the distal margins of dorsal soft rays. More pigment is present on the dorsal fin than on the anal fin. Dorsal spines are darkly pigmented and small scattered chromatophores are visible in the interradial membranes.* On TN specimens, the caudal spot appears wedge-shaped or triangular. A submarginal band of pigment in caudal fin is evident to the naked eye in TN* but at 34.0 mm SL (about 41 mm TL) in AL the submarginal band of caudal fin is barely defined to the unaided eye (Figure 96B).[55] At 35.7 mm SL (about 43 mm TL) secondary melanistic pigment is visible midlaterally; the adjacent primary bars remain dark (Figure 97B).[75]

>45.0 mm SL (about 54 mm TL). In life, have white iridescent caudal fin margins.[55]

50.0–65.0 mm SL (about 60–79 mm TL). Lateral pigment as described above. Caudal spot is elongate and smaller than eye diameter. Submarginal band of caudal fin is dark and distinct (Figure 96C).[55]

70.0 mm SL (about 85 mm TL). Caudal band begins to expand posteriad and disappear anteriad.[55]

Midlateral secondary bars of melanistic pigment forming; adjacent primary bars remain dark (Figure 97C).[75]

75.7 mm SL (about 92 mm TL). Lateral pigment little changed. Caudal spot remains smaller than eye diameter. Caudal band becomes diffuse (Figure 96D).[55]

120.0–169.0 mm SL (about 145–204 mm TL). Primary and secondary dorsal and midlateral melanistic bars become confluent and indistinguishable (Figure 97D).[75]

Lateral stripe is distinctly present in juvenile *Micropterus punctulatus henshalli* until 120 mm; until 169 mm for *M. p. punctulatus*.[55]

Larger juveniles. Unpigmented areas on the outer margins of the caudal fin lobes are retained by some juveniles but eventually disappear with growth.[55]

Young spotted bass <150 mm have a tricolored tail which is orange-yellow basally, the middle section a vertical, dusky bar, and the distal third whitish.[49,61] Lateral band is usually solid, tending to break up into vertical bars. Caudal spot is large and prominent. Spots on lower sides less well developed in large young but may be absent in smallest young.[49] Three dusky bars are generally present on the cheeks and opercles. Eyes usually reddish.[63]

TAXONOMIC DIAGNOSIS OF YOUNG SPOTTED BASS

Similar species: other centrarchids, especially other *Micropterus* species; also superficially similar to elassomatids.

For comparisons of important morphometric and meristic data, developmental benchmarks, and pigmentary characters between spotted bass and all other centrarchids and elassomatids that occur in the Ohio River drainage, see Tables 4 and 5.

For characters used to distinguish between young basses and the young of other centrarchid genera and *Elassoma*, see provisional key to genera (pages 28–30).

For identification of young basses in the Ohio River drainage, see "Provisional Key to Young of *Micropterus* Species" in genus introduction (pages 251–252).

Characters such as broken lateral band of blotches, prominent basicaudal spot, and tricolor caudal fin of spotted bass were not considered by some as reliable for taxonomic separation of spotted and largemouth bass fry or fingerlings (16–30 mm TL), particularly after preservation in 10% formalin and fading had ensued. Branching of pyloric caeca in largemouth bass as opposed to non-branching of the caeca in spotted bass is a reliable character for distinguishing the species in most instances, but branching or non-branching is difficult to determine for individuals less than 20 mm TL. However, at larger sizes reliable identification of the young of these two species can be obtained by counting pyloric caeca; spotted bass have 10–13 (usually 11–12) compared to 20–33 (usually 24) for largemouth bass.[9,95]

ECOLOGY OF EARLY LIFE PHASES

Occurrence and Distribution (Figure 98)

Eggs. Male spotted bass are reported to defend their nests during egg development.[11,61] They begin fanning the eggs immediately after departure of the female, and thereafter constantly defend them against predators.[11] In AL ponds, male spotted bass were never observed guarding their nests.[46]

Larvae. After hatching, larvae remain in nests for 3–8 days.[39,61] In MO, spotted bass fry disperse from the nest 8–9 days after fertilization if water temperature is >20°C.[61] Male spotted bass guard schools of larvae as long as 4 weeks,[11] or desert the fry about the time they leave the nest.[61]

When spotted bass fry emerged from nests in Bull Shoals Reservoir, AR, they formed compact schools that were constantly guarded and apparently guided into areas of cover by parent males for periods up to 4 weeks. The fry reached lengths as great as 30 mm before the school dispersed.[11] Male spotted bass in hatchery ponds exhibit less parental care.[39] In AL brood ponds, male spotted bass were never observed guarding the young, which scattered shortly after swim-up.[46]

Larval spotted bass (6–15 mm) were collected with drift nets in July 1976 and May 1977 from the New River, VA; estimates of total drift were 125,000 in 1976 and 55,000 in 1977.[35]

Swim-up spotted bass larvae were collected from an east TN stream on 5/16/1986; larvae were found around boulders in slow-moving water 21.1°C and 30–60 cm deep.* Spotted bass spawned and hatched in circular tanks swam up at 6.0–6.5 mm TL.*

Juveniles. In Normandy Reservoir, TN, young spotted bass 125–170 mm TL, collected with electrofishing gear in the spring, came from coves and off of gravel habitat; fish >200 mm came from rubble or riprap habitats or mixed habitats. In the fall, spotted bass captured with electrofishing gear were <200 mm TL; smaller fish (<125 mm) once again were collected from coves and gravel habitats.[22]

Young spotted bass (16–27 mm TL) were collected from gravel bank habitat of the Tennessee River with backpack electrofishing gear in early May from water 19°C; young 18–35 mm were also collected from slough habitat in late May (water temperature 21.5°C).[88] In the Tennessee River, in north

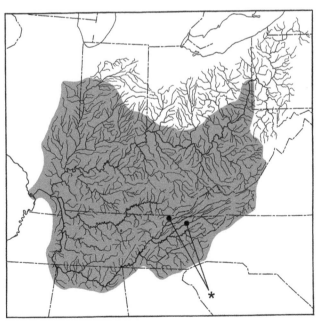

Figure 98 General distribution of spotted bass in the Ohio River system (shaded area), and areas where early life history information has been collected (circles). Asterisk indicates original data and information collected by TVA.

AL, spotted bass <100 mm TL were more common in habitats with fine substrate. Distribution of larger fish was closely associated with proximity to the river channel; substrate size seemed not to be a factor.[88]

The effects of the trophic state of six AL reservoirs displaying a wide range of limnological and morphological characteristics were described and compared to age-0 population characteristics of largemouth and spotted bass. Density and biomass of age-0 spotted bass were not associated with chlorophyll–a concentrations. Higher biomass of young spotted bass was weakly associated with shorter-retention reservoirs with more stable water characterized by higher ionic strength. Biomass increased with trophic state; biomass averaged 197, 139, and 92 g/ha in eutrophic, mesotrophic, and oligotrophic systems, respectively. Spotted bass density did not vary as much as largemouth density among trophic states.[23]

Regression models used to define relationships between reservoir operations and abundance of young fish indicated that seasonal surface area, storage volume, and ratio of inflow to release volume were significant variables for predicting abundance of young spotted bass in southeastern reservoirs.[36] In Bull Shoals Reservoir, AR, biomass of small spotted bass increased in high-water years, but it was so dwarfed by increases in largemouth bass biomass that its percent contribution to total biomass gradually declined; biomass was significantly correlated with hectare-days of flooding during spawning and post-spawning periods. Flooding of vegetation in Bull Shoals Reservoir apparently alters the trophic base because age-0 spotted bass grow faster despite the higher densities and switch to a more productive diet of fish earlier than they do in drier years (an abundance of suitable forage is apparently critical).[87,95] Accelerated growth and an increased complexity of habitat as refuge from predation are suggested as important factors for increased survival, especially during the first summer of life. Biomass of age-0 spotted bass in August is a fairly reliable indicator of future year-class strength.[87]

In Norris Reservoir, TN, young black basses 40–131 mm TL (probably age-0 fish) including spotted bass were preyed upon only by other black basses.[15]

Early Growth
Growth of young spotted bass across its range is variable, but generally slow with TL at age 1 ranging from 53 to 339 mm and reported average at age 4 ranging from 142 to 459 mm (Table 109). Average growth in streams is slower than in reservoirs and maximum growth is smaller (Figure 99).[39,65,67] This relationship is shown for OK[33,64,76] and MO[34,54,61] reservoirs and streams in Table 110 and also holds true for southern waters. Boschung and

Table 109

Central 50% and range of mean of means of TL (mm) for young spotted bass related to age summarized by Carlander (1977) from numerous reports from various locations.[56]

Age	No. of Fish	TL (mm)	
		Central 50%	Range
0	289+	86–109	60–155
1	640+	147–178	53–339
2	878+	190–277	61–387
3	819	244–280	117–424
4	216+	261–377	142–459

Table 110

Reports of average growth for young spotted bass from several locations.

Location	TL (mm) at Each Annulus			
	1	2	3	4
Pickwick Reservoir, AL[29]	118	199	262	306
Lewis Smith Reservoir, AL[54]	107	264	350	391
Norris Reservoir, TN (1939)[89]	155	290	356	429
Norris Reservoir, TN (1936–1945)[30]	124	262	335	378
Norris Reservoir, TN (1936–1973)[91]	121	256	330	378
Cherokee Reservoir, TN[31]	94	218	284	—
Dale Hollow Reservoir, TN[32]	108	192	251	—
Center Hill Reservoir, TN[a]	170	264	354	406
Claytor Lake, VA[b]	104	198	282	340
W. Kerr Scott Reservoir, NC[c,12]	98	182	235	280
Sardis Lake, MS[d]	152	285	343	386
Granada Lake, MS[d]	147	279	345	391
Lake Wappapello, MO[34]	132	259	315	345
Lake Taneycomo, MO[54]	102	213	267	300
MO streams[61]	86	183	254	292
Grand Lake, OK[33]	104	213	300	356
8 OK lakes[76]	114	186	252	375
OK streams[64]	84	165	224	277
Lake Fort Smith, AR[14]	132	203	257	289

[a] From Hargis, H.L. (1965), cited in reference 29.

[b] From Rosebury, D.A. (1950), cited in reference 29.

[c] Introduced population.

[d] From Towery, B.A. (1964), cited in reference 39.

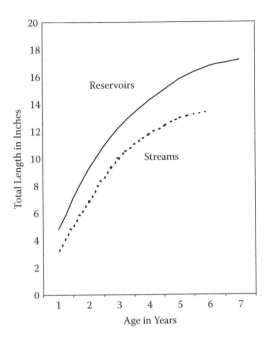

Figure 99 Comparison of growth of northern spotted bass in streams and reservoirs. Unweighted average growth is calculated from data for 1,910 reservoir fish from seven states and 1,115 stream fish from four states. (Reprinted from Figure 1, reference 39, with publisher's permission.)

In KS, young spotted bass were 64–127 mm at age 1, 165–178 mm by age 2, and 218 mm at age 3.[50] In southern IL, average calculated TL at first annulus is about 79 mm; 8 age-1 spotted bass had a mean TL of 173 mm.[2]

At the northern edge of its range in the middle Wabash River, IN, growth of spotted bass is slow.[77] Growth was rapid in the White River, IN; ranges and average SL (mm) for fish collected in July, August, and September were 17–51 (38.5), 33–147 (60.4), and 61–280 (99.7).[74] In OH, YOY in October were 38–100 mm (abundantly fed hatchery fish reached a maximum length of 170 mm); at age 1, 76–180 mm.[49]

Growth of young spotted bass was variable among 7 KY reservoirs,[6,78] but relatively slow, with a maximum calculated TL of only 320 mm at age 4 (Table 111).

Larval spotted bass from a small stream in east TN were 8.5 mm TL on May 16 and grew to about 14 mm in 30 days when cultured in an aquarium (water temperature 20.5–23.3°C) and fed brine shrimp. In another study, larval spotted bass spawned in large circular tanks grew from about 6.0 to 6.5 mm at swim-up to as large as 47 mm (21–47 mm) in about 60 days (water temperature 20.5–23.3°C); bass were fed brine shrimp daily and, as they grew larger, their diet was supplemented with ground trout chow, tubifex worms, and krill fish food.*

Construction of Norris Dam on the Clinch River, TN, was completed in 1936. Growth rates of spotted bass in Norris Reservoir were determined on several occasions during the period 1936–1973 (Table 110).[30,89,91] Regression analysis of TL vs. year indicated that all age groups of spotted bass grew slower as the reservoir aged.[91]

Mayden (2004) tabulated average annual growth of spotted bass based on reports of growth from 8 reservoirs in KY, TN, MS, and AL and 3 streams in KY and AL and reported the following comparisons for the first 4 years: age 1, 82 mm TL in streams compared to 122 mm in reservoirs; age 2, 157 mm compared to 233 mm; age 3, 228 mm compared to 304; and age 4, 278 mm compared to 385 mm.[65]

Table 111

Calculated growth (ranges in mm TL) of young spotted bass from six KY reservoirs during the period 1981–1983[78] and from Cave Run Lake during the period 1985–1994.[6]

	Age			
Reservoir	1	2	3	4
Barkley Reservoir[78]	102–125	178–193	246–295	297–318
Kentucky Reservoir[78]	91–119	185–211	262–287	320
Barren River Lake[78]	66–163	127–193	218–254	282–290
Dewey Lake[78]	69–91	137–158	191–203	234–249
Green River Lake[78]	69–104	140–208	203–251	246–300
Nolin River Lake[78]	53–127	119–223	155–264	188–287
Cave Run Lake[6]	119–126	128–183	221–234	249–277

Age-0 spotted bass grew at rates of 0.4–1.0 mm/day in Normandy Reservoir, TN; first-year growth was not directly affected by the timing of threadfin or gizzard shad hatching.[21]

In the Tennessee River, AL, young spotted bass were 16–27 mm TL in early May, 35–42 mm in late May, and about 100 mm in December.[88]

In AL reservoirs, growth and condition of spotted bass are positively related to reservoir trophic state.[17,23,86] Daily growth rates of age-0 spotted bass were higher in more shallow, productive, short-retention reservoirs than in deeper, long-retention, less productive reservoirs; growth rates were higher in eutrophic than in oligotrophic and mesotrophic systems.[23,86] Young spotted bass grew 16–22% faster in eutrophic reservoirs than in lower trophic states.[23] They reached 304 mm (12 inches) in 2.43 years in eutrophic reservoirs compared to 2.77 years in oligomesotrophic reservoirs.[86]

Age-1 Alabama spotted bass *Micropterus punctulatus henshalli* from 10 AL reservoirs averaged 151 mm TL (119–214 mm); age 2, 251 mm (227–283 mm); age 3, 336 mm (317–359 mm); and age 4, 394 mm (359–427 mm).[17]

Growth was rapid in AL brood ponds.[55] *Micropterus punctulatus punctulatus* hatched in early May, were 21–28 mm TL by early June when stocked into earthen rearing ponds, where they grew to an average of 189 mm and 66.7 g after 199 days. *M. p. henshalli* hatched April 10 were 11–12 mm TL on April 25, and 19–26 mm by May 8; fry 25–33 mm stocked into rearing ponds on May 9 grew to an average of 136 mm and 19.5 g after 222 days. Rearing ponds containing both subspecies were abundantly stocked with *Pimephales promelas* fry for forage.[46] After 18 months, survival of spotted bass fingerlings stocked at an average of 25 mm TL in combination with bluegill and fathead minnows was 84% and the range in size was 229–330 mm and 182–427 g.[47]

Female Alabama spotted bass live longer and attain larger sizes than males, but growth is approximately the same until the third year of life.[54] Growth rates of male and female spotted bass in AR are similar for the first 4 years. Average calculated growth at each annulus for males and females from Lake Fort Smith, AR, are reported as follows:[14]

Age 1:	Males	132
	Females	128
Age 2:	Males	203
	Females	203
Age 3:	Males	257
	Females	262
Age 4:	Males	289
	Females	296

Length–weight relationships for young spotted bass captured in OK lakes[76] are presented in Table 112. Length–weight relationships for spotted bass from AL (Table 113 and Table 114) were similar.[20,29] See Carlander (1977) for additional information on the growth of spotted bass.[56]

Feeding Habits

Crayfish, fish, and other aquatic insects are important food items for young spotted bass.[7,8,14,18]

Six spotted bass 19–122 mm TL, from a LA river, fed mostly on small crustaceans, insects, and fish fry; fish occurred in the stomach of a 43-mm specimen.[13] In the New River, VA, the diets of young spotted and smallmouth bass were very similar. Small bass (40–80 mm TL) preyed heavily on mayflies and stoneflies and other aquatic insects. Terrestrial insects, crayfish, and fish became increasingly important for young bass 81–150 mm TL.[18]

In OK, the most important food for spotted bass in size ranges from <125 to >201 mm SL was crayfish, representing the greatest total volume of food (63.6%) in all length groups and appearing at a greater frequency in stomachs (54.4%) than any other food organism. Forage fishes were the second most utilized food both in total volume (18.4%) and total frequency (34.5%) followed by aquatic invertebrates other than crayfish (15.3% total volume and 33.9% total frequency). Aquatic invertebrates other than crayfish were used to the greatest extent by the 126- to 150-mm-length group.[7]

No difference was found between the diets of smaller and larger sizes (151–357 mm TL) of spotted bass in two WV reservoirs. Crayfish were the most frequently encountered food followed by fish and fish remains; insects were also present.[8]

In the White River, IN, there were differences in diet related to size, but Corixidae (water boatmen) were important in all of the three size categories (<40, 40–69, and ≥70 mm SL) studied, making up about 25% of the total volume of food eaten. Chironomid larvae were also important in all three size groups, present in 30% or more of the stomachs examined and constituting 20.2 and 29% of the total volume of food eaten by spotted bass <40 and ≥70 mm SL. Ephemerid naiads were also an important food item for spotted bass <70 mm SL. Fish became increasingly important with size, representing 13.2, 20.8, and 25.7% volume, respectively, in the three size groups.[74]

Seasonal feeding of young spotted bass in Lake Fort Smith, AR, has been documented. Immature fish fed very little in the winter, only 27% of those examined contained food. Feeding frequency (81%) and mean diet weight (0.208 g) increased sharply in the spring; major food items and their associated percent of total weight of diet were fish (85.5%),

Table 112

Statewide length–weight relationships of young
spotted bass captured from OK lakes
during the period 1972–1973.

TL (mm)	Weight (g)	TL (mm)	Weight (g)	TL (mm)	Weight (g)
25	0.1	100	11.0	175	65.0
30	0.2	105	12.9	180	71.0
35	0.4	110	14.9	185	72.4
40	0.6	115	17.2	190	84.3
45	0.8	120	19.7	195	91.5
50	1.2	125	22.4	200	99.1
55	1.6	130	25.4	205	107.1
60	2.2	135	28.6	210	115.6
65	2.8	140	32.1	215	124.5
70	3.5	145	35.8	220	133.9
75	4.4	150	39.9	225	143.8
80	5.4	155	44.3	230	154.1
85	6.6	160	48.9	235	165.0
90	7.9	165	53.9	240	176.3
95	9.4	170	59.3	245	188.2
				250	200.6

Source: Table Q, page 122 (in part), reference 76.

Table 113

Length–weight relationships of
young spotted bass collected in AL
during the period 1949–1964.

Average Total Length (mm)	Number	Average Empirical Weight (g)
25	33	0.5
51	6,420	1.4
76	4,214	4.2
102	1,665	9.5
127	483	22.7
152	356	39.5
178	254	63.6
203	173	99.9
254	147	136.2
279	84	195.2
305	71	258.8
330	53	363.2

Source: Constructed from data presented in
unnumbered table on page 67, reference 20.

Table 114

Calculated weights at selected
lengths of young spotted bass
from Pickwick Reservoir, AL.

TL (mm)	Weight (g)		
	Males	Females	All
50	2	1	1
100	13	11	10
150	48	42	39
200	116	107	103
250	233	222	216
300	411	403	397

Source: Constructed from data presented
in Table 7 (in part), reference 29.

insects (13.2%), crayfish (1.1%), and fish eggs (0.2%). Eighty-one percent of the immature spotted bass examined in the summer contained food, but the summer mean diet weight was low (0.074 g). Fishes contributed 86.7% of the diet and insects 11.2%. Highest frequency of feeding (84.6%) and mean diet weight (0.287 g) was exhibited in the fall; fish, crayfish, and insects contributed 92.1, 5.4, and 2.5% of the total weight of diet, respectively. In this study, feeding was also characterized by sex.[14]

The principal food of YOY spotted bass during the filling of Beaver Reservoir, AR, changed from entomostracans to insects to fishes as the bass increased in size.[4]

Important food items for spotted bass <100 mm TL from the Tennessee River, north AL, were aquatic insects (41% frequency of occurrence) and malacostracans (34%); entomostracans were also consumed. Spotted bass became almost completely piscivorous after reaching 100 mm TL; between 201 and 400 mm TL, fish occurred in 95% of the stomachs examined; crayfish (6%) were also consumed.[88]

In AL ponds, young spotted bass stocked in combination with bluegill and fathead minnows had little apparent effect on the bluegill population during the first summer after stocking, but virtually eliminated fathead minnows by mid-August.[47]

See Carlander (1977) for additional information on the feeding habits of spotted bass.[56]

LITERATURE CITED

1. MacCrimmon, H.R. and W.H. Robbins. 1975.
2. Lewis, W.M. and D. Elder. 1953.
3. Fajen, O.F. 1975c.
4. Hodson, R.G. and K. Strawn. 1969.
A 5. Tillman, J.S. and C.S. Guy. 1997.
6. Buynak, G.L. 1995.
7. Scalet, C.G. 1977.
8. Lewis, G.E. 1976.
9. Applegate, R.L. 1966.
10. Pierce, P.C. and M.J. Van Den Avyle. 1997.
11. Vogele, L.E. 1975a.
A 12. Simpson, J.A. et al. 1988.
13. Ryan, P.W. et al. 1970.
14. Olmsted, L.L. 1974.
15. Raburn, S.W. et al. 2003.
16. Layher, W.G. et al. 1987.
17. DiCenzo, V.J. et al. 1995.
18. Scott, M.C. and P.L. Angermeier. 1998.
19. Tillma, J.S. et al. 1998.
A 20. Swingle, W.E. 1965.
21. Sammons, S.M. et al. 1999.
22. Sammons, S.M. and P.W. Bettoli. 1999.
23. Greene, J.C. and M.J. Maceina. 2000.
24. Sammons, S.M. and P.W. Bettoli. 2000.
25. Maceina, M.J. and D.R. Bayne. 2001.
26. Bayne, D.R. et al. 2002.
27. Horton, T.B. et al. 2004.
28. Long, J.M. and W.L. Fisher. 2005.
29. Hubert, W.A. 1976a.
30. Stroud, R.H. 1948.
31. Stroud, R.H. 1949.
32. Range, J.D. 1973.
33. Jenkins, R.M. 1953a.
34. Patriarche, M.H. 1953.
35. Potter, W.A. et al. 1978.
36. Reinert, T.R. et al. 1996.
37. Buynak, G.L. et al. 1989.
38. Horton, T.B. and C.S. Guy. 2002.
39. Vogele, L.E. 1975b.
40. Cady, E.R. 1945.

41. Dendy, J.S. 1945.
42. Dendy, J.S. 1946.
43. Haslbauer, O.F. 1945.
44. Funk, J.L. 1957.
45. Vogele, L.E. and W.C. Rainwater. 1975.
46. Smitherman, R.O. and J.S. Ramsey. 1972.
47. Smitherman. R.O. 1975.
48. Miller, R.J. 1975.
49. Trautman, M.B. 1981.
50. Cross, F.B. 1967.
51. Jenkins, R.M. 1975.
52. Funk, J.L. 1975b.
53. Fajen, O.F. 1975a.
54. Webb, J.F. and W.C. Reeves. 1975.
55. Ramsey, J.S. and R.O. Smitherman. 1971.
56. Carlander, K.D. 1977.
57. Dendy, J.S. 1948.
58. Pearson, W.D. and L.A. Krumholz. 1984.
59. Gammon, J.R. 1998.
60. Burr, B.M. and M.L. Warren, Jr. 1986.
61. Pflieger, W.L. 1975b.
62. Robison, H.W. and T.M. Buchanan. 1988.
63. Etnier, D.A. and W.C. Starnes. 1993.
64. Finnell, J.C. et al. 1956.
65. Boschung, H.T., Jr. and R.L. Mayden. 2004.
66. Menhinick, E.F. 1991.
67. Ross, S.T. 2001.
68. Pearson, W.D. and B.J. Pearson. 1989.
69. Van Hassel, J.H. et al. 1988.
70. Reash, R.J. and J.H. Van Hassel. 1988.
71. Coutant, C.C. 1977.
72. Baker, J.A. et al. 1991.
73. Eschmeyer, R.W. 1944.
74. Whitaker, J.O., Jr. 1974.
75. Mabee, P.M. 1995.
A 76. Mense, J.B. 1976.
A 77. Gammon, J.R. 1991.
A 78. Axon, J.R. 1984.
A 79. EA Engineering, Science, and Technology. 1994.

A 80. Environmental Science and Engineering, Inc. 1995.

A 81. EA Engineering, Science, and Technology. 2004.

82. Clay, W.M. 1975.

83. Smith, P.W. 1979.

84. Mettee, M.F. et al. 2001.

85. Jenkins, R.E. and N.M. Burkhead. 1994.

86. Maceina, M.J. et al. 1996.

87. Ploskey, G.R. et al. 1996.

88. Janssen, F.W. 1992.

89. Eschmeyer, R.W. 1940.

90. Koppelman, J.B. 1994.

91. Chance, C.J. et al. 1975.

92. Thomas, J.A. et al. 2004.

93. Cherry, D.S. et al. 1975.

94. Hurst, H.N. 1969.

95. Applegate, R.L. et al. 1966.

* Original observations on spawning times, early development, culture, early life ecology, and early growth come from unpublished TVA data and from brood stock and subsequent progeny captured from the Holston River in east TN and also from field collections by R. Wallus of larval spotted bass from Tackett Creek, Campbell County, TN. Specimens are curated at the Division of Fishes, Aquatic Research Center, Indiana Biological Survey, Bloomington, IN (catalogue numbers: TV2615, DS64; TV3100).

OTHER IMPORTANT LITERATURE

Bennett, G.W. 1965.

Hargis, H.L. 1965.

Howland, J.D. 1932.

Rosebery, D.A. 1950.

Towery, B.A. 1963. (In Appendix)

Towery, B.A. 1964. (In Appendix)

LARGEMOUTH BASS

Micropterus salmoides (Lacepede)

J. Fred Heitman and Robert Wallus

Micropterus: small fin — apparently in reference to a torn fin on the type specimen which appeared as a separate fin; *salmoides*: from salmo, a name often formerly applied to this species in the southern United States.

Largemouth bass is one of the most important game fishes in the United States.[22] Populations have flourished since the early 20th century due to the creation of reservoirs where habitats generally favor this species over related species. Few species have provided as popular and dependable sport fisheries as the largemouth bass, and still fewer have the proven potential for intensive management within natural and man-made waters.[131] Socially, bass fishing has increased from moderate interest in the 1950s to cult status by the 21st century. As a game fish, largemouth bass is one of the most sought after warm water fishes in America. Annually, thousands of anglers participate in bass fishing tournaments organized from the local to the national level.

RANGE

Largemouth bass was the first of the black basses to be recorded in North America, taken from northern FL during the French Ribault expedition of 1562. Early records seldom differentiate between species of black basses, so determination of the native distribution is difficult at best. Species differentiation was established in the early 19th century.[14] Most reports indicate that largemouth bass are native to the central and southeastern United States,[1,87] including the Great Lakes, St. Lawrence and Mississippi Basins,[20] the Ohio River Valley,[88,234] and the Gulf and south Atlantic slopes,[20] with original range from northeastern Mexico to FL, much of the Mississippi River, north to southern QU and ON, and on the Atlantic slope north only to central SC.[259]

While native to most of the eastern United States, largemouth bass has been extensively stocked in all of the lower 48 states[1,69,87,88,134] and elsewhere in the world,[14] including Puerto Rico,[2,3,113] Guatemala,[135] Austria, Finland, France, Hungary, Czechoslovakia, Germany, Italy, Holland, Spain, Sweden, England, Russia, Philippines, Japan, Cuba, South Africa, Natal, West Cameroon, Madagascar, and Belgium.[22] Early successes in the spread of this species in the United States stimulated the U.S. Fish Commission to try to augment the supply of wild fish with fish from a more controlled source. A bass hatchery for this purpose was established about 1886 at Wythville, VA.[149] These extensive introductions throughout the United States and the world have often been made as management options more politically than biologically based. Successful introductions of largemouth bass are seldom described or measured and the impact on native fish is seldom addressed.[88]

There are two distinct subspecies of *Micropterus salmoides*: northern largemouth bass *M. s. salmoides* and the Florida largemouth bass *M. s. floridanus*.[152,254] The extent of the original distribution of Florida largemouth bass is thought to be peninsular FL,[20] but this species has been extensively introduced outside its native range in the United States as part of management efforts to improve sport fish stocks, primarily through changes in the genetic composition of local northern largemouth bass populations resulting from hybridization.[79,102,103,105,107,111,114,131,198,256] Meristic data were originally used to distinguish between the two subspecies;[125] however, because of the overlap in their ranges and hybridization, external meristics often lead to confusion in discriminating between the two.[6,30,105,111] Recently, use of genetic markers has proved a reliable way to distinguish between the subspecies.[61,63,108] Examination of mitochondrial DNA allows distinction between not only the northern and Florida subspecies but also within northern largemouth bass between northern and southern latitudes in North America.[254] Generally it is agreed that Florida largemouth bass require a longer growing season to reach the trophy-size status desired by anglers and the northern limit of introductions for management purposes would include only the southern portion of the Ohio River drainage including waters in the southern Tennessee River valley.

The reader is referred to MacCrimmon and Robbins (1975) for an excellent description of the native distribution of all black basses and an historical review of the spread of largemouth bass through-out the United States.[14]

HABITAT AND MOVEMENT

Largemouth bass prefer more eutrophic conditions than other members of the genus. They often select microhabitats that contain significant cover, both woody and/or soft vegetative. While they prefer a gravel substrate they are frequently found in many other areas.[163]

Historically, largemouth bass inhabited the lower reaches of rivers and streams.[140] In these habitats, waters tended to be more eutrophic and slower moving. Other *Micropterus* species inhabited other stream reaches. In Ozark streams largemouth bass replaced smallmouth bass when stream temperatures increased and pool area increased.[174]

Largemouth bass reaches relatively high levels of abundance in rivers and streams in parts of its range,[137,140] yet little is known about its population dynamics in streams or rivers, as most work has been done on lentic populations.[140] From 15 streams in the midwest and southern United States, the standing crop of largemouth bass ranged from 0 to 5 kg/ha with a mean of 3.6 kg/ha. In streams where largemouth bass was the predominant predator, there was a median of 53 other species; 18% sunfish, 17% minnows, 7% darters, and 5% suckers, and largemouth bass made up 1% of the total number of individuals, minnows 44%, and sunfish 16%.[137]

There are three distinct stream communities in which largemouth bass have been commonly found. They occur as the only bass species in low gradient streams in the Atlantic and Gulf coastal plains. In these systems, they are the major predator in a complex community of sunfishes, shads, and many estuarine and marine species. Minnows and darters are not typically abundant in these systems. Largemouth was the only bass present in prairie streams in the central United States. These systems typically contained relatively few sunfishes and minnows but carp, carpsuckers, and freshwater drum were likely abundant. Largemouth bass were not abundant in these systems, and it was suggested that those individuals present may have escaped from stocked ponds in the watershed. Largemouth bass may be present in systems where either spotted bass or smallmouth bass predominate. In these situations largemouth bass make up a small part of the total biomass of the fish community.[137]

In the Shoals Reach of the Tennessee River, AL, largemouth bass occupied a wider range of habitat types than did smallmouth bass or spotted bass. Adult largemouth bass were most frequently associated with shallow boulder areas with little current,

but deep, turbulent areas also were frequented.[200] In the lower Mississippi River watershed, largemouth bass are abundant in oxbow lakes and borrow pits and commonly found in sloughs and seasonal floodplain waters.[219]

Since the late 1930s, large reservoirs have become increasingly abundant and this habitat has allowed largemouth bass populations to flourish. Particularly in the South and Southern Plains, reservoir construction has provided millions of acres of excellent largemouth bass habitat. Population characteristics of largemouth bass in lakes and reservoirs are influenced by both biotic and abiotic environmental factors. Increased density of largemouth bass typically leads to reduced growth, size structure, and condition, particularly in bodies of water <40 ha.[52]

There is a tendency for largemouth bass populations to exhibit a boom-and-bust response in newly created reservoirs. During the initial years of impoundment, there is an abundance of cover and excess nutrients that stimulate fish production. During the early period (up to about 5 years) after dam construction, fish growth and recruitment are remarkable. After the initial surge of nutrients and the deterioration of soft vegetative cover, reservoirs begin to produce year-classes that are of a more typical size, and largemouth bass will exhibit more typical growth rates. The large initial year-classes produced during the boom will carry forward throughout the life of those year-classes and continue to provide angling opportunities for up to 15 years after impoundment. Subsequent year-classes exhibit typical fluctuations in strength based on a variety of factors.[53,146,147,157,258]

In reservoirs largemouth bass are generally associated with littoral areas,[240] and mostly associated with coves,[54,56,200,267] although increasingly it seems that adults may inhabit offshore locales (little is known about these subpopulations). They occur in most inshore habitats[240] and tend to be found more often in the upper ends of coves and away from wind-swept beaches without structure.[38] They are usually found in areas associated with some type of cover and, generally, populations are positively influenced by increased vegetative cover.[26,53,135] In three southern reservoirs, largemouth bass were more abundant during all seasons in habitats with coarse woody debris when compared to developed and undeveloped habitats.[165] However, if cover is too dense, bass populations may suffer. In experimental ponds, juvenile and adult largemouth bass preferred medium interstice (150 mm) structure over small (40 mm) or large (350 mm) interstices.[12]

Largemouth bass are successful in large southern lakes that are shallow enough for rooted vegetation to grow in all parts of the lake. In these desirable habitats, largemouth bass is a principal predator in the system as well as a major game species for anglers. In northern lakes, typically too deep for vast areas of aquatic plants, largemouth bass is not the major game fish species and is restricted to the weedy littoral areas along the shoreline.[135]

In general, largemouth bass tend to prefer and thrive in more eutrophic conditions than do other *Micropterus* spp.[172,176] In AL, largemouth bass are more abundant in eutrophic (chlorophyll *a* >8 mg/m³) reservoirs than are spotted bass. They also are more abundant than spotted bass in shallower, generally shorter retention reservoirs with higher conductivities and chlorophyll *a* than in deeper, less productive reservoirs with generally longer retention and lower conductivities. Chlorophyll *a* was the strongest determinant of largemouth bass density and biomass.[170] Reservoir sport fish populations decline when water quality becomes more oligotrophic.[24] This effect was observed for largemouth bass populations in AL reservoirs, although catch rates did not differ between oligotrophic and eutrophic reservoirs.[25] Electrofishing catch rates of largemouth bass in Cave Run Lake, KY, were higher in the upper portion of the reservoir, which was classified as eutrophic, than in mid-lake or lower lake areas, which were classified as mesotrophic and oligotrophic, respectively.[176] Oligotrophication due to reduced total phosphorus of West Point Reservoir, GA, resulted in a restructuring of black bass populations with fewer, smaller, and less robust largemouth bass.[155]

Growth, natural mortality, latitude, elevation, average air temperature, and degree-days exceeding 10°C were summarized for 698 populations of largemouth bass in North America and used in simulations to determine the effects of fishing on populations of varying productivity. Length at age and growth equation parameters varied greatly among populations and resulted in ages at quality length (300 mm) ranging from 1 to 10 years. Natural mortality also varied widely, but the conditional annual rate averaged 35%. Natural mortality was significantly correlated with latitude, mean air temperature, and degree-days exceeding 10°C. Age at quality length, used as an index of growth rate, was positively correlated with latitude and elevation and negatively correlated with mean air temperature and degree-days exceeding 10°C. Simulations suggested that the effects of exploitation and minimum length limits on yield, harvest, catch rate of stock-length fish (≥200 mm), proportional stock density, and biomass vary substantially with population productivity and that management options increase with population productivity.[269]

Positive correlations were found between standing crops of largemouth bass and outlet depth, length of growing season, and total dissolved solids in 171 U.S. reservoirs. With regard to other species, largemouth bass populations were positively related to gizzard shad and some sunfish, as might be expected, but also to coexisting predators such as spotted bass, crappie, white bass, spotted gar, and chain pickerel. These correlations should be negative if interspecific competition were a controlling factor. Largemouth bass populations also were positively influenced by omnivorous catfishes, bottom feeding carp, and zooplankton feeding bigmouth buffalo. There was also a positive correlation with total standing crop, suggesting that relatively high basic fertility overrides the importance of species associations. Populations of largemouth bass were negatively related to populations of smallmouth bass.[134]

Largemouth bass inhabited warmer water than any other game fish in Norris Reservoir, TN. During the summer, most largemouth bass were reported from the epilimnion in water only slightly, if at all, cooler than the surface water. As the epilimnion thickened, they moved to deeper water, which seems to indicate that water temperature is an important factor affecting their vertical distribution. They not only tolerated water temperatures of 26.7°C, but appeared to prefer such conditions to cooler water. They were normally found above the thermocline, and only a small portion of the population was required to make adjustments in depth distribution as a result of oxygen depletion.[221] They were collected from water depths ranging from 0.3 to 22.3 m; average depth of capture ranged from 4.2 to 5.8 m. About half of the largemouth bass collected with gill nets fished immediately above the bottom came from the lower third of the gill nets, suggesting an affinity for that stratum of water immediately above the floor of the reservoir.[222,223]

Temperature is undoubtedly the environmental factor that most consistently influences the lives and habits of largemouth bass (Table 115). It controls the reproductive cycle, determines the rate of feeding and metabolism, determines swimming performance and growth rate, and influences distribution in lakes and rivers. Fish actively select to avoid conditions that approach lethal levels, for temperature or other pollutants. In cold climates largemouth bass overwinter in deep water, which is the warmest available. In east Tennessee as the spring sun begins to warm the surface layers, largemouth bass move into the shallows and are generally found in nearly the warmest water available. They stay there until the water temperature rises above 27°C. When seasonal or daily temperatures at the surface rise above 27°C, fish that have cooler water below them in a stratified lake swim deeper, following the water that stays near 27°C as long as oxygen is available.[1,43]

Table 115

Important water temperatures (°C) in the life history of largemouth bass (LMB).[143]

Activity	Temperature	Notes
Spawning	18.9–20	
Incubation	15.25	Below this temperature males desert the nest and the eggs die
Feeding (maximum)	20–27	
Growth range	10–35.5	
Growth optimum	27	
Max swimming speed	25–30	
Incipient lethal temp	36.5	
Preferred temp (adults)	27	
Preferred temp (young)	30–32	
Maximum temp	36.5	Northern LMB
Maximum temp	34	Florida LMB

Largemouth bass are often associated with warmer waters than other related species. The upper critical thermal maximum temperature for northern largemouth bass, Florida largemouth bass, and their reciprocal hybrids was 39.2°C and increased with acclimation temperature.[11] In another study, laboratory testing of preferred temperatures of northern largemouth bass, Florida largemouth bass, and their F_1 hybrids indicated a linear relationship between acclimation temperature and preferred temperature. For all acclimation temperatures (seven temperatures ranging from 8 to 32°C), except 32°C, all stocks preferred temperatures higher or nearly equal to those to which they were acclimated. All stocks acclimated at 32°C preferred areas of their test chambers with lower temperatures.[32]

Small largemouth bass seem to prefer warmer water than do adults. Young bass reportedly stay in the shallows longer in the summer than adults. Winter attraction of largemouth bass to power plant discharges is common.[143] Largemouth bass distribution was affected by high temperatures in a cooling reservoir. Adult largemouth bass were found in more desirable — although very high (up to 35°C) — temperature ranges than juvenile largemouth bass, which were found in water 36–39°C. In winter, bass were found throughout the reservoir, but in summer were restricted to shallow areas of slightly cooler water temperatures.[148]

Largemouth bass have greater oxygen requirements for survival and growth than do many other species of fish. They cannot remain in low oxygen environments, such as the hypolimnion, for extended periods of time. Bass require more dissolved oxygen (DO) at higher temperatures and the water holds less oxygen so there is greater stress on the bass. They strongly avoided water with DO <1.5 ppm; the avoidance reaction decreased sharply when the DO was 3 ppm, and ceased when the DO was 6 ppm.[144] Largemouth bass tolerate DO concentrations of 0.92 mg/l at 25°C, 1.19 mg/l at 30°C,[144] and 1.4 mg/l at 25°C[19] and 35°C,[144] but avoid concentrations >3.0 mg/l at 25°C.[19] When the DO was lowered slowly, largemouth bass could tolerate lower DO concentrations — 0.83 ppm at 25°C and 1.23 ppm at 35°C.

Though sensitive to dissolved oxygen levels, largemouth bass are tolerant of a wide range of environmental conditions. In a laboratory study all fish lived 30 days at a pH of 4.5, but none survived more than 9 days at pH 4.0.[33,144,145] They can tolerate un-ionized NH_3 at a concentration of 1.68 mg/l and NO_3 at 480 mg/l.[19] They are tolerant of phosphorus and turbidity[163] (101,000 mg/l).[144] Largemouth bass continued feeding when pentachlorophenal concentrations were 88 µg/l.[4] They are reported from salinities of 13.4 ppt[144] to 24.4 ppt.[22] This species is also considered tolerant of watershed disturbance and shoreline disturbance based on sampling in 169 lakes in the northeastern United States.[163] Effects of elevated CO_2 levels are not well understood in largemouth bass.[144]

Latitude was not a good predictor of largemouth bass naturalization in WY where populations were not expected to become established in lentic environments at altitudes greater than 1,900 feet above mean sea level and without at least a 100-day growing season. [175]

Mean standing crop of largemouth bass varies greatly by location. Where largemouth bass, smallmouth bass, and spotted bass were all present, largemouth bass made up 66%, spotted bass 22%, and smallmouth bass 12% of the black bass standing crop. Black basses typically make up about one-third of the predator standing crop. In 173 reservoirs the mean standing crop of largemouth bass was 10.0 kg/ha with a maximum of 59.1 kg/ha.[134] Standing crop data for largemouth bass in natural lakes are limited, but in several studies it ranged from 3.36 to 74.48 kg/ha, and in smaller northern lakes the standing crop was 1.3 to 3.36 kg/ha.[135]

Numbers of largemouth bass in OH ponds ranged from 250/ha to 1,668/ha with an average of 472/ha. The standing crop of largemouth bass in "balanced" KY ponds averaged 55 kg/ha; 41 kg/ha in "unbalanced" ponds. In midwestern ponds, largemouth bass standing crop ranged from 3.4 to 106.7 kg/ha totaling 3–3,203 individuals/ha. Population averages were 20 bass and 39.3 kg/ha. Densities in OK ponds ranged from 2 to 359 kg/ha and averaged 49 kg/ha. In an AL community lake (3.95 ha), there were 1,569 fish from 51 mm to 55.8 cm TL that weighed a total 111.04 kg.[134,136]

At all life stages, largemouth bass generally exhibit little movement; however, there tends to be two types of largemouth bass, those that are essentially solitary and closely associated with structural elements and a small portion of the population that will move significant distances.[87,135] The reasons that some individuals tend to move more than others are not known.

The mean home-range size for largemouth bass in Cedar Lake, IL, during June–December 1980 and May–July 1981 was 0.18–2.07 ha. Two of nine largemouth bass had two separate home ranges and one bass had two primary occupation areas within its home range.[266]

The relative motility of largemouth bass in an AL reservoir was determined based on tissue polychlorinated biphenyl (PCB) residues. Largemouth bass 223–585 mm TL (N = 121) moved an average of 17.2 km (0–45.6 km) away from the area of exposure to the PCBs.[268]

In the Great Lakes, largemouth bass may move short distances inshore or into marshes to spawn.[205] As they mature in northern lakes, they move to deeper waters, but not to the depths that are typically inhabited by smallmouth bass.[135]

Largemouth bass caught in bass tournaments on the Hudson and St. Lawrence Rivers, NY, moved little after being caught and released.[23]

Largemouth bass are often stocked with bluegill or other sunfish in small farm ponds throughout the United States. Particularly in the South, the bass–bluegill combination has proven to be an easily sustainable community in small impoundments. In a 10-year study of 38 OK farm ponds, it was determined that in ponds stocked at an approximate 1:1 ratio of bluegill or redear sunfish to largemouth bass and in quantities of somewhat less than 75 fish of each species per acre, the bass–bluegill combination is capable of providing more satisfactory fish populations.[206]

Largemouth bass virus (LMBV) was first discovered in 1991 in Florida.[48] LMBV has been cited as a potential cause of numerous fish kills in the eastern United States and has been linked to declines in the catch of memorable-sized largemouth bass.[95] In five AL reservoirs, LMBV was most common in largemouth bass that ranged from 250 to 400 mm. The virus was most common in adults and seldom was found in juveniles <100 mm TL and was rare in fish >500 mm. Infected fish generally had a lower relative weight and grew more slowly after age 3 than uninfected fish. While LMBV could not definitively be linked to changes in largemouth bass population metrics in five AL reservoirs, the virus appeared to be linked to the phenomenon.[95] A good review of LMBV may be found in the work of Grizzle and Brunner (2003)[49] and an excellent review of the diseases of largemouth bass is given by Sullivan (1975).[141]

DISTRIBUTION AND OCCURRENCE IN THE OHIO RIVER SYSTEM

Largemouth bass is native to the entire Ohio River drainage which is located in the heart of its original area of distribution.[14,20] This species has been collected from throughout the Ohio River since the early 1800s,[202] and it is commonly found in the river from Pittsburg to Cairo.[203] It probably became the most abundant of the black basses in the river before 1950, following the completion of the original set of 46 navigation dams in 1937.[203] Its abundance has increased as the river and its tributaries have become more impounded. Being typically a species that prefers backwater areas of larger streams and rivers, the largemouth bass has thrived in the multitude of impoundments that have been created in the last 75 years. As a result of the impoundments, the species is currently more abundant in the Ohio River drainage than at any time in history.[201–203, 225, 251, 252]

Largemouth bass was never found in great abundance in the upper Ohio River in the late 1950s, but was among the most commonly collected species during the period 1973–1985. It was one of the dominant species in electrofishing catches made near five electric generating plants between ORM 494 and ORM 54. It was statistically more abundant in the upper reaches of the river.[201] During the period 1981–1993, largemouth bass were present all years in collections made near the W.H. Sammis electric generating plant (ORM 54). They occurred in samples in low numbers from 1981 to 1987 and thereafter were generally common in samples.[251,252]

During the period 1981–2003, largemouth bass were present all years but generally in low numbers in collections made near Cardinal electric generating plant (22 years) at ORM 77 and the Kyger Creek Plant (17 years) at ORM 260. They were also present in collections all 23 years at Tanners Creek Plant (ORM 494), where they were collected in low numbers from 1981 to 1985; thereafter, they were generally common in collections.[250] At TVA's Shawnee Plant (ORM 946), largemouth bass were present in collections in low numbers all years during the period 1987–1992.[251] Based on lockchamber surveys (1957–2001) largemouth bass have increased in abundance in the Ohio River since 1957.[225]

The Florida subspecies, while not native to the river system, has been widely introduced, particularly in the southern portion of the drainage.[79,102,103,105,107,111,114,131,198]

SPAWNING

Location
Male largemouth bass clear nest depressions of leaves, silt, and other debris prior to spawning.[26] A distinct nest may or may not be excavated and largemouth bass nests are much less conspicuous than those of smallmouth bass. Only a layer of sediment may be swept off the nesting area. The nests of the largemouth are the property of individual males.[242,243] In some cases males make "test" nests, which may or may not be used later by largemouth bass or other species.[191]

Largemouth bass have been reported to build nests and spawn in a wide variety of locales such as protected littoral areas in lakes or tributaries, including marshes, bays, harbors, sloughs, lagoons, and creek mouths.[46,205] Nests are usually built near or among vegetation or near physical structures[26,46,76,191,197,205] such as logs, stumps,[26,46,76,205] the base of ledges, or around the base of large boulders.[191] Spawning substrates include gravel,[26,46,54,57,205,239,243,245] rock,[46,57,205] rubble/cobble,[26,57,191] clay,[46,54,205] sand,[46,54,57,205,243,245] mud,[46,205,239] detritus,[46,205] vegetation,[46,67,199,205,239,242,243] organic debris,[57,245] bedrock,[57] silt,[54,57,144,188] roots,[57,242,243] filamentous algae mats,[57] spatterdock rhizomes,[67] and fallen leaves.[242,243] Some reports indicate that spawn-

ing seldom occurs over silt[26,57,144,188] or mud,[46] or, if it does, the soft substrate is excavated down to firm bottom.[57,205] Areas selected for spawning generally are also primary nursery habitats.[46] In the Great Lakes, largemouth bass may move short distances inshore or into marshes to spawn.[205]

In Lake Mead, NV, largemouth bass nests were found on a variety of substrates including bedrock, tamarisk roots, sessile filamentous algae mats, rocks, rubble, and gravel. The typical nest consisted of 2.5% rubble, 31.1% course gravel, 41.5% fine gravel, 1.4% sand, 7.4% silt, and 0.2% organic debris. No nests were observed in sand or silt; however, many nests were observed from which 1 inch of silt or sand had been removed from typical composite substrate. The range of slope gradients was 3–52°, but nest failures were common in steep gradients. Nests were not successful and did not produce fry when exposed to wind and wave action.[57]

In Little Rock Lake, WI, largemouth bass nests were primarily (82%) on the macrophyte, *Myriophyllum tenellum*; the remainder of nests were on some combination of logs, rocks, and gravel.[199] In Lake Orange, FL, Florida largemouth bass, in the absence of traditional substrate, spawned on spatterdock rhizomes, emergent grasses, and smartweed. Although all nests were associated with vegetation, firm bottom or above-bottom substrate appeared to be important in nest selection.[67]

In hatchery conditions largemouth bass successfully spawned on nylon mat materials.[179]

In 69 of 74 spawning events, Florida largemouth bass in raceways selected fibrous spawning mats over rock, gravel, and black fibrous synthetic spawning mats.[104]

Most spawning males constructed their nests near physical structure. Parental males were more aggressive when nesting near complex structure, largely because they faced intrusion by more nest predators compared to individuals that nested near simple structure or areas without structure.[197] Male largemouth bass utilized artificial logs equally with natural logs in an Ozark reservoir. Mating, hatching, and nesting success were equally high for broods located near supplemental logs and naturally occurring logs[76] and near boulders and pilings.[22]

Reports of depths of largemouth bass nests are variable, although spawning is usually reported in relatively shallow water. Reported depths range from 0.1 to 5.5 m.[13,46,54,56,57,191,197] Nests are usually 30–50 cm wide[13,54] or one to two times the length of the male.[54,245] Normally nests are grouped near the shoreline or in a specific cove. They are reported within 2.1–2.4 m of shore[54] and may be 2 m[54] to 6 m[19] apart.

Several instances of shallow cave nesting are noted. In these cases, the male bass seemed to select a defensible position that offered limited approach with good visibility.[57]

Season

Largemouth bass have been reported to spawn in all months except September and October (Table 116) and the season appears dependent upon latitude. Spawning begins earlier in the southern latitudes, as early as November in Puerto Rico[256] and December in southern Florida,[51] and spawning dates are progressively later at more northern latitudes, as late as August in the Great Lakes.[205,243] Generally, in the Ohio River valley, largemouth bass spawning occurs from early April through early June.[148,170,200,204,260-262]

Table 116

Reported spawning times for largemouth bass from various locations.

Location	Spawning Time
Puerto Rico[94,256]	November–June
CA[22,150]	Mid-March–late May
UT/AZ[191]	Mid-April–mid-June
Great Lakes[205,238,243]	April–August
MD[247]	May–July
Delaware Bay[245]	April–June
NY[46]	Mid-May–mid-June
OK[22,124,138,234]	April 15–June 10
AR[26,53,120,239]	April–May
IL[22]	Late April–June
SC[22]	March–April
GA[126]	Late March–mid-May
MS[39]	Mid-March
AL[170,200,224,260]	Late March–mid-June
VA[20,261]	May and June
TN[262]	Late April–June
FL (southern)[51]	Begins in December

Spawning may take place over a period of several weeks[153] or longer (Table 117). A lengthy spawning season can cause bimodal length frequencies that may ultimately affect year-class strength.[13,51,114] The duration of the spawning season was shorter in the northern United States than in the south.[54] Largemouth bass tended to spawn over a longer period in AL reservoirs than did spotted bass. They spawned later in oligotrophic reservoirs than spotted bass but in mesotrophic and eutrophic reservoirs spawning of both species occurred at the same time.[170]

Table 117

The duration of largemouth bass spawning seasons reported from several locations.

Location	Duration of Spawning Season
Puerto Rico[94,256]	90+ days
Puerto Rico[83]	About 180 days
UT/AZ[191]	About 60 days
NY[40]	26 days
TX[30,41]	68–75 days
OK[111]	30 days
OK[138]	53 days
AR[120]	31 days
AR[53,56]	About 45 days
IL[6,43,68]	36–51 days
FL, south[51]	4–6 months
FL, north[51]	2–3 months
FL, central[51]	2–4 months
MS[39]	60–71 days

Northern largemouth bass spawned earlier in experimental ponds in IL than did Florida largemouth bass.[6] Also, northern largemouth bass grew larger (length and weight) than Florida largemouth bass. There was significant interbreeding between the subspecies, particularly when species-specific spawning acts overlapped in time. Northern largemouth bass spawned earlier in Lake Aquilla, TX, than did Florida largemouth bass[30] and, in another report, northern largemouth bass spawned earlier and grew faster than did resident-hatched Florida largemouth bass during the first year of life.[107] It has also been reported that Florida largemouth bass may spawn earlier than northern largemouth bass.[7,8]

Once initiated, the spawning act can be interrupted by changes in environmental factors or climatological conditions.[138,150] Comparison of calculated spawning dates with records of average wind velocity, air temperature, and water turbidity indicated that spawning success was greatest during short intervals when weather was stable. Passage of frontal systems apparently disrupted spawning of largemouth bass in Lake Carl Blackwell, OK, and probably increased mortality of bass embryos and larvae.[138] Disjunctive spawning caused bimodal length frequency of YOY largemouth bass in Millerton Lake, CA.[150] In Puerto Rico, declining water levels caused spawning to be terminated.[94]

Temperature

Largemouth bass spawn over a wide range of water temperatures (12.7 to ≥25.5°C, Table 118).

Table 118

Reports of spawning temperatures for largemouth bass from several locations.

Location	Temperature (°C)
Puerto Rico[143]	≥25.5
CA[150]	16.0
UT/AZ[191]	14.4
Great Lakes[205]	14.4
NY[46]	16.7
OK[87,124,138]	12.7–21.0
AR[56,239]	13.0–15.0
AL[200,224]	15.0–16.0
FL[22,51]	14.7–15.6

Typically, nest construction begins when water temperature rises to about 15.0–15.5°C[19,239] and spawning begins at about 18.3°C.[19,153] Because of many abiotic factors, defining a precise temperature range for spawning may not be realistic. Courtship behavior precedes actual egg deposition, so the actual spawning temperature will depend on the duration of courtship, time for "building" a nest, and the rates of temperature change while these activities are underway.[239]

In some Puerto Rico reservoirs that support self-sustaining largemouth bass populations,[143] the water never cools below 25.5°C, which is well above most reports of spawning temperatures in the United States (Table 118). Largemouth bass in a cooling reservoir in SC spawned earlier than normal — from December to March.[148]

A temperature decline of only a few degrees is especially crucial when temperature at the time of spawning is <18.3°C.[138] A sudden drop in water temperature may cause the male to abandon its nest.[54,138,143]

Fecundity

It is difficult to count the number of eggs that a female bass produces annually because, like all sunfishes, their ovaries contain many ova in various stages of development.[22,138] The adult female largemouth bass, prior to spawning, may contain as many as 109,000 ova in various stages of development or as few as 2,000.[140] Ova >0.75 mm are generally considered mature.[22] The fecundity of individual females is not related to counts of eggs in bass nests because not all eggs in a female are spawned in one spawning period.[140]

Prior to spawning, up to 10% of the body weight of a female largemouth bass may be egg mass.[239] Reports of the number of eggs per body weight also vary with a range from 4,000 to 80,000 eggs per pound (8,800–176,000 per kg) given in one report,[239] and about 2,000–7,000 eggs per pound of body weight (mean = 4,000 eggs) was also reported.[19] In ME, fecundity was better correlated with age than with length or weight and fecundity declined after age 7.[22] In Lake Fort Smith, AR, fecundity estimates for largemouth bass ranged from 2,942 to 30,709 per female with a mean of 10,464. Total length, SL, weight, age, and condition factor were all significantly correlated with fecundity.[239]

In Lake Aquilla, TX, higher relative survival rates were evident for Florida largemouth bass and hybrids than for northern largemouth bass due to a size-dependent fecundity advantage for the larger Florida largemouth bass females apparent by age 3.[30]

A 305-mm largemouth bass will have about 17,000 eggs, while a similar-sized smallmouth bass will have about 4,500 eggs.[138]

Sexual Maturity

The embryonic history of germ cells of largemouth bass has been described.[212] Sexual maturity of largemouth bass is more related to size than age with both genders reaching maturity at about 250 mm TL.[22,54,166] Typically this would be after two to three growing seasons but could be at age 1.[22,54,139,166,239] In Lake Fort Smith, AR, 2, 43, and 100% of largemouth bass were mature at ages 1, 2, and 3, respectively.[239] Average calculated total lengths for ages 1, 2, and 3 largemouth were 138, 244, and 310 mm.[239] In Bull Shoals Reservoir, AR, some males and females were sexually mature at age 1. All males were mature at age 2. Females matured by about 300 mm TL.[192]

In Lewis Smith Reservoir, AL, no age-1 fish were sexually mature (168-mm males, 201-mm females), by age 2 most fish of both sexes were sexually mature (295-mm males, 282-mm females), and by age 3 all fish were sexually mature (356-mm males, 358-mm females).[142]

Spawning Act

Male largemouth bass prepare and defend their nests for one to several days prior to spawning.[22,26,199] The spawning act is relatively simple. After nest construction, the male circles the nest until attracting a female. When a female joins the male, they circle the nest side by side for a while with the male on the outside.[54,56,245] Their circles may be 1.2–1.5 m in diameter.[56] While circling, the male will nip or nudge the female frequently in the opercular region or on the underside of the body near the vent or tail.[56] Actual spawning may be preceded by a "false spawning" during which the female quivers convulsively but releases no eggs.[56] Color of both sexes is enhanced during circling and spawning.[56,57] When spawning occurs the male and female tilt onto their sides, with their vents in close proximity, and release sperm and eggs.[19,54,245] The female, quivering convulsively, releases a thin stream of about 30–50 eggs and the male discharges a thin, whitish stream of milt about 1 inch in length.[56,245] The female lays several hundred eggs during a spawning act.[54] The male fans the eggs with his pectoral fins which helps to mix the milt and eggs and to disperse the fertilized eggs onto the substrate.[245] After spawning, the female moves away, returning later to the same or a different nest.[54,56,57] The nest of an individual male may contain eggs from several females.[19,54] A male largemouth bass may accept several females in succession[56] or receive the same female a second time after a considerable interval. There are reports of males spawning with two females over a nest at the same time. This was inefficient and often resulted in the eggs being more or less scattered about.[242] Nests may contain 5,000–43,000 eggs.[54]

Gonad analysis indicates that multiple partial spawns by individual fish occur.[16,83,169] The release of mature ova by female largemouth bass may occur in a few hours, but repeated spawns over longer periods may occur due to environmental perturbations such as temperature changes or changes in water level.[39,44,45] Generally, females spawn once per year but multiple spawns are common, and one female in Watts Bar Reservoir, TN, was determined to have spawned three separate times in one season (personal communication, D. Harris, 1070 Lela Way, Seymour, TN 37865).

After spawning, the male largemouth bass maintains the nest and guards the incubating eggs from predators.[13,26,199,242,245] He hovers over the nest fanning the eggs with his pectoral fins. This fanning behavior prevents the eggs from being smothered by layers of sediment[242,245] and moves water over the eggs.[22,54,145] Fry will remain in the nest until their yolk sac is absorbed, usually 1–2 weeks, depending on temperature, before they become free swimming.[22,56,57] Once the fry leave the nest, they form a dense school that is guarded by the male for up to 2 weeks. The male usually guards the nest or the school of fry until the fry disperse,[19,22,54,56,57,191,199] which is reported to occur when they are 12–32 mm long. The presence of the male guarding the nest was essential for a successful nesting event[22] in Lake Mead, NV,[57] in Lake Powell, UT/AZ,[191] and in Little Rock Lake, WI.[199] Successful nests were guarded for 15 days and unsuccessful nests were guarded for 3 days.[199] When anglers removed male bass from 34 nests, all nests failed.[22]

Largemouth bass spawning events appeared to be associated with rising water or relatively stable water levels. Spawning ceased when water levels dropped sharply (>0.05 m/day) in Puerto Rican reservoirs.[94] In Millerton Lake, CA, largemouth bass were unsuccessful spawning when lake levels fluctuated during the spawning season.[150] Also, a drop in water temperature may result in desertion of the nest by the male, making the eggs more susceptible to predation, siltation, and suffocation (which also may accompany cool, windy periods in spring),[138,143] and the spread of fungus.[143]

Ovarian development suggests that larger largemouth bass spawn earlier than smaller ones.[224] In one report, the average 500-mm female started spawning 9–12 days earlier than did a 300-mm female.[166]

Golden shiner Notemogonus chrysleucas and lake chubsucker Erimyzon sucetta are sometimes allowed by the male largemouth bass to lay their eggs in his nest, after which he will guard the eggs.[54,135] One study indicated that excessive utilization by golden shiners (>20%) may have reduced largemouth bass hatch.[135] This may be why it is often suggested that golden shiners not be stocked with largemouth bass, even though the shiners provide excellent forage.

Largemouth bass failed to spawn in eutrophic lakes with overcrowded sunfish populations. Though the adults were capable of spawning, they did not and the eggs were eventually reabsorbed.[188] Largemouth bass also failed to spawn in association with dense populations of sunfish (>336 kg/ha) in controlled ponds.[180]

Largemouth bass failed to reproduce in more than 10–12% seawater[144] and no successful reproduction occurred at 0.5% salinity. pH must be between 5 and 10 for largemouth bass to successfully spawn. Reproduction occurred with turbidity of up to 83 ppm but not at 348–612 ppm.[22]

Males do not eat while on the nest.[22,54,56,124,256] This includes the pre-spawn, spawning, and post-spawning periods.

Even though there are conflicting reports of which subspecies spawns first, there is a significant overlap in the spawning times of northern largemouth bass and Florida largemouth bass that allows for hybridization. The more the spawning periods overlap the greater the likelihood that hybridization will occur.[30] In most cases Florida largemouth bass soon hybridize with native northern largemouth bass, producing intergrade hybrids (F_1 and F_x).[114]

Florida largemouth bass introduced into AL[158] and OK lakes[111] integrated with native northern largemouth bass populations. Florida largemouth bass were stocked into 82 reservoirs in OK between 1970 and 1986. Of 251 bass collected from 34 reservoirs, 93% contained Florida largemouth bass alleles.[102]

Subspecific alleles of Florida largemouth bass were present in 28 of 30 populations (93%), and were found in >50% of the fish from 8 (27%) reservoirs. Reservoirs in the southern portion of OK that were stocked with fingerlings >100 mm TL at rates >25/ha for several consecutive years had the highest degrees of introgression of Florida largemouth bass alleles into the existing bass populations.[111]

It is well documented that largemouth bass hybridize with other *Micropterus* species. In central TX, largemouth bass, Florida largemouth bass, smallmouth bass, and Guadalupe bass have hybridized extensively. One bass examined from this area had genetic markers from at least three species. The hybrids and back crosses were all viable.[37]

EGGS

Description
Fertilized eggs are spherical,[54,245] and demersal and adhesive;[238,243,245,247] light yellow, orange, or cream in color.[122,243] Reports of egg diameter, after water hardening, range from 1.4 to 2.0 mm,[54,236,242,243,245,247] with averages reported at 1.5 mm[242] to 1.6 mm.[236] Egg diameter increases with the size of the fish.[54] The shell consists of two layers, the outer of which is transparent, about three times as thick as the inner, rough on its outer surface, and very adhesive.[242] The yolk is granular in appearance and the perivitelline space is narrow[242,245] (0.16–0.24 mm wide).[245] Within the yolk is a single large, amber-colored oil globule 0.52–0.55 mm in diameter[245] or about one-third the diameter of the yolk.[242]

Incubation
Incubation time for largemouth bass eggs is temperature dependent. Generally, eggs take longer to hatch in cooler water and a shorter time in warmer water (Table 119).[19,22,54,56,57,62,191,205,242,243] In the typical spawning range of 18.3–23.9°C, incubation time is about 3–4 days. Incubation temperature affects egg survival in a parabolic way, with the best survival near 20°C and progressively poorer survival as temperature deviates from this point.[22] Egg survival declines with sudden thermal shocks and the eggs are most sensitive within the first day of fertilization.[143] Cold shock can cause eggs to not hatch, but a short-duration cold shock (~3 h) followed by a quick return to the acclimated temperature caused no adverse impacts on hatching success.[22]

Development
See Figure 100.

Events in the development of largemouth bass eggs related to age (time in hours after fertilization) incubated at 23–26°C:

Table 119
Reported incubation times for largemouth bass eggs at various temperatures.

Temperature (°C)	Incubation Time (days)
10.0	13.2[22,54]
12.5	9.8[22]
13.0–15.5	5.4[57]
15.0	6.8[22]
15.5–18.3	4.7[57]
15.6–16.8	4.0[243]
16.8–19.6	3.0[243]
17.5	4.8[22]
17.5	4.0[22]
18.0	2.3[54]
18.3	10.0[19,54]
18.3–21.1	3.1[57]
18.9	5.0[243]
20.0	2.87[22]
21.1–24.0	3.1[57]
21.7–23.9	2.0[243]
22.0	1.9[22]
22.2	2.0[243]
22.5	2.87[22]
23.0–25.0	2.0–3.0[22]
26.7	5.0[19,54]
27.5–28.0	2.04[54]
30.0	1.5[22]
Ambient	4.0[191]
Ambient	2.0–15.0[205]
Ambient	3.0–4.0[242]
Ambient	2.0–6.0[56]

0.25 h — water-hardened; blastodisc forming; chorion no longer adhesive
0.50 h — blastodisc flattened
0.58 h — blastodisc swollen to form cup
0.93 h — first cleavage furrow evident
1.00 h — 2-cell stage
1.25 h — 4-cell stage
1.50 h — 8-cell stage
2.00 h — 16-cell stage
7.00 h — periblast evident
8.00 h — blastopore, germ ring, and embryonic shield evident
11.00 h — yolk plug stage

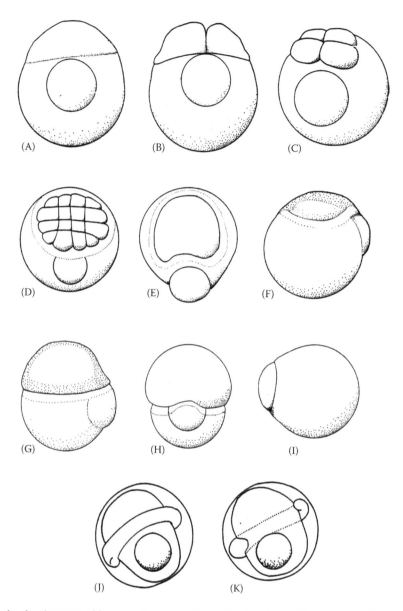

Figure 100 Events in the development of largemouth eggs. A–K: (A) Fertilized egg, blastodisc developed (35 min old); (B) 2–cell stage; (C) 4–cell stage; (D) 16–cell stage; (E) periblast formed; (F) same as E, lateral view; (G) blastoderm 1/3 over yolk; (H) blastoderm beyond equator; (I) yolk plug stage; (J) embryo, dorsal view; (K) embryo, ventral view. (A–I reprinted from Figures 1–6 and 10–12, reference 244; J–K reprinted from Figure 18, reference 242.)

12.00 h — periblast thickened in region of posterior pole

13.00 h — a ridge evident in median line of embryonic shield

14.00 h — head forming

16.00 h — blastopore nearly closed; 4 somites formed; notochord evident

24.00 h — olfactory bulbs, optic vesicles, and otic capsules formed or forming; 10 somites

26.00 h — optic vesicles invaginated; constriction between midbrain and hindbrain evident; 16 somites; Kupffer's vesicle formed

30.00 h — tail free

32.00 h — lens forming

42.00 h — caudal region compressed laterally; finfold formed

47.00 h — hatching[244]

Additional descriptions of largemouth bass egg development are reported.[263,274]

YOLK–SAC LARVAE

See Figure 101.

Size Range

Reports of hatching size for largemouth bass vary and include: 4.2 mm TL;[242] 3.6–4.1 mm TL;[245]

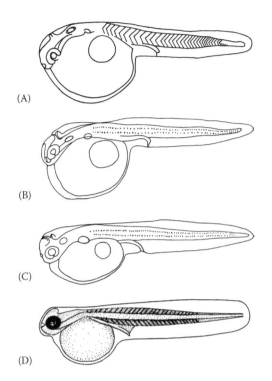

Figure 101 Development of young largemouth bass. A–D, yolk-sac larvae: (A) 3.4 mm TL; (B) 4.1 mm TL; (C) 5.0 mm TL; (D) 5.4 mm TL. (A–C reprinted from Figures 31 and 32, reference 244; D reprinted from Figure 2, reference 236, with publisher's permission.)

2.7–4.3 mm TL (average 3.5 mm).[22, 247] Yolk is no longer visible by 8 mm TL and is probably absorbed[242] or remnants of yolk are present on specimens up to 10 mm TL or larger.*

Myomeres

6.2–6.7 mm TL (n = 5). Preanal 14–15; postanal 16–17; total 31–32.*

7.2–7.6 mm TL (n = 5). Preanal 15–17; postanal 15–16; total 31–32.*

8.1–8.7 mm TL (n = 5). Preanal 16–19; postanal 13–15; total 31–32.*

9.0–9.8 mm TL (n = 5). Preanal 15–17; postanal 15–17; total 32–33.*

Other reports:

3.0–6.0 mm TL. Preanal 11; postanal 23–25; total 34–36.[243,244]

5.0–8.0 mm TL. Preanal 11; postanal 17–19; total 29–30.[243,244,247]

Morphology

3.6–4.1 mm TL (newly hatched). Body is robust, tadpole-like, with a large round yolk sac. Head is strongly deflected downward over anterior margin of the yolk sac. A single oil globule is present in the posterior portion of the yolk sac.[245]

4.2 mm TL (at hatching). Head barely extends beyond yolk sac; only slight cranial flexure.[242]

4.75 mm TL. Head extends somewhat beyond the yolk sac; parietal flexure formed so that tip of head points downward.[242]

5.3 mm TL. Head directed nearly forward; mouth cavity formed, but no jaws.[242]

5.6 mm TL. Lower jaw not yet visible; ventral surface of the yolk sac very adhesive.[242]

6.2–6.3 mm TL. Yolk sac shape variable, depending on amount of yolk remaining; a single oil globule 0.7 mm in diameter is still visible on some individuals posterior to the forming air bladder.* Head and mouth larger;* lower jaw developed and free from yolk sac, not reaching to end of snout;[242] upper jaw formed;[243,244] opercle developing;*[243,244] developing gill filaments visible.*[242]

6.5–7.2 mm TL. Yolk reduced and pear-shaped; air bladder prominent and filled[242,244] above and in front of oil globule; hind gut very prominent posterior to yolk; jaws of equal length;[242] mouth functional.[243,244] Opercle covers gills by 6.7 mm.*

7.2–8.0 mm TL. Varying amounts of yolk remain,* yolk nearly absorbed on some individuals; oil globule no longer present;*[242] air bladder now as large as original oil globule;[242] digestive tract functional;* gut coiling evident,* massively coiled by 8 mm;[247] intestine very large and visible through body wall.[242] Head and mouth large. Opercular development nearing completion or complete; branchiostegal rays visible.*

8.1–8.7 mm TL. Some specimens up to 8.4 mm have highly visible amounts of yolk remaining but by 8.7 mm little yolk remains as well as sometimes a few scattered oil droplets.*

9.0–9.8 mm TL. A small amount of yolk remains in all specimens with small oil droplets in some.*

Morphometry
See Table 120.

3.0–6.0 mm TL. As percent TL: PreAL 62; GD 41; ED 7–8.[243,244]

5.0–8.0 mm TL. As percent TL: PreAL 44–49; GD 20–23; ED 6–9.[243,244]

Table 120

Morphometric data for southern stock of young largemouth bass. Data are presented as mean percent total length (TL) and head length (HL) for selected intervals of TL (N = sample size).

Length Range (mm TL)	TL Grouping											
	6.2–6.7	7.2–7.6	8.1–8.7	9–9.8	10–10.8	11.4–11.9	12.5–12.7	13.1–13.8	14.1–14.9	15.3–15.9	16.1–16.4	19.6–22.4
N	5	5	5	5	5	3	4	5	5	4	2	6
Length as percent TL												
SL	96.3	96.3	94.6	91.3	89.2	87.5	85.9	84.3	83.4	83.4	81.2	82.6
PreAL	49.6	52.0	52.4	50.9	51.7	51.9	51.8	50.8	51.0	51.5	51.4	51.6
GD	20.6	22.1	21.9	21.5	22.4	23.1	21.6	21.3	21.6	21.9	21.8	22.1
Depth at anus	13.4	13.5	14.9	13.8	15.1	16.0	16.9	16.8	17.1	17.1	17.2	17.6
CPD	5.9	5.6	7.2	7.0	8.0	8.3	8.7	8.9	8.9	8.8	8.9	9.0
ED	10.7	10.5	10.9	11.3	10.8	10.8	10.9	10.9	10.7	10.7	10.2	10.2
HL	22.3	24.5	25.6	28.7	28.4	29.3	28.8	29.4	28.9	29.7	30.2	31.1
Diameter as percent HL												
ED	47.9	42.8	42.6	39.3	38.1	36.9	37.9	37.2	36.9	36.0	33.7	32.9

6.2–6.7 mm TL. Body tapers to a narrow caudal peduncle; greatest depth is measured at yolk sac or sometimes at the head.*

7.2–7.6 mm TL. Greatest depth measured at the posterior margin of the head on most specimens.*

Fin Development

3.6–4.1 mm TL (newly hatched). Wide finfolds.[245]

4.75 mm TL. Pectoral fin bud present as a low ridge.[242]

5.3 mm TL. Pectoral fins more prominent.[242]

5.6 mm TL. Pectoral fins attached at an angle of about 45°.[242]

6.2–6.7 mm TL. Median finfold originates between myomeres 4–6 and is present along the dorsum and around urostyle proceeding ventrally and anteriorly to posterior margin of the yolk sac; urostyle straight. Hypural differentiation begins at about 6.7 mm. Pectoral fins fan-like in shape;* in life, capable of vigorous movement.[242] Pelvic fins are not visible.*

7.0–7.2 mm TL. Caudal rays begin forming.[236]

7.2–8.0 mm TL. Median finfold is still present; elevated in the regions of future dorsal and anal fins.* Caudal fin ray development complete by 8.0 mm

or[236] caudal, anal, and dorsal fins differentiating;*,[242] pterygiophores visible at the bases of developing dorsal and anal fins on most individuals.* Urostylar flexion not yet apparent[242] or flexion begins between 7.2 and 7.6 mm and is very evident by 7.6 mm.* Hypural development evident and rays are visible ventrally in developing caudal fin.* Formation of rays in pectoral fins is visible by 7.6 mm.*

8.1–8.7 mm TL. Median finfold is still present but low in profile on caudal peduncle dorsally and ventrally and still visible ventrally anterior to anus. Profiles of soft dorsal and anal fins are pronounced but insertions are not yet clearly defined; fin bases are well developed with pterygiophores visible on all specimens; dorsal and anal fins with 0–8 developing rays but no spines. Caudal fin is rounded, flexion is obvious on all specimens; 8–16 developing rays with up to 8 segmented rays on some individuals. Pelvic buds are visible on some individuals by 8.4 mm TL.*

9.0–9.9 mm TL. Predorsal finfold limited to area of developing spinous dorsal fin; at 9.7–9.8 mm, developing spines are visible. Insertion of dorsal and anal fins nearly defined and finfold on caudal peduncle is much reduced, as is ventral finfold anterior to the anus. Bases of dorsal and anal fins are well developed. Dorsal fin has 9–11 pterygiophores and 6–9 developing soft rays. Anal fin has 10–11 pterygiophores and 8–10 developing soft rays. Urostyle is flexed at about 45° and caudal fin is squared off but becomes bilobed on larger specimens. Caudal fin has 12–17 developing rays, 4–12 segmented. Pelvic fins are flap-like and small.* Also reported that by 9.7–9.9 mm soft fin ray development is complete in all fins,[236] or soft rays are differentiated in dorsal, caudal, pectoral, and anal fins by 9.7 mm.[245]

Pigmentation

3.6–4.1 mm TL (newly hatched). Eyes are not pigmented; little if any body pigment.[245]

4.2 mm TL (at hatching). No pigment.[242]

4.75 mm TL. No pigment in eye or elsewhere.[242]

5.0 mm TL. Pigmentation evident on the ventral and dorsal aspects of the air bladder.[243]

5.3 mm TL. Eye slightly pigmented; no pigment elsewhere.[242,244]

5.6 mm TL. Eye very black and becoming iridescent; a few melanophores scattered along junction of body and yolk sac and on the yolk sac.[242] Melanophores are present on the head, dorsum, and along the median myosypta.[243,244]

6.2–6.7 mm TL. Eyes are dark*,[245] or iridescent (6.3 mm).[242] Dorsally, chromatophores are scattered on the snout and the head is heavily pigmented*,[245] from about the middle of the eyes posteriorly onto the nape and on the body over the yolk sac.* Scattered pigment is present over the developing air bladder and dorsolaterally on the yolk on some.*,[242] Internal pigment extends anterio-ventrally from the dorsum of the air bladder forming a "V" pattern at the isthmus on some individuals.* Lightly pigmented along the digestive tract and abdomen.[245]

7.2–8.0 mm TL. Pigmentation has increased and is visible over whole body and yolk sac.*,[242] Dorsally, the head is generally covered with pigment from the snout to the occiput except for a small area between the eyes. Chromatophores are present along the dorsum from the occiput to the caudal fin. Pigment is scattered on the remaining yolk sac, abdominal cavity, and laterally and ventrally on the gut. Laterally a bar of pigment begins forming on the opercle posterior to the eye. A lateral stripe is present along the median myosepta on most individuals, extending from the operculum to the caudal fin. This stripe consists of scattered chromatophores in the region of the abdominal cavity and gut narrowing to a single or double row of pigment from the gut/anus region to the caudal fin. On some specimens, small scattered chromatophores are barely visible laterally along the caudal peduncle. Internally, heavy pigment covers the air bladder and extends over the dorsal margin of the gut on some individuals. Ventral pigmentation varies with scattered pigment on the lower jaw of some specimens; small scattered chromatophores are present on some individuals posterior to the anus on the caudal peduncle; pigment is scattered to concentrated mid-ventrally on the stomach or yolk sac.*

8.1–8.7 mm TL. Pigment as described above with more small scattered pigment laterally on some individuals.*

9.0–9.8 mm TL. Pigmentation much as before with the following noticeable changes: dorsally, the tip of the snout is darkly pigmented; more pigment is present ventrally and internally on remaining yolk and stomach; pigment over air bladder and gut is as described before; many small scattered chromatophores are visible laterally on the body from between the dorsal and anal fins onto the caudal peduncle. Lateral pigment along the body's midline is pronounced and easily discernible on all specimens. This lateral stripe now consists of a narrow row of concentrated small chromatophores present from the opercle to the caudal fin. Anteriorly, this stripe arches over the air bladder and appears to connect with the well-defined bar of pigment on the opercle posterior to the eye (this apparent connection is due to a concentrated patch of pigment internally just below the occiput and just anterior to the air bladder). The lateral stripe widens on the caudal peduncle as it nears the base of the caudal fin. On larger specimens this lateral pigment continues across the developing hypural plate onto the middle 2–3 caudal fin rays, extending to about the middle of the rays.*

It is also reported that by 9.6 mm the caudal and anal fins are pigmented[242] and that a broad, dark, lateral stripe begins to form at 9.7 mm.[236]

POST YOLK–SAC LARVAE

See Figures 102 and 103.

Size Range

Phase begins at 8 mm TL[242] or at about 10 mm TL (very small remnants of yolk may persist in some specimens between 10 and 11 mm TL).* Transition to juvenile phase is at 12–16 mm TL[236,245] or 15–16 mm TL.*,[236]

Myomeres

10.0–10.8 mm TL (n = 5). Preanal 16–17; postanal 15–16; total 32.*

11.4–11.9 mm TL (n = 3). Preanal 17–18; postanal 14–15; total 31–32.*

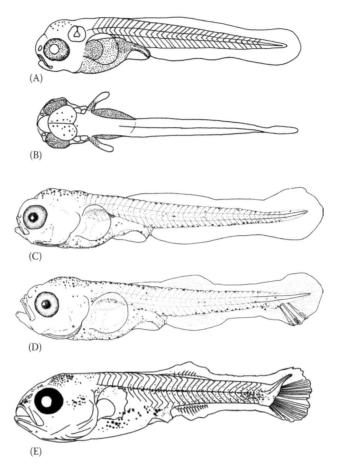

(A)

(B)

(C)

(D)

(E)

Figure 102 Development of young largemouth bass. A–B, post yolk-sac larva, 6.2 mm TL: (A) lateral view; (B) dorsal view. C–E, post yolk-sac larvae: (C) 6.6 mm TL; (D) 8.4 mm TL; (E) 9.6 mm TL. (A–B reprinted from Figure 13, reference 275; C and D reprinted from Figures 2 and 3, reference 246, with author's permission; E reprinted from Figure 20, reference 276, with permission of Progress Energy Service Company, LLC.)

12.5–12.7 mm TL (n = 4). Preanal 17; postanal 15; total 32.*

Post yolk-sac larvae are also reported with 11 preanal and 19 postanal myomeres.[243]

Morphology
10.2 mm TL. Body is elongate, deepest at the head and thoracic region. The gut is massive and coiled.[236,243,245,246] Maxillary extends past middle of the eye.[236,245]

Morphometry
See Table 120.

Fin Development
10.0–10.8 mm TL. Profiles of the spinous and soft dorsal and anal fins are established. Remaining median finfold consists of small remnants near the base of the caudal fin, dorsally and ventrally; ventrally, preanal finfold is almost absorbed.*,[236] Spinous dorsal fin is visible, low in profile, with 2–8 developing spines; soft dorsal fin has 8–11 developing rays visible. Anal fin has one spine

apparent on some individuals and 8–12 developing soft rays. Hypural plate is well defined. Caudal fin appearance becomes bilobed with 15–17 rays, 11–15 segmented. Rays are developing in the pectoral fins. Pelvic fins are still small and flap-like, but ray differentiation begins.*

11.0 mm TL. Tail becomes homocercal.[236]

11.4–11.9 mm TL. Median finfold completely absorbed on some individuals or small remnants still present. Urostyle still extends beyond hypural plate; caudal fin bilobed with 14–17 segmented rays. Spinous dorsal fin is still low in profile, about 0.1 mm high; 8–9 spines visible. Soft dorsal fin well-formed with 10–12 rays. Anal fin has one apparent spine and 11 soft rays. Pelvic fins are still flap-like.*

12.5–12.7 mm TL. Only small remnants of median finfold remain near the dorsal and ventral base of the caudal fin. Urostyle still extends beyond the hypural plate; caudal fin has the adult complement of 17 segmented rays. Dorsal spines 8–10, soft rays 12; anal spines 2–3, soft rays 11.*

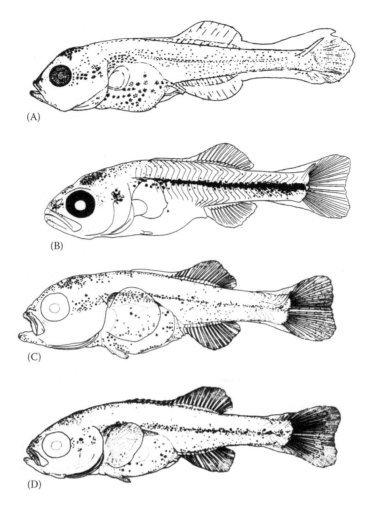

Figure 103 Development of young largemouth bass. A–D, post yolk-sac larvae: (A) 10.2 mm TL from CA; (B) 11.5 mm TL from SC; (C) 12.1 mm TL from LA; (D) 14.7 mm TL from LA. (A reprinted from Figure 2, reference 236, with publisher's permission; B reprinted from Figure 20, reference 276, with permission of Progress Energy Service Company, LLC; C and D reprinted from Figures 4 and 5, reference 246, with author's permission.)

13.1–13.8 mm TL. Median finfold completely absorbed. Adult complement of fin elements present in median fins. Dorsal fin has 10 spines and 12 rays; anal fin has 3 spines and 11 rays; caudal fin has 17 rays. At 13.8 mm, pectoral fin has about 13 soft rays. Pelvic fins are 1.0–1.1 mm long; developing rays are clearly visible by 13.8 mm.*

14.1–14.9 mm TL. Median fins have the adult complement of fin elements. Pectoral and pelvic fin development is not quite complete. At 14.9 mm, pectoral fins have 10–11 rays and pelvic fins have 4–5 developing rays. Morphology of spinous dorsal fin becomes elevated on some individuals.*

Pigmentation

Note: Conversions from SL to TL are based on the formula TL = 1.22 SL, provided by Carlander (1977).[22]

10.2 mm TL. Dark blotch of pigment on dorsal surface of head near posterior margin of eye; tip of

snout darkly pigmented; small melanophores scattered over much of body; large, stellate melanophores scattered laterally on gut region and on abdomen; broad lateral stripe forming behind eye and obvious over gut (descriptions based on an illustration in figure 2, reference 236).[236]

11.8–11.9 mm TL. Pigmentation is generally as before with heavy pigment on tip of snout. Lateral stripe wider — widest over air bladder and gut and near caudal fin base. Oval blotches of pigmentation form along lateral stripe.*

>10 mm SL (about 12.2 mm TL[22]). Primary bars of melanistic pigment are initially visible as 10–12 mid-lateral patches of melanophores, often difficult to distinguish because of a deep lateral stripe of pigment along the horizontal myosepta (Figure 104A).[253]

12.5–12.7 mm TL. Lateral stripe wider and prominent — widest at the posterior margin of the caudal peduncle. Stripe now appears as a series of lateral blotches. On some individuals, pigment extends

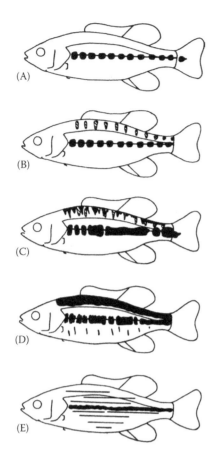

Figure 104 Development of young largemouth bass — schematic drawings showing ontogeny of melanistic pigment patterns of young largemouth bass. (A) 19 mm SL, representative of fish >10 mm SL; (B) 24.8 mm SL, representative of fish 20–30 mm SL; (C) 30.9 mm SL; (D) 85.0 mm SL; (E) 85.0 mm SL. (A–E reprinted from Figure 4, reference 253, with publisher's permission.)

across the hypural plate onto the bases of the middle caudal fin rays.*

13.1–13.8 mm TL. Pigment becomes concentrated at the base of the middle caudal fin rays — appears as a caudal spot on some individuals.*

14.1–14.9 mm TL. Pigmentation is generally as before, except that a caudal spot is now distinct on most individuals.*

15.5 mm TL. A broad dark lateral stripe extends from the snout through the eye to the caudal fin. Numerous stellate melanophores are present on the ventral surface of the gut and on top of the head. Snout is darkly pigmented.[236]

JUVENILES

See Figure 105.

Size Range
12–16 mm TL[236,245] or 15–16 mm TL* to about 250 mm TL.[22,54,166]

Morphology
15.3–15.9 mm TL. Look like little bass.*

16.8 mm SL (about 20.5 mm TL). A few scales are present on the posterior portion of the lateral line.[123]

A study was conducted to determine which scales from four regions of the body of largemouth bass were best at predicting fish length at the time of scale formation. Using a regression intercept method of back-calculation, scales from the caudal peduncle region predicted largemouth bass were 19–25 mm TL at beginning squamation, which is in line with other published results. Scales from the pectoral region consistently estimated a larger size at the beginning of scale formation than that reported in the literature.[18]

36.0 mm TL. Squamation complete.[22,264]

75.0 mm TL. Body elongate; mouth very large and oblique, maxilla extending to back of eye, lower jaw projecting.[238]

Larger juveniles. Scales in lateral line 55–73;[20,21,151,260,262] branchiostegal rays 6–7; gill rakers 6–8; pyloric caeca 24–28;[21] vertebrae 30–32.[21,262]

Morphometry
See Table 120.

75.0 mm TL. As percent TL: SL 70; HL 28; GD 19.5; ED 4.8.[238]

Fins
15.3–15.9 mm TL. Median finfold is completely absorbed and adult complement of fin elements is present in all fins. Spinous dorsal fin has 10 spines and profile is elevated (about 0.3 mm high); soft dorsal fin has 12–13 rays; anal fin has 3 spines and 11–12 soft rays. Caudal fin has 17 rays.*

Larger juveniles. Area between spiny and soft dorsal fin deeply notched;*,21,86,238 dorsal spines 9–11,*,20,86,87,238,247,257,262,265,266 soft dorsal rays 12–14;*,6,20,87,238,247,257,260,262,265,266 anal spines 3,*,20,21,86,151,238,247 soft rays 10–12;*,20,21,86,151,238,247,260,262 pectoral fin rays 13–15 in Canada,[21] usually 14–15 in the South;[20,86,151,260,262] pelvic fin has 1 spine and 5 rays.[21]

Pigmentation
(*Note:* Conversions from SL to TL are based on the formula TL = 1.22 SL, provided by Carlander, 1999.[22])

15.3–16.4 mm TL. Lateral stripe is wide, dark, and prominent from the snout, through the eye, across the opercle and now obvious to the naked eye

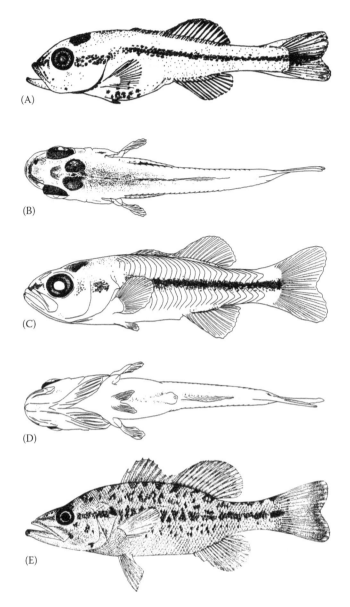

Figure 105 Development of young largemouth bass. (A) Juvenile, 15.5 mm TL from CA. B–D, juvenile, 16.4 mm TL from NC: (B) dorsal view; (C) lateral view; (D) ventral view. (E) Juvenile, 79.0 mm SL OH. (A reprinted from Figure 2, reference 236, with publisher's permission; B–D reprinted from Figure 54, reference 277, with permission of Progress Energy Service Company, LLC; E reprinted from Figure 162, reference 55, with publisher's permission.)

along the body to the base of the caudal fin and extending beyond onto the middle rays of the caudal fin. On some individuals this pigment is present on about the proximal half of the middle two caudal fin rays.*

20 mm TL. In life, the lower surface of the body and throat have become silvery and a black stripe has appeared laterally and runs through the eye, while the rest of the fish is gray or greenish gray.[242]

19.6–22.4 mm TL. Lateral stripe is continuous, wide, and essentially uninterrupted on unscaled young (Figure 106A).[123] Lateral blotches along the stripe are visible on some individuals.* Caudal spot

is a well-formed circle of pigment about 0.5 mm in diameter and visible to the naked eye.* The caudal spot may be continuous with or separated from lateral stripe.*,[123]

20.0–30.0 mm SL (about 24–37 mm TL). Primary melanistic bars remain as short midlateral blotches; they do not become dorsoventrally elongated. At about 30 mm TL, melanophores begin to form secondary bars along the dorsal midline that continue to develop dorsolaterally (Figure 104B).[253]

30.9–35.0 mm SL (about 38–43 mm TL). Secondary bars of pigment terminate ventrally at the trunk canal (Figure 104C).[253] Lateral stripe begins to

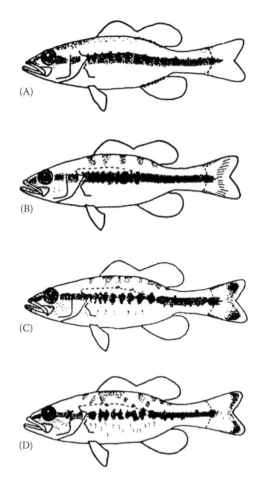

Figure 106 Development of young largemouth bass — details of pigmentation for largemouth bass from AL. Body proportions are constant in the drawing outlines and do not reflect proportional differences between sizes. (A) 16.8 mm SL; (B) 31.6 mm SL; (C) 45.9 mm SL; (D) 57 mm SL. (Reprinted from Figure 1, reference 123, with publisher's permission.)

appear disrupted in scaled fish. Caudal spot distinct and elongated posteriorly. Submarginal band on caudal fin becomes visible in some specimens (Figure 106B).[123]

35.0–60.0 mm SL (about 43–73 mm TL). Caudal spot distinct and elongated posteriorly. Submarginal band on caudal fin is apparent, but never intensely developed, usually indistinct anteriad (Figure 106C and 106D).[123]

75.0 mm TL. In life, greenish above and silvery below. A prominent dark lateral stripe is present. Three oblique stripes are apparent across cheek and opercles posterior to eye. Very small black chromatophores are abundant on top and sides of body, darker and more numerous above the lateral stripe, and arranged more heavily on outer margins of scales. Belly is white. All fins except ventrals are sprinkled with chromatophores.[238]

85.0 mm SL (about 104 mm TL). The number of blotches comprising the lateral stripe has approximately

doubled, remaining confluent as a broad midlateral stripe (Figure 104D). On some individuals, especially larger adults, longitudinal stripes are formed along scale rows (Figure 104E).[253]

> 75 mm TL. The lateral stripe becomes fainter in larger juvenile specimens, especially those from turbid streams.[123]

TAXONOMIC DIAGNOSIS OF YOUNG LARGEMOUTH BASS

Similar species: other centrarchids, especially *Micropterus* species; also superficially similar to elassomatids.

For comparisons of important morphometric and meristic data, developmental benchmarks, and pigmentary characters between largemouth bass and all other centrarchids and elassomatids that occur in the Ohio River drainage, see Tables 4 and 5.

For characters used to distinguish between young basses and the young of other centrarchid genera and *Elassoma*, see provisional key to genera (pages 27–29).

For identification of young basses in the Ohio River drainage, see "Provisional Key to Young of *Micropterus* Species" in genus introduction (pages 249–250).

Characters such as broken lateral band of blotches, prominent basicaudal spot, and tricolor caudal fin of spotted bass were not considered by some as reliable for taxonomic separation of spotted and largemouth bass fry or fingerlings (16–30 mm TL), particularly after preservation in 10% formalin and fading had ensued. Branching of pyloric caeca in largemouth bass as opposed to non-branching of the caeca in spotted bass is a reliable character for distinguishing the species in most instances, but branching or non-branching is difficult to determine for individuals less than 20 mm TL. However, at larger sizes reliable identification of the young of these two species can be obtained by counting pyloric caeca; spotted bass have 10–13 (usually 11–12) compared to 20–33 (usually 24) for largemouth bass.[229]

ECOLOGY OF EARLY LIFE PHASES

Occurrence and Distribution (Figure 107)

Eggs. During the spawning act, male and female largemouth bass swim in a circle around the nest releasing eggs and sperm over the nest site. Most

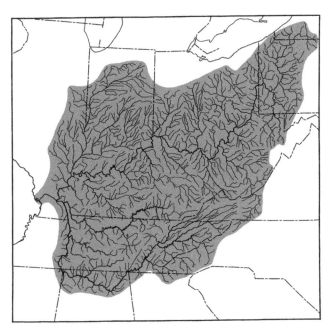

Figure 107   General distribution of largemouth bass in the Ohio River system (shaded area).

eggs fall into the nest; however, some fall outside the perimeter of the nest and these eggs usually die from lack of attention from the male. Eggs adhere to roots and stones on bottom of nest.[54,57,205,238]

After spawning, the male largemouth bass maintains the nest and, during incubation, guards the eggs from predators.[13,26,199,242,245] He hovers over the nest fanning the eggs with his pectoral fins. This fanning behavior prevents the eggs from being smothered by layers of sediment[242,245] and moves water over the eggs.[22,54,145]

An environmental variable that has an impact on largemouth bass embryo survival is wind-induced wave action that destroys the nest.[191] Egg hatching may also be adversely affected by lower dissolved oxygen levels.[22] Oxygen levels high enough to sustain adult life may be too low for embryonic development.[144]

Density-independent sources of mortality, such as water level and temperature, affected largemouth bass recruitment primarily through embryo mortality.[29] Incubation temperatures affect egg survival in a parabolic way, with the best survival near 20°C and progressively poorer survival as temperature deviates from this point. Egg survival declines with sudden thermal shocks and the eggs are most sensitive within the first day of fertilization.[143] In laboratory experiments regarding water temperature and largemouth bass egg mortality: (1) the sensitivity of bass embryos to temperature changes decreased gradually as development proceeded; (2) mortality was somewhat greater after sudden increases in temperature than after gradual changes; (3) mortality increased gradually from a

minimum at 20°C toward the low and high extremes of the range of temperatures tested.[22]

A drop in water temperature (which also may accompany cool, windy periods in spring) often results in desertion of the nest by the male, making the eggs more susceptible to predation, siltation, suffocation,[138,143] and the spread of fungus.[57,143]

Many largemouth bass eggs are lost to predation by other fishes, particularly sunfishes.[19,26,145,191] Sometimes entire nests are lost.[19,145,191] Sunfishes consumed large quantities of bass eggs in May. In one report, all longear, green, and bluegill sunfish examined in early May had fed on black bass eggs.[118] Common carp *Cyrinus carpio* was often present in the vicinity of largemouth bass nests in Lake Mead, NV. Carp would disturb the nest with their caudal fin on one pass and then suck up eggs and organic matter on a second pass by the nest. Several well-defended nests were observed to be surrounded with loose carp scales.[57] Yellow perch *Perca flavecens* preyed on largemouth bass embryos but were not observed preying on fry. Nest guarding males "yawned" upon being approached by a group of yellow perch. The male would nip the first few yellow perch. When several perch attacked the nest the male moved off to the side and abandoned the nest.[199] Dragonfly nymphs, pond snails, and backswimmers are also reported to eat bass eggs.[145]

In a study of the effects of habitat alteration and spring angling on the nesting success of largemouth bass in five MI lakes, it was determined that although local habitat characteristics are likely to be important factors affecting nesting success, lakewide features of lakes are also important and help explain large-scale patterns in nesting success that would be missed if only local habitat characteristics or single lakes were considered. Understanding the ecology and management of the black basses is challenged by the disconnect between the effects of fishing and habitat on individual nests and ultimate population-level effects. The authors of this study suggest that quantification of the magnitude of anthropogenic and natural effects at the whole-lake scale is needed. Their findings demonstrated that dwelling density around lakes warrants more attention and study.[271]

Yolk-sac larvae. At hatching largemouth bass have no mouth and must lie on their side in the nest until the air bladder begins to inflate. They obtain all nutrition from the yolk sac.[54] At 20°C, the mouth forms 192 h after fertilization.[54]

The yolk sac of recently hatched largemouth bass (<6.5 mm TL) is adhesive. Sediment or bottom scum will adhere to the surface of the yolk sac forming a coating that fits the yolk sac like a shell and has considerable coherence. This shell weights the young bass down and holds them in place, so that, although they may make vigorous movements with their tails and pectoral fins, they do not progress. In the laboratory, when sediment or ooze was pipetted onto a young largemouth bass, there was at once a vigorous wriggling movement that threw off the sediment. At the same time the shell of adherent particles on the ventral surface of the yolk sac keeps them from sinking deeper into the ooze of the bottom. The adhesive yolk sac and the wriggling movements protect the young largemouth from being smothered by sediment.[242]

Largemouth bass fry remain in the nest until their yolks are absorbed, usually 1–2 weeks, and then disperse when they are between 12 and 32 mm long. The male usually guards the nest until the fry disperse.[19,22,54,56,57,191,199,205,238] Three to eight days after hatching,[236,242] yolk-sac larvae 6.0–6.5 mm TL are able to break away from their attachment to the bottom and swim up from nest bottom.[22,39,51,54,236,242] At this size the yolk sac shows little trace of adhesiveness.[242] At 7.3–8.0 mm TL, the oil is gone and the yolk nearly absorbed. The air bladder is large and the young largemouth bass are swimming free above the bottom of the nest.[242] Largemouth bass fry are free-swimming in 240 h and have absorbed their yolk by 312 h. The fry must eat within 6 days of becoming free-swimming or they will die.[54]

At temperatures (°C) of:	Sac fry remain in the nest:[57]
13.0–15.5	7.2 days
15.5–18.3	7.2 days
18.3–21.1	6.7 days
21.1–24.0	6.0 days

Water temperature ranged from 13.0 to 21.0°C during the first week of largemouth bass hatching in Normandy Reservoir, TN. Hatching duration time was 26–42 days. Hatching began as early as April 1 and as late as May 6 over a 5-year period. The first day of hatching was positively related to the first day water levels reached full pool in the reservoir, leading the authors to believe that spawning time was not related to water temperature.[173]

The reproductive success of largemouth bass was studied after 56 days exposure to varying concentrations of paper mill effluents. After the exposure period, the largemouth bass were placed in hatchery ponds and allowed to spawn. Those exposed to higher effluent concentrations produced yolk-sac larvae with a higher incidence of abnormalities and a decrease in weights. Abnormalities included yolk-sac edema, abnormal vertebral column, and abnormal head shape.[10]

Post yolk-sac larvae. After yolk absorption, largemouth bass larvae remain over or near the nest until they disperse when they are reported to be from 12 to 32 mm long (usually 20–25 mm). The male usually guards the nest and/or fry until the fry disperse.[19,22,54,56,57,191,199,205,238] Larvae remain in the nest after hatching for upward of 2 weeks.[238] After rising from the bottom, largemouth bass larvae hover over the nest like a swarm, which soon forms a dense school. The school then leaves the nest, guarded by the male parent. The male herds the school into shallow shoreline areas where he apparently keeps them close to shore by swimming back and forth in the arc of a circle outside them. The male remains with the school until the young are at least 20 mm TL[242] or 12–32 mm (see above). In Lake Mead, NV, several swarms or schools of larvae in a small area often merge, with one guardian male assuming responsibility for the entire group.[57]

The microhabitat use patterns of largemouth bass larvae in a small Arkansas reservoir were studied. After becoming free-swimming, the larvae initially stayed directly over the substratum of their natal nests, but within 24 h they moved, usually less than a meter away from the nest. The larvae hovered <0.5 m above the substratum. While the larvae tended to remain within a few meters of the nest and at the same depth, it was not possible to predict which direction they would move. Normally they moved parallel to the shoreline. They swam into nearby structure to escape threats and predators.[26]

The first week of May saw the maximum predation of largemouth bass 4–10 mm TL by bluegill and green sunfish. Bass larvae made up 41% of the diet of green sunfish >100 mm and 36% of larger bluegill. During the second week of May, largemouth bass larvae were only 1% of the food eaten by green sunfish, but they continued to be present in the diets of bluegill and longear sunfish.[120]

In Orange Lake, FL, larval largemouth bass were collected primarily in panic grasses and floating/emergent vegetation, beginning in March.[209]

Largemouth bass larvae were collected in low numbers from the Tennessee River channel and the discharge canal at TVA's Colbert Steam Plant during the period April–June 1973.[210] Larval largemouth bass were collected with light traps from aquatic vegetation habitats in a backwater lake of the upper Mississippi River. Exact time of occurrence was not reported, but samples were collected from late May to mid-August.[220] Larval largemouth bass and spotted bass were collected in Lake Texoma, OK, with ichthyoplankton trawls in shallow water, usually at night.[234]

Weekly survival of young largemouth bass in their first summer was positively related to reservoir water level.[173]

Juveniles. Because largemouth bass is such an important sport fish, considerable effort has been expended in the study of its YOY ecology. It is generally accepted that year-class strength is largely determined late in the first summer for largemouth bass. However, the fate of smaller individuals that over-winter to age 1 is poorly understood.

Diverse factors such as floods, predation, and inadequate food undoubtedly take a tremendous toll of YOY bass in streams.[140]

In the first year of life, largemouth bass in reservoirs are largely confined to areas of cover in shallows.[207,240] It should be noted, however, that YOY will inhabit the best available cover for their changing needs throughout the first year. Young largemouth bass spend their first summer in sheltered, littoral, weedy areas near the spawning grounds and move offshore in the fall. Habitats over substrates of vegetation, sand, mud, detritus, and occasionally stone and rubble are preferred, usually in water >1.8 m deep, but also sometimes at depths of 4.2–6.0 m around structure.[205]

In three NY lakes, over 99% of the age-0 largemouth bass observed were in vegetated habitats; none were observed in areas of cobble substrate. Among age-0 smallmouth bass, 94% were observed over cobble substrates and only 3% were observed in vegetated habitats.[273]

In the Chipola River, FL, age-0 largemouth bass were associated with areas of reduced velocity and those with higher-than-average amounts of woody debris.[270]

In the Tennessee River, AL, young largemouth bass <200 mm TL were most often found in shallow off-channel habitats with fine substrates, little current velocity, and woody cover. They were seldom found in deeper, rocky, main channel habitats frequented by smallmouth and spotted basses. Young largemouth bass (16–27 mm TL) were first collected in early May in backpack electrofishing samples from gravel bank habitat in water 19°C; at the end of May, young largemouth bass 18–35 mm were collected from slough habitat (water temperature 21.5°C).[200]

Young largemouth bass were more abundant in cove habitats than in either sloughs or in main lake sites in four MS reservoirs. Shorelines of coves and sloughs were more protected from wind and wave action than main lake sites. These sheltered areas may have allowed greater nesting success and survival of the YOY. During high water years, the number of fish in cove and slough habitats increased 2- and 4-fold, respectively, whereas in the main lake habitats the increase was 10- to 22-fold. The disproportionate increase in the main lake suggests that flooding of terrestrial vegetation may have reduced the deleterious effects of wave action and nutrient paucity on the reproductive success and early survival.[168] Juvenile largemouth bass collected with shoreline seining in Lake Texoma, OK, were most abundant in small coves where the water was calm and usually clear. Abundance in seine collections was lowest on wave-swept shores where turbidity was often high and no vegetation was present.[234]

Maximum productivity of juvenile largemouth bass occurred in coves with moderate (~20% aerial coverage) vegetative cover.[72] Standing crop of largemouth bass was reported highest in vegetated areas as opposed to non-vegetated areas. Bass YOY were present in 7 of 10 collections from vegetated areas, but in only 1 of 10 from non-vegetated areas. Observations indicated that highest largemouth bass concentrations seemed to be in areas of flooded Bermuda grass, wheat, and/or aquatic vegetation. Vegetated areas of smart weed did not seem to hold as many largemouth bass as other types of vegetation.[42] Abundance of age-0 largemouth bass was positively related to hydrilla coverage in two Florida lakes. In contrast, native aquatic plant coverage was not related to age-0 largemouth bass abundance, perhaps because coverage by native plants was relatively constant from year to year.[78] In Lake Erie, juvenile largemouth bass were restricted to weedy areas along the lake shore and to quieter tributaries and ponds.[238] In Lake Mead, NV, yearling largemouth bass were associated with the presence of aquatic plants.[57] In a study of vegetated and un-vegetated areas of Guntersville Reservoir, AL, densities and biomass estimates for age-0 largemouth bass were greater in the vegetated areas of the reservoir.[233]

In Lake Opinicon, ON, age-0 and age-1 largemouth bass occupy inshore open waters.[240] Young-of-year tend to be associated with gravel substrates and cover. Often YOY can be found in water that is only slightly deeper than their body depth. Being in water this shallow may be a means to escape predation by juvenile bass. Juvenile largemouth bass are most commonly associated with complex physical structure and use simple structures such as horizontal logs or rock piles. Juveniles were never observed in areas without structure.[26]

Age-0 largemouth bass have patchy distributions which are positively related to specific landscape features. In Jordan Lake, NC, the distribution of YOY was positively related to slope and gravel substrata. Small patches (<20 m in length) of high-sloping gravel habitat accounted for some of the highest abundances of age-0 largemouth bass.[72] Age-0 largemouth bass were not distributed equally in embayments and a significant difference in abundance between embayments could be an important concern for management efforts.[278]

Young largemouth bass will readily utilize artificial substrates. In Jordan Lake, NC, artificial habitat substrates of gravel were placed in areas that historically held few YOY bass. Young bass actively selected for gravel patches of artificial substrate. Over the 4 years of study, more YOY bass were collected over the artificial habitats than over natural habitats and the number increased through time.[194] Complex habitat embayments provided the extent of landscape for age-0 largemouth bass in this reservoir. If complex structure is present, YOY will not leave, and those areas of artificial habitat produce more bass than areas with less complex habitat.[186,194] In a small Tennessee impoundment YOY largemouth bass were more abundant near wooden shipping pallets than along shorelines lacking cover.[13]

Feeding habitat of juvenile largemouth bass significantly overlapped with juvenile *Morone* in Lake Texoma, OK/TX. However, spatial segregation prevented potential competition between the species. Largemouth bass tended to be found more often in the upper ends of coves and away from wind-swept beaches without structure.[38] Juvenile largemouth bass have been observed schooling at the edge of lakes and ponds but the reason for this is not known.[56] In large reservoirs they have been observed with pelagic schools of white bass and hybrids chasing shad. In such situations the largemouth bass are usually found at the edges and below the moronid schools (personal observation of J. Fred Heitman).

Year-class strength of largemouth bass is highly variable between years.[53,138,139,143,145–147,150] Adult stock abundance had little effect on bass year-class strength, but stock-specific characteristics such as hatching time and growth influence juvenile survival.[29,138,139,143,145] Mortality rates are very high during their first year.[29,52] Year-class strength can be attributed to high mortality — often 99% or more — between egg fertilization and the end of the first

few weeks on life,[145] and survival to 1 year is generally, if not always, <0.5%.[140] Total annual mortality of largemouth bass is highly variable.[52] In 22 NE lakes, it ranged from 10 to 56% with a mean of 30%.[52] A significant variability of natural mortality rates for largemouth bass in large natural lakes was reported. Various studies indicate that the natural mortality rate ranged from 12 to 47% per year. Natural mortality of largemouth bass increased with age.[139]

Extreme year-to-year variations in abundance of YOY are common in large reservoirs. Estimates of YOY present in the fall, in Lake Carl Blackwell, OK, were independent of the number of brood fish present during the spawning season. Year-class strength may be established in the first few weeks of life due to abiotic factors;[138,139,143,145] however, it is more likely that biotic factors have an influence on year-class strength later in the summer.

Mortality after the first year is generally relatively low until the fish reach maturity. In general, if a largemouth bass reaches yearling status, its chances of becoming sexually mature are excellent, whereas there are a multitude of factors that affect survival of YOY. When they have reached a harvestable size by anglers, then mortality increases greatly and is dependent on local conditions. Mortality of largemouth bass in Beaver Lake, AR, was highest between ages 2 and 3.[147]

Environmental factors that have been shown to affect recruitment of largemouth bass in reservoirs can be grouped into two categories: hydrology and weather.[183] Extreme year-to-year variations in abundance of largemouth YOY are common in large reservoirs. In some reservoirs, primarily on the plains, passage of weather fronts disrupted spawning by causing males to abandon their nests. The reasons for nest abandonment were related to wind-induced wave action and/or sharp decline in water temperature. Although some nests are abandoned, spawning usually continues after passage of a weather front. Subsequent weather fronts might also cause nest abandonment. The potential result is a local largemouth bass population experiencing disjunctive rather than continuous spawning and hatching. Later in the year this disjunctive spawning will be related to bimodal length distribution of YOY and probable disparities in survival to age 1 of the YOY fish. Year-class strength, under these conditions, may have been established in the first few weeks of life with the most critical environmental influences appearing to be a sharp decline in temperature and the effect of wind action;[138,143,145,150] if spawning is disjunctive there is also significant mortality of YOY throughout the summer due to the loss of the later-spawned fish.[150]

Over-winter survival rates did not vary among years, despite TL differences of up to 50 mm. Little evidence was found to suggest size-selective over-winter mortality. Mortality rates of young largemouth bass from school dispersal through the yearling stage did not differ among years, suggesting that variations in cohort dynamics did not act to override initial differences in relative year-class strength.[66] No evidence of over-winter mortality was detected in Skiatook Reservoir, OK.[183]

The hypothesis that post-winter abundance of age-0 largemouth bass is strongly related to pre-winter size in IL lakes was not supported by the data. It appeared as if post-winter abundance of largemouth bass was dictated by the abundance of age-0 fish at the end of the first growing season.[195] Northern largemouth bass survived better in IL than did Florida largemouth bass and their reciprocal crosses were intermediate. While all stocks exhibited increased over-winter mortality as winters were more severe, the trend was more pronounced in Florida largemouth bass. There was no statistical difference in growth of the stocks.[36]

Over-winter survival of YOY Florida largemouth bass was significantly less than that of northern largemouth bass in VA[114] and OK.[18,111] Differential over-winter mortality occurred among phenotypes with the percentage of Florida largemouth bass declining by two-thirds, while intergrade bass increased proportionally in Briery Creek Lake, VA. First-year growth and relative weight did not vary consistently among phenotypes.[114] In Boomer Lake, OK, a thermally enriched reservoir, over-winter survival of Florida largemouth bass was low, particularly after a cold winter. During the cold winter Florida largemouth bass were largely found only in the thermal effluent of the power plant.[129] In Dripping Springs Lake, OK, a new impoundment, over-winter survival of hybrids was greater than Florida largemouth bass, which was less than northern largemouth bass.[132]

Abundance of Florida largemouth bass stocked in small OK impoundments was generally higher in the southeastern portion of the state as opposed to the northwestern portion, which follows a trend of decreasingly cold climate across the state from northwest to southeast. There was no statistical relationship between Florida largemouth bass over-winter survival and heating degree days. Estimates of over-winter survival ranged from 0–68% with an overall mean of 25%. Stocked Florida smallmouth bass made up 15% of the year-class in the fall but only 5% in the following spring.[105] In OK reservoirs the mean number of days with sub-freezing air temperatures at a given location was one of the most influential independent variables that determined the success of Florida largemouth bass introductions.[111]

Whole body lipid content of young largemouth bass increased from 3.5% of dry body weight in fall to 5.9% in December and decreased to 3.3% in March. All young had equal amounts of stored lipids by late fall, but by March the surviving small young had lower lipid levels than large young, suggesting that during winter, smaller fish spent their reserves at a faster rate. Protein content averaged 74.6% of the dry body weight and neither differed among months nor changed with fish weight. Mortality of the smallest young was accelerated in the late fall and winter, coinciding with decreases in lipid stores and rapid drops in water temperature.[58] During the first winter of life, loss of energy reserves as a function of low feeding activity and scarce prey may contribute to high mortality of age-0 largemouth bass.[64]

Juvenile largemouth bass prefer warmer water temperatures than adults,[143] with preferred temperatures of 26.5–32°C reported[237] and their presence in water 36–39°C also reported.[148] Below-normal temperatures can adversely affect anatomical structures, feeding behavior, or size of post-larval bass in ways that predispose them to increased mortality from starvation or from deformity per se.[145] The maximum swimming speed of largemouth bass fingerlings and adults declines at temperatures above 30°C.[143]

Yearling northern largemouth bass and Florida largemouth bass were evaluated for cold shock. The estimated 96-h median tolerance for northern largemouth bass was about 6°C and for Florida largemouth bass about 8.5°C. At all temperatures where fish died the Florida largemouth bass died first.[130]

In Lake Kissimmee, FL, heavily vegetated areas that had dissolved oxygen (DO) concentrations of <2 mg/l were not considered as suitable habitat for YOY largemouth bass.[196] In low DO environments, the appetite of young largemouth bass is suppressed.[144] Swimming ability of juvenile largemouth bass in the lab was markedly reduced when DO was below 5–6 ppm. At DO >6 ppm, swimming speed was independent of DO levels.[144]

Largemouth bass fingerlings increased their oxygen consumption about ten times as temperature increased from 5 to 30°C. Increased oxygen demand by the tissues was accommodated by an increase in opercular movement. A fivefold increase in opercular activity was needed to keep largemouth bass blood saturated with oxygen at 32–35°C.[144]

YOY largemouth bass acclimated in water that was 0 and 5% salinity, preferred 0%.[34]

When 100-mm largemouth bass were exposed to chromium they "coughed" more frequently and there was widespread destruction of the intestinal lining.[144]

Water level fluctuation is one of several abiotic factors that influence recruitment of largemouth bass in reservoirs of the southeastern United States.[138,145] Weekly survival of largemouth bass in their first summer was positively related to reservoir water level.[173] Density-independent factors such as climate (air and water temperatures and wind action)[138,145] and water level fluctuations have an effect on recruitment. In contrast, density-dependent factors (limited carrying capacity, reduced forage availability, and predator success) can have their effects, also. Further, there are large variations geographically in YOY largemouth bass densities between systems and years that, in turn, influence each of the above factors.[110,150]

Hydrology affects largemouth bass year-class strength. Largemouth bass are more numerous and exhibit better growth when lake levels are high.[144] In Skiatook Reservoir, OK, increased water levels during the spawning season and distribution of nutrients throughout the reservoir via water releases after the spawning period enhanced largemouth bass production in the reservoir.[183] Hatching success was disrupted during rapid water level rises and drops in two IL reservoirs. Peak hatching success occurred when water levels were relatively stable. Recruitment of largemouth bass was considerably higher in higher water years than in low water years, and appeared to be unrelated to water level fluctuations during the spawning season.[68]

Annual catch rates of age-1 largemouth bass were significantly related to estimated mean structural habitat availability the previous season. Total structural habitat availability declined sharply at lower water levels. Coarse substrates (gravel, cobble, and boulder) were more common at higher water levels and at lower levels the primary habitat was fine substrates (clays). Water level appears to affect largemouth bass population size partially through availability of habitat.[184]

In Bull Shoals Reservoir, AR, largemouth bass standing crops were highly correlated with water inflow, seasonal mean lake elevations, and standing crop of YOY threadfin shad and gizzard shad. None of these variables was correlated in Beaver Reservoir, AR, during the same period.[146] In Beaver and Bull Shoals Reservoirs[147] and also in Hugo Reservoir, OK,[85] holding water in the flood pool during summer enhanced recruitment of largemouth bass. In Bull Shoals Reservoir, cover and survival of YOY were positively related. Flooding of terrestrial vegetation was positively correlated to YOY abundance, particularly abundance from June 1 to mid-August. Duration of flooding after largemouth bass schools break up about June 1 is highly correlated with the number of YOY surviving the first summer of life.[147] Although the availability of cover appeared to influence profoundly the early survival of YOY in Bull Shoals Reservoir, rate of growth for largemouth bass may be an even more important factor to annual survival and year-class strength.[53]

Year-class strength of largemouth bass was fixed by late summer or early fall in Normandy Reservoir, TN. Catch of age-1 bass in early spring was related to the number of days the reservoir was at or above full stage while the fish were age 0. Largemouth bass produced in wet or intermediate years were more than twice as abundant at age 3 than fish produced during dry years. There was often no relation between YOY abundance in mid-summer (peak abundance) and late summer. However, late fall abundance was related to abundance in the following spring.[156] Decreasing water levels in late summer reduce shoreline cover available to young largemouth bass, exposing them to increased rates of predation and reducing their feeding efficiency.[53,73,156]

Growth and survival of reservoir YOY largemouth bass are positively affected by reaching full pool in the reservoir early in the spring and maintaining that level for at least 90 days.[173] Predictive models and correlations were used to emphasize the importance of high, stable water conditions on the spawning and nursery stages of largemouth bass. Maintaining high, stable water during spring and summer followed by a late summer/fall drawdown to allow re-vegetation and oxidation of organic and mineral matter, thus providing nutrients upon re-flooding, may be the best management plan for largemouth bass in southeastern reservoirs.[187] There was a trend for increased YOY largemouth bass abundance in summer seining when lake levels on April 15 were at or above full pool in Lake Eufaula, OK.[17]

Significantly positive relationships among abundance of YOY largemouth bass and days of littoral flooding during spawning and nursery seasons, draw downs during re-vegetation season in the previous calendar year, combinations thereof, and water levels during these seasons indicated that recruitment was influenced by water level fluctuations in Grand Lake, OK. A difference between re-vegetation and nursery seasons elicited favorable overall conditions for recruitment of largemouth in the lake.[110]

In four mainstream impoundments of the Tennessee River, year-class strength of largemouth bass was inversely related to average June–July discharge and positively associated with retention (reservoir volume/discharge). Thus, weak year-classes were produced when early summer conditions were wet and stronger year-classes were produced when early summer conditions were dry. Late summer aquatic plant abundance and water level fluctuations during April–May, while spawning was occurring, were not related to largemouth bass recruitment in these studies. The authors speculated that higher discharges and faster flushing rates were associated with reduced production at lower trophic levels and poorer survival of young largemouth bass.[74]

When comparing population level patterns across systems, the density of largemouth bass adults (above some minimum threshold) had little impact on recruitment strength.[27] There was no relationship between YOY abundance and the number or weight of spawners, or the number of nests.[199] One of the biotic factors that could influence which YOY actually contribute to the year-class may be the fact that larger largemouth bass (>356 mm) spawn 2 weeks earlier than smaller largemouth bass.[161]

It appears that earlier-spawned largemouth bass grow larger in their first year than do later-spawned individuals and that the larger individuals may have a higher survival rate than the smaller, later-spawned fish.[39] Early swim-up larvae are critical to year-class production at age 1 because these fish have a growth advantage over later swim-up larvae. This advantage is carried through the first year of life if the prey of young bass becomes too large to be eaten by the later-spawned YOY, resulting in smaller bass eating a lower caloric value prey (i.e., insects) and, as a result, exhibiting slower growth. The length advantage attained by the earlier-spawned fish may make them the primary contributor to both age-0 largemouth bass present in the fall and age-1 fish present the following spring.[171] The slower-growing YOY are more susceptible to predation and less likely to survive to age 1. Larger adult male largemouth bass spawn earlier than smaller males, providing competitive advantage to the progeny of the larger males.[166,171]

Nutrient alteration and impact on phytoplankton by oligotrophication could favor young spotted bass, and eutrophication of low productivity waters may select for young largemouth bass. Oligotrophication will probably reduce black bass growth rates, condition, and size of fish caught by anglers, and probably will increase the proportion of spotted bass in these fisheries.[170]

Piscivory on age-0 shad by age-0 largemouth bass is more likely to occur in eutrophic than in oligotrophic or mesotrophic reservoirs. Eutrophic impoundments have high chlorophyll *a* values and high larval gizzard shad and threadfin shad densities. Density and growth of larval shad and age-0 largemouth bass increased with chlorophyll *a* in Kentucky impoundments. Duration of occurrence for larvae of both species of shad was positively related to chlorophyll *a*. Eutrophic reservoirs contained larval shad that were ≤40% of the mean age-0 TL of largemouth bass (i.e., the size at which the shad would be vulnerable to predation) in late June–July, whereas larval shad were generally not collected in late June or July in oligotrophic or mesotrophic impoundments.[172] Age-0 largemouth bass were more abundant in AL reservoirs when concentrations of chlorophyll *a* were >8 mg/m^3.[155]

Largemouth bass exposed to exhaustive exercise, air exposure (i.e., hooking by anglers), culling, live-well conditions, and weigh-in procedures were evaluated for spawning success and survival of their young. Age-0 largemouth bass produced from parents subjected to stress were smaller and weighed less than control fish not subjected to stress; the stressed fry had a later swim-up date than controls.[81]

Largemouth bass populations in Lake DeGray, AR, were dominated numerically by age-1 and age-2 fish (73–96%).[90] In West Point Reservoir, AL/GA, abundance of YOY largemouth was highest in June (n = 611/ha) and declined steadily through September (n = 33/ha). There was no statistical difference in number or weight of YOY largemouth bass between the upper, middle, and lower reaches of the reservoir. Changes in number were related to mortality by YOY rather than changes in distribution.[182]

Young-of-year largemouth bass in Lake Mead, NV, were preyed upon by yearling largemouth bass, green sunfish, bluegill, and carp.[57] Size-dependent cannibalism during summer acted to remove smaller, later-hatched largemouth bass. Early hatching may enhance survival during larval and early juvenile stages in southern systems.[65] Overwinter survival of largemouth bass is sometimes size related, with smaller fish exhibiting higher mortalities than larger individuals.[71,89] Abundance of YOY largemouth bass declined from 542 young/ha in June (mean length = 37 mm) down to 12/ha in March (mean length = 149 mm). During the period September–January, smaller YOY decreased in abundance from 176 to 9/ha while the mean length remained at 60 mm; while fish in the larger mode decreased from 76 to 27/ha and increased in length from 108 to 154 mm.[58]

Juvenile largemouth bass had increasing numbers of parasitic taxa as size increased; the number of parasites, both endoparasites and ectoparasites, also increased with size. Parasites were found in 73% of largemouth bass <40 mm TL.[35,135]

The ecological neighborhood of YOY largemouth bass in Lake Jordan, NC, was relatively small (shoreline of an embayment), yet differed based on time of day and habitat characters. About 93% of individuals were captured within 1 m of cover.[194] Most young largemouth bass move very little and mass migrations of YOY or yearling largemouth bass between embayments are unlikely. Since adult largemouth bass generally move little, only if there is significant movement of the fish between age 1 and sexual maturity would there be much movement beyond the locale where the bass are spawned. A small portion of the largemouth bass do move substantial distances.[167] Fingerling largemouth bass in Lake Mead, NV, frequently migrated between adjacent coves and into deeper water.[57] Similar YOY behavior of largemouth bass was reported in the Arkansas River.[88] Water level fluctuation had no detectable effect on movements.[167]

A study of diel activity patterns assessed by comparing day and night catches in activity traps in a variety of lowland hardwood wetland habitats indicated that juvenile largemouth bass were primarily night-active.[214] In a central Piedmont reservoir in NC, offshore movement of YOY largemouth during daylight hours was observed. Smaller bass tended to remain inshore during the day, suggesting an ontogenetic shift in habitat use, possibly in response to predation pressure. The authors also suggest that some diel differences in distribution of YOY largemouth may be related to habitat characteristics.[232]

While native largemouth bass tend to move little, stocked bass seem to offer additional insights to movement in reservoirs. Generally, stocked YOY largemouth exhibited some movement along the shoreline but most individuals stayed relatively close to the stocking location. Often, stocked fish served as a forage base for existing predators.[84] In pools of the Arkansas River, stocked YOY largemouth bass exhibited some movements between the release site and nearby coves but most stayed in the area where they were stocked.[88] About 31% of recaptured northern largemouth bass, Florida largemouth bass, and hybrid largemouth bass in Chickamauga Lake, TN, had moved >600 m from the stocking site, indicating rapid dispersal by some fish.[198] Largemouth bass YOY stocked into Jordan Lake, NC, dispersed an average of 1,100 m, then established a limited area home range similar to native fish.[255]

Early Growth

Growth in young largemouth bass may be broken into two discussion periods: growth during the first growing season and subsequent growth to maturity. In general, growth in either period is largely dependent on the availability of suitable forage. In the Ohio River drainage largemouth bass can usually be expected to reach 250 mm, the size of sexual maturity, by the end of the second growing season. If that size is not attained then local conditions are inhibiting growth.

Historically, scales have been used to determine age and growth of largemouth bass,[22,75,119,127,138,178] but in OK, scale reading was determined an inaccurate method for aging this species.[93] More recently, counting growth rings on otoliths has been determined to be a more accurate method for aging largemouth bass, particularly in more southern portions of its range.[75,88,162,177] Use of sagitta otoliths has become well-established as a means of determining the daily growth rates of YOY[6,31,170,198] as well as a means of aging juveniles and adults,[86,88] although this technique was found invalid for aging largemouth bass in tropical regions.[9]

Growth of largemouth bass fry was directly related to the amount of food in their stomachs during the first 2 weeks after rising from the nest. Near the end of the first growing season growth was directly related to the ratio of large to small organisms in their stomachs.[22]

Largemouth bass tend to live longer in northern regions, but their growth rates are slower,[192] while faster-growing and shorter-lived bass are typical of southern locales.[22,131] However, variations in growth in all regions is so high that regional differences are probably not significant.[22] Length at age and growth equation parameters for 698 populations of largemouth bass in North America varied greatly among populations and resulted in ages at quality length (300 mm) ranging from 1 to 10 years.[269] Growth of largemouth bass in the North is likely limited by cooler water temperatures and short growing seasons, while populations in the south may be limited by year-round warm water temperatures and extended reproductive drain (the energetics associated with spawning causes a reduction in growth and a reduced life expectancy).[256]

First-year growth of largemouth bass is extremely important in determining which fish survive to reach age 1. Larvae spawned by larger adults had more daily growth rings on the otoliths than larvae from smaller-sized adults, indicating that larger, older bass spawned earlier than smaller bass.[39] One author suggests that largemouth bass populations with larger brood fish may produce bigger YOY, better prepared to become piscivorous when prey fish begin to hatch,[224] but another study indicated that length of larvae was independent of length of adults (larger largemouth bass did not produce larger larvae). Rather, positive correlations between larval age and length indicate that the initial length advantage was the result of an earlier spawning date.[39] Largemouth bass exhibited bimodal length frequency by mid-summer in wet years but maintained unimodal length in dry years. When bimodal length distributions were the case, earlier-hatched fish grew faster than later-hatched fish.[173]

Earlier spawned largemouth bass have a head start in growth, an advantage that is carried through the first year because earlier spawned fish have available more prey, a longer growing season,[6,27,29,39] and the advantage of being able to utilize larger-bodied prey, acquiring increased metabolic efficiency.[207] Earlier-hatched largemouth bass consume a higher percentage of fish in their diet than do later-hatched bass. Earlier-hatched bass have extended access to prey fish as they grow with them, whereas the later-hatched largemouth bass have a shorter time within which prey fish are of a suitable size for consumption.[62]

If older and larger largemouth bass tend to spawn earlier than younger individuals,[39] populations dominated by large individuals may be characterized by early hatches and rapidly growing and ultimately more successful offspring in natural lakes; however, this phenomenon was not apparent in reservoirs.[27] Diet and hatch time strongly affected growth, with subsequent effects on recruitment variability through size-dependent mortality of juveniles.[29]

Growth rates differed for groups of large- and small-sized age-0 largemouth bass in Kentucky Lake, KY. There was a pivotal period in mid-July when there was a divergence in growth rates. Growth rate of the smaller age-0 bass slowed dramatically, while the growth rate of the larger young increased. During this period, the large bass were consuming more fish and fewer insects and zooplankton by weight than were the small bass. The variation of growth rates resulted in a multi-modal length frequency distribution by the end of the summer; growth of the small fish stopped but the large fish continued to grow. Difference in growth between the two groups was related to the quality of diet.[185]

Growth of young largemouth bass in Lake Blackshear, GA, was highly variable within a year-class. Some individuals were as long as 160 mm TL at the end of the first year, while others were only 70 mm. It was evident that some YOY gained a competitive advantage allowing them to grow relatively quickly through the first year of life. This advantage was lost somewhat during the second growing season as evidenced by a weak correlation between TL at first annulus and second annulus formation.[126]

In years with extensive flooding, YOY largemouth bass in Bull Shoals Reservoir, AR, exhibited bimodal length frequency in August samples. This bimodal growth may have been due to differences in the feeding habits of the YOY. Some young bass began eating fish as early as mid-June, when they were not significantly larger than other YOY bass that were eating plankton and insects. Growth of the bass that ate fish greatly accelerated during the remainder of the summer. Stomachs of the bass that ate fish contained more than seven times the relative food volume than those of bass that fed on invertebrates throughout the summer. There appeared to be a highly positive relationship between the number of fast-growing fish in the first summer and the number of yearling largemouth bass the following spring. In the spring the yearling bass had a uni-modal length frequency indicating nearly 100% mortality of the slower growing bass. The authors concluded that to some extent there must exist a relationship between the growth of these fast-growing fish-eating bass and the availability of properly sized and numbers of forage fish.[53]

Growth of young largemouth bass also varies depending on numerous environmental[6,29,173] and genetic[6] variables. First-year growth of largemouth bass in Lake Carl Blackwell was positively correlated to the average annual lake level; in the second

Table 121

Central 50% and range of mean of means of TL (mm) for young largemouth bass related to age summarized by Carlander (1977) from numerous reports from various locations.[22]

Age (years)	Age Month	TL (mm)		
		N	Central 50%	Range
0	April	975	19–52	12–52
	June	3,897 +	30–48	13–156
	July	6,682 +	46–74	23–203
	August	3,007 +	56–97	18–251
	September	1,009 +	71–137	36–315
	Oct.–Dec.	6,047 +	79–165	38–351
1		10,604 +	150–226	28–406
2		7,742 +	218–300	71–517
3		6,070 +	270–350	119–511
4		4,359 +	310–396	165–559

year the relationship was reversed.[190] Warmer autumn air temperatures in NE were related to increased largemouth bass recruitment, possibly by allowing for increased growth.[52]

Carlander (1977) provides an extensive summary of data relating TL to age (Table 121) and back-calculated length estimates at age for largemouth bass.[22] He concludes that generally growth is slower in the North and faster in the South and that variation within regions is extensive (Table 122), even within a state (Tables 122, 123, and 124). There is apparently a tremendous influence by local environmental variables on growth rates of largemouth bass despite any inherent growth potential;[100,109] largemouth in the South grow larger than those in the North.[22,139]

A comparison of reports of growth for largemouth bass from stream and reservoir populations shows variable but comparable growth at age 1, but reservoir largemouth bass are generally larger thereafter. Reports of calculated lengths for age-1 largemouth bass from reservoirs range from 89–187 mm TL, from streams 64–187 mm; for age 2, 170–347 from reservoirs and 116–245 from streams; for age 3, 236–386 from reservoirs and 164–338 from streams (Tables 122 and 123).

The findings of early researchers comparing first- and second-year growth of northern and Florida subspecies of largemouth bass were inconclusive and conflicting.[6] This problem was probably due to overlapping meristic counts between the two

Table 122

Reports of age and growth calculated for northern largemouth bass from various locations.

Location	TL (mm) at Age:			
	1	2	3	4
Lake Wappapello, MO[215]	137	277	338	409
Beaver Reservoir, AR[192]	152	277	333	—
Ball Shoals Reservoir, AR[192]	176	297	377	—
Lake Ft. Smith, AR[239]	138	244	310	353
OK, state average[17]	139	216	278	336
Lake Eufaula, OK[17]	144	250	322	385
Upper Spavinaw Lake, OK[97]	185	320	386	450
Lower Spainaw Lake, OK[97]	140	249	323	394
Lake Carl Blackwell, OK[138]	162	347	401	440
Lake Carl Blackwell, OK[193]	128	283	374	—
Lake Carl Blackwell, OK[190]	140	279	369	—
Cain Lake, TX[128]	193	280	297	317
OH, state average[22]	89	178	257	—
St. Mary's Lake, OH[22]	89	170	236	279
Vesuvius Lake, OH[22]	89	178	249	297
Meander Lake, OH[22]	91	190	251	—
Lake Alma, OK[22]	102	198	257	—
Lake Blackshear, GA[126]	106	253	350	418
KY Lakes[22]	150	292	—	—
Kentucky Lake, KY[22]	109	213	300	—
Lake Barkley, KY (2003)[a]	178	279	335	—
Lake Barkley, KY (2006)[a]	188	284	343	—
Ohio River, Meldahl Pool, KY[b]	170	259	317	—
Ohio River, Markland Pool, KY[b]	152	246	307	—
Unweighted mean TN, KY[22]	153	268	—	—
Norris Reservoir, TN[50]	136	178	262	—
Norris Reservoir, TN[248]	175	315	373	409
White Oak Lake, TN[22]	102	236	330	—
Dale Hollow Reservoir, TN[22]	124	227	302	—
Chickamauga Reservoir, TN[211]	103	212	269	327
Lewis Smith Res., AL (males)[142]	168	295	356	383
Lewis Smith Res., AL (females)[142]	201	282	358	410

[a] G.N. Jackson. Personal communication, KY Department of Fish and Wildlife Resources.

[b] H.D. Henley. Personal communication, KY Department of Fish and Wildlife Resources.

subspecies that confounded accurate identifications. The confusion was later corrected by the use of electrophoretic evaluation of fish that could positively differentiate between each of the subspecies

and their intergrades. Conflicts within the earlier works were likely due to misidentification of the intergrades. When growth rates of Florida largemouth bass and northern largemouth bass are compared there is often little if any difference between the subspecies (Tables 122 and 124), although there appears to be a general trend that first-year growth of Florida largemouth bass is somewhat better below 35° N latitude.[114] Total lengths of Florida largemouth bass from Lake Cain, TX, ranged from 182 to 198 mm at age 1, 269–287 mm at age 2, 301–344 mm at age 3, and 344–350 mm at age 4, growth comparable to the growth reported for FL populations of the subspecies (Table 124).

In OK, statewide average growth calculated for the period 1972–1973 was reported as follows: 133 mm TL at age 1 (35 lakes); 231 mm at age 2 (34 lakes); 308 mm at age 3 (29 lakes); and 386 mm at age 4 (19 lakes).[218] These growth rates were comparable to growth reported for the period 1952–1963 (Table 125).

In the Tennessee River, AL, young largemouth bass were 16–27 mm TL in early May and 35–42 mm in late May; average TL at the end of the first growing season was 109 mm.[200] In OH, YOY in October were 51–140 mm TL and up to 200 mm in hatcheries; at about 1 year, young largemouth bass were 76–190 mm TL.[55]

TL (mm) ranges for YOY and yearling largemouth bass in Lake DeGray, AR, during the period May through September 1977–1979 were reported as follows:[91]

Month	TL	
	YOY	Yearlings
May	38–60	86–239
June	38–86	86–264
August	38–132	190–290
September	86–239	140–315

Growth of young largemouth bass in KY lakes, calculated during the period 1981–1983, was variable among year-classes and among lakes (Table 126).[249] Ranges of calculated TL for young largemouth bass collected from Cave Run Lake, KY, during the period 1985–1994 were as follows: age 1, 132–152 mm, age 2, 196–231 mm, age 3, 239–297 mm, and age 4, 300–343 mm.[230]

During the first 10 years after impoundment (1936–1946) of Norris Reservoir, average calculated TL at the end of each year of life for young largemouth bass was 175 mm at age 1, 315 mm at age 2, 373 mm at age 3, and 409 mm at age 4 (Table 122). Growth varied considerably from year to year,[55,248]

Table 123

Reports of age and growth for stream populations
of largemouth bass.

Location	TL (mm) at Age:			
	1	2	3	4
Cape Fear River, NC[181]	98	182	241	284
Chowan River, NC[181]	123	215	277	320
Pasquotank River, NC[181]	136	223	296	352
Tar–Pamlico River, NC[181]	121	217	298	362
KS streams[217]	64	116	164	285
KY streams[22]	114	231	338	—
MS National Forest streams (N = 8)[261]	187	245	257	296
Tennessee/Tombigee waterway dwst. of Aberdeen Dam[265]	149	230	291	362
Tennessee/Tombigee waterway dwst. of Columbus Dam[265]	149	228	300	326

Table 124

Reports of age and growth for Florida largemouth bass from several FL locations.

Sex	Age	TL (mm) from:							
		Lake George[100]	Lake Poinsett[100]	Washington & Sawgrass[100]	Weir Lake[100]	Suwanne Lake[100]	Kissimmee Lake[100]	Lake George[107]	Henderson Lake[107]
Male	1	176	179	171	169	148	176	175	169
	2	266	281	267	262	240	265	267	269
	3	314	322	333	305	292	309	331	343
	4	337	351	354	334	328	331	362	382
Female	1	189	203	182	169	151	178	172	166
	2	301	317	290	270	252	283	269	297
	3	368	381	370	334	325	355	388	349
	4	424	430	431	388	373	413	428	400

Sex	Age	TL (mm) from:					
		Newnans Lake[107]	Orange Lake[107]	Sante Fe Lake[107]	6 Lakes Mean of Means[100]	4 Vegetated Lakes[116]	4 Unvegetated Lakes[116]
Male	1	193	182	138	170	204	166
	2	302	244	269	264	264	281
	3	343	308	319	313	310	336
	4	411	348	352	339	327	378
Female	1	180	143	178	179	210	295
	2	295	275	258	286	288	383
	3	380	312	343	356	361	417
	4	470	416	407	410	448	495

Table 125

Statewide average length (mm TL)/weight (g) relationships related to age for young largemouth bass in OK waters during the period 1952–1963 with slowest and fastest growth rates also recorded.

Age (years)	Slowest Growth		Average Growth		Fastest Growth	
	TL	Wt	TL	Wt	TL	Wt
1	63.5	—	139.7	31.8	284.5	304.2
2	124.5	22.7	246.4	199.8	391.2	894.4
3	170.2	59.0	317.5	454.0	510.2	2,129.3
4	264.2	249.7	378.5	808.1	553.7	2,773.9

Source: Constructed from data presented in Table 1, reference 208.

Table 126

Age and growth determinations of young largemouth bass from KY lakes. Calculations are based on various year-classes during the period 1972–1982.

Location	TL (mm) Range at Age:			
	1	2	3	4
Barkley Lake (1983)	114–135	196–216	272–292	323–353
Kentucky Lake (1983)	109–125	196–236	259–323	320–381
Rough River Lake (1983)	89–137	158–224	257–290	318–345
Nolin River Lake (1983)	140–170	221–244	282–307	325–356
Mauzy Lake (1981)	122–147	193–246	246–358	338–422
Carpenter Lake (1982)	114–180	193–259	264–305	297–353
Kingfisher Lakes (1983)	104–145	157–224	244–285	295–333
Washburn Lake (1981)	117–165	165–246	221–300	264–373
Barren River Lake (1981–83)	86–97	157–180	234–259	284–330
Briggs Lake (1983)	122–145	218–252	297–312	363–378
Green River Lake (1981, 1983)	66–140	127–213	188–269	267–323
Marion County Lake (1983)	127–178	218–267	287–356	406
Shanty Hollow Lake (1983)	102	183	244	315
Spurlington Lake (1981, 1983)	137–175	241–251	307–310	373–376
Dewey Lake (1983)	76–137	168–239	241–297	284–338

Source: Constructed from data presented in Tables 1, 3, 5, 8, 12–17, 21, 24, 33, 37, 40, 47, 52, 65, 72, 73, and 78, reference 249.

with growth in the latter years of the period nearly as good as during the first 3 years of impoundment.[248]

In Jordan Lake, NC, in 1990, stocked YOY largemouth bass grew at an average daily rate of 0.18 mm, while native YOY grew at a daily rate of 0.14 mm, not a statistically significant difference. In 1991, stocked bass grew at a rate of 0.3 mm/day compared with 0.2 mm/day for the natives; this was statistically significant.[250] During the period June 13–July 5 young bass grew at a rate of 0.6 mm/day and between July 5 and July 25 growth was at a rate of 0.65 mm/day.[112] In Normandy Reservoir, TN, young largemouth bass growth ranged from 0.677 to 0.744 mm/day.[173] In eutrophic AL reservoirs, daily growth of young largemouth bass was about 0.7 mm/day compared to only about 0.55 mm/day in mesotrophic and oligotrophic reservoirs.[170]

The rate of growth for YOY largemouth bass in Florida was highly variable between lakes and time of hatching. Growth was slowest in early-spawned groups and tended to be higher in later-spawned groups. This probably was due to ambient water temperatures. Growth rates ranged from 0.43 to 0.82 mm/day.[51]

Early growth rate was much faster in Puerto Rico. Daily growth of Florida largemouth bass in Lucchetti Reservoir averaged 1.26 mm/day and Florida largemouth bass × northern largemouth bass hybrids averaged 1.23 mm/day; no statistical difference. This growth rate was consistent up to about 275 mm TL. At age 1 the growth rate slowed to 0.25 mm/day and by age 2 the growth rate was only 0.06 mm/day.[164,169]

A study of the timing of spawning, hatching, and growth of largemouth bass in Florida lakes at different latitudes showed that hatching periodicity influenced growth and survival of YOY bass. Fish that were hatched early (December–January) in south FL exhibited slow growth (0.40 mm/day) and high mortality probably because of low and variable water temperatures in January and February. Later-hatched fish in the same lakes had faster growth (0.5 mm/day) and higher survival than early-hatched fish. While bass in northern FL lakes hatched later than bass in southern FL lakes, the northern lake largemouth bass had rapid growth that nearly compensated for the difference in size by age 1. Age-1 bass ranged in length from 120 to 250 mm in south FL, 100–250 mm in the central lakes, and 70–250 mm in the northern lakes. Most YOY bass exceeded 100 mm TL by fall and grew throughout the winter.[51]

In Lewis Smith Reservoir, AL, female largemouth bass lived longer than males but the rate of growth for the sexes was similar through age 4 (Table 122).

Annulus formation began in April and continued through June.[142]

Female largemouth bass tend to be larger than males at all ages except age 1.[22,189] In some FL lakes male Florida largemouth bass grew faster than did females; however, by the end of their second year, females were longer than males. Also, females lived longer than males.[111] Female Florida largemouth bass through age 3+ and males age 4+ were significantly larger in un-vegetated lakes than in vegetated lakes (Table 124).[116]

In systems where sunfish were the dominant prey, the growth of YOY largemouth bass was largely regulated by the availability of late-hatched sunfish.[27] In systems dominated by gizzard shad, growth of YOY largemouth bass is regulated by complex ecosystem-level interactions that vary along environmental gradients. Total phosphorus affects the success of YOY gizzard shad abundance, thereby mediating its role as a competitor or prey.[27] Another study suggested that first-year growth of largemouth bass was not directly affected by the timing of threadfin shad or gizzard shad hatching.[173]

Largemouth bass growth initially improved in CA reservoirs after threadfin shad were introduced as a forage species:[150]

	TL (mm) at Age:			
	1	2	3	4
Before threadfin shad	107	216	318	389
After threadfin shad	150	272	358	409

Formulas of length–weight relationships from several locations are presented in Table 127 and combined data from various locations relating ranges of average weight to TL ranges of young largemouth bass are presented in Table 128.

Growth data from OK (Tables 125 and 129) and AL (Table 130) seem to fit within the ranges presented in Table 128, but young largemouth bass from Chautauqua Lake, IL, appear to be heavier at comparable sizes (Table 131).

Condition factors are a measurement of well-being or plumpness in fish. Condition factors (K) of largemouth bass vary among locations and populations but are often reported to increase with age and growth. Selected reports of K for young largemouth bass are presented in Table 132. Condition factors of YOY largemouth bass were higher during May–July, then declined in August and September.[22] Oftentimes condition factors will improve again in the fall after summer stresses are relieved. For largemouth bass 61–260 mm TL in Lake Eufaula, OK, K ranged from 1.10 to 1.21 and the

Table 127

Length–weight relationships reported for largemouth bass from several locations (w = weight; L = length; TL = total length; SL = standard length).

Location	TL (mm)	Relationship
FL canals[98]	150–559	log w = −5.469 + 3.250 log TL
Blue Cypress Lake, FL[98]	216–627	log w = −5.587 + 3.285 log TL
Cain Lake, TX (FLMB)[128]	—	log 10w = −5.4339 + 3.3375 log L
Cain Lake, TX (NLMB)[128]	—	log 10w = −4.9409 + 3.1349 log L
Cain Lake, TX (hybrids)[128]	—	log 10w = −5.7019 + 3.4491 log L
Lewis Smith Reservoir, AL (males)[142]	—	log w = 5.3904 + 3.1818 log L
Lewis Smith Reservoir, AL (females)[142]	—	log w = −5.6085 + 3.2689 log L
Lewis Smith Reservoir, AL (immature)[142]	—	log w = 4.5008 + 2.7931 log L
AL[235]	51–254	log w = −4.8 + 2.96 log TL
AL[235]	275–533	log w = −5.26 + 3.16 log TL
Norris Reservoir, TN[248]	—	log w = −4.8776 + 3.115 log SL
Cumberland Lake, KY[22]	127–518	log w = −5.562 + 3.274 log TL

Table 128

Reports of weight (g) related to TL (mm) of young largemouth bass from combined data from various locations, summarized by Carlander (1977).[22]

TL (mm)	Weight (g)	Location
25–50	1.5	AL
51–75	2–5	AL, OK, TN
76–101	6–9	AL, AR, OK, TN
102–126	13–20	AL, CA, CO, MI, MN, OK, PA, TN, UT
127–151	23–37	AL, KY, MO, OK, TN
152–177	42–55	AL, IN, KY, OH, TN
178–202	71–100	AL, AR, IL, IN, KY, MO, OH, OK, TN
203–228	102–150	AL, AR, IL, IN, KY, MO, OH, TN
229–253	150–202	AL, AR, IL, IN, KY, MO, OH, OK, TN

Table 129

Length–weight relationships of young largemouth bass in OK waters (1952–1963).

Total Length (mm)	Weight (g)
157.5	47
177.8	69
203.2	107
228.6	157
254.0	221
279.4	301
304.8	399
330.2	517
355.6	658

Source: Constructed from data presented in Table 2, reference 208.

OK statewide average was 1.18 to 1.94.[47] The reader is referred to Carlander (1977)[22] for a more complete listing of reported largemouth bass condition factors.

Small (100–300 mm TL) Florida largemouth bass displayed lower W_r (the weight of an individual compared to a universal standard weight used for that species) values than small northern largemouth bass. Differences in W_r were not apparent between 301 and 500 mm TL. Relative weights for small F_1 hybrids were intermediate between those of the two subspecies in Lake Aquilla, TX. There was no evidence that lower temperatures in Lake

Table 130

Length and weight relationships of young largemouth bass collected in AL during the period 1949–1964.

TL (mm)	N	Average Empirical Weight (g)
51	190	2.0
76	1,083	6.4
102	1,554	12.3
127	1,283	25.4
152	839	45.4
178	391	77.2
203	234	109.0
229	202	149.8
254	208	208.8
279	119	308.7
305	121	395.0
330	70	490.3
356	66	612.9
381	33	758.2
406	27	935.2

Source: Constructed from data presented in unnumbered table on page 37, reference 235.

Table 131

Average observed lengths and weights of young largemouth bass in Lake Chautauqua, IL, during the period 1950–1959.

Age	N	Average TL (mm)	Average Weight (g)
1	62	163	68
2	252	249	232
3	120	312	477
4	41	363	767

Source: Constructed from data presented in Table 10, reference 227.

Aquilla when compared to FL caused a decline in Florida largemouth bass condition.[31]

Numbers and weight of YOY and yearling largemouth bass per hectare in DeGray Lake, AR, are presented in Table 133.

Growth is influenced mainly by available food, water temperature, and number of other bass present.[139] In 22 Nebraska sandhill lakes, it took

Table 132

Selected reports of condition factors (K) for young largemouth bass.[22]

Standard Length (SL)		
SL Range, Age, or Development Stage	Location	K (SL)
—	Norris Lake, TN	1.90
41–394 mm	OH ponds (unbalanced)	1.99
41–419 mm	OH ponds (balanced)	2.12
200–299 mm	Clinch River, TN	2.19
—	Douglas Reservoir, TN	2.28
—	Cherokee Reservoir, TN	2.30
Age 2	Chickamauga Reservoir, TN	2.47
—	Norris Reservoir, TN	2.48
Yearlings	Chickamauga Reservoir, TN	2.63

Total Length (TL)		
TL Range or Age	Location	K (TL)
51 mm	AL	1.56
Age 0	IL ponds	1.24
Age 0	Lake Carl Blackwell, OK	1.24
76–254 mm	AL	1.27
102–445 mm	IL	1.36
Age 1	IL ponds	1.18
Age 1	Lake Carl Blackwell, OK	1.41
Age 2	IL ponds	1.27

Table 133

Number of YOY and yearling largemouth bass per hectare, with average weight (kg) in parentheses, found in DeGray Lake, AR, during the period May–September, 1977–1979.[91]

	Year Class		
Month	1977	1978	1979
YOY			
May	393 (0.1)	—	—
June	408 (0.5)	605 (1.0)	213 (0.3)
August	939 (4.9)	1,915 (9.1)	451 (4.6)
September	802 (10.2)	439 (4.4)	114 (2.6)
Yearlings			
May	104 (4.4)	53 (2.8)	34 (2.0)

largemouth bass an average of 3.6 years to reach 300 mm TL,[52] whereas that length is typically achieved about 1 year sooner in southern populations. Bass stopped growing in summer due to high water temperatures in a power plant cooling reservoir.[148]

Growth seems to begin for largemouth bass at about 10°C, but may begin sooner for smaller fish. When summer water temperatures rise above the growth optimum, largemouth bass growth rate slows. Maximum growth was observed at about 27°C. In southern locales, warmer water and a longer growing season result in faster growth.[143]

Juvenile largemouth bass in the laboratory gained weight progressively slower as dissolved oxygen (DO) levels decreased. Efficiency of food conversion was lessened when DO levels were <4 ppm. The principal reason for reduction in growth at lowered DO is a reduced appetite. Digestion requires additional oxygen and in a low DO environment little extra DO is available, so appetite is suppressed.[144]

Growth for age-1 largemouth bass in Lake Kissimmee, FL, increased after a habitat enhancement project removed dense shoreline vegetation; also, catch per unit effort (CPUE) increased after enhancement activities.[80]

At time of stocking, mean length of stocked and wild YOY largemouth bass in Arkansas River pools did not differ, but mean length was significantly different between pools at age 0 (181 and 172 mm) and at age 1 (204 and 182 mm).[69,88]

Initially, in a new OK reservoir, both Florida largemouth bass and northern largemouth bass were longer and heavier than hybrids, but after 3 years, Florida largemouth bass and hybrids were longer and heavier than northern largemouth bass.[132] In a thermally enriched lake in OK, growth of YOY Florida largemouth bass was greater than that of northern largemouth bass.[129] Florida largemouth bass stocked in small OK impoundments were significantly longer than northern largemouth bass (mean TL of 139 vs. 128; P = 0.001), but had significantly poorer body condition (W_r = 88 vs. 93; P = 0.0007).[105]

The ratio of TL/SL for largemouth bass in Norris Reservoir, TN, declined from 1.236 for specimens 50–69 mm to 1.174 for fish >450 mm.[22] Carlander (1977) calculated that TL = 1.22 SL for largemouth bass up to 200 mm TL and then 1.215 SL up to 380 mm.[22] In Lewis Smith Reservoir, AL, for largemouth bass measured in inches, SL = 0.832 TL and FL = 0.958 TL.[142]

The oldest known largemouth bass was reported from NY to be 23+ years old.[70] Conversely, largemouth bass in Puerto Rico seldom live beyond

about age 4.[94,256] Female largemouth bass of both subspecies tend to live longer and grow to a larger size than do males.[100,107]

Feeding Habits

It is important that newly hatched largemouth bass quickly begin feeding or they soon become too weak to effectively forage. Laboratory experiments have shown that: (1) energy reserves in the yolk are adequate to meet a larva's energy requirements until the time of yolk absorption; (2) swimming ability of larvae denied food started to decline sharply on the fourth or fifth day after the larvae became free-swimming; (3) at an average temperature of 19.4°C, 7 days after swim-up starved larvae reached a "point of no return," after which most could not feed successfully, even when abundant food was made available.[145]

Young largemouth bass progress through a pattern of diet shifts from phytoplankton and zooplankton to insects to crustaceans (Table 134) to fish (Table 135) as they increase in size.[51,53,56,62,87,94,99,113,115–117,154,160,185,191,216,239] All of these changes in diet occur in their first growing season. Their ability to successfully advance through these dietary changes can, in fact, significantly influence their chances of survival to age 1.[5] Small largemouth bass fry eat more often than larger bass because they pass food through their stomachs in about 3 h.[54]

As soon as larval largemouth bass swim up from the bottom, they begin to feed.[191] A stable zooplankton population at the time of yolk absorption is important.[106] In Lake Powell no food was found in the stomachs of largemouth bass fry up to 8 mm TL.[191] Largemouth bass larvae at 10 mm TL are reported to feed on crustaceans.[242]

In hatchery ponds, 5-mm largemouth bass begin feeding on small midge larvae, cladocerans, and copepods or on immature forms.[121] Cladocerans consumed by 5-mm largemouth bass averaged 0.36 mm in length, copepods 0.45 mm, and midges 0.84 mm. Largemouth bass 10 mm showed only a slight increase in the size of prey selected. There was a general increase in the length of midge larvae with an increase of largemouth bass length up to 30 mm, after which average length of consumed midge larvae decreased for larger largemouth bass. The greatest number of organisms occurred in 20-mm largemouth bass and averaged 21.8 copepods, 6.3 cladocerans, and 3.3 midge larvae along with other prey per stomach.[121]

Generally, by about 15 mm, largemouth bass begin eating insects, with the size of insects eaten increasing as the young bass grow. Largemouth as small as 22 mm TL have been found with fish in their stomachs, but it is generally reported that at about 30 mm

Table 134

Reported invertebrate prey consumed by young largemouth bass (LMB)
from various locations.

LMB Length or Developmental Stage	Location	Food
8 mm	—	Zooplankton[22]
≤10 mm	Weiss Reservoir, AL[60]	Zooplankton
<15 mm	West Point Lake, AL[60]	Zooplankton
<15 mm	Hatchery Ponds, AL[121]	Copepods, Cladocerans
6–20 mm	Puerto Rico Res.[94]	Rotifers, Cladocerans, Copepods, Ostracods
21–90 mm	L. Blackshear, GA[126]	Microcrustaceans, cyclopods, cladocerans, amphipods, *Ephemeroptera*, tendipedids
30–50 mm	ON[207]	Cladocerans
51–70 mm	ON[207]	Odonates, *Ephemeroptera*, crayfish
>40 mm	—	Aquatic insects, dragonflies, mayflies, damselflies
50–100 mm	Smith Mt. Lake, VA[77]	Insects
<100 mm	Tenn. River, AL[200]	Insects, entomostracans, malacostracans
<100 mm	OK[226]	Insects
<100 mm	TX[160]	Aquatic insects, amphipods, *Paleomonetes* spp.
95–130 mm	Lake Opinicon, ON[240]	Decapods, zygopterans
>100 mm	OK streams[228]	Crayfish
Age 0	VA ponds[82]	Insects
Age 0	West Point Lake, AL[99]	*Bosmina, Daphnia, Cyclops, Diaptomus*, odonates, cladocerans, *Ephemeroptera*, copepods, chironomids
YOY	Lake Jordan, NC[112]	Zooplankton
Immature	Lake Ft. Smith, AR[239]	*Bosmina, Daphnia*, amphipods, ostracods, *Ceriodaphnia*, copepods, *Diaphanosoma*, hemipterans, ephemeropterans, corixids, dipterans, coleopterans, hymenopterans, megalopterans

they begin to become piscivorous with the degree of piscivory usually increasing with length. By around 75 mm, fish may be the predominant food in their diet and usually after largemouth bass exceed about 100 mm TL, they are largely piscivorous. However, the degree of piscivory may be influenced by prey availability.[5,53,60,62,94,99,107,112,113,115–117,126,154,191,216]

In Lucchetti Reservoir, Puerto Rico, the trophic stages of young largemouth bass follow the general trend of zooplanktivory (<40 mm), insectivory (46–60 mm), and piscivory (>60 mm).[113]

In the Chipola River, FL, largemouth bass <59 mm TL consumed primarily grass shrimp (>80% by weight). At 60 mm TL, largemouth bass consumed mostly grass shrimp (>50% by weight) and fish (about 27–28%). Fish increased in importance as food with growth of the young largemouth bass. At 80 mm TL, largemouth bass still consumed mostly grass shrimp (about 35%) and fish (about

58%) and at 100 mm fish accounted for about 70% of the food consumed and grass shrimp about 20%. At lengths of 120–180 mm, largemouth bass ate predominantly other fish and grass shrimp disappeared from the diet. A shift in the diet of young largemouth bass occurred at lengths ≥200 mm. Between 200 and 300 mm they continued to eat other fish but consumed greater amounts of crayfish (60–80%). At 350 mm, crayfish amounted to about 90% of the food consumed.[270]

Sometimes a significant portion of largemouth bass YOY will move more slowly to a fish diet and continue to consume invertebrate prey (Table 134).[6,20,113] One study found that insectivory was high in situations of low fish prey abundance and was accompanied by an increased occurrence of empty stomachs.[113] Young bass that move more slowly to a fish diet grow slower than do their piscivorous cohorts. By late summer there is often a bimodal

Table 135

Reports of piscivory by young largemouth bass
related to size or development of the bass.

Length or Developmental Stage	Location	Prey
>20 mm	—	Fish[22]
>20 mm	Reservoirs[134]	Fish
>25 mm	West Point Lake, AL[99]	Fish
30–150 mm	West Point L., GA/AL[133]	Bluegill preferred
>38 mm	Jordon Lake, NC[112]	Fish
>43 mm	Lake Blackhear, GA[126]	Fish
>80 mm	ON[207]	Fish
95–130 mm	Lake Opinicon, ON[240]	Fish
>100 mm	Tennessee River, AL[200]	Fish
>100 mm	OK streams[228]	Fish
>150 mm	West Point L., GA/AL[133]	Shad preferred
>200 mm	OK[226]	Sunfish, silversides
Age 0	Smith Mt. Lake, VA[77]	Sunfish, darters
Immature	Lake Ft. Smith, AR[239]	Sunfish, darters
YOY	Lake Kissimmee, FL[196]	Fish
YOY	L. Carl Blackwell, OK[124]	Gizzard shad

or even a multimodal length frequency distribution of YOY, the smaller individuals having preyed largely on insects, while the larger bass became more piscivorous. In Bull Shoals Reservoir, AR, some YOY largemouth bass converted to fish in mid-June and these bass grew faster for the remainder of the summer than did bass that ate insects. First-year survival was also greater for the fish-eaters than the insect-eaters.[53]

The ability of YOY largemouth bass to shift and to remain piscivorous should be determined by prey availability, which tends to decrease as the season progresses. Since earlier-hatched bass are larger than those hatched later, and the onset of piscivory is directly related to bass size, earlier-hatched bass should have a greater ability to become and remain piscivorous, which would presumably lead to higher growth rates, improved condition, and greater survival. In a study of diets and daily ages of largemouth bass in a NC reservoir, earlier-hatched bass were piscivorous at a younger age and maintained a higher level of piscivory later in the growing season than did later-hatched bass. Growth rates were higher for earlier-hatched bass but only during their first 85 days of life; growth rates were not different for bass older than 85 days. Hatching date did not appear to affect bass survival or condition. No explanation was apparent for the

failure of greater growth and condition of earlier-hatched bass to result in greater survival.[231]

The earlier an individual shifts to fish prey, the earlier it consumes more energy per predatory act (due to the mass of the prey). Largemouth bass at about 100 mm TL ate prey that was about 35% of their TL, while bass at about 200 mm TL ate prey at about 25% of their TL. Both size groups selected prey that was smaller than the theoretical maximum size prey they might consume.[101]

While structural complexity of habitat has been shown to enhance survival of young largemouth bass, it appears that bass in these situations exhibit slower growth because they tend to prey on insects longer than largemouth bass in habitats less structurally complex. In less complex habitats, young largemouth bass prey on fish of a smaller size.[160]

In LA, copepods occurred in 30% of the 30 young (12–34 mm TL) largemouth bass stomachs examined and were 52% of the food items consumed, but were less than 1% of total volume consumed. Fish accounted for 40% of the total volume of the diet but only two of the young bass had fish in their stomachs.[213]

In Ridge Lake, IL, during the period 1987–1989, the diets of largemouth bass <200 mm TL did not differ among years, and differences in diets of larger bass were due to declines in crayfish volume over time (30% to 5% from 1987 to 1989). Age-0 bluegills made up 89% of the diet of age-0 largemouth bass, 25% for age-1 largemouth bass, and <9% for larger fish. The proportion of age-1 and older bluegill and crayfish increased with increasing largemouth bass size. Other invertebrate and vertebrate fauna were consumed but individually made up only small portions of the diet (<5% by volume).[272]

In a southwestern reservoir, juvenile largemouth bass <150 mm consumed a variety of prey items including fish and invertebrates. By number and weight corixids were the most numerous prey (1.27 items and 5.1 mg per fish) and baetid mayflies also were important. Copepods were not found and chironomids were rare in stomachs. Among fish prey, the most important species was inland silversides *Menidia beryllina*.[38]

In Puerto Rico reservoirs, largemouth bass ≤50 mm consumed plankton (12% of total number of food items), insects (53%) and fish (41%); largemouth bass ≥51 mm ate primarily fish and insects. Small amounts of plankton were present in stomachs of largemouth bass up to 200 mm long.[94]

In Aquilla Lake, TX, juvenile largemouth bass >150 mm TL were piscivorous. There was no evidence of subspecific (northern largemouth bass, Florida largemouth bass, F_1, F_x) niche segregation for feeding, as dietary composition was similar among largemouth bass phenotypes.[107]

The feeding habits of juvenile largemouth bass were studied in the intake (cool) and discharge (warm) arms of Lake Sangchris, a central IL lake that provides condenser cooling water to an electric generating station. Bass from both arms of the lake fed essentially on the same food items, aquatic insects, fishes, and zooplankton. Aquatic insects were the most important food resource in both areas followed by fishes. Bass 18–50 mm fed heavily on zooplankton. Chironomids were also frequently eaten by small bass but were utilized to some extent by bass as large as 130 mm. While fish were used as food by largemouth bass as small as 22 mm, they were not a consistently important item until bass reached a length of 90–100 mm. The extended growing season in the discharge arm of the lake allowed juvenile largemouth bass to make major length and weight gains, and become predominantly piscivorous during their first growing season.[241]

Cannibalism is documented for young largemouth bass. Yearling largemouth bass preyed upon YOY largemouth bass in Lake Mead[57] and juvenile largemouth bass eat smaller largemouth bass in Bull Shoals Reservoir, AR.[120] Young largemouth bass become cannibalistic when they reach a length of about 120 mm TL. By the time they are mature, they are usually not cannibalistic.[57,120,121,124,138,145,149,150,199]

As a general rule largemouth bass can consume cylindrical-shaped prey, such as minnows, up to 50% of their own length, but they usually do not consume prey that is >40% of their length. They may consume deep-bodied prey, such as sunfishes, up to 35% of their length, but they usually consume such deep-bodied prey ≤25% of their own body length. Preference for prey fish such as shad, fathead minnows, and silversides would be expected over sunfishes.[15,96,101,117,124,153,159] In experimental ponds stocked with Florida largemouth bass and blue tilapia *Tilapia aurea*, bass rarely ate tilapia >27% of the bass TL.[92]

Largemouth bass usually eat their prey head first, except decapods which are consumed tail first.[96,117,126,159] They eat larger individual food items as they themselves increase in size so that in lotic or lentic environments larger largemouth bass consume larger prey.[124,133,140] Young-of-year largemouth bass tend to have some food items in their stomach up to 90% of the time but an increase in piscivory may increase the number of empty stomachs. Apparently the increased energy obtained from a fish diet as compared to a largely insect diet more than offsets the lack of food. Several studies have found that the diet of YOY largemouth bass overlaps those of other reservoir species such as moronids or shad but that spatial or temporal differences prevent any significant degree of competition.[38,60,112]

Young largemouth bass consumed their body weight in food in 15 days as compared to 36 days for gar. Bass eating forage fish consumed 2.2% of their body weight per day. At 10°C, largemouth bass weighing 4.5 g ate 2.4–5.5% of their weight daily; at 20°C they ate 17.4% of body weight per day. At 10°C, bass weighing 9 g ate 3.4–4.6% of their body weight daily; at 20°C consumption increased to 11.1–14.0% of their body weight per day. Largemouth bass weighing 200–400 g were fed gizzard shad equivalent to 3.0–4.1% of their body weight — the average time to evacuate the stomach was 20 h at 27°C and 30 h at 18°C. Evacuation time of actively feeding larval largemouth bass was 354.6 min at 10.41°C, and that of larvae not actively feeding was 638.5 min.[22] Gastric evacuation rates for juvenile largemouth bass (87–157 mm and 15–70 g, wet weight) at 26°C fed 2% of dry body weight were curvilinear, declining substantially within the first 2.5 h after meal consumption; a small portion of the meal was present 10 h postfeeding.[59]

The feeding rate of young largemouth bass is greatly influenced by water temperature. When water temperature is reduced below 10°C, age-0 largemouth bass will continue to consume prey until the temperature drops below 6°C.[28] Between 10 and 20°C the feeding rate increases rapidly with temperature rise. Feeding occurs at a uniformly high rate to at least 27°C, above which feeding rate seems to decline slightly. Annual growth rings suggest that small fish may begin feeding at lower temperatures than larger fish.[143] Young largemouth bass introduced into aquaria in November ceased feeding when water temperature was dropped to below 12°C but resumed feeding when temperature was raised to over 13°C; they responded this way repeatedly.[207]

Appetite of young largemouth bass is suppressed in a low dissolved oxygen environment.[144]

LITERATURE CITED

1. Waters, D.S. et al. 2005.
2. Heidinger, R.C. 1976.
3. Erdman, D.S. 1984.
4. Brown, J. A. et al. 1987.
5. Wicker, A.M. and W.E. Johnson. 1987.
6. Isley, J.J. et al. 1987.
7. Hunsaker, D. and R.W. Crawford. 1964.
8. Chew, R.L. 1975.
9. Neal, J.W. et al. 1997.
10. Sepulveda, M.S. et al. 1999.
11. Fields, R. et al. 1987.
12. Johnson, D.L. et al. 1988.
A 13. Heitman, J.F. 2004.
14. MacCrimmon, H.R. and W.H. Robbins. 1975.
15. Shelton, W.L. et al. 1979.
16. Gran, J.E. 1995.
A 17. Heitman, J.F. 1980.
18. Jackson, D.C. and R.V. Kilambi. 1983.
19. Tidwell, J.H. et al. 2000.
20. Jenkins, R.E. and N.M. Burkhead. 1993.
21. Scott, W.B. and E.J. Crossman. 1973.
22. Carlander, K.D. 1977.
23. Stang, D.L. et al. 1996.
24. Ney, J.J. 1996.
25. Maceina, M.J. et al. 1996.
26. Annett, C. et al. 1996.
27. Garvey, J.E. et al. 2002.
28. Garvey, J.E. et al. 1998.
29. Parkos, J.J. III and D.H. Wahl. 2002.
30. Maceina, M.J. et al. 1988.
31. Maceina, M.J. and B.R. Murphy. 1988.
32. Koppelman, J.B. et al. 1988.
33. McCormick, J.H. et al. 1989.
34. Meador, M.R. and W.E. Kelso. 1989.
35. Fischer, S.A. and W.E. Kelso. 1990.
36. Phillipp, D.P. and G.S. Whitt. 1990.
37. Morizot, D.C. et al. 1991.
38. Matthews, W.J. et al. 1992.
39. Goodgame, L.S. and L.E. Miranda. 1993.
40. Schmidt, R.C. and M.C. Fabrizio. 1980.
41. Isley, J.J. and R.L. Noble. 1987.
A 42. Heitman, J.F. 1982.
43. Miller, S.J. and T. Storck. 1982.
44. Kramer, R.H. and L.L. Smith. 1960.
45. Timmons, T.J. et al. 1980.
46. Nack, S.B. et al. 1993.
A 47. Wright, F.G. 1977.
48. Grizzle, J.M. et al. 2002.
49. Grizzle, J.M. and C.J. Brunner. 2003.
50. Eschmeyer, R.W. 1948.
A 51. Rogers, M. and M.S. Allen. 2005.
52. Paukert, C.P. and D.W. Willis. 2004.
53. Aggus, L.R. and G.V. Elliott. 1975.
54. Heidinger, R.C. 1975.
55. Trautman, M.B. 1981.
56. Miller, R.J. 1975.
57. Allan, R.C. and J. Romero. 1975.
58. Miranda, L.E. and W.D. Hubbard. 1994a.
59. Hayward, R.S. and M.E. Bushman. 1994.
60. Hirst, S.C. and D.R. DeVries. 1994.
61. Nedbal, M.A. and D.P. Philipp. 1994.
62. Phillips, J.M. et al. 1995.
63. Gelwick, F.P. et al. 1995.
64. Wright, R.A. et al. 1999.
65. Pine, W.E. III et al. 2000.
66. Jackson, J.R. and R.L. Noble. 2000.
67. Bruno, N.A. et al. 1990.
68. Kohler, C.C. et al. 1993.
69. Heitman, N.E. et al. 2006.
70. Green, D.M. and R.C. Heidinger. 1994.
71. Miranda, L.E. and W.D. Hubbard. 1994b.
72. Miranda, L.E. and L.L. Pugh. 1997.
73. Phillips, J.M. et al. 1997.
74. Maceina, M.J. and P.W. Bettoli. 1998.
75. Long, J.M. and W.L. Fisher. 2001.
76. Hunt, J. and C.A. Annett. 2002.
77. Sutton, T.M. and J.J. Ney. 2002.
78. Tate, W.B. et al. 2003.
79. Buckmeier, D.L. et al. 2003.
80. Allen, M.S. et al. 2003.
81. Ostrand, K.G. et al. 2004.
82. Brenden, T.O. and B.R. Murphy. 2004.
83. Waters, D.S. and R.L. Noble. 2004.
84. Buckmeier, D.L. et al. 2005.
85. Boxrucker, J.C. et al. 2005.
86. Mettee, M.F. et al. 2001.
87. Miller, R.J. and H.W. Robison. 1973.
88. Heitman, E. 2005.
89. Boxrucker, J. 1983.
90. Dewey, M.R. and T.E. Moen. 1982.
91. Dewey, M.R. et al. 1981.
92. Shafland, P.L. and J.M. Pestrak. 1982.
93. Wigtail, G.W. 1982.
94. Neal, J.W. et al. 2006.
95. Maceina, M.J. and J.M. Grizzle. 2006.
96. Lawrence, J.M. 1958.

97. Jackson, S.W., Jr. 1957.
98. Herke, W.H. 1960.
99. Timmons, T.J. 1984.
100. Porak, W. et al. 1987.
101. Goldstein, R.M. 1994.
102. Horton, R.A. and E.R. Gilliland. 1994.
103. Cofer, L.M. 1994.
104. Isaac, J., Jr. and V.H. Staata. 1993.
105. Gilliland, E.R. 1992.
106. Young, C.H. and S.A. Flickinger. 1989.
107. Maceina, M.J. and B.R. Murphy. 1989.
108. Phillipp, D.P. et al. 1983.
109. Schramm, H.L. and D.C. Smith. 1988.
110. Fisher, W.L. and A.V. Zale. 1992.
111. Gilliland, E.R. and J. Whitaker. 1989.
112. Jackson, J.R. et al. 1991.
113. Alicea, A.R. et al. 1997.
114. Hoover, R.S. et al. 1997.
115. Lilyestrom, C.G. and T.N. Churchill. 1997.
116. Cailteux, R.L. et al. 1997.
117. Snow, J.R. 1962.
118. Applegate, R.L. et al. 1967.
119. Prather, E.E. 1967.
120. Mullan, J.W. and R.L. Applegate. 1968.
121. Rogers, W.A. 1967.
122. Snow, J.R. 1971.
123. Ramsey, J.S. and R.O. Smitherman. 1971.
124. Zweiacker, P.L. and R.C. Summerfelt. 1974.
125. Buchanan, J.P. 1973.
126. Pasch, R.W. 1975.
127. Prentice, J.A. and B.G. Whiteside. 1975.
128. Inman, C.R. et al. 1977.
129. Rieger, P.W. and R.C. Summerfelt. 1977.
130. Cichra, C.E. et al. 1981b.
131. Smith, R.P. and J.L. Wilson. 1981.
132. Wright, G.L. and G.W. Wigtil. 1981.
133. Timmons, T.J. and O. Pawaputanon. 1981.
134. Jenkins, R.M. 1975.
135. Carlander, K.D. 1975.
136. Hackney, P.A. 1975.
137. Funk, J.L. 1975b.
138. Summerfelt, R.C. 1975.
139. Latta, W.C. 1975.
140. Fajen, O.F. 1975b.
141. Sullivan, J.R. 1975.
142. Webb, J.F. and W.C. Reeves. 1975.
143. Coutant, C.C. 1975.
144. Bulkley, R.V. 1975.
145. Eipper, A.W. 1975.
146. Rainwater, W.C. and A. Houser. 1975.
147. Houser, A. and W.C. Rainwater. 1975.
148. Siler, J.R. and J.P. Clugston. 1975.
149. Snow, J.R. 1975.
150. von Geldern, C., Jr. and D.F. Mitchell. 1975.
151. Ross, S.T. 2001.
152. Bailey, R.M. and C.L. Hubbs. 1949.
153. Snow, J.R. 1969.
154. Hodson, R.G. and K. Strawn. 1969.
155. Maceina, M.J. and D.R. Bayne. 2001.
156. Sammons, S.M. and P.W. Bettoli. 2000.
157. Novinger, G.D. 1987.

158. Dunham, R.A. et al. 1992.
159. Stoeckel, J.N. and R.C. Heidinger. 1992.
160. Bettoli, P.W. et al. 1992.
161. Miranda, L.E. and R.J. Muncy. 1987a.
162. Schramm, H.L. et al. 1992.
163. Whittier, T.R. and R.M. Hughes. 1998.
164. Neal, J.W. and R.L. Noble. 2002.
165. Barwick, D.H. 2004.
166. Miranda, L.E. and R.J. Muncy. 1987b.
167. Copeland, J.R. and R.L. Noble. 1994.
168. Meals, K.O. and L.E. Miranda. 1991.
169. Ozen, O. and R.L. Noble. 2005.
170. Greene, J.C. and M.J. Maceina. 2000.
171. Buynak, G.L. et al. 1999.
172. Allen, M.S. et al. 1999.
173. Sammons, S.M. et al. 1999.
174. Zweifel, R.D. et al. 1999.
175. Hubert, W.A. 1988.
176. Buynak, G.L. et al. 1989.
177. Crawford, S. et al. 1989.
178. Muncy, R.J. 1966.
179. Chastain, G.A. and J.R. Snow. 1966.
180. Smith, S.L. and J.E. Crumpton. 1977.
181. Guier, C.R. et al. 1978.
182. Timmons, T.J. et al. 1979.
183. Long, J.M. and W.L. Fisher. 2002.
184. Neal, J.W. et al. 2002.
185. Dreves, D.P. and T.J. Timmons. 2002.
186. Jackson, J.R. et al. 2001.
187. Reinert, T.R. et al. 1996.
188. Chew, R.L. 1973.
189. Holbrook, J.A. et al. 1973.
190. Zweiacker, P.L. et al. 1972.
191. Miller, K.D. and R.H. Kramer. 1971.
192. Bryant, H.E. and A. Houser. 1971.
193. Zweiacker, P.L. and B.E. Brown. 1971.
194. Irwin, E.R. et al. 2002.
195. Fuhr, M.A. et al. 2002.
196. Allen, M.S. and K.I. Tugend. 2002.
197. Hunt, J. et al. 2002.
198. Hoffman, K.J. and P.W. Bettoli. 2005.
199. Swenson, W.A. 2002.
200. Janssen, F.W. 1992.
201. Van Hassel, J.H. et al. 1988.
202. Pearson, W.D. and B.J. Pearson. 1989.
203. Pearson, W.D. and L.A. Krumholz. 1984.
204. Gammon, J.R. 1998.
205. Goodyear, C.S. et al. 1982.
206. Gasaway, C.R. 1968.
207. Keast, A. 1970.
A 208. Houser, A. and M.G. Bross. 1963.
209. Conrow, R. and A.V. Zale. 1985.
210. Wrenn, W.B. 1976.
A 211. Tennessee Valley Authority. 1978.
212. Johnston, P.M. 1951.
213. Wallace, R.K., Jr. 1981.
214. Stewart, E.M. and T.R. Finger. 1985.
215. Patriarche, M.H. 1953.
216. Reid, S.M. et al. 1999.
217. Delp, J.G. et al. 2000.
A 218. Mense, J.B. 1976.

219. Baker, J.A. et al. 1991.
220. Dewey, M.R. and C.A. Jennings. 1992.
221. Dendy, J.S. 1945.
222. Dendy, J.S. 1946.
223. Haslbauer, O.F. 1945.
224. Miranda, L.E. 1987.
225. Thomas, J.A. et al. 2004.
A 226. Oklahoma Department of Wildlife Conservation. 1992.
227. Starrett, W.C. and A.W. Fritz. 1965.
228. Scalet, C.G. 1977.
229. Applegate, R.L. 1966.
230. Buynak, G.L. 1995.
231. Phillips, J.M. et al. 1993.
232. Irwin, E.R. and R.L. Noble. 1993.
233. Rider, S.J. and M.J. Maceina. 1993.
234. Taber, C.A. 1964.
A 235. Swingle, W.E. 1965.
236. Meyer, F.A. 1970.
237. Coutant, C.C. 1977.
238. Fish, M.P. 1932.
239. Olmsted, L.L. 1974.
240. Keast, A. 1978b.
241. Sule, M.J. 1981.
242. Reighard, J. 1906.
243. Tin, H.T. 1982.
244. Carr, M.H. 1942.
245. Wang, J.C.S. and R.J. Kernehan. 1979.
246. Conner, J.V. 1979.
247. Lippson, A.J. and R.L. Moran. 1974.
248. Stroud, R.H. 1948.
A 249. Axon, J.R. 1984.
A 250. EA Engineering, Science, and Technology. 2004
A 251. EA Engineering, Science, and Technology. 1994.
A 252. Environmental Science and Engineering, Inc. 1995.

253. Mabee, P.M. 1995.
254. Lutz–Carrillo, D.J. et al. 2006.
255. Jackson, J.R. et al. 2002.
256. Neal, J.W. and R.L. Noble. In Press.
257. Miller, R.J. and H.W. Robison. 2004.
258. Kimmel, B.L. and A.W. Groeger. 1986.
259. Lee, D.S. et al. 1980.
260. Boschung, H.T., Jr. and R.L. Mayden. 2004.
A 261. Shewmake, J.W. and D.C. Jackson. 2004.
262. Etnier, D.A. and W.C. Starnes. 1993.
263. Chew, R.L. 1974.
264. Carter, B.T. 1967.
265. Marler, B.J. 1990.
266. Fish, P.A. and J. Savitz. 1983.
267. Long, J.M. and W.L. Fisher. 2005.
268. Bayne, D.R. et al. 2002.
269. Beamesderfer, R.C.P. and J.A. North. 1995.
270. Wheeler, A.P. and M.S. Allen. 2003.
271. Wagner, T. et al. 2006.
272. Santucci, V.J., Jr. and D.H. Wahl. 2003.
273. Olson, M.H. et al. 2003.
274. Hardy, J.D., Jr. 1978.
275. Taber, C.A. 1969.
A 276. McGowan, E.G. 1984.
A 277. McGowan, E.G. 1988.
278. Irwin, E.R. et al. 1997.

 * Original morphometric and meristic data and descriptions of early development were taken from developmental series of larvae and juveniles cultured by TVA from eggs and yolk-sac larvae provided by Owens Mill Fish Hatchery and Cohutta National Fish Hatchery, GA. All specimens are housed at the Division of Fishes, Aquatic Research Center, Indiana Biological Survey, Bloomington, IN (catalogue numbers: TV738; TV758; TV3339).

GENUS
Pomoxis Rafinesque

Robert Wallus

The genus *Pomoxis* contains two widely distributed species, white crappie *P. annularis* and black crappie *P. nigromaculatus*. Both species are native to eastern North America and both are popular sport and food fishes (Etnier and Starnes, 1993; Boschung and Mayden, 2004). The crappies have a paradoxical place in fishery management. They are palatable fishes that are readily caught and capable of sustaining heavy fishing pressure; but the size normally attained is marginal in terms of angler acceptability. Regulating crappie populations to maintain them at favorable levels for angling has proved to be a difficult task (Cross, 1967).

Distinguishing adult characteristics for *Pomoxis* include: an oblong body, very deep and strongly compressed; an anterior dorsal profile that is long and sloping; concave napes; a projecting lower jaw; well-developed supramaxilla; maxilla that reach beyond the pupil of the eye; opercles with two flat projections; 5–7 anal fin spines; 5–8 dorsal fin spines in a continuous dorsal fin; teeth on tongue, vomer, and palatine bones; and contrasting black and white coloration (Etnier and Starnes, 1993; Boschung and Mayden, 2004).

TAXONOMY AND SYSTEMATICS OF GENUS POMOXIS

Hypotheses concerning the systematic relationships of *Pomoxis* have variously grouped the genus with *Centrarchus* and *Archoplites* (Bailey, 1938; Smith and Bailey, 1961; Branson and Moore, 1962; Mabee, 1988). Wainwright and Lauder (1992) and Mabee (1993) hypothesized that *Centrarchus* is a sister group to *Archoplites* and together they form a sister clade to *Pomoxis*. Branson and Moore (1962) hypothesized a close relationship between *Pomoxis* and *Centrarchus* on the basis of acoustico-lateralis system and select meristics. Mok (1981) considers *Pomoxis* to be sister to *Centrarchus* based on kidney morphology, and Eaton (1956) used olfactory variation data to place *Pomoxis* in an unresolved trichotomy with *Centrarchus* and *Archoplites*. Chang (1988) and Roe et al. (2003) also consider *Pomoxis* and *Centrarchus* to be sister groups.

Taxonomy of Larvae and Early Juveniles
Black and white crappies are superficially very similar throughout most of their larval development. There is no single character that will reliably distinguish the two species throughout all larval phases. Success in identification requires the use of different characters depending on size or phase of development (Chatry, 1977). Siefert (1969b) compared the development of several morphological and meristic characteristics of larval black and white crappies from Missouri River main stem reservoirs. His methods for identifying crappie larvae in the 5–16 mm TL range are presented in Table 136.

Although the meristic trends identified by Siefert (1969b) were evident in LA crappie, Chatry (1977) and Conner (1979) noted that there were certain inconsistencies and variations of details. Siefert noted that the reliability of myomere count differences dissipated with decreasing specimen length. Conner (1979) pointed out that many crappie larvae collected during ichthyoplankton sampling were within this small size group. Chatry (1977) found that regional inconsistencies in allometric development and meristic characters emphasizes the importance of considering geographic variation when developing identification criteria for crappies. His observations of allometric development for crappies in southeastern LA were contrary to those of Siefert (1969b) for Missouri River crappies. Siefert found that black crappies consistently attained fin structures at a smaller size than white crappies.

Table 136

Summary of characteristics for distinguishing between white and black crappie larvae >5 mm TL from Missouri River mainstem reservoirs, SD.

TL (mm)	Character	White Crappie	Black Crappie
5.0–6.5	Postanal myomeres	≤19	≥21
6.0–16.0	Total myomeres	30–31 (usually 30)	32
>16.0	Dorsal fin spines	5–6	7

Source: Constructed from Table 3 in Siefert (1969b).

Chatry reported greater variability and overlap in myomere counts than did Siefert, indicating that Siefert's widely used characters (Table 136) for distinguishing crappie species (Anjard, 1974; Hardy, 1978) were largely unreliable in southeastern LA. Recent studies indicate that hybridization and water temperatures during egg development can influence myomere numbers and thereby interfere with identification of larval crappies. Higher temperatures cause the anus to develop more posteriorly in larval crappie. A recent study by Spier and Ackerson (2004) concluded that because myomere counts overlap so much and because hybridization and natal water temperatures can induce variation in myomere numbers, myomere counts are not useful for discriminating between larval black and white crappie.

A summary of the taxonomic criteria developed by Chatry (1977) for identifying larval crappies in southeastern LA follows.

Characters were not discovered for identifying crappie larvae <3.5 mm TL. For fully straightened yolk-sac larvae, prior to the appearance of the air bladder (3.5–4.0 mm TL), species recognition is usually possible based on the position of the oil globule. The oil globule is usually positioned in the posterior half of the yolk sac in white crappie larvae, while centrally or anteriorly placed in black crappies. Position of the oil globule occasionally varies, so identification using this character is tentative. Chatry (1977) notes that crappie larvae <4.0 mm are seldom collected during conventional ichthyoplankton sampling. The following key for identifying crappie larvae >4.0 mm TL was developed from information presented in Chatry's thesis.

Key for Identification of Crappie Larvae and Early Juveniles from Southeastern LA

1a. Larvae 4–13 mm TL...2
1b. Larvae >13 mm TL...4

2a. Air bladder visible but not sufficiently distinct to afford an accurate eye to air bladder measurement ....*Pomoxis* spp.
2b. Air bladder sufficiently distinct to afford an accurate eye to air bladder measurement.................3

3a. Distance from the posterior margin of the eye to the anterior-most portion of the air bladder is >15% TL... white crappie *P. annularis*
3b. Distance from the posterior margin of the eye to the anterior-most portion of the air bladder is <15% TL... black crappie *P. nigromaculatus*

4a. Larvae 13–16 mm TL...*Pomoxis* spp.
4b. Larvae and early juveniles >16 mm TL...5

5a. Length of dorsal fin base is equal to or included in the distance from the eye to the dorsal fin origin (dorsal fin length/eye to fin distance is ≥1)..................................... white crappie *P. annularis*
    Note: In the case of larger juveniles and adults, such a specimen would be identified as a black crappie.
5b. Length of dorsal fin base is greater than the eye to dorsal fin length (dorsal fin length/eye to fin distance is <1) .... black crappie *P. nigromaculatus*

Table 137

Comparison of developmental changes in black and white crappie larvae from southeastern LA. Standard deviation (SD^{50}) represents actual size interval that structure(s) appear in more than 50% of specimens examined; numbers in parentheses are actual percentages having structures in question.

Structure	No. of Fish	Minimum TL (mm) of First Appearance	Maximum TL at Which Structure(s) Had Not Appeared	SD^{50}
White Crappie				
Air bladder	44	3.65	3.91	3.90–3.99 (86)
Caudal fin rays	22	7.31	7.50	7.00–7.99 (56)
Complete caudal fin	41	9.32	10.61	10.00–10.99 (80)
Dorsal fin rays	41	9.84	10.35	10.00–10.99 (80)
5 dorsal spines	34	13.20	15.69	15.00–15.99 (90)
Black Crappie				
Air bladder	84	3.41	3.74	3.60–3.69 (93)
Caudal fin rays	39	7.74	8.00	8.00–8.99 (95)
Complete caudal fin	12	10.96	11.55	11.00–11.99 (83)
Dorsal fin rays	6	10.18	10.86	10.00–10.99 (66)
5 dorsal spines	18	12.59	14.52	14.00–14.99 (85)

Source: Taken (in part) from Table 4 in Chatry (1977).

In addition to the diagnostic characters presented above, Chatry (1977) suggested that allometric development can aid in discriminating the two crappie species. The size that certain developmental milestones occur depends upon the species (Table 137). For example, white crappies tend to have a complete caudal fin ray, a complete caudal fin, and a complete dorsal fin ray at smaller sizes than black crappies. The air bladder and the fifth dorsal fin spine tend to appear at smaller sizes in black crappies.

Large Juveniles and Adult Taxonomy

Regional works (Scott and Crossman, 1973; Smith, 1979; Trautman, 1981; Becker, 1983; Robison and Buchanan, 1988; Etnier and Starnes, 1993; Jenkins and Burkhead, 1994; Ross, 2001; Boschung and Mayden, 2004) provide criteria for the identification of large juveniles and adult crappies. The number of dorsal spines (6 or fewer for white crappie and 7 or more for black crappie) has been determined to be a fairly reliable character, though it is often noted that occasional overlaps in these counts occur. The length of the dorsal fin base into nape length is also often used as an identification tool. One fairly consistent meristic skeletal difference (requires clearing and staining) between black and white crappies is in the number of interneurals, usually 7 or more for white crappie and 6 or fewer for black crappie (Hofstetter et al., 1958; Chatry, 1977).

Natural hybridization by black and white crappies is common and reported from hatcheries (Dunham et al., 1994), ponds, newly constructed reservoirs (Trautman, 1981), and older reservoirs (Dunham et al., 1994; Smith et al., 1994; Travnichek et al., 1996; and Travnichek et al., 1997). Black crappies, white crappies, and their hybrids have overlapping morphological characteristics, which makes it difficult to identify the parentals and their hybrids in the field. Smith et al. (1995), using allozyme examination to verify identifications, found that spine counts and nape measures are not reliable for distinguishing crappie species or their hybrids in AL reservoirs where the two species coexist. In Weiss Lake, 43% of the crappies were misidentified when both characteristics were used. In two other AL lakes, 11–13% of crappies were misidentified when dorsal spine counts were used.

Note: In light of the regional inconsistencies in allometric development and meristic characters of young crappies (Chatry, 1977) and the unreliability of myomere counts as a taxonomic tool (Spier and Ackerson, 2004), identification of crappie larvae to species is tentative, at best. Even when geographic variation is taken into account, developing identification criteria for crappie larvae may not be possible due to the effects of adult hybridization. Black crappies, white crappies, and their hybrids have overlapping morphological characteristics, which makes it difficult to identify the parentals and their hybrids (Smith et al., 1995). In areas where the identification of adult crappies is questionable, attempts to identify larvae of the two species is ineffectual. Biologists who do not have a clear understanding of the population dynamics and reproductive biology of the two species in the water body of concern should consider identifying all young crappies as *Pomoxis* spp.

LITERATURE CITED

Anjard, C.A. 1974.
Bailey, R.M. 1938.
Becker, G.C. 1983.
Boschung, H.T., Jr. and R.L. Mayden. 2004.
Branson, B.A. and G.A. Moore. 1962.
Chang, C.H. 1988.
Chatry, M.F. 1977.
Conner, J.V. 1979
Cross, F.B. 1967.
Dunham, R.A. et al. 1994.
Eaton, T.H., Jr. 1956.
Etnier, D.A. and W.C. Starnes. 1993.
Hardy, J.D., Jr. 1978.
Hofstetter, A.M. et al. 1958.
Jenkins, R.E. and N.M. Burkhead. 1994.
Mabee, P.M. 1988.

Mabee, P.M. 1993.
Mok, H.K. 1981.
Robison, H.W. and T.M. Buchanan. 1988.
Roe, K.J. et al. 2002.
Ross, S.T. 2001.
Scott, W.B. and E.J. Crossman. 1973.
Siefert, R.E. 1969b.
Smith, C.L. and R.M. Bailey. 1961.
Smith, P.W. 1979.
Smith, S.M. et al. 1994.
Smith, S.M. et al. 1995.
Spier, T.W. and J.R. Ackerson. 2004.
Trautman, M.B. 1981.
Travnichek, V.H. et al. 1996.
Travnichek, V.H. et al. 1997.
Wainwright, P.C. and G.V. Lauder. 1992.

WHITE CRAPPIE

Pomoxis annularis Rafinesque

Robert Wallus

Pomoxis, Greek: poma, "lid, cover" and oxys, "sharp," alluding to the opercles ending in two flat points instead of an ear flap; *annularis*, Latin: "having rings," probably in reference to the vague vertical bars on the body.

RANGE

Native range was restricted to the freshwaters of east central North America from southern ON and southwestern NY west of the Appalachians, south to the Gulf coast, and west to TX, SD, and southern MN. Widely introduced elsewhere in the United States.[1]

HABITAT AND MOVEMENT

White crappie has wide ecological tolerances, occurring in virtually all types of water except very small streams and ponds,[24] and is tolerant to a wide variety of habitats[19,73,124] including turbidity[1,19,46,57,114,124] and siltation.[1,19] This species is often found in ponds,[1,19,23,25,46,57,124] lakes,[1,16,19,23,25,46,57,124] or impoundments[19,25,46,57,124] of more than 2 ha (5 acres), especially in those where competition and predation from sunfishes and black basses are slight.[19]

White crappie also inhabit sloughs,[19,25,46,57,63,73,124] bayous,[19,25,46,57,73,124] oxbows,[19,25,46,57,63,124] borrow pits,[63] swamps,[16] seasonal floodplain habitat,[63] and sluggish pools of small[1,16,23,46,57] to large streams[1,16,19,23,25,46,57,63,73,124] of moderate[19,46,57,124] or low gradient flow.[1,19,23,124] They are found over both soft and hard bottoms[124] of mud,[19,57] clay,[19,57] sand,[19,57] gravel,[19,57] rubble,[57] and silt;[57] frequently found in association with aquatic vegetation[19,46,124] or congregated around submerged brush, logs, stumps, and tree roots.[19,46,57,124]

White crappie is reported by some as a schooling species,[57] but it is suggested by others that it does not school, but congregates, in loose aggregations, about submerged structures such as trees, boat docks, and other suitable cover.[21]

White crappie attains great abundance in lowland lakes, navigation pools of large rivers, and man-made impoundments. In reservoirs it frequents areas around brush piles, standing timber, rock ledges, or other cover.[15,21] In flowing streams, it avoids noticeable current and is common in slow-moving pools and backwaters over a variety of substrates around cover such as log jams, woody debris, vegetation, and undercut banks; it is seldom taken in stream channels.[15,16,21] It avoids streams that are excessively turbid and those kept continuously cool by spring inflow.[21]

Many reports suggest that white crappie are more tolerant of turbid water than black crappie.[1,101] Though it can tolerate severe turbidity levels, it appears to avoid them.[19,21,53] Linear regression analysis of growth and length data from stocked juvenile and adult crappies showed that growth of both species was similar across a range of turbidities in IL ponds. However, in larger bodies of water, crappies might be indirectly influenced by factors that are correlated with turbidity, such as submerged vegetation.[101]

White crappies are found at almost any depth where food is available.[46] They appear to be influenced by light penetration, generally occurring at greater depths when turbidity is low and nearer the surface as turbidity increases.[8] In mainstem Missouri River reservoirs, white crappie were most often captured in the water depths of 6–15 m.[41]

Vertical distribution of white crappies in the Buncombe Creek arm of Lake Texoma was similar in open water and near-shore habitats, although the depth of capture was deeper in open water than near shore. Crappies >254 mm TL (10 inches) occurred at greater depths than smaller fish. Crappies <254 mm were taken in open water more commonly than near the shoreline, while larger fish were more commonly taken near the shore.[8]

Mean depth of capture of white crappies was deepest (7.9 m) during January and nearest the surface (5.2 m) in April. In general they were distributed more or less evenly from the surface to the bottom during the fall (September, October, and November); were concentrated near the bottom in winter (December, January, and February); were nearer the surface and more widely distributed during spring (March, April, and May); and were deeper in distribution during the summer (June, July, and August).[8]

In reservoirs, white crappie occur in open areas and may show a pattern of vertical migration in which they move closer to the surface at night. In summer they seek out cool water and are located near or below the thermocline. Because water below the thermocline is often devoid of oxygen, they tend to be forced into narrow zones around the thermocline where there is enough oxygen in cooler water.[116]

In Lake Carl Blackwell, OK, catch rate of white crappies was negatively correlated with water depth and positively correlated with biomass of major forage items (mayflies and gizzard shad). As much as 47.6% of the variability in catch rate was accountable to depth of water, organic content of sediment, and particle diameter and 59% of the variability accounted biomass of gizzard shad, biomass of chironomids, biomass of ephemerids, water depth, and organic content.[9]

Field and laboratory experiments were conducted to investigate possible competitive effects of introducing inland silversides and threadfin shad on native white crappie populations in Thunderbird Reservoir, OK. Findings indicated that the introduced planktivores competed with white crappie and could possibly be a contributing factor to poor growth rates seen in the crappie population.[108] In a study examining the effects of gizzard and threadfin shad on zooplankton abundance and reproduction of white crappie in a TX pond, the authors suggest that presence of the shad species may decrease the density and biomass of YOY white crappies, but increase their mean size.[80]

In a later study on Thunderbird Reservoir, introduction of a predator (saugeye = walleye × sauger hybrid) for over abundant intermediate-size white crappie and a prey species (threadfin shad) to facilitate crappie dietary shift from invertebrates to fish was evaluated. Results indicated a decline in the density of intermediate-size crappies (130–199 mm TL) and an increase in the catch of larger crappies (≥200 mm). Crappie growth rates improved, but remained below desired levels. Threadfin shad were seldom taken by the crappie and recommendations of the study were to continue stocking the predator but to discontinue stocking the shad.[109]

Though local anglers and fishing guides from Weiss Lake, AL, had voiced concerns that natural reproduction of striped bass in the lake had negatively impacted the crappie fishery (both species), subsequent study indicated that the crappie population of the lake was little changed and that no negative impact was indicated.[110]

The proportional stock density (PSD) of white crappie populations in 8 small OK impoundments during 1984 was inversely related to the largemouth bass PSD and directly related to largemouth bass density. Reduction in crappie numbers through predation by largemouth bass was hypothesized to be the determining factor in establishing favorable population structure of crappies in small impoundments.[28]

White crappie populations with diverse size distributions might possibly be achieved in flood prevention lakes in TX by proper management of largemouth bass/bluegill populations. In 30 flood prevention lakes in north-central TX, total numbers of white crappie were not significantly correlated to total numbers of largemouth bass or bluegill. However, numerous relationships existed between white crappie and largemouth bass/bluegill size structure variables. Largemouth bass/bluegill lakes dominated by intermediate-size bluegill (100–159 mm TL) supported large numbers and high proportions of stunted white crappies (100–199 mm TL) and low proportions of large white crappies (≥200 mm TL).[106]

White crappie was found at dissolved oxygen concentrations as low as 3.3 mg/l, but was rarely found at 2.0 mg/l.[8,53]

Studies of white crappie movement or migration showed that upstream and downstream movements were at random or perhaps haphazard. Crappie have a strong positive reaction to water currents and this may be a factor in their movement patterns.[46,50]

Returns of tagged white crappies in Lake Texoma showed that some individuals moved considerable distances (up to about 23 miles), but the majority remain in a limited area; about 73% of those tagged remained within 2 miles of their release point. It was concluded that migrations or horizontal movements did not significantly influence the vertical distribution of white crappies.[8]

White crappie migrate inshore and ascend tributaries of the Great Lakes to spawn. Females usually return to deeper water after spawning.[22] In the spring in MO, spawning fish are found in relatively shallow water near the upper ends of coves; later they move to deeper water, commonly occurring at depths of 4.6 m or more.[21]

Ultrasonic telemetry was used to determine monthly and diel movements of white crappie in a SD glacial lake. Movement was highest in May and, other than in early spring, movement did not vary significantly among months. White crappie occur in shallow water significantly more during June and October and were nearest to shore in April, June, and October and farthest from shore in August. Home range varied from <0.1 to 85 ha with a median of 15.8 ha and differed significantly among months. Movement differed significantly among diel periods for May, June, and August. Lake depth at fish location was significantly different among diel periods for April, May, June, and September. Distance from shore differed significantly among diel periods for all months except July and October. Weak relationships were noted between environmental variables and movement.[82]

A seasonal cycle of distribution of white crappie occurred in two OK reservoirs related primarily to changes in temperature and dissolved oxygen. Lake Arbuckle was stratified by midsummer of both years of this study. Stratification in this reservoir appeared to force white crappie into the thermocline but anoxic conditions excluded them from the hypolimnion. In Eufaula Reservoir, white crappies were distributed deeper when surface water temperatures increased, but their distribution in depth was not limited by anoxic water. White crappie were found nearer the surface in the fall when surface water cooled.[105]

White crappie often presents a dilemma for fishery managers, especially in small lakes and farm ponds. In these habitats, they frequently overpopulate, resulting in stunted populations dominated by a single year-class. These management problems are so severe that some biologists have recommended that the species not be stocked into small impoundments.[78] Juvenile mortality in white crappie populations is seldom excessive; more often, too many small fish of a year-class survive and deplete their food supply before attaining a size acceptable to anglers. Such "stunting" does occur in large lakes but is more likely to take place in small ones.[119]

Survival of F_1 hybrid crappie in ponds was low, especially beyond age 1. Recruitment was also low; none of the stocked hybrid populations was able to hold their initial densities. Because their recruitment is low and they have growth rates similar to their parent species, hybrid crappie were recommended as a viable alternative to stocking white or black crappie in small impoundments or ponds.[126]

Techniques for producing non-reproductive triploid crappie, utilizing temperature shock soon after artificial fertilization, have been developed to expand options available to crappie fishery managers.[78,79] Triploid white crappie may prove to be valuable management tools, especially useful for stocking small lakes and farm ponds where overcrowding and stunting offer severe management problems.[78]

DISTRIBUTION AND OCCURRENCE IN THE OHIO RIVER SYSTEM

The Falls of the Ohio River is the type locality for white crappie.[15,64] The species is common and has been reported throughout the length of the Ohio River since 1800.[18,19,64,65] In 1984 it was considered the third most abundant centrarchid (after bluegill and longear sunfish) in the river,[18,64] and had demonstrated no obvious changes in distribution since 1970.[64] Lockchamber surveys, during the period 1957–2001, showed an increase in abundance in the Ohio River during the period of study,[62] and collections by the Ohio River Ecological Research Program, during the period 1981–2003, have yielded white crappies in low numbers at most sites most years.[121–123]

White crappie is naturally distributed in all the main drainage systems of IN and widely introduced into natural and artificial lakes of the state.[16] Though not very abundant in the Wabash River, it is a regular component of electrofishing catch.[66] It is common in the Ohio River drainages of IL,[24] and distributed and common throughout the flowing and static waters of OH, where it is widely introduced throughout the state.[19] White crappie is generally distributed and common throughout KY, where it has been widely introduced into reservoirs statewide.[15,17]

Because it avoids mountain streams, white crappie was probably not indigenous to the upper Tennessee and Big Sandy drainages of VA or the entire New River drainage.[115] It is present in the Tennessee River drainages of western NC[113] and throughout the Tennessee River system of TN,[114] AL,[117,118] and MS.[116]

SPAWNING

Location

Depth of water and substrates are variable where white crappies build their nests. They are reported to spawn under overhanging banks, cliffs, ledges, and tree roots[6,7,124] and over a variety of submerged structures including brush, trees, logs, and roots;[6,7,67,124] among beds of aquatic plants and over substrates such as rocks, boulders, mud, clay, sand, gravel, and vegetation; and they will spawn in turbid water.[6,7,124] Nests are found in deeper water than those of other sunfishes, 5–61 cm,[53,71] and a positive correlation between spawning depths and water clarity has been reported.[53]

In reservoirs, nests are usually located in coves, protected bays, and shallow island areas that are protected from wave action.[2,21,34,67,114] Many nests are sometimes concentrated in the same cove. Nests are often constructed near cover or overhanging banks[21,114] over substrates such as fine gravel or finely divided plant roots to which fertilized eggs attach.[21] Spawning has been reported in ponds at water depths of 0.9 m or less over clay, silt, gravel, and sod. The authors felt that nests at these depths in unprotected areas of Lewis and Clark Lake would be destroyed by wave action or minor water-level fluctuations.[2]

Observations of nesting white crappies in two IL lakes were consistent in these respects: the nests were on hard bottom (clay or gravel); nest depressions were lacking or difficult to see; eggs were attached to aquatic or terrestrial plants; most of the nests were in colonies; and nesting fish remained in fixed positions.[3]

On May 21, 1956, 60–70 white crappies were observed guarding nests in Lake of the Woods, IL. Nests were found 0.6–9.1 m from shore and at depths of 0.1–1.5 m. Most nests were on hard clay, a few were on gravel; none were on sand. Nests were usually arranged in colonies; of over 150 nests observed, only three solitary nests were found. The two largest colonies observed were circular in shape. One was in a bed of *Elodea* on a steep slope and contained 35 nests in water 0.5–1.2 m deep. Another colony of about 50 nests was observed on a gentle slope where the water averaged about 0.6 m deep and where the bottom was dotted with clumps of algae. Nests in the circular colonies were spaced 0.5–0.6 m apart. Other colonies were observed that were linear in shape and included 3–15 nests. These linear colonies lay parallel with the lake margin or were parallel with some of the margins of boat docks. Nests in these linear colonies were spaced 0.6–1.2 mm apart. There was usually some kind of plant growth in or near the nests; eggs were found attached to such plant materials as filamentous algae, *Elodea*, and tree leaves. Nests found in a bed of *Elodea* were 2–2.5 cm in diameter with most of the plant growth removed from the bottom. Some of the nests were exposed to full sunlight, others were shaded part of the day.[3] White crappie in Lake of the Woods, IL, usually did not return to the same areas to spawn in subsequent years.[3]

Observations of white crappie nests and spawning were made in pens built in ponds at a fish hatchery in SD. Water depths over nests ranged from 20 to 97 cm. Well-defined nest depressions were not observed but average nest diameter was about 30 cm. No substrate preference for nesting was apparent, but fish appeared to select areas near objects or bottom vegetation. Substrates used included gravel, rock pile in a pen corner, a hay mat, a sod clump, silt swept to natural pond bottom, and natural pond bottom near a submerged tree.[35]

Season

The spawning season for white crappie varies geographically,[52] beginning as early as March[6,22,25,72,116] and lasting into July.[2,22,34,46,67]

In OH and the Great Lakes region, spawning occurs from late March into July[6,22,72] with peaks in May and early June.[6,22]

In WI and in Missouri River reservoirs in ND and SD, spawning is reported from May to July.[34,35,46,67] In ND and SD peak spawning occurs in the middle of June.[34,67]

In Lake Texoma, OK, spawning occurs earlier in the year for older fish than for younger ones. Spawning period is April and May, with peak in late April and early May.[52]

In MO, spawning is reported from the second week of April to early June[21] and also from mid-May to mid-July.[2]

White crappie spawn from April to June in IL,[3,24,50] the Delaware Bay region,[75] VA,[115] NC,[113] and TN.[114]

Spawning occurs from March to May in MS with variations in timing largely dependent on water temperatures.[25,116]

Temperature

Reports of spawning temperatures for white crappie range from 14 to 25°C,[2,3,21,22,35,52,115] with most reports between about 16 and 20°C.[2,3,21,35,115]

White crappie spawning peaks at 16.1–20°C in the Great Lakes,[22] 15.6–20.6°C in MO,[2,21] 15.6–25°C in Lake Texoma, OK,[52] and 17.8–20°C in IL.[3]

Fecundity

In Lewis and Clark Lake, SD, white crappies 211 and 316 mm TL had 22,880 and 194,100 eggs, respectively.[34]

In Lake Texoma, OK, the number of mature eggs produced by female white crappies ranged from 25,600 to 91,700 and averaged 53,000. Fecundity varied with size of the female and the degree that the ovaries were spent.[52]

In Buckeye Lake, OH, the number of eggs in mature female white crappie varied with the age and size of the fish. A female 145 mm TL weighing 43.3 g had a total of 1,908 eggs of which 970 were mature. A 330-mm female weighed 581.1 g and had a total of 325,677 eggs, of which 213,213 were mature.[6]

Individual egg quality, measured as ovary energy density, increased with TL for white crappie from seven OH reservoirs in 1999 and from three in 2000. Among these same individuals, egg quality increased with maternal condition factor (measured as residual wet mass for a given length) in 1999 but not in 2000. In 2000, somatic energy density (an improved measure of condition) was estimated; egg quality increased with somatic energy density, but somatic energy density was also strongly correlated with maternal length. No determination could be made concerning whether length or condition was the primary factor influencing the quality of white crappie eggs.[95]

In three AL impoundments, crappie fecundity (both species) and fish length were positively correlated. Both ovary weight (OW) and GSI were significant predictors of total fecundity (OW: fecundity = 3,996 (OW) + 323.1, r^2 = 0.75, n = 465; GSI: fecundity = 20,585 (GSI) – 23,795, r^2 = 0.54, n = 289; both P = 0.0001).[92]

Sexual Maturity

Sexual maturity for white crappie is usually reached in the second or third year of life.[6,7,21,50,52,59,114,115]

In Lewis and Clark Lake, SD, white crappie mature some years at age 2, but age at maturity ranges from age 2 to 5. White crappies as small as 180 mm TL were reported sexually mature.[34] In Lake Oahe, 72% of male white crappies were sexually mature at age 2, only about 6% of the females were mature at age 2. At age 3, about 75% of the males and 80% of the females were mature; all males were mature at age 4 as were 98% of the females.[59]

In Lake Texoma, OK, sexual maturity of white crappie is reached in the second or third year of life but may be more related to growth than to age. Fish 2 years old and >178 mm TL were mature, while those 2 years old and <178 mm TL were immature.[52]

In OH, the smallest white crappie reported to spawn was about 145 mm TL.[6] The smallest ripe female observed in IL was 142 mm TL.[50]

In Kentucky Lake, male white crappies were reported sexually mature at 190.5 mm TL and females at 236 mm. One age-1 female and 3 age-1 males were found sexually mature.[5]

Spawning Act

The bottom of the typical white crappie nest was swept enough to remove loose sediment but not enough to leave a visible depression. Occasionally fish were observed guarding an area that had not been swept clean or that had not been kept clean.[3]

Observations of spawning white crappies were made in pens constructed in a hatchery pond in SD. Prior to spawning, some males exhibited territorial behavior aggressively chasing intruders from a defended area that did not always correspond to a later nesting area. Most spawning took place in nests established by a male less than 24 h earlier. Nest preparation consisted of short, vigorous periods when sediment was swept out of the nest with fin and body movements. Occasionally a female exhibited the same nest-sweeping movements as the male immediately before and during a spawning run. A female, after being repulsed from a male's territory a number of times, would eventually stop retreating from the territory when chased, and would be accepted by the male. After circling the nest several times, the female positioned herself beside the male facing the same direction. They remained this way motionless for a few seconds and then the sides of their bodies touched, which precipitated slow movement forward and upward with bodies quivering. The female slid under the male in the process, pushing him up and to the side, causing the pair to move in a curve as sex products were presumably emitted. During their movement, the male exerted a steady pressure on the female's abdomen. Each spawning act lasted 2–5 s, with most 4 s. Intervals between acts ranged from one-half hour to 20 min. The maximum spawning acts in one spawning run was 50. In one observation, an intruding male quickly positioned himself on the unoccupied side of the female and participated in the act; he was subsequently chased from the area by the defending male. On another occasion, a female spawned with a second male guarding a nearby nest. The first male chased the female away from the second nest and the second male chased the invading male back to his own territory. One male successfully nested twice (12 days apart) in different areas of the pen. Females sometimes remained in the nest area with the male between spawning acts and participated in sweeping the bottom; more frequently they swam to deeper water.[35] The female does not void all her eggs during one spawning act, but may ovulate over a period of time.[6]

Males exhibit aggressive nest defense during incubation of the eggs,[22,35,50,56] which lasts from 43 to 103 h.[35] A guarding male remains in a relatively fixed position over the nest; males were not observed circling the nest and were never observed leaving the nest.[3] The average time from the start of hatching to the departure of broods from the nests was 95 h (51–162).[35] After departure of the larvae from the nest, the guarding male moved into deeper water, with no further indication of parental care.[35]

Natural hybridization with black crappie is common[84,94,96] and has been reported from hatcheries,[84] ponds, newly constructed reservoirs,[19] and older reservoirs.[83,84,87,88,96] An intergeneric hybrid between white crappie and flier *Centrarchus macropterus* was seined from a borrow pit in Clinton County, IL, in 1970.[51]

EGGS

Description

Unfertilized white crappie eggs vary from 0.82 to 0.9 mm in diameter; perivitelline space is 0.05 mm in diameter. The yolk is granular and contains a large oil globule near the center.[6]

Fertilized white crappie eggs are adhesive, demersal,[2,6,21,50,71] and colorless[20,71] or amber to pale yellow in color.[20,73] The adhesive egg capsule is initially highly elastic; at hatching, it dissolves into a sticky mass.[6] Reports of fertilized egg diameters range from 0.82 to 0.92 with an average of 0.89 mm.[7,52]

Incubation

White crappie eggs hatch in 2–4 days at normal spawning temperatures, with shorter incubation periods at warmer temperatures.[2,21]

White crappie eggs are reported to hatch in 93 h at 14.4°C, 43–51 h at 18.3–19.4°C, and 42 h at 22.8°C;[35] also a report of 24–27.5 h at 21.1–22.2°C.[6] Highest mortalities occurred for eggs incubated at the lowest water temperatures with corresponding extended incubation periods. Lowest egg mortality was exhibited in water temperatures of 18.9–19.4°C.[35]

Development

See Figure 108.

Following are stages of development of white crappie eggs related to time after fertilization when incubated at 21.2–22.2°C:[6]

    5 min — blastodisc formed
    50 min — 2-cell stage
    55 min — 4-cell stage
    60 min — 5- to 8-cell stage
    1.25 h — morula
    2.25 h — late morula
    3.25 h — blastoderm to equator of egg
    5.25 h — blastoderm below equator of egg;
        embryo distinct
    9.00 h — somites forming
    16.25 h — 20–22 somites; blastopore not yet
        closed
    23.00 h — tail free on embryo; heart beat estab-
        lished

YOLK-SAC LARVAE

See Figures 109 and 110.

Size Range

White crappie are reported to hatch at 1.2–2.0 mm,[6] 2.1 mm,[125] and 2.56 mm TL.[58]

Yolk sac is absorbed by 4.0[6]–4.6 mm TL,[35] but traces of yolk material are present in some individuals at 5.0 mm.[125]

Myomeres

In Missouri River: preanal 10–11; postanal 19–20; total 28–31.[58]

In LA: preanal 10–13, usually 11–12; postanal 18–23, usually 19–20; total 29–33, usually 31–32.[125]

Morphology

1.2–2.0 mm TL (at hatching). Poorly developed; movements are described as quite feeble. The head is still attached to the yolk. From external view the circulatory system is indistinct except in the heart region and beneath the gills.[6] Larva is coiled about the yolk.[6,125] The egg membrane dissolves into a sticky mass at hatching to which the yolk-sac larvae are attached head on.[6]

1.75–3.0 mm TL. Body is tadpole-like in shape with a large, oval yolk sac.[75] Body straightens.[6,125] Bulbous yolk is granular with a single large posteriorly[125] and dorsally[75] placed oil globule.[75,125] Heart action is much stronger and the circulation of the blood can now be followed all over the body.[6] At 3.0 mm the head is no longer deflected;[75] brain is well developed; digestive tube is visible from mouth region to vent; jaws are not yet evident; gill clefts appear in the pharynx; optic and otic capsules are well developed; semicircular canals are present in the otic capsules; the heart is active; the auricle, ventricle, and ventral aorta are well defined; there is no evidence of an air bladder.[6]

3.5 mm TL. Oil globule is generally in the posterior-most portion of the remaining yolk. The first skeletal structure, the cartilaginous cleithrum, appears.[125]

3.7 mm TL. Brain plexuses are evident; pharynx and mouth developing; jaws present but not moveable.[6] Air bladder becomes apparent.[125]

3.9 mm TL. Jaws well developed and moveable; gill arches and gill filaments are present; air bladder is now visible dorsal to the posterior end of the yolk sac; alimentary canal begins to fold in the region of the yolk sac.[6]

4.0–5.0 mm TL. Discrete bulbous mass of yolk absorbed; remnants of yolk may still persist on some individuals.[125]

Morphometry

Chatry (1977) provided statistics describing regressions of body measurements on total length for young white crappies from southeastern LA (Table 138).[125] Plots of the following regressions can also be found in Chatry's thesis:

    Eye diameter on TL
    Distance from eye to air bladder on TL
    Air bladder length on TL
    Dorsal fin length on TL
    Distance from eye to dorsal fin origin on TL
    Dorsal fin length/distance from eye to dorsal
        fin origin on TL[125]

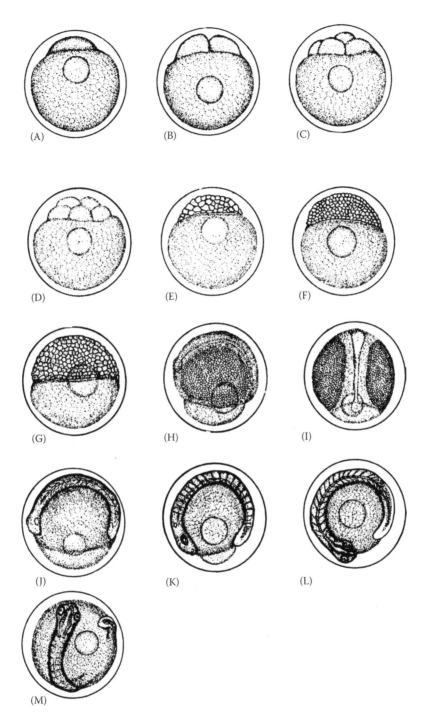

Figure 108 Development of white crappie eggs from OH related to time after fertilization when incubated at 21.2–22.2°C.[6] A. 5 min; B. 50 min; C. 55 min; D. 1 h; E. 1.25 h; F. 2.25 h; G. 3.25 h; H. 5.25 h; I. Same as H, dorsal view; J. 9 h; K. 16.25 h; L. 21.25 h; M. 23 h. (A–M reprinted from Figures 3a–3m, reference 6.)

Fin Development

1.2–2.0 mm TL (at hatching). Tail movement is pronounced.[6] At 1.75 mm, the median finfold originates dorsally near the middle of the yolk, is present around the urostyle, and is continuous ventrally to the vent.[6]

3.0 mm TL. Median finfold extends from the head around the urostyle to the vent; no rays visible in caudal fin; pectoral fins have not formed.[6]

3.7 mm TL. Pectoral fins visible; fin rays appearing in caudal fin.[6]

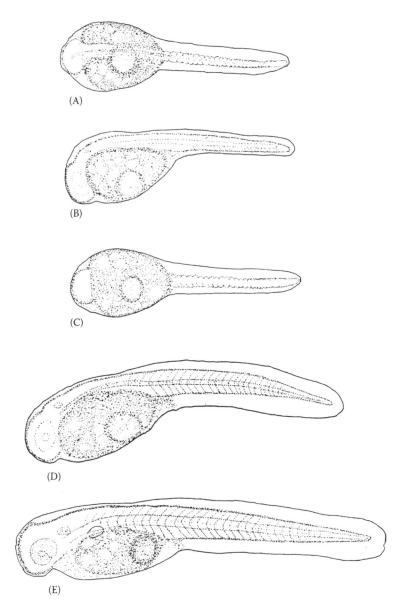

Figure 109 Development of young white crappie. A–C. Yolk-sac larva, 2.25 mm TL: (A) dorsal view; (B) lateral view; (C) ventral view. D–E. Yolk-sac larvae: (D) 3.03 mm TL; (E) 3.47 mm TL. (A–C reprinted from Figure 3, reference 125; D–E reprinted from Figure 4, reference 125.)

3.9 mm TL. Pectoral fins with rays.[6,124]

4.2 mm TL. Fin rays begin to appear in dorsal and ventral finfold.[6]

Note: This statement by Morgan (1954)[6] is questioned by Hardy (1978).[124]

Pigmentation
Eye apparently not pigmented at hatching[6] and possibly not throughout phase.[124]

3.0–4.0 mm TL. Eyes gradually become pigmented.[125] At 3.7 mm TL, pigment is visible on air

bladder.[124] At 3.9 mm TL, pigment spots are visible along the hypaxial muscles between the anus and caudal peduncle and on the dorsal surface of the air bladder.[6]

POST YOLK-SAC LARVAE

See Figures 110 and 111.

Size Range
4.0[6,58,125] to about 20 mm[125] or >20.0 mm TL (length at which 6th dorsal spine is evident).[58,124]

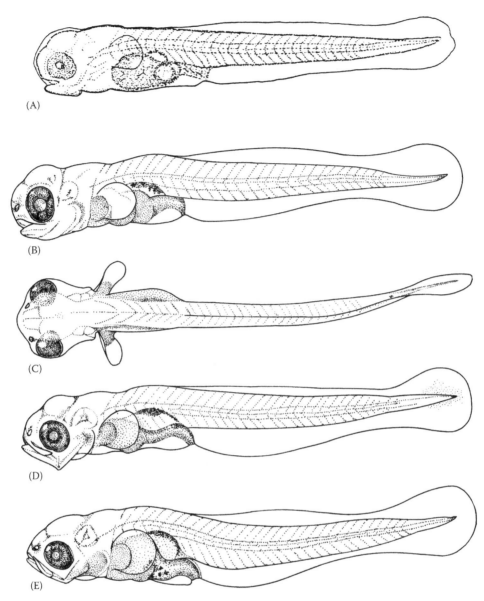

Figure 110 Development of young white crappie. (A) Yolk-sac larva, 4.04 mm TL. B–C. Post yolk-sac larva, 4.3 mm TL: (B) lateral view; (C) dorsal view. D–E. Post yolk-sac larvae: (D) 4.8 mm TL; (E) 6.0 mm TL. (A reprinted from Figure 6, reference 125; B–E reprinted from Figure 16, reference 111.)

Table 138

Statistics used to describe regressions of body measurements on TL for young white crappie from southeastern LA.[125]

Body Measurement	mm TL (X)	N	Regression Equation	R²
Eye diameter	3.12–20.90	383	$Y = 0.258 + 0.0036X$	0.982
Distance from eye to air bladder	3.66–16.22	296	$Y = -0.445 + 0.312X - 0.011X^2$	0.966
Air bladder length	3.65–20.90	304	$Y = 0.085 + 0.011X^2$	0.974

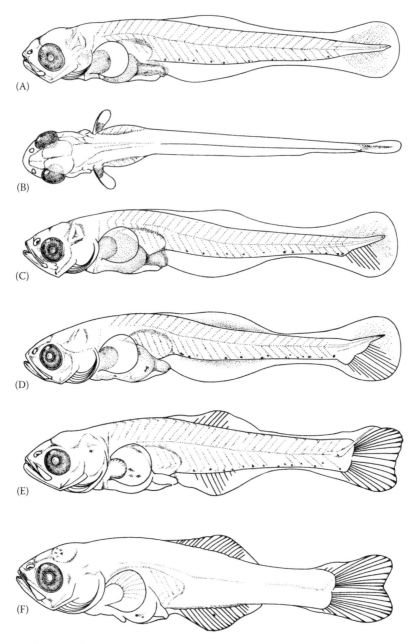

Figure 111 Development of young white crappie. A–B. Post yolk-sac larva, 7.2 mm TL: (A) lateral view; (B) dorsal view. C–F. Post yolk-sac larvae: (C) 8.3 mm TL; (D) 9.3 mm TL; (E) 11.2 mm TL; (F) 13.5 mm TL. (A–F reprinted from Figure 16, reference 111.)

Myomeres

In Lewis and Clark Lake, Missouri River:

4.0–7.49 mm TL. Preanal: mean 10.3–12.0, mode 10–12; postanal: mean 18.6–19.7, mode 18–20; total: mean 29.6–30.8, mode 29–31.[58]

7.5–9.49 mm TL. Preanal: mean 11.2–12, mode 11–12; postanal: mean 18.4–19.1, mode 18–19; total: mean 30.0–30.6, mode 30.[58]

9.5–16.49 mm TL. Preanal: mean 11.6–13.0, mode 12–13; postanal: mean 17.8–18.6, mode 17–18; total: mean 29.8–31.0, mode 29–31.[58]

In LA:

4.0–7.5 mm TL. Preanal 10–13, usually 10–12; postanal 17–22, usually 19–21; total 28–33, usually 30–33.[125]

7.3–9.3 mm TL. Preanal 11–13, usually 11–12; postanal 17–22, usually 18–20; total 29–33, usually 30–32.[125]

9.3–16.0 mm TL. Preanal 11–14, usually 12–13; postanal 17–20, usually 17–18; total 29–33, usually 30–31.[125]

Morphology

4.0–7.2 mm TL. Gut well coiled at 4.0 mm.[6] Jaws and gill arches are well developed by 6.0 mm,[6] otherwise external morphology changes little.[75,125]

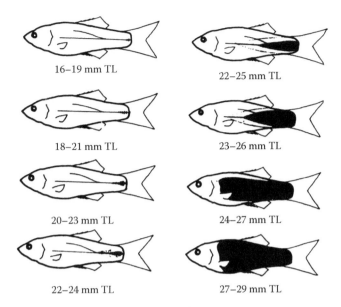

16–19 mm TL

22–25 mm TL

18–21 mm TL

23–26 mm TL

20–23 mm TL

24–27 mm TL

22–24 mm TL

27–29 mm TL

Figure 112 Early development of young white crappie — sequence of scale development. (Reprinted from Figure 1, reference 4.)

7.5 mm TL. Shape of the mouth and head has changed. The premaxillaries are well defined and protractile. The head tapers toward the snout and the cleithrum becomes externally evident. Chondrification of the premaxillary, maxillary, dentary, frontal, supraoccipital, and branchiostegals has begun; the cleithrum, dentary, and maxillaries are ossified. Coincident with the ossification of the head parts is the formation of cartilaginous neural and haemal spines, and hypurals.[125]

8.3 mm TL. Snout more elongate; mouth upturned; intestine short and coiled.[75]

9.5 mm TL. Vertebral column ossified.[125]

11.0–13.0 mm TL. Interneurals appear and complete formation.[125] White crappie this size and larger are deep-bodied with extreme lateral compression.[75]

16.0–19.0 mm TL. Pockets of scales first appear on the caudal peduncle and form progressively anterior and ventral to the lateral line with growth of the fish (Figure 112).[4]

Morphometry

See information presented in Yolk-Sac Larvae section (page 360) and Table 138.

4.0–13.0 mm TL. Distance from the posterior margin of the eye to the anterior-most portion of the air bladder is >15% TL.[125]

>11.0 mm TL. Body is deepest at origin of the dorsal fin. Base of dorsal fin is shorter than the base of the anal fin.[75]

>16 mm TL. The ratio of dorsal fin length to the distance from the eye to the dorsal fin is less than 1.0.[125]

The gape of larval white crappie increased linearly with fish size in Clark and Stonelick Lakes, OH; relationships did not differ between lakes (Table 139).[89]

Fin Development

6.0 mm TL. There are now indentations in the margins of the dorsal and ventral finfolds indicating beginning differentiation of the dorsal and anal fins. Pectoral fins are well developed. Caudal fin beginning to take the shape of the adult homocercal tail. No evidence of the pelvic fins.[6]

7.0–11.5 mm TL. Minimum length at first appearance of caudal fin rays variously indicated as 4.0 mm or less (questioned),[6] 7.2 mm,[75] 7.3 mm (complete caudal ray visible),[125] 8.3 mm,[111] or 11–11.5 mm.[58] In LA, a full complement of principal caudal fin rays was first observed at 9.3 mm.[125] Urostyle is slightly flexed at 8.3 mm[111] and flexion is completed by 9.5 mm[125] or 11.2 mm.[111] In LA, dorsal and anal fin pterygiophores and rays first appear at about 8.6 mm.[125] Other reports indicate later development with mesenchyme visible in region of future dorsal and anal fins at 9.3 mm.[111,124] Also, first dorsal rays reported to appear at 9.0–11.5 mm[58,71,125] and other reports indicate that the first incipient rays appear in dorsal and anal fins at 11–11.5 mm.[58,75] At 11.0–11.5 mm, the caudal fin becomes bilobed and the first dorsal spine appears.[125]

12.4 mm TL. In LA, pelvic fins appear and rays are visible in pectoral fins.[125]

Table 139

Linear regression of gape size on fish length for young white crappie from Clark and Stonelick Lakes, Ohio. The equation is of the form: gape = (slope × length) + intercept. Parenthetical values are SEs.

Lake	TL Range (mm)	N	Slope (SE)	Intercept (SE)	r^2
Clark	4.4–23.4	58	0.154 (0.004)	–0.419 (0.055)	0.94
Stonelick	4.9–21.4	39	0.163 (0.005)	–0.404 (0.060)	0.96
Clark and Stonelick	4.4–23.4	97	0.158 (0.004)	–0.417 (0.043)	0.95

Source: Constructed from data presented in Table 1, reference 89.

13.5–14.0 mm TL. Remnant of preanal finfold still visible;[6,124] dorsal and anal spines evident;[58,71] pectoral fins not well developed until 13.5 mm; pelvic buds evident at 13.5 mm.[111,124]

14.5 mm TL. In LA, all median fin ray elements completed.[125]

16.0 mm TL. 5–6 dorsal fin spines.[58,125]

20.0 mm TL. Full complement of dorsal spines present.[125]

Pigmentation

Pigment patterns apparently vary geographically.[124]

4.1–4.6 mm TL. Body transparent.[71]

In OH:

4.0–6.0 mm TL. Air bladder with large, widely and evenly spaced chromatophores; a heavy row of chromatophores, about one per myomere, along ventrum between anus and tip of tail; a similar, less pronounced, row just beneath notochord from region of air bladder to a point about one-fourth TL from tip of tail; eye apparently unpigmented.[6,124]

In OK:

4.3 mm TL. Eye dark; dorsum of air bladder densely pigmented, otherwise the body appears clear.[111,124]

7.2 mm TL. A ventral row of widely spaced chromatophores, about one every other myomere, is present between the anus and the tail.[111,124]

11.2 mm TL. Ventral pigment as above but number of chromatophores has increased; air bladder still pigmented above but pigment less dense than in earlier stages.[111,124]

13.5 mm TL. Ventral pigment apparently absent; a few melanophores visible over brain.[111,124]

In LA:

4.0–7.5 mm TL. Eyes become heavily pigmented. Pigmentation present on the dorsal wall of the air bladder and in a ventrolateral row.[125]

7.3–9.3 mm TL. Ventrolateral pigment is present on about 50% of individuals; head and visceral mass pigment virtually absent.[125]

13.0–16.0 mm TL. Pigment is mostly limited to the ventrolateral region and on the visceral mass.[125]

JUVENILES

See Figure 113.

Size Range

>20 mm[124] to as small as 142 mm TL (in an apparently normal population).[50]

Maturity reportedly reached as small as 109 mm in a stunted population.[19]

Morphology

22.0–24.0 mm TL. Imbrication of scales begins.[4]

27.0 mm TL. Some individuals fully scaled (Figure 112).[4]

Larger juveniles. White crappie have 34–45 lateral line scales;[20,114,117] 30–33 total vertebrae;[20,114,125] 17–19 caudal vertebrae;[125] 10 thin pyloric caeca;[2] 7 branchiostegal rays; and 25–32 long and slender gill rakers.[114]

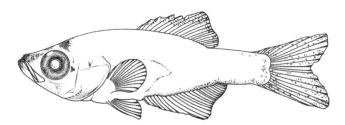

Figure 113 Young white crappie juvenile, 25.5 mm TL. (Reprinted from Figure 16, reference 111.)

Morphometry

See Table 138.

The gape of young white crappie through early juvenile development increased linearly with fish size in Clark and Stonelick Lakes, OH; relationships did not differ between lakes (Table 139).[89]

25.5 mm TL. Body depth conspicuously less than in adult, about 4.5 times in TL.[111,124]

Fins

Reports of white crappie dorsal fin spine counts range from 4 to 8 (extreme),[19] but are more often 5–7,[20,76,114,117,125] and usually 5–6 in the south central United States.[19,20,25,114–117,125] Reports of dorsal fin soft rays range from 13 to 16,[114] but are usually 13–15.[20,25,76,114–117]

Anal fin spines range from 4[76] to 7,[20,114] but are usually 6–7.[20,25,114–117] Reports of soft rays in the anal fin range from 17 to 19,[25,116] most often 16–18.[20,76,114,115,117]

Pectoral fin rays are usually 13,[20,115] or 14–16.[117] Pelvic fins have 1 spine and 5 soft rays.[20]

Pigmentation

25.5 mm TL. Pigment over head, snout, and lower jaw; a row of pigment along ventrum from middle of anal fin to tail; a few chromatophores below dorsal fin and scattered chromatophores on caudal peduncle; pigment on membranes of caudal fin and distal half of dorsal fin.[111,124]

19.7–85.0 mm SL. Development of primary and secondary bars of melanistic pigment in white crappies is very similar to that described for black crappies[132] (see black crappie species account, page 389).

52.0 mm TL. A series of narrow vertical bands are present on the body and several broad bands on the caudal peduncle; small punctations are visible on the lower sides of the body, abdomen, and caudal penduncle.[124]

Large juveniles have light-centered primary bars of melanistic pigment interspersed with narrow intercalated secondary bars.[132]

"Smallest young" are described as sometimes faintly spotted, and "small young" as having dorsal, anal, and caudal fins plain or inconspicuously marked.[19] Young with much less black pigment than adults.[6]

TAXONOMIC DIAGNOSIS OF YOUNG WHITE CRAPPIE

Similar species: other centrarchids, especially black crappie and flier.

For characters used to distinguish young crappie from the young of other centrarchid genera, see the provisional key to genera (pages 27–29).

See "Taxonomic Diagnosis" section in flier species account (page 85) for discussions concerning differences between young flier and crappie.

For taxonomic discussions about black and white crappie, see information presented in the introduction to this chapter (page 351–354).

Also, see comments from literature presented below.

White crappie vs. other centrarchids. White crappie larvae have little pigmentation except for melanophores on the surface of the air bladder and they lack the anal melanophore characteristic of *Lepomis* spp. A coil forms in the gut at 3–4 mm TL, or prior to yolk absorption, in contrast to *Lepomis* spp., where the first coil does not form until at least 4.5–6.0 mm. Preanal length is always less than postanal length. White crappie larvae are smaller, have a more laterally compressed body, sparser pigmentation, and a less massive gut than *Micropterus* spp. The gut does not extend conspicuously beyond the air bladder as does the gut of *Lepomis* spp. Pigmentation is sparse into the juvenile phase. White crappie juveniles can be distinguished from the black basses and sunfishes in that the spinous dorsal fin originates above or only slightly anterior to the vent.[76,112]

White crappie vs. black crappie: At lengths ≥11 mm TL, the body is more elongate in appearance for white crappie than for black crappie. Caudal peduncle is longer and narrower for young white crappie than for young black crappie.[75]

ECOLOGY OF EARLY LIFE PHASES

Occurrence and Distribution (Figure 114)

Eggs. White crappie eggs are protected by the parent male during incubation.[35] The adhesive and demersal eggs become attached to the nest bottom, to surrounding objects such as twigs and vegetation, and to each other. The tolerance limit of eggs to foreign material is not known. Turbidity may represent a mortality factor.[2]

Across years (1994 and 1996–1998) in Normandy Reservoir, TN, white crappies hatched as early as April 13 and as late as June 6. Hatch date and length-frequency distributions were unimodal in all years. The first hatch date and mean hatch date of white crappies were positively correlated with the first day the reservoir achieved full pool. Hatching began when water temperatures ranged from 14 to 17°C, and peak hatching occurred at water temperatures ranging from 20 to 22°C. The warmest water temperature observed during hatching was 26°C.[90]

Yolk-sac larvae. Newly hatched yolk-sac larvae do not pop out of the egg membrane. At hatching, the membrane dissolves into a sticky mass to which the yolk-sac larvae remain attached at the head for some time. The egg membranes stick together forming one large mass, and as many as 50 larvae have been observed attached to a mass at one time.[6]

After hatching, larvae remain on the nest for 2–4 days during which time they are protected by the parent male. Larvae are 4.1–4.6 mm long when they leave the nest and no schooling of free-swimming larvae was observed.[2,35,124] After they leave the nest, they are no longer protected by the male.[35]

In the Buncombe Creek arm of Lake Texoma, OK, yolk-sac larval white crappies were more abundant in samples from shallow water stations which were apparently closer to spawning areas. They evenly distributed vertically in the daytime but showed a higher concentration near the surface at night.[111]

Post yolk-sac larvae. White crappie larvae move from shallow, protected nursery areas into deeper water as their size increases.[2]

In the Buncombe Creek arm of Lake Texoma, OK, white crappie larvae 5–10 mm TL were fairly abundant in surface and mid-water collections. They were more abundant in bottom hauls in the daytime and almost absent from mid-water trawls. At night, white crappie this size were evenly distributed in the water column. Larvae ≥10.5 mm were mostly collected in cove areas, were caught primarily in bottom trawls, and were slightly more abundant there in night collections. Very few larval white crappie were collected near the bottom in deep water far from shore; those that moved into open water stayed at higher levels in the water

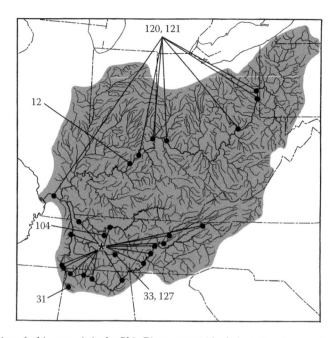

Figure 114 General distribution of white crappie in the Ohio River system (shaded area) and areas where early life history information has been collected (circles). Numbers indicate appropriate references. Asterisk indicates TVA collection localities.

column. White crappie larvae were more abundant in shallow areas in April and then moved from these spawning areas into deeper water beginning in late May.[111]

In the upper Mississippi River, larval white crappies occur in ichthyoplankton collections in early to mid-May. They are most abundant in backwaters and demonstrate a significant drift into the main channel at dusk.[73,74]

In a 2-year study of entrainment of ichthyoplankton associated with hydropower generation at Pensacola Dam on Grand Lake O' the Cherokees, OK, white crappie larvae were dominant in collections near the intakes. In 1988 they constituted 15% of the total larval catch and in 1989 composed the majority (54.5%) of the larvae collected near the intakes. The estimated entrainment of white crappie larvae was almost 3 million in 1988 and almost 40 million in 1989.[81]

There is no evidence of spawning in the tidal waters of Delaware Bay, but young white crappie are commonly collected from tidal spillpools of ponds and specimens were also collected from the western part of the C & D Canal. Specimens collected in the spillpools were probably spawned in upstream impoundments. A few young white crappie have been reported from waters with salinities as high as 2.0 ppt.[75]

Larval white crappies were collected in low numbers with light traps from channel, tupelo, and oak habitats of the Cache River, AR.[37,55]

Ichthyoplankton collections made in the Ohio River from ORM 569 to 572 yielded crappie larvae. Both black and white crappie larvae were present, but most larvae were identified as white crappie. Crappie first appeared in ichthyoplankton tows in June at water temperature of 17.2°C. Surface drift was indicated for 55% of the crappie larvae collected, but they occurred in mid-depth and bottom samples as well.[27]

Larval crappies were consistently collected during ichthyoplankton surveys at sites (ORM 54, 77, 260, 494, and 946) along the Ohio River. Composition was assumed evenly distributed between black and white crappies since both species are common throughout the length of the river. Larval crappies were rarely abundant in ichythyoplankton tows and no apparent trends in density were evident between locations or survey years (1981–1993).[120,121] Peak densities typically occurred from mid-May to July. Initial appearance of crappies occurred from early May to mid-June for tow samples and from late May to July for seine collections. Water temperatures at initial appearance were extremely variable between locations ranging from 14 to 33.7°C.[120]

During the period 1973–1986, larval crappies were frequent and often abundant in TVA ichthyoplankton samples taken from the Tennessee River system. Both black and white crappie are present in most of the waters sampled by TVA, but, at the time, crappie larvae were not routinely identified to species. Crappies normally began to appear in ichthyoplankton samples in early April but the earliest record was March 22, 1976 at the Bellefonte Nuclear Plant site on Guntersville Reservoir, AL. They occurred in samples as late as early August. Specimens ranged from 3 to >30 mm TL, but most were in the 4–10 mm range.*

Peak densities of larval crappies (<20 mm TL, both species assumed present) increased progressively during each year for the first 3 years following impoundment (1976–1978) of Normandy Reservoir, Duck River, TN. Crappie larvae first appeared in collections on April 26, 1976 and were the most abundant taxon in April 1977. In 1978 crappie larvae constituted 49% of the total ichthyoplankton catch and the periods of greatest larval density (May 10 and May 24) were dominated (90%) by crappie larvae (780 and 760/1000 m³, respectively).[33]

Crappie larvae (both species assumed present) showed increases in relative abundance in ichthyoplankton samples each consecutive season during the first 3 years of impoundment (1976–1978) of Little Bear Creek Reservoir, AL. Crappie larvae first appeared in samples in mid-April and remained dominant in collections until early May. Mean water temperature associated with first occurrence was about 15.2°C.[31]

In a 4-year study (1996–1999) in Walker County Lake, AL, white crappie larvae were the first larvae collected (along with golden shiners) in limnetic larval fish collections made in 1996 (April 11). Gizzard shad were the first larvae collected in 1997 (March 20), white crappies and gizzard shad were first in 1998 (April 9), and white crappies were the first in 1999 (April 1). Surface water temperatures associated with these collections were 16°C in 1996, 19.3°C in 1997, 18.4°C in 1998, and 16.6°C in 1999. In years when larval gizzard shad appeared in collections before larval white crappies, larval white crappie densities were low. During 1996 and 1999 (years when densities of larval white crappies were the highest), white crappies appeared in samples 1–2 weeks prior to the first collection of gizzard shad larvae. Early in life both species are obligate planktivores, therefore an earlier timing of first appearance may have contributed to increased growth and survival of larval white crappies in the absence of larval shad through reduced competition for zooplankton.[103]

During the period 1998–2002 in J. Percy Priest Reservoir, TN, both larval tow-net and fall trap-net samples effectively sampled age-0 black and white crappies and accurately indexed year-class strength. The density of larval crappies in May was correlated to age-0 catches in fall trap nets, thus density of larval crappies was positively related to year-class strength. However, fall trap-net samples of age-0 crappies provided a more precise and cost-effective index of year-class strength than larval tow-net samples. May discharge through the dam was an accurate predictor of age-0 abundance in fall trap nets. No relation was evident in this study between mean water temperatures over sample periods and abundance of age-0 crappie.[104]

Larval sampling of white and black crappies over a few weeks in spring in Normandy Reservoir, TN, can accurately demonstrate the presence or absence of strong year-classes much earlier and with less effort than traditional sampling techniques such as fall gillnetting or cove sampling. Larval crappie (both species) recruited to a 1 × 2-m neuston net at 9 mm TL; few over 15 mm were collected.[127]

Juveniles. Fingerling white crappies have been collected with seines from weedy habitats,[16] and juveniles are recorded along beaches in the Delaware River.[124]

Young white crappies are often observed over open water of considerable depth[21] and one report indicates that, at 50–60 mm, white crappies leave shallow water and move to deeper areas.[2] In Lewis and Clark Lake, young white crappies emigrate from spawning-nursery areas into the open reservoir during their first summer of life.[34]

Mid-water trawl studies in a KS reservoir recorded white crappies 80–170 mm TL. Young crappies were distributed throughout most depths during the cool times of the year but were grouped more vertically during warmer months, in part because of the anaerobic hypolimnion extending up to 5-m depth.[47]

Fall trap net sampling provides a good index of white crappie year-class strength in TN reservoirs.[97] A comparison of catch statistics for trap nets, electrofishing, and gill nets in OK indicated that fall trap nets were the only gear that collected substantial numbers of age-0 white crappies.[107] In another study, trap nets with 13-mm mesh caught significantly more white crappies <130 mm TL than did trap nets with 16-mm mesh. Catch rates of white crappies ≥130 mm were similar for both mesh sizes.[129]

In TX ponds, the presence of gizzard or threadfin shad was associated with decreased total number and biomass of YOY white crappies, a shift in white crappie size distributions toward larger individuals, and increased mean weight in some size classes.[80]

The presence of gizzard shad in 0.1-ha ponds in AL did not negatively affect age-0 white crappie. At a given size, age-0 white crappie weighed more in ponds with gizzard shad than in ponds without them.[85]

In Chautauqua Lake, NY, abundance of age-0 white crappies was positively associated with the biomass of walleye prey and negatively associated with that of walleye, demonstrating that the availability of alternative prey affects walleye predation on white crappie. The mortality of white crappies from the fall of age 0 to the fall of age 1 was not related to winter severity, predation, or prey abundance.[98]

In laboratory experiments, severe winter conditions strongly influenced survival of age-0 white crappies. Only 47% of all white crappies survived the severe winter conditions, whereas 97% survived the mild winter conditions. Mortality was not influenced by size, feeding level, or energy depletion. Osmoregulatory failure may have occurred during exposure to temperatures colder than 4°C for at least 1 week, suggesting that the availability of oxygenated water with temperatures >4°C during winter may be critical to the survival of age-0 white crappies. In the northern portion of their range, winter temperatures may account for some of the variability in recruitment common to white crappie populations.[93]

Maximum reported salinity for young white crappie is 2 ppt.[75] Reported temperature range for young white crappie is 1.2–19.5°C.[124]

In TN, reservoir hydrology appeared to have a stronger effect on white crappie recruitment in tributary storage reservoirs than in mainstem Tennessee River impoundments, possibly because recruitment was more variable in tributary systems.[99]

Crappie (both species) recruitment was also related to hydrology in most AL reservoirs studied. In reservoirs where water levels fluctuate, a rising or higher-than-average water level in the months preceding and during the spawning period may increase crappie year-class production. Hydrologic manipulation of these reservoirs could increase reproductive success of white crappie. If erratic crappie recruitment in mainstem reservoirs, where water levels do not fluctuate, is due to variable and uncontrolled hydrologic conditions dictated by climate, then management options to increase crappie year-class strength would be limited.[102]

In new reservoirs, strong year classes of white crappie are produced during the first year of impoundment.[31–34,40] YOY white crappies were abundant during the filling of Lake Oahe, an upper Missouri River storage reservoir, but general abundance declined after the reservoir filled.[40] White crappie juveniles occurred in ichthyoplankton samples during the first 2 years of impoundment of Normandy Reservoir, Duck River, TN. Size range was 20–55 mm TL in 1976 and 113–143 mm TL in 1977.[33] In Little Bear Creek Reservoir, AL, crappie reproduction (both species were assumed present in the reservoir, though only white crappie appeared in cove population collections during the period of study) was good during the first 3 years of impoundment as evidenced by increased densities of crappie larvae each year. However, crappie year-class strengths for the second and third years were relative failures, which was probably associated with increased numbers of predators, especially yearling largemouth bass.[31]

The conclusion that the establishment of a strong year-class of crappie is not dependent on a prolific spawn (as measured by larval abundance) was also supported by study of larval fish populations during the first 3 years of impoundment of Normandy Reservoir on the Duck River, TN. The strongest year-class of crappie (both species present) was observed during the first year, even though greater larval abundance was observed (5 times) the third year of the study.[33]

White crappies produce poor year-classes in dry years and strong year-classes during wet years in Normandy Reservoir, TN. Responses were variable during intermediate years.[130]

Early Growth

Carlander (1977) reviewed and summarized numerous reports on the growth of white crappie. His average calculated growth rates related to age of white crappie in the United States are: 78 mm TL at age 1; 158 mm at age 2; 213 mm at age 3; and 257 mm at age 4. Regional growth varies greatly with ranges of 65–100 mm at age 1, 138–200 mm at age 2, 192–263 at age 3, and 232–310 at age 4 (Table 140). The combined data of numerous empirical reports of growth also show great variability in white crappie growth related to age (Table 141).[14]

Growth rates of white crappie vary greatly between populations and among years, even regionally, as demonstrated by reports of calculated length by age for populations in OK (Table 142), KY (Table 143), and in the Tennessee and Cumberland River systems (Table 144).

Etnier and Starnes (1993) reported that the growth of individuals in those TN reservoirs having relatively unconstrained growth rates averaged 65–75 mm TL during the first year; subsequent year-classes averaged 180, 240, and 285, respectively.[114] These calculations are generally comparable to others presented for Kentucky, Barkley, Douglas, and Hiwassee Reservoirs in the Tennessee River system (Table 144). First-year growth calculated for Kentucky Reservoir white crappies in 1952 and 1983 and in Barkley Reservoir in 1983 greatly exceeded the average range reported by Etnier and Starnes. Calculated growth in Cherokee Reservoir (1944–1945) was much lower the first year but exceeded their averages in years 2 and 3 (Table 144).

Table 140

Mean calculated TL (mm) at each annulus for young white crappie from regional waters of the United States.[14]

Location	Age			
	1	2	3	4
Western waters	65	148	202	241
Northern waters	66	138	193	238
Central waters	74	146	192	232
South central waters	84	166	226	263
AR, OK waters	84	162	210	254
Southeast waters	74	146	205	263
Gulf states	100	200	263	310
Mean of regions	78	158	213	257

In Lewis and Clark Lake, SD, YOY white crappies reached a maximum length (15–90 mm TL) by late August, and by late fall were up to 105 mm. Average calculated TL (mm) at ages 1–4 were 79, 171, 231, and 256, respectively.[34]

Growth rates of young white crappie were greater in Upper Lake Spavinaw, OK, after 2 years of impoundment compared to Lower Lake Spavinaw (impounded 31 years). Calculated TL for white crappie from Upper Lake Spavinaw was 236 mm at age 2 compared to 208 mm for crappie from Lower Lake Spavinaw.[43] Another report suggests that stratification of OK lakes and the resultant loss of fish habitat retard the growth of young white crappie.[38]

In MO reservoirs, white crappie are about 99 mm TL or larger at the end of their first year and about 259 mm or larger by age 3. In MO streams, a 3-year-old white crappie averages about 206 mm TL.[21] White crappie grew faster downstream than upstream in the 170-mile-long Salt River, MO:[11]

	TL (mm) at age			
	1	2	3	4
Upstream	64	122	175	211
Downstream	84	165	201	—

In OH, YOY white crappie in October were 25–97 mm long; at about 1 year, 30–130 mm long.[19] In Buckeye Lake, OH, average calculated TLs (mm) at ages 1–4 were 57, 106, 149, and 194, respectively.[6]

Growth of young white crappie is rapid in MS; larvae are reported 63–76 mm in mid-June with the average about 38 mm.[25]

Table 141

Mean of means and range of TL (mm) related to age calculated and summarized by Carlander (1977) from the combined data of numerous empirical reports of growth for white crappie.[14]

Age	N	TL (mm)	
		Mean of Means	Range
Age 0 — June	464+	28	5–58
Age 0 — July	2,722+	43	12–91
Age 0 — August	2,085+	65	13–150
Age 0 — September	287	80	26–120
Age 0 — October	408+	91	33–251
Age 0 — November	—	118	102–123
Age 0 — December	4,569	140	112–168
Age 0 — combined	444+	82	30–142
Age 1	16,184+	151	58–310
Age 2	19,773+	189	102–366
Age 3	8,952+	221	122–387
Age 4	2,229+	253	130–388

Table 142

Reports of average calculated TL (mm) related to age for young white crappie from OK waters.

Location	Age			
	1	2	3	4
Grand Lake (1942–1952)[13]	89	147	218	262
Canton Reservoir (1950s)[13]	104	198	264	—
*Canton Reservoir (1960s)[69]	76	203	282	318
*Keystone Reservoir (1966)[12]	112	154	191	252
*Keystone Reservoir (1967)[12]	111	142	188	221
Lake Texoma[13]	97	142	185	226
Lake Texoma (12 yrs)[52]	88	152	208	257
*Wister Reservoir[13]	104	201	269	330
*Fort Gibson Reservoir[13]	160	236	287	—
*Tenkiller Reservoir[13]	127	279	315	—
Cimarron River[39]	56	122	168	241
OK state average (1961)[39]	74	150	198	249

* Impounded 3 years or less.

Growth rate of YOY white crappie in a LA lake was calculated as 0.229 ± 0.127 mm per day with a TL of approximately 76 mm reached during their first growing season.[10]

Table 143

Calculated growth (ranges in mm TL) related to age of young white crappie from six KY reservoirs during 1983.

Location	Age			
	1	2	3	4
Rough River Lake	79–81	155	—	—
Nolin River Lake	84–89	163	224	—
Barren River Lake	58–97	157–180	231	—
Green River Lake	137	—	—	—
Shanty Hollow Lake	69	163	—	—
Spurlington Lake	99	231	—	—

Source: Constructed from data presented in Tables 6, 10, 26, 43, 69, and 75 in reference 77.

In Normandy Reservoir, TN, age-0 white crappie grew more slowly (0.58 mm/day) than white crappie in MS reservoirs (0.69 mm/day),[91] in Weiss Lake, AL (0.72 mm/day),[87] or in small AL ponds (0.69 mm/day).[85] In Normandy Reservoir within each year (1994 and 1996–1998), earlier-hatched fish grew at slower rates than later-hatched fish, probably due to warmer water temperatures experienced by the later-hatched fish. However, year-classes that

Table 144

Comparison of reports of calculated lengths (mm TL) related to age of young white crappie from Kentucky, Barkley, and other reservoirs in the Tennessee and Cumberland River systems.

Reservoir	Age			
	1	2	3	4
Kentucky (1952)[5]	118	200	265	301
Kentucky (1982)[45]	75	163	234	296
Kentucky (1983)[77]	91–112[a]	163–191	216–244	269–292
Kentucky (1998)[97]	65[b]	175	235	270
Barkley (1982)[45]	78	164	230	280
Barkley (1983)[77]	86–114[c]	150–183	208–229	244–257
Barkley (1998)[97]	75[b]	150	205	260
Cherokee (1944–45)[48]	38	221	295	—
Douglas (1943–44)[48]	74	185	234	—
Hiwassee (1942–43)[48]	61	173	241	241

[a] mm TL ranges converted from lengths presented in inches in Table 4, reference 77.

[b] mean TL approximated from graphs presented in Figure 1, reference 97.

[c] mm TL ranges converted from lengths presented in inches in Table 2, reference 77.

hatched earlier grew faster on average. Early growth did not appear to be affected by the density of white crappie larvae or by spring or summer water level events but may have been influenced by hatch date and zooplankton density.[90]

During the period 1992–1994, age-0 crappie were collected from Weiss Lake, AL, and identified to species and their hybrids with starch gel electrophoresis. Hatching dates and early growth were determined from otoliths. Growth rate of F_1 hybrids (0.86 mm/day) was significantly higher than those of black crappie (0.67 mm/day) and white crappie (0.72 mm/day). Back-calculated TLs for age-0 white crappies at 10, 20, and 30 days were 3.9, 7.9, and 13 mm, respectively.[87]

In three AL impoundments, back-calculated lengths of white crappie larvae at 4 days were directly related to chlorophyll a concentration and inversely related to larval density.[92]

The capacity of white crappie to grow in ponds may be severely constrained by summer water temperatures, particularly in the middle and southern United States. Also, the warm and annually variable summer temperature and dissolved oxygen

regimes of many U.S. impoundments may substantially underlie the slow growth rates and among-year variation in size structure and recruitment that are characteristic of crappie populations in these environments.[86] In a KY reservoir, water temperature at the 3-m depth explained significant portions of variability in growth of both age-1 and age-2 white crappies over a 5-year (1989–1993) period of study. Higher temperatures in July and August resulted in slower instantaneous growth rates.[128]

Stocking of saugeye (sauger × walleye hybrids) in Thunderbird Reservoir, OK, increased the predation on white crappie and elicited a density-dependent growth response. After the saugeye population became established, the mean length of white crappie ages 1, 2, 3, and 5 increased.[100] Another report indicated that in 8 small impoundments in OK in 1984, mean lengths of age-2 white crappie were directly related to largemouth bass densities.[28]

Growth rate of age-2 and age-3 white crappies increased significantly in Dale Hollow Reservoir, TN, after the introduction of threadfin shad, though they showed no significant increase in growth at age 1.[131]

Average yearly growth of young male and female white crappie is about the same.[57] Only small differences were found in Lake Decatur, IL:[50]

| | Average TL (mm) | |
	Males	Females
Age 1	211	208
Age 2	244	244
Age 3	252	262
Age 4	264	297

Length–weight relationships of young white crappie collected in OK waters during the 1960s are presented in Table 145.[54] Length–weight relationships for Canton Reservoir, OK, white crappie during the same time period were very similar to the state averages, average weight differing by no more than 2 g for any average length.[69] The OK relationships were also comparable to those of white crappie collected from AL during the period 1949–1964[61] (Table 145).

Growth of white crappie in IA farm ponds was slow; three growing seasons were necessary to produce a fish of harvestable size (6 inches, 152 mm TL). Length–weight relationships of young white crappie from IA farm ponds (Table 146)[49] appear comparable to those from OK[54] and AL[61] (Table 145).

In OK, the use of the otolith method for aging white crappie is recommended over the scale method.[29]

Overcrowding and stunting are commonly reported for white crappie populations.[114] Poor growth has been documented for small ponds and reservoirs, but good growth is usually documented in large reservoirs (M.A. Colvin, personal communication, retired Resource Scientist, MO Department of Conservation).

Feeding Habits

White crappie fry begin feeding on copepods, rotifers, and algae.[30,53] As they grow they feed on a variety of zooplankton and small insects;[30,53,114] at about 120 mm, planktonic insects become a predominant item in the diet.[30,53] Young white crappie prey progressively more on fishes as they mature.[53,114]

In three AL impoundments, the biomass of young crappie (both species) diets was dominated by crustacean zooplankton, although diets were numerically dominated by rotifers. Larvae 4–14 mm TL strongly selected the smallest prey available in all lakes.[92]

In Lewis and Clark Lake, zooplankton was the total diet of white crappie larvae 4.5–15 mm TL and the most important food of larger YOY.[34] In another report from Lewis and Clark Lake, bottom fauna

Table 145

Comparison of length–weight relationships of young white crappie collected in AL during the period 1949–1964 and in OK during the 1960s.

| | Weight (g) | |
TL (mm)	OK [54]	AL [61]
25	—	0.2
51	—	1.7
76	4.0	4.5
102	11.0	12.3
127	23.0	25.9
152	42.0	45.4
178	70.0	68.1
203	110.0	113.5
229	164.0	158.9
254	234.0	236.1

Source: Constructed from data presented in Table 4, reference 54 and unnumbered table on page 78, reference 61.

organisms were unimportant food items for white crappie 11–100 mm TL. Common zooplankton in the diet included *Daphnia, Cyclops, Diaptomus,* and *Leptadora. Cyclops* was selected most by crappie up to 30 mm and became unimportant as growth progressed; the other three genera increased in importance with fish growth with order of priority as listed above.[2]

In Clear Lake, IA, young crappies 5–11 mm (possibly a mix of both species) primarily consumed copepod nauplii; adult *Cyclops* were also important in the diet.[26]

In a hatchery pond, nauplii were the most important food items of white crappie larvae at the initiation of active feeding. At 6–7.9 mm TL, a sharp increase in the selection of *Cyclops* occurred. *Daphnia* was a significant food item for larvae 10–13.9 mm TL.[35]

The diet of white crappies 20–75 mm SL from a MN lake was dominated by entomostraca. Chironomids were more important for larger fish within this range than for smaller fish.[30]

In Buckeye Lake, OH, white crappies 51–76 mm TL consumed mostly copepods (89.1%) and some chironomids (10.9%).[6] In another OH report, mean prey size of white crappie in three reservoirs continued to increase with fish size; however, these white crappie continued to consume prey that were

Table 146

Length–weight relationships of young white crappie from IA farm ponds.

AverageTL (mm)	N	Average Weight (g)
97	2	10.4
107	6	17.3
119	18	23.2
130	11	29.5
147	1	35.9
175	2	59.5
183	2	87.6
185	39	77.2
196	59	92.2
198	2	114.0
206	4	99.0
208	4	136.7
218	3	154.4
231	1	183.3
264	1	355.0
269	1	276.0
285	1	343.2

Source: Constructed from data presented in Table 8, reference 49.

smaller than other available prey, even when they were no longer gape-limited.[89]

The food of 8 white crappies 25–165 mm SL from the White River, IN, consisted of the following: chironomid larvae (48.7% volume, 62.5% frequency), ephemerid naiads (27.5, 50.0), Corixidae (13.1, 25.0), and Trichoptera (10.6, 25.0).[36]

White crappies 50–100 mm TL collected in September, November, and December from Bull Shoals Reservoir, AR, consumed *Cyclops*, *Daphnia*, *Bosmina*, and *Diaptomus*. *Bosmina* was of greatest overall importance during the 3-month period. White crappies >100 mm consumed mostly *Cyclops* and *Daphnia* in September, *Bosmina* in November, and immature chironomids and *Cyclops* in December. In July, the diet was mostly shad and ephemeropteran nymphs. Shad was the sole food in August.[60] In Beaver and Bull Shoals Reservoirs, AR, the primary food items of yearling and older white crappie up to about 175 mm TL were fishes and zooplankton in the summer and fall. During the other seasons, the diet was composed of zooplankton and chironomids. Larger fish in this size range tended to concentrate on fishes as food during the winter. White crappie >175 mm consume mostly fishes, although zooplankton (especially *Chaoborus*) and chironomids were taken in large numbers when forage species were scarce.[60]

In 1977, white crappies ≤200 mm TL from Thunderbird Reservoir, OK, fed primarily on invertebrates; those >200 mm ate gizzard shad and sunfishes.[70] During the period 1985–1991, white crappies <200 mm fed almost exclusively on zooplankton and insects, but some individuals in the 161–199 mm TL size range were piscivorous and primarily ate sunfish. Insects continued to be important in the diet of white crappies ≥200 mm, constituting up to 80% frequency of occurrence; frequency of fish in stomachs ranged from 0 to 42% in the larger crappie.[109] In another report from Thunderbird Reservoir, white crappies 40–200 mm TL positively selected for *Daphnia* (the dominant cladoceran during most of the growing season) during all months (March–November 1987) except July. *Bosmina* were generally not found in the diet and patterns of preference for copepods were inconsistent.[108]

Young white crappie feed mainly during daylight hours.[34,47,68] In a KS reservoir, they rarely fed at night and fed entirely on zooplankton during the day. *Daphnia*, even when relatively sparse in zooplankton samples, comprised a significant portion of the diet. Diaptomid copepods, though of similar size to *Daphnia*, were relatively rare in the diet.[47]

In Conwingo Reservoir, most of the food eaten by young white crappie over a 24-h period was eaten during daylight hours, mainly morning and early afternoon. Seasonal food studies indicated that *Daphnia* were important in the diet June through October. *Cyclops* was important September through April. Chironomid larvae and pupae were mostly eaten in April and May. *Bosmina*, *Leptadora*, *Alona*, *Diaptomus*, amphipods, mayfly nymphs, and algae were also eaten seasonally in small quantities.[68]

The preferred diet of age 1 and older white crappies in Lewis and Clark Lake was zooplankton followed by insects, especially *Hexagenia*. White crappie growth after the first season of life appeared related to the population density of *Hexagenia*. Young white crappie fed mainly during daylight hours.[34]

The principal diet of young white crappie in Rend Lake, IL, is zooplankton and insects. Few crappies utilize fish as a major portion of their diet until age 5. At ages 2–6, white crappie feeding on fish have back-calculated total lengths greater than those crappie of the same age feeding on invertebrates.[44]

The feeding rate of white crappies 100–150 mm on *Daphnia pulex* (laboratory study) was significantly reduced when turbidity was elevated from 80 to 160 NTU (Nepholometer Turbidity units). These tests showed no significant difference in the feeding rates of black and white crappie at the turbidity levels tested.[42]

LITERATURE CITED

1. Lee, D.S. et al. 1980.
2. Nelson, W.R. et al. 1968.
3. Hansen, D.F. 1965.
4. Siefert, R.E. 1965.
A 5. Carter, E.R. 1953.
6. Morgan, G.D. 1954.
7. Hansen, D.F. 1943.
8. Grinstead, B.G. 1969.
9. Summerfelt, R.C. 1971.
10. Lambou, V.W. 1958.
11. Purkett, C.A., Jr. 1958.
12. Al-Rawi, T.R. and D.W. Toetz. 1972.
13. Jenkins, R.M. 1953b.
14. Carlander, K.D. 1977.
15. Burr, B.M. and M.L. Warren, Jr. 1986.
16. Gerking, S.D. 1945.
17. Clay, W.M. 1975.
18. Pearson, W.D. and L.A. Krumholz. 1984.
19. Trautman, M.B. 1981.
20. Scott, W.B. and E.J. Crossman. 1973.
21. Pflieger, W.L. 1975b.
22. Goodyear, C.S. et al. 1982.
23. Smith, C.L. 1985.
24. Smith, P.W. 1979.
25. Cook, F.A. 1959.
26. Bulkley, R.V. et al. 1976.
27. Simon, T.P. 1986.
28. Boxrucker, J. 1987.
29. Boxrucker. J. 1986.
30. Nurnberger, P.K. 1930.
31. Scott, E.M., Jr. and J.P. Buchanan. 1979.
32. Walburg, C.H. 1976.
33. Buchanan, J.P. and E.M. Scott. 1979.
34. Siefert, R.E. 1969a.
35. Siefert, R.E. 1968.
36. Whitaker, J.O., Jr. 1974.
37. Killgore, K.J. and J.A. Baker. 1996.
38. Gebhart, G.E. and R.C. Summerfelt. 1978.
39. Linton, T.L. 1961.
40. June, F.C. 1976.
41. Benson, N.G. 1968.
42. Barefield, R.L. and C.D. Ziebell. 1986.
43. Jackson, S.W., Jr. 1957.
44. Heidinger, R.C. et al. 1985.
45. Parrish, D.L. et al. 1986.
46. Schneberger, E. 1972.
47. O'Brien, W.J. et al. 1984.
48. Stroud, R.H. 1949.
49. Moorman, R.B. 1957.
50. Hansen, D.F. 1951.
51. Burr, B.M. 1974.
52. Whiteside, B.G. 1964.
53. Edwards, E.A. et al. 1982b.
A 54. Houser, A. and M.G. Bross. 1963.
55. Baker, J.A. and K.J. Killgore. 1994.
56. Breder, C.M., Jr. and D.E. Rosen. 1966.
57. Becker, G.C. 1983.
58. Siefert, R.E. 1969b.
59. Nelson, W.R. 1974.
60. Ball, R.L. 1972.
A 61. Swingle, W.E. 1965.
62. Thomas, J.A. et al. 2004.
63. Baker, J.A. et al. 1991.
64. Pearson, W.D. and L.A. Krumholz. 1984.
65. Pearson, W.D. and B.J. Pearson. 1989.
66. Gammon, J.R. 1998.
67. June, F.C. 1977.
68. Mathur, D. and T.W. Robbins. 1971.
69. Lewis, S.A. et al. 1971.
A 70. Oklahoma Department of Wildlife Conservation. 1982.
71. Tin, H.T. 1982.
72. Morgan, G.D. 1951a.
73. Holland-Bartels, L.E. et al. 1990.
74. Holland, L.E. and J.R. Sylvester. 1983.
75. Wang, J.C.S. and R.J. Kernehan. 1979.
76. Anjard, C.A. 1974.
A 77. Axon, J.R. 1984.
78. Baldwin, N.W. et al. 1990.
79. Parsons, G.R. 1993.
80. Guest, W.C. et al. 1990.
81. Travnichek, V.H. et al. 1993.
82. Guy, C.S. et al. 1994.
83. Smith, S.M. et al. 1994.
84. Dunham, R.A. et al. 1994
85. Pope, K.L. and D.R. DeVries. 1994.
86. Hayward, R.S. and E. Arnold. 1996.
87. Travnichek, V.H. et al. 1996.
88. Travnichek, V.H. et al. 1997.
89. DeVries, D.R. et al. 1998.
90. Sammons, S.M. et al 2001.
91. Hammers, B.E. 1990.
92. Dubuc, R.A. and D.R. DeVries. 2002.
93. McCollum, A.B. et al. 2003.
94. Spier, T.W. and J.R. Ackerson. 2004.
95. Bunnell, D.B. et al. 2005.
96. Smith, S.M. et al. 1995.
97. Sammons, S.M. et al. 2002a.
98. McKeown, P.E. and S.R. Mooradian. 2002.
99. Sammons, S.M. et al. 2002b.
100. Boxrucker, J. 2002a.
101. Spier, T.W. and R.C. Heidinger. 2002.
102. Maceina, M.J. 2003.
103. Irwin, B.J. et al. 2003.
104. St. John, R.T. and W.P. Black. 2004.
105. Gebhart, G.E. and R.C. Summerfelt. 1975.
106. Cichra, C.E. et al. 1981a.
107. Boxrucker, J. and G. Ploskey. 1988.
108. Crowl, T.A. and J. Boxrucker. 1988.
109. Boxrucker, J. 1992.
110. Slipke, J.W. et al. 2000.
111. Taber, C.A. 1969.
112. Conner, J.V. 1979.
113. Menhinick, E.F. 1991.
114. Etnier, D.A. and W.C. Starnes. 1993.
115. Jenkins, R.E. and N.M. Burkhead. 1994.

116. Ross, S.T. 2001.
117. Boschung, H.T., Jr. and R.L. Mayden. 2004.
118. Mettee, M.F. et al. 2001.
119. Cross, F.B. 1967.
A 120. Environmental Science and Engineering, Inc. 1992.
A 121. Environmental Science and Engineering, Inc. 1995.
A 122. EA Engineering, Science, and Technology. 1994.
A 123. EA Engineering, Science, and Technology. 2004.
124. Hardy, J.D., Jr. 1978.
125. Chatry, M.F. 1977.

126. Hooe, M.L. et al. 1994.
127. Sammons, S.M. and P.W. Bettoli. 1998.
128. Hale, R.S. 1999.
129. Jackson, J.J. and D.L. Bauer. 2000.
130. Sammons, S.M. and P.W. Bettoli. 2000.
131. Range, J.D. 1973.
132. Mabee, P.M. 1995.

* Discussions of spatio-temporal distribution of crappie larvae in the Tennessee River system are based on ichthyoplankton data collected by TVA during the period 1973–1986.

OTHER IMPORTANT LITERATURE

Boxrucker, J. 2002b.
Boxrucker, J. and E. Irwin. 2002.
Bister, T.J. et al. 2002.
Dorr, B. et al. 2002.
Hurley, K.L. and J.J. Jackson. 2002.
Isermann, D.A. et al. 2002.

BLACK CRAPPIE

Pomoxis nigromaculatus (Lesueur)

Robert Wallus

Pomoxis, Greek: poma, "lid, cover" and oxys, "sharp," alluding to the opercles ending in two flat points instead of an ear flap; *nigromaculatus*, Latin: "black-spotted."

RANGE

Black crappie has been so widely transplanted that reconstruction of its native range is difficult. The natural distribution was restricted to the fresh waters, and rarely brackish waters, of eastern and central North America.[33] The species was apparently native from southern MB and QU south along the Atlantic slope to FL, along the Gulf Coast to central TX, north to ND and eastern MT, and east to the Appalachians.[13,33,35,63,124] It has been widely introduced elsewhere in North America[13,35] and also in Europe.[110]

HABITAT AND MOVEMENT

Black crappie reportedly prefer clearer, deeper, and cooler water than white crappie,[13,33,35,63] and the range of black crappie extends farther north;[63] otherwise the biology of black crappie is similar to that of white crappie.[16,17,70] Black crappie are much more abundant in natural lakes with clear waters and vegetation, and the species also does well in some less turbid reservoirs.[70] In MO reservoirs, the black crappie is noticeably more abundant in the arms fed by the clearer streams.[17]

Though black crappie are reportedly less tolerant of turbidity and siltation than white crappie,[16,17,35,37,38] linear regression analysis of growth and length data from stocked juvenile and adult crappies showed that growth of both species was similar across a range of turbidities in IL ponds. In larger bodies of water, crappie might be indirectly influenced by factors that are correlated with turbidity, such as submerged vegetation.[84]

Abundant cover is necessary for growth and reproduction. Common daytime habitat is shallow water in dense submerged aquatic vegetation[13,16,17,33,35,37,38,66,78,110,124,126] and other abundant cover in the form of standing timber, submerged trees, brush, or other objects.[13,16,17,33,124,126] In reservoirs, they may be associated with inundated terrestrial vegetation,[67] but they have also been recorded from areas with essentially no vegetation.[110] Fallen trees are often placed along drawdown zones in reservoirs to serve as "crappie attractors."[71]

Black crappie are often found in quiet, clear,[13,16,17,33,35,37,38,78] warm waters such as large ponds, shallow areas of lakes,[13,16] pools or backwaters of streams, rivers, big rivers, and wetland subsystems.[16,33,35,66,78] They are also reported from streams, creeks (including tidal-fresh creeks), oxbows, lakes, ponds, impoundments, spill pools, lagoons, and bayous.[66,110] They occur over a variety of sub-strates including: sand,[16,33,35,78,110] mud,[16,78] gravel,[16,78,110] silt,[78,110] rubble,[78,110] boulders,[78,110] clay,[78,110] muck,[33,35,110] hardpan,[78] organic debris,[16] and detritus.[78]

Because of their preference for low velocities[16,124] it is assumed that black crappie prefer quiet, sluggish rivers with high percentage of pools, backwaters, and cut-off areas.[124] They are absent from high gradient streams (>2 m/km) and are common in base or low gradient streams (<0.5 m/km).[35] They are not found in montane streams.[71]

Black crappie occupy middle to upper sections of the water column;[67] maximum reported depth is 25 m.[110] Data on depth distribution of various fish species in Norris Reservoir, TN, were collected with gill nets from March to October 1943. Netting was confined to cleared areas, not attractive habitat to black crappie.[107] Depth distribution, through time, of the black crappies collected was as follows:

Date	Number Collected	Depth
March and April	4	6.1–13.7 m
May and early June	7	Upper 3 m
Mid-June	3	4.6–8.8 m
August	1	3.4 m
August	26	7.3–11.3 m
Early October	5	1.2–24.7 m

In Missouri River mainstem reservoirs, adult black crappie were commonly netted in water 0–6 m deep, while white crappie were most commonly collected in the 6- to 15-m zone. Similar distribution patterns for crappie were reported from Clear Lake, IA.[49]

The recorded temperature range for black crappie is 4–32.5°C.[110] Coutant (1977) summarized reports of temperature preference data: large northern (WI) black crappie preferred water temperatures of 27–30°C; upper avoidance temperature for small black crappies in a laboratory was about 30°C and lower avoidance temperatures were 25.5–26.5°C. In other laboratory studies, preferred temperatures of adult black crappie varied little between seasons with a range of 20.5 to 22.2°C.[20]

In 90% of the streams where adult black crappie were found in the Mississippi Valley and along the east coast, the mean weekly summer (July and August) temperatures were 23–32°C, with a mean of about 26°C.[124] In summer, black crappie in Norris Reservoir were found in water 21–27°C with dissolved oxygen content above 1.5 ppm.[108]

In backwater lakes of the upper Mississippi River during winter, when dissolved oxygen (DO) was above 2 mg/L, black crappie selected areas with no detectable current and water temperatures greater >1°C. Areas with water temperature <1°C and current velocity >1 cm/s were avoided. If DO fell below 2 mg/L, they moved to areas with higher DO, despite water temperatures of 1°C and lower current velocities of 1 cm/s.[42]

Black crappie were rarely found in brackish water in Canada,[33] but are reported from waters that varied in salinity from 1.32 to 14.9 ppt.[110,124] They tolerate acidic waters and enter Dismal Swamp and occasionally tidal freshwaters of VA.[71]

Black crappie was classed as a species of slightly alkaline eutrophic waters with limits of 900 ppm total alkalinity, 250 ppm carbonates, and 200 ppm potassium and sodium.[63,93]

Black crappie population characteristics differ among ecosystems. In SD, natural lakes typically had populations with low density, unstable recruitment, fast growth rates, and high condition factors. Small impoundments (≤40 ha) had populations with high density, more stable recruitment, slow growth, and low condition factors. Population characteristics in large impoundments (>40 ha) were typically intermediate between those of natural lakes and small impoundments.[43]

Density of black crappie in FL lakes was highly variable but positively related to zooplankton abundance. Density was unrelated to lake surface area, chlorophyll *a* concentrations, chlorophyll *a* values adjusted to include nutrients encompassed in macrophytes, or macrophyte abundance. Florida lakes most likely to sustain high catch rates of harvestable-sized black crappie were relatively large (>100 ha), contained high levels of chlorophyll *a*, and had relatively high black crappie densities (>200 fish/ha). Small (<10 ha), oligotrophic (chlorophyll *a* concentrations <3 mg/L) lakes exhibited low black crappie occurrences. Growth or density of black crappie was not strongly affected by the presence or density of aquatic macrophytes.[52]

Though local anglers and fishing guides from Weiss Lake, AL, had voiced concerns that natural reproduction of striped bass in the lake had negatively impacted the crappie (both species) fishery, subsequent study indicated that the crappie population of the lake was little changed between 1990 and 1999 and that no negative impact was indicated.[92]

Black crappie is a gregarious species that travels in discrete, small,[110] or moderately large[33] schools;[33,78,110] during the day a school may be located near submerged cover in shallow (1–3 m deep) water, but during the evening the schools move into deeper water where they remain within 1–3 m of the surface.[78] In fall after the temperature falls, black crappie move to deeper water.[110]

Before spawning, black crappie form into schools and move into shallow water,[36,65] where they are caught by anglers in large numbers around heavy cover.[65] Movements also include migrations; individuals living in large bodies of water may migrate several kilometers. Evidence of winter movement under the ice in northern populations is noted.[78]

In Lake Chautauqua, IL, black crappie were more mobile than white crappie. In a period of 121 days, one tagged black crappie (183 mm TL) left the lake and moved 35 miles up the Illinois River. Another tagged specimen was netted 3 miles from the point of release in the lake 147 days after being tagged.[125] Black crappie moved considerably in Norris Reservoir, TN, with a range of distance traveled of 0–20 miles[102,104] and average movements of 5.4[102] and about 6.5 miles[104] reported.

Black crappie and white crappie are popular sport fishes, but because of erratic recruitment they are difficult to manage and are not recommended for stocking in small impoundments or ponds. Some authors suggest that, because the recruitment of hybrid crappies is low and their growth rates similar to the parent species, they provide a viable alternative for stocking into small water bodies.[41]

DISTRIBUTION AND OCCURRENCE IN THE OHIO RIVER SYSTEM

Black crappie have been reported from throughout the length of the Ohio River since 1800.[9,10] Data from lock chamber surveys in the Ohio River indicate no significant change in black crappie abundance during the period 1957–2001,[46] and another report indicated no obvious change in black crappie distribution in the river between 1970 and 1984.[9]

Ohio River Ecological Research Program collections at five electric generating plants on the Ohio River (W.H. Sammis Plant, ORM 54; Cardinal Plant, ORM 76; Kyger Creek Plant, ORM 260; Tanners Creek Plant, ORM 494; Shawnee Steam Plant ORM 946) yielded black crappie most years. During the period 1981–1993, black crappie were collected in very low numbers at each site most years, usually accounting for less than 1% of total catch with all gears.[121,122] Occurrence in electrofishing catches during the period 1981–2003 was sporadic and usually very low.[123]

Black crappie are present in the Allegheny River system of NY.[38] They are not very abundant in the Wabash River, IN[11] and IL,[37] but they are a regular component of electrofishing catches.[11]

Black crappie may have been almost or entirely absent from the Ohio River drainage of OH prior to 1925. Since then, however, fairly stable populations have occurred in those ponds, lakes, and impoundments where competition from white crappie was not too severe, where the waters were usually clear, and where there was an abundance of submerged aquatic vegetation and the sandy or muck bottoms.[35]

Black crappie are reported statewide and in all drainages of KY[15] where they are found occasionally in reservoirs, ponds, and rivers. They are less common than white crappie.[16] In VA, the species probably is not indigenous to the upper portions of the Tennessee and Big Sandy drainages, nor to any of the New River drainage, based on the lack of early records and avoidance by the species of montane streams.[71]

Black crappie are present in the Tennessee River drainage of NC,[64] TN,[70] AL,[65,66] and MS.[67] In TN, they occur sporadically statewide and do relatively well in less turbid reservoirs of the Tennessee and Cumberland Rivers. They are uncommon in small tributary streams. Recent studies show that black crappie may be supplanting white crappie in abundance in TN reservoirs that have developed extensive growths of aquatic vegetation.[70,128]

In the late 1980s, the Tennessee Wildlife Resources Agency began supplemental stocking of crappie into reservoirs in an attempt to improve crappie fisheries. The majority of the crappie stocked in TN are a morphological variant of the black crappie, the black-nosed crappie, which is characterized by a black predorsal stripe.[81,82]

SPAWNING

Location

Black crappie spawn in freshwater.[117] Males build nests[8,33,36,38,117,124] in shallow (0.3–0.6 m) areas near shore[36,66,67,110] over substrates of gravel,[33,66,110] sand,[33,38,66,110] clay,[66,110] or mud,[33,66,110] most often associated with some kind of vegetated cover[33,38,66,67,110,117] or beneath undercut banks.[33,66,110]

In the Great Lakes, spawning occurs in sheltered near-shore areas, including bays, harbors, marshes, sloughs, lagoons, creek mouths, and lower reaches of rivers; usually near vegetation beds or plant roots.[36]

Males clear shallow depressions or just clear a section of the bottom in water 2.5–6.1 m deep, where there is some vegetation. Nesting areas may be cleared in the protection of an undercut bank. Nests are 20–38 cm in diameter and colonial but 1.5–1.8 m apart.[33,38,78]

One nest was observed that was a small circular area on hard clay substrate that had been cleared of silt. This nest was in 10 cm of water and was 30 cm from an overhanging ledge.[109]

In Lake Oahe, SD and ND, black crappie most commonly nested under overhanging ledges on soft substrate in the upper reaches of protected embayments.[45]

Season

Black crappie spawn early, constructing nests over the period from late February to July,[66,71,117] with earliest spawning in the south.[66,71] Northern populations generally spawn during May and June,[8,12,33,36,38,45,78,109,112,119] but in colder seasons spawning may be delayed into July.[8,33,36,38,78]

Spawning occurs from early May to early July in IL and IN[119] and from late March to early June in Buckeye Lake, OH.[99]

The reproductive season for black crappie in the southeastern United States is from late February to early May.[30,64–67,71] Spawning occurs mostly in April in VA,[71] early April through early June in NC,[64] April and May in AL,[65] and February to March in MS.[67] In a FL lake, hatching occurred over a 12-week period from early March to mid-May, based on daily otolith rings.[30] Off-season spawning is also reported from Orange Lake, FL; black crappies <60 mm TL were observed spawning in December.[54]

Temperature

In Canada, black crappie spawn at temperatures of 19–20°C.[33,110] In the Great Lakes, reports of spawning temperature range from 16.1 to 26.1°C.[36] In WI, activity may begin at 14.4°C, but favorable spawning temperatures range from 17.8 to 20°C.[63,78]

Water temperatures during the spawning season in Buckeye Lake, OH, reportedly ranged from 4.4 to 15.6°C,[99] but records of spawning at a minimum temperature of 4.4°C[99] and a maximum of 28°C[117] are questioned.[110]

Fecundity

Development of black crappie ovaries takes place in the fall and fish are ready to spawn early in the year, with little further development.[99] This species is highly fecund, with numbers of eggs variously reported between 3,000 and 188,000.[63,71,78,110]

In Canada, 3- to 4-year-old black crappie females 195–230 mm had 26,700–65,520 eggs with an average of 37,796.[33] Other reports summarized by Becker (1983) include: in IN, 3- to 4-year-old black crappie averaged 33,700 and 41,900 eggs, respectively; Lake Wingra, WI, females 163–180 mm TL had an average of 11,400 eggs (7,900–19,000).[78]

In three AL impoundments, black crappie fecundity (both species) and fish length were positively correlated. Both ovary weight (OW) and GSI were significant predictors of total fecundity (OW: fecundity = 3.996 (OW) + 323.1, r^2 = 0.75, n = 465; GSI: fecundity = 20,585 (GSI) – 23,795, r^2 = 0.54, n = 289; both P = 0.0001).[7]

Sexual Maturity

Sexual maturity of black crappie reportedly occurs in the second to fourth year in northern populations[3,33,110] and generally at age 2 or 3 elsewhere.[66,71,94,124]

In Lake Oahe during the period 1963–1968, more than 50% of male and female black crappie were mature at age 2, 83% of males and 79% of females were mature at age 3, and all males and females were mature at age 4.[94]

Average calculated TL for mature age-2 black crappie from Lake Oahe was 167 mm for males and 176 mm for females.[94] Total length at maturity for black crappie is also reported from 175 to 200 mm[124] and at about 150 mm for males.[110]

Spawning Act

As temperature approaches 18°C in the spring, male black crappie (usually smaller males first)[63] move into the shallows of river backwaters or littoral areas in lakes and reservoirs to establish territories and construct nests.[33,63,124] Spawning behavior is very similar to that of white crappies,[33,63,70] except that nests are more associated with vegetation where possible,[63,70] and black crappie may have a preference for cleaner substrates.[70] Nests are usually built after attracting females to the area, and although each male establishes and guards a territory, the nests are often close together.[63]

Female black crappie probably spawn with different males in more than one nest[33,78] and may exude eggs several times during the spawning period.[78]

Males guard the nest, guard and fan the eggs until they hatch, and guard the young for a short time.[33]

Normally, a male black crappie guarding his nest is quite belligerent and will attack intruders or objects within his territory. He guards the nest and protects the young until they start to feed.[78]

Black crappie and white crappie co-occur within drainages in many parts of their ranges, suggesting that the two species occupy discrete niches. However, under perturbed ecological conditions such as those created by impoundments, effective ecological isolation may break down resulting in hybridization.[6] Natural hybridization between black crappie and white crappie occurs,[3,33,35,71] and is documented in reservoirs of the southeastern United States.[4]

Extensive natural hybridization between black crappie and white crappie occurs in Weiss Lake, AL.[2,89] First-generation (F_1) hybrids have faster annual growth than black crappie and white crappie in this reservoir.[89] Factors that may have contributed to hybridization include failure to identify specific mates due to low water clarity, short and overlapping spawning times, and fluctuating water levels. Also, Weiss Lake is located at the interface of sympatric and allopatric crappie populations and this has been associated with interspecific hybridization among other fishes.[2]

EGGS

Description

Fertilized, water-hardened black crappie eggs are whitish,[33,110] adhesive,[8,33,36,110] demersal,[8,33,36,110] and slightly less than 1 mm in diameter.[33,110] Two hundred fertilized, water-hardened eggs from NC averaged 0.93 ± 0.082 mm.[24]

Incubation

Adhesive eggs incubate for 3–10 days on vegetation above the nest bottom or on roots at the bottom of the nest.[36] Incubation is reported in 3–5 days at ambient temperatures[33] and also in 48.1–67.8 (mean 57.5) h at 18.3°C.[63]

Development

No information.

YOLK-SAC LARVAE

See Figure 115.

Size Range

Black crappie hatch at 2.3 mm TL in the Missouri River[111] and at 2.0 mm TL in LA.[115] Hatching lengths of 4–5 mm are also reported.[8]

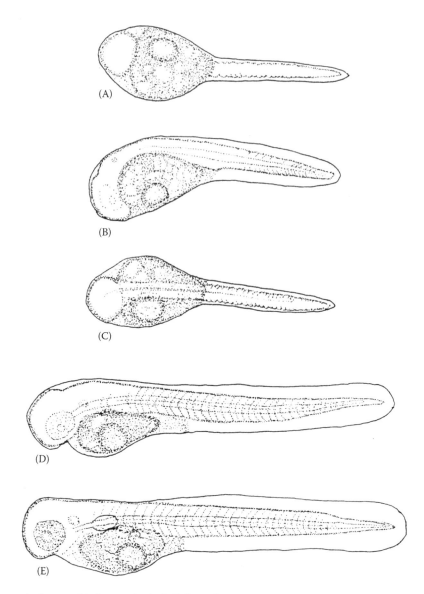

Figure 115 Development of young black crappie. A–C. Yolk-sac larva, 2.23 mm TL: (A) ventral view; (B) lateral view; (C) dorsal view. D–E. Yolk-sac larvae: (D) 3.15 mm TL; (E) 3.42 mm TL. (A–C reprinted from Figure 3, D–E from Figure 4, reference 115.)

In the Missouri River population, yolk is absorbed before fish reach 5 mm TL.[111] In LA, the yolk is absorbed by 3.7 mm, but traces of yolk are sometimes found in individuals up to 5.0 mm.[115]

Myomeres

Natal water temperature can induce variation in myomere numbers; higher temperatures cause the anus to development more posteriorly, influencing the number of preanal and postanal myomeres.[90]

In Missouri River:

2.5–4.9 mm TL. Preanal mode 10–11; postanal mode 19–21; total mode 29–31.[111]

In LA:

2.0–3.7 mm TL. Preanal 10–13, usually 11–12; postanal 19–23, usually 20–21; total 31–34, usually 31–33.[115]

Morphology

2.0–3.7 mm TL. At hatching (2.0 mm), body is coiled around the yolk sac; yolk is bulbous and granular with a single oil globule either centrally located or anteriorly placed. Body is straightened by 2.25 mm[115] and gut is coiled at less than 3.0 mm.[114] At about 3.5 mm, the air bladder becomes apparent; measurement of eye to air bladder distance becomes possible at this size. The first skeletal structure, the cartilaginous cleithrum, appears at about 3.5 mm followed closely by chondrification of the premaxillary, maxillary, dentary, frontal, supraoccipital, and branchiostegals. At 3.7 mm, a discrete, bulbous mass of yolk is no longer discernible.[115]

3.7–5.0 mm TL (in LA). If present, only traces of yolk remain.[115]

Table 147

Morphometric data for young black crappie from Erickson Lake, WI.
Measurements are expressed as range of lengths (mm).

| | TL (mm) Groupings | | | | | | | |
N	4.5–5.0 10	7.5–8.0 10	8.0–8.5 10	10.5–11.0 10	11.0–11.5 10	13.5–14.0 10	14.0–14.5 10	16.5–17.0 10
Character								
HL	0.6–0.8	1.3–1.5	1.5–1.75	2.15–2.5	2.25–2.5	3.1–3.5	3.3–3.6	3.9–4.5
ED	0.25–0.35	0.4–0.5	0.45–0.55	0.7–0.8	0.75–0.9	0.95–1.05	1.0–1.1	1.3–1.4
PreAL	1.6–1.85	2.9–3.3	3.15–3.6	4.15–4.7	4.4–4.85	5.3–5.7	5.6–6.1	6.7–7.1
GBD	0.6–0.8	1.15–1.3	1.35–1.5	1.75–2.25	2.0–2.3	2.5–3.05	2.8–3.1	3.3–3.7
SI[a]	0.5–0.65	0.65–0.8	0.5–0.8	0.4–0.8	0.5–0.75	0.45–0.75	0.5–0.6	0.5–0.8
ABL[b]	0.3–0.45	0.75–0.95	0.85–1.1	1.4–1.6	1.45–7.75	2.0–2.5	2.3–2.6	3.0–3.4
ABH[c]	0.15–0.25	0.4–0.55	0.45–0.6	0.6–0.75	0.65–0.8	0.85–1.0	0.9–1.2	1.3–1.5

[a] Straight intestine; [b] air bladder length; [c] air bladder height.

Source: Constructed from data presented in Table II, reference 112.

Table 148

Statistics used to describe regressions of body measurements on TL
for young black crappie from southeastern LA.[115]

Body Measurement	mm TL (X)	N	Regression Equation	R^2
Eye diameter	3.00–19.67	252	$Y = 0.261 + 0.0042X^2$	0.979
Distance from eye to air bladder	3.41–15.87	205	$Y = -0.180 + 0.191X - 0.004X^2$	0.981
Air bladder length	3.42–19.67	211	$Y = 0.120 + 0.011X^2$	0.976

4.5–5.0 mm TL. Usually a single oil globule is present; rarely several oil globules. Oil is usually positioned anteriorly on the right side of the future stomach.[110,112]

Mouth and operculum are functional by 4.8 mm.[22]

Morphometry

Morphometric measurements of young black crappie from Erickson Lake, WI, are presented in Table 147.

Chatry (1977) provided statistics for describing regressions of body measurements on TL for young black crappie from southeastern LA (Table 148). Plots of the following regressions can also be found in Chatry's thesis:[115]

Eye diameter on TL
Distance from eye to air bladder on TL
Air bladder length on TL
Dorsal fin length on TL

Distance from eye to dorsal fin origin on TL
Dorsal fin length/distance from eye to dorsal fin origin on TL[115]

Fin Development

2.0–3.7 mm TL. Median finfold present dorsally, continuous around urostyle, and present ventrally from tip of urostyle anteriorly to posterior margin of the yolk.[115]

4.5–5.0 mm TL. Origin of dorsal finfold is just above anus or slightly anterior to this point; finfold is continuous around urostyle and present ventrally to the intestine.[112]

Pigmentation

3.0–4.0 mm TL. Eyes gradually become pigmented.[115]

4.5–5.0 mm TL. Transparent.[110] Pigment first evident over air bladder, on sides of head, on anterior portion of stomach, and mid-ventrally.[110,112]

POST YOLK-SAC LARVAE

See Figures 116 and 117.

Size Range

Yolk is absorbed at 3.7–5.0 mm TL.[111,115]

In LA, black crappie are often 24–26 mm TL before the 7th and/or 8th dorsal fin spines are added.[115]

Myomeres

In Missouri River:

5.0–9.9 mm TL. Preanal mode 10–13; postanal mode 19–21; total mode 31–32.[111]

10.0–16.5 mm TL. Preanal mode 13–14 (usually 14); postanal mode 18–19 (usually 18); total mode 32.[111]

In WI:

8.0–11.0 mm TL. Preanal 11–12; postanal 19–22; total 31–34.[112]

11.0–14.0 mm TL. Preanal 12–13; postanal 18–19; total 31–32.[112]

14.0–17.0 mm TL. Preanal 13–14; postanal 16–17; total 30–31.[112]

In LA:

3.7–8.0 mm TL. Preanal 9–13, usually 10–12; postanal 19–23, usually 21–22; total 30–35, usually 31–33.[115]

7.7–11.0 mm TL. Preanal 10–13, usually 11–13; postanal 19–23, usually 20–21; total 31–35, usually 32–33.[115]

> 11.0 mm TL. Preanal 11–13, usually 11–12; postanal 19–21, usually 19–20; total 31–34, usually 32.[115]

Morphology

3.7–8.0 mm TL. Except for an increase in TL and the proportional increase in size of various structures, the external morphology changes little. The most obvious morphological change is in the shape of

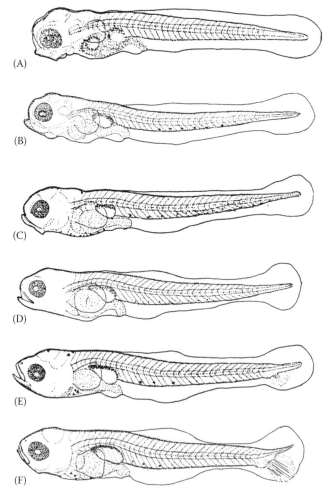

(A)

(B)

(C)

(D)

(E)

(F)

Figure 116 Development of young black crappie. A–F. Post yolk-sac larvae: (A) 3.99 mm TL; (B) 4.85 mm TL; (C) 5.97 mm TL; (D) 6.98 mm TL; (E) 7.95 mm TL; (F) 8.8 mm TL. (A–B reprinted from Figure 6, C–D from Figure 8, and E–F from Figure 9, reference 115.)

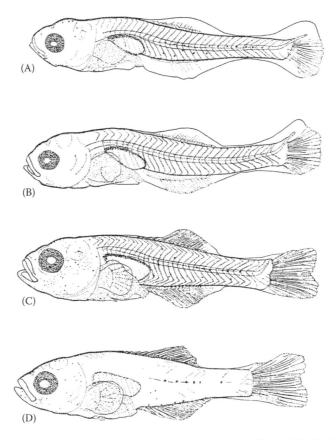

Figure 117 Development of young black crappie. A–D. Post yolk-sac larvae: (A) 9.96 mm TL; (B) 11.01 mm TL; (C) 12.42 mm TL; (D) 14.73 mm TL. (A and B reprinted from Figures 10 and 11, reference 115; C and D reprinted from Figures 13 and 14, reference 115.)

the mouth and head. The premaxillaries become well defined and protractile. The head tapers toward the snout and the cleithrum becomes externally evident.[115]

8.5 mm TL. The cleithrum, dentary, and maxillaries are ossified. Coincident with the ossification of the head parts is the formation of cartilaginous neural and haemal spines, hypurals, and caudal rays.[115]

16.0–26.0 mm TL. Scales first evident at 16 mm at the base of the caudal fin (Figure 118), but mean length for first appearance of scales is 17.7 mm. At 20–22 mm TL, a band of scales is present from the base of the caudal fin, ventral to and along median myosepta, anteriorly to a point above the base of the pectoral fins (Figure 118).[113] By 26.0 mm, scales are formed laterally over most of body, present anteriorly to the base of the pectoral fins, to the opercle, and along the dorsum almost to origin of spinous dorsal fin (Figure 118).[113]

Morphometry
See Tables 147 and 148 and information presented in Yolk-Sac Larvae section (page 383).

4.0–13.0 mm TL. The distance from the posterior margin of the eye to the anterior-most portion of the air bladder is <15% TL.[115]

> 16.0 mm TL. Length of the base of the dorsal fin is greater than the distance from the eye to the dorsal fin origin.[115]

Fin Development
7.7–8.0 mm TL. In LA, the first complete caudal fin ray occurs; some 8.0-mm individuals were found without caudal fin rays.[115] At 8.0 mm, the median finfold has broad dorsal and ventral depressions in area of future caudal peduncle; finfold is opaque in areas of future caudal, dorsal and anal fins; pectoral fins are evident, but without rays; and urostyle flexion is evident.[110,112]

8.5–9.0 mm TL. First incipient fin rays visible in caudal fin.[111,115]

9.4–10.9 mm TL. First pterygiophores and incipient fin rays form in soft dorsal fin and anal fin.[111,115] At 10.8 mm the urostyle is completely flexed.[115]

11.0–12.0 mm TL. Finfold no longer continuous between dorsal and caudal fins and caudal and anal fins; pelvic buds first evident.[110,112,115] Notochord has bent dorsally and rays have developed in the entire caudal fin.[112] Rays are visible in dorsal, anal,[112] and pectoral[115] fins. First spines appear in dorsal and anal fins.[111,115] Vertebral column is ossified by 11.2 mm.[115]

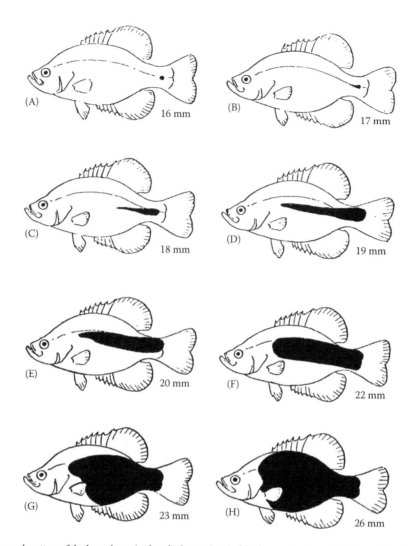

Figure 118 Sequence and pattern (blackened area) of scale formation in black crappie from OK. (Reprinted from Figure 2, reference 113, with publisher's permission.)

In LA, between 11.0 and 11.6 mm, a full complement of principal caudal rays is present; at 11.0 mm, no rays are visible in pectoral fins.[115] Interneurals appear at about 12.0 mm.[115]

12.4 mm TL. Caudal fin is bilobed.[115]

13.0 mm TL. Basipterygia of the pelvic fins usually formed.[115]

14.0–15.0 mm TL. Incipient rays visible in pectoral fins and pelvic fins.[110,112] Formation of interneurals is completed.[115] Five dorsal spines are present in 85% of black crappies examined.[11] By 14.7 mm all median fin ray elements are completed.[115]

17.0 mm TL. Remnant of the median finfold is still present just anterior to the anus.[112]

24.0–26.0 mm TL. Seventh and/or eighth dorsal spines added (in LA), completing fin development.[115]

Pigmentation

3.7–8.0 mm TL. Eyes are heavily pigmented. Pigmentation is present on the dorsal wall of the air bladder, in a ventrolateral row, and occasionally on the head and the visceral mass. Air bladder pigmentation is more intense on black crappie larvae than on white crappie.[115] In northern populations at lengths less than 8.0 mm, a prominent subsurface melanophore is visible between dorsal myomere bundles just posterior to head.

8.0–11.0 mm TL. At 8.0 mm, chromatophores are present on both sides of myomeres below the horizontal myosepta, and in some individuals, a recognizable ventrolateral line of pigment is visible.

A few chromatophores develop in the future caudal fin and the ventrolateral line of pigment becomes well developed.[110,112] In LA, ventrolateral pigment occurs on most individuals; head and visceral pigment intensifies.[115]

11.0–14.0 mm TL. A double row of pigment is present along base of anal fin; scattered melanophores are present from anal fin base to the caudal fin. A lateral pigment line is usually visible. The ventrolateral pigment line is less conspicuous than in smaller individuals and is sometimes absent. There is usually no pigment posterior to the head or above the horizontal myoseptum. Melanophores are present on top and sides of head, upper and lower jaws, and on sides of stomach.[110,112]

12.0–16.0 mm TL. In LA, melanophores are often scattered over the entire body and on the median fins.[115]

14.0–17.0 mm TL. A double row of pigment lines is present at base of dorsal fin, a few scattered melanophores are present posterior to the dorsal fin, and in some individuals, pigment is visible on the dorsal and anal fins.[110,112]

JUVENILES

Size Range

In LA, black crappie are 24–26 mm TL before the 7th dorsal spine appears.[115]

Sexual maturity is reported at about 150 mm TL for males.[110] In the Missouri River, possibly mature age-2 males and females averaged 167 and 176 mm TL, respectively.[94]

Morphology

Branchiostegal rays 7;[33,70] lateral line scales 33–44;[33,38,65,66,70,71,78] vertebrae 31–34;[33,38,70] pyloric caeca 8;[33] gill rakers 27–32, long and slender.[38,70]

In six rivers in the south central United States, reported counts for caudal vertebrae were 18–20, usually 18–19; total vertebrae counts were reported as 32–34, usually 32–33; interneurals were usually 6.[115]

Reports of patterns and size ranges for scale development are varied.[113,116] As described above (Figure 118), in OK, first scales appear at the base of the caudal fin and subsequent development proceeds anteriorly just ventral to the lateral line with lateral extensions ventrally and dorsally; squamation is nearly complete by 26 mm TL (Figure 118) and completely scaled black crappie averaged 30 mm.[113]

In CT, scales first appear at larger sizes than in OK and a different pattern of scale development is reported. Squamation begins at 28 mm TL and is complete by 40.0 mm. At 28–30 mm, scales are apparent posterior to the dorsal and anal fins and anterior to the pelvic fins (Figure 119). Scale formation then extends posteriorly to the base of the caudal fin and anteriorly along the dorsal and ventral midlines with lateral extension to the lateral line. Simultaneously, scale development proceeds posteriorly and dorsally from the region in front of the pelvic fins. Lateral line scales appear initially at the base of the caudal fin and in the most anterior region of the lateral line (30–32 mm, Figure 119). Body scale development was completed by 36 mm, head scales by 38 mm, and opercle scales by 40 mm. The largest individual without scales was 34 mm and the smallest fully scale fish was 37 mm.[116]

Morphometry

The length of the dorsal fin base is greater than the distance from the eye to the dorsal fin origin.[115]

Fins

The spines and rays of the black crappie dorsal fin are completely joined and so graduated as to appear as one fin;[33,78] dorsal spines 6–9, usually 7–8; dorsal fin soft rays 14–18, usually 14–16. Anal fin has 5–7 graduated spines and 16–19 soft rays. Pectoral fins are elongate with 13–15 rays. Pelvic fins are thoracic, originating under the pectoral fins, with 1 spine and 5 rays.[33–35,38,66,67,70,71,78]

In six rivers in the south central United States black crappie usually had 7–8 (7–10) dorsal spines.[115] In VA[71] and TN,[70] dorsal spines usually number 7–8 (rarely 6).

Pigmentation

19.7–85 mm SL. Narrow primary and secondary bars of melanistic pigment develop. There appears to be much variability in the position of the first-appearing primary bars. No single bar could be recognized as the first to appear, but the primary bars are added from the center of the body bi-directionally (Figures 120A–C). Secondary bars develop between some of the primary bars (Figure 120D); the order and position of secondary bar formation are apparently random. The primary bars become broken dorsoventrally and distorted laterally like the secondary bars, to produce the characteristically mottled appearance of adult black crappies (Figure 120E and 120F).[129]

47.5 mm TL. Lateral pigment arranged in indefinite vertical bands.[110]

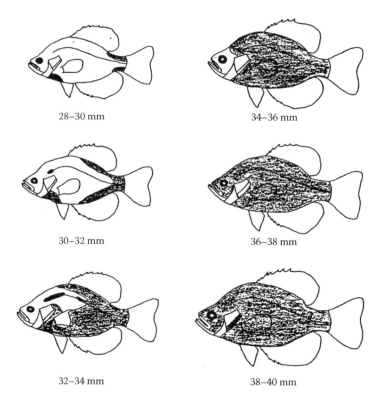

28–30 mm 34–36 mm

30–32 mm 36–38 mm

32–34 mm 38–40 mm

Figure 119 Sequence and pattern (blackened area) of scale formation in black crappie from CT. (Reprinted from Figure 1, reference 116, with publisher's permission.)

TAXONOMIC DIAGNOSIS OF YOUNG BLACK CRAPPIE

Similar species: other centrarchids, especially white crappie and flier.

For characters used to distinguish young crappie from the young of other centrarchid genera, see the provisional key to genera (pages 27–29).

See "Taxonomic Diagnosis" section in flier species account (page 85) for discussions concerning differences between young fliers and crappies.

For taxonomic discussions about black and white crappie, see information presented in the introduction to this chapter (page 351–354).

ECOLOGY OF EARLY LIFE PHASES

Occurrence and Distribution (see Figure 121)

Eggs. Black crappie eggs incubate while attached to aquatic plants, sometimes 5–10 cm above the bottom of the nest; also attach to fine roots on the nest bottom.[36,110] The male parent guards and fans the eggs until they hatch.[33]

Yolk-sac larvae. Black crappie yolk-sac larvae remain in the nest guarded by the male parent until they swim up.[33,110] In Lake Opinicon, ON, black crappie fry began free-swimming at about 4.8 mm TL.[22]

Post yolk-sac larvae. Black crappie larvae migrate from their nest area into shallow inshore waters soon after hatching.[110,126] In the Great Lakes, young black crappie are found close to spawning grounds in sheltered near-shore areas, including shores, bays, creek mouths, and weed beds over substrates of mud, sand, silt, and vegetation and also in the pelagic zone.[36]

An understanding of larval distribution is essential for developing an appropriate sampling design to monitor larval abundances. A SD study designed to assess spatial differences for black crappie larvae in an impoundment and a natural lake showed that spatially stratified sampling sites more effectively collected black crappie larvae in an impoundment, whereas random sites were probably more appropriate in natural lakes.[44]

In a Great Lakes coastal wetland, black crappie larvae constituted a major component of ichthyoplankton samples in April and May. Though generally occurring in river channel collections prior to their appearance in bayous, black crappie

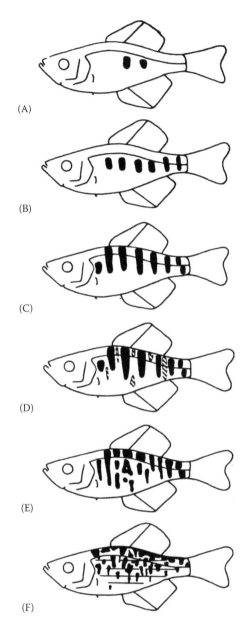

(A)

(B)

(C)

(D)

(E)

(F)

Figure 120 Development of young black crappie — schematic drawings showing melanistic pigment patterns of young black crappie. (A) 19.7 mm SL; (B) 17.3 mm SL; (C) 24.5 mm SL; (D) 24.5 mm SL; (E) 26.0 mm SL; (F) 85 mm SL. (A–F reprinted from Figure 11, reference 129, with publisher's permission.)

larvae were much more important components of bayou collections than those from river channels.[27]

Larval black crappie occur in ichthyoplankton collections from the upper Mississippi River from early to mid-May. They are most abundant in backwaters but tend to drift into the main channel at dusk.[12]

Larval fish were collected from river channel and flooded, bottomland tupelo and oak forest habitats in a study of fish use of flooded wetlands in the Cache River system, AR. Three gears were used:

light traps, ichthyoplankton nets, and a pump. Black crappie larvae were collected in low numbers and usually occurred in light trap samples from oak forest habitat.[47,50]

Ichthyoplankton collections made in the Ohio River from ORM 569 to 572 yielded crappie larvae. Both black and white crappie larvae were present, but most larvae were identified as white crappie. Crappie first appeared in ichthyoplankton tows in June at water temperature of 17.2°C. Surface drift was indicated for 55% of the crappie larvae collected, but they occurred in mid-depth and bottom samples as well.[39]

Larval crappie were consistently collected during ichthyoplankton surveys at sites (ORM 54, 77, 260, 494, and 946) along the Ohio River. Composition was assumed evenly distributed between black and white crappie since both species are common throughout the length of the river. Larval crappie were rarely abundant in ichthyoplankton tows and no apparent trends in density were evident between locations or survey years (1981–1993).[120,121] Peak densities typically occurred from mid-May to July. Initial appearance of crappie occurred from early May to mid-June for tow samples and from late May to July for seine collections. Water temperatures at initial appearance were extremely variable between locations ranging from 14 to 33.7°C.[120]

During the period 1973–1986, larval crappie were frequent and often abundant in TVA ichthyoplankton samples taken from the Tennessee River system. Both black and white crappie are present in most of the waters sampled by TVA, but, at the time, crappie larvae were not routinely identified to species. Crappie normally began to appear in ichthyoplankton samples in early April but the earliest record was March 22, 1976 at the Bellefonte Nuclear Plant site on Guntersville Reservoir, AL. They occurred in samples as late as early August. Specimens ranged from 3 to >30 mm TL, but most were in the 4- to 10-mm range.*

Crappie larvae (both species assumed present) showed increases in relative abundance in ichthyoplankton samples each consecutive season during the first 3 years of impoundment (1976–1978) of Little Bear Creek Reservoir, AL. Crappie larvae first appeared in samples in mid-April and remained dominant in collections until early May. Mean water temperature associated with first occurrence was about 15.2°C.[18]

Peak densities of larval crappie (<20 mm TL, both species assumed present) increased progressively during each year for the first 3 years following impoundment (1976–1978) of Normandy Reservoir, Duck River, TN. Crappie larvae first appeared in collections on April 26, 1976 and were the most abundant taxon in April 1977. In 1978 crappie larvae

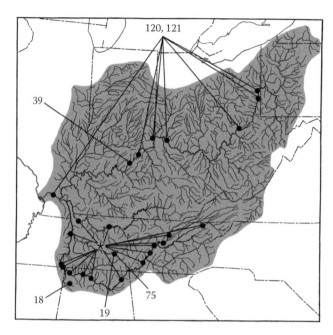

Figure 121 General distribution of black crappie in the Ohio River system (shaded area) and areas where early life history information has been collected (circles). Numbers indicate appropriate references. Asterisk indicates TVA collection locations.

constituted 49% of the total ichthyoplankton catch and the periods of greatest larval density (May 10 and May 24) were dominated (90%) by crappie larvae (780 and 760/1000 m³, respectively).[19]

The collection of larval white and black crappie over a few weeks in spring in Normandy Reservoir, TN, accurately demonstrated the presence or absence of strong year-classes much earlier and with less effort than traditional sampling techniques such as fall gillnetting or cove sampling. Larval crappie (both species) recruited to a 1 × 2–m neuston net at 9 mm TL; few over 15 mm were collected.[75]

Black crappie larvae were collected in low numbers in passive drift-net sampling in the James River, a shallow prairie river in SD.[29]

Black crappie larvae, though dominant in collections from the pelagic region of a WI lake, were never very abundant in open water. Small-mesh nets caught smaller black crappie and in greater numbers than larger mesh nets, but the larger mesh nets caught larger black crappie.[112]

In Orange Lake, FL, during the period June 1983 to June 1984, larval fish collections were made using a conical tow net and a floating light trap. Black crappie were collected in greatest numbers in late winter and spring (late February through April). Black crappie larvae 2–8 mm TL were collected in open water and in vegetated zones dominated by panic grass and floating-emergent vegetation. Relative abundance was highest near floating-emergent vegetation. Black crappie larvae and juveniles 9–40 mm were also collected in low numbers from panic grass and floating-emergent habitats; none in this size range were collected in open water. No crappie larvae were collected in vegetated zones dominated by hydrilla.[32]

Juveniles. Black crappie juveniles (>25 mm TL) are most abundant in shallow, vegetated areas with cover and food;[59,124] also in the lower courses of creeks and at the mouths of streams.[110] In Lake Opinicon, ON, age-0 and age-1 black crappie were found in shallower water than older fish.[126]

Age-0 black crappie had pelagic life history phases during the period 1988–1990 in Lake Mendota, WI, as determined by purse-seine sampling. Young black crappie were caught at all sites across the pelagic transect. The duration of their pelagic residence varied from 2 to 6 weeks among years. Concentrations at deeper stations were observed early in their pelagic residence with concentrations at the shallowest station toward the end of their pelagic residence. When age-0 black crappie disappeared from the pelagic zone, they appeared in fyke net and gill net collections in the littoral zone.[28]

Young-of-the-year black crappie were common in seine collections from Lake Oahe, SD, during the period 1966–1974, but as they got larger they moved into deeper water.[51]

Trap nets were evaluated as sampling gear for black crappie in Lake Wylie, NC and SC. Compared with creel survey, cove sampling with rotenone, and spring electrofishing, trap nets were the most efficient and cost-effective gear for capturing crappie. More small black crappie (<250 mm TL) were captured in nets with 1.9-cm bar mesh than in nets with 2.5-cm bar mesh.[61] In another study, comparison of efficiencies of 13- and 16-mm mesh trap nets showed that the catch rate of black crappie <130 mm TL was similar for both mesh sizes. The 16-mm mesh caught significantly more black crappie ≥130 mm than the smaller mesh.[72]

In two FL lakes, both trawls and trap nets were effective in collecting black crappie <150 mm TL, but trawls collected significantly more adult fish (>250 mm TL).[53] The efficiency for capturing young black crappie by two sizes of otter trawls was also compared in two FL lakes. A large (31.8-mm knotted stretch mesh) trawl was less effective than a smaller trawl of similar design (6.8-mm knotless stretch mesh) for catching juvenile black crappie. Total length of captured black crappie from the two lakes averaged 157 and 143 mm in the large trawl; specimens averaged 135 and 116 in the smaller trawl. The smaller trawl also caught more fish <100 mm.[74]

During the period 1998–2002 in J. Percy Priest Reservoir, TN, both larval tow-net and fall trap-net samples effectively sampled age-0 black and white crappie and accurately indexed year-class strength. Larval densities of crappie in May were correlated to age-0 catches in fall trap nets, thus density of larval crappie was positively related to year-class strength. However, fall trap-net samples of age-0 crappie provided a more precise and cost-effective index of year-class strength than larval tow-net samples. May discharge through the dam was an accurate predictor of age-0 abundance in fall trap nets. No relation was evident in this study between mean water temperatures over sample periods and abundance of age-0 crappie.[76]

In TN, reservoir hydrology appeared to have a stronger effect on black crappie recruitment in tributary storage reservoirs than in mainstem Tennessee River impoundments, possibly because recruitment was more variable in tributary systems.[80]

Crappie recruitment (both species) was related to hydrology in most AL reservoirs studied. In reservoirs where water levels fluctuate, a rising or higher-than-average water level in the months preceding and during the spawning period may increase crappie year-class production. Hydrologic manipulation of these reservoirs could increase reproductive success of black crappie. If erratic crappie recruitment in mainstream reservoirs, where water levels do not fluctuate, is due to variable and uncontrolled hydrologic conditions dictated by climate, then management options to increase crappie year-class strength would be limited.[77]

Black crappie produced poor year-classes in dry years and strong year-classes in wet years in Normandy Reservoir, TN. Responses were variable during intermediate years.[73]

In three FL lakes, age-0 black crappie abundance and size during fall were not predicted solely by food availability, stock abundance, or water clarity, suggesting that these factors interacted to influence recruitment through the first year. Juvenile abundance of black crappie in early summer was positively related to stock abundance during the previous fall among lakes and also to juvenile abundance in the fall, suggesting that year-class strength was set by early summer in all three lakes. There was no relationship between crustacean zooplankton density and black crappie size across lakes.[40]

Black crappie and white crappie were introduced into Chautauqua Lake, NY, in the late 1920s; abundance of both species rapidly increased and an intensive recreational fishery developed. Both species remained at high levels of abundance until the 1970s. However, the catch of age-2 crappie declined after 1978. Trawling data during the period 1978–1999 indicated that year-class strength of black crappie and white crappie was synchronized and fixed between the fall of age 0 and fall of age 1. Age-0 crappie abundance was positively associated with the biomass of walleye prey and negatively associated with the biomass of walleye, showing that the availability of alternative prey affects walleye predation on crappie. The mortality of both species from fall of age 0 to fall of age 1 was not related to winter severity, predation, or prey abundance. Low abundance of age-2 crappie and lack of any significant relation between adult stock abundance and age-0 abundance indicated that angler removal of adult crappie from Lake Chautauqua was an unlikely cause of poor recruitment.[79]

The introduction of saugeye (sauger × walleye hybrids) into Richmond Lake, SD, probably reduced the abundance and intraspecific competition of small black crappie (<200 mm), which resulted in increased growth and improved size structure for black crappie. Black crappie >200 mm were not found in saugeye stomachs during the study.[83]

In AL ponds, YOY largemouth bass significantly reduced YOY black crappie when the crappies were small enough to be eaten. When crowded, YOY of the two species competed for aquatic insects as food.[55]

Black crappie juveniles showed a capacity to osmoregulate and survive for up to 30 days at pH 4.5 and above but lost osmoregulatory control at pH 4.0 and died after 16 days. However, when exposure to pH 4.0 was halted and pH was raised to 7.0 before death, blood osmolality returned to near-initial values within 15 days. Histomorphological responses were observed at a less acidic pH and shorter time of exposure than those associated with loss of osmoregulatory control, suggesting that changes in gill morphology may be more useful as an early indicator of potentially lethal acid stress. Recovery of normal blood osmolarity and gill morphology after short-term exposure to pH 4.0 suggests that juvenile black crappie may tolerate short episodes of exposure to potentially lethal pH.[1]

Black crappie juveniles have been recorded from water having surface salinity of 0.2 and bottom salinity of 12.7 ppt.[110]

Early Growth

No consistent difference in growth rate occurs between sexes.[23,63,71,94] The average calculated TL (mm) at the end of the year for young male and female black crappies from Lake Oahe, SD, during the period 1963–1968 was reported as follows:[94]

	Age			
	1	2	3	4
Males	88	167	225	260
Females	85	176	230	273

Carlander (1977) summarized regional reports of growth for black crappie (Table 149). Reports of growth for age-1 black crappie varied little (74–85 mm TL) among regions, but growth was variable for older fish. Greatest growth rates were reported from the southern waters of KY and TN.[63]

In northern populations of black crappie, growth varies widely with size of population and with the size and productivity of habitat.[33] In WI, YOY black crappie were 25 mm long at the end of June, 45 mm in July, 51 mm in August, and 64 mm at the end of September.[78]

State-average calculated TL related to age of young black crappie from OK, during the period 1964–1968, was about 79 mm at age 1, 160 mm at age 2, 208 mm at age 3, and 252 mm at age 4.[97] Average length of black crappie from the Arkansas River, OK, at the end of each of the first 3 years was comparable to the state average, but the state averages exceeded the corresponding averages for the river fish in all years of life.[98]

Average observed lengths and weights by age of young black crappie collected during the period 1950–1959 from Lake Chautauqua, IL, are presented below:[125]

Age	N	Average TL (mm)	Average Wt (g)
2	835	168	73
3	1,022	213	159
4	776	244	241

In OH, YOY black crappie in October were 25–76 mm; around 1 year 28–110 mm.[35]

From 53 populations of rivers, ponds, or reservoirs in VA, the range of mean lengths at age 3 is 132–270 mm TL, with an unweighted grand mean of 203 mm. The indicated growth rates are generally typical of the species elsewhere in the eastern and central United States (Table 149).[63,71]

Total length related to age for young black crappie from Kentucky Reservoir and other KY waters in 1983 is presented in Table 150.[118] Mean TL (mm) of age-1 to age-4 black crappie from Kentucky Reservoir in 1998 were 80, 160, 225, and 230,[85] respectively,

Table 149

Calculated mean of means TL (mm) at each annulus for young black crappie from regional waters of the United States.

Region	Mean Calculated TL at Each Annulus			
	1	2	3	4
Western waters	85	156	206	234
Northern waters	74	139	187	217
Central waters	80	147	188	220
AR, OK waters	76	158	212	246
KY, TN waters	81	199	251	298
DE, MD, NC, SC waters	75	148	197	242
FL, TX waters	78	151	206	246
Mean of regional means	78	157	207	243

Source: Constructed from data summarized by Carlander (1977).[63]

Table 150

Calculated TL (mm) related to age for young black crappie from KY waters in 1983.

Location	Age			
	1	2	3	4
Kentucky Lake	76–135	140–203	198–246	305
Barren River Lake	66–132	168	—	—
Marion County Lake	137	—	—	—
Rough River Lake tailwater	76	125	178	—
Nolin River Lake tailwater	74–91	221	—	—

Source: Constructed from data presented in Tables 4, 7, 27, 50, and 63 in reference 118.

which appeared similar to earlier reports on growth in KY and TN waters (Table 149 and 150) with the exception of age-4 fish. Mean total lengths (mm) of black crappie collected from Barkley Reservoir compared to Kentucky Reservoir in 1998 were similar for age-1 fish (85 mm TL), but smaller for older fish — 150 mm at age 2, 190 mm at age 3, and 210 mm at age 4. Lengths from this report are approximated and based on this author's interpretation of graphs presented in Figure 1, reference 85.

Growth of young black crappie in the 1940s was variable among tributary reservoirs of the Tennessee River (Table 151).[100,101]

In TN reservoirs, black crappie average about 80 mm TL after 1 year's growth with wide variation between populations; subsequent year-classes averaged about 200, 250, and 300 mm TL (Table 149).[63,70] These sizes considerably exceed estimates for northern populations.[23,70]

Table 151

Average calculated total lengths (TL) attained at successive ages by black crappie during the 1940s in Norris,[100] Cherokee, Douglas, and Hiwassee Reservoirs.[101]

Reservoir	Average Calculated TL (mm) at Age:			
	1	2	3	4
Norris, TN	64	234	292	322
Cherokee, TN	46	183	262	—
Douglas, TN	119	178	206	—
Hiwassee, NC	74	181	259	292

In three AL impoundments, back-calculated lengths of crappie (both species) at 4 days were directly related to chlorophyll *a* concentration and inversely related to larval density.[7] In AL ponds, young black crappie grew faster when their numbers were reduced by bass predation.[55]

In Weiss Lake, AL, young black crappie grew at a rate of 0.67 mm/day. Back-calculated lengths at 10, 20, and 30 days were higher for F_1 hybrids compared to black crappie. The faster growth of F_1 hybrids starts just after hatching and continues throughout life.[89] The calculated growth rate for young black crappie in a LA lake was 0.229 ± 0.076 mm/day. Young crappie were about 76 mm TL at the end of their first growing season.[14] In MS reservoirs, young black crappie grew at a rate of about 0.73 mm/day.[86,88] In two SD lakes, young black crappie grew at rates of about 1.03 and 1.11 mm/day.[87,88]

In a FL lake, age-0 black crappie that hatched in March had lower daily growth rates (0.72 mm/day) than those hatched later (May) in the spawning season (0.82 mm/day), higher instantaneous mortality rates, and lower survival to the end of the first summer. The mean daily growth rate was positively related to water temperature, which increased over the hatching season.[30]

In a study of 60 natural FL lakes, black crappie length at age 3 was quadratically related to fish density but was unrelated to trophic indicators, lake surface area, or macrophyte abundance.[52]

Growth rates for the 1980 year-class of black crappie in the St. Johns River, FL, were determined from collections made during the period 1981–1987. Although back calculated lengths were overestimated or underestimated in all cases, differences between actual and back-calculated lengths were not significant (Table 152).[62]

Table 152

Actual mean TL (mm) compared to back-calculated TL for black crappie collected from the St. Johns River, FL, during the period 1981–1987.[62]

Actual Age	N	Actual Mean TL	Back-Calculated TL
1	101	163	143–181
2	178	215	208–237
3	364	249	252–284
4	80	307	298–308

Black crappies <60 mm TL observed in Orange Lake, FL, in December were presumed the result of off-season spawning.[54]

An abundant year-class of black crappie developed in Lewis and Clark Lake, SD, the year following impoundment. During the period 1954–1967, growth of black crappie in the lake was slow compared with most other waters and was slower the first few years than 8–10 years after impoundment. No difference in growth rate was observed between sexes. No differences were detected in length–weight relationships between samples collected from different areas of the lake, nor between sexes.[23] Conversely, black crappie appeared to grow rapidly during the early years of impoundment of Norris Reservoir, TN (1938–1940).[105,106] The following growth data for black crappie collected in 1939 and 1940 are reported:

Age	N	Year(s) Collected	Average TL (mm)
1	48	1940	114
2	172	1939–1940	226–292
3	36	1939–1940	320–343
4	3	1940	343

Using these fish, the calculated TL at age 1 ranged from 71–170 mm; age 2 ranged from 218–290.[106]

Greenwood Lake, IN, also produced a very large crop of young black crappie during the year after the lake began to fill. These fish grew rapidly, initially, averaging 114, 166, and 190 mm TL their first 3 years. Rate of growth decreased after the lake was filled.[103]

Carlander (1977) observed that length–weight relationships vary considerably for black crappie and that it was difficult to distinguish any regional differences. He stated that, in general, weights at most lengths seemed less in the south than in the north.

Table 153

Mean of means and range in weight (g) related to TL (mm) for young black crappie from the combined data of numerous reports summarized by Carlander (1977).

TL (mm)	N	Weight (g) Means of Means	Weight (g) Range
15–24	22	0.06	0.02–0.08
25–50 (AL)	295	0.4	0.3–0.9
25–50 (OK)	93	0.55	0.1–1.4
51–75	3,775+	3	0.5–6
76–101	2,999+	15	2–76
102–126	7,695+	19	9–34
127–151	2,561+	37	9–71
152–177	1,149+	62	28–122
178–202	887+	97	45–169
203–228	1,770+	143	56–268
229–253	2,515+	210	77–340
254–278	1,849+	289	104–454

A summary of the combined data from numerous reports relating lengths and weights of young black crappie is presented in Table 153.[63]

The relationship of age, length, and weight for young black crappie from Lake of the Woods, ON, in 1949 is reported as follows:[33]

Age	FL (mm)	Wt (g)
1	112	45.4
2	168	90.8
3	213	181.6
4	254	317.8

In Lake Oahe during the period 1963–1968, the length–weight relationship for 318 black crappie 95–394 mm TL was expressed by the formula: log W = –5.25247 + 3.19758 log L (W = weight in grams; L = total length in millimeters).[94]

Length–weight relationships reported for young black crappie from OK waters, during the period 1952–1963 (Table 154),[95] vary little from those reported for black crappie collected from AL waters during the period 1949–1964 (Table 155).[55] Growth in weight in AL ponds, in 1970, was similar to the state average (Table 155).

Optimum temperature for growth was reported to be 22–25°C, with no growth recorded below 11°C or above 30°C.[124]

Table 154

Length–weight relationships of young black crappie from OK waters for the period 1952–1963.

TL (mm)	Average Weight (g)
76	5
102	11
127	24
152	44
178	74
203	116
229	173
254	245

Source: Constructed from data presented in Table 6, reference 95.

Table 155

Average empirical weight (g) related to TL (mm) of young black crappie collected from AL waters during the period 1949–1964[96] compared to the average calculated weight of young black crappie from AL ponds in 1970.[55]

Mean TL	Average State Empirical Weight	Average Calculated Pond Weight
25	0.4	0.4
51	1.8	1.8
76	4.1	5.9
102	14.1	13.6
127	22.7	27.2
152	36.3	45.4
178	72.6	72.6
203	104.4	108.0

Sources: Constructed from data presented in unnumbered table on page 7, reference 96, and in Table 6, reference 55.

Dense hydrilla (*Hydrilla verticillata*) coverage and/or reduced shad abundance appeared to restrict black crappie growth in Lake Baldwin, FL. Mean calculated total lengths of young black crappie were significantly higher after the disappearance of hydrilla from the lake. Black crappie did not reach harvestable size (>228 mm TL) until age 3 and 4 during periods of extensive hydrilla coverage and reduced threadfin shad abundance. Following removal of hydrilla and the reestablishment

of the shad population, harvestable-size black crappie were present by age 2 and the growth of older fish increased.[58]

Feeding Habits

Body morphology channels each fish species toward an ecological role different from that of other species. Black crappie have large numbers of long gill rakers (25–29 rays on the first arch compared to 10–12 rays for other centrarchids) that function as a sieve to retain small food organisms in the pharynx when water is ejected through the gill slits. The combination of a gibbous body, a high number of gill rakers, and a somewhat enlarged mouth mark the black crappie as a mid-water feeder.[126,127]

Black crappie are carnivorous in their feeding habits,[78] and the size of young black crappie at the onset of exogenous feeding influences early feeding behavior.[22] In Ontario waters, black crappie are largely nocturnal feeders, concentrating by day along shady areas of shoreline where aquatic vegetation extends through the water column, dispersing at night to feed.[127]

Some authors report that the food habits of black crappie are similar to those of white crappie, with young feeding on microcrustaceans and insects and with forage fishes becoming progressively more important in the diet as the fish grow and mature.[33,56,70,91,124] However, reports on the importance of fish in the diet of young black crappie vary. Microcrustacea and other plankton may remain important in the diet of juveniles and adults; the numerous long, slender gill rakers possessed by black crappie are apparently well adapted to plankton feeding.[60,70]

Prey selection patterns as a function of prey size did not differ for young black crappie or white crappie between northern temperate natural lakes and reservoirs, in spite of dramatic differences in zooplankton size structure. Studies for both species across a wide geographical area and in systems that differ in a variety of ways suggest that larval crappie consume larger prey with increasing gape size, but continue to select prey below their gape limitation.[5]

Sampling regimes and differences in prey availability have possibly affected reports of seasonal diets of young black crappie. Though some reports indicated that black crappies <80 mm generally did not eat fish,[26] one study suggested that fish fry may be an important prey.[25] Subsequent findings showed that the dominant prey of age-0 black crappies in the Ottawa River, ON, was copepods and that fish larvae occasionally constituted up to 88% of stomach content for black crappies 15–30 mm long. The effect of increasing fish size was not evident for age-0 black crappie except that chaoborids were only consumed by fish >50 mm. Consumption of

Chaoborus larvae during daylight hours implied that young black crappie made feeding forays into the hypolimnion. Strong positive, curvilinear relationships between mean prey weight and fish weight were observed.[26]

Black crappie show a change in food habits with growth. Young fish feed mainly on microcrustaceans and midges. At 60–115 mm TL, they continue to feed on small crustaceans and midges along with small insect larvae and pupae, but rarely on fishes. At larger sizes, they begin to regularly include fishes in their diet; the actual size at which fishes begin to constitute a major part of the diet varies among areas. By 160 mm TL, black crappie feed primarily on small fishes such as minnows and sunfishes. The switch to fish as prey is related to prey availability and to energetics, as large black crappie cannot consume enough zooplankton to sustain positive growth.[67–69] Another study suggests that small black crappies (<200 mm TL) benefit from, but are not dependent upon, the availability of fish forage.[91]

In Lake Opinicon, Ontario, very young black crappie are dependent on larval and small cyclopoid copepods, the only abundant prey of sufficiently small size. In June, larval black crappies 4–12 mm TL fed primarily on nauplii and *Cyclops*. Older black crappie, 95–110 mm, fed primarily on *Chaoboras*, chironomids, and fish (86% by volume, cumulative); cladocerans, dipterans, ephemeropterans, zygopterans, and anisopterans were also present in their stomachs.[33,126]

In WI, black crappie larvae had a diet dominated by copepods and selected progressively larger prey as fish length increased. Gape size predicted only 8% of the variability in mean prey size ingested by young black crappie; they fed on smaller zooplankton sizes than the average size available in the lake even though their gapes were large enough to allow ingestion of average size zooplankton.[31] The food of 140 black crappies 165–200 mm long, taken between February and October in a WI lake, was composed of cladocerans (33% by volume), chironomid larvae and pupae (24%), amphipods (11%), fish (9%), ephemeropteran nymphs (6%), copepods (5%), adult chironomids (4%), and Odonata nymphs (2%); other food items, each amounting to less than 1% by volume, included hemipteran nymphs and adults, caddisfly larvae, grasshoppers, beetles, ostracods, mites, snails, leeches, algae, plant material, silt, and debris.[78]

In the early impoundment period of Beaver Reservoir, AR, when terrestrial invertebrates were available in large numbers, young black crappie consumed large numbers of earthworms and some terrestrial arthropods.[59]

In Beaver and Bull Shoals Reservoirs, AR, the primary food items of yearling and older black crappie up to about 175 mm TL were fishes and zooplankton in the summer and fall. During other seasons, the diet was composed of zooplankton and immature chironomids. Larger members of this size range tended to concentrate on fish through the winter. Black crappie >175 mm TL consumed fish, zooplankton, chironomids, and benthic invertebrates, especially insects. Feeding on insects was greatest in March and April and probably May. Larger black crappie consumed fewer fish and more benthic insects than white crappie, and black crappies >200 mm consumed more zooplankton than white crappies in this size range.[59]

Food of larval black crappie from Keowee Reservoir, SC, was determined in 1973 and 1976, before and after commercial power generation by a nuclear power plant. Black crappie larvae 5.0–10.9 mm TL consumed primarily cladocerans (*Diaphanosoma* spp.) and copepod nauplii. Although water temperatures were higher in 1976, food of the larvae appeared unchanged by operation of the nuclear power plant.[57]

In three AL impoundments, the biomass of young crappie (both species) diets was dominated by crustacean zooplankton, although diets were numerically dominated by rotifers. Larvae 4–14 mm TL strongly selected the smallest prey available in all lakes.[7]

In AL ponds, black crappie larvae 16–72 mm TL fed mainly on zooplankton but consumed more aquatic insects as their sizes increased (86–110 mm TL). In one pond on July 8, 1970, zooplankton was present in all stomachs of 6 black crappie 85–100 mm TL, but in low volume. Bluegill fry occurred in all six stomachs and averaged 63% of the volume.[55]

In a FL lake, common prey of black crappie larvae included calanoid copepods, *Daphnia* and *Bosmina* spp., and cyclopoid copepods.[30]

Dense hydrilla probably retards crappie feeding efficiency.[58] Black crappies >100 mm TL primarily consume aquatic insect larvae, other benthic invertebrates, and fish.[56,58–60] Hydrilla harbors large numbers of small forage fishes and macroinvertebrates, including aquatic insects; however, these may be inaccessible to predation. One author suggests that young black crappies <106 mm TL are nocturnal, mid-water filter feeders primarily consuming crustacean zooplankton, *Chaoborus*, and chironomid larvae, which would make the physical structure of hydrilla communities restrictive to their feeding habits.[58]

In a 2-year food habit study in a 50-acre fishing impoundment in CA, black crappies (75–200 mm TL) consumption of stocked threadfin shad was only 17% by volume; arthropods, mainly cladocerans and insects, made up the remainder of their diets.[21]

The feeding rate of black crappies 100–150 mm on *Daphnia pulex* (laboratory study) was significantly reduced when turbidity was elevated from 80 to 160 NTU (Nepholometer Turbidity units). These tests showed no significant difference in the feeding rates of black crappie and white crappie at the turbidity levels tested.[48]

LITERATURE CITED

1. McCormick, J.H. et al. 1989.
2. Smith, S.M. et al. 1994.
3. Dunham, R.A. et al. 1994.
4. Travnichek, V.H. et al. 1997.
5. DeVries, D.R. et al. 1998.
6. Epifanio, J.M. et al. 1999.
7. Dubuc, R.A. and D.R. DeVries. 2002.
8. Faber, D.J. 1984a.
9. Pearson, W.D. and L.A. Krumholz. 1984.
10. Pearson, W.D. and B.J. Pearson. 1989.
11. Gammon, J.R. 1998.
12. Holland–Bartels, L.E. 1990.
13. Lee, D.S. et al. 1980.
14. Lambou, V.W. 1958.
15. Clay, W.M. 1975.
16. Burr, B.M. and M.L. Warren, Jr. 1986.
17. Pflieger, W.L. 1975b.
18. Scott, E.M. and J.P. Buchanan. 1979.
19. Buchanan, J.P. and E.M. Scott. 1979.
20. Coutant, C.C. 1977.
21. McConnell, W.J. and J.H. Gerdes. 1964.
22. Brown, J.A. and P.W. Colgan. 1985.
23. Vanderpuye, C.J. and K.D. Carlander. 1971.
24. Merriner, J.V. 1971.
25. Hanson, J.M. and S.U. Qadri. 1979.
26. Hanson, J.M. and S.U. Qadri. 1984.
27. Chubb, S.L. and C.R. Liston. 1986.
28. Post, J.R. et al. 1995.
29. Muth, R.T. and J.C. Schmulbach. 1984.
30. Pine, W.E., III and M.S. Allen. 2001.
31. Schael, D.M. et al. 1991.
32. Conrow, R. et al. 1990.
33. Scott, W.B. and E.J. Crossman. 1973.
34. Cook, F.A. 1959.
35. Trautman, M.B. 1981.
36. Goodyear, C.D. et al. 1982
37. Smith, P.W. 1979.
38. Smith, C.L. 1985.
39. Simon, T.P. 1986.
40. Dockendorf, K.J. and M.S. Allen. 2005.
41. Hooe, M.L. et al. 1994.
42. Knights, B.C. et al. 1995.
43. Guy, C.S. and D.W. Willis. 1995.
44. Pope, K.L. and D.W. Willis. 1998a.
45. June, F.C. 1977.
46. Thomas, J.A. et al. 2004.
47. Baker, J.A. and K.J. Killgore. 1994.
48. Barefield, R.L. and C.D. Ziebell. 1986.
49. Benson, N.G. 1968.
50. Killgore, K.J. and J.A. Baker. 1996.
51. June, F.C. 1976.
52. Allen, M.S. et al. 1998.
53. Allen, M.S. et al. 1999.
54. Nagid, E.J. et al. 2004.
55. Tucker, W.H. 1973.
56. Ager, L.A. 1976.
57. Barwick, D.H. 1978.
58. Maceina, M.J. and J.V. Shireman. 1982.
59. Ball, R.L and R.V. Kilambi. 1973.
60. Keast, G.K. 1968.
61. McInerny, M.C. 1988.
62. Crumpton, J.E. et al. 1988.
63. Carlander, K.D. 1977.
64. Menhinick, E.F. 1991.
65. Mettee, M.F. et al. 2001.
66. Boschung, H.T., Jr. and R.L. Mayden. 2004.
67. Ross, S.T. 2001.
68. Reid, G.R. 1949.
69. Ellison, D.G. 1984.
70. Etnier, D.A. and W.C. Starnes. 1993.
71. Jenkins, R.E. and N.M. Burkhead. 1994.
72. Jackson, J.J. and D.L. Bauer. 2000.
73. Sammons, S.M. and P.W. Bettoli. 2000.
74. Pine, W.E. III. 2000.
75. Sammons, S.M. and P.W. Bettoli. 1998.
76. St. John, R.T. and W.P. Black. 2004.
77. Maceina, M.J. 2003.
78. Becker, G.C. 1983.
79. McKeown, P.E and S.R. Mooradian. 2002.
80. Sammons, S.M. et al. 2002a.
81. Buchanan, J.P. and H.E. Bryant. 1973.
82. Isermann, D.A. et al. 2002.
83. Galinat, G.F. et al. 2002.
84. Spier, T.W. and R.C. Heidinger. 2002.
85. Sammons, S.M. et al. 2002b.
86. Hammers, B.E. 1990.
87. Pope, K.L. and D.W. Willis. 1998b.
88. Sammons, S.M. et al. 2001.
89. Travnichek, V.H. et al. 1996.
90. Spier, T.W. and J.R. Ackerson. 2004.
91. Jackson, J.R. and S.L. Bryant. 1993.
92. Slipke, J.W. et al. 2000.
93. McCarraher, D.B. 1971.
94. Nelson, W.R. 1974.
A 95. Houser, A. and M.G. Bross. 1963.
A 96. Swingle, W.E. 1965.
97. Lewis, S.A. et al. 1971.
98. Linton, T.L. 1961.
99. Morgan, G.D. 1951a.
100. Stroud, R.H. 1948.
101. Stroud, R.H. 1949.
102. Chance, C.J. 1958.
103. Johnson, W.L. 1945.
104. Eschmeyer, R.W. 1942.

105. Eschmeyer, R.W. 1940.
106. Eschmeyer, R.W. and A.M. Jones. 1941.
107. Cady, E.R. 1945.
108. Dendy, J.S. 1945.
109. Wang, J.C.S. and R.J. Kernehan. 1979.
110. Hardy, J.D., Jr. 1978
111. Siefert, R.E. 1969b.
112. Faber, D.J. 1963.
113. Ward, H.C. and E.M. Leonard. 1952.
114. Tin, H.T. 1982.
115. Chatry, M.F. 1977.
116. Cooper, J.A. 1971.
117. Lippson, A.J. and R.L. Moran. 1974.
A 118. Axon, J.R. 1984.
119. Breder, C.M., Jr. and D.E. Rosen. 1966.
A 120. Environmental Science and Engineering, Inc. 1992.

A 121. Environmental Science and Engineering, Inc. 1995.
A 122. EA Engineering, Science, and Technology. 1994.
A 123. EA Engineering, Science, and Technology. 2004.
124. Edwards, E.A. et al. 1982.
125. Starrett, W.C. and A.W. Fritz. 1965.
126. Keast, A. 1978b.
127. Keast, A. 1970.
128. McDonough, T.A. and J.P. Buchanan. 1991.
129. Mabee, P.M. 1995.

* Discussions of spatio-temporal distribution of crappie larvae in the Tennessee River system are based on ichthyoplankton data collected by TVA during the period 1973–1986.

OTHER IMPORTANT LITERATURE

Boxrucker, J. and E. Irwin. 2002.
Bister, T.J. et al. 2002.
Dorr, B. et al. 2002.
Hurley, K.L. and J.J. Jackson. 2002.
Isermann, D.A. et al. 2002.

BIBLIOGRAPHY

Adams, C.C. and T.L. Hankinson. 1928. The ecology and economics of Oneida Lake fish. New York State College of Forestry, Syracuse University Bulletin, Roosevelt Wildlife Annals 1:236–548.

Aday, D.D., R.J.H. Hoxmeier, and D.H. Wahl. 2003. Direct and indirect effects of gizzard shad on bluegill growth and population size structure. *Transactions of the American Fisheries Society* 132:47–56.

Ager, L.A. 1976. Monthly food habits of various size groups of black crappie in Lake Okeechobee. *Proceedings of the Twenty-Ninth Annual Conference of the Southeastern Association of Game and Fish Commissioners* 29:336–342.

Aggus, L.R. and G.V. Elliott. 1975. Effects of cover and food on year-class strength of largemouth bass. Pages 317–322 in R. H. Stroud and H. Clepper, editors. *Black Bass Biology and Management.* Sport Fishing Institute, Washington, D.C., USA.

Ali, A.B. and D.R. Bayne. 1985. Food related problems in bluegill populations of West Point Reservoir. *Proceedings of the Thirty-Ninth Annual Conference of the Southeastern Association of Fish and Wildlife Agencies* 39:207–216.

Alicea, A.R., R.L. Noble, and T.N. Churchill. 1997. Trophic dynamics of juvenile largemouth bass in Lucchetti Reservoir, Puerto Rico. *Proceedings of the Fifty-First Annual Conference of the Southeastern Association of Fish and Wildlife Agencies* 51:149–158.

Allan, R.C. and J. Romero. 1975. Underwater observations of largemouth bass spawning and survival in Lake Mead. Pages 104–112 in R.H. Stroud and H. Clepper, editors. *Black Bass Biology and Management.* Sport Fishing Institute, Washington, D.C., USA.

Allen, M.S., J.C. Greene, F.J. Snow, M.J. Maceina, and D.R. DeVies. 1999. Recruitment of largemouth bass in Alabama reservoirs: relations to trophic state and larval shad occurrence. *North American Journal of Fisheries Management* 19:67–77.

Allen, M.S., M.M. Hale, and W.E. Pine III. 1999. Comparison of trap nets and otter trawls for sampling black crappie in two Florida lakes. *North American Journal of Fisheries Management* 19:977–983.

Allen, M.S., M.V. Hoyer, and D.E. Canfield, Jr. 1998. Factors related to black crappie occurrence, density, and growth in Florida lakes. *North American Journal of Fisheries Management* 18:864–871.

Allen, M.S. and K.I. Tugend. 2002. Effects of a large-scale habitat enhancement project on habitat quality for age-0 largemouth bass at Lake Kissimmee, Florida. Pages 265–276 in D.P. Phillip and M.S. Ridgway, editors. *Black Bass: Ecology, Conservation, and Management.* American Fisheries Society, Symposium 31, Bethesda, Maryland, USA.

Allen, M.S., K.I. Tugend, and M.J. Mann. 2003. Largemouth bass abundance and angler catch rates following a habitat enhancement project at Lake Kissimmee, Florida. *North American Journal of Fisheries Management* 23:845–855.

Al-Rawi, T.R. and D.W. Toetz. 1972. Growth of white crappie and gizzard shad in Lake Keystone, Oklahoma. *Proceedings of the Oklahoma Academy of Science* 52:1–5.

Anderson, R.O. and R.M. Neumann. 1996. Length, weight, and associated structural indices. Pages 447–482 in B.R. Murphy and D.W. Willis, editors. *Fisheries Techniques,* 2nd edition. American Fisheries Society, Bethesda, Maryland, USA.

Anderson, R.O. and A.S. Weithman. 1978. The concept of balance for coolwater fish populations. Pages 371–381 in R.L. Kendall, editor. *Selected Coolwater Fishes of North America.* American Fisheries Society, Special Publication 11, Bethesda, Maryland, USA.

Angermeier, P.L. 1982. Resource seasonality and fish diets in an Illinois stream. *Environmental Biology of Fishes* 7:251–264.

Angermeier, P.L. 1985. Spatio-temporal patterns of foraging success for fishes in an Illinois stream. *American Midland Naturalist* 114:342–359.

Anjard, C.A. 1974. Centrarchidae Sunfishes. Pages 178–195 in A.J. Lippson and R.L. Moran, editors. *Manual for the Identification of Early Developmental Stages of Fishes of the Potomac River Estuary.* Environmental Technological Center, Martin Marietta Corporation, Baltimore, Maryland, USA.

Annett, C., J. Hunt, and E.D. Dibble. 1996. The complete bass: habitat use patterns of all stages of the life cycle of largemouth bass. Pages 306–314 in L.E. Miranda and D.R. DeVries, editors. *Multidimensional Approaches to Reservoir Fisheries Management.* American Fisheries Society, Symposium 16, Bethesda, Maryland, USA.

Applegate, R.L. 1966. Pyloric caeca counts as a method for separating the advanced fry and fingerlings of largemouth and spotted bass. *Transactions of the American Fisheries Society* 95:226.

Applegate, R.L., J.W. Mullan, and D.L. Morais. 1966. Food and growth of six centrarchids from shoreline areas of Bull Shoals Reservoir. *Proceedings of the Twentieth Annual Conference of the Southeastern Association of Game and Fish Commissioners* 20:469–482.

Armstrong, J.G. III. 1969. A Study of the Spring Dwelling Fishes of the Southern Bend of the Tennessee River. Master's thesis. University of Alabama, Tuscaloosa, Alabama, USA.

Ashley, K.W. and R.T. Rachels. 1998. Changes in redbreast sunfish population characteristics in the Black and Lumbee Rivers, North Carolina. *Proceedings of the Fifty-Second Annual Conference of the Southeastern Association of Fish and Wildlife Agencies* 52:29–38.

Auer, N.A. 1982. *Identification of Larval Fishes of the Great Lakes Basin with Emphasis on the Lake Michigan Drainage.* Great Lakes Fishery Commission, Special Publication 82-3, Ann Arbor, Michigan, USA.

Austen, D.J. and D.J. Orth. 1985. Food utilization by riverine smallmouth bass in relation to minimum length limits. *Proceedings of the Thirty-Ninth Annual Conference of the Southeastern Association of Fish and Wildlife Agencies* 39:97–107.

Austen, D.J. and D.J. Orth. 1988. Evaluation of a 305-mm minimum-length limit for smallmouth bass in the New River, Virginia and West Virginia. *North American Journal of Fisheries Management* 8:231–239.

Avila, V.L. 1976. A field study of nesting behavior of male bluegill sunfish (*Lepomis macrochirus* Rafinesque). *The American Midland Naturalist* 96:195–206.

Avise, J.C. and M.H. Smith. 1977. Gene frequency comparisons between sunfish (Centrarchidae) populations at various stages of evolutionary divergence. *Systematic Zoology* 26:319–335.

Avise, J.C., D.O. Straney, and M.H. Smith. 1977. Biochemical genetic of sunfish. IV. Relationships of centrarchid genera. *Copeia* 1977:250–258.

Axelrod, H.P. and L.P. Schultz. 1971. *Handbook of Tropical Aquarium Fishes.* T.F.H. Publications, Inc., Jersey City, New Jersey, USA.

Axelrod, H.P. and S.R. Shaw. 1967. *Breeding Aquarium Fishes. Book 1.* T.F.H. Publications, Inc., Hong Kong, China.

Bade, E. 1931. Pages 361–820 in Das Susswasser-Aquarium. Die Flora und Fauna des Susswassers und ihre Pflege im Zimmer-Aquarium [in German]. *2. Teil, die Susswasser-Fauna, 5th ed.* Fritz Pfenningstorff, Berlin.

Bailey, J.R. and J.A. Oliver. 1939. The fishes of the Connecticut watershed. Pages 150–189 in Biological Survey of the Connecticut Watershed. *New Hampshire Fish and Game Department Survey Report 4.*

Bailey, R.M. 1938. A Systematic Revision of the Centrarchid Fishes, with a Discussion of Their Distribution, Variations, and Probable Interrelationships. Doctoral dissertation, University of Michigan, Ann Arbor, Michigan, USA.

Bailey, R.M., J.E. Fitch, E.S. Herald, E.A. Lachner, C.C. Lindsey, C.R. Robins, and W.B. Scott. 1970. *A List of Common and Scientific Names of Fishes from the United States and Canada*, 3rd ed. American Fisheries Society Special Publication, Number 6, Bethesda, Maryland, USA.

Bailey, R.M. and C.L. Hubbs. 1949. The black basses (*Micropterus*) of Florida, with description of a new species. *Occasional Papers of the Museum of Zoology University of Michigan* 516:1–43.

Bailey, R.M., H.E. Winn, and C.L. Smith. 1954. Fishes from the Escambia River, Alabama and Florida, with ecologic and taxonomic notes. *Proceedings of the Academy of Natural Sciences, Philadelphia* 106:109–164.

Bain, M.B. and L.A. Helfrich. 1983. Role of male parental care in survival of larval bluegills. *Transactions of the American Fisheries Society* 112:47–52.

Baker, C.L. 1939. Additional fishes of Reelfoot Lake. *Journal Tennessee Academy of Science* 14:46–53.

Baker, J.A. and K.J. Killgore. 1994. Use of a flooded bottomland hardwood wetland by fishes in the Cache River system, Arkansas. *United States Army Corps of Engineers Wetlands Research Program Technical Report WRP-CP-3*, Waterways Experiment Station, Vicksburg, Mississippi, USA.

Baker, J.A., K.J. Killgore, and R.L. Kasul. 1991. Aquatic habitats and fish communities in the lower Mississippi River. *Aquatic Sciences* 3:313–356.

Baldwin, N.W., C.A. Busack, and K.O. Meals. 1990. Induction of triploidy in white crappie by temperature shock. *Transactions of the American Fisheries Society* 119:438–444.

Balkenbush, P.E. and W.L. Fisher. 1999. Population characteristics and management of black bass in eastern Oklahoma streams. *Proceedings of the Fifty-Third Annual Conference of the Southeastern Association of Fish and Wildlife Agencies* 53:130–143.

Ball, R.L. 1972. The Feeding Ecology of the Black Crappie, *Pomoxis nigromaculatus*, and the White Crappie, *Pomoxis annularis*, in Beaver Reservoir. Master's thesis. The University of Arkansas, Fayetteville, Arkansas, USA.

Ball, R.L. and R.V. Kilambi. 1973. The feeding ecology of the black and white crappies in Beaver Reservoir, Arkansas, and its effect on the relative abundance of the crappie species. *Proceedings of the Twenty-Sixth Annual Conference of the Southeastern Association of Game and Fish Commissioners* 26:577–590.

Balon, E. 1959. Spawning of *Lepomis gibbosus* (Linne 1758) acclimatized in the backwaters of the Danube and its development during the embryonic period. *Acta of the Zoology Society Bohemoslovenicae* 23:1–22.

Balon, E.K. 1975. Reproductive guilds of fishes: a proposal and definition. *Journal of the Fisheries Research Board of Canada* 32:821–864.

Balon, E.K. 1981. Addition and amendments to the classification of reproductive styles in fishes. *Environmental Biology of Fishes* 6:377–389.

Barans, C.A. and R.A. Tubb. 1973. Temperatures selected seasonally by four fishes from western Lake Erie. *Journal of the Fisheries Research Board of Canada* 30:1697–1703.

Barclay, L.E. 1985. Responses of stream fishes to flooding and drought. Abstract. 65th Annual Meeting of the American Society of Ichthyologists and Herpetologists. University of Tennessee, Knoxville, Tennessee, USA.

Barefield, R.L. and C.D. Ziebell. 1986. Comparative feeding ability of small white and black crappie in turbid water. *Iowa State Journal of Research* 61:143–146.

Barlow, J.R., Jr. 1980. Geographic Variation in *Lepomis megalotis* (Rafinesque) Osteichthyes: Centrarchidae. Doctoral dissertation. Texas A&M University, College Station, Texas, USA.

Barney, R.L. and B.J. Anson. 1920. Life history and ecology of the pygmy sunfish, *Elassoma zonatum*. *Ecology* 1:241–256.

Barney, R.L. and B.J. Anson. 1923. Life history and ecology of the orangespotted sunfish, *Lepomis humilis*. *Appendix XV, report of the U.S. Commission of Fisheries* 1922:1–16. Document 938.

Barrett, P.J. and O.E. Maughan. 1994. Habitat preferences of introduced smallmouth bass in a central Arizona stream. *North American Journal of Fisheries Management* 14:112–118.

Barwick, D.H. 1978. Food of larval black crappies in relation to electrical power generation, Keowee Reservoir, South Carolina. *Proceedings of the Thirty-Second Annual Conference of the Southeastern Association of Fish and Wildlife Agencies* 32:485–489.

Barwick, D.H. 2004. Species richness and centrarchid abundance in littoral habitats of three southern U.S. reservoirs. *North American Journal of Fisheries Management* 24:76–81.

Barwick, D.H. and P.R. Moore. 1983. Abundance and growth of redeye bass in two South Carolina Reservoirs. *Transactions of the American Fisheries Society* 112:216–219.

Bass, D.G., Jr. and V.G. Hitt. 1973. Sport fishing ecology of the Suwannee and Santa Fe Rivers, Florida. Report III. Life history aspects of the redbreast sunfish, *Lepomis auritus* (Linnaeus), in Florida. *Florida Game and Fresh Water Fish Commission* 1973:1–34.

Bass, D.G., Jr. and V.G. Hitt. 1975. Ecological aspects of the redbreast sunfish, *Lepomis auritus*, in Florida. *Proceedings of the Twenty-Eighth Annual Conference of the Southeastern Association of Game and Fish Commissioners* 28:296–307.

Bauer, B.H. 1980. *Lepomis marginatus* (Holbrook). Dollar sunfish. Page 599 in Lee, D.S. et al., editors. *Atlas of North American Freshwater Fishes*. Publication 1980-12, North Carolina State Museum of Natural History, Raleigh, North Carolina, USA.

Baylis, J.R., D.D. Wiegmann, and M.H. Hoff. 1993. Alternating life histories of smallmouth bass. *Transactions of the American Fisheries Society* 122:500–510.

Bayne, D.R., E. Reutebuch, and W.C. Seesock. 2002. Relative motility of fishes in a southeastern reservoir based on tissue polychlorinated biphenyl residues. *North American Journal of Fisheries Management* 22:122–131.

Beamesderfer, R.C.P. and J.A. North. 1995. Growth, natural mortality, and predicted response to fishing for largemouth bass and smallmouth bass populations in North America. *North American Journal of Fisheries Management* 15:688–704.

Bean, T.H. 1903. Catalogue of the fishes of New York. *New York State Museum Bulletin* 9:1–784.

Beard, T.D. 1982. Population dynamics of young-of-the-year bluegill. *Technical Bulletin Number 127*, Wisconsin Department of Natural Resources, Madison, Wisconsin, USA.

Becker, G.C. 1976. *Environmental Status of the Lake Michigan Region. Volume 17. Inland Fishes of the Lake Michigan Drainage Basin*. ANL/ES-40 Environmental Control Technology and Earth Sciences. Argonne National Laboratory, Argonne, Illinois, USA.

Becker, G.C. 1983. *Fishes of Wisconsin*. The University of Wisconsin Press, Madison, Wisconsin, USA.

Beckman, W.C. 1952. *Guide to the Fishes of Colorado*. University Colorado Museum Leaflet 11:1–110.

Beitinger, T.L. and J.J. Magnuson. 1979. Growth rates and temperature selection of bluegill, *Lepomis macrochirus*. *Transactions of the American Fisheries Society* 108:378–382.

Belk, M.C. 1994. Effects of presence of predators on age at maturity in bluegill sunfish, *Lepomis macrochirus*: a phenotypic response. (Abstract) *Bulletin of the Ecological Society of America* 75:13.

Belk, M.C. and L.S. Hales, Jr. 1993. Predator-induced differences in growth and reproduction of bluegills (*Lepomis macrochirus*). *Copeia* 1993:1034–1044.

Bennett, G.W. 1965. The environmental requirements of centrarchids with special reference to largemouth bass, smallmouth bass, and spotted bass. Pages 156–160 in R.A. Taft, editor. *Biological Problems in Water Pollution, Transactions of the 1962 Seminar*. Technical Report 999-Wp-25, Sanitary Engineering Center, U.S. Public Health Service, Washington, D.C., USA.

Benoit, D.A. 1975. Chronic effects of copper on survival, growth, and reproduction of the bluegill (*Lepomis macrochirus*). *Transactions of the American Fisheries Society* 104:353–358.

Benson, N.G. 1968. Review of fishery studies on Missouri River main stem reservoirs. *United States Fish and Wildlife Service Bureau of Sport Fisheries and Wildlife Research Report Number 71*, Washington, D.C., USA.

Bermingham, E. and J.C. Avise. 1986. Molecular zoogeography of freshwater fishes in the southeastern United States. *Genetics* 113:939–965.

Bernard, G. and M.G. Fox. 1997. Effects of body size and population density on overwinter survival of age-0 pumpkinseeds. *North American Journal of Fisheries Management* 17:581–590.

Berra, T.M. and G.E. Gunning. 1972. Seasonal movement and home range of the longear sunfish, *Lepomis megalotis* (Rafinesque) in Louisiana. *The American Midland Naturalist* 88:368–375.

Bertoldt, W. 1949. *Elassoma evergladei. The Aquarium* 18:112–113.

Bettoli, P.W. 1979. Age, Growth, and Food Habits of Fishes Common to the New River, Tennessee. Master's thesis. Tennessee Technological University, Cookeville, Tennessee, USA.

Bettoli, P.W., M.J. Maceina, R.L. Noble, and R.K. Betsill. 1992. Piscivory in largemouth bass as a function of aquatic vegetation abundance. *North American Journal of Fisheries Management* 12:509–512.

Bettoli, P.W., M.J. Maceina, R.L. Noble, and R.K. Betsill. 1993. Response of a reservoir fish community to aquatic vegetation removal. *North American Journal of Fisheries Management* 13:110–124.

Bevelhimer, M.S. 1995. Smallmouth bass habitat use and movement patterns with respect to reservoir thermal structure. *Proceedings of the Forty-Ninth Annual Conference of the Southeastern Association of Fish and Wildlife Agencies* 49:240–249.

Bevelhimer, M.S. 1996. Relative importance of temperature, food, and physical structure to habitat choice by smallmouth bass in laboratory experiments. *Transactions of the American Fisheries Society* 125:274–283.

Bietz, B.F. 1981. Habitat availability, social attraction and nest distribution patterns in longear sunfish (*Lepomis megalotis peltastes*). *Environmental Biology of Fishes* 6:193–200.

Bister, T.J., D.W. Willis, A.D. Knapp, and T.R. St. Sauver. 2002. Evaluation of a 23-cm minimum length limit for black and white crappies in a small South Dakota impoundment. *North American Journal of Fisheries Management* 22:1364–1368.

Blackwell, B.G. and M.L. Brown. 2000. A comparison of fish distributions in simple and complex lake basins. *Journal of Freshwater Ecology* 15:353–362.

Bodensteiner, L.R. and W.M. Lewis. 1994. Downstream drift of fishes in the upper Mississippi River during winter. *Journal of Freshwater Ecology* 9:45–56.

Böhlke, J.E. 1956. A new pygmy sunfish from southern Georgia. *Notulae Naturae* 294:1–11.

Boschung, H.T., Jr. 1992. Catalog of freshwater and marine fishes of Alabama. *Bulletin of the Alabama Museum of Natural History* 1992:1–206.

Boschung, H.T., Jr. and R.L. Mayden. 2004. *Fishes of Alabama*. Smithsonian Books, Washington, D.C., USA.

Bosley, T.R. and J.V. Conner. 1984. Geographic and temporal variation in numbers of myomeres in fish larvae from the lower Mississippi River. *Transactions of the American Fisheries Society* 113:238–242.

Boulenger, G.A. 1895. *A Catalogue of the Fishes of the British Museum. Volume 1, 2nd ed.* Order of Trustees, London, England.

Boxrucker, J. 1983. First year growth and survival of stocked largemouth bass in a small Oklahoma impoundment. *Proceedings of the Annual Conference of the Southeastern Association of Fish and Wildlife Agencies* 36:369–376.

Boxrucker, J. 1986. A comparison of the otolith and scale methods for aging white crappies in Oklahoma. *North American Journal of Fisheries Management* 6:122–125.

Boxrucker, J. 1987. Largemouth bass influence on size structure of crappie populations in small Oklahoma impoundments. *North American Journal of Fisheries Management* 7:273–278.

Boxrucker, J. 1992. Results of concomitant predator and prey stockings as a management strategy in combating stunting in an Oklahoma crappie population. *Proceedings of the Forty-sixth Annual Conference of the Southeastern Association of Fish and Wildlife Agencies* 46:327–336.

Boxrucker, J. 2002a. Improved growth of a white crappie population following stocking of saugeyes (sauger × walleye): a top-down, density-dependent growth response. *North American Journal of Fisheries Management* 22:1425–1437.

Boxrucker, J. 2002b. Rescinding a 254-mm minimum length limit on white crappies at Ft. Supply Reservoir, Oklahoma: the influence of variable recruitment, compensatory mortality, and angler dissatisfaction. *North American Journal of Fisheries Management* 22:1340–1348.

Boxrucker, J. and E. Irwin. 2002. Challenges of crappie management continuing into the 21st century. *North American Journal of Fisheries Management* 22:1334–1339.

Boxrucker, J. and G. Ploskey. 1988. Gear and seasonal biases associated with sampling crappie in Oklahoma. *Proceedings of the Forty-Second Annual Conference of the Southeastern Association of Fish and Wildlife Agencies* 42:89–97.

Boxrucker, J.C., G.L. Summers, and E.R. Gilliland. 2005. Effects of the extent and duration of seasonal flood pool inundation on recruitment of threadfin shad, white crappies, and largemouth bass in Hugo Reservoir, Oklahoma. *North American Journal of Fisheries Management* 25:709–716.

Boyd, J.A., B.M. Burr, L.M. Page, and P.W. Smith. 1975. A study of threatened and/or unique fishes within the boundaries of Shawnee National Forest of southern Illinois. Pages 1–29 in *Those on the Brink of Doom: A Study of Rare Fishes in the Shawnee National Forest*. Illinois Natural History Survey, Champaign, Illinois, USA.

Boyer, R.L. 1969. Aspects of the Behavior and Biology of the Longear Sunfish, *Lepomis megalotis* (Rafinesque) in Two Arkansas Reservoirs. Master's thesis. Oklahoma State University, Stillwater, Oklahoma, USA.

Boyer, R.L. and L.E. Vogele. 1971. Longear sunfish behavior in two Ozark reservoirs. Pages 13–25 in G.E. Hall, editor. *Reservoir Fisheries and Limnology.* American Fisheries Society Special Publication Number 8, Bethesda, Maryland, USA.

Bozek, M.A., P.H. Short, C.J. Edwards, M.J. Jennings, and S.P. Newman. 2002. Habitat selection of nesting smallmouth bass *Micropterus dolomieu* in two north temperate lakes. *American Fisheries Society Symposium* 31:135–148.

Branson, B.A. and G.A. Moore. 1962. The lateralis components of the acoustico-lateralis system of the sunfish family Centrarchidae. *Copeia* 1962:1–108.

Breder, C.M., Jr. 1936. The reproductive habits of the North American sunfishes (family Centrarchidae). *Zoologica* 21:1–49.

Breder, C.M., Jr. and R.F. Nigrelli. 1935. The influence of temperature and other factors on the winter aggregations of the sunfish, *Lepomis auritus*, with critical remarks on the social behavior of fishes. *Ecology* 16:33–47.

Breder, C.M., Jr. and A.C. Redmond. 1929. The blue-"spotted" sunfish. A contribution to the life history and habits of *Enneacanthus*, with notes on other Lepominae. *Zoologica* 9:379–401.

Breder, C.M., Jr. and D.E. Rosen. 1966. *Modes of Reproduction in Fishes.* Natural History Press, Garden City, New York, USA.

Brenden, T.O. and B.R. Murphy. 2004. Experimental assessment of age-0 largemouth bass and juvenile bluegill competition in a small impoundment in Virginia. *North American Journal of Fisheries Management* 24:1058–1070.

Brice, J.J. 1898. A manual of fish culture, based on the methods of the United States Commission of Fish and Fisheries. *Appendix to United States Commission Fisheries Report* 23:1–340.

Brown, J.A. and P.W. Colgan. 1981. The use of lateral-body bar markings in identification of young-of-year sunfish (*Lepomis*) in Lake Opinicon, Ontario. *Canadian Journal of Zoology* 59:1852–1855.

Brown, J.A. and P.W. Colgan. 1982. The inshore vertical distribution of young-of-year *Lepomis* in Lake Opinicon, Ontario. *Copeia* 1982:958–960.

Brown, J.A. and P.W. Colgan. 1985. Interspecific differences in the ontogeny of feeding behaviour in two species of centrarchid fish. *Zeitschrift für Tierspsychol.* 70:70–80.

Brown, J.A., P.H. Johansen, P.W. Colgan, and R.A. Mathers. 1987. Impairment of early feeding behavior of largemouth bass by pentachlorophenol exposure: a preliminary assessment. *Transactions of the American Fisheries Society* 116:71–78.

Bruno, N.A., R.W. Gregory, and H.L. Schramm. 1990. Nest sites used by radio-tagged largemouth bass in Orange Lake, Florida. *North American Journal of Fisheries Management* 10:80–84.

Bryant, H.E. and A. Houser. 1971. Population estimates of largemouth bass in Beaver and Bull Shoals Reservoirs. Pages 349–357 in G. E. Hall, editor. *Reservoir Fisheries and Limnology.* Special Publication Number 8, American Fisheries Society, Washington, D.C., USA.

Bryant, H.E. and T.E. Moen. 1980. Food of bluegill and longear sunfish in Degray Reservoir, Arkansas, 1976. *Arkansas Academy of Science Proceedings* 34:31–33.

Buchanan, J.P. 1973. Separation of the subspecies of largemouth bass, *Micropterus salmoides salmoides* and *M. s. floridanus* and intergrades by use of meristic characteristics. *Proceedings of the Annual Conference of the Southeastern Association of Fish and Wildlife Agencies* 27:608–619.

Buchanan, J.P. and H.E. Bryant. 1973. The occurrence of a predorsal stripe in the black crappie, *Pomoxis nigromaculatus. Journal of the Alabama Academy of Science* 44:293–297.

Buchanan, J.P. and E.M. Scott, Jr. 1979. Larval fish populations during the first three years of impoundment in Normandy Reservoir. *Proceedings of the Thirty-Third Annual Conference of the Southeastern Association of Fish and Wildlife Agencies* 33:648–659.

Buckmeier, D.L., R.K. Betsill, and J.W. Schlechte. 2005. Initial predation of stocked fingerling largemouth bass in a Texas reservoir and implications for improving stocking efficiency. *North American Journal of Fisheries Management* 25:652–659.

Buckmeier, D.L., J.W. Schlechte, and R.K. Betsill. 2003. Stocking fingerling largemouth bass to alter genetic composition: efficacy and efficiency of three stocking rates. *North American Journal of Fisheries Management* 23:523–529.

Bulkley, R.V. 1975. Chemical and physical effects on the centrarchid basses. Pages 286–293 in R.H. Stroud and H. Clepper, editors. *Black Bass Biology and Management.* Sport Fishing Institute, Washington, D.C., USA.

Bulkley, R.V, V.L. Spykermann, and L.E. Inmon. 1976. Food of pelagic young of walleyes and five cohabiting fish species in Clear Lake, Iowa. *Transactions of the American Fisheries Society* 105:77–82.

Bulow, F.J., M.A. Webb, W.D. Crumby, and S.S. Quisenberry. 1988. Effectiveness of a fish barrier dam in limiting movement of rough fishes from a reservoir into a tributary stream. *North American Journal of Fisheries Management* 8:273–275.

Bunnell, D.B., M.A. Scantland, and R.A. Stein. 2005. Testing for evidence of maternal effects among individuals and populations of white crappie. *Transactions of the American Fisheries Society* 134:607–619.

Burdick, G.E., H.J. Dean, and E.J. Harris. 1958. Toxicity of cyanide to brown trout and smallmouth bass. *New York Fish and Game Journal* 5:133–163.

Burns, J.R. 1976. The reproductive cycle and its environmental control in the pumpkinseed, *Lepomis gibbosus* (Pisces: Centrarchidae). *Copeia* 1976:449–455.

Burr, B.M. 1974. A new intergeneric hybrid combination in nature: *Pomoxis annularis* × *Centrarchus macropterus*. *Copeia* 1974:269–271.

Burr, B.M. 1977. The bantam sunfish, *Lepomis symmetricus*: systematics and distribution, and life history in Wolf Lake, Illinois. *Illinois Natural History Survey Bulletin* 31:437–466.

Burr, B.M., K.M. Cook, D.J. Eisenhour, K.R. Piller, W.J. Poly, R.W. Sauer, C.A. Taylor, E.R. Atwood, and G.L. Seegert. 1996. Selected Illinois fishes in jeopardy: new records and status evaluations. *Transactions of the Illinois State Academy of Science* 89:169–186.

Burr, B.M. and R.L. Mayden. 1979. Records of fishes in western Kentucky with additions to the known fauna. *Transactions of the Kentucky Academy of Science* 40:58–67.

Burr, B.M. and M.L. Warren, Jr. 1986. *A Distributional Atlas of Kentucky Fishes*. Kentucky Nature Preserves Commission Scientific and Technical Series Number 4, Frankfort, Kentucky, USA.

Burr, B.M. and M.L. Warren, Jr. 1990. Records of nine endangered, threatened, or rare Kentucky fishes. *Transactions of the Kentucky Academy of Science* 51:188–190.

Burr, B.M., M.L. Warren, Jr., and K.S. Cummings. 1988. New distributional records of Illinois fishes with additions to the known fauna. *Transactions of the Illinois Academy of Science* 81:163–170.

Burr, B.M., M.L. Warren, Jr., G.K. Weddle, and R.R. Cicerello. 1990. Records of nine endangered, threatened, or rare Kentucky fishes. *Transactions of the Kentucky Academy of Science* 51:188–190.

Buss, D.G. 1974. Orangespotted sunfish *Lepomis humilis*. Pages 721 in A.J. McClane, editor. *McClane's New Standard Fishing Encyclopedia and International Angling Guide*. Holt Rinehart & Winston, New York, New York, USA.

Butler, M.J. IV. 1988. In situ observations of bluegill (*Lepomis macrochirus* Raf.) foraging behavior: the effects of habitat complexity, group size, and predators. *Copeia* 1988:939–944.

Buynak, G.L. 1995. Evaluation of a differential-black bass size limit regulation on largemouth and spotted bass populations at Cave Run Lake. Kentucky Department of Fish and Wildlife Fisheries Bulletin Number 96.

Buynak, G.L., L.E. Kornman, A. Surmont, and B. Mitchell. 1989. Longitudinal differences in electrofishing catch rates and angler catches of black bass in Cave Run Lake, Kentucky. *North American Journal of Fisheries Management* 9:226–230.

Buynak, G.L., L.E. Kornman, A. Surmont, and B. Mitchell. 1991. Evaluation of a smallmouth bass stocking program in a Kentucky reservoir. *North American Journal of Fisheries Management* 11:293–297.

Buynak, G.L. and B. Mitchell. 2002. Response of smallmouth bass to regulatory and environmental changes in Elkhorn Creek, Kentucky. *North American Journal of Fisheries Management* 22:500–508.

Buynak, G.L., B. Mitchell, D. Bunnell, B. McLemore, and P. Rister. 1999. Management of largemouth bass at Kentucky and Barkley Lakes, Kentucky. *North American Journal of Fisheries Management* 19:59–66.

Buynak, G.L. and H.W. Mohr, Jr. 1978. Larval development of the redbreast sunfish (*Lepomis auritus*) from the Susquehanna River. *Transactions of the American Fisheries Society* 107:600–604.

Buynak, G.L. and H.W. Mohr, Jr. 1979. Larval development of rock bass from the Susquehanna River. *The Progressive Fish-Culturist* 41:39–42.

Buynak, G.L. and H.W. Mohr, Jr. 1981. Small-scale culture techniques for obtaining spawns from fish. *The Progressive Fish-Culturist* 43:38–39.

Byrd, I.B. 1952. Depth distribution of the bluegill, *Lepomis macrochirus* Rafinesque, in farm ponds during summer stratification. *Transactions of the American Fisheries Society* 1952:162–170.

Cady, E.R. 1945. Fish distribution, Norris Reservoir, Tennessee, 1943. I. Depth distribution of fish in Norris Reservoir. *Journal of the Tennessee Academy of Science* 20:103–114.

Cahn, A.R. 1927. An ecological study of southern Wisconsin fishes. *Illinois Biology Monogram* 11:1–151.

Cailteux, R.L., W.F. Porak, S. Crawford, and L.L. Connor. 1997. Differences in largemouth bass food habits and growth in vegetated and unvegetated North-Central Florida lakes. *Proceedings of the Fiftieth Annual Conference of the Southeastern Association of Fish and Wildlife Agencies* 50:201–211.

Caldwell, R.D., H.T. Odum, T.R. Heller, Jr., and F.H. Berry. 1955. Populations of spotted sunfish and Florida largemouth bass in a constant temperature spring. *Transactions of the American Fisheries Society* 85:120–134.

Carbine, W.F. 1939. Observations on the spawning habits of centrarchids fishes in Deep Lake, Oakland County, Michigan. *Transactions of the North American Wildlife Conference* 4:275–287.

Cargnelli, L.M. and M.R. Gross. 1996. The temporal dimension in fish recruitment: birth date, body size, and size-dependent survival in a sunfish (bluegill: *Lepomis macrochirus*). *Canadian Journal of Fisheries and Aquatic Science* 53:360–367.

Carlander, K.D. 1975. Community relations of bass in large natural lakes. Pages 125–130 in R.H. Stroud and H. Clepper, editors. *Black Bass Biology and Management*. Sport Fishing Institute, Washington, D.C., USA.

Carlander, K.D. 1977. *Handbook of Freshwater Fishery Biology, Volume Two*. The Iowa State University Press, Ames, Iowa, USA.

Carlander, K.D. 1982. Standard intercepts for calculating lengths from scale measurements for some centrarchid and percid fishes. *Transactions of the American Fisheries Society* 111:332–336.

Carr, A.F. 1936. A key to the freshwater fishes of Florida. *Proceedings of the Florida Academy of Science* 1:72–86.

Carr, A.F., Jr. 1939. Notes on the breeding habits of the warmouth bass. *Proceedings of the Florida Academy of Science* 4:108–112.

Carr, M.H. 1942. The breeding habits, embryology and larval development of the largemouth bass in Florida. *Proceedings of the New England Zoological Club* 20:43–77.

Carr, M.H. 1946. Notes on the breeding habits of the eastern stumpnocker, *Lepomis punctatus punctatus* (Cuvier). *Quarterly Journal Florida Academy of Science* 9:101–106.

Carter, B.T. 1967. Growth of three centrarchids in Lake Cumberland, Kentucky. *Kentucky Fisheries Bulletin* 44:30.

Carver, D.C. 1967. Distribution and abundance of the centrarchids in the recent delta of the Mississippi River. *Proceedings of the Twentieth Annual Conference of the Southeastern Association of Game and Fish Commissioners* 20:390–404.

Cashner, R.C. 1974. A Systematic Study of the Genus *Ambloplites*, with Comparisons to Other Members of the Tribe Ambloplitini (Pisces: Centrarchidae). Doctoral dissertation. Tulane University, New Orleans, Louisiana, USA.

Cashner, R.C. and R.E. Jenkins. 1982. Systematics of the Roanoke bass, *Ambloplites cavifrons*. *Copeia* 1982:581–594.

Cashner, R.C. and R.D. Suttkus. 1997. *Ambloplites constellatus*, a new species of rock bass from the Ozark uplands of Arkansas and Missouri with a review of western rock bass populations. *American Midland Naturalist* 98:147–161.

Casselman, J.M., D.M. Brown, J.A. Hoyle, and T.H. Eckert. 2002. Effects of climate and global warming on year-class strength and relative abundance of smallmouth bass in eastern Lake Ontario. *American Fisheries Society Symposium* 31:73–90.

Catchings, E.D. 1979. Age and growth of redeye bass in Shoal and Little Shoal Creeks, Alabama. *Proceedings of the Thirty-Second Annual Conference Southeastern Association of Fish and Wildlife Agencies* 32:380–390.

Cathey, H.J. 1973. A Study of Redeye Bass, *Micropterus coosae*, Smallmouth Bass, *Micropterus dolomieui*, and Rock Bass, *Ambloplites rupectris* in Roaring River, Tennessee. Master's thesis. Tennessee Technological University, Cookeville, Tennessee, USA.

Chable, A.C. 1947. A Study of the Food Habits and Ecological Relationships of the Sunfishes of Northern Florida. Master's thesis. University of Florida, Gainesville, Florida, USA.

Champion, M.J. and G.S. Whitt. 1976. Differential gene expression in multilocus isozyme systems of the developing green sunfish. *Journal of Experimental Zoology* 196:263–281.

Chance, C.J. 1958. History of fish and fishing in Norris — a TVA tributary reservoir. *Proceedings of the Twelfth Annual Conference of the Southeastern Association of Game and Fish Commissioners* 12:126.

Chance, C.J., A.O. Smith, J.A. Holbrook II, and R.B. Fitz. 1975. Norris Reservoir: a case history in fish management. Pages 399–407 in R.H. Stroud and H. Clepper, editors. *Black Bass Biology and Management*. Sport Fishing Institute, Washington, D.C., USA.

Chang, C.H. 1988. Systematics of the Centrarchidae (Perciformes: Percoidei) with Notes on the Haemal and Axial Character Complex. Doctoral dissertation. The City University of New York, New York, New York, USA.

Chastain, G.A. and J.R. Snow. 1966. Nylon mats as spawning sites for largemouth bass, *Micropterus salmoides*, Lac. *Proceedings of the Nineteenth Annual Conference of the Southeastern Association of Fish and Wildlife Agencies* 19:405–408.

Chatry, M.F. 1977. Comparative Developmental Morphology of the Crappies, *Pomoxis annularis* Rafinesque and *P. nigromaculatus* (LeSueur). Master's thesis. Louisiana State University, Baton Rouge, Louisiana, USA.

Cherry, D.S., K.L. Dickson, and J. Cairns, Jr. 1975. Temperatures selected and avoided by fish at various acclimation temperatures. *Journal of the Fisheries Research Board of Canada* 32:485–491.

Chew, R.L. 1973. The failure of largemouth bass, *Micropterus salmoides floridanus* (LeSueur), to spawn in eutrophic, over-crowded environments. *Proceedings of the Twenty-Sixth Annual Conference of the Southeastern Association of Fish and Wildlife Agencies* 26:306–319.

Chew, R.L. 1974. Early life history of the Florida largemouth bass. *Florida Game and Freshwater Fish Commission Bulletin* 7:1–76.

Chew, R.L. 1975. The Florida largemouth bass. Pages 450–458 in H. Clepper, editor. *Black Bass Biology and Management*. Sport Fishing Institute, Washington, D.C., USA.

Childers, W.F. 1965. Hybridization of Four Species of Sunfishes (Centrarchidae). Doctoral thesis. University of Illinois, Urbana, Illinois, USA.

Childers, W.F. 1967. Hybridization of four species of sunfishes (Centrarchidae). *Illinois Natural History Survey Bulletin* 29:159–214.

Childers, W.F. and G.W. Bennett. 1961. Hybridization between three species of sunfish (*Lepomis*). *Illinois Natural History Survey, Biological Notes No. 46*.

Christie, W.J. and H.A. Regier. 1973. Temperature as a major factor influencing reproductive success of fish — two examples. *Journal of the International Council for the Exploration of the Sea* 164:208–218.

Chubb, S.L. 1985. Spatial and Temporal Distribution and Abundance of Larval Fishes in Pentwater Marsh, a Coastal Wetland on Lake Michigan. Master's thesis. Michigan State University, East Lansing, Michigan, USA.

Chubb, S.L. and C.R. Liston. 1986. Density and distribution of larval fishes in Pentwater Marsh, a coastal wetland on Lake Michigan. *Journal of Great Lakes Research* 12:332–343.

Cichra, C.E., W.H. Neill, and R.L. Noble. 1981. Differential resistance of northern and Florida largemouth bass to cold shock. *Proceedings of the Thirty-Fourth Annual Conference of the Southeastern Association of Fish and Wildlife Agencies* 34:19–24.

Cichra, C.E., R.L. Noble, and B.W. Farquhar. 1981. Relationships of white crappie populations to largemouth bass and bluegill. *Proceedings of the Thirty-Fifth Annual Conference of the Southeastern Association of Fish and Wildlife Agencies* 35:416–423.

Cincotta, D.A. and J.R. Stauffer, Jr. 1984. Temperature preference and avoidance studies of six North American freshwater fish species. *Hydrobiologia* 109:173–177.

Clark, E.W. 1937. *Elassoma evergladei* in Marion County, Florida. *The Aquarium* 5:250–252.

Clark, F.W. and M.H.A. Keenleyside. 1967. Reproductive isolation between the sunfish *Lepomis gibbosus* and *L. macrochirus*. *Journal of the Fisheries Research Board of Canada* 24:495–514.

Clark, P.W. 1990. Behavioral Responses of Spawning Bluegill, *Lepomis macrochirus*, and Redear Sunfish, *Lepomis microlophus*, to Herbicide Applications. Master's thesis. Tennessee Technological University, Cookeville, Tennessee, USA.

Clay, W.M. 1975. *The Fishes of Kentucky*. Kentucky Department of Fish and Wildlife Resources, Frankfort, Kentucky, USA.

Coahran, D.A. 1990. Effects of Liming on Plankton and Young-of-the-Year Bluegill Growth in Flat Top Lake, West Virginia. Master's thesis. Virginia Polytechnic Institute and State University, Blacksburg, Virginia, USA.

Coble, D.W. 1975. Smallmouth bass. Pages 21–33 in R.H. Stroud and H. Clepper, editors. *Black Bass Biology and Management*. Sport Fishing Institute, Washington, D.C., USA.

Cofer, L.M. 1994. Evaluation of a trophy bass length limit on Lake Fuqua, Oklahoma. *Proceedings of the Forty-Seventh Annual Conference of the Southeastern Association of Fish and Wildlife Agencies* 47:702–710.

Cofer, L.M. 1995. Invalidation of the Wichita spotted bass, *Micropterus punctulatus wichitae*, subspecies theory. *Copeia* 1995:487–490.

Cole, M.B. and J.R. Moring. 1997. Relation of adult size to movements and distribution of smallmouth bass in a central Maine lake. *Transactions of the American Fisheries Society* 126:815–821.

Cole, V.W. 1951. Contributions to the Life History of the Redear Sunfish (*Lepomis microlophus*) in Michigan Waters. Master's thesis. Michigan State College, East Lansing, Michigan, USA.

Coleman, R.M. and R.U. Fischer. 1991. Brood size, male fanning effort and the energetics of a nonshareable parental investment in bluegill sunfish, *Lepomis macrochirus* (Teleostei: Centrarchidae). *Ethology* 87:177–188.

Colgan, P.W., W.A. Nowell, and N.W. Stokes. 1981. Spatial aspects of nest defense by pumpkinseed sunfish (*Lepomis gibbosus*): stimulus features and an application of catastrophe theory. *Animal Behavior* 29:433–442.

Colle, D.E., J.V. Shireman, W.T. Haller, J.C. Joyce, and D.E. Canfield, Jr. 1987. Influence of hydrilla on harvestable sport-fish populations, angler use, and angler expenditures at Orange Lake, Florida. *North American Journal of Fisheries Management* 7:410–417.

Conley, J.M. 1966. Ecology of the Flier, *Centrarchus macropterus* (Lacepede), in Southeast Missouri. Master's thesis. University of Missouri, Columbia, Missouri, U.S.A.

Conley, J.M. and A. Witt, Jr. 1966. The origin and development of scales in the flier, *Centrachus macropterus* (Lacepede). *Transactions of the American Fisheries Society* 95:433–434.

Conner, J.V. 1979. Identification of larval sunfishes (Centrarchidae, Elassomatidae) from southern Louisiana. Pages 17–52 in R.D. Hoyt, editor. *Proceedings of the Third Symposium on Larval Fish*. Western Kentucky University, Bowling Green, Kentucky, USA.

Conrow, R. and A.V. Zale. 1985. Early life history stages of fishes of Orange Lake, Florida: an illustrated identification manual. Florida Cooperative Fish and Wildlife Unit, Technical Report Number 15:1–45.

Conrow, R., A.V. Zale, and R.W. Gregory. 1990. Distributions and abundances of early life stages of fishes in a Florida lake dominated by aquatic macrophytes. *Transactions of the American Fisheries Society* 119:521–528.

Cook, F.A. 1959. *Freshwater Fishes in Mississippi*. Mississippi Game and Fish Commission, Jackson, Mississippi, USA.

Cook, K.M. 1994. Fishes of the Clear Creek Drainage, Illinois: Faunal Composition, Biotic Integrity, and Historical Overview. Master's thesis. Southern Illinois University, Carbondale, Illinois, USA.

Cooner, R.W. and D.R. Bayne. 1982. Diet overlap in redbreast and longear sunfishes from small streams of east central Alabama. *Proceedings of the Thirty-Sixth Annual Conference of the Southeastern Association of Fish and Wildlife Agencies* 36:106–114.

Cooper, J.A. 1971. Scale development as related to growth of juvenile black crappie, *Pomoxis nigromaculatus* LeSueur. *Transactions of the American Fisheries Society* 100:570–572.

Copeland, J.R. and R.L. Noble. 1994. Movements by young-of-year and yearling largemouth bass and their implications for supplemental stocking. *North American Journal of Fisheries Management* 14:119–124.

Copp, G.H. and B. Cellot. 1988. Drift of embryonic and larval fishes, especially *Lepomis gibbosus* (L.), in the upper Rhone River. *Journal of Freshwater Ecology* 4:419–424.

Couey, F.M. 1935. Fish food studies of a number of northeastern Wisconsin lakes. *Transactions of the Wisconsin Academy of Science, Arts, and Letters* 29:131–172.

Coutant, C.C. 1975. Responses of bass to natural and artificial temperature regimes. Pages 272–285 in R.H. Stroud and H. Clepper, editors. *Black Bass Biology and Management.* Sport Fishing Institute, Washington, D.C., USA.

Coutant, C.C. 1977. Compilation of temperature preference data. *Journal of the Fisheries Research Board of Canada* 34:739–745.

Crawford, S., W.S. Coleman, and W.F. Porak. 1989. Time of annulus formation in otoliths of Florida largemouth bass. *North American Journal of Fisheries Management* 9:231–233.

Cross, F.B. 1950. Effects of sewage and of a headwaters impoundment on the fishes of Stillwater Creek in Payne County, Oklahoma. *American Midland Naturalist* 43:128–145.

Cross, F.B. 1967. *Handbook of Fishes of Kansas.* Museum of Natural History, University of Kansas, Lawrence, Kansas, USA.

Crossman, E.J., J. Houston, and R.R. Campbell. 1996. The status of the warmouth, *Chaenobryttus gulosus,* in Canada. *Canadian Field-Naturalist* 110:495–500.

Crowl, T.A. and J. Boxrucker. 1988. Possible competitive effects of two introduced planktivores on white crappie. *Proceedings of the Forty-Second Annual Conference of the Southeastern Association of Fish and Wildlife Agencies* 42:185–192.

Crumpton, J.E., M.M. Hale, and D.J. Renfro. 1988. Growth history of black crappie Spawned 1980, St. Johns River, Florida. *Proceedings of the Forty-Second Annual Conference of the Southeastern Association of Fish and Wildlife Agencies* 42:107–111.

Crutchfield, J.U., Jr., T.E. Thompson, and J.M. Swing. 2003. Ecological changes following an alewife introduction in an oligotrophic reservoir: a case history. *Proceedings of the Fifty-Seventh Annual Conference of the Southeastern Association of Fish and Wildlife Agencies* 57:44–58.

Curd, M.R. 1959. The morphology of abnormal lateral-line canals in the centrarchid fish *Lepomis humilis* (Girard). *Proceedings of the Oklahoma Academy of Science* 39:359–427.

Danylchuk, A.J. and M.G. Fox. 1994. Age and size-dependent variation in the seasonal timing and probability of reproduction among mature female pumpkinseed, *Lepomis gibbosus. Environmental Biology of Fishes* 39:119–127.

Davis, J.R. 1972. The spawning behavior, fecundity rates, and food habits of the redbreast in southeastern North Carolina. *Proceedings of the Twenty-Fifth Annual Conference of the Southeastern Association of Game and Fish Commissioners* 25:556–560.

Dawley, R.M. 1987. Hybridization and polyploidy in a community of three sunfish species (Pisces: Centrarchidae). *Copeia* 1987:326–335.

Delp, J.G., J.S. Tillma, M.C. Quist, and C.S. Guy. 2000. Age and growth of four centrarchid species in southeastern Kansas streams. *Journal of Freshwater Ecology* 15:475–478.

DeMont, D.J. 1982. Use of *Lepomis macrochirus* Rafinesque nests by spawning *Notemigonus crysoleucas* (Mitchill) (Pisces: Centrarchidae and Cyprinidae). *Brimleyana* 8:61–63.

Dendy, J.S. 1945. Fish distribution, Norris Reservoir, Tennessee, 1943. II. Depth distribution of fish in relation to environmental factors, Norris Reservoir. *Journal of the Tennessee Academy of Science* 20:114–135.

Dendy, J.S. 1946. Further studies of depth distribution of fish, Norris Reservoir, Tennessee. *Journal of the Tennessee Academy of Science* 21:94–104.

Dendy, J.S. 1948. Predicting depth distribution of fish in three TVA storage type reservoirs. *Transactions of the American Fisheries Society* 75:65–71.

DeParker, W.D. 1971. Preliminary studies on sexing adult largemouth bass by means of an external characteristic. *The Progressive Fish-Culturist* 33:54–55.

Desselle, W.J., M.A. Poirrier, J.S. Rogers, and R.C. Cashner. 1978. A discriminate function analysis of sunfish (*Lepomis*) food habits and feeding niche segregation in the Lake Ponchartrain, Louisiana estuary. *Transactions of the American Fisheries Society* 107:713–719.

DeVries, D.R., M.T. Bremigan, and R.A. Stein. 1998. Prey selection by larval fishes as influenced by available zooplankton and gape limitation. *Transactions of the American Fisheries Society* 127:1040–1050.

DeVries, D.R. and R.A. Stein. 1990. Manipulating shad to enhance sport fisheries in North America: an assessment. *North American Journal of Fisheries Management* 10:209–223.

Dewey, M.R. 1992. Effectiveness of a drop net, a pop net, and an electrofishing frame for collecting quantitative samples of juvenile fishes in vegetation. *North American Journal of Fisheries Management* 12:808–813.

Dewey, M.R. and C.A. Jennings. 1992. Habitat use by larval fishes in a backwater lake of the upper Mississippi River. *Journal of Freshwater Ecology* 7:363–372.

Dewey, M.R. and T.E. Moen. 1982. Population dynamics of largemouth bass in DeGray Lake, Arkansas. *Proceedings of the Thirty-Fifth Annual Conference of the Southeastern Association of Fish and Wildlife Agencies* 35:430–437.

Dewey, M.R., T.E. Moen, and H.E. Bryant. 1981. Seasonal biomass of selected centrarchids in Degray Lake, Arkansas, 1977–79. *Proceedings of the Thirty-Fifth Annual Conference of the Southeastern Association of Fish and Wildlife Agencies* 35:438–442.

Dewey, M.R., W.B. Richardson, and S.J. Zigler. 1997. Patterns of foraging and distribution of bluegill sunfish in a Mississippi River backwater: influence of macrophytes and predation. *Ecology of Freshwater Fish* 6:8–15.

DeWoody, J.A., D.E. Fletcher, M. Mackiewicz, S.D. Wilkins, and J.C. Avise. 2000. The genetic mating system of spotted sunfish (*Lepomis punctatus*): mate numbers and the influence of male reproductive parasites. *Molecular Ecology* 9:2119–2128.

Dibble, R.D. 1991. Intra-guild habitat partitioning in centrarchids from clear water Ozark lakes. *Bulletin of the Ecological Society of America* 72:100 (abstract).

DiCenzo, V.J., M.J. Maceina, and W.C. Reeves. 1995. Factors related to growth and condition of the Alabama subspecies of spotted bass in reservoirs. *North American Journal of Fisheries Management* 15:794–798.

DiConstanzo, C.J. 1957. Growth of bluegill, *Lepomis macrochirus*, and pumpkinseed, *L. gibbosus*, of Clear Lake, Iowa. *Iowa State College Journal of Science* 32:19–34.

Dimond, W.F. and T.W. Storck. 1985. Abundance, spatiotemporal distribution, and growth of bluegill and redear sunfish fry in a 0.6-ha pond. *Journal of Freshwater Ecology* 3:93–102.

Dimond, W.F., T.W. Storck, and K.C. Kruse. 1985. A turbidity-related delay in bluegill (*Lepomis macrochirus*) reproduction and some size-dependent differences in the spatial distribution of bluegill fry. *Transactions of the Illinois Academy of Science* 78:49–56.

Doan, K.H. 1940. Studies of the smallmouth bass. *Journal of Wildlife Management* 4:241–266.

Dockendorf, K.J. and M.S. Allen. 2005. Age-0 black crappie abundance and size in relation to zooplankton density, stock abundance, and water clarity in three Florida lakes. *Transactions of the American Fisheries Society* 134:172–183.

Dominey, W.J. 1981. Maintenance of female mimicry as a reproductive strategy in bluegill sunfish (*Lepomis macrochirus*). *Environmental Biology of Fishes* 6:59–64.

Dominey, W.J. 1983. Mobbing in colonially nesting fishes, especially the bluegill, *Lepomis macrochirus*. *Copeia* 1983:1086–1088.

Dorr, B., I.A. Munn, and K.O. Meals. 2002. A socioeconomic and biological evaluation of current and hypothetical crappie regulations in Sardis Lake, Mississippi: an integrated approach. *North American Journal of Fisheries Management* 22:1376–1384.

Dreves, D.P. and T.J. Timmons. 2002. Relationships between diet and growth of age-0 largemouth bass in a Kentucky Lake embayment. *Proceedings of the Fifty-Fifth Annual Conference of the Southeastern Association of Fish and Wildlife Agencies* 55:175–193.

Dubuc, R.A. and D.R. DeVries. 2002. An exploration of factors influencing crappie early life history in three Alabama impoundments. *Transactions of the American Fisheries Society* 131:476–491.

Duffy, J.J. and D.C. Jackson. 1994. Bluegill populations associated with water lily, water shield, and pondweed stands: management implications. *Proceedings of the Forty-Eighth Annual Conference of the Southeastern Association of Fish and Wildlife Agencies* 48:566–574.

Dunham, R.A., K.G. Norgren, L. Robison, R.O. Smitherman, T. Steeger, D.C. Peterson, and M. Gibson. 1994. Hybridization and biochemical genetics of black and white crappies in the southeastern USA. *Transactions of the American Fisheries Society* 123:141–149.

Dunham, R.A., C.J. Turner, and W.C. Reeves. 1992. Introgression of the Florida largemouth bass genome into native populations in Alabama public lakes. *North American Journal of Fisheries Management* 12:494–498.

Dupuis, H.M.C. and M.H.A. Keenleyside. 1988. Reproductive success of nesting male longear sunfish (*Lepomis megalotis peltastes*). *Behavioral Ecology and Sociobiology* 23:109–116.

Eaton, J.G., W.A. Swenson, T.D. McCormick, T.D. Simonson, and K.M. Jensen. 1992. A field and laboratory investigation of acid effects on largemouth bass, rock bass, black crappie, and yellow perch. *Transactions of the American Fisheries Society* 121:644–658.

Eaton, T.H. 1953. Pygmy sunfishes and the meaning of pygmism. *Journal of the Elisha Mitchell Society* 69:98–99.

Eaton, T.H., Jr. 1956. Notes on the olfactory organs in Centrarchidae. *Copeia* 1956:196–199.

Echelle, A.F. and A.A. Echelle. 2002. Reproductive behavior in banded pygmy sunfish, *Elassoma zonatum* (Elassomatidae), with comments on implications for relationships of the genus. *Universidad Autónoma de Nuevo León/Facultad de Ciencias Biológicas*, Monterrey, Mexico 2002:79–86.

Eddy, S. 1969. *How to Know the Freshwater Fishes*. W.C. Brown Company, Dubuque, Iowa, USA.

Eddy, S. and T. Surber. 1943. *Northern Fishes with Special Reference to the Upper Mississippi Valley*, 3rd edition. The University of Minnesota Press, Minneapolis, Minnesota, USA.

Eddy, S. and T. Surber. 1947. *Northern Fishes with Special Reference to the Upper Mississippi Valley*. The University of Minnesota Press, Minneapolis, Minnesota, USA.

Eddy, S. and J.C. Underhill. 1974. *Northern Fishes*. University of Minnesota Press, Minneapolis, Minnesota, USA.

Edwards, E.A., D.A. Krieger, M. Bacteller, and O.E. Maughan. 1982a. Habitat suitability index models: black crappie. United States Department of the Interior, Fish and Wildlife Service. FWS/OBS-82/10.6. Washington, D.C., USA.

Edwards, E.A., D.A. Krieger, G. Gebhart, and O.E. Maughan. 1982b. Habitat suitability index models: white crappie. United States Fish and Wildlife Service, FWS/OBS-82/10.7, Washington, D.C., USA.

Ehlinger, T.J. 1991. Allometry and analysis of morphometric variation in bluegill, *Lepomis macrochirus*. *Copeia* 1991:347–357.

Eipper, A.W. 1975. Environmental influences on the mortality of bass embryos and larvae. Pages 295–305 in R.H. Stroud and H. Clepper, editors. *Black Bass Biology and Management*. Sport Fishing Institute, Washington, D.C., USA.

Ellison, D.G. 1984. Trophic dynamics of a Nebraska black crappie and white crappie population. *North American Journal of Fisheries Management* 4:355–364.

Elrod, J.H., W.N. Bush, B.L. Griswold, C.P. Schneider, and D.R. Wolfert. 1981. Food of white perch, rock bass, and yellow perch in eastern Lake Ontario. *New York Fish and Game Journal* 28:191–207.

Emery, A.R. 1973. Preliminary comparisons of day and night habitats of freshwater fish in Ontario lakes. *Journal of the Fisheries Research Board of Canada* 30:761–774.

Engelbrecht, R.S., R.G. Blankenbaker, F.C. Campbell, W.J. Kilgour, L.N. Clausing, N. DeBenedictis, W.M. Abbitt, Jr., L.C. Hansbarger, R.C. Armstrong, C. McGoy, K.B. Floyd, R.D. Hoyt, and S. Timbrook. 1984. Chronology of appearance and habitat partitioning by stream larval fishes. *Transactions of the American Fisheries Society* 113:217–223.

Epifanio, J.M., M. Hooe, D.H. Buck, and D.P. Philipp. 1999. Reproductive success and assortative mating among *Pomoxis* species and their hybrids. *Transactions of the American Fisheries Society* 128:104–120.

Erdman, D.S. 1984. Exotic fishes of Puerto Rico. Pages 162–176 in W.R. Courtenay and J.R. Stauffer, Jr., editors. *Distribution, Biology, and Management of Exotic Fishes*. Johns Hopkins University Press, Baltimore, Maryland, USA.

Ernest, J.R. 1960. The Embryology and Larval Development of the Northern Rock Bass, *Ambloplites rupestris rupestris* (Rafinesque). Master's thesis. Ohio State University, Columbus, Ohio, USA.

Eschmeyer, R.W. 1940. Growth of fishes in Norris Lake, Tennessee. *Tennessee Academy of Science* 15:329–341.

Eschmeyer, R.W. 1942. The catch, abundance, and migration of game fishes in Norris Reservoir, Tennessee, 1940. *The Tennessee Academy of Science* 17:90–115.

Eschmeyer, R.W. 1944. *Norris Lake Fishing 1944*. Division of Game and Fish, Department of Conservation, State of Tennessee, Nashville, Tennessee, USA.

Eschmeyer, R.W. 1948. Growth of fishes in Norris Lake, Tennessee. *Proceedings of the Tennessee Academy of Science* 15:329–341.

Eschmeyer, R.W. and A.M. Jones. 1941. The growth of game fishes in Norris Reservoir during the first five years of impoundment. *Transactions of the Sixth North American Wildlife Conference* 6:222–240.

Etnier, D.A. 1968. Reproductive success of natural populations of hybrid sunfish in three Minnesota lakes. *Transactions of the American Fisheries Society* 97:466–471.

Etnier, D.A. and W.C. Starnes. 1993. *The Fishes of Tennessee*. The University of Tennessee Press, Knoxville, Tennessee, USA.

Everhart, W.H. 1949. Body length of the smallmouth bass at scale formation. *Copeia* 1949:110–115.

Evermann, B.W. and H.W. Clark. 1920. *Lake Maxinkuckee, a Physical and Biological Survey*. Indiana Department of Conservation Publication 7:1–660.

Faber, D.J. 1963. Larval Fish from the Pelagic Region of Two Wisconsin Lakes. Doctoral dissertation. University of Wisconsin, Madison, Wisconsin, USA.

Faber, D.J. 1967. Limnetic larval fish in northern Wisconsin lakes. *Journal of the Fisheries Research Board of Canada* 24:927–937.

Faber, D.J. 1982. Fish larvae caught by a light-trap at littoral sites in Lac Henry, Quebec, 1979 and 1980. *Proceedings of the Fifth Annual Larval Fish Conference* 5:42–46. Louisiana Cooperative Fisheries Research Unit, Baton Rouge, Louisiana, USA.

Faber, D.J. 1984a. Water babies, larval fishes of Ottawa and vicinity. Part III. Larval fishes of shallow water species. Ottawa Field Naturalist's Club Publication "*Trail and Landscape*" 18:273–283. Available through interlibrary loan: Library Services Division, Natural Museums of Canada, 2086 Walkley Road, Ottawa, Ontario, Canada.

Faber, D.J. 1984b. Water babies, larval fishes of Ottawa and vicinity. Part I. Distribution and phrenology of baby fish in lakes and ponds. Ottawa Field Naturalist's Club Publication "*Trail and Landscape*" 18:84–92. Available through interlibrary loan: Library Services Division, Natural Museums of Canada, 2086 Walkley Road, Ottawa, Ontario, Canada.

Faber, D.J. 1985. Water babies, larval fishes of Ottawa and vicinity. Part IV. Larval fishes of deep water species. Ottawa Field Naturalist's Club Publication "*Trail and Landscape*" 19:18–25. Available through interlibrary loan: Library Services Division, Natural Museums of Canada, 2086 Walkley Road, Ottawa, Ontario, Canada.

Fajen, O.F. 1975a. Population dynamics of bass in rivers and streams. Pages 195–203 in R.H. Stroud and H. Clepper, editors. *Black Bass Biology and Management*. Sport Fishing Institute, Washington, D.C., USA.

Fajen, O.F. 1975b. The standing crop of smallmouth bass and associated species in Courtois Creek. Pages 240–249 in R.H. Stroud and H. Clepper, editors. *Black Bass Biology and Management*. Sport Fishing Institute, Washington, D.C., USA.

Fajen, O.F. 1975c. The establishment of spotted bass fisheries in some northern Missouri streams. *Proceedings of the 29th Annual Conference of the Southeastern Association of Game and Fish Commissioners* 29:28–35.

Farris, D.A. 1959. A change in the early growth rates of four larval marine fishes. *Limnology and Oceanography* 4:29–36.

Fayram, A.H. and T.H. Sibley. 2000. Impact of predation by smallmouth bass on sockeye salmon in Lake Washington, Washington. *North American Journal of Fisheries Management* 20:81–89.

Fields, R., S.S. Lowe, C. Kaminski, G.S. Whitt, and D.P Phillip. 1987. Critical and chronic thermal maxima of northern and Florida largemouth bass and their reciprocal F₁ and F₂ hybrids. *Transactions of the American Fisheries Society* 116:856–863.

Finger, T.R. 1982. Fish community-habitat relations in a central New York stream. *Journal of Freshwater Ecology* 1:343–352.

Finnell, J.C., R.M. Jenkins, and G.E. Hall. 1956. The fishery resources of the Little River system. McCurtain County, Oklahoma. Report Number 55, Oklahoma Fisheries Research Laboratory, Norman. 81 pp.

Fischer, S.A. and W.E. Kelso. 1988. Potential parasite-induced mortality in age-0 bluegills in a floodplain pond of the lower Mississippi River. *Transactions of the American Fisheries Society* 117:565–573.

Fischer, S.A. and W.E. Kelso. 1990. Parasite fauna development in juvenile bluegills and largemouth bass. *Transactions of the American Fisheries Society* 119:877–894.

Fish, M.P. 1932. Contributions to the early life histories of sixty-two species of fishes from Lake Erie and its tributaries. *United States Bureau of Fisheries Bulletin* 47:293–398.

Fish, P.A. and J. Savitz. 1983. Variations in home ranges of largemouth bass, yellow perch, bluegills, and pumpkinseeds in an Illinois lake. *Transactions of the American Fisheries Society* 112:147–153.

Fisher, W.L. and A.V. Zale. 1992. Effects of water level fluctuations on abundance of young-of-year largemouth bass in a hydropower reservoir. *Proceedings of the Forty-Fifth Annual Conference of the Southeastern Association of Fish and Wildlife Agencies* 45:422–431.

Fitz, R.B. 1966. Distribution records of the pumpkinseed (*Lepomis gibbosus*) and redbreast sunfish (*Lepomis auritus*) in Tennessee. *Journal of the Tennessee Academy of Science* 41:98.

Fleener, G.G. 1975. Harvest of smallmouth bass and associated species in Courtois Creek. Pages 250–256 in R.H. Stroud and H. Clepper, editors. *Black Bass Biology and Management*. Sport Fishing Institute, Washington, D.C., USA.

Flemer, D.A. and W.S. Woolcott. 1966. Food habits and distribution of the fishes of Tachahoe Creek, Virginia, with special emphasis on the bluegill, *Lepomis m. macrochirus* Rafinesque. *Chesapeake Science* 7:75–89.

Flickinger, S.A., R.J. Anderson, and S.J. Puttmann. 1975. Intensive culture of smallmouth bass. Pages 373–379 in R.H. Stroud and H. Clepper, editors. *Black Bass Biology and Management*. Sport Fishing Institute, Washington, D.C., USA.

Floyd, K.B., R.D. Hoyt, and S. Timbrook. 1984. Chronology of appearance and habitat partitioning by stream larval fishes. *Transactions of the American Fisheries Society* 113:217–223.

Forbes, S.A. 1880. Various papers on the food of fishes. *Bulletin of the Illinois State Laboratory of Natural History* 1:18–65.

Forbes, S.A. and R.E. Richardson. 1909. *The Fishes of Illinois*. State Laboratory of the Natural History Survey of Illinois, Champaign, Illinois, USA.

Forbes, S.A. and R.E. Richardson. 1920. *The Fishes of Illinois*. Illinois Natural History Survey, Urbana, Illinois, USA.

Forney, J.L. 1974. Interactions between yellow perch abundance, walleye predation and survival of alternate prey in Oneida Lake, New York. *Transactions of the American Fisheries Society* 103:15–24.

Fowler, H.W. 1920. Notes on New Jersey, Pennsylvania and Virginia fishes. *Proceedings of the Academy of Natural Science of Philadelphia.* 1919:292–300.

Fowler, H.W. 1936. Notes on Florida fishes 1931–32. *The Fish Culturist* 15:16–17.

Fowler, H.W. 1945. A study of the fishes of the southern Piedmont and Coastal Plain. *Academy of Natural Science Philadelphia*, Monograph 7:1–408.

Fox, M.G., J.E. Claussen, and D.P. Philipp. 1997. Geographic patterns in genetic and life history variation in pumpkinseed populations from four east-central Ontario watersheds. *North American Journal of Fisheries Management* 17:543–556.

Fox, M.G. and A. Keast. 1991. Effect of overwinter mortality on reproductive life history characteristics of pumpkinseed (*Lepomis gibbosus*) populations. *Canadian Journal of Fisheries and Aquatic Science* 48:1792–1799.

Freeman, M.C. 1993. Movements by little fish in a big stream. *Abstracts of the First Mid-Year Technical Meeting, Southern Division American Fisheries Society* 1993:23–24.

French, J.R.P. III. 1988. Effect of submersed aquatic macrophytes on resource partitioning in yearling rock bass (*Ambloplites rupestris*) and pumpkinseeds (*Lepomis gibbosus*) in Lake St. Clair. *Journal of Great Lakes Research* 14:291–300.

French, J.R.P. III and M.N. Morgan. 1995. Preference of redear sunfish on zebra mussels and rams-horn snails. *Journal of Freshwater Ecology* 10:49–55.

Fuhr, M.A., D.H. Wahl, and D.P. Phillip. 2002. Fall abundance of age-0 largemouth bass is more important than size in determining age-1 year-class strength in Illinois. Pages 91–99 in D.P. Phillip and M.S. Ridgway, editors. *Black Bass: Ecology, Conservation, and Management.* American Fisheries Symposium 31, Bethesda, Maryland, USA.

Funk, J.L. 1955. Movement of stream fishes in Missouri. *Transactions of the American Fisheries Society* 85:39–57.

Funk, J.L. 1957. Movement of stream fishes in Missouri. *Transactions of the American Fisheries Society* 85:39–57.

Funk, J.L. 1975a. Evaluation of the smallmouth bass population and fishery in Courtois Creek. Pages 257–269 in R.H. Stroud and H. Clepper, editors. *Black Bass Biology and Management.* Sport Fishing Institute, Washington, D.C., USA.

Funk, J.L. 1975b. Structure of fish communities in streams which contain bass. Pages 140–153 in R.H. Stroud and H. Clepper, editors. *Black Bass Biology and Management.* Sport Fishing Institute, Washington, D.C., USA.

Funk, J.L. and W.L. Pflieger. 1975. Courtois Creek, a smallmouth bass stream in the Missouri Ozarks. Pages 224–230 in R.H. Stroud and H. Clepper, editors. *Black Bass Biology and Management.* Sport Fishing Institute, Washington, D.C., USA.

Galinat, G.F., D.W. Willis, B.G. Blackwell, and M.J. Hubers. 2002. Influence of a saugeye (sauger × walleye) introduction program on the black crappie population in Richmond Lake, South Dakota. *North American Journal of Fisheries Management* 22:1416–1424.

Gallagher, R.P. and J.V. Conner. 1983. Comparison of two ichthyoplankton sampling gears with notes on microdistribution of fish larvae in a large river. *Transactions of the American Fisheries Society* 112:280–285.

Gammon, J.R. 1998. *The Wabash River Ecosystem.* Indiana University Press, Bloomington, Indiana, USA.

Garren, D.A., J.J. Ney, and P.E. Bugas, Jr. 2001. Seasonal movement and distribution of smallmouth bass in a Virginia impoundment. *Proceedings of the Fifty-Fifth Annual Conference of the Southeastern Association of Fish and Wildlife Agencies* 55:165–174.

Garvey, J.E. and R.A. Stein. 1998. Competition between larval fishes in reservoirs: the role of relative timing of appearance. *Transactions of the American Fisheries Society* 127:1021–1039.

Garvey, J.E., R.A. Stein, R.A. Wright, and M.T. Bremigan. 2002. Exploring ecological mechanisms underlying largemouth bass recruitment along environmental gradients. Pages 7–24 in D.P. Phillip and M.S. Ridgway, editors. *Black Bass: Ecology, Conservation, and Management.* American Fisheries Society, Symposium 31, Bethesda, Maryland, USA.

Garvey, J.E., R.A. Wright, and R.A. Stein. 1998. Overwinter growth and survival of age-0 largemouth bass. *Canadian Journal of Fisheries and Aquatic Sciences* 55:2414–2424.

Gasaway, C.R. 1968. Comparison of bass-bluegill and bass-redear sunfish stocking in Oklahoma farm ponds. *Proceedings of the Oklahoma Academy of Science* 47:397–406.

Geaghan, J.P. and M.T. Huish. 1981. Evaluation of condition factor of fliers in three North Carolina swamp streams. Pages 163–167 in L.A. Krumholz, editor. *The Warmwater Streams Symposium.* American Fisheries Society, Bethesda, Maryland, U.S.A.

Gebhart, G.E. and R.C. Summerfelt. 1975. Factors affecting the vertical distribution of white crappie (*Pomoxis annularis*) in two Oklahoma reservoirs. *Proceedings of the Twenty-Eighth Annual Conference of the Southeastern Association of Game and Fish Commissioners* 28:355–366.

Gebhart, G.E. and R.C. Summerfelt. 1978. Seasonal growth rates of fishes in relation to conditions of lake stratification. *Proceedings of the Oklahoma Academy of Science* 58:6–10.

Gelwick, F.P., E.R. Gilliland, and W.J. Matthews. 1995. Introgression of the Florida largemouth bass genome into stream populations of northern largemouth Bass in Oklahoma. *Transactions of the American Fisheries Society* 124:550–562.

George, E.L. and W.F. Hadley. 1979. Food and habitat partitioning between rock bass (*Ambloplites rupestris*) and smallmouth bass (*Micropterus dolomieui*) young of the year. *Transactions of the American Fisheries Society* 108: 253–261.

Gerald, J.W. 1970. Species Isolating Mechanisms in the Genus *Lepomis.* Doctoral dissertation. University of Texas, Austin, Texas, USA.

Gerald, J.W. 1971. Sound production during courtship in six species of sunfish (Centrarchidae). *Evolution* 25:75–87.

Gerber, G.P. 1987. Movements and Behavior of Smallmouth Bass, *Micropterus dolomieui*, and Rock Bass, *Ambloplites rupestris*, in Southcentral Lake Ontario and Two Tributaries. Master's thesis. State University of New York, Brockport, New York, USA.

Gerber, G.P. and J.M. Haynes. 1988. Movements and behavior of smallmouth bass, *Micropterus dolomieui*, and rock bass, *Ambloplites rupestris*, in southcentral Lake Ontario and two tributaries. *Journal of Freshwater Ecology* 4:425–440.

Gerking, S.D. 1945. Distribution of the fishes of Indiana. *Investigations of Indiana Lakes and Streams* 3:1–137.

Gerking, S.D. 1953. Evidence for the concepts of home range and territory in stream fishes. *Ecology* 34:347–365.

Germann, J.F., L.E. McSwain, D.R. Holder, and C.D. Swanson. 1975. Life history of the warmouth in the Suwannee River and Okefenokee Swamp, Georgia. *Proceedings of the Twenty-Eighth Annual Conference of the Southeastern Association of Game and Fish Commissioners* 28:259–278.

Ghent, A.W. and B. Grinstead. 1965. A new method of assessing contagion, applied to a distribution of redear sunfish. *Transactions of the American Fisheries Society* 94:135–142.

Gilbert, C.R. 2004. Family Elassomatidae Jordan 1877 — pygmy sunfishes. *California Academy of Sciences Annotated Checklists of Fishes* 33:1–5.

Gilliland, E.R. 1992. Experimental stocking of Florida largemouth bass into small Oklahoma reservoirs. *Proceedings of the Forty-Sixth Annual Conference of the Southeastern Association of Fish and Wildlife Agencies* 46:487–494.

Gilliland, E., R. Horton, B. Hysmith, and J. Moczygemba. 1991. Smallmouth bass in Lake Texoma, a case history. *First International Smallmouth Bass Sysposium* 1991:136–142.

Gilliland, E.R. and J. Whitaker. 1989. Introgression of Florida largemouth bass introduced into northern largemouth bass populations in Oklahoma reservoirs. *Proceedings of the Forty-Third Annual Conference of the Southeastern Association of Fish and Wildlife Agencies* 43:182–190.

Goddard, K. and A. Mathis. 1997. Microhabitat preferences of longear sunfish: low light intensity versus submerged cover. *Environmental Biology of Fishes* 49:495–499.

Goldstein, R.M. 1994. Size selection of prey by young largemouth bass. *Proceedings of the Forty-Seventh Annual Conference Southeastern Association of Fish and Wildlife Agencies* 47:596–604.

Goldstein, R.M. and T.P. Simon. 1999. Toward a united definition of guild structure for feeding ecology of North American freshwater fishes. Pages 123–202 in T.P. Simon, editor. *Assessing the Sustainability and Biological Integrity of Water Resources Using Fish Communities.* CRC Press, Boca Raton, Florida, USA.

Goodgame, L.S. and L.E. Miranda. 1993. Early growth and survival of age-0 largemouth bass in relation to parental size and swim-up time. *Transactions of the American Fisheries Society* 122:131–138.

Goodyear, C.S., T.A. Edsall, D.M. Ormsby Dempsey, G.D. Moss, and P.E. Polowski. 1982. *Atlas of the Spawning and Nursery Areas of Great Lakes Fishes. Volume 13: Reproductive Characteristics of Great Lakes Fishes.* United States Fish and Wildlife Service, FWS/OBS-82/52, Washington, D.C., USA.

Gotceitas, V. and P. Colgan. 1987. Selection between densities of artificial vegetation by young bluegills avoiding predation. *Transactions of the American Fisheries Society* 116:40–49.

Gotceitas, V. and P. Colgan. 1990. The effects of prey availability and predation risk on habitat selection by juvenile bluegill sunfish. *Copeia* 1990:409–417.

Gran, J.E. 1995. Gonad Development and Spawning of Largemouth Bass in a Tropical Reservoir. Master's thesis. North Carolina State University, Raleigh, North Carolina, USA.

Green, D.M. and R.C. Heidinger. 1994. Longevity record for largemouth bass. *North American Journal of Fisheries Management* 14:464–465.

Greene, J.C. and M.J. Maceina. 2000. Influence of trophic state on spotted bass and largemouth bass spawning time and age-0 population characteristics in Alabama reservoirs. *North American Journal of Fisheries Management* 20:100–108.

Greenwood, P.H., D.E. Rosen, S.H. Weitzman, and M.S. Myers. 1966. Phyletic studies of teleostean fishes with a provisional classification of living forms. *Bulletin of the American Museum of Natural History* 131:341–455.

Gregory, R.S. and P.M. Powles. 1985. Chronology, distribution, and sizes of larval fish sampled by light traps in macrophytic Chemung Lake. *Canadian Journal of Zoology* 63:2569–2577.

Gregory, R.S. and P.M. Powles. 1988. Relative selectivities of Miller High-Speed samplers and light traps for collecting ichthyoplankton. *Canadian Journal of Fisheries and Aquatic Science* 45:993–998.

Grinstead, B.G. 1969. The vertical distribution of the white crappie in the Buncombe Creek arm of Lake Texoma. *Oklahoma Fishery Research Laboratory, Bulletin Number 3.* University of Oklahoma, Norman, Oklahoma, USA.

Grinstead, B.G. 1975. Response of bass to removal of competing species by commercial fishing. Pages 436–449 in R.H. Stroud and H. Clepper, editors. *Black Bass Biology and Management.* Sport Fishing Institute, Washington, D.C., USA.

Grizzle, J.M., I. Altinok, W.A. Fraser, and R. Francis-Floyd. 2002. First isolation of largemouth bass virus. *Diseases of Aquatic Organisms* 50:233–235.

Grizzle, J.M. and C.J. Brunner. 2003. Review of largemouth bass virus. *Fisheries* 28:10–14.

Gross, M.R. and W.A. Nowell. 1980. The reproductive biology of rock bass, *Ambloplites rupestris* (Centrachidae), in Lake Opinicon, Ontario. *Copeia* 1980:482–494.

Grossman, G.D., R.E. Ratajczak, Jr., and M.K. Crawford. 1995. Do rock bass (*Ambloplites rupestris*) induce microhabitat shifts in mottled sculpins (*Cottus bairdi*)? *Copeia* 1995:343–353.

Guest, W.C., R.W. Drenner, S.T. Threlkeld, F.D. Martin, and J.D. Smith. 1990. Effects of gizzard shad and threadfin shad on zooplankton and young-of-year white crappie production. *Transactions of the American Fisheries Society* 119:529–536.

Guier, C.R., W.G. Miller, A.W. Mullis, and L.E. Nichols. 1978. Comparison of growth rates and abundance of largemouth bass in selected North Carolina coastal rivers. *Proceedings of the Thirty-Second Annual Conference of the Southeastern Association of Fish and Wildlife Agencies* 32:391–400.

Gunning, G.E. and W.M. Lewis. 1955. The fish population of a spring-fed swamp in the Mississippi bottoms of southern Illinois. *Ecology* 36:552–558.

Gunning, G.E. and C.R. Shoop. 1963. Occupancy of home range by longear sunfish, *Lepomis m. megalotis* (Rafinesque), and bluegill, *Lepomis m. macrochirus* Rafinesque. *Animal Behaviour* 11:325–330.

Gunning, G.E. and R.D. Suttkus. 1990. Species dominance in two river populations of sunfishes (Pisces: Centrarchidae): 1966 to 1988. *The Southwestern Naturalist* 35:346–348.

Guy, C.S. and D.W. Willis. 1990. Structural relationships of largemouth bass and bluegill populations in South Dakota ponds. *North American Journal of Fisheries Management* 10:338–343.

Guy, C.S. and D.W. Willis. 1995. Population characteristics of black crappies in South Dakota waters: a case for ecosystem-specific management. *North American Journal of Fisheries Management* 15:745–765.

Guy, C.S., D.W. Willis, and J.J. Jackson. 1994. Biotelemetry of white crappies in a South Dakota lake. *Transactions of the American Fisheries Society* 123:63–70.

Gwinner, H.R. 1973. A Study of the Redeye Bass, *Micropterus coosae*, Smallmouth Bass, *Micropterus dolomieui*, and Rock Bass, *Ambloplites rupestris* in Spring Creek, Tennessee. Master's thesis. Tennessee Technological University, Cookeville, Tennessee, USA.

Gwinner, H.R., H.J. Cathey, and F.J. Bulow. 1975. A study of two populations of introduced redeye bass, *Micropterus coosae* Hubbs and Bailey. *Journal of the Tennessee Academy of Science* 50:102–105.

Hackney, P.A. 1975. Bass populations in ponds and community lakes. Pages 131–139 in R.H. Stroud and H. Clepper, editors. *Black Bass Biology and Management*. Sport Fishing Institute, Washington, D.C., USA.

Hale, R.S. 1999. Growth of white crappies in response to temperature and dissolved oxygen conditions in a Kentucky reservoir. *North American Journal of Fisheries Management* 19:591–598.

Hall, D.J. and E.E. Werner. 1977. Seasonal distribution and abundance of fishes in the littoral zone of a Michigan lake. *Transactions of the American Fisheries Society* 106:545–555.

Hambrick, P.S., C.H. Hocutt, M.T. Masnik, and J.H. Wilson 1973. Additions to the West Virginia ichthyofauna, with comments on the distribution of other species. *Journal of the West Virginia Academy of Science* 45:58–60.

Hammers, B.E. 1990. Comparison of Methods for Estimating Age and Growth of Crappie in Mississippi. Master's thesis. Mississippi State University, Starkville, Mississippi, USA.

Hankinson, T.L. 1908. A biological survey of Walnut Lake, Michigan. *State Biology Report, Geological Survey, Michigan* 1907:153–288.

Hankinson, T.L. 1919. Notes on life-histories of Illinois fish. *Transactions of the Illinois State Academy of Science* 17:132–150.

Hansen, D.F. 1943. On nesting of the white crappie, *Pomoxis annularis*. *Copeia* 1943:259–260.

Hansen, D.F. 1951. Biology of the white crappie in Illinois. *Bulletin of the Illinois Natural History Survey* 25:211–265.

Hansen, D.F. 1965. Further observations on nesting of the white crappie, *Pomoxis annularis*. *Transactions of the American Fisheries Society* 94:182–184.

Hanson, J.M. and S.U. Qadri. 1979. Seasonal growth, food, and feeding habits of young-of-the-year black crappie in the Ottawa River. *Canadian Field-Naturalist* 93:232–238.

Hanson, J.M. and S.U. Qadri. 1984. Feeding ecology of age 0 pumpkinseed (*Lepomis gibbosus*) and black crappie (*Pomoxis nigromaculatus*) in the Ottawa River. *Canadian Journal of Zoology* 62:613–621.

Hardy, J.D., Jr. 1978. *Development of Fishes in the Mid-Atlantic Bight. An Atlas of Egg, Larval and Juvenile Stages. Vol. III. Aphredoderidae through Rachycentridae*. United States Fish and Wildlife Service. FWS/OBS-78/12.

Hargis, H.L. 1965. Age and Growth of *Micropterus salmoides, Micropterus dolomieui*, and *Micropterus punctulatus* in Center Hill Reservoir, Tennessee. Master's thesis. Tennessee Polytechnic Institute, Cookeville, Tennessee, USA.

Harlan, J.R. and E.B. Speaker. 1969. *Iowa Fishes and Fishing*. Iowa State Conservation Commission, Des Moines, Iowa, USA.

Harrel, R.C., B.J. Davis, T.C. Dorris. 1967. Stream order and species diversity of fishes in an intermittent Oklahoma stream. *American Midland Naturalist* 78:428–436.

Hart, T.F., Jr. and R.G. Werner. 1987. Effects of prey density on growth and survival of white sucker, *Catostomus commersoni*, and pumpkinseed, *Lepomis gibbosus*, larvae. *Environmental Biology of Fishes* 18:41–50.

Harvey, B.C. 1991. Interactions among stream fishes: predator-induced habitat shifts and larval survival. *Oecologia* 87:29–36.

Haslbauer, O.F. 1945. Fish distribution, Norris Reservoir, Tennessee, 1943. III. Relation of the bottom to fish distribution, Norris Reservoir. *Journal of the Tennessee Academy of Science* 20:135–138.

Hatch, J.T., M.C. Whiteside, and J.A. Schuldt. 1990. Selectivity in larval fish predation: a Lake Itasca case study. (Abstract.) *Proceedings of the 14th Larval Fish Conference*. Duke Marine Laboratory, Beaufort, North Carolina, USA.

Hatleli, D.C. 1996. Ecological Study of Irogami Lake, Wisconsin, with Emphasis on Mitigating Bluegill Spawning Habitat. Master's thesis. University of Wisconsin, Stevens Point, Wisconsin, USA.

Hay, O.P. 1883. On a collection of fishes from eastern Mississippi. *Proceedings of the U.S. National Museum* 3:488–515.

Hayward, R.S. and E. Arnold. 1996. Temperature dependence of maximum daily consumption in white crappie: implications for fisheries management. *Transactions of the American Fisheries Society* 125:132–138.

Hayward, R.S. and M.E. Bushman. 1994. Gastric evacuation rates for juvenile largemouth bass. *Transactions of the American Fisheries Society* 123:88–93.

Heidinger, R.C. 1975. Life history and biology of the largemouth bass. Pages 11–20 in R.H. Stroud and H. Clepper, editors. *Black Bass Biology and Management*. Sport Fishing Institute, Washington, D.C., USA.

Heidinger, R.C. 1976. Synopsis of biological data on the largemouth bass, *Micropterus salmoides* (Lacepede) 1802. FAO (Food and Agriculture Organization of the United Nations) Fisheries Synopsis 115.

Heidinger, R.C., B. Tetzlaff, and J. Stoeckel. 1985. Evidence of two feeding subpopulations of white crappie (*Pomoxis annularis*) in Rend Lake, Illinois. *Journal of Freshwater Ecology* 3:133–139.

Heinrich, O. 1921. *Centrarchus macropterus*, seine Zucht und Pflege im Zimmer- und Freilandbecken. *Wochenschr. Aquar.-Terrarienk.* 18:109–110.

Heitman, E. 2005. An Evaluation of Population Dynamics and Contribution of Stocked Largemouth Bass in Pools of the Arkansas River. Master's thesis. University of Arkansas, Pine Bluff, Arkansas, USA.

Heitman, N.E., C.L. Racey, and S.E. Lochmann. 2006. Stocking contribution and growth of largemouth bass in pools of the Arkansas River. *North American Journal of Fisheries Management* 26:175–179.

Helfman, G.S. 1981. Twilight activities and temporal structure in a freshwater fish community. *Canadian Journal of Fisheries and Aquatic Sciences* 38:1405–1420.

Helfrich, L.A., K.W. Nutt, and D.L. Weigmann. 1991. Habitat selection by spawning redbreast sunfish in Virginia streams. *Rivers* 2:138–147.

Hellier, T.R., Jr. 1967. The fishes of the Santa Fe River system. *Bulletin of the Florida State Museum of Biological Sciences* 11:1–37.

Hendricks, M.L., J.R. Stauffer, Jr., and C.H. Hocutt. 1983. The zoogeography of the fishes of the Youghiogheny River system, Pennsylvania, Maryland, and West Virginia. *American Midland Naturalist* 110:145–164.

Herke, W.H. 1960. Comparison of the length–weight relationship of several species of fish from two different, but connected, habitats. *Proceedings of the Thirteenth Annual Conference of the Southeastern Association of Fish and Wildlife Agencies* 13:299–313.

Hester, F.E. 1970. Phylogenetic relationships of sunfishes as demonstrated by hybridization. *Transactions of the American Fisheries Society* 99:100–104.

Hickman, G.D. and M.R. Dewey. 1973. Notes on the upper lethal temperature of the duskystripe shiner, *Notropis pilsbryi*, and the bluegill, *Lepomis macrochirus*. *Transactions of the American Fisheries Society* 102:838–840.

Hildebrand, S.F. and I.L. Towers. 1928. Annotated list of fishes collected in the vicinity of Greenwood, Mississippi, with descriptions of three new species. *Bulletin of the United States Bureau of Fisheries* 43:105–136.

Hile, R. 1931. The rate of growth of fishes of Indiana. *Investigations of Indiana Lakes and Streams* 2:9–55.

Hile, R. 1941. Age and growth of the rock bass, *Ambloplites rupestris* (Rafinesque), in Nebish Lake, Wisconsin. *Transactions of the Wisconsin Academy of Science, Arts, and Letters* 33:189–337.

Hile, R. 1942. Growth of rock bass, *Ambloplites rupestris* (Rafinesque), in five lakes of Northeastern Wisconsin. *Transactions of the American Fisheries Society* 71: 131–143.

Hinch, S.G. and N.C. Collins. 1991. Importance of diurnal and nocturnal nest defense in the energy budget of male smallmouth bass: insights from direct video observations. *Transactions of the American Fisheries Society* 120:657–663.

Hiranvat, S. 1975. Age and growth of the redbreast sunfish in the proposed West Point Reservoir area, Alabama and Georgia. *Bulletin of the Georgia Academy of Science* 33:106–116.

Hirst, S.C. and D.R. DeVries. 1994. Assessing the potential for direct feeding interactions among larval black bass and larval shad in two southeastern reservoirs. *Transactions of the American Fisheries Society* 123:173–181.

Hocutt, C.H., R.F. Denoncourt, and J.R. Stauffer, Jr. 1978. Fishes of the Greenbreier River, West Virginia, with drainage history of the central Appalachians. *Journal of Biogeography* 5:59–80.

Hocutt, C.H. and E.O. Wiley (editors). 1986. *The Zoogeography of North American Freshwater Fishes*. John Wiley & Sons, New York, New York, USA.

Hodgson, J.R., D.E. Schindler, and X. He. 1998. Homing tendency of three piscivorous fishes in a north temperate lake. *Transactions of the American Fisheries Society* 127:1078–1081.

Hodson, R.G. and K. Strawn. 1969. Food of young-of-the-year largemouth and spotted bass during the filling of Beaver Reservoir, Arkansas. *Proceedings of the Twenty-Second Annual Conference of the Southeastern Association of Game and Fish Commissioners* 22:510–516.

Hoffman, G.C., C.L. Milewski, and D.W. Willis. 1990. Population characteristics of rock bass in three northeastern South Dakota lakes. *Prairie Naturalist* 22:33–40.

Hoffman, J.M. 1955. Age and Growth of the Green Sunfish, *Lepomis cyanellus* Rafinesque, in the Niangua Arm of the Lake of the Ozarks. Master's thesis. University of Missouri, Columbia, Missouri, USA.

Hoffman, K.J. and P.W. Bettoli. 2005. Growth, dispersal, mortality, and contribution of largemouth bass stocked into Chickamauga Lake, Tennessee. *North American Journal of Fisheries Management* 25:1518–1527.

Hofstetter, A.M., C.F. Dineen, and P.S. Stokely. 1958. Skeletal differences between white and black crappies. *Transactions of the American Microscopic Society* 77:19–21.

Holbrook, J.A., D. Johnson, and J.P. Strzemienski. 1973. Management implications of bass fishing tournaments. *Proceedings of the Twenty-Sixth Annual Conference of the Southeastern Association of Fish and Wildlife Agencies* 26:320–324.

Holder, D.R. 1970. A study of fish movements from the Okefenokee Swamp into the Suwanee River. *Proceedings Annual Conference Southeastern Association of Game and Fish Commission* 24:591–608.

Holland, L.E. 1986. Distribution of early life history stages of fishes in selected pools of the Upper Mississippi River. *Hydrobiologia* 136:121–130.

Holland, L.E. and M.L. Huston. 1985. Distribution and food habits of young-of-the-year fishes in a backwater lake of the upper Mississippi River. *Journal of Freshwater Ecology* 3:81–91.

Holland, L.E. and J.R. Sylvester. 1983. Distribution of larval fishes related to potential navigation impacts on the upper Mississippi River, Pool 7. *Transactions of the American Fisheries Society* 112:293–301.

Holland-Bartels, L.E., S.K. Littlejohn, and M.L. Huston. 1990. *A Guide to Larval Fishes of the Upper Mississippi River.* United States Fish and Wildlife Service, National Fisheries Research Center, LaCrosse, Wisconsin, USA.

Hooe, M.L., D.H. Buck, and D.H. Wahl. 1994. Growth, survival, and recruitment of hybrid crappies stocked in small impoundments. *North American Journal of Fisheries Management* 14:137–142.

Hoover, R.S., J.J. Ney, E.M. Hallerman, and W.B. Kittrell, Jr. 1997. Comparison of Florida, northern, and intergrade juvenile largemouth bass in a Virginia reservoir. *Proceedings of the Fifty-First Annual Conference of the Southeastern Association of Fish and Wildlife Agencies* 51:192–198.

Horton, R.A. and E.R. Gilliland. 1994. Monitoring trophy largemouth bass in Oklahoma using a taxidermist network. *Proceedings of the Forty-Seventh Annual Conference of the Southeastern Association of Fish and Wildlife Agencies* 47:679–685.

Horton, T.B. and C.S. Guy. 2002. Habitat use and movement of spotted bass in Otter Creek, Kansas. *American Fisheries Society Symposium* 31:161–171.

Horton, T.B., C.S. Guy, and J.S. Pontius. 2004. Influence of time interval on estimations of movement and habitat use. *North American Journal of Fisheries Management* 24:690–696.

Houser, A. 1963. Loss in weight of sunfish following aquatic vegetation control using the herbicide Silvex. *Proceedings of the Oklahoma Academy of Science* 43:232–237.

Houser, A. and W.C. Rainwater. 1975. Production of largemouth bass in Beaver and Bull Shoals Lakes. Pages 310–316 in R.H. Stroud and H. Clepper, editors. *Black Bass Biology and Management.* Sport Fishing Institute, Washington, D.C., USA.

Houston, J. 1990. Status of redbreast sunfish, *Lepomis auritus*, in Canada. *Canadian Field-Naturalist* 104:64–68.

Howland, J.W. 1932. The spotted or Kentucky black bass in Ohio. *Bulletin of the Bureau of Scientific Research, Division of Conservation, Ohio Department of Agriculture* 1:1–19.

Hoyt, R.D., S.E. Neff, and V.H. Resh. 1979. Distribution, abundance, and species diversity of fishes of the Upper Salt River drainage, Kentucky. *Transactions of the Kentucky Academy of Science* 40:1–20.

Hubbs, C.L. 1920. Notes on hybrid sunfishes. *Aquatic Life* 5:101–103.

Hubbs, C.L. 1922. Variations in the number of vertebrae and other meristic characters of fishes correlated with the temperature of water during development. *The American Naturalist* 56:360–372.

Hubbs, C.L. 1943. Terminology of early stages of fishes. *Copeia* 1943:260.

Hubbs, C.L. 1955. Hybridization between fish species in nature. *Systematic Zoology* 4:1–20.

Hubbs, C.L. and E.R. Allen. 1943. Fishes of Silver Springs, Florida. *Quarterly Journal of the Florida Academy of Science* 6:110–130.

Hubbs, C.L. and R.M. Bailey. 1938. *The Small-Mouthed Bass.* Cranbrook Institute of Science, Bulletin Number 10, Cranbrook, Michigan, USA.

Hubbs, C.L. and R.M. Bailey. 1940. A revision of the black basses (*Micropterus* and *Huro*) with descriptions of four new forms. *Miscellaneous Publications Museum of Zoology University of Michigan* 48:1–63.

Hubbs, C.L. and L.C. Hubbs. 1931. Experimental verification of natural hybridization between distinct genera of sunfishes. *Papers of the Michigan Academy of Science, Arts, and Letters* 15:427–437.

Hubbs, C.L. and K.F. Lagler. 1958. *Fishes of the Great Lakes Region.* Cranbrook Institute of Science. Cranbrook, Michigan, USA.

Hubbs, C.L. and K.F. Lagler. 1964. *Fishes of the Great Lakes Region with a New Preface.* University of Michigan Press, Ann Arbor, Michigan, USA.

Hubert, W.A. 1976a. Age and growth of three black bass species in Pickwick Reservoir. *Proceedings of the Twenty-Ninth Annual Conference of the Southeastern Association of Game and Fish Commissioners* 29:126–134.

Hubert, W.A. 1976b. Estimation of the fecundity of smallmouth bass, *Micropterus dolomieui* Lacepede, found in the Wilson Dam tailwaters, Alabama. *Journal of the Tennessee Academy of Science* 51:142–144.

Hubert, W.A. 1988. Altitude as the determinant of distribution of largemouth bass and smallmouth bass in Wyoming. *North American Journal of Fisheries Management* 8:386–387.

Huck, L.L. and G.E. Gunning. 1967. Behavior of the longear sunfish, *Lepomis megalotis* (Rafinesque). *Tulane Studies in Zoology* 14:121–131.

Huckins, C.J.F. 1997. Functional linkages among morphology, feeding performance, diet, and competitive ability in molluscivorous sunfish. *Ecology* 78:2401–2414.

Hudson, R.G. and F.E. Hester. 1976. Movements of the redbreast sunfish in Little River, near Raleigh, North Carolina. *Proceedings of the Twenty-Ninth Annual Conference of the Southeastern Association of Game and Fish Commissioners* 29:325–329.

Hudson, W.F. and F.J. Bulow. 1984. Relationships between squamation chronology of the bluegill, *Lepomis macrochirus* Rafinesque, and age-growth methods. *Journal of Fish Biology* 24:459–469.

Hunsaker, D. and R.W. Crawford. 1964. Preferential spawning behavior of largemouth bass, *Micropterus salmoides*. *Copeia* 1964:240–241.

Hunt, J. and C.A. Annett. 2002. Effects of habitat manipulation on reproductive success of individual largemouth bass in an Ozark reservoir. *North American Journal of Fisheries Management* 22:1201–1208.

Hunt, J., N. Bacheler, D. Wilson, E. Videan, and C.A. Annett. 2002. Enhancing largemouth bass spawning: behavioral and habitat considerations. Pages 277–290 in D.P. Phillip and M.S. Ridgway, editors. *Black Bass: Ecology, Conservation, and Management*. American Fisheries Symposium 31, Bethesda, Maryland, USA.

Hunter, J.R. 1963. The reproductive behavior of the green sunfish, *Lepomis cyanellus*. *Zoologica: New York Zoological Society* 48:13–24.

Hunter, J.R. and A.D. Hasler. 1965. Spawning association of the redfin shiner, *Notropis umbratilis*, and the green sunfish, *Lepomis cyanellus*. *Copeia* 1965:265–281.

Hurley, K.L. and J.J. Jackson. 2002. Evaluation of a 254-mm minimum length limit for crappies in two southeast Nebraska reservoirs. *North American Journal of Fisheries Management* 22:1369–1375.

Hurst, H., G. Bass, and C. Hubbs. 1975. The biology of the Guadalupe, Suwannee, and redeye basses. Pages 47–53 in R.H. Stroud and H. Clepper, editors. *Black Bass Biology and Management*. Sport Fishing Institute, Washington, D.C., USA.

Hurst, H.N. 1969. Comparative Life History of the Redeye Bass, *Micropterus coosae* Hubbs and Bailey, and the Spotted Bass, *Micropterus p. punctulatus* (Rafinesque) in Halawakee Creek, Alabama. Master's thesis. Auburn University, Auburn, Alabama, USA.

Hutton, G.D. 1982. Comparative Developmental Morphology of Larvae and Early Juveniles of Orangespotted Sunfish (*Lepomis humilis* (Girard)) and Bluegill (*Lepomis macrochirus* Rafinesque) from Southeastern Louisiana. Master's thesis. Louisiana State University, Baton Rouge, Louisiana, USA.

Illinois Natural History Survey. 1971. Where do all the little fishes go? *The Illinois Natural History Survey Reports*, Number 106.

Ingram, W.M. and E.P. Odum. 1941. Nests and behavior of *Lepomis gibbosus* (Linnaeus) in Lincoln Pond, Rensselaerville, New York. *American Midland Naturalist* 26:182–193.

Inman, C.R., R.C. Dewey, and P.P. Durocher. 1977. Growth comparisons and catchability of three largemouth bass strains. *Proceedings of the Thirtieth Annual Conference of the Southeastern Association of Fish and Wildlife Agencies* 30:40–47.

Innes, W.T. 1969. *Exotic Aquarium Fishes*. T.F.H. Publications, Inc., Jersey City, New Jersey, USA.

Inslee, T.D. 1975. Increased production of smallmouth bass fry. Pages 357–361 in R.H. Stroud and H. Clepper, editors. *Black Bass Biology and Management*. Sport Fishing Institute, Washington, D.C., USA.

Irwin, B.J., D.R. DeVries, and G.W. Kim. 2003. Responses to gizzard shad recovery following selective treatment in Walker County Lake, Alabama, 1996–1999. *North American Journal of Fisheries Management* 23:1225–1237.

Irwin, E.R., J.R. Jackson, and R.L. Noble. 2002. A Reservoir landscape for age-0 largemouth bass. Pages 61–71 in D.P. Phillip and M.S. Ridgway, editors. *Black Bass: Ecology, Conservation, and Management*. American Fisheries Symposium 31, Bethesda, Maryland, USA.

Irwin, E.R. and R.L. Noble. 1993. Diel inshore-offshore movements of young-of-year largemouth bass in a central Piedmont reservoir. *Abstracts of the First Mid-Year Technical Meeting, Southern Division American Fisheries Society* 1993:23–24.

Irwin, E.R., R.L. Noble, and J.R. Jackson. 1997. Distribution of age-0 largemouth bass in relation to shoreline landscape features. *North American Journal of Fisheries Management* 17:882–893.

Isaac, J., Jr. and V.H. Staata. 1993. Florida largemouth bass raceway spawning substrate evaluation. *Proceedings of the Forty-Sixth Annual Conference of the Southeastern Association of Fish and Wildlife Agencies* 46:453–457.

Isermann, D.A., P.W. Bettoli, S.M. Sammons, and T.N. Churchill. 2002. Initial poststocking mortality, oxytetracycline marking, and year-class contribution of black-nosed crappies stocked into Tennessee Reservoirs. *North American Journal of Fisheries Management* 22:1399–1408.

Isermann, D.A., S.M. Sammons, P.W. Bettoli, and T.N. Churchill. 2002. Predictive evaluation of size restrictions as management strategies for Tennessee reservoir crappie fisheries. *North American Journal of Fisheries Management* 22:1349–1357.

Isley, J.J. and R.L. Noble. 1987. Use of daily otolith rings to interpret development of length distribution of young largemouth bass. Pages 475–482 in R.C. Summerfelt and G.E. Hall, editors. *Age and Growth of Fish*. Iowa State University Press, Ames, Iowa, USA.

Isley, J.J., R.L. Noble, J.B. Koppelman, and D.P. Phillip. 1987. Spawning period and first-year growth of northern, Florida and intergrade stocks of largemouth bass. *Transaction of the American Fisheries Society* 116:757–762.

Jackson, D.A. 2002. Ecological effects of *Micropterus* introductions: the dark side of black bass. *American Fisheries Society Symposium* 31:221–232.

Jackson, D.C. and R.V. Kilambi. 1983. Selection of scales for growth analysis of largemouth bass. *Arkansas Academy of Science* 37:36–37.

Jackson, J.J. and D.L. Bauer. 2000. Size structure and catch rates of white crappie, black crappie, and bluegill in trap nets with 13-mm and 16-mm mesh. *North American Journal of Fisheries Management* 20:646–650.

Jackson, J.R. and S.L. Bryant. 1993. Impacts of a threadfin shad winterkill on black crappie in a North Carolina reservoir. *Proceedings of the Forty-Seventh Annual Conference of the Southeastern Association of Fish and Wildlife Agencies* 47:511–519.

Jackson, J.R. and R.L. Noble. 2000. First-year cohort dynamics and overwinter mortality of juvenile largemouth bass. *Transactions of the American Fisheries Society* 129:716–726.

Jackson, J.R., R.L. Noble, and J.R Copeland. 2002. Movements, growth, and survival of individually-marked fingerling largemouth bass supplementally stocked into a North Carolina reservoir. Pages 677–689 in D.P. Phillips and M.S. Ridgway, editors. *BlackBass: Ecology, Conservation, and Management*. American Fisheries Symposium 31, Bethesda, Maryland, USA.

Jackson, J.R., R.L. Noble, E.R. Irwin, and S.L. Van Horn. 2001. Response of juvenile largemouth bass to habitat enhancement through addition of artificial substrates. *Proceedings of the Fifty-Fourth Annual Conference of the Southeastern Association of Fish and Wildlife Agencies* 54:46–58.

Jackson, J.R., J.M. Phillips, R.L. Noble, and J.A. Rice. 1991. Relationship of planktivory by shad and diet shifts by young of year largemouth bass in a southern reservoir. *Proceedings of the Forty-Fourth Annual Conference of the Southeastern Association of Fish and Wildlife Agencies* 44:114–125.

Jackson, S.W., Jr. 1957. Comparison of the age and growth of four fishes from Lower and Upper Spavinaw Lakes, Oklahoma. *Proceedings of the Eleventh Annual Conference of the Southeastern Association of Game and Fish Commissioners* 11:232–249.

Jackson, W.D. and G.L. Harp. 1973. Ichthyofaunal diversification and distribution in an Ozark stream in northcentral Arkansas. *Arkansas Academy of Science Proceedings* 27:42–33.

Jacobs, R.P. 1979. Natural Hybridization of Sunfish (*Lepomis*) in Hall's Pond, Connecticut. Master's thesis. University of Connecticut, Storrs, Connecticut, USA.

Jandebeur, T.E. 1979. Distribution, Life History, and Ecology of the Spring Pygmy Sunfish, *Elassoma* species. Master's thesis. University of Alabama, Tuscaloosa, Alabama, USA.

Janssen, F.W. 1992. Ecology of Three Species of Black Bass in the Shoals Reach of the Tennessee River and Pickwick Reservoir, Alabama. Master's thesis. Auburn University, Auburn, Alabama, USA.

Jayne, B.C. and G.V. Lauder. 1994. Comparative morphology of the myomeres and axial skeleton in four genera of centrarchid fishes. *Journal of Morphology* 220:185–205.

Jenkins, R.E. and N.M. Burkhead. 1994. *Freshwater Fishes of Virginia*. American Fisheries Society, Bethesda, Maryland, USA.

Jenkins, R.M. 1953a. Growth histories of the principal fishes in Grand Lake (O' the Cherokees), Oklahoma, through fifteen years of impoundment. Oklahoma Fisheries Research Laboratory Report Number 34.

Jenkins, R.M. 1953b. An eleven-year growth history of white crappie in Grand Lake, Oklahoma. *Proceedings of the Oklahoma Academy of Science* 34:40–47.

Jenkins, R.M. 1975. Black bass crops and species associations in reservoirs. Pages 114–124 in R.H. Stroud and H. Clepper, editors. *Black Bass Biology and Management*. Sport Fishing Institute, Washington, D.C., USA.

Jennings, M.J. 1991. Sexual Selection, Reproductive Strategies, and Genetic Variation in the Longear Sunfish (*Lepomis megalotis*). Doctoral dissertation. University of Illinois at Urbana–Champaign, Urbana, Illinois, USA.

Jennings, M.J. and D.P. Philipp. 1992. Genetic variation in the longear sunfish (*Lepomis megalotis*). *Canadian Journal of Zoology* 70:163–168.

Jennings, M.J. and D.P. Philipp. 1994. Biotic and abiotic factors affecting survival of early life history intervals of a stream-dwelling sunfish. *Environmental Biology of Fishes* 39:153–159.

Johnson, D.L., R.A. Beaumier, and W.E. Lynch, Jr. 1988. Selection of habitat structure interstice size by bluegills and largemouth bass in ponds. *Transactions of the American Fisheries Society* 117:171–179.

Johnson, G.D. 1984. Percoidei: development and relationships. Pages 464–498 in Moser, H.G. et al., editors. *Ontogeny and Systematics of Fishes*. Special Publication Number 1, American Society of Ichthyology and Herpetology, Lawrence, Kansas, USA.

Johnson, G.D. and C. Patterson. 1993. Percomorph phylogeny: a survey of acanthomorphs and a new proposal. *Bulletin of Marine Science* 52:554–626.

Johnson, J.H. 1983. Summer diet of juvenile fish in the St. Lawrence River. *New York Fish and Game Journal* 30:91–99.

Johnson, J.H. and D.S. Dropkin. 1993. Diel variation in diet composition of a riverine fish community. *Hydrobiology* 271:149–158.

Johnson, J.H. and D.S. Dropkin. 1995. Diel feeding chronology of six fish species in the Juniata River, Pennsylvania. *Journal of Freshwater Ecology* 10:11–18.

Johnson, S.L. 1993. Cover choice by bluegills: orientation of underwater structure and light intensity. *Transactions of the American Fisheries Society* 122:148–154.

Johnson, W.L. 1945. Age and growth of the black and white crappies of Greenwood Lake, Indiana. *Investigations of Indiana Lakes and Streams* 2:297–324.

Johnston, C.E. 1994a. The benefit to some minnows of spawning in the nests of other species. *Environmental Biology of Fishes* 40:213–218.

Johnston, C.E. 1994b. Nest association in fishes: evidence for mutualism. *Behavioral Ecology and Sociobiology* 35:379–383.

Johnston, C.E. and E.B. Smithson. 2000. Movement patterns of stream-dwelling sunfishes: effects of pool size. *Journal of Freshwater Ecology* 15:565–566.

Johnston, P.M. 1951. The embryonic history of the germ cells of the largemouth bass, *Micropterus salmoides salmoides* (Lacepede). *Journal of Morphology* 88:471–542.

Jones, P.W., F.D. Martin, and J.D. Hardy, Jr. 1978. *Development of Fishes of the Mid-Atlantic Bight, an Atlas of Egg, Larval, and Juvenile Stages. Volume 1. Acipenseridae through Ictaluridae*. United States Fish and Wildlife Service, Biological Services Program, FWS/OBS-78/12, Washington, D.C., USA.

Jones, T.C. and W.H. Irwin. 1965. Temperature preferences by two species of fish and the influence of temperature on fish distribution. *Proceedings of the Sixteenth Annual Conference of the Southeastern Association of Game and Fish Commissioners* 16:323–333.

Jones, W.J. and J.M. Quattro. 1999. Phylogenetic affinities of pygmy sunfishes (*Elassoma*) inferred from mitochondrial DNA sequences. *Copeia* 1999:470–474.

Jordan, D.S. 1877. Contributions to North American ichthyology based primarily on the collections of the U.S. National Museum. II. Notes on the Cottidae, Etheostomatidae, Percidae, Centrarchidae, Aphredoderidae, Dorysomatidae, and Cyprinidae, with revisions of the genera and descriptions of new or little known species. *Bulletin of the U.S. National Museum* 10:1–68.

Jordan, D.S. 1884. List of fishes collected in Lake Jessup and Indian River, Florida by R.E. Earll, with descriptions of two new species. *Proceedings of the U.S. National Museum* 7:322–324.

Jordan, D.S. and B.W. Evermann. 1896a. The fishes of North and Middle America. *Bulletin of the U.S. National Museum* 47:1–1240.

Jordan, D.S. and B.W. Evermann. 1896b. Descriptions of a new species of shad (*Alosa alabamae*) from Alabama. In B.W. Evermann, *Report to the United States Commission on Fisheries* (1895) 21:203–205.

Jordan, D.S. and B.W. Evermann. 1923. *American Food and Game Fishes. A Popular Account of All the Species Found in America North of the Equator, with Keys for Ready Identification, Life Histories and Methods of Capture*. Doubleday, Page and Company, New York, New York, USA.

Jordan, D.S., B.W. Evermann, and H.W. Clark. 1930. *Checklist of the Fishes and Fishlike Vertebrates of North and Middle America North of the Northern Boundary of Venezuela and Colombia*. Report of the U.S. Fish Commission for 1928 Appendix 10:1–670.

Jordan, D.S. and C.H. Gilbert. 1883. Synopsis of the fishes of North America. *Bulletin of the U.S. National Museum* 16:1–68.

June, F.C. 1976. Changes in young-of-the-year fish stocks during and after filling of Lake Oahe, an upper Missouri River storage reservoir, 1966–74. *United States Fish and Wildlife Service Technical Paper Number 87*.

June, F.C. 1977. Reproductive patterns in seventeen species of warmwater fishes in a Missouri River reservoir. *Environmental Biology of Fishes* 2:285–296.

Kassler, T.W., J.B. Koppelman, T.J. Near, C.B. Dillman, J.M. Levengood, D.L. Swofford, J.L. VanOrman, J.E. Claussen, and D.P. Philipp. 2002. Molecular and morphological analyses of the black basses: implications for taxonomy and conservation. *American Fisheries Society Symposium* 31:291–322.

Kay, L.K., R. Wallus, and B.L. Yeager. 1994. *Reproductive Biology and Early Life History of Fishes in the Ohio River Drainage. Volume 2: Catostomidae*. Tennessee Valley Authority, Chattanooga, Tennessee, USA.

Kaya, C.M. and A.D. Hasler. 1972. Photoperiod and temperature effects on the gonads of green sunfish, *Lepomis cyanellus* (Rafinesque), during the quiescent, winter phase of its annual sexual cycle. *Transactions of the American Fisheries Society* 101:270–275.

Keast, A. 1970. Food specializations and bioenergetic interrelations in the fish fauna of some small Ontario waterways. Pages 377–411 in J.H. Steel, editor. *Marine Food Chains*. Oliver and Boyd, Edinburgh, England.

Keast, A. 1978a. Feeding interrelations between age-groups of pumpkinseed (*Lepomis gibbosus*) and comparisons with bluegill (*L. macrochirus*). *Journal of the Fisheries Research Board of Canada* 35:12–27.

Keast, A. 1978b. Trophic and spatial interrelationships in the fish species of an Ontario temperate lake. *Environmental Biology of Fishes* 3:7–31.

Keast, A. 1980. Food and feeding relationships of young fish in the first weeks after the beginning of exogenous feeding in Lake Opinicon, Ontario. *Environmental Biology of Fishes* 5:305–314.

Keast, A. 1985. Planktivory in a littoral-dwelling lake fish association: prey selection and seasonality. *Canadian Journal of Fisheries and Aquatic Sciences* 42:1114–1126.

Keast, A. and J. Eadie. 1984. Growth in the first summer of life: a comparison of nine co-occurring fish species. *Canadian Journal of Zoology* 62:1242–1250.

Keast, A. and L. Welch. 1968. Daily feeding periodicity, food uptake rates, and dietary changes with hour day in some lake fishes. *Journal of the Fisheries Research Board of Canada* 25:1133–1144.

Keast, G.K. 1968. Feeding biology of the black crappie, *Pomoxis nigromaculatus*. *Journal of the Fisheries Research Board of Canada* 25:285–297.

Keenleyside, M.H.A. 1972. Intraspecific intrusions into nests of spawning longear sunfish (Pisces: Centrarchidae). *Copeia* 1972:272–278.

Keenleyside, M.H.A. 1978. Reproductive isolation between pumpkinseed (*Lepomis gibbosus*) and longear sunfish (*L. megalotis*) (Centrarchidae) in the Thames River, southwestern Ontario. *Journal of the Fisheries Research Board of Canada* 35:131–135.

Kerr, S.R. 1966. Thermal Relations of Young Smallmouth Bass, *Micropterus dolomieui* Lacepede. Master's thesis. Queen's University, Kingston, Ontario, Canada.

Kieffer, J.D. and P.W. Colgan. 1993. Foraging flexibility in pumpkinseed (*Lepomis gibbosus*): influence of habitat structure and prey type. *Canadian Journal of Fisheries and Aquatic Science* 50:1699–1705.

Kieffer, J.D., M.R. Kubacki, F.J.S. Phelan, D.P. Philipp, and B.L. Tufts. 1995. Effects of catch-and-release angling on nesting male smallmouth bass. *Transactions of the American Fisheries Society* 124:70–76.

Kilby, J.D. 1955. The fishes of two Gulf coastal marsh areas of Florida. *Tulane Studies in Zoology* 2:175–247.

Killgore, K.J. and J.A. Baker. 1996. Patterns of larval fish abundance in a bottomland hardwood wetland. *Wetlands* 16:288–295.

Kimmel, B.L. and A.W. Groeger. 1986. Limnological and ecological changes associated with reservoir aging. Pages 103–109 in G.E. Hall and M.J. Van Den Avyle, editors. *Reservoir Fisheries Management, Strategies for the 80s*. American Fisheries Society, Bethesda, Maryland, USA.

Kindschi, G.A., R.D. Hoyt, and G.J. Overmann. 1979. Some aspects of the ecology of larval fishes in Rough River Lake, Kentucky. *Proceedings of the Third Symposium on Larval Fish* 1979:139–166.

Knights, B.C., B.L. Johnson, and M.B. Sandheinrich. 1995. Responses of bluegills and black crappies to dissolved oxygen, temperature, and current in backwater lakes of the upper Mississippi River during winter. *North American Journal of Fisheries Management* 15:390–399.

Kohler, C.C., R.J. Sheehan, and J.J. Sweatman. 1993. Largemouth bass hatching success and first-winter survival in two Illinois reservoirs. *North American Journal of Fisheries Management* 13:125–133.

Kolander, T.D., D.W. Willis, and B.R. Murphy. 1993. Proposed revision of the standard weight (W_s) equation for smallmouth bass. *North American Journal of Fisheries Management* 13:398–400.

Koppelman, J.B. 1994. Hybridization between smallmouth bass, *Micropterus dolomieu*, and spotted bass, *M. punctulatus*, in the Missouri River system, Missouri. *Copeia* 1994:204–210.

Koppelman, J.B. and G.P. Garrett. 2002. Distribution, biology, and conservation of the rare black bass species. *American Fisheries Society Symposium* 31:333–341.

Koppelman, J.B., G.S. Whitt, and D.P. Phillipp. 1988. Thermal preferenda of northern, Florida and reciprocal F_1 hybrid largemouth bass. *Transactions of the American Fisheries Society* 117:238–244.

Kraai, J.E. and C.R. Munger. 2000. Three decades of managing largemouth bass, smallmouth bass, and walleye in Meredith Reservoir, Texas. *Proceedings of the Fifty-Fourth Annual Conference of the Southeastern Association of Fish and Wildlife Agencies* 54:3–17.

Kramer, R.H. and L.L. Smith. 1960. First year growth of the largemouth bass, *Micropterous salmoides* (Lacepede), and some related ecological factors. *Transactions of the American Fisheries Society* 89:222–233.

Krumholz, L.A. 1949. Rates of survival and growth of bluegill yolk fry stocked at different intensities in hatchery ponds. *Transactions of the American Fisheries Society* 76:190–203.

Krumholz, L.A. 1950. Further observations on the use of hybrid sunfish in stocking small ponds. *Transactions of the American Fisheries Society* 79:112–124.

Kudrna, J.J. 1965. Movement and homing of sunfishes in Clear Lake. *Proceedings of the Iowa Academy of Science* 72:263–271.

Kuehne, R.A. 1962. A classification of streams, illustrated by fish distribution in an eastern Kentucky creek. *Ecology* 43:608–614.

Kwak, T.J. 1988. Lateral movement and use of floodplain habitat by fishes of the Kankakee River, Illinois. *The American Midland Naturalist* 120:241–249.

Kwak, T.J., M.J. Wiley, L.L. Osborne, and R.W. Larimore. 1992. Application of diel feeding chronology to habitat suitability analysis of warmwater stream fishes. *Canadian Journal of Fisheries and Aquatic Sciences* 49:1417–1430.

Lagler, K.E., J.E. Bardack, and R.R. Miller. 1962. *Ichthyology.* John Wiley & Sons, New York, New York, USA.

Lambert, T.R. 1980. Status of redeye bass, *Micropterus coosae*, in the South Fork Stanislaus River, California. *California Fish and Game* 1980:240–242.

Lambou, V.W. 1958. Growth rate of young-of-the-year largemouth bass, black crappie, and white crappie in some Louisiana lakes. *The Proceedings of the Louisiana Academy of Sciences* 21:63–69.

Lane, C.E., Jr. 1954. Age and growth of the bluegill, *Lepomis m. Macrochirus* (Rafinesque), in a new Missouri impoundment. *Journal of Wildlife Management* 18:358–365.

Langhurst, R.W. and D.L. Schoenike. 1990. Seasonal migration of smallmouth bass in the Embarrass and Wolf Rivers, Wisconsin. *North American Journal of Fisheries Management* 10:224–227.

Langlois, T. H. 1954. *The Western End of Lake Erie and Its Ecology.* J.W. Edwards, Publisher, Inc., Ann Arbor, Michigan, USA.

Larimore, R.W. 1952. Home pools and homing behavior of smallmouth black bass in Jordan Creek. *Illinois Natural History Survey Biological Note Number 28.*

Larimore, R.W. 1957. Ecological life history of the warmouth (Centrarchidae). *Illinois Natural History Survey Bulletin* 27:1–83.

Larimore, R.W. 1975. Visual and tactile orientation of smallmouth bass fry under floodwater conditions. Pages 323–332 in R.H. Stroud and H. Clepper, editors. *Black Bass Biology and Management.* Sport Fishing Institute, Washington, D.C., USA.

Larimore, R.W. 2002. Temperature acclimation and survival of smallmouth bass fry in flooded warmwater streams. *American Fisheries Society Symposium* 31:115–122.

Larimore, R.W. and P.W. Smith. 1963. The fishes of Champaign County, Illinois, as affected by 60 years of stream changes. *Illinois Natural History Survey Bulletin* 28:299–382.

Latta, W.C. 1975. Dynamics of bass in large natural lakes. Pages 175–181 in R.H. Stroud and H. Clepper, editors. *Black Bass Biology and Management.* Sport Fishing Institute, Washington, D.C., USA.

Lauder, G.V. 1983a. Neuromuscular patterns and the origin of trophic specialization in fishes. *Science* 219:1235–1237.

Lauder, G.V. 1983b. Functional and morphological basis of trophic specialization in sunfishes (Teleostei, Centrarchidae). *Journal of Morphology* 178:1–21.

Laughlin, D.R. and E.E. Werner. 1980. Resource partitioning in two coexisting sunfish: pumpkinseed (*Lepomis gibbosus*) and northern longear sunfish (*Lepomis megalotis peltastes*). *Canadian Journal of Fisheries and Aquatic Sciences* 37:1411–1420.

Lawrence, J.M. 1958. Estimated sizes of various forage fishes largemouth bass can swallow. *Proceedings of the Eleventh Annual Conference of the Southeastern Association of Fish and Wildlife Agencies* 11:220–225.

Layher, W.G., O.E. Maughan, and W.D. Warde. 1987. Spotted bass habitat suitability related to fish occurrence and biomass and measurements of physicochemical variables. *North American Journal of Fisheries Management* 7:238–251.

Layzer, J.B. and M.D. Clady. 1991. Microhabitat and diet segregation among coexisting young-of-year sunfishes (Centrarchidae). Pages 99–108 in R.D. Hoyt, editor. *Larval Fish Recruitment and Research in the Americas: Proceedings of the Thirteenth Annual Conference.* NOAA Technical Report NMFS 95.

Leary, J.L. 1912. Brief notes on pond culture at San Marcos, Texas. *Transactions of the American Fisheries Society* 41:149–151.

Lee, D.S. and B.M. Burr. 1985. Observations on life history of the dollar sunfish, *Lepomis marginatus* (Holbrook). *Association of Southeastern Biologist Bulletin* 32:58 (abstract).

Lee, D.S., C.R. Gilbert, C.H. Hocutt, R.E. Jenkins, D.E. McAllister, and J.R. Stauffer, Jr. 1980. *Atlas of North American Freshwater Fishes.* Publication 1980-12, North Carolina State Museum of Natural History, Raleigh, North Carolina, USA.

Lemke, A.E. 1977. Optimum temperature for growth of juvenile bluegills. *The Progressive Fish-Culturalist* 39:55–58.

Lemly, A.D. 1980. Effects of a larval parasite on the growth and survival of young bluegill. *Proceedings of the Thirty-Fourth Annual Conference of the Southeastern Association of Fish and Wildlife Agencies* 34:263–274.

Lemly, A.D. 1985. Suppression of native fish populations by green sunfish in first-order streams of Piedmont North Carolina. *Transactions of the American Fisheries Society* 114:705–712.

Lemly, A.D. and J.F. Dimmick. 1982. Growth of young-of-the-year and yearling centrarchids in relation to zooplankton in the littoral zone of lakes. *Copeia* 1982:305–321.

Leslie, J.K. and J.E. Moore. 1985. Ecology of young-of-the-year fish in Muscote Bay (Bay of Quinte), Ontario. *Canadian Technical Report of Fisheries and Aquatic Sciences Number 1377.*

Lester, N.P., W.I. Dunlop, and C.C. Willox. 1996. Detecting changes in the nearshore fish community. *Canadian Journal of Fisheries and Aquatic Sciences* 53:391–402.

Levine, D.S., A.G. Eversole, and H.A. Loyacano. 1986. Biology of redbreast sunfish in beaver ponds. *Proceedings of the Fortieth Annual Conference of the Southeastern Association of Fish and Wildlife Agencies* 40:216–226.

Lewis, G.E. 1976. Summer and fall food of spotted bass in two West Virginia reservoirs. *The Progressive Fish-Culturist* 38:175–176.

Lewis, S.A., K.D. Hopkins, and T.F. White. 1971. Average growth rates and length–weight relationships of sixteen species of fish in Canton Reservoir. Oklahoma. *Oklahoma Fishery Research Laboratory Bulletin Number 9*, Norman, Oklahoma, USA.

Lewis, W.M. and D. Elder. 1953. The fish population of the headwaters of a spotted bass stream in southern Illinois. *Transactions of the American Fisheries Society* 82:193–202.

Lewis, W.M. and T.S. English. 1949. The warmouth, *Chaenobryttus coronarius* (Bartram), in Red Haw Hill Reservoir, Iowa. *Iowa State College Journal of Science* 23:317–322.

Lilyestrom, C.G. and T.N. Churchill. 1997. Diet and movement of largemouth bass and butterfly peacocks in La Plata Reservoir, Puerto Rico. *Proceedings of the Annual Conference Southeastern Association of Fish and Wildlife Agencies* 50:192–200.

Linton, T.L. 1961. A study of fishes of the Arkansas and Cimarron Rivers in the area of the proposed Keystone Reservoir. *Oklahoma Fishery Research Laboratory Report Number 81*, Norman, Oklahoma, USA.

Lippson, A.J. and R.L. Moran. 1974. *Manual for Identification of Early Developmental Stages of Fishes of the Potomac River Estuary*. Power Plant Siting Program of the Maryland Department of Natural Resources Publication PPSP-MP-13. Baltimore, Maryland, USA.

Livingstone, A.C. 1987. Habitat Use and Food Habits of Young-of-Year Centrarchids in the Jacks Fork River. Master's thesis. University of Missouri–Columbia, Columbia, Missouri, USA.

Lobb, M.D. III and D.J. Orth. 1991. Habitat use by an assemblage of fish in a large warmwater stream. *Transactions of the American Fisheries Society* 120:65–78.

Long, J.M. and W.L. Fisher. 2001. Precision and bias of largemouth, smallmouth, and spotted bass age estimation from scales, whole otoliths, and sectioned otoliths. *North American Journal of Fisheries Management* 21:636–645.

Long, J.M. and W.L. Fisher. 2002. Environmental influences on largemouth bass recruitment in a southern Great Plains reservoir. *Proceedings of the Fifty-Fifth Annual Conference of the Southeastern Association of Fish and Wildlife Agencies* 55:146–155.

Long, J.M. and W.L. Fisher. 2005. Distribution and abundance of black bass in Skiatook Lake, Oklahoma, after introduction of smallmouth bass and a liberalized harvest regulation on spotted bass. *North American Journal of Fisheries Management* 25:49–56.

Lönnberg, E. 1894. List of fishes observed and collected in South Florida. *Öfversigt af Kongl. Vetenskaps-Akademiens Förhandlingar* 8:123–124.

Lotrich, V.A. 1973. Growth, production, and community composition of fishes inhabiting a first-, second-, and third-order stream of eastern Kentucky. *Ecological Monographs* 43:377–397.

Lukas, J.A. and D.J. Orth. 1993a. Reproductive ecology of redbreast sunfish *Lepomis auritus* in a Virginia stream. *Journal of Freshwater Ecology* 8:235–244.

Lukas, J.A. and D.J. Orth. 1993b. Factors affecting nesting success of smallmouth bass in a regulated Virginia stream. *Proceedings of Southern Division American Fisheries Society First Mid-Year Technical Meeting* 1993:18 (Abstract).

Lukas, J.A. and D.J. Orth. 1995. Factors affecting nesting success of smallmouth bass in a regulated Virginia stream. *Transactions of the American Fisheries Society* 124:726–735.

Lutz-Carrillo, D.J., C.C. Nice, T.H. Bonner, M.R.J. Forstner, and L.T. Fries. 2006. Admixture analysis of Florida largemouth bass and northern largemouth bass using microsatellite loci. *Transactions of the American Fisheries Society* 135:779–791.

Lux, F.E. 1960. Notes on first-year growth of several species of Minnesota fish. *The Progressive Fish-Culturist* 22:81–82.

Lynch, W.E., Jr. and D.L. Johnson. 1989. Influences of interstice size, shade, and predators on the use of artificial structures by bluegills. *North American Journal of Fisheries Management* 9:219–225.

Lyons, J. and P. Kanehl. 2002. Seasonal movements of smallmouth bass in streams. *American Fisheries Society Symposium* 31:149–160.

Mabee, P.M. 1987. Phylogenetic Change and Ontogenetic Interpretation in the Family Centrarchidae (Perciformes: Centrarchidae). Doctoral dissertation. Duke University, Durham, North Carolina, USA.

Mabee, P.M. 1988. Supraneural and predorsal bones in fishes development and homologies. *Copeia* 1988:827–838.

Mabee, P.M. 1993. Phylogenetic interpretations of ontogenetic change: sorting out the actual and artifactual in an empirical case study of centrarchid fishes. *Zoological Journal of the Linnean Society* 107:175–291.

Mabee, P.M. 1995. Evolution of pigment pattern development in centrarchid fishes. *Copeia* 1995:586–607.

MacCrimmon, H.R. and W.H. Robbins. 1975. Distribution of the black basses in North America. Pages 56–66 in R.H. Stroud and H. Clepper, editors. *Black Bass Biology and Management*. Sport Fishing Institute, Washington, D.C., USA.

Maceina, M.J. 2003. Verification of the influence of hydrologic factors on crappie recruitment in Alabama reservoirs. *North American Journal of Fisheries Management* 23:470–480.

Maceina, M.J. and D.R. Bayne. 2001. Changes in the black bass community and fishery with oligotrophication in West Point Reservoir, Georgia. *North American Journal of Fisheries Management* 21:745–755.

Maceina, M.J., D.R. Bayne, A.S. Hendricks, W.C. Reeves, W.P. Black, and V.J. DiCenzo. 1996. Compatibility between water clarity and quality black bass and crappie fisheries in Alabama. *American Fisheries Society Symposium* 16:296–305.

Maceina, M.J. and P.W. Bettoli. 1998. Variation in largemouth bass recruitment in four mainstream impoundments of the Tennessee River. *North American Journal of Fisheries Management* 18:998–1003.

Maceina, M.J. and J.M. Grizzle. 2006. The relation of largemouth bass virus to largemouth bass population metrics in five Alabama lakes. *Transactions of the American Fisheries Society* 135:545–555.

Maceina, M.J. and B.R. Murphy. 1988. Variation in the weight-to-length relationship among Florida and northern largemouth bass and their intraspecific F_1 hybrid. *Transactions of the American Fisheries Society* 117:232–237.

Maceina, M.J. and B.R. Murphy. 1989. Florida, northern, and hybrid largemouth bass feeding characteristics in Aquilla Lake, Texas. *Proceedings of the Forty-Second Annual Conference of the Southeastern Association of Fish and Wildlife Agencies* 42:112–119.

Maceina, M.J., B.R. Murphy, and J.J. Isley. 1988. Factors regulating Florida largemouth bass stocking success and hybridization with northern largemouth bass in Aquilla Lake, Texas. *Transactions of the American Fisheries Society* 117:221–231.

Maceina, M.J. and J.V. Shireman. 1982. Influence of dense *Hydrilla* infestation on black crappie growth. *Proceedings of the Thirty-Sixth Annual Conference of the Southeastern Association of Fish and Wildlife Agencies* 36:394–402.

MacRae, P.S.D. and D.A. Jackson. 2001. The influence of smallmouth bass (*Micropterus dolomieu*) predation and habitat complexity on the structure of littoral zone fish assemblages. *Canadian Journal of Aquatic Science* 58:342–351.

Mansueti, A.J. 1962. Early development of the yellow perch, *Perca flavescens*. *Chesapeake Science* 3:46–66.

Mansueti, A.J. and J.D. Hardy, Jr. 1967. *Development of Fishes of the Chesapeake Bay Region: An Atlas of Egg, Larval, and Juvenile Stages*. Natural Resources Institute, University of Maryland, Baltimore, Maryland, USA.

Mansueti, R.J. 1962. Eggs, larvae, and young of the hickory shad, *Alosa mediocris*, with comments on its ecology in the estuary. *Chesapeake Science* 3:173–205.

Mansueti, R.J. 1964. Eggs, larvae, and young of the white perch, *Roccus americanus*, with comments on its ecology in the estuary. *Chesapeake Science* 5:3–45.

Mantini, L., M.V. Hoyer, J.V. Shireman, and D.E. Canfield, Jr. 1992. Annulus validation, time of formation, and mean length at age of three sunfish species in north central Florida. *Proceedings of the Forty-Sixth Annual Conference of the Southeastern Association of Fish and Wildlife Agencies* 46:357–367.

Marler, B.J. 1990. Distribution and Abundance Patterns of Largemouth Bass in Stream Reaches below Aberdeen and Columbus Dams, Tennessee-Tombigbee Waterway. Master's thesis. Mississippi State University, Mississippi State, Mississippi, USA.

Martin, F.D. 1968. Some Factors Influencing Penetration into Rivers by Fishes of the Genus *Cyprinodon*. Doctoral thesis. University of Texas, Austin, Texas, USA.

Martinez, G.J. 1980. A Study of Age, Growth, and Food Habitats of Certain Game Fishes in Cordell Hull Reservoir, Tennessee. Master's thesis. Tennessee Technological University, Cookeville, Tennessee, USA.

Mathur, D. and T.W. Robbins. 1971. Food habits and feeding chronology of young white crappie, *Pomoxis annularis* Rafinesque, in Conwingo Reservoir. *Transactions of the American Fisheries Society* 100:307–311.

Mathur, D., R.M. Schutsky, E.J. Purdy, Jr., and C.A. Silver. 1981. Similarities in acute temperature preferences of freshwater fishes. *Transactions of the American Fisheries Society* 110:1–13.

Matthews, W.J. 1986. Fish faunal structure in an Ozark stream: stability, persistence and a catastrophic flood. *Copeia* 1986:388–397.

Matthews, W.J., F.P. Gelwick, and J.J. Hoover. 1992. Food of and habitat use by juveniles of species of *Micropterus* and *Morone* in a southwestern reservoir. *Transactions of the American Fisheries Society* 121:54–66.

May, E.B. and C.R. Gasaway. 1967. A preliminary key to the identification of larval fishes of Oklahoma with particular reference to Canton Reservoir, including a selected bibliography. Oklahoma Fishery Research Laboratory Bulletin Number 5, Contribution Number 164 of the Oklahoma Fishery Research Laboratory, Norman, Oklahoma, USA.

Mayden, R.L. 1993. *Elassoma alabamae*, a new species of pygmy sunfish endemic to the Tennessee River drainage of Alabama (Teleostei: Elassomatidae). *Bulletin of the Alabama Museum of Natural History* 16:1–14.

Mayden, R.L., B.M. Burr, L.M. Page, and R.R. Miller. 1992. The native freshwater fishes of North America. Pages 827–863 in R.L. Mayden, editor. *Systematics, Historical Ecology, and North American Freshwater Fishes*. Stanford University Press, Stanford, California, USA.

Mayer, F. 1934. *Elassoma evergladei* Jordan (the pygmy sunfish). *The Aquarium* 3:101–102.

McCarraher, D.B. 1971. Survival of some freshwater fishes in the alkaline eutrophic waters of Nebraska. *Journal of the Fisheries Research Board of Canada* 28:1811–1814.

McCarraher, D.B., M.L. Madsen, and R.E. Thomas. 1971. Ecology and fishery management of McConaughy Reservoir, Nebraska. Pages 299–311 in G.E. Hall, editor. *Reservoir Fisheries and Limnology.* Special Publication Number 8, American Fisheries Society, Washington, D.C., USA.

McClane, A.J. 1978. *Field Guide to Freshwater Fishes of North America.* Holt, Rinehart, & Winston, New York, New York, U.S.A.

McClendon, D.D. and C.F. Rabeni. 1987. Physical and biological variables useful for predicting population characteristics of smallmouth bass and rock bass in an Ozark stream. *North American Journal of Fisheries Management* 7:46–56.

McCollum, A.B., D.B. Bunnell, and R.A. Stein. 2003. Cold, northern winters: the importance of temperature to overwinter mortality of age-0 white crappies. *Transactions of the American Fisheries Society* 132:977–987.

McConnell, W.J. and J.H. Gerdes. 1964. Threadfin shad, *Dorosoma petenense,* as food of yearling centrarchids. *California Fish and Game* 50:170–175.

McCormick, J.H., K.M. Jensen, and R.L. Leino. 1989. Survival, blood osmolality, and gill morphology of juvenile yellow perch, rock bass, black crappie, and largemouth bass exposed to acidified soft water. *Transactions of the American Fisheries Society* 118:386–399.

McDonough, T.A. and J.P. Buchanan. 1991. Factors affecting abundance of white crappies in Chickamauga Reservoir, Tennessee. *North American Journal of Fisheries Management* 11:513–524.

McElroy, T.C., K.L. Kandl, J. Garcia, and J.C. Trexler. 2003. Extinction-colonization dynamics structure genetic variation of spotted sunfish (*Lepomis punctatus*) in the Florida Everglades. *Molecular Ecology* 12: 355–368.

McHugh, J.J. 1990. Response of bluegills and crappies to reduced abundance of largemouth bass in two Alabama impoundments. *North American Journal of Fisheries Management* 10:344–351.

McInerny, M.C. 1988. Evaluation of trapnetting for sampling black crappie. *Proceedings of the Forty-Second Annual Conference of the Southeastern Association of Fish and Wildlife Agencies* 42:98–106.

McKeown, P.E. and S.R. Mooradian. 2002. Factors influencing recruitment of crappies in Chautauqua Lake, New York. *North American Journal of Fisheries Management* 22:1385–1392.

McLane, W.M. 1955. The Fishes of the St. Johns River System. Doctoral dissertation. University of Florida, Gainesville, Florida, USA.

McNeill, A.J. 1995. An overview of the smallmouth bass in Nova Scotia. *North American Journal of Fisheries Management* 15:680–687.

Meador, M.R. and W.E. Kelso. 1989. Behavior and movements of largemouth bass in response to salinity. *Transactions of the American Fisheries Society* 118:409–415.

Meals, K.O. and L.E. Miranda. 1991. Variability in abundance of age-0 centrarchids among littoral habitats of flood control reservoirs in Mississippi. *North American Journal of Fisheries Management* 11:298–304.

Medford, D.W. and B.A. Simco. 1971. The fishes of the Wolf River, Tennessee and Mississippi. *Journal of the Tennessee Academy of Science* 46:121–123.

Meinken, H. Not dated. Leiferung 92. *Family Centrarchidae. Ambloplites rupestris* (Rafinesque). Blatt 940–941 in M. Holly, H. Meinken, and A. Rachow, *Die Aquarienfische in Wort und Bild* [in German]. Alfred Kernan Verlag, Stuttgart. (Loose-leaf, pagination open.)

Menhinick, E.F. 1991. *The Freshwater Fishes of North Carolina.* North Carolina Wildlife Resources Commission, Raleigh, North Carolina, USA.

Merriner, J.V. 1971. Egg size as a factor in intergeneric hybrid success of centrarchids. *Transactions of the American Fisheries Society* 100:29–32.

Merriner, J.V. 1971b. Development of intergeneric centrarchid hybrid embryos. *Transactions of the American Fisheries Society* 100:611–618.

Mettee, M.F. 1974. A Study on the Reproductive Behavior, Embryology, and Larval Development of the Pygmy Sunfishes of the Genus *Elassoma.* Doctoral dissertation. University of Alabama, Tuscaloosa, Alabama, USA.

Mettee, M.F., P.E. O'Neil, and J.M. Pierson. 1996. *Fishes of Alabama and the Mobile Basin.* Oxmoor House, Inc., Birmingham, Alabama, USA.

Mettee, M.F., P.E. O'Neil, and J.M. Pierson. 2001. *Fishes of Alabama and the Mobile Basin.* 2nd edition. Oxmoor House, Inc., Birmingham, Alabama, USA.

Mettee, M.F. and J.J. Pulliam. 1986. Reintroduction of an undescribed species of *Elassoma* into Pryor Branch, Limestone County, Alabama. *Southeastern Fishes Council Proceedings* 4:14–15.

Mettee, M.F. and J.S. Ramsey. 1986. Spring pygmy sunfish, *Elassoma* species. Pages 4–5 in R.H. Mount, editor. *Vertebrate Animals of Alabama in Need of Special Attention.* Alabama Agricultural Experimental Station, Auburn University, Auburn, Alabama, USA.

Mettee, M.F. and C. Scharpf. 1998. Reproductive biology, embryology, and larval development of four species of pygmy sunfish. *American Currents* 24:1–10.

Meyer, F.A. 1970. Development of some larval centrarchids. *The Progressive Fish-Culturist* 32:130–136.

Michaletz, P.H. and J.L. Bonneau. 2005. Age-0 gizzard shad abundance is reduced in the presence of macrophytes: implications for interactions with bluegills. *Transactions of the American Fisheries Society* 134:149–159.

Miller, H.C. 1964. The behavior of the pumpkinseed sunfish, *Lepomis gibbosus* (Linneaus), with notes on the behavior of other species of *Lepomis* and the pigmy sunfish, *Elassoma evergladi*. *Behaviour* 22:88–151.

Miller, K.D. and R.H. Kramer. 1971. Spawning and early life history of largemouth bass (*Micropterus salmoides*) in Lake Powell. Pages 73–83 in G.E. Hall, editor. *Reservoir Fisheries and Limnology*. Special Publication Number 8, American Fisheries Society, Washington, D.C., USA.

Miller, R.J. 1975. Comparative behavior of centrarchid basses. Pages 85–94 in R.H. Stroud and H. Clepper, editors. *Black Bass Biology and Management*. Sport Fishing Institute, Washington, D.C., USA.

Miller, R.J. and H.W. Robison. 1973. *The Fishes of Oklahoma*. Oklahoma State University Press, Stillwater, Oklahoma, USA.

Miller, R.J. and H.W. Robison. 2004. *Fishes of Oklahoma*, 2nd edition. University of Oklahoma Press, Norman, Oklahoma, USA.

Miller, S.J. and T. Storck. 1982. Daily growth rings in otoliths of young-of-the-year largemouth bass. *Transactions of the American Fisheries Society* 111:527–530.

Miner, J.G. and R.A. Stein. 1993. Interactive influence of turbidity and light on larval bluegill (*Lepomis macrochirus*) foraging. *Canadian Journal of Fisheries and Aquatic Science* 50:781–788.

Miner, J.G. and R.A. Stein. 1996. Detection of predators and habitat choice by small bluegills: effects of turbidity and alternative prey. *Transactions of the American Fisheries Society* 125:97–103.

Minns, C.K., J.R.M. Kelso, and W. Hyatt. 1978. Spatial distribution of nearshore fish in the vicinity of two thermal generating stations, Nanticoke and Douglas Point, on the Great Lakes. *Journal of the Fisheries Research Board of Canada* 35:885–892.

Miranda, L.E. 1987. Fish community changes in the upper Tombigbee River following connection with the Tennessee River, with special reference to largemouth bass–prey interactions in relation to spawning sequence and prey availability. *Dissertation Abstracts International Part B: Science and Engineering* 48:1–12.

Miranda, L.E. and W.D. Hubbard. 1994a. Length-dependent winter survival and lipid composition of age-0 largemouth bass in Bay Springs Reservoir, Mississippi. *Transactions of the American Fisheries Society* 123:80–87.

Miranda, L.E. and W.D. Hubbard. 1994b. Winter survival of age-0 largemouth bass relative to size, predators, and shelter. *North American Journal of Fisheries Management* 14:790–796.

Miranda, L.E. and R.J. Muncy. 1987a. Spawning sequence of largemouth bass, bluegill, and gizzard shad. *Proceedings of the Forty-First Annual Conference of the Southeastern Association of Fish and Wildlife Agencies* 41:197–204.

Miranda, L.E. and R.J. Muncy. 1987b. Recruitment of young-of-year largemouth bass in relation to size structure of parental stock. *North American Journal of Fisheries Management* 7:131–137.

Miranda, L.E. and L.L. Pugh. 1997. Relationship between vegetation cover and abundance, size, and diet of juvenile largemouth bass during winter. *North American Journal of Fisheries Management* 17:601–610.

Mischke, C.C. and J.E. Morris. 1997. Out-of-season spawning of sunfish *Lepomis* spp. in the laboratory. *The Progressive Fish-Culturist* 59:297–302.

Mittelbach, G.G. 1984. Predation and resource partitioning in two sunfishes (Centrarchidae). *Ecology* 65:499–513.

Mok, H.K. 1981. The phylogenetic implications of centrarchid kidneys. *Bulletin Institute Zoology, Academia Sinica* 20:59–67.

Moody, D.P. 1979. Separation of *Lepomis punctatus* larvae from selected other *Lepomis* species. *American Society of Zoologists* 19:948 (Abstract).

Moore, G.A. 1962. Fishes. Pages 21–165 in Blair, W.F., editor. *Vertebrates of the United States*. McGraw-Hill Book Company, New York, New York, USA.

Moore, G.A. and F.B. Cross. 1950. Additional Oklahoma fishes with validation of *Poecilichthys parvipinnis* (Gilbert and Swain). *Copeia* 1950:139–148.

Moore, G.A. and M.E. Sisk. 1963. The spectacle of *Elassoma zonatum* Jordan. *Copeia* 1963:346–350.

Moore, W.G. 1942. Field studies on the oxygen requirements of certain fresh-water fishes. *Ecology* 23:319–329.

Moorman, R.B. 1957. Reproduction and growth of fishes in Marion County, Iowa, farm ponds. *Iowa State College Journal of Science* 32:71–88.

Morgan, G.D. 1951a. A comparative study of the spawning periods of the bluegills, *Lepomis macrochirus*, the black crappie, *Pomoxis nigromaculatus* (LeSeuer), and the white crappie, *Pomoxis annularis* (Rafinesque), of Buckeye Lake, Ohio. *Journal of the Scientific Laboratories, Denison University* 42:112–118.

Morgan, G.D. 1951b. The life history of the bluegill sunfish, *Lepomis macrochirus*, of Buckeye Lake (Ohio). *Journal of the Scientific Laboratories, Denison University* 42:21–59.

Morgan, G.D. 1954. The life history of the white crappie (*Pomoxis annularis*) of Buckeye Lake, Ohio. *Ohio Journal of the Scientific Labs, Denison University* 43:113–144.

Morizot, D.C., S.W. Calhoun, L.L. Clepper, M.E. Schmidt, J.H. Williamson, and G.J. Carmichael. 1991. Multispecies hybridization among native and introduced centrarchid basses in central Texas. *Transactions of the American Fisheries Society* 120:283–289.

Moss, D.D. and D.C. Scott. 1961. Dissolved-oxygen requirements of three species of fish. *Transactions of the American Fisheries Society* 90:377–393.

Mount, R.H., ed. 1984. *Vertebrate Wildlife of Alabama.* Alabama Agricultural Experimental Station, Auburn University, Auburn, Alabama, USA.

Moynan, K.M. 1989. The Effects of Acid and Water Hardness on Bluegill Embryo-Larvae Determined by Laboratory and On-Site Toxicity Tests. Master's thesis. Virginia Polytechnic Institute and State University, Blacksburg, Virginia, USA.

Mraz, D. and E.L. Cooper. 1957. Reproduction of carp, largemouth bass, bluegills, and black crappies in small rearing ponds. *Journal of Wildlife Management* 21:127–133.

Mullan, J.W. and R.L. Applegate. 1968. Centrarchid food habits in a new and old reservoir during and following bass spawning. *Proceedings of the Twenty-First Annual Conference of the Southeastern Association of Fish and Wildlife Agencies* 21:332–342.

Muller, R. and F.E.J. Fry. 1976. Preferred temperature of fish: a new method. *Journal of the Fisheries Research Board of Canada* 33:1815–1817.

Muncy, R.J. 1966. Aging and growth of largemouth bass, bluegill, and redear sunfish from Louisiana ponds of known stocking history. *Proceedings of the Nineteenth Annual Conference of the Southeastern Association of Game and Fish Commissioners* 19:343–349.

Musick, J.A. 1972. Fishes of the Chesapeake Bay and the adjacent Coastal Plain. Pages 175–212 in M.L. Wass, editor. *A Checklist of the Biota of Lower Chesapeake Bay.* Virginia Institute of Marine Science Special Scientific Report 65.

Muth, R.T. and J.C. Schmulbach. 1984. Downstream transport of fish larvae in a shallow prairie river. *Transactions of the American Fisheries Society* 113:224–230.

Nachstedt, J. and H. Tusche. 1954. *Breeding Aquarium Fishes* (translated from the German by William Vorderwinkler). Aquarium Stock Company, New York, New York, USA.

Nack, S.B., D. Bunnell, D.M. Green, and J.L. Forney. 1993. Spawning and nursery habitats of largemouth bass in the tidal Hudson River. *Transactions of the American Fisheries Society* 122:208–216.

Nagid, E.J., K.J. Dockendorf, and R.W. Hujik. 2004. Off-season spawning of black crappie in a north-central Florida lake. *North American Journal of Fisheries Management* 24:299–302.

Nakamura, N., S. Kasahare, and T. Yada. 1971. Studies on the usefulness of the bluegill sunfish, *Lepomis macrochirus* Rafinesque, as an experimental standard animal. II. On the developmental stages and growth from egg through one year (in Japanese, English subtitle and summary). *Journal of the Faculty of Fisheries and Animal Husbandry Hiroshima University* 10:139–151.

Naughton, G.P., D.H. Bennett, and K.B. Newman. 2004. Predation on juvenile salmonids by small-mouth bass in the lower Granite Reservoir system, Snake River. *North American Journal of Fisheries Management* 24:534–544.

Neal, J.W., N.M. Bacheler, R.L. Noble, and C.G. Lilyestrom. 2002. Effects of reservoir drawdown on available habitat: implications for a tropical largemouth bass population. *Proceedings of the Fifty-Fifth Annual Conference of the Southeastern Association of Fish and Wildlife Agencies* 55:156–164.

Neal, J.W. and R.L. Noble. 2002. Growth, survival, and site fidelity of Florida and intergrade largemouth bass stocked in a tropical reservoir. *North American Journal of Fisheries Management* 22:528–536.

Neal, J.W. and R.L. Noble. In Press. A bioenergetics-based approach to explain largemouth bass Size in tropical reservoirs. *Transactions of the American Fisheries Society.*

Neal, J.W., R.L. Noble, A.R. Alicea, and T.N. Churchill. 1997. Invalidation of otolith aging techniques for tropical largemouth bass. *Proceedings of Fifty-First Annual Conference of the Southeastern Association of Fish and Wildlife Agencies* 51:159–165.

Neal, J.W., R.L. Noble, and C.G. Lilyestrom. 2006. Evaluation of the ecological compatibility of butterfly peacock cichlids and largemouth bass in Puerto Rico reservoirs. *Transactions of the American Fisheries Society* 135:288–296.

Near, T.J., D.I. Bolnick, and P.C. Wainwright. 2005. Fossil calibrations and molecular divergence time estimates in centrarchid fishes (Teleostei: Centrarchidae). *Evolution* 59:1768–1782.

Nedbal, M.A. and D.P. Philipp. 1994. Differentiation of mitochondrial DNA in largemouth bass. *Transactions of the American Fisheries Society* 123:460–468.

Nelson, J.S. 1994. *Fishes of the World.* 3rd edition. John Wiley & Sons, New York, New York, USA.

Nelson, J.S., E.J. Crossman, H. Espinosa-Perez, L.T. Findley, C.R. Gilbert, R.N. Lea, and J.D. Williams. 2004. *Common and Scientific Names of Fishes from the United States, Canada, and Mexico.* American Fisheries Society, Special Publication 29, Bethesda, Maryland, USA.

Nelson, M.N. and A.D. Hasler. 1942. The growth, food, distribution and relative abundance of the fishes of Lake Geneva, Wisconsin, in 1941. *Transactions of the Wisconsin Academy of Science, Arts, and Letters* 34:137–148.

Nelson, W.R. 1974. Age, growth, and maturity of thirteen species of fish from Lake Oahe during the early years of impoundment, 1963–68. Technical Papers (Number 77) of the United States Fish and Wildlife Service, Washington, D.C., USA.

Nelson, W.R., R.E. Siefert, and D.V. Swedberg. 1968. Studies of the early life histories of reservoir fishes. *Reservoir Fisheries Research Symposium.* Southern Division of the American Fisheries Society. University of Georgia Press, Athens, Georgia, USA.

Neophitou, C. and A.J. Giapis. 1994. A study of the biology of pumpkinseed (*Lepomis gibbosus* (L.)) in Lake Kerkini (Greece). *Journal of Applied Ichthyology* 10:123–133.

Neves, R.J. 1975. Factors affecting fry production of smallmouth bass (*Micropterus dolomieui*) in South Branch Lake, Maine. *Transactions of the American Fisheries Society* 104:83–87.

Ney, J.J. 1996. Oligotrophication and its discontents: effects of reduced nutrient loading on reservoir fisheries. Pages 285–295 in L.E. Miranda and D.R. DeVries, editors. *Multidimensional Approaches to Reservoir Fisheries Management.* American Fisheries Society Symposium 16.

Noltie, D.B. 1986. A method for measuring reproductive success in the rock bass (*Ambloplites rupestris*), with applicability to other substrate brooding fishes. *Journal of Freshwater Ecology* 3:319–323.

Noltie, D.B. 1989. Status of the orangespotted sunfish, *Lepomis humilis*, in Canada. *Canadian Field Naturalist* 104:69–86.

Noltie, D.B. and M.H.A. Keenleyside. 1987. The breeding behavior of stream dwelling rock bass, *Ambloplites rupestris* (Centrachidae). *Biological Behavior* 12:196–206.

Novinger, G.D. 1987. Evaluation of a 15.0-inch minimum length limit on largemouth and spotted bass catches at Table Rock Lake, Missouri. *North American Journal of Fisheries Management* 7:260–272.

Nurnberger, P.K. 1930. The plant and animal food of the fishes of Big Sandy Lake. *Transactions of the American Fisheries Society* 1930:253–259.

O'Brien, W.J., B. Loveless, and D. Wright. 1984. Feeding ecology of young white crappie in a Kansas reservoir. *North American Journal of Fisheries Management* 4:341–349.

Olmsted, L.L. 1974. The Ecology of Largemouth Bass (*Micropterus salmoides*) and Spotted Bass (*Micropterus punctulatus*) in Lake Fort Smith, Arkansas. Doctoral dissertation. University of Arkansas, Fayetteville, Arkansas, USA.

Olson, M.H. and B.P. Young. 2003. Patterns of diet and growth in co-occurring populations of largemouth bass and smallmouth bass. *Transactions of the American Fisheries Society* 132:1207–1213.

Olson, M.H., B.P. Young, and K.D. Blinkoff. 2003. Mechanisms underlying habitat use of juvenile largemouth bass and smallmouth bass. *Transactions of the American Fisheries Society* 132:398–405.

O'Neil, P.E. 2004. Fishes. Pages 60–100 in R.E. Mirarchi, editor. *Alabama Wildlife. Volume 1. A Checklist of Vertebrates and Selected Invertebrates: Aquatic Mollusks, Fishes, Amphibians, Reptiles, Birds, and Mammals.* The University of Alabama Press, Tuscaloosa, Alabama, USA.

Ongarato, R.J. and E.J. Snucins. 1993. Aggression of guarding male smallmouth bass (*Micropterus dolomieui*) towards potential brood predators near the nest. *Canadian Journal of Zoology* 71:437–440.

O'Rear, R.S. 1970. A growth study of redbreast, *Lepomis auritus* (Gunther), and bluegill, *Lepomis macrochirus* (Rafinesque), populations in a thermally influenced lake. *Proceedings of the Twenty-Third Annual Conference of the Southeastern Association of Game and Fish Commissioners* 23:545–553.

Orth, D.J. and T.J. Newcomb. 2002. Certainties and uncertainties in defining essential habitats for riverine smallmouth bass. *American Fisheries Society Symposium* 31:251–264.

Osenberg, C.W., G.G. Mittelbach, and F.C. Wainright. 1992. Two-stage life histories in fish: the interaction between juvenile competition and adult performance. *Ecology* 73:255–267.

Osenberg, C.W., E.E. Werner, G.G. Mittelbach, and D.J. Hall. 1988. Growth patterns in bluegill (*Lepomis microchirus*) and pumpkinseed (*L. gibbosus*) sunfish: environmental variation and the importance of ontogenetic niche shifts. *Canadian Journal of Fisheries and Aquatic Science* 45:17–26.

Ostrand, K.G., S.J. Cooke, and D.H. Wahl. 2004. Effects of stress on largemouth bass reproduction. *North American Journal of Fisheries Management* 24:1038–1045.

Ozen, O. and R.L. Noble. 2005. Assessing age-0 year-class strength of fast growing largemouth bass in a tropical reservoir. *North American Journal of Fisheries Management* 25:163–170.

Page, L.M. and B.M. Burr. 1991. *A Field Guide to Freshwater Fishes of North America North of Mexico.* Houghton Mifflin Company, Boston, Massachusetts, USA.

Pajak, P. and R.J. Neves. 1987. Habitat suitability and fish production: a model evaluation for rock bass in two Virginia streams. *Transactions of the American Fisheries Society* 116:839–850.

Paller, M.H., J.B. Gladden, and J.H. Heuer. 1992. Development of the fish community in a new South Carolina reservoir. *American Midland Naturalist* 128:95–114.

Panek, F.M. and C.R. Cofield. 1978. Fecundity of bluegill and warmouth from a South Carolina blackwater lake. *The Progressive Fish-Culturist* 40:67–68.

Paragamian, V.L. and M.J. Wiley. 1987. Effects of variable stream flows on growth of smallmouth bass in the Maquoketa River, Iowa. *North American Journal of Fisheries Management* 7:357–362.

Parkos, J.J. III and D.H. Wahl. 2002. Towards an understanding of recruitment mechanisms in largemouth bass. Pages 25–45 in D.P. Phillip and M.S. Ridgway, editors. *Black Bass: Ecology, Conservation, and Management*. American Fisheries Society, Symposium 31, Bethesda, Maryland, USA.

Parrish, D.L., T.D. Forsythe, and T.J. Timmons. 1986. Growth and condition comparisons of white crappie in two similar reservoirs. *Journal of the Tennessee Academy of Science* 1986:92–95.

Parsons, G.R. 1993. Comparisons of triploid and diploid white crappies. *Transactions of the American Fisheries Society* 122:237–243.

Parsons, J.W. 1954. Growth and habits of the redeye bass. *Transactions of the American Fisheries Society* 83:202–211.

Parsons, J.W. and E. Crittenden. 1959. Growth of redeye bass in Chipola River, Florida. *Transactions of the American Fisheries Society* 68:191–192.

Partridge, D.G. and D.R. DeVries. 1999. Regulation of growth and mortality in larval bluegills: implications for juvenile recruitment. *Transactions of the American Fisheries Society* 128:625–638.

Pasch, R.W. 1975. Some relationships between food habits and growth of largemouth bass in Lake Blackshear, Georgia. *Proceedings of the Twenty-Eighth Annual Conference of the Southeastern Association of Fish and Wildlife Agencies* 28:307–321.

Patriarche, M.H. 1953. The fishery in Lake Wappapello, a flood-control reservoir on the St. Francis River, Missouri. *Transactions of the American Fisheries Society* 82:242–254.

Patton, T.M. and W.A. Hubert. 1996. Water temperature affects smallmouth bass and channel catfish in a tailwater stream on the Great Plains. *North American Journal of Fisheries Management* 16:124–131.

Paukert, C.P. and D.W. Willis. 2004. Environmental influences on largemouth bass *Micropterus salmoides* in shallow Nebraska lakes. *Fisheries Management and Ecology* 11:345–352.

Paukert, C.P., D.W. Willis, and J.A. Klammer. 2002. Effects of predation and environment on quality of yellow perch and bluegill populations in Nebraska sandhill lakes. *North American Journal of Fisheries Management* 22:86–95.

Pearse, A.S. 1921. Distribution and food of the fishes of three Wisconsin lakes in summer. *University of Wisconsin Studies in Science* 3:1–61.

Pearson, W.D. and L.A. Krumholz. 1984. Distribution and status of Ohio River fishes. ORNL/Sub/79-7831/1. Oak Ridge National Laboratory, Oak Ridge, Tennessee, USA.

Pearson, W.D. and B.J. Pearson. 1989. Fishes of the Ohio River. *The Ohio Journal of Science* 89:181–187.

Peek, F. 1966. Age and growth of the smallmouth bass *Micropterus dolomieui* Lacepede in Arkansas. *Proceedings of the Nineteenth Annual Conference of the Southeastern Association of Game and Fish Commissioners* 19:422–431.

Pert, E.J., D.J. Orth, and M.J. Sabo. 2002. Lotic-dwelling age-0 smallmouth bass as both resource specialists and generalists: reconciling disparate literature reports. *America Fisheries Society Symposium* 31:185–189.

Peterson, D.C. and A.I. Myhr III. 1977. Ultrasonic tracking of smallmouth bass in Center Hill Reservoir, Tennessee. *Proceedings of the Thirty-First Annual Conference of the Southeastern Association of Fish and Wildlife Agencies* 31:618–624.

Peterson, J.T. and C.F. Rabeni. 1996. Natural thermal refugia for temperate warmwater stream fishes. *North American Journal of Fisheries Management* 16:738–746.

Peterson, M.S., N.J. Musselman, J. Francis, G. Habron, and K. Dierolf. 1993. Lack of salinity selection by freshwater and brackish populations of juvenile bluegill, *Lepomis macrochirus* Rafinesque. *Wetlands* 13:194–199.

Pflieger, W.L. 1966. Reproduction of the smallmouth bass (*Micropterus dolomieui*) in a small Ozark stream. *The American Midland Naturalist* 76:410–418.

Pflieger, W.L. 1975a. Reproduction and survival of the smallmouth bass in Courtois Creek. Pages 231–239 in R.H. Stroud and H. Clepper, editors. *Black Bass Biology and Management*. Sport Fishing Institute, Washington, D.C., USA.

Pflieger, W.L. 1975b. *The Fishes of Missouri*. Missouri Department of Conservation, Jefferson City, Missouri, USA.

Pflieger, W.L. 1997. *The Fishes of Missouri*. Second edition. Missouri Department of Conservation, Jefferson City, Missouri, USA.

Phillipp, D.P., W.F. Childers, and G.S. Whitt. 1983. A biochemical genetic evaluation of the northern and Florida subspecies of largemouth bass. *Transactions of the American Fisheries Society* 112:1–20.

Philipp, D.P., C.A. Toline, M.F. Kubacki, D.B.F. Philipp, and F.J.S. Phelan. 1997. The impact of catch-and-release angling on the reproductive success of smallmouth bass and largemouth bass. *North American Journal of Fisheries Management* 17:557–567.

Phillipp, D.P. and G.S. Whitt. 1990. Survival and growth of northern, Florida, and reciprocal F_1 hybrid largemouth bass in central Illinois. *Transactions of the American Fisheries Society* 120:58–64.

Phillips, J.M., J.R. Jackson, and R.L. Noble. 1993. The effect of hatching date on survival, age-specific piscivory and growth rates in young of year largemouth bass. *Abstracts of the First Mid-Year Technical Meeting, Southern Division American Fisheries Society* 1993:23.

Phillips, J.M., J.R. Jackson, and R.L. Noble. 1995. Hatching date influence on age-specific diet and growth of age-0 largemouth bass. *Transactions of the American Fisheries Society* 124:370–379.

Phillips, J.M., J.B. Jackson, and R.L. Noble. 1997. Spatial variation in abundance of age-0 largemouth bass among reservoir embayments. *North American Journal of Fisheries Management* 17:894–901.

Pierce, P.C. and M.J. Van Den Avyle. 1997. Hybridization between introduced spotted bass and smallmouth bass in reservoirs. *Transactions of the American Fisheries Society* 126:939–947.

Pierson, J.M. 1990. Status of endangered, threatened, and special concern freshwater fishes in Alabama. *Journal of the Alabama Academy of Science* 61:106–116.

Pine, W.E. III. 2000. Comparison of two otter trawls of different sizes for sampling black crappies. *North American Journal of Fisheries Management* 20:819–821.

Pine, W.E. III and M.S. Allen. 2001. Differential growth and survival of weekly age-0 black crappie cohorts in a Florida lake. *Transactions of the American Fisheries Society* 130:80–91.

Pine, W.E. III, S.A. Ludsin, and D.R. DeVries. 2000. First-summer survival of largemouth bass cohorts: is early spawning really best? *Transactions of the American Fisheries Society* 129:504–513.

Pipas, J.C. and F.J. Bulow. 1998. Hybridization between redeye bass and smallmouth bass in Tennessee streams. *Transactions of the American Fisheries Society* 127:141–146.

Ploskey, G.R., J.M. Nestler, and W.M. Bivin. 1996. Predicting black bass reproductive success from Bull Shoals Reservoir hydrology. *American Fisheries Society Symposium* 16:422–441.

Pope, K.L., M.L. Brown, and D.W. Willis. 1995. Proposed revision of the standard weight (W_s) equation for redear sunfish. *Journal of Freshwater Ecology* 10:129–134.

Pope, K.L. and D.R. DeVries. 1994. Interactions between larval white crappie and gizzard shad: quantifying mechanisms in small ponds. *Transactions of the American Fisheries Society* 123:975–987.

Pope, K.L. and D.W. Willis. 1998a. Larval black crappie distribution: implications for sampling impoundments and natural lakes. *North American Journal of Fisheries Management* 18:470–474.

Pope, K.L. and D.W. Willis. 1998b. Early life history and recruitment of black crappie (*Pomoxis nigromaculatus*) in two South Dakota waters. *Ecology of Freshwater Fish* 7:56–68.

Popiel, S.A., A. Perez-Fuentetaja, D.J. McQueen, and N.C. Collins. 1996. Determinants of nesting success in the pumpkinseed (*Lepomis gibbosus*): a comparison of two populations under different risks from predation. *Copeia* 1996:649–656.

Porak, W., W.S. Coleman, and S. Crawford. 1987. Age, growth, and mortality of Florida largemouth bass utilizing otoliths. *Proceedings of the Fortieth Annual Conference of the Southeastern Association of Fish and Wildlife Agencies* 40:206–215.

Post, J.R., L.G. Rudstam, and D.M. Schael. 1995. Temporal and spatial distribution of pelagic age-0 fish in Lake Mendota, Wisconsin. *Transactions of the American Fisheries Society* 124:84–93.

Potter, W.A., K.L. Dickson, and L.A. Nielsen. 1978. Larval sport fish drift in the New River. *Proceedings of the Annual Conference of the Southeastern Association of Fish and Wildlife Agencies* 32:672–679.

Powles, P.M., D.R. Vandeloo, and B. Clancy. 1980. Some features of larval rock bass, *Ambloplites rupestris* (Rafinesque), development in central Ontario. Pages 36–44 in L.A. Fuiman, editor. *Proceedings of the Fourth Annual Larval Fish Conference.* FWS/OBS-80/43. U.S. Fish and Wildlife Service, National Power Plant Team, Ann Arbor, Michigan, USA.

Poyser, W.A. 1919. Notes on the breeding habits of the pygmy sunfish. *Aquatic Life* 4:65–69.

Prather, E.E. 1967. A Note on the accuracy of the scale method in determining the ages of largemouth bass and bluegill from Alabama waters. *Proceedings of the Twentieth Annual Conference of the Southeastern Association of Fish and Wildlife Agencies* 20:483–486.

Prentice, J.A. and B.G. Whiteside. 1975. Validation of aging techniques for largemouth bass and channel catfish in central Texas farm ponds. *Proceedings of the Twenty-Eighth Annual Conference of the Southeastern Association of Fish and Wildlife Agencies* 28:414–428.

Probst, W.E., C.F. Rabeni, W.G. Covington, and R.E. Marteney. 1984. Resource use by stream-dwelling rock bass and smallmouth bass. *Transactions of the American Fisheries Society* 113: 283–294.

Purkett, C.A., Jr. 1958. Growth of fishes in the Salt River, Missouri. *Transactions of the American Fisheries Society* 87:116–130.

Putnam, J.H., C.L. Pierce, and D.M. Day. 1995. Relationships between environmental variables and size-specific growth rates of Illinois stream fishes. *Transactions of the American Fisheries Society* 124:252–261.

Putman, J.H. II and D.R. DeVries. 1993. Growth of young-of-year bluegill during limnetic-littoral migrations. (Abstract.) *Bulletin of the Ecological Society of America* 74:400–401.

Pyron, M. and C.M. Taylor. 1993. Fish community structure of Oklahoma Gulf Coastal Plains. *Hydrobiologia* 257:29–35.

Raborn, S.W., L.E. Miranda, and M.T. Driscoll. 2003. Modeling predation as a source of mortality for piscivorous fishes in a southeastern U.S. reservoir. *Transactions of the American Fisheries Society* 132:560–575.

Radomski, P. and T.J. Goeman. 2001. Consequences of human lakeshore development on emergent and floating-leaf vegetation abundance. *North American Journal of Fisheries Management* 21:46–61.

Rafinesque, C.S. 1817. First decade of new North American fishes. *American Monthly Magazine and Critical Reviews* 14:120–121.

Rainwater, W.C. and A. Houser. 1975. Relation of physical and biological variables to black bass crops. Pages 306–309 in R.H. Stroud and H. Clepper, editors. *Black Bass Biology and Management.* Sport Fishing Institute, Washington, D.C., USA.

Ramsey, J.S. 1975. Taxonomic history and systematic relationships among species of *Micropterus.* Pages 67–75 in R.H. Stroud and H. Clepper, editors. *Black Bass Biology and Management.* Sport Fishing Institute, Washington, D.C., USA.

Ramsey, J.S. 1976. Freshwater fishes. Pages 53–65 in H. Boschung, editor. *Endangered and Threatened Plants and Animals of Alabama.* Bulletin of the Alabama Museum of Natural History, Number 2:1–92.

Ramsey, J.S. and R.O. Smitherman. 1971. Development of color pattern in pond-reared young of five *Micropterus* species of southeastern U.S. *Proceedings of the Twenty-Fifth Annual Conference of the Southeastern Association of Game and Fish Commissioners* 25:348–356.

Raney, E.C. 1965. Some pan fishes of New York — rock bass, crappies, and other sunfishes. *The Conservationist* 19:21–35.

Range, J.D. 1973. Growth of five species of game fishes before and after introduction of threadfin shad into Dale Hollow Reservoir. *Proceedings of the Twenty-Sixth Annual Conference of the Southeastern Association of Game and Fish Commissioners* 26:510–518.

Reash, R.J. and J.H. Van Hassel. 1988. Distribution of upper and middle Ohio River fishes, 1973–1985. II. Influence of zoogeographic and physico-chemical tolerance factors. *Journal of Freshwater Ecology* 4:459–476.

Redmon, W.L. and L.A. Krumholz. 1978. Age, growth, condition, and maturity of sunfishes of Doe Run, Meade County, Kentucky. *Transactions of the Kentucky Academy of Science* 39:60–73.

Reed, H.D. and A.H. Wright. 1909. The vertebrates of the Cayuga Lake basin, New York. *Proceedings American Philosophical Society* 48:370–459.

Reed, M.S. and C.F. Rabeni. 1989. Characteristics of an unexploited smallmouth bass population in a Missouri Ozark stream. *North American Journal of Fisheries Management* 9:420–426.

Reed, R.J. 1971. Underwater observations of the population density and behavior of pumpkinseed, *Lepomis gibbosus* (Linnaeus), in Craneberry Pond, Massachusetts. *Transactions of the American Fisheries Society* 100:350–353.

Reid, G.R. 1949. Food of the black crappie, *Pomoxis nigromaculatus* (LeSueur), in Orange Lake, Florida. *Transactions of the American Fisheries Society* 79:145–154.

Reid, N. and P.M. Powles. 1982. Decline and cessation in fall feeding of 0 and age 1-year-old *Lepomis gibbosus* in central Ohio. *Proceedings of the Fifth Annual Larval Fish Conference* 5:16–19. Louisiana Cooperative Fisheries Research Unit, Baton Rouge, Louisiana, USA.

Reid, S.M., M.G. Fox, and T.H. Whillans. 1999. Influence of turbidity on piscivory in largemouth bass (*Micropterus salmoides*). *Canadian Journal of Fisheries and Aquatic Sciences* 56:1362–1369.

Reighard, J. 1906. The breeding habits, development and propagation of the black bass (*Micropterus dolomieu* Lacepede and *Micropterus salmoides* Lacepede). *Bulletin of the Michigan Fish Commission*, Number 7.

Reighard, J.E. 1915. An ecological reconnaissance of the fishes of Douglas Lake, Cheboygan County, Michigan, in midsummer. *United States Bureau of Fisheries Bulletin* 33:215–249.

Reinert, T.R., G.R. Ploskey, and M.J. Van Den Avyle. 1996. Effects of hydrology on black bass reproductive success in four southeastern reservoirs. *Proceedings of the Forty-Ninth Annual Conference of the Southeastern Association of Fish and Wildlife Agencies* 49:47–57.

Rettig, J.E. 1998. Variation in species composition of the larval assemblage in four southwest Michigan lakes: using allozyme analysis to identify larval sunfish. *Transactions of the American Fisheries Society* 127:661–668.

Reutter, J.M. and C.E. Herdendorf. 1976. Thermal discharge from a nuclear power plant: predicted effects on Lake Erie fish. *Ohio Journal of Science* 76:39–45.

Rice, L.A. 1942. The food of seventeen Reelfoot Lake fishes in 1941. *Journal of the Tennessee Academy of Science* 17:4–13.

Richardson, R.E. 1913. Observations on the breeding habits of fishes at Havana, Illinois, 1910 and 1911. *Illinois State Laboratory of Natural History Bulletin* 9:405–416.

Richardson, R.E. 1942. The rate of growth of bluegill sunfish in lakes of northern Indiana. *Investigations of Indiana Lakes and Streams* 2:161–214.

Richmond, N. 1940. Nesting of the sunfish, *Lepomis auritus* (Linnaeus), in tidal waters. *Zoologica* 25:329–330.

Ricker, W. 1948. Hybrid sunfish for stocking small ponds. *Transactions of the American Fisheries Society* 75:84–96.

Ricker, W.E. 1945. Abundance, exploitation and mortality of the fishes in two lakes. *Investigations of Indiana Lakes and Streams* 2:345–448.

Rider, S.J. and M.J. Maceina. 1993. The use of a catch-depletion method to estimate abundance and biomass of age-0 largemouth bass in vegetated and unvegetated regions of Lake Guntersville, Alabama. *Abstracts of the First Mid-Year Technical Meeting, Southern Division American Fisheries Society* 193:32.

Ridgway, M.S. and B.J. Shuter. 1996. Effects of displacement on the seasonal movements and home range characteristics of smallmouth bass in Lake Opeongo. *North American Journal of Fisheries Management* 16:371–377.

Ridgway, M.S. and B.J. Shuter. 1997. Predicting the effects of angling for nesting male smallmouth bass on production of age-0 fish with an individual-based model. *North American Journal of Fisheries Management* 17:568–580.

Ridgway, M.S., B.J. Shuter, T.A. Middel, and M.L. Gross. 2002. Spatial ecology and density-dependent processes in smallmouth bass: the juvenile transition hypothesis. *American Fisheries Society Symposium* 31:47–60.

Rieger, P.W. and R.C. Summerfelt. 1977. An evaluation of the introduction of Florida largemouth bass into an Oklahoma reservoir receiving a heated effluent. *Proceedings of the Thirtieth Annual Conference of the Southeastern Association of Fish and Wildlife Agencies* 30:48–57.

Roberts, F.L. 1964. A chromosome study of twenty species of Centrarchidae. *Journal of Morphology* 115:401–418.

Robins, C.R., R.M. Bailey, C.E. Bond, J.R. Brooker, E.A. Lachner, R.N. Lea, and W.B. Scott. 1991. *A List of Common and Scientific Names of Fishes from the United States and Canada*. American Fisheries Society Publication 20.

Robison, H.W. 1975. New distributional records of fishes from the lower Ouachita River system in Arkansas. *Arkansas Academy of Science Proceedings* 29:54–56.

Robison, H.W. and T.M. Buchanan. 1988. *Fishes of Arkansas*. The University of Arkansas Press, Fayetteville, Arkansas, USA.

Rodeheffer, I.A. 1940. The movements of marked fish in Douglas Lake, Michigan. *Papers of the Michigan Academy of Science, Arts, and Letters* 26:265–280.

Roe, K.J., P.M. Harris, and R.L. Mayden. 2002. Phylogenetic relationships of the genera of North American sunfishes and basses (Percoidei: Centrarchidae) as evidenced by the mitochondrial cytochrome *b* gene. *Copeia* 2002:897–905.

Roell, M.J. and D.J. Orth. 1993. Trophic basis of production of stream-dwelling smallmouth bass, rock bass, and flathead catfish in relation to invertebrate bait harvest. *Transactions of the American Fisheries Society* 122:46–62.

Rogers, W.A. 1967. Food habits of young largemouth bass (*Micropterus salmoides*) in hatchery ponds. *Proceedings of the Twenty-First Annual Conference of the Southeastern Association of Fish and Wildlife Agencies* 21:543–553.

Rohde, F.C. and R.G. Arndt. 1987. Two new species of pygmy sunfishes (Elassomatidae, *Elassoma*) from the Carolinas. *Proceedings of the Academy of Natural Science of Philadelphia* 139:65–85.

Rosebery, D.A. 1950. Game Fisheries Investigation of Clayton Lake, a Mainstream Impound of New River, Pulaski County, Virginia, with Emphasis on *Micropterus punctulatus* (Rafinesque). Doctoral dissertation. Virginia Polytechnic Institute and State University, Blacksburg, Virginia, USA.

Rosebery, D.A. and R.R. Bowers. 1952. Under the cover of Lake Drummond. *Virginia Wildlife* 13:21–23.

Ross, S.T. 2001. *Inland Fishes of Mississippi*. University of Mississippi Press, Jackson, Mississippi, USA.

Ruhr, C.E. 1957. Effect of stream impoundment in Tennessee on the fish populations of tributary streams. *Transactions of the American Fisheries Society* 86:144–157.

Runyun, S. 1961. Early development of the clingfish, *Gobiesox strumosus* Cope. *Chesapeake Science* 2:113–141.

Ryan, P.M. and H.H. Harvey. 1977. Growth of rock bass, *Ambloplites rupestris*, in relation to the morphoedaphic index as an indicator of an environmental stress. *Journal of the Fisheries Research Board of Canada* 34:2079–2088.

Ryan, P.W., J.W. Avault, Jr., and R.O. Smitherman. 1970. Food habits and spawning of spotted bass in Tchefuncte River, southeastern Louisiana. *The Progressive Fish-Culturist* 1970:162–167.

Sabat, A.M. 1994a. Costs and benefits of parental effort in a brood-guarding fish (*Ambloplites rupestris*, Centrarchidae). *Behavioral Ecology* 5:195–201.

Sabat, A.M. 1994b. Mating success in brood-guarding male rock bass, *Ambloplites rupestris*: the effect of body size. *Environmental Biology of Fishes* 39:411–415.

Sabo, M.J. and W.E. Kelso. 1991. Relationship between morphometry of excavated floodplain ponds along the Mississippi River and their use as fish nurseries. *Transactions of the American Fisheries Society* 120:552–561.

Sabo, M.J. and D.J. Orth. 1994. Temporal variation in microhabitat use by age-0 smallmouth bass in the North Anna River, Virginia. *Transactions of the American Fisheries Society* 123:733–746.

Sabo, M.J. and D.J. Orth. 1996. Absence of size-selective mortality among larval smallmouth bass in a Virginia stream. *Transactions of the American Fisheries Society* 125:920–924.

Sachdev, S.C. 1993. Molecular Phylogenetic Analysis of *Ambloplites* Populations. Master's thesis. University of Missouri, Columbia, Missouri, USA.

Sammons, S.M. and P.W. Bettoli. 1998. Larval sampling as a fisheries management tool: early detection of year-class strength. *North American Journal of Fisheries Management* 18:137–143.

Sammons, S.M. and P.W. Bettoli. 1999. Spatial and temporal variation in electrofishing catch rates of three species of black bass (*Micropterus* spp.) from Normandy Reservoir, Tennessee. *North American Journal of Fisheries Management* 19:454–461.

Sammons, S.M. and P.W. Bettoli. 2000. Population dynamics of a reservoir fish community in response to hydrology. *North American Journal of Fisheries Management* 20:791–800.

Sammons, S.M., P.W. Bettoli, and V.A. Greear. 2001. Early life history characteristics of age-0 white crappies in response to hydrology and zooplankton densities in Normandy Reservoir, Tennessee. *Transactions of the American Fisheries Society* 130:442–449.

Sammons, S.M., P.W. Bettoli, D.A. Isermann, and T.N. Churchill. 2002a. Recruitment variation of crappies in response to hydrology of Tennessee reservoirs. *North American Journal of Fisheries Management* 22:1393–1398.

Sammons, S.M., D.A. Isermann, and P.W. Bettoli. 2002b. Variation in population characteristics and gear selection between black and white crappies in Tennessee reservoirs: potential effects on management decisions. *North American Journal of Fisheries Management* 22:863–869.

Sammons, S.M., L.G. Dorsey, P.W. Bettoli, and F.C. Fiss. 1999. Effects of reservoir hydrology on reproduction by largemouth bass and spotted bass in Normandy Reservoir, Tennessee. *North American Journal of Fisheries Management* 19:78–88.

Sandow, J.T., Jr., D.R. Holder, and L.E. McSwain. 1975. Life history of the redbreast sunfish in the Satilla River, Georgia. *Proceedings of the Twenty-Eighth Annual Conference of the Southeastern Association of Game and Fish Commissioners* 28:279–295.

Santucci, V.J., Jr. and D.H. Wahl. 2003. The effects of growth, predation, and first-winter mortality on recruitment of bluegill cohorts. *Transactions of the American Fisheries Society* 132:346–360.

Saunders, R., M.A. Bozek, C.J. Edwards, M.J. Jennings, and S.P. Newman. 2002. Habitat features affecting smallmouth bass *Micropterus dolomieu* nesting success in four northern Wisconsin lakes. *American Fisheries Society Symposium* 31:123–134.

Savino, J.F., E.A. Marschall, and R.A. Stein. 1992. Bluegill growth as modified by plant density: an exploration of underlying mechanisms. *Oecologia* 153–160.

Savitz, J., P.A. Fish, and R. Weszely. 1983. Habitat utilization and movement of fish as determined by radio-telemetry. *Journal of Freshwater Ecology* 2:165–174.

Savitz, J. and G. Funk. 2001. Population size and recruitment of smallmouth bass in Calumet Harbor of Lake Michigan. *Journal of Freshwater Ecology* 16:317–319.

Scalet, C.G. 1977. Summer food habits of sympatric stream populations of spotted bass, *Micropterus punctulatus*, and largemouth bass, *M. salmoides* (Osteichthyes: Centrarchidae). *The Southwest Naturalist* 21:493–501.

Schael, D.M., L.G. Rudstam, and J.R. Post. 1991. Gape limitation and prey selection in larval yellow perch (*Perca flavescens*), freshwater drum (*Aplodinotus grunniens*), and black crappie (*Pomoxis nigromaculatus*). *Canadian Journal of Fisheries and Aquatic Science* 48:1919–1925.

Scheidegger, K.J. 1990. Larval Fish Assemblage Composition and Microhabitat Used in Two Southeastern Rivers. Master's thesis. Auburn University, Auburn, Alabama, USA.

Schmidt, R.C. and M.C. Fabrizio. 1980. Daily growth rings on otoliths for aging young-of-the-year largemouth bass from wild populations. *The Progressive Fish-Culturist* 42:78–80.

Schneberger, E. 1972. *White Crappie Life History, Ecology and Management*. Publication 244, Wisconsin Department of Natural Resources, Madison, Wisconsin, USA.

Schneider, J.C. 1999. Dynamics of quality bluegill populations in two Michigan lakes with dense vegetation. *North American Journal of Fisheries Management* 19:97–109.

Schoffman, R.J. 1939. Age and growth of redear sunfish in Reelfoot Lake. *Journal of the Tennessee Academy of Science* 14:61–71.

Schoffman, R.J. 1940. Age and growth of the black and white crappie, the warmouth bass, and the yellow bass in Reelfoot Lake. *Report of the Reelfoot Lake Biological Station* 4:22–42.

Schoffman, R.J. 1952. Growth of the bluegills and crappies in Reelfoot Lake, Tennessee. *Journal of the Tennessee Academy of Science* 27:15–26.

Schramm, H.L., Jr. and K.J. Jirka. 1989. Epiphytic macroinvertebrates as a food resource for bluegills in Florida lakes. *Transactions of the American Fisheries Society* 118:416–426.

Schramm, H.L., S.P. Malvestuto, and W.A. Hubert. 1992. Evaluation of procedures for back-calculation of lengths of largemouth bass aged by otoliths. *North American Journal of Fisheries Management* 12:604–608.

Schramm, H.L. and D.C. Smith. 1988. Differences in growth rate between sexes of Florida largemouth bass. *Proceedings of the Forty-First Annual Conference of the Southeastern Association of Fish and Wildlife Agencies* 41:76–84.

Schreer, J.F. and S.J. Cooke. 2002. Behavioral and physiological responses of smallmouth bass to a dynamic thermal environment. *American Fisheries Society Symposium* 31:191–203.

Schwartz, F.J. 1964. Natural salinity tolerances of some freshwater fishes. *Underwater Naturalist* 2:13–15.

Scott, D.C. 1949. A study of a stream population of rock bass, *Ambloplites rupestris*. *Investigations of Indiana Lakes and Streams* 3:169–234.

Scott, E.M., Jr. and J.P. Buchanan. 1979. Larval fish populations in Little Bear Creek Reservoir during the first three years of impoundment. *Proceedings of the Thirty-Third Annual Conference of the Southeastern Association of Fish and Wildlife Agencies* 33:660–672.

Scott, M.C. and P.L. Angermeier. 1998. Resource use by two sympatric black basses in impounded and riverine sections of the New River, Virginia. *North American Journal of Fisheries Management* 18:221–235.

Scott, W.B. and E.J. Crossman. 1973. *Freshwater Fishes of Canada*. Fisheries Research Board of Canada, Bulletin 184. Ottawa, Ontario, Canada.

Sepulveda, M.S., D.S. Ruessler, B.P. Quinn, N.D. Denslow, S.E. Holm, and T.S. Gross. 1999. Effects of paper mill effluents on reproductive success of Florida largemouth bass (*Micropterus salmoides*). Presented at the SETAC 20th Annual Meeting, Philadelphia, Pennyslvania, USA.

Serns, S.L. 1972. Age, Growth and Condition of Bluegill Sunfish, *Lepomis macrochirus* Rafinesque, in Four Heated Reservoirs in Texas. Master's thesis. Texas A&M University, College Station, Texas, USA.

Shafland, P.L. and J.M. Pestrak. 1982. Predation on blue tilapia by largemouth bass, in experimental ponds. *Proceedings of the Thirty-Fifth Annual Conference of the Southeastern Association of Fish and Wildlife Agencies* 35:443–448.

Shannon, E.H. 1967. Geographical distribution and habitat requirements of the redbreast sunfish *Lepomis auritus* in North Carolina. *Proceedings of the Twentieth Annual Conference of the Southeastern Association of Game and Fish Commissioners* 20: 319–323.

Shao, B. 1997. Nest association of pumpkinseed, *Lepomis gibbosus*, and golden shiner, *Notemigonus crysoleucas*. *Environmental Biology of Fishes* 50:41–48.

Shelton, W.L., W.D. Davies, T.A. King, and T.J. Timmons. 1979. Variation in the growth of the initial year class of largemouth bass in West Point Reservoir, Alabama and Georgia. *Transactions of the American Fisheries Society* 108:142–149.

Shortt, L.R. 1956. A new pygmy sunfish. *The Aquarium* 25:133–135.

Shoup, D.E., R.E. Carlson, and R.T. Heath. 2003. Effects of predation risk and foraging return on the diel use of vegetated habitat by two size-classes of bluegills. *Transactions of the American Fisheries Society* 132:590–597.

Shute, P. and J.R. Shute. 1991. Sharing the nest. *Tropical Fish Hobbyist* 1991 (December): 76–93.

Shuter, B.J. and P.E. Ihssen. 1991. Chemical and biological factors affecting acid tolerance of smallmouth bass. *Transactions of the American Fisheries Society* 120:23–33.

Shuter, B.J. and M.S. Ridgway. 2002. Bass in time and space: operational definitions of risk. *American Fisheries Society Symposium* 31:235–249.

Siefert, R.E. 1965. Early scale development in the white crappie. *Transactions of the American Fisheries Society* 94:182.

Siefert, R.E. 1968. Reproductive behavior, incubation and mortality of eggs, and postlarval food selection in the white crappie. *Transactions of the American Fisheries Society* 97:252–259.

Siefert, R.E. 1969a. Biology of the white crappie in Lewis and Clark Lake. *Technical Papers of the Bureau of Sport Fisheries and Wildlife* 22:1–16.

Siefert, R.E. 1969b. Characteristics for separation of white and black crappie larvae. *Transactions of the American Fisheries Society* 98:326–328.

Siefert, R.E. 1972. First food of larval yellow perch, white sucker, bluegill, emerald shiner, and rainbow smelt. *Transactions of the American Fisheries Society* 101:219–225.

Siler, J.R. and J.P. Clugston. 1975. Largemouth bass under conditions of extreme thermal stress. Pages 333–341 in R.H. Stroud and H. Clepper, editors. *Black Bass Biology and Management.* Sport Fishing Institute, Washington, D.C., USA.

Simon, T.P. 1985. Descriptions of Larval Percidae Inhabiting the Upper Mississippi River Basin (*Osteichthyes: Ethostomatini*). Master's thesis. University of Wisconsin, Lacrosse, Wisconsin, USA.

Simon, T.P. 1986. Variation in seasonal, spatial, and species composition of main channel ichthyoplankton abundance, Ohio River miles 569–572. *Transactions of the Kentucky Academy of Science* 47:19–26.

Simon, T.P. 1999. Assessment of Balon's reproductive guilds with application to Midwestern North American freshwater fishes. Pages 97–121 in T.P. Simon, editor. *Assessing the Sustainability and Biological Integrity of Water Resources Using Fish Communities.* CRC Press, Boca Raton, Florida, USA.

Simon, T.P., R.L. Dufour, and B.E. Fisher. 2005. Changes in the biological integrity of fish assemblages in the Patoka River drainage as a result of anthropogenic disturbance from 1888 to 2001. Pages 383–398 in J.N. Rinne, R.M. Hughes, and B. Calamusso, editors. *Historical Changes in Large River Fish Assemblages of the Americas.* American Fisheries Society Symposium 45, Bethesda, Maryland, USA.

Simon, T.P. and R. Wallus. 2003. *Reproductive Biology and Early Life History of Fishes in the Ohio River Drainage. Volume 3: Ictaluridae — Catfish and Madtoms.* CRC Press, Boca Raton, Florida, USA.

Simon, T.P. and R. Wallus. 2006. *Reproductive Biology and Early Life History of Fishes in the Ohio River Drainage. Volume 4: Percidae.* CRC Press, Boca Raton, Florida, USA.

Simonson, T.D. and W.A. Swenson. 1990. Critical stream velocities for young-of-year smallmouth bass in relation to habitat use. *Transactions of the American Fisheries Society* 119:902–909.

Sizemore, D.R. and W.M. Howell. 1990. Fishes of springs and spring-fed creeks of Calhoun County, Alabama. *Southeastern Fishes Council Proceedings* 22:1–6.

Slipke, J.W., M.J. Maceina, V.H. Travnichek, and K.C. Weathers. 1998. Effects of a 356-mm minimum length limit on the population characteristics and sport fishery of smallmouth bass in the Shoals Reach of the Tennessee River, Alabama. *North American Journal of Fisheries Management* 18:76–84.

Slipke, J.W., S.M. Smith, and M.J. Maceina. 2000. Food habits of striped bass and their influence on crappie in Weiss Lake, Alabama. *Proceedings of the Fifty-Fourth Annual Conference of the Southeastern Association of Fish and Wildlife Agencies* 54:88–96.

Slipp, J.W. 1943. The rock bass, *Ambloplites rupestris*, in Washington state. *Copeia* 1943(2):132.

Smale, M.A. and C.F. Rabeni. 1995a. Hypoxia and hyperthermia tolerances of headwater stream fishes. *Transactions of the American Fisheries Society* 124:698–710.

Smale, M.A. and C.F. Rabeni. 1995b. Influences of hypoxia and hyperthermia on fish species composition in headwater streams. *Transactions of the American Fisheries Society* 124:711–725.

Smith, C.L. 1985. *The Inland Fishes of New York State*. New York State Department of Environmental Conservation, Albany, New York, USA.

Smith, C.L. and R.M. Bailey. 1961. Evolution of the dorsal-fin supports of percoid fishes. *Papers of the Michigan Academy of Science, Arts, and Letters* 46:345–363.

Smith, G.R. and J.G. Lundberg. 1972. The Sand Draw fish fauna. *Bulletin of the American Museum of Natural History* 148:40–54.

Smith, H.M. 1907. The Fishes of North Carolina. *North Carolina Geology Economic Survey*. E.M. Uzzell and Company, Raleigh, North Carolina, USA.

Smith, H.M. and B.A. Bean. 1899. List of fishes known to inhabit the waters of the District of Columbia and vicinity. *United States Commissioners Fisheries Bulletin* 18(1898):179–187.

Smith, M.W. 1952. Limnology and trout angling in Charlotte County lakes, New Brunswick. *Journal of the Fisheries Research Board of Canada* 8:383–452.

Smith, P.W. 1968. An assessment of changes in the fish fauna of two Illinois rivers and its bearing on their future. *Transactions of the Illinois State Academy of Science* 61: 31–45.

Smith, P.W. 1979. *The Fishes of Illinois*. University of Illinois Press, Urbana, Illinois, USA.

Smith, P.W., A.C. Lopinot, and W.L. Pflieger. 1971. A distributional atlas of upper Mississippi River fishes. *Illinois Natural History Survey Biological Notes* 73:1–20.

Smith, R.J.F. 1969. Control of prespawning behaviour of sunfish (*Lepomis gibbosus* and *L. megalotis*). I. Gonadal androgen. *Animal Behaviour* 17:279–285.

Smith, R.P. and J.L. Wilson. 1981. Growth comparison of two subspecies of largemouth bass in Tennessee ponds. *Proceedings of the Thirty-Fourth Annual Conference of the Southeastern Association of Fish and Wildlife Agencies* 34:25–30.

Smith, S.L. and J.E. Crumpton. 1977. Interrelationship of vegetative cover and sunfish population density in suppressing spawning in largemouth bass. *Proceedings of the Thirty-First Annual Conference of the Southeastern Association of Fish and Wildlife Agencies* 31:318–321.

Smith, S.M., M.J. Maceina, and R.A. Dunham. 1994. Natural hybridization between black crappie and white crappie in Weiss Lake, Alabama. *Transactions of the American Fisheries Society* 123:71–79.

Smith, S.M., M.J. Maceina, V.H. Travnichek, and R.A. Dunham. 1995. Failure of quantitative phenotypic characteristics to distinguish black crappie, white crappie, and their first-generation hybrid. *North American Journal of Fisheries Management* 15:121–125.

Smith, S.M., J.S. Odenkirk, and S.J. Reeser. 2005. Smallmouth bass recruitment variability and its relation to stream discharge in three Virginia rivers. *North American Journal of Fisheries Management* 25:1112–1121.

Smith, W.E. 1975. Breeding and culture of two sunfish, *Lepomis cyanellus* and *L. megalotis*, in the laboratory. *The Progressive Fish-Culturist* 37:227–229.

Smitherman, R.O. 1975. Experimental species associations of basses in Alabama ponds. Pages 76–84 in R.H. Stroud and H. Clepper, editors. *Black Bass Biology and Management*. Sport Fishing Institute, Washington, D.C., USA.

Smitherman, R.O. and J.S. Ramsey. 1972. Observations on spawning and growth of four species of basses (*Micropterus*) in ponds. *Proceedings of the Twenty-Fifth Annual Conference of the Southeastern Association of Game and Fish Commissioners* 25:357–365.

Smithson, E.B. and C.E. Johnston. 1999. Movement patterns of stream fishes in Ouachita Highlands stream: an examination of the restricted movement paradigm. *Transactions of the American Fisheries Society* 128:847–853.

Smith-Vaniz, W.F. 1968. *Freshwater Fishes of Alabama*. Auburn University Agricultural Experimental Station, Auburn, Alabama, USA.

Snow, H., A. Ensign, and J. Klingbiel. 1966. The bluegill, its life history, ecology, and management. Wisconsin Department of Conservation, Publication 230, Madison, Wisconsin, USA.

Snow, J.R. 1962. Forage fish preference and growth rate of largemouth bass fingerlings under experimental conditions. *Proceedings of the Fifteenth Annual Conference of the Southeastern Association of Fish and Wildlife Agencies* 15:303–313.

Snow, J.R. 1969. Some progress in the controlled culture of the largemouth bass, *Micropterus salmoides* (Lac.). *Proceedings of the Twenty-Second Annual Conference of the Southeastern Association of Fish and Wildlife Agencies* 22:380–387.

Snow, J.R. 1971. Fecundity of largemouth bass, *Micropterus salmoides* Lacepede receiving artificial food. *Proceedings of the Twenty-Fourth Annual Conference of the Southeastern Association of Fish and Wildlife Agencies* 24:550–559.

Snow, J.R. 1975. Hatchery propagation of the black basses. Pages 344–356 in R.H. Stroud and H. Clepper, editors. *Black Bass Biology and Management.* Sport Fishing Institute, Washington, D.C., USA.

Snucins, E.J. and B.J. Shuter. 1991. Survival of introduced smallmouth bass in low-pH lakes. *Transactions of the American Fisheries Society* 120:209–216.

Snyder, D.E. 1976. Terminologies for intervals of larval fish development. In J. Boreman, editor. *Great Lakes Fish Egg and Larvae Identification.* FWS/OBS-76/23. Office of Biological Services, United States Fish and Wildlife Service, Washington, D.C., USA.

Sowa, S.P. and C.F. Rabeni. 1995. Regional evaluation of the relation of habitat to distribution and abundance of smallmouth bass and largemouth bass in Missouri streams. *Transactions of the American Fisheries Society* 124:240–251.

Speir, H.J. 1969. Comparison of Rock Bass, *Ambloplites rupestris*, Populations in Obed River and Spring Creek, Tennessee. Master's thesis. Tennessee Technological University, Cookeville, Tennessee, USA.

Spier, T.W. and J.R. Ackerson. 2004. Effect of temperature on the identification of larval black crappies, white crappies, and F$_1$ hybrid crappies. *Transactions of the American Fisheries Society* 133:789–793.

Spier, T.W. and R.C. Heidinger. 2002. Effect of turbidity on growth of black crappie and white crappie. *North American Journal of Fisheries Management* 22:1438–1441.

Sprugel, G., Jr. 1955. The growth of green sunfish (*Lepomis cyanellus*) in Little Wall Lake, Iowa. *Iowa State College Journal of Science* 29:707–719.

St. John, R.T. and W.P. Black. 2004. Methods for predicting age-0 crappie year-class strength in J. Percy Priest Reservoir, Tennessee. *North American Journal of Fisheries Management* 24:1300–1308.

Stang, D.L., D.M. Green, R.M. Klindt, T.L. Chiotti, and W.M. Miller. 1996. Black bass movememnts after release from fishing tournaments in four New York waters. Pages 163–171 in L.E. Miranda and D.R. DeVries, editors. *Multidimensional Approaches to Reservoir Fisheries Management.* American Fisheries Society Symposium 16.

Stark, W.J. and A.A. Echelle. 1998. Genetic structure and systematics of smallmouth bass, with emphasis on Interior Highlands populations. *Transactions of the American Fisheries Society* 127:393–416.

Starrett, W.C. and A.W. Fritz. 1965. A biological investigation of the fishes of Lake Chautauqua, Illinois. *Illinois Natural History Survey Bulletin* 29:1–104.

Stauffer, J.R., Jr., K.L. Dickson, J. Cairns, Jr., and D.S. Cherry. 1976. The potential and realized influences of temperature on the distribution of fishes in the New River, Glen Lyn, Virginia. *Wildlife Monographs* 50:5–40.

Steele, B.D. 1981. Commensal spawning behavior of the rosefin shiner, *Notropis ardens*, in nests of the longear sunfish, *Lepomis megalotis. Environmantal Biology of Fishes* 6:138 (abstract).

Stegman, J.L. 1958. The Fishes of Kincaid Creek, Illinois. Master's thesis. Southern Illinois University, Carbondale, Illinois, USA.

Stegman, J.L. 1959. The fishes of Kincaid Creek, Illinois. *Transactions of the Illinois State Academy of Science* 52: 25–32.

Steinhart, G.B., E.A. Marschall, and R.A. Stein. 2004. Round goby predation on smallmouth bass offspring in nests during simulated catch-and-release angling. *Transactions of the American Fisheries Society* 133:121–131.

Sterba, G. 1967. *Freshwater Fishes of the World.* The Pet Library, Ltd., New York, New York. USA.

Stevenson, F., W.T. Momot, and F.J. Svoboda III. 1969. Nesting success of the bluegill, *Lepomis macrochirus* Rafinesque, in a small Ohio farm pond. *The Ohio Journal of Science* 69:347–355.

Stewart, E.M. and T.R. Finger. 1985. Diel activity patterns of fishes in lowland hardwood wetlands. *Transactions, Missouri Academy of Science* 19:5–9.

Stoeckel, J.N. and R.C. Heidinger. 1992. Relative susceptibility of inland and brook silversides to capture by largemouth bass. *North American Journal of Fisheries Management* 12:499–503.

Stone, U.B., D.G. Pasko, and R.M. Roecker. 1954. A study of Lake Ontario–St. Lawrence River smallmouth bass. *New York Fish and Game Journal* 1:1–26.

Storr, J.F., P.J. Hadden-Carter, J.M. Myers, and A.G. Smythe. 1983. Dispersion of rock bass along the south shore of Lake Ontario. *Transactions of the American Fisheries Society* 112:618–628.

Stroud, R.H. 1948. Growth of the basses and black crappie in Norris Reservoir, Tennessee. *Journal of the Tennessee Academy of Science* 23:31–99.

Stroud, R.H. 1949. Rate of growth and condition of game and pan fish in Cherokee and Douglas Reservoirs, Tennessee, and Hiwassee Reservoir, North Carolina. *Journal of the Tennessee Academy of Science* 24:60–74.

Sule, M.J. 1981. First-year growth and feeding of largemouth bass in a heated reservoir. Pages 520–535 in R.W. Larimore and J.A. Tranquilli, editors. *The Lake Sangchris Study: Case History of an Illinois Cooling Lake. Illinois Natural History Survey Bulletin* 32. Champaign, Illinois, USA.

Sullivan, J.R. 1975. Some diseases of the black basses. Pages 95–103 in R.H. Stroud and H. Clepper, editors. *Black Bass Biology and Management.* Sport Fishing Institute, Washington, D.C., USA.

Summerfelt, R.C. 1971. Factors influencing the horizontal distribution of several fishes in an Oklahoma Reservoir. Pages 425–439 in G.E. Hall, editor. *Reservoir Fisheries and Limnology.* American Fisheries Society Special Publication Number 8.

Summerfelt, R.C. 1975. Relationship between weather and year-class strength of largemouth bass. Pages 166–174 in R.H. Stroud and H. Clepper, editors. *Black Bass Biology and Management.* Sport Fishing Institute, Washington, D.C., USA.

Suski, C.D., F.J.S. Phelan, M.F. Kubacki, and D.P. Philipp. 2002. The use of sanctuaries for protecting nesting black bass from angling. *American Fisheries Society Symposium* 31:371–378.

Suski, C.D. and D.P. Philipp. 2004. Factors affecting the vulnerability to angling of nesting male largemouth and smallmouth bass. *Transactions of the American Fisheries Society* 133:1100–1106.

Suski, C.D., J.H. Svec, J.B. Ludden, F.J.S. Phelan, and D.P. Philipp. 2003. The effect of catch-and-release angling on the parental care behavior of male smallmouth bass. *Transactions of the American Fisheries Society* 132:210–218.

Suttkus, R.D. 1955. Age and growth of a small-stream population of "stunted" smallmouth black bass, *Micropterus dolomieui* (Lacepede). *New York Fish and Game Journal* 2:83–91.

Sutton, T.M. and J.J. Ney. 2002. Trophic resource overlap between age-0 striped bass and largemouth bass in Smith Mountain Lake, Virginia. *North American Journal of Fisheries Management* 22:1250–1259.

Swann, D.L., L.E. Rider, and F.J. Bulow. 1991. Age, growth, and summer foods of four centrarchid species in a Big South Fork National River and Recreation Area stream fish community. *Journal of the Tennessee Academy of Science* 66:23–28.

Swenson, W.A. 2002. Demographic changes in a largemouth bass population following closure of the fishery. Pages 627–637 in D.P. Phillip and M.S. Ridgway, editors. *Black Bass: Ecology, Conservation, and Management.* American Fisheries Society Symposium 31, Bethesda, Maryland, USA.

Swenson, W.A., B.J. Shuter, D.J. Orr, and G.D. Heberling. 2002. The effects of stream temperature and velocity on first-year growth and year-class abundance of smallmouth bass in the upper Mississippi River. Pages 101–113 in D.P. Phillip and M.S. Ridgway, editors. *Black Bass: Ecology, Conservation, and Management.* American Fisheries Society Symposium 31, Bethesda, Maryland, USA.

Swing, J.M. and E.G. McGowan. 1987. Characteristics for distinguishing flier (*Centrarchus macropterus*) and black crappie (*Pomoxis nigromaculatus*) larvae. *The Journal of the Elisha Mitchell Society* 103:81–87.

Swingle, H.S. 1956. Appraisal of methods of fish population study. IV. Determination of balance in farm fish ponds. *Transactions of the Twenty-First North American Wildlife Conference* 21:298–318.

Taber, C. 1964. Spectacular development in the pygmy sunfish *Elassoma zonatum*, with observations on spawning habitats. *Proceedings of the Oklahoma Academy of Science* 45:73–78.

Taber, C.A. 1969. The Distribution and Identification of Larval Fishes in the Buncombe Creek Arm of Lake Texoma with Observations on Spawning Habits and Relative Abundance. Doctoral dissertation. University of Oklahoma, Norman, Oklahoma, USA.

Tate, W.B., M.S. Allen, R.A. Myers, E.J. Nagid, and J.R. Estes. 2003. Relation of age-0 largemouth bass abundance to *Hydrilla* coverage and water level at Lochloosa and Orange Lakes, Florida. *North American Journal of Fisheries Management* 23:251–257.

Taubert, B.D. 1977. Early morphological development of the green sunfish, *Lepomis cyanellus*, and its separation from other larval *Lepomis* species. *Transactions of the American Fisheries Society* 106:445–448.

Tebo, L.B. and E.G. McCoy. 1964. Effect of sea-water concentration on the reproduction and survival of largemouth bass and bluegills. *Progressive Fish-Culturist* 26:99–106.

Thomas, J.A., E.B. Emery, and K.H. McCormick. 2004. Detection of temporal trends in Ohio River fish assemblages based on lock chamber surveys (1957–2001). *American Fisheries Society Symposium* 2004:1–19.

Thompson, H.E. 1941. A short trip to Florida. *The Fish Culturist* 20:19–20.

Tidwell, J.H., S.D. Coyle, and T.A. Woods. 2000. Species profile largemouth bass. *Southern Regional Aquaculture Center Publication Number 722.*

Tillma, J.S., C.S. Guy, and C.S. Mammoliti. 1998. Relations among habitat and population characteristics of spotted bass in Kansas streams. *North American Journal of Fisheries Management* 18:886–893.

Timmons, T.J. 1984. Comparative food habits and habitat preference of age-0 largemouth bass and yellow perch in West Point Lake, Alabama–Georgia. *Proceedings of the Thirty-Eighth Annual Conference of the Southeastern Association of Fish and Wildlife Agencies* 38:302–312.

Timmons. T.J. and O. Pawaputanon. 1981. Relative size relationship in prey selection by largemouth bass in West Point Lake, Alabama–Georgia. *Proceedings of the Thirty-Fourth Annual Conference of the Southeastern Association of Fish and Wildlife Agencies* 34:248–252.

Timmons, T.J., W.L. Shelton, and W.D. Davies. 1979. Sampling reservoir fish populations with rotenone in littoral areas. *Proceedings of the Thirty-Second Annual Conference of the Southeastern Association of Fish and Wildlife Agencies* 32:474–485.

Timmons, T.J., W.L. Shelton, and W.D. Davies. 1980. Differential growth of largemouth bass in West Point Reservoir, Alabama and Georgia. *Transactions of the American Fisheries Society* 109:176–186.

Tin, H.T. 1982. Family Centrarchidae, sunfishes. Pages 524–580 in N.A. Auer, editor. *Identification of Larval Fishes of the Great Lakes Basin with Emphasis on the Lake Michigan Drainage*. Special Publication 82-3, Great Lakes Fishery Commission, Ann Arbor, Michigan, USA.

Todd, B.L. and C.F. Rabeni. 1989. Movement and habitat use by stream-dwelling smallmouth bass. *Transactions of the American Fisheries Society* 118:229–242.

Toetz, D.W. 1965. Factors Affecting the Survival of Bluegill Sunfish Larvae. Doctoral dissertation. Indiana University, Bloomington, Indiana, USA.

Tomcko, C.M. and R.B. Pierce. 2001. The relationship of bluegill growth, lake morphometry, and water quality in Minnesota. *Transactions of the American Fisheries Society* 130:317–321.

Tomcko, C.M. and R.B. Pierce. 2005. Bluegill recruitment, growth, population size structure, and associated factors in Minnesota lakes. *North American Journal of Fisheries Management* 25:171–179.

Trautman, M.B. 1957. *The Fishes of Ohio with Illustrated Keys*. Ohio State University Press, Columbus, Ohio, USA.

Trautman, M.B. 1981. *The Fishes of Ohio*. Ohio State University Press, Columbus, Ohio, USA.

Travnichek, V.H., M.J. Maceina, and R.A. Dunham. 1996. Hatching time and early growth of age-0 black crappies, white crappies, and their naturally produced F_1 hybrids in Weiss Lake, Alabama. *Transactions of the American Fisheries Society* 125:334–337.

Travnichek, V.H., M.J. Maceina, M.C. Wooten, and R.A. Dunham. 1997. Symmetrical hybridization between black crappie and white crappie in an Alabama reservoir based on analysis of the cytochrome-*b* gene. *Transactions of the American Fisheries Society* 126:127–132.

Travnichek, V.H., A.V. Zale, and W.L. Fisher. 1993. Entrainment of ichthyoplankton by a warmwater hydroelectric facility. *Transactions of the American Fisheries Society* 122:709–716.

Truitt, R.V., B.A. Bean, and H.W. Fowler. 1929. The Fishes of Maryland. *Master's Conservative Department Bulletin Number 3*.

Tucker, W.H. 1973. Food habits, growth, and length–weight relationships of young-of-the-year black crappie and largemouth bass in ponds. *Proceedings of the Twenty–Sixth Annual Conference of the Southeastern Association of Game and Fish Commissioners* 26:565–577.

Tyus, H.M. 1969. Spawning of rock bass in North Carolina during 1968. *The Progressive Fish-Culturist* 32:25.

Ulrey, L., C. Risk, and W. Scott. 1938. The number of eggs produced by some of our common freshwater fishes. *Investigations of Indiana Lakes and Streams* 1:73–77.

VanArnum, C.J.G., G.L. Buynak, and J.R. Ross. 2004. Movement of smallmouth bass in Elkhorn Creek, Kentucky. *North American Journal of Fisheries Management* 24:311–315.

Vanderkooy, K.E., C.F. Rakoinski, and R.W. Heard. 2000. Trophic relationships of three sunfishes (*Lepomis* spp.) in an estuarine bayou. *Estuaries* 23:621–632.

Vanderpuye, C.J. and K.D. Carlander. 1971. Age, growth and condition of black crappie, *Pomoxis nigromaculatus* (Le Sueur), in Lewis and Clark Lake, South Dakota, 1954 to 1967. *Iowa State Journal of Science* 45:541–555.

Van Hassel, J.H., R.J. Reash, H.W. Brown, J.L. Thomas, and R.C. Matthews, Jr. 1988. Distribution of upper and middle Ohio River fishes, 1973–1985. I. Associations with water quality and ecological variables. *Journal of Freshwater Ecology* 4:441–458.

Vogele, L.E. 1975a. Reproduction of spotted bass, *Micropterus punctulatus*, in Bull Shoals Reservoir, Arkansas. *Technical Papers of the U.S. Fish and Wildlife Service; Number 84*.

Vogele, L.E. 1975b. The spotted bass. Pages 34–45 in R.H. Stroud and H. Clepper, editors. *Black Bass Biology and Management*. Sport Fishing Institute, Washington, D.C., USA.

Vogele, L.E. and W.C. Rainwater. 1975. Use of brush shelters as cover by spawning black basses (*Micropterus*) in Bull Shoals Reservoir. *Transactions of the American Fisheries Society* 104:264–269.

von Geldern, C., Jr. and D.F. Mitchell. 1975. Largemouth bass and threadfin shad in California. Pages 436–449 in R.H. Stroud and H. Clepper, editors. *Black Bass Biology and Management*. Sport Fishing Institute, Washington, D.C., USA.

Wagner, T., A.K. Jubar, and M.T. Bremigan. 2006. Can habitat alteration and spring angling explain largemouth bass nest success? *Transactions of the American Fisheries Society* 135:843–852.

Wainwright, P.C. 1996. Ecological explanation through functional morphology: the feeding biology of sunfishes. *Ecology* 77:1336–1343.

Wainwright, P.C. and G.V. Lauder. 1992. The evolution of feeding biology in sunfishes (Centrarchidae). Pages 472–491 in R.L. Mayden, editor. *Systematics, Historical Ecology, and North American Freshwater Fishes*. Stanford University Press, Stanford, California, USA.

Walburg, C.H. 1976. Changes in the fish populations of Lewis and Clark Lake, 1956–1974, and their relation to water management and the environment. United States Fish and Wildlife Service Research Report Number 79.

Wallace, R.K., Jr. 1981. An assessment of diet-overlap indexes. *Transactions of the American Fisheries Society* 110:72–76.

Wallus, R. and T.P. Simon. 2006. *Reproductive Biology and Early Life History of Fishes in the Ohio River Drainage. Volume 5: Aphredoderidae through Cottidae, Moronidae, and Sciaenidae.* CRC Press, Boca Raton, Florida, USA.

Wallus, R., T.P. Simon, and B.L. Yeager. 1990. *Reproductive Biology and Early Life History of Fishes in the Ohio River Drainage. Volume 1: Acipenseridae through Esocidae.* Tennessee Valley Authority, Chattanooga, Tennessee, USA.

Walsh, S.J. and B.M. Burr. 1984. Life history of the banded pygmy sunfish, *Elassoma zonatum* Jordan (Pisces, Centrarchidae) in western Kentucky. *Bulletin of the Alabama Museum of Natural History* 8:31–52.

Walters, D.A., W.E. Lynch, Jr., and D.L. Johnson. 1991. How depth and interstice size of artificial structures influence fish attraction. *North American Journal of Fisheries Management* 11:319–329.

Walters, J.P. and J.R. Wilson. 1996. Intraspecific habitat segregation by smallmouth bass in the Buffalo River, Arkansas. *Transactions of the American Fisheries Society* 125:284–290.

Wang, J.C.S. and R.J. Kernehan. 1979. *Fishes of the Delaware Estuaries, A Guide to the Early Life Histories.* Ecological Analysts, Towson, Maryland, USA.

Ward, H.C. 1953. *Know Your Oklahoma Fishes.* Oklahoma Game and Fish Department, Oklahoma City, Oklahoma, USA.

Ward, H.C. and E.M. Leonard. 1952. Order of appearance of scales in the black crappie, *Pomoxis nigromaculatus. Proceedings of the Oklahoma Academy of Science* 33:138–140.

Warren, M.L., Jr. 2004. Fishes. Pages 184–185 in R.E. Mirarchi, R.E. Garner, M.F. Mettee, and P.E. O'Neil, editors. *Alabama Wildlife, Volume 2, Imperiled Aquatic Mollusks and Fishes.* The University of Alabama Press, Tuscaloosa, Alabama, USA.

Warren, M.L., Jr., S.J. Walsh, H.L. Bart, Jr., R.C. Cashner, D.A. Etnier, B.J. Freeman, B.R. Kuhajda, R.L. Mayden, H.W. Robison, S.T. Ross, and W.C. Starnes. 2000. Diversity, distribution, and conservation status of the native freshwater fishes of the southern United States. *Fisheries* 25:7–29.

Warren, M.L., Jr. 1992. Variation of the spotted sunfish *Lepomis punctatus* complex (Centrarchidae): meristics, morphometrics, pigmentation, and species limits. *Bulletin of the Alabama Museum Natural History* 12:1–47.

Warren, M.L., Jr. and R.R. Cicerello. 1982. New records, distribution, and status of ten rare fishes in the Tradewater and lower Green rivers, Kentucky. *Southeastern Fishes Council Proceedings* 3:1–7.

Warren, M.L., Jr., R.R. Cicerello, C.A. Taylor, E. Laudermilk, and B.M. Burr. 1991. Ichthyofaunal records and range extensions for the Barren, Cumberland, and Tennessee River drainages. *Transactions of the Kentucky Academy of Science* 52:17–20.

Waters, D.S. and R.L. Noble. 2004. Spawning season and nest fidelity of largemouth bass in a tropical reservoir. *North American Journal of Fisheries Management* 24:1240–1251.

Waters, D.S., R.L. Noble, and J.E. Hightower. 2005. Fishing and natural mortality of adult largemouth bass in a tropical reservoir. *Transactions of the American Fisheries Society* 134:563–571.

Waters, T.F., J.P. Kaehler, T.J. Polomis, and T.J. Kwak. 1993. Production dynamics of smallmouth bass in a small Minnesota stream. *Transactions of the American Fisheries Society* 122:588–598.

Weathers, K.C. and M.B. Bain. 1992. Smallmouth bass in the Shoals Reach of the Tennessee River: population characteristics and sport fishery. *North American Journal of Fisheries Management* 12:528–537.

Weaver, L.A. and G.C. Garman. 1994. Urbanization of a watershed and historical changes in a stream fish assemblage. *Transactions of the American Fisheries Society* 123:163–172.

Weaver, M.J., J.J. Magnuson, and M.K. Clayton. 1996. Habitat heterogeneity and fish community structure: inferences from north temperate lakes. *American Fisheries Society Symposium* 16:335–346.

Weaver, M.J., J.J. Magnuson, and M.K. Clayton. 1997. Distribution of littoral fishes in structurely complex macrophytes. *Canadian Journal of Fisheries and Aquatic Science* 54:2277–2289.

Webb, J.F. and W.C. Reeves. 1975. Age and growth of Alabama spotted bass and northern largemouth bass. Pages 204–215 in R.H. Stroud and H. Clepper, editors. *Black Bass Biology and Management.* Sport Fishing Institute, Washington, D.C., USA.

Webb, P.W. 1984. Body and fin form and strike tactics of four teleost predators attaching fathead minnow (*Pimephales promelas*) prey. *Canadian Journal of Fisheries and Aquatic Sciences* 41:157–165.

Webster, D.A. 1942. The life histories of some Connecticut fishes. Pages 122–227 in L. Thorpe, editor. A Fishery Survey of Important Connecticut Lakes. *Bulletin Connecticut Geology Natural History Survey.*

Webster, D.A. 1948. Relation of temperature to survival and incubation of the eggs of smallmouth bass (*Micropterus dolomieui*). *Transactions of the American Fisheries Society* 75:43–47.

Weddle, G.K. 1986. Fishes of the Green River drainage in Taylor County, Kentucky. *Transactions of the Kentucky Academy of Science* 47:37–42.

Wehrly, K.E., M.J. Wiley, and P.W. Seelbach. 2003. Classifying regional variation in thermal regime based on stream fish community patterns. *Transactions of the American Fisheries Society* 132:18–38.

Welker, M.T., C.L. Pierce, and D.H. Wahl. 1994. Growth and survival of larval fishes: roles of competition and zooplankton abundance. *Transactions of the American Fisheries Society* 123:703–717.

Werner, R.G. 1966. Ecology and Movements of Bluegill Sunfish Fry in a Small Northern Indiana Lake. Doctoral dissertation. Indiana University, Bloomington, Indiana, USA.

Werner, R.G. 1967. Intralacustrine movements of bluegill fry in Crane Lake, Indiana. *Transactions of the American Fisheries Society* 96:416–420.

Werner, R.G. 1969. Ecology of limnetic bluegill (*Lepomis macrochirus*) fry in Crane Lake, Indiana. *The American Midland Naturalist* 81:164–181.

West, J.L. and F.E. Hester. 1966. Intergeneric hybridization of centrarchids. *Transactions of the American Fisheries Society* 95:280–288.

Wheeler, A.P. and M.S. Allen. 2003. Habitat and diet partitioning between shoal bass and largemouth bass in the Chipola River, Florida. *Transactions of the American Fisheries Society* 132:438–449.

Whitaker, J.O., Jr. 1974. Foods of some fishes from the White River at Petersburg, Indiana. *Indiana Academy of Science* 1974:491–499.

Whitehurst, D.K. 1976. Movement and Trapping Success of Fish in an Eastern North Carolina Swamp Stream. Master's thesis. North Carolina State University, Raleigh, North Carolina, USA.

Whiteside, B.G. 1964. Biology of the White Crappie, *Pomoxis annularis*, in Lake Texoma, Oklahoma. Master's thesis. Oklahoma State University, Stillwater, Oklahoma, USA.

Whittier, T.R. and R.M. Hughes. 1998. Evaluation of fish species tolerances to environmental stressors in lakes in the northeastern United States. *North American Journal of Fisheries Management* 18:236–252.

Whitworth, W.R., P.L. Barrien, and W.T. Keller. 1968. *Freshwater Fishes of Connecticut*. Bulletin Connecticut Geology Natural History Survey.

Wicker, A.M. and M.T. Huish. 1982. Morphology of bluegill (*Lepomis macrochirus*), chain pickerel (*Esox niger*) and yellow perch (*Perca flavescens*) spermatozoa, as determined by scanning electron microscopy. *Copeia* 1982:955–957.

Wicker, A.M. and W.E. Johnson. 1987. Relationships among fat content, condition factor, and first-year survival of Florida largemouth bass. *Transactions of the American Fisheries Society* 116:264–271.

Wiebe, A.H. 1931. Notes in the exposure of several species of fish to sudden changes in hydrogen-ion concentration of the water and to an atmosphere of pure oxygen. *Transactions of the American Fisheries Society* 61:216–224.

Wiener, J.G. and W.R. Hanneman. 1982. Growth and condition of bluegills in Wisconsin lakes: effects of population density and lake pH. *Transactions of the American Fisheries Society* 111:761–767.

Wigtail, G.W. 1982. Lack of consistency of age and growth analysis in Oklahoma. *Proceedings of the Thirty-Fifth Annual Conference of the Southeastern Association of Fish and Wildlife Agencies* 35:579–584.

Wilbur, R.L. 1969. The redear sunfish in Florida. *Fisheries Bulletin of the Florida Game and Freshwater Fish Commission* 5:1–64.

Winkelman, D.L. 1994. Growth, Lipid Use, and Behavioral Choice in Two Prey Fishes, Dollar Sunfish (*Lepomis marginatus*) and Mosquitofish (*Gambusia holbrooki*), under Varying Predation Pressure. Doctoral dissertation. University of Georgia, Athens, Georgia, USA.

Winkelman, D.L. 1996. Reproduction under predatory threat: trade-offs between nest guarding and predator avoidance in male dollar sunfish (*Lepomis marginatus*). *Copeia* 1996:845–851.

Witt, A., Jr. and R.C. Marzolf. 1954. Spawning and behavior of longear sunfish, *Lepomis megalotis megalotis*. *Copeia* 1954:188–190.

Witt, L.A. 1970. The fishes of the Nemaha Basin, Nebraska. *Transactions of the Kansas Academy of Sciences* 73:70–88.

Wood, R., R.H. Macomber, and R.K. Franz. 1956. Trends in fishing pressure and catch, Allatoona Reservoir, Georgia, 1950–53. *Journal of the Tennessee Academy of Science* 31:215–223.

Wrenn, W.B. 1976. Preliminary assessment of larval fish entrainment, Colbert Steam Plant, Tennessee River. *Thermal Ecology II, Proceedings of Thermal Ecology Symposium*, Augusta, Georgia, April 1975. Available as CONF-750425 from NTIS, United States Department of Commerce, Springfield, Virginia, USA.

Wrenn, W.B. 1984. Smallmouth bass reproduction in elevated temperature regimes at the species' native southern limit. *Transactions of the American Fisheries Society* 113:295–303.

Wrenn, W.B. and K.L. Grannemann. 1980. Effects of temperature on bluegill reproduction and young-of-the-year standing stocks in experimental ecosystems. Pages 703–714 in J.P. Giesy, Jr., editor. *Microcosms in Ecological Research*. Published by United States Technical Information Center, U.S. Department of Energy, Symposium Series 52 (CONF-781101).

Wright, G.L. and G.W. Wigtil. 1981. Comparison of growth, survival, and catchability of Florida, northern, and hybrid largemouth bass in a new Oklahoma reservoir. *Proceedings of the Thirty-Fourth Annual Conference of the Southeastern Association of Fish and Wildlife Agencies* 34:31–38.

Wright, R.A., J.E. Garvey, A.H. Fullerton, and R.A. Stein. 1999. Predicting how winter affects energetics of age-0 largemouth bass: how do current models fare? *Transactions of the American Fisheries Society* 128:603–612.

Yeager, B.L. 1981. Early development of the longear sunfish, *Lepomis megalotis* (Rafinesque). *Journal of the Tennessee Academy of Science* 56:84–88.

Young, C.H. and S.A. Flickinger. 1989. Zooplankton production and pond fertilization for largemouth bass fingerling production. *Proceedings of the Forty-Second Annual Conference of the Southeastern Association of Fish and Wildlife Agencies* 42:66–73.

Zanden, M.J.V. and J.B. Rasmussen. 2002. Food web perspectives on studies of bass populations in north-temperate lakes. *American Fisheries Society Symposium* 31:173–184.

Zorn, T.G. and P.W. Seelbach. 1995. The relation between habitat availability and the short-term carrying capacity of a stream reach for smallmouth bass. *North American Journal of Fisheries Management* 15:773–783.

Zweiacker, P.L. and B.E. Brown. 1971. Production of a minimal largemouth bass population in a 3000-acre turbid Oklahoma reservoir. Pages 481–493 in G.E. Hall, editor. *Reservoir Fisheries and Limnology.* Special Publication Number 8, American Fisheries Society, Washington, D.C., USA.

Zweiacker, P.L. and R.C. Summerfelt. 1974. Seasonal variation in food and diel periodicity in feeding of northern largemouth bass, *Micropterus salmoides* (Lacepede), in an Oklahoma reservoir. *Proceedings of the Twenty-Seventh Annual Conference of the Southeastern Association of Fish and Wildlife Agencies* 27:579–591.

Zweiacker, P.L., R.C. Summerfelt, and J.N. Johnson. 1972. Largemouth bass growth in relationship to annual variations in mean pool elevations in Lake Carl Blackwell, Oklahoma. *Proceedings of the Twenty-Sixth Annual Conference of the Southeastern Association of Fish and Wildlife Agencies* 26:530–540.

Zweifel, R.D., R.S. Hayward, and C.F. Rabeni. 1999. Bioenergetics insight into black bass distribution shifts in Ozark Border Region streams. *North American Journal of Fisheries Management* 19:192–197.

APPENDIX

Axon, J.R. 1984. Annual performance report for State-wide Fisheries Management Project, Part I of III, Subsection I: Lakes and Tailwaters Research and Management. Commonwealth of Kentucky D-J Project Number F-50, Segment 6. Department of Fish and Wildlife Resources, Frankfort, Kentucky, USA.

Banner, A. and J.A. Van Arman. 1973. Thermal effects on eggs, larvae, and juveniles of bluegill sunfish. *EPA report: EPA-R3-73-041.* Project 18050 GAB, National Water Quality Laboratory, 6201 Congdon Boulevard, Duluth, Minnesota, USA.

Brown, E.H., Jr. 1960. Little Miami River headwaterstream investigations. (The fisheries in relation to land-use improvement and to hydrological conditions.) Final Report Dingell-Johnson Project F-1-R, Division of Wildlife, Ohio Department of Natural Resources, Columbus, Ohio, USA.

Carter, E.R. 1953. Growth rates of the white crappie *Pomoxis annularis* in Kentucky Lake. Dingle-Johnson Project F-2-R Report, Division of Fisheries, Department of Fish and Wildlife Resources, Frankfort, Kentucky, USA.

Carver, D.M. 1976. Early life history of the bluegill, *Lepomis macrochirus.* Student summer report, 1975 (Reference Number 76-46). University of Maryland Center for Environmental and Estuarine Studies, Chesapeake Biological Laboratory, Solomons, Maryland, USA.

Darr, D.P. and G.R. Hooper. 1991. Spring pygmy sunfish population monitoring. Unpublished final report for the Alabama Department of Conservation and Natural Resources, Montgomery, Alabama, USA.

EA Engineering, Science, and Technology. 1994. *1992 Ohio River Ecological Research Program, Adult and Juvenile Fishes.* Report submitted to American Electric Power Service Corporation, Ohio Edison Company, Ohio Valley Electric Corporation, and Tennessee Valley Authority. EA Engineering, Science, and Technology, Deerfield, Illinois, USA.

EA Engineering, Science, and Technology. 2004. *Final Report 2003 Ohio River Ecological Research Program.* Report submitted to Electric Power Research Institute, American Electric Power Service Corporation, Ohio Valley Electric Corporation, and Buckeye Power, Inc. EA Engineering, Science, and Technology, Deerfield, Illinois, USA.

Enamait, E.C. and R.M. Davis. 1982. Biological surveys of lakes, ponds, and impoundments. Maryland Department of Natural Resources, Federal Aid in Sport Fish Restoration, Project F-29-R-6, Jobs 1-2, January 1–December 31, 1981. Progress Report, Annapolis, Maryland, USA.

Environmental Science and Engineering, Inc. 1992. Ohio River Ecological Research Program, Analysis of Long-Term Larval Fish Data 1978-1990. Report prepared for American Electric Power Service Corporation, Ohio Edison Company, Ohio Valley Electric Corporation, Cincinnati Gas and Electric Company, and Tennessee Valley Authority. ESE Number 591-1065-0300, Environmental Science and Engineering, Inc., St. Louis, Missouri, USA.

Environmental Science and Engineering, Inc. 1995. *1993 Ohio River Ecological Research Program.* Report submitted to American Electric Power Service Corporation, Ohio Edison Company, and Ohio Valley Electric Corporation. ESE Number 593-1041-0600, Environmental Science and Engineering, Inc., St. Louis, Missouri, USA.

Gammon, J.R. 1991. *The Environment and Fish Communities of the Middle Wabash River.* A report for Eli Lilly and Company, Indianapolis, Indiana, and PSI Energy, Plainfield, Indiana, USA.

Heitman, J.F. 1980. Study to evaluate the fishery of Lake Eufaula, Oklahoma: Annual Performance report. Oklahoma Department of Wildlife Conservation, Oklahoma City, Oklahoma, USA.

Heitman, J.F. 1982. Study to evaluate the fishery of Lake Eufaula, Oklahoma: Annual Performance report. Oklahoma Department of Wildlife Conservation, Oklahoma City, Oklahoma, USA

Heitman, J.F. 2004. Annual Report on the Lakes at Cumberland Mountain Sand in Coffee and Grundy Counties, Tennessee. Unpublished Data.

Hill, J.E. and C.E. Cichra. 2005. Biological Synopsis of Five Selected Florida Centrarchid Fishes with an Emphasis on the Effects of Water Level Fluctuations. Special Publication SJ2005-SP3, St. Johns River Water Management District, Palatka, Florida.

Hogue, J.J., Jr., R. Wallus, and L.K. Kay. 1976. *Preliminary Guide to the Identification of Larval Fishes in the Tennessee River.* Tennessee Valley Authority, Norris, Tennessee, USA.

Houser, A. and M.G. Bross. 1963. Average growth rates and length–weight relationships for fifteen species of fish in Oklahoma waters. *Report Number 85, Oklahoma Fishery Research Laboratory*, Norman, Oklahoma, USA.

Huish, M.T. and G.B. Pardue. Ecological studies of one channelized and two unchannelized wooded coastal swamp streams in North Carolina. U.S. Department of the Interior, Fish and Wildlife Service, OBS-78/85, Raleigh, North Carolina.

Indiana Division of Fish and Wildlife. 1996. Population characteristics of smallmouth bass in southern Lake Michigan, 1995–1996. Indiana Division of Fish and Wildlife, Indianapolis, Indiana, USA.

Jenkins, R., R. Elkin, and J.C. Finnell. 1955. Growth rates of six sunfishes in Oklahoma. *Oklahoma Fisheries Research Laboratory Report* 49:1–73, Norman, Oklahoma, USA.

Jones, A.R. 1970. Inventory and classification of streams in the Licking River drainage. *Kentucky Fisheries Bulletin*, No. 42:63 pp.

Jones, A.R. 1973. Inventory and classification of streams in the Kentucky River drainage. *Kentucky Fisheries Bulletin*, No. 56:108 pp.

Lewis, S.A., K.D. Hopkins, and T.F. White. 1971. Average growth rates and length–weight relationships of sixteen species of fish in Canton Reservoir, Oklahoma. *Bulletin Number 9, Oklahoma Fisheries Research Laboratory*, Norman, Oklahoma, USA.

McGowan, E.G. 1984. *An Identification Guide for Selected Larval Fishes from Robinson Impoundment, South Carolina.* Biological Unit, Carolina Power and Light Company, New Hill, North Carolina, USA.

McGowan, E.G. 1988. *An Illustrated Guide to Larval Fishes from Three North Carolina Piedmont Impoundments.* Biological Unit, Carolina Power and Light Company, New Hill, North Carolina, U.S.A.

Mense, J.B. 1976. Growth and length–weight relationships of twenty-one reservoir fishes in Oklahoma. *Bulletin Number 13, Oklahoma Fishery Research Laboratory*, Norman, Oklahoma, USA.

North Carolina Wildlife Resource Commission. 1983. *Some North Carolina Freshwater Fishes.* Raleigh, North Carolina, U.S.A.

Oklahoma Department of Wildlife Conservation. 1982. Mississippi silversides evaluation. Fish Research and Surveys for Oklahoma Lakes and Reservoirs, Job Number 1, Federal Aid Project Number F-37-R. *Contribution Number 200 of the Oklahoma Fishery Research Laboratory*, Norman, Oklahoma, USA.

Oklahoma Department of Wildlife Conservation. 1992. Growth responses of white crappie and recruitment of largemouth bass following the introduction of threadfin shad in Thunderbird Reservoir. Fish Research and Surveys for Oklahoma Lakes and Reservoirs, Job Number 15, Federal Aid Project Number F-37-R, Oklahoma Fishery Research Laboratory, Norman, Oklahoma, USA.

Peterka, J.J. and J.S. Kent. 1976. Dissolved oxygen, temperature, survival of young at fish spawning sites. Grant Number R801976; EPA-600/3-76-113. Environmental Research Laboratory, Duluth, Minnesota, USA.

Rogers, M. and M.S. Allen. 2005. Hatching duration, growth and survival of age-0 largemouth bass along a latitudinal gradient of Florida lakes. Final Report, Florida Fish and Wildlife Conservation Commission, Tallahassee, Florida, USA.

Shewmake, J.W. and D.C. Jackson. 2004. Fisheries resources assessment in small streams on Mississippi National Forests. Project Completion Report. Mississippi State University, Department of Wildlife and Fisheries, Box 9690, Mississippi State, Mississippi, USA.

Simon, T.P. 1994. Survey of remaining populations of the bantam sunfish (*Lepomis symmetricus* Forbes) in the West Fork White River. Indiana Non-Game Grant Report. Division of Fish and Wildlife, Indiana Department of Natural Resources, Indianapolis, Indiana, U.S.A.

Simpson, J.A., J.H. Mickey, Jr., and J.C. Borawa. 1988. Investigation of largemouth and spotted bass populations in W. Kerr Scott Reservoir. Federal Aid in Fish Restoration Final Report, Project Number: F-24-12. Division of Boating and Inland Fisheries, North Carolina Wildlife Resources Commission, Raleigh, North Carolina, USA.

Smith, P. and J. Kaufman. 1982. Age and growth of Virginia's freshwater fishes. Virginia Commission of Game and Inland Fisheries, Richmond, Virginia, U.S.A.

Snow, H. 1969. Comparative growth of eight species of fishes in thirteen northern Wisconsin lakes. *Wisconsin Department of Natural Resources Research Report* 46.

Swingle, W.E. 1965. Length–weight relationships of Alabama fishes. Auburn University Zoology–Entomology Department Series Number 3. Auburn Agricultural Experiment Station, Auburn University, Auburn, Alabama, USA.

Tennessee Valley Authority. 1978. Preoperational fisheries report for the Sequoyah Nuclear Plant. Tennessee Valley Authority, Chattanooga, Tennessee, USA.

Tillman, J.S. and C.S. Guy. 1997. Evaluation of spotted bass populations in Kansas streams. Federal Aid in Sport Fish Restoration Final Report, Project Number: F-27-R. Kansas Cooperative Fish and Wildlife Research Unit, Kansas State University, Manhattan, Kansas, USA.

Towery, B.A. 1963. Age and size at maturity of largemouth bass. Completion Report F-6-R (Job 1). Mississippi Game and Fish, Jackson, Mississippi, USA.

Towery, B.A. 1964. Age and size at maturity of spotted bass. Completion Report F-6-R (Job 1). Mississippi Game and Fish, Jackson, Mississippi, USA.

Vessel, M.F. and S. Eddy. 1941. A preliminary study of the egg production of certain Minnesota fishes. *Minnesota Bureau of Fisheries Research, Investigation Report* 26:1–26.

Wojtalik, T.A. 1970. Summary of temperature literature on largemouth bass, *Micropterus salmoides* (Lacepede), smallmouth bass (*Micropterus dolomieu* Lacepede), carp (*Cyprinus carpio* Linnaeus), channel catfish, *Ictalurus punctatus* (Rafinesque), and bluegill, *Lepomis macrochirus* (Rafinesque). Internal Tennessee Valley Authority report. Environment Biology Branch, Muscle Shoals, Alabama, USA.

Wright, F.G. 1977. Lake Eufaula, Performance Report, Job Number 2. Oklahoma Department of Wildlife Conservation F-15-R-14. Oklahoma City, Oklahoma, USA.

Plate 1 Photographs of juvenile centrarchids that occur in the Ohio River drainage: (A) flier *Centrarchus macropterus*, 27 mm SL, Isom Lake, TN; (B) redbreast sunfish *Lepomis auritus*, 28 mm SL, Holston River system, TN; (C) green sunfish *L. cyanellus*, 32 mm SL, Potomac River system, VA; (D) pumpkinseed *L. gibbosus*, 38 mm SL, Potomac River system, VA; (E) warmouth *L. gulosus*, 50 mm SL, Hatchie River system, TN; (F) orangespotted sunfish *L. humilis*, 40 mm SL, Running Reelfoot Bayou, TN. (A reprinted from Plate 186d and B–F reprinted from Figure 130a–e in Etnier and Starnes, 1993, with permission of senior author and photographer, R.T. Bryant.)

Plate 2   Photographs of juvenile *Lepomis* spp. that occur in the Ohio River drainage: (A) bluegill *L. macrochirus*, 29 mm SL, farm pond, Knox County, TN; (B) dollar sunfish *L. marginatus*, 31 mm SL, Hatchie River system, TN; (C) longear sunfish *L. megalotis*, 30 mm SL, lower Tennessee River system, TN; (D) redear sunfish *L. microlophus*, 29 mm SL, farm pond, Blount County, TN. (E) redspotted sunfish *L. miniatus*, 27 mm SL, Tennessee River system, AL; (F) bantam sunfish *L. symmetricus*, 40 mm SL, Lake Isom, TN. (A–F reprinted from Figure 130f–k in Etnier and Starnes, 1993, with permission of senior author and photographer, R.T. Bryant.)